Festschrift Masatoshi Fukushima

In Honor of Masatoshi Fukushima's Sanju

INTERDISCIPLINARY MATHEMATICAL SCIENCES*

Published

Vol. 10: Ordinal and Relational Clustering (with CD-ROM)
Melvin F. Janowitz

Vol. 11: Advances in Interdisciplinary Applied Discrete Mathematics
eds. Hemanshu Kaul & Henry Martyn Mulder

Vol. 12: New Trends in Stochastic Analysis and Related Topics: A Volume in Honour of Professor K D Elworthy
eds. Huaizhong Zhao & Aubrey Truman

Vol. 13: Stochastic Analysis and Applications to Finance: Essays in Honour of Jia-an Yan
eds. Tusheng Zhang & Xunyu Zhou

Vol. 14 Recent Developments in Computational Finance: Foundations, Algorithms and Applications
eds. Thomas Gerstner & Peter Kloeden

Vol. 15 Recent Advances in Applied Nonlinear Dynamics with Numerical Analysis: Fractional Dynamics, Network Dynamics, Classical Dynamics and Fractal Dynamics with Their Numerical Simulations
eds. Changpin Li, Yujiang Wu & Ruisong Ye

Vol. 16 Hilbert–Huang Transform and Its Applications (2nd Edition)
eds. Norden E Huang & Samuel S P Shen

Vol. 17 Festschrift Masatoshi Fukushima In Honor of Masatoshi Fukushima's Sanju
eds. Zhen-Qing Chen, Niels Jacob, Masayoshi Takeda & Toshihiro Uemura

Vol. 18 Global Attractors of Non-Autonomous Dynamical and Control Systems (2nd Edition)
by David N Cheban

*For the complete list of titles in this series, please go to
http://www.worldscientific.com/series/ims

Interdisciplinary Mathematical Sciences – Vol. 17

$$(\mathcal{E}, \mathfrak{F}) \quad \Longleftrightarrow \quad \mathbf{M} = (X_t, \mathbb{P}_x)$$

$$A_t^{[u]} = \widetilde{u}(X_t) - \widetilde{u}(X_0) = M_t^{[u]} + N_t^{[u]}$$

Festschrift Masatoshi Fukushima

In Honor of Masatoshi Fukushima's Sanju

Editors

Zhen-Qing Chen
University of Washington, USA

Niels Jacob
Swansea University, UK

Masayoshi Takeda
Tohoku University, Japan

Toshihiro Uemura
Kansai University, Japan

World Scientific

NEW JERSEY • LONDON • SINGAPORE • BEIJING • SHANGHAI • HONG KONG • TAIPEI • CHENNAI

Published by

World Scientific Publishing Co. Pte. Ltd.
5 Toh Tuck Link, Singapore 596224
USA office: 27 Warren Street, Suite 401-402, Hackensack, NJ 07601
UK office: 57 Shelton Street, Covent Garden, London WC2H 9HE

Library of Congress Cataloging-in-Publication Data
Festschrift Masatoshi Fukushima : in honor of Masatoshi Fukushima's Sanju / edited by Zhen-Qing Chen (University of Washington, USA), Niels Jacob (Swansea University, UK), Masayoshi Takeda (Tohoku University, Japan), Toshihiro Uemura (Kansai University, Japan).
pages cm. -- (Interdisciplinary mathematical sciences ; volume 17)
Includes bibliographical references.
ISBN 978-9814596527 (hard cover : alk. paper)
1. Mathematical analysis. 2. Stochastic analysis. 3. Probabilities. I. Fukushima, Masatoshi, 1935– II. Chen, Zhen-Qing. III. Jacob, Niels. IV. Takeda, Masayoshi. V. Uemura, Toshihiro.
QA371.F38 2014
519.2--dc23

2014034313

British Library Cataloguing-in-Publication Data
A catalogue record for this book is available from the British Library.

Printed in Singapore

Preface

Compiling a Festschrift and presenting it at a larger international conference is arguably the most personal way the academic community can express its appreciation to one of its members. More than five decades, Professor Masatoshi Fukushima has contributed to our discipline and is still surprising us with highly original contributions and deep insights. He has long been recognised as one of the towering figures in stochastic analysis who has shaped, and in fact initiated, the probabilistic part of the theory of Dirichlet spaces. His scholarship is much admired, as have his mentorship of the younger generation and his contributions to the international mathematical community.

The overwhelming and cordial responses we have received when inviting contributions for the Festschrift Masatoshi Fukushima to be published on the occasion of his Sanju is further testimony of Professor Fukushima's standing in our community. The Festschrift contains 26 articles, some being surveys, some being cutting-edge research contributions, as well as an update of Professor Fukushima's bibliography published a few years ago in "Selecta Masatoshi Fukushima". We would like to take the opportunity to thank again all colleagues for their contributions, and in particular for collaborating with us smoothly. We also want to thank those who helped us as referees, and special thanks go to K. Yamazaki, A. Arranz-Carreño and K. Evans for their tex-nical support. Further thanks go to Ms. Tan Rok Ting of World Scientific Publishing Co. Pte. Ltd. for her constant and truly generous cooperation and also to the publisher for the uncomplicated collaboration.

We consider it as a great honour to present to Professor Fukushima this Festschrift on behalf of his colleagues, students and friends, as a tribute to his mathematical achievements and as gratitude for his friendship.

Zhen-Qing Chen (Seattle)
Niels Jacob (Swansea)
Masayoshi Takeda (Sendai)
Toshihiro Uemura (Osaka)

Contents

Part 1

Professor Fukushima's Work

Chapter 1

The mathematical work of Masatoshi Fukushima - An Essay

Zhen-Qing Chen, Niels Jacob, Masayoshi Takeda and Toshihiro Uemura

Over more than five decades Professor Fukushima has made remarkable contributions to mathematics, especially to stochastic analysis and its applications to science, in particular to mathematical physics. For many of his colleagues, students and friends it becomes an excellent opportunity to express their appreciation of himself and his work by celebrating his Sanju, the 80th birthday in the Japanese tradition, with an international conference and by publishing a Festschrift in his honour. Naturally such a Festschrift should contain a scientific account of Professor Fukushima's oeuvre. When making here an attempt, i.e. writing "un essai" in the classical meaning, to present such an account, we must be aware that we are not discussing a completed oeuvre. Professor Fukushima is continuing to make highly original research contributions with deep insights. Obviously it is pre-matured to describe or guess the lasting impact these latest contributions will have. However it is fair to say that some of his earlier works are meanwhile considered as to belong to the "classical canon" of stochastic analysis, and hence we will concentrate more on those contributions. But we will see how some of his latest research indeed connect with his earlier work in a natural way.

Professor Fukushima started his graduate studies in 1959 as a student of Kiyoshi Itô at Kyoto University. For anyone who is interested in the history of probability theory it is a temptation to write about the amazing changes in the field during the three decades between the publication of the works of P. Lévy [24], J. L. Doob [8], K. Itô [19] and [20], and S. Kakutani [22], and the publication of the monograph of K. Itô and H. P. McKean [21]. This period also includes the contributions of the E. B. Dynkin school, and we refer to the monographs [10] - [11], and for us quite important the work of W. Feller, in particular [12] and [13], as well as the contributions of G. A. Hunt [16] - [18]. (Clearly, this list is not exhausting.) However we have already some good, although brief, account of this period, especially when having in mind the developments in Japan, see D. W. Stroock and S. R. S. Varadhan [29], H. Föllmer [15], and in particular to M. Fukushima [F39] as well as [F49]. We should also mention the work of Kôsaku Yosida combining one-parameter operator semigroup theory with probability theory and potential theory on which S. Watanabe

[30] gave a readable account.

For Professor Fukushima, early in his career, the problem to understand Feller's boundary theory for one-dimensional diffusion processes in higher dimensions, say Brownian motion in bounded domains, took an important place, see [F1] and [F2], and in some sense this is a topic that still occupies him after 50 years, see for example [F44]. Already in these early papers the idea was to use Dirichlet spaces as introduced by A. Beurling and J. Deny [2] and [3]. Feller's work in one dimension could use the fact that (in a rough formulation) integrable functions with an integrable distributional derivative are absolutely continuous. Clearly this does not apply in higher dimensions, but the results of the paper by J. Deny and J.-L. Lions [7], see also J. L. Doob [9], provided a type of substitute, i.e. allowing to pass to quasi everywhere fine continuous versions. In [7] many technical tools originated from classical potential theory, by which we mean really the theory of harmonic functions and the Newton potential, and thus the method had a natural limitation. In the paper [F2] Professor Fukushima introduced already a significant change into the theory of Dirichlet spaces as initiated by A. Beurling and J. Deny: the concept of "Dirichlet spaces related to $L^2(X;m)$" was introduced, where X is a certain Hausdorff space with some finite measure m. Although this looks at first glance as technical or minor, today this is the standard definition of (symmetric) Dirichlet spaces (with some more general space X and measure m), once we add regularity, see the comments to [F3] below.

The papers [F3] and [F4] are Professor Fukushima's early seminal contribution in which he laid the foundation of what is nowadays understood as the modern theory of Dirichlet forms. Most of this work was done while he was a post doctoral fellow at the University of Illinois, Urbana - Champaign, working in the group of J. L. Doob. In [F3] parts of the quadratic form based potential theory of symmetric Dirichlet forms are studied, and in particular "regular representations" are discussed. The basic problem was to find for a given L^2 - related Dirichlet space an isometric version on a locally compact metric space that contains "sufficiently many" regular (continuous) elements. Now the ground was prepared for [F4], the construction of a strong Markov process associated to a regular Dirichlet space. The ideas picked up from the Deny - Lions paper [7], and employed already in [F2], were combined with the regular representation result. Using the natural capacity in a regular Dirichlet space it is possible to obtain quasi-continuous refinements of elements and this makes it eventually possible to construct the corresponding Hunt process (up to an invariant negligible set). This result was immediately recognised as a breakthrough and Professor Fukushima was invited to give a talk at the ICM in Helsinki 1978.

The construction of the Hunt process associated with a Dirichlet form we find now in the monograph [F43] and its predecessors [F10] and [F23] follows the modification by M. L. Silverstein which appeared in [27], not the paper [F4]. Professor Fukushima always points to the important contributions of Martin Silverstein to the theory of Dirichlet forms [27] and the boundary theory of symmetric Markov processes [28], as did Martin Silverstein appreciated his work. Martin Silverstein was among the first to recognize the importance of the regular representation of Dirichlet forms considered by Professor Fukushima, see [27], Section 2. Maybe in times without citation index, impact factors, etc., it was much easier to add to academic standards also civilized standards in human relations, respect for the work of colleagues, and a certain intellectual honesty.

The construction of a Hunt process being associated with a regular Dirichlet form was already a big step forward, but immediately several question arose, for example:

1. Itô's stochastic calculus is an analysis on path level. How can we derive a corresponding theory for symmetric Markov processes which may not be semi-martingales?
2. The processes are constructed up to a set of zero capacity. Is it possible to remove such exceptional set?
3. Can this approach to stochastic processes help to treat problems in science, especially mathematical physics?
4. Is it eventually possible to extend Feller's boundary theory to higher dimensions?

Itô calculus changed its face under the influence of P. A. Meyer and his school. The basic objects for a "reasonable" stochastic calculus were identified as semi-martingales, but the Hunt processes constructed with the help of Dirichlet forms are in general not semi-martingales. The breakthrough was the paper [F8] where the "Fukushima decomposition" was introduced. Using versions of elements in the Dirichlet space, additive functionals were introduced with the help of the associated Hunt process and it was proved that every such additive functional admits a decomposition into a martingale and an additive functional of zero energy. Then a stochastic calculus for additive functionals was established. Meanwhile the Fukushima decomposition and the related stochastic calculus is a core element of stochastic analysis with many applications in the studies of diffusions in infinite dimensions or on fractals, and in the study of Markov processes in random environment, just to mention some areas.

One of the hardest problem is to construct "nice" versions of the process associated with a Dirichlet form. The best solution would be to get rid of any exceptional set. Several papers of Professor Fukushima's, including [F13, F16, F21], are devoted to this problem. Arguably the most influential one is that jointly with H. Kaneko [F16]. Here (r,p)-capacities, originally introduced by P. Malliavin [26], were taken up with the aim to get version up to a set of (r,p)-capacity zero. These capacities are monotone in r and in p , and it may happen that for r or p large the only set of

(r, p)-capacity zero is the empty set. H. Kaneko [23] has used this to study diffusion processes in an L^p-setting.

Many of the non-Japanese colleagues of Professor Fukushima may not know about the book [F9] in Japanese co-authored with the physicist K. Ishii titled "Natural Phenomena and Stochastic Processes" first published in 1980. Influenced by the work of K. Ishii, Professor Fukushima turned to the study of the spectra of random Schrödinger operators and related topics, see [F6] written with H. Nagai and S. Nakao, [F7] with S. Nakao, as well as [F18] with S. Nakao and M. Takeda. Subsequently he continued to publish on mathematical physics and was in closed contact with colleagues in Bielefeld, including S. Albeverio, W. Karwowski and L. Streit, see for example [F11]. His interest and work had much impact on the later work of S. Kotani.

Within our essay we can not discuss all of Professor Fukushima's contributions, so we mention only briefly his work on ergodic theorems [F5, F12], on problems related to large deviation, for example [F15] with M. Takeda, on stochastic analysis on fractals [F20, F22, F27] in collaboration with T. Shima and M. Takeda, on Dirichlet forms related to complex analysis of several variables [F17] and [F14, F19] with M. Okada, or on BV functions, capacitary inequalities and the "geometry" of sets [F28] - [F32], including joint works with M. Hino or T. Uemura, or finally, already more related to the boundary theory of Markov processes, the study of reflecting diffusions with M. Tomisaki [F24] - [F26].

However we want and we have to return to the beginning of his career and the attempt to extend Feller's theory. Time changes of Markov processes are closely related to their boundary theory. Motivated by the Douglas integral, which was the main tool of J. Douglas to give a solution to the Plateau problem and that also characterizes the trace of reflecting Brownian motion in the unit disk on its boundary, Professor Fukushima with P. He and J. Ying in [F33] studied time changes of symmetric diffusions and the role played by the Feller measure. They showed that the jumping and killing measures for the time-changed process are bounded below by the Feller measure and the supplementary Feller measure, respectively. A complete solution to this problem is obtained in a joint work of Professor Fukushima with Z.-Q. Chen and J. Ying [F35], where not only the jumping and killing measures of the time-changed process are identified, but also the complete characterization of the Beurling - Deny decomposition, see also the extension in [F36].

The boundary theory for one-dimensional diffusions is well understood thanks to the fundamental works of W. Feller, K. Itô, and H. P. McKean. Much less is known for the boundary theory of multi-dimensional diffusions and of Markov processes with discontinuous sample paths. In his work with H. Tanaka [F34] Professor Fukushima studied the one-point extension of absorbing symmetric diffusions using excursion theory. In a series of joint works with Z.-Q. Chen [F38, F40, F41] and with Z.-Q. Chen and J. Ying [F37], he has further studied one-point extensions of general symmetric Markov processes, as well as strong Markov processes in weak duality, and their characterization. Moreover, many-point extensions of strong Markov processes were also studied. These works provide us with a satisfactory picture of the boundary theory for symmetric Markov processes with finitely many boundary points, and more generally, for strong Markov processes having weak duals, with finitely many boundary points. An important tool used in this study is the reflected Dirichlet space first introduced by M. L. Silverstein [27] in 1974. Very recently in joint work with Z.-Q. Chen [F48] one point extension theory was used to give a direct construction of Walsh's Brownian motion. This construction uses Dirichlet form methods. The Dirichlet form characterization of this process together with the results in [4] and [6] gives the analytic characterization of harmonicity and subharmonicity for Walsh's Brownian motion which recovers the main result of the recent paper of P. Fitzsimmons and K. E. Kutler [14]. Using Dirichlet forms, in [F42]and [F45] Professor Fukushima has revisited Feller's boundary theory for one-dimensional diffusions providing us with an elegant characterization. In the joint monograph [F44] with Z.-Q. Chen many of these and more results are presented. In particular this monograph covers the potential theory for regular symmetric Dirichlet forms as well as quasi-regular symmetric Dirichlet forms initiated by S. Albeverio, Z.-M. Ma and M. Röckner, see [1] and [25], in a unified and systematic way.

A further new area of Professor Fukushima's latest research is the study of the interaction between probability theory and complex analysis, more precisely Komatu - Loewner evolution in multiply connected planar domains. Here we refer to the recent papers [F46] with Z.-Q. Chen and S. Rohde, as well as [F50], and [5]. A special case of the many-point extension of absorbing Brownian motion, called Brownian motion with darning, played an important role in such study.

References to publications of M. Fukushima

[F1] *A construction of reflecting barrier Brownian motions for bounded domains.* Osaka J. Math. **4** (1967), 183 – 215.

[F2] *On boundary conditions for multi-dimensional Brownian motions with symmetric resolvent densities.* J. Math. Soc. Japan **21** (1969), 58 – 93.

[F3] *Regular representations of Dirichlet spaces.* Trans. Amer. Math. Soc. **155** (1971), 455 – 473.

[F4] *Dirichlet spaces and strong Markov processes.* Trans. Amer. Math. Soc. **162** (1971), 185 – 224.

[F5] *Almost polar sets and an ergodic theorem.* J. Math. Soc. Japan **26** (1974), 17 – 32.

[F6] *On an asymptotic property of spectra of a random difference operator* (with H. Nagai and S. Nakao). Proc. Japan Acad. **51** (1975), 100 – 102.

[F7] *On the spectra of the Schrödinger operator with a white Gaussian noise potential* (with S. Nakao). Z. Wahrscheinlichkeitstheorie und verw. Gebiete **37** (1976/77), 267 – 274.

[F8] *A decomposition of additive functionals of finite energy.* Nagoya Math. J. **74** (1979), 137 – 168.

[F9] (with K. Ishii) *Natural Phenomena and Stochastic Processes (Japanese).* Nippon-Hyoron-Sha Co., Ltd., Tokyo 1980, (enlarged ed. 1996.)

[F10] *Dirichlet Forms and Markov Processes.* North-Holland, Amsterdam, and Kodansha, Tokyo, 1980.

[F11] *Capacity and quantum mechanical tunnelling* (with S. Albeverio, W. Karwowski, and L. Streit). Commun. Math. Phys. **81** (1981), 501 – 513.

[F12] *A note on irreducibility and ergodicity of symmetric Markow processes. In : Stochastic Processes and Quantum Theory and Statistical Physics*, Marseille 1981. Lecture Notes in Physics 173, Springer Verlag, Berlin 1982, 200 – 207.

[F13] *Basic properties of Brownian motion and a capacity on the Wiener space.* J. Math. Soc. Japan **36** (1984), 161 – 176.

[F14] *On conformal martingale diffusions and pluripolar sets* (with M. Okada). J. Funct. Anal. **55** (1984), 377 – 388.

[F15] *A transformation of a symmetric Markov process and the Donsker-Varadhan theory* (with M. Takeda). Osaka J. Math. **21** (1984), 311 – 326.

[F16] (r,p)*-capacities for general Markovian semigroups* (with H. Kaneko). In : Infinite Dimensional Analysis and Stochastic Processes, Bielefeld 1983. Research Notes in Mathematics **124**, Pitman, Boston MA 1985, 41– 47.

[F17] *On the continuity of plurisubharmonic functions along conformal diffusions.* Osaka J. Math. **23** (1986), 69 – 75.

[F18] *On Dirichlet forms with random data - recurrence and homogenization* (with S. Nakao and M. Takeda). In : Stochastic Processes - Mathematics and Physics II, Bielefeld 1985. Lecture Notes in Math. 1250, Springer Verlag, Berlin 1987, 87 – 97.

[F19] *On Dirichlet forms for plurisubharmonic functions* (with M. Okada). Acta Math **159** (1987), 171 – 213.

[F20] *On a spectral analysis for the Sierpinski gasket* (with T. Shima). Potential Analysis **1** (1992), 1 – 35.

[F21] *Two topics related to Dirichlet forms : Quasi-everywhere convergence and additive functionals* In : Dirichlet Forms, Varenna 1992. Lecture Notes in Math. 1563, Springer Verlag, 1993, 21 – 53.

[F22] *On discontinuity and tail behaviours of the integrated density of states for nested pre-fractals* (with T. Shima). Commun. Math. Phys. **163** (1994), 461 – 471.

[F23] (with Y. Oshima and M. Takeda) *Dirichlet Forms and Symmetric Markov Processes.* Walter de Gruyter Verlag, Berlin 1994.

[F24] *Reflecting diffusions on Lipschitz domains with cusps - analytic construction and Skorohod representation* (with M. Tomisaki). Potential Analysis **4** (1995),

377 – 408.

[F25] *Construction and decomposition of reflecting diffusions on Lipschitz domains with Hölder cusps* (with M. Tomisaki). Probab. Theory Related Fields **106** (1996), 521 – 557.

[F26] *On decomposition of additive functionals of reflecting Brownian motions* (with M. Tomisaki). In : Ito's Stochastic Calculus and Probability Theory. Springer Verlag, Tokyo 1996, 51 – 61.

[F27] *Large deviations and related LIL's for Brownian motion on nested fractals* (with T. Shima and M. Takeda). Osaka J. Math. **36** (1999), 497 – 537.

[F28] *BV functions and distorted Ornstein - Uhlenbeck processes over abstract Wiener space.* J. Funct. Anal. **174** (2000), 227 – 249.

[F29] *On the space of BV functions and a related stochastic calculus in infinite dimensions* (with M. Hino). J. Funct. Anal. **183** (2001), 245 – 268.

[F30] *On Sobolev and capacitary inequalities for contractive Besov spaces over d-sets* (with T. Uemura). Potential Analysis **18** (2003), 59 – 77.

[F31] *Capacitary bounds of measures and ultracontactivity of time changed processes* (with T. Uemura). J. Math. Pures Appl. **82** (2003), 553 – 77.

[F32] *On spectral synthesis for contractive p-norms and Besov spaces* (with T. Uemura). Potential Analysis **20** (2004), 195 – 206.

[F33] *Time changes of symmetric diffusions and Feller measures* (with P. He and J. Ying). Ann. Probab. **32** (2004), 3138– 3166.

[F34] *Poisson point processes attached to symmetric diffusions* (with H. Tanaka). Ann. Inst. Henri Poincaré Probab. Stat. **41** (2005), 419 – 459.

[F35] *Traces of symmetric Markov processes and their characterizations* (with Z.-Q. Chen and J. Ying). Ann. Probab. **34** (2006), 1052 – 1102.

[F36] *Entrance law, exit system and Lévy system of time changed processes* (with Z.-Q. Chen and J. Ying). Illinois J. Math. **50** (2006), 269 – 312.

[F37] *Extending Markov processes in weak duality by Poisson point processes of excursions* (with Z.-Q. Chen and J. Ying). The Abel Symposium 2005. Springer Verlag, Berlin 2007, 153 – 196.

[F38] *On Feller's boundary problem for Markov processes in weak duality* (with Z.-Q. Chen) J. Funct. Anal. **252** (2007), 710 – 733.

[F39] *On the works of Kiyoshi Itô and stochastic analysis.* Japanese J. Math. **2** (2007), 45 – 53.

[F40] *One-point extensions of Markov processes by darning* (with Z.-Q. Chen). Probab. Theory Related Fields 141 (2008), 61 – 112.

[F41] *Flux and lateral conditions for symmetric Markov processes* (with Z.-Q. Chen). Potential Analysis **29** (2008), 241 – 269.

[F42] *From one dimensional diffusions to symmetric Markov processes* Stochastic Processes Appl. **120** (2010), 590 – 604.

[F43] (with Y. Oshima and M. Takeda) *Dirichlet Forms and Symmetric Markov Processes.* Second revised and extended editions. Walter de Gruyter Verlag,

Berlin 2011.

[F44] (with Z.-Q. Chen) *Symmetric Markov Processes, Time Change, and Boundary Theory.* Princeton University Press, Princeton NJ, 2012.

[F45] *On general boundary conditions for one-dimensional diffusions with symmetry.* J. Math Soc. Japan **66** (2014), 289 – 316.

[F46] *Chordal Komatu-Loewner equation and Brownian motion with darning in multiply connected domains* (with Z.-Q. Chen and S. Rohde). Trans. Amer. Math. Soc. (to appear).

[F47] *On Villat's kernels and BMD Schwarz kernels in Komatu - Loewner equations* (with H. Kaneko). Stochastic Analysis and Applications 2014, Springer Proceedings in Mathematics. (to appear)

[F48] *One - point reflection* (with Z.-Q. Chen). Stochastic Processes and their Applications. (to appear)

[F49] *Feller's contributions to the one - dimensional diffusion theory and beyond.* In: Selected Works of William Feller, 2 vols. Eds. R. Schilling, Z. Vondracek, W. Woyczynski. Springer Verlag. (to appear)

[F50] *Stochastic Komatu - Loewner evolution and BMD domain constant* (with Z.-Q. Chen), arXiv:1410.8257 [math PR] (preprint)

Further References

[1] Albeverio, S., and Ma, Z.-M., *Necessary and sufficient conditions for the existence of m-perfect processes associated with Dirichlet forms.* In: Séminaire de Probabilités. XXV. Lecture Notes in Mathematics, Vol.1485, Springer Verlag, Berlin 1991, 374–406.

[2] Beurling, A., and Deny, J., *Espace de Dirichlet I. Le cas élémentaire.* Acta Math. **99** (1958), 203 – 224.

[3] Beurling, A., and Deny, J., *Dirichlet spaces.* Proc. Nat. Acad. Sci. U.S.A. **45** (1959), 208 – 215.

[4] Chen, Z.-Q., *On notions of harmonicity.* Proc. Amer. Math. Soc. **137** (2009), 3497 – 3510.

[5] Chen, Z.-Q., *Brownian Motion with Darning.* Lecture notes for talks given at RIMS, Kyoto University, 2012.

[6] Chen, Z.-Q., and Kuwae, K., *On subharmonicity for symmetric Markov processes.* J. Math. Soc. Japan **64** (2012), 1181 – 1209.

[7] Deny, J., and Lions, J.-L., *Les espaces du type de Beppo Levi.* Ann. Inst. Fourier (Grenoble) **5** (1953/54), 305 – 370.

[8] Doob, J.L., *Stochastic processes depending on a continuous parameter.* Trans. Amer. Math. Soc. **42** (1937), 107 – 140.

[9] Doob, J.L., *Boundary properties of functions with finite Dirichlet integrals.* Ann. Inst. Fourier (Grenoble) **12** (1962), 573 – 621.

[10] Dynkin, E.B., *Markov Processes.* Vol. I. Springer Verlag, Berlin 1965.

[11] Dynkin, E.B., *Markov Processes.* Vol. II. Springer Verlag, Berlin 1965.

[12] Feller, W., *Zur Theorie der stochastischen Prozesse (Existenz-und Eindeutigkeitssätze).* Math. Ann. **113** (1936), 113 – 160.

[13] Feller, W., *The parabolic differential equations and the associated semi-groups of transformations.* Ann. of Math. (2) **55** (1952), 468 – 519.

[14] Fitzsimmons, P. J., and Kuter, K.E., *Harmonic functions on Walsh's Brownian motion.* Stochastic Processes Appl. **124** (2014), 2228 – 2248.

[15] Föllmer, H., *On Kiyosi Itô's work and its impact.* In : International Congress of Mathematicians, Vol. I. Madrid 2006. European Mathematical Society, Zürich 2007, 109 – 123.

[16] Hunt, G.A., *Markov processes and potentials I.* Illinois J. Math. **1** (1957), 44 – 93.

[17] Hunt, G.A.,*Markov processes and potentials II.* Illinois J. Math. **1** (1957), 316 – 369.

[18] Hunt, G.A., *Markov processes and potentials III.* Illinois J. Math. **2** (1958), 151– 213.

[19] Itô, K., *On stochastic processes. I.* (Infinitely divisible laws of probability), (doctoral thesis) , Jap. J. Math. **18** (1942), 261 – 301.

[20] Itô, K., *Differential equations determining a Markoff process.* (In Japanese). J. Pan-Japan Math. Coll. **1077** (1942), 1352 – 1400. English translation in: Kiyosi Itô. Selected Papers (Eds. D.W. Stroock and S.R.S. Varadhan.) Springer Verlag, Berlin 1986, 42 – 75.

[21] Itô, K., and McKean, H.P., *Diffusion Processes and Their Sample Paths.* Springer Verlag, Berlin 1965.

[22] Kakutani, S., *Two-dimensional Brownian motion and harmonic functions.* Proc. Imp. Acad. Tokyo 20 (1944), 706 - 714.

[23] Kaneko, H., *On (r,p) - capacities for Markov processes.* Osaka J. Math. **23** (1986), 325 – 336.

[24] Lévy, P., *Théorie de l'Addition des Variables Aléatoires.* Gauthier - Villars, Paris 1937.

[25] Ma, Z.-M., and Röckner, M., *Introduction to the theory of (nonsymmetric) Dirichlet forms.* Universitext. Springer Verlag, Berlin 1992.

[26] Malliavin, P., *Implicit functions in finite co-rank on the Wiener space.* In: Stochastic Analysis, Katata / Kyoto 1982. North - Holland Publisher, Amsterdam 1984, 369 – 386.

[27] Silverstein, M.L., *Symmetric Markov Processes.* Lecture Notes in Mathematics, Vol. **426**. Springer Verlag, Berlin 1974.

[28] Silverstein, M.L., *Boundary Theory for Symmetric Markov Processes.* Lecture Notes in Mathematics, Vol. **516**, Springer Verlag, 1976.

[29] Stroock, D.W., and Varadhan, S.R.S., *Introduction. In: Kiyosi Itô. Selected Papers* (Eds. D.W. Stroock and S.R.S. Varadhan.) Springer Verlag, Berlin 1986, vii - xii.

[30] Watanabe, S., *Markov Processes (Comments), and Potential Theory (Comments)*. In: Kôsaku Yosida. Collected Papers. Ed. K. Itô. Springer Verlag, Berlin 1992, 489 – 491 and 563 – 564.

M. Fukushima with S. Watanabe and H. Kunita in London, 1980.

M. Fukushima at study, 2011.

M. Fukushima with N. Jacob and Z.-Q. Chen at SAA-Osaka, 2010.

Chapter 2

Bibliography of Masatoshi Fukushima

Research Papers (English)

[R1] On Feller's kernel and the Dirichlet norm, *Nagoya Math. J.*, **24** (1964) 167–175

[R2] Resolvent kernels on a Martin space, *Proc. Japan Acad.*, **41** (1965), 260–263

[R3] On spectral functions related to birth and death processes, *J. Math. Kyoto Univ.*, **5** (1966), 151–161

[R4] On a class of Markov processes taking values on lines and the central limit theorem, (with M. Hitsuda), *Nagoya Math. J.*, **30** (1967), 47–56.

[R5] A construction of reflecting barrier Brownian motions for bounded domains, *Osaka J. Math.*, **4** (1967), 183–215

[R6] On boundary conditions for multi-dimensional Brownian motions with symmetric resolvent densities, *J. Math. Soc. Japan*, **21** (1969), 58–93

[R7] On Dirichlet spaces and Dirichlet rings, *Proc. Japan Acad.*, **45** (1969), 433–436

[R8] Regular representations of Dirichlet spaces, *Trans. Amer. Math. Soc.*, **155** (1971), 455–473

[R9] Dirichlet spaces and strong Markov processes, *Trans. Amer. Math. Soc.*, **162** (1971), 185–224

[R10] On transition probabilities of symmetric strong Markov processes, *J. Math. Kyoto Univ.*, **12** (1972), 431–450

[R11] On the generation of Markov processes by symmetric forms, in *Proceedings of the Second Japan-USSR Symposium on Probability Theory*, (Kyoto, 1972), 46–79, Lecture Notes in Math., **330**, Springer, Berlin, 1973

[R12] Almost polar sets and an ergodic theorem, *J. Math. Soc. Japan*, **26** (1974), 17–32

[R13] On the spectral distribution of a disordered system and the range of a random walk, *Osaka J. Math.*, **11** (1974), 73–85

[R14] Local property of Dirichlet forms and continuity of sample paths, *Z. Wahrscheinlichkeitstheorie und Verw. Gebiete*, **29** (1974), 1–6

[R15] On an asymptotic property of spectra of a random difference operator (with H Nagai and S. Nakao), *Proc. Japan Acad.*, **51** (1975), 100–102

[R16] Asymptotic properties of the spectral distributions of disordered systems, in *International Symposium on Mathematical Problems in Theoretical Physics*, (Kyoto Univ., Kyoto, 1975), 224–227, Lecture Notes in Phys., **39**, Springer, Berlin, 1975

[R17] On spectra of the Schrödinger operator with a white Gaussian noise potential (with S. Nakao), *Z. Wahrscheinlichkeitstheorie und Verw. Gebiete*, **37** (1976/77), 267–274

[R18] Potential theory of symmetric Markov processes and its applications, in *Proceedings of the Third Japan-USSR Symposium on Probability Theory*, (Tashkent, 1975), 119–133, Lecture Notes in Math., **550**, Springer, Berlin, 1976

[R19] On an L^p-estimate of resolvents of Markov processes, *Publ. Res. Inst. Math. Sci.*, **13** (1977/78), 277–284

[R20] A decomposition of additive functionals of finite energy, *Nagoya Math. J.*, **74** (1979), 137–168

[R21] On additive functionals admitting exceptional sets, *J. Math. Kyoto Univ.*, **19** (1979), 191–202

[R22] Dirichlet spaces and additive functionals of finite energy, in *Proceedings of the International Congress of Mathematicians*, (Helsinki, 1978), 741–747, Acad. Sci. Fennica, Helsinki, 1980

[R23] A generalized stochastic calculus in homogenization, in *Quantum fields–algebras, processes*, (Proc. Sympos., Univ. Bielefeld, Bielefeld, 1978), 41–51, Springer, Vienna, 1980

[R24] On a stochastic calculus related to Dirichlet forms and distorted Brownian motions, in *New stochasitic methods in physics, Phys. Rep.*, **77** (1981), 255–262

[R25] On a representation of local martingale additive functionals of symmetric diffusions, in *Stochastic integrals*, (Proc. Sympos. Durham, 1980), 110–118, Lecture Notes in Math., **851**, Springer, Berlin, 1981

[R26] Capacity and quantum mechanical tunneling (with S. Albeverio, W. Karwowski and L. Streit), *Comm. Math. Phys.*, **81** (1981), 501–513

[R27] On asymptotics of spectra of Schrödinger operators, in *Statistical and physical aspects of Gaussian processes*, (Saint-Flour, 1980), 335–347, Colloq. Internat. CNRS, **307**, CNRS, Paris, 1981

[R28] On absolute continuity of multidimensional symmetrizable diffusions, in *Functional analysis in Markov processes*, (Katata/Kyoto, 1981), 146–176, Lecture Notes in Math., **923**, Springer, Berlin-New York, 1982

[R29] A note on irreducibility and ergodicity of symmetric Markov processes, in *Stochastic processes in quantum theory and statistical physics*, (Marseille, 1981), 200–207, Lecture Notes in Phys., **173**, Springer, Berlin, 1982

[R30] Markov processes and functional analysis, in *Proc. International Math. Conf. Singapore,* 187-202, Eds. L.H. Chen, Y.B. Ng, M.J. Wicks, North Holland, 1982

[R31] Capacitary maximal inequalities and an ergodic theorem, in *Probability theory and mathematical statistics*, (Tbilisi, 1982), 130–136, Lecture Notes in Math., **1021**, Springer, Berlin, 1983

[R32] Basic properties of Brownian motion and a capacity on the Wiener space, *J. Math. Soc. Japan* **36** (1984), 161-176

[R33] On conformal martingale diffusions and pluripolar sets (with M. Okada), *J. Funct. Anal.*, **55** (1984), 377–388.

[R34] A transformation of a symmetric Markov process and the Donsker-Varadhan theory (with M. Takeda), *Osaka J. Math.*, **21** (1984), 311–326

[R35] A Dirichlet form on the Wiener space and properties on Brownian motion, in *Théorie du potentiel*, (Orsay, 1983), 290–300, Lecture Notes in Math., **1096**, Springer, Berlin, 1984

[R36] Energy forms and diffusion processes, in *Mathematics + physics.* Vol. **1**, 65–97, World Sci. Publishing, Singapore, 1985

[R37] (r,p)-capacities for general Markovian semigroups (with H. Kaneko), in *Infinite-dimensional analysis and stochastic processes*, (Bielefeld, 1983), 41–47, Res. Notes in Math., **124**, Pitman, Boston, MA, 1985

[R38] On the continuity of plurisubharmonic functions along conformal diffusions, *Osaka J. Math.*, **23** (1986), 69–75

[R39] A stochastic approach to the minimum principle for the complex Monge-Ampère operator, in *Stochastic processes and their applications*, (Nagoya, 1985), 38–50, Lecture Notes in Math., **1203**, Springer, Berlin, 1986

[R40] Reversibility of solutions to martingale problems (with D. W. Stroock), in *Probability, statistical mechanics, and number theory*, 107–123, Adv. Math. Suppl. Stud., **9**, Academic Press, Orlando, FL, 1986

[R41] On recurrence criteria in the Dirichlet space theory, in *From local times to global geometry, control and physics*, (Coventry, 1984/85), 100–110, Pitman Res. Notes Math. Ser., **150**, Longman Sci. Tech., Harlow, 1986

[R42] On Dirichlet forms with random data—recurrence and homogenization (with S. Nakao and M. Takeda), in *Stochastic processes—mathematics and physics, II*, (Bielefeld, 1985), 87–97, Lecture Notes in Math., **1250**, Springer, Berlin, 1987

[R43] On Dirichlet forms for plurisubharmonic functions (with M. Okada), *Acta Math.*, **159** (1987), 171–213

[R44] A note on capacities in infinite dimensions, in *Probability theory and mathematical statistics*, (Kyoto, 1986), 80–85, Lecture Notes in Math., **1299**, Springer, Berlin, 1988

[R45] On holomorphic diffusions and plurisubharmonic functions, in *Geometry of random motion*, (Ithaca, N.Y., 1987), 65–78, Contemp. Math., **73**, Amer. Math. Soc., Providence, RI, 1988

[R46] On two classes of smooth measures for symmetric Markov processes, in *Stochastic analysis*, (Paris, 1987), 17–27, Lecture Notes in Math., **1322**, Springer, Berlin, 1988

[R47] On the skew product of symmetric diffusion processes (with Y. Oshima), *Forum Math.*, **1** (1989), 103–142

[R48] Capacities on Wiener space: tightness and invariance (with S. Albeverio, W. Hansen, Z.-M. Ma and M. Roĉkner), *C. R. Acad. Sci. Paris Se'r. I Math.*, **312** (1991), 931–935.

[R49] On quasi-supports of smooth measures and closability of pre-Dirichlet forms (with Y. LeJan), *Osaka J. Math.*, **28** (1991), no. 4, 837–845.

[R50] On the closable parts of pre-Dirichlet forms and the fine supports of underlying measures (with K. Sato and S. Taniguchi), *Osaka J. Math.*, **28** (1991), 517–535.

[R51] An invariance result for capacities on Wiener space (with S. Albeverio, W. Hansen, Z.-M. Ma and M. Roc̆kner), *J. Functional Analysis*, **106** (1992), 35–49

[R52] On $(r,2)$-capacities for a class of elliptic pseudo differential operators (with N. Jacob and H. Kaneko), *Math. Ann.*, **293** (1992), 343–348.

[R53] Dirichlet forms, diffusion processes and spectral dimensions for nested fractals, in *Ideas and methods in mathematical analysis, stochastics, and applications*, (Oslo, 1988), 151–161, Cambridge Univ. Press, Cambridge, 1992

[R54] (r,p)-capacities and Hunt processes in infinite dimensions, in *Probability theory and mathematical statistics*, (Kiev, 1991), 96–103, World Sci. Publishing, River Edge, NJ, 1992

[R55] On a spectral analysis for the Sierpiński gasket (with T. Shima), *Potential Analysis*, **1** (1992), 1–35.

[R56] Two topics related to Dirichlet forms: quasi-everywhere convergences and additive functionals, in *Dirichlet forms* (Varenna, 1992), 21–53, G. Dell'Antonio and U. Mosco (Eds), Lecture Notes in Math., **1563**, Springer, Berlin, 1993

[R57] On discontinuity and tail behaviours of the integrated density of states for nested pre-fractals (with T. Shima), *Comm. Math. Phys.*, **163** (1994), 461–471.

[R58] On a strict decomposition of additive functionals for symmetric diffusion processes, *Proc. Japan Acad. Ser. A Math. Sci.*, **70** (1994), 277–281.

[R59] On a decomposition of additive functionals in the strict sense for a symmetric Markov processes, in *Dirichlet forms and stochastic processes,* Z. Ma, M.Roeckner, J.Yan (Eds), Walter de Gruyter, (1995), 155-169

[R60] Reflecting diffusions on Lipschitz domains with cusps—analytic construction and Skorohod representation (with M. Tomisaki), in *Potential theory and degenerate partial differential operators* (Parma), *Potential Analysis*, **4** (1995), 377–408.

[R61] Construction and decomposition of reflecting diffusions on Lipschitz domains with Ho"lder cusps (with M. Tomisaki), *Probab. Theory Related Fields*, **106** (1996), 521–557

[R62] On decomposition of additive functionals of reflecting Brownian motions (with M. Tomisaki), in *Itô's stochastic calculus and probability theory,* N. Ikeda, S. Watanabe, M. Fukushima, H. Kunita (Eds), Springer, Tokyo, 1996, 51–61

[R63] Dirichlet forms, Caccioppoli sets and the Skorohod equation, in *Stochastic Differential and Difference equations,* Csiszar Michaletzky (Eds.), Birkhauser 1997, 56-66

[R64] Distorted Brownian motions and BV functions, in *Trends in probability and related analysis*, (Taipei, 1996), 143–150, N-R.Shieh (Ed), World Sci. Publishing, River Edge, NJ, 1997

[R65] Large deviations and related LIL's for Brownian motions on nested fractals (with T. Shima and M. Takeda), *Osaka J. Math.*, **36** (1999), 497–537.

[R66] On semi-martingale characterizations of functionals of symmetric Markov processes, *Electron. J. Probab.*, **4** (1999), Paper 18, 1-32, http://www.math.washington.edu/~ejpecp

[R67] *BV* functions and distorted Ornstein Uhlenbeck processes over the abstract Wiener space, *J. Funct. Analysis.* **174** (2000), 227–249.

[R68] On limit theorems for Brownian motions on unbounded fractal sets, in *Fractal geometry and stochastics, II* (Greifswald/Koserow, 1998), 227–237, *Progr. Probab.*, S. Graf (Ed), **46**, Birkhäuser, Basel, 2000.

[R69] On Ito's formulae for additive functionals of symmetric diffusion processes, in *Stochastic processes, physics and geometry: new interplays, I* (Leipzig, 1999), 201–211, F. Getsztesy (Ed), *CMS Conf. Proc.*, **28**, Amer. Math. Soc., Providence, RI, 2000.

[R70] On the space of BV functions and a related stochastic calculus in infinite dimensions (with M. Hino), *J. Funct. Analysis*, **183** (2001), 245–268.

[R71] Dynkin games via Dirichlet forms and singular control of one-dimensional diffusions (with M. Taksar), *SIAM J. Control Optim.*, **41** (2002), 682–699

[R72] On Sobolev and capacitary inequalities for contractive Besov spaces over *d*-sets (with T. Uemura), *Potential Analysis*, **18** (2003), 59-77

[R73] Capacitary bounds of measures and ultracontractivity of time changed processes (with T. Uemura), *J. Math. Pures Appl.*, **82** (2003), 553-572

[R74] A note on regular Dirichlet subspaces (with J. Ying), *Proc. Am. Math. Soc.*, **131** (2003) 1607-1610 (2003); erratum ibid. **132** (2004), 1559

[R75] On spectral synthesis for contractive p-norms and Besov spaces (with T. Uemura), *Potential Analysis*, **20** (2004), 195-206

[R76] Function spaces and symmetric Markov processes, in *Stochastic analysis and related topics in Kyoto, Advanced Studies in Pure Mathematics*, **41** (2004), 75-89, Math. Soc. Japan

[R77] Time changes of symmetric diffusions and Feller measures (with P. He and J. Ying), *Ann. Probab.*, **32** (2004) 3138-3166

[R78] On regular Dirichlet subspaces of $H^1(I)$ and associated linear diffusions (with X. Fang and J. Ying), *Osaka J. Math.*, **42** (2005), 27-41

[R79] Poisson point processes attached to symmetric diffusions (with H. Tanaka), *Ann. Inst. Henri Poincaré Probab. Stat.*, **41** (2005), 419-459

[R80] Traces of symmetric Markov processes and their characterizations (with Z.-Q. Chen and J. Ying), *Ann. Probab.*, **34** (2006), 1052-1102

[R81] Entrance law, exit system and Lévy system of time changed processes (with Z.-Q. Chen and J. Ying), *Illinois J. Math.*, **50** (2006), 269-312

[R82] On Feller's boundary problem for Markov processes in weak duality (with Z.-Q. Chen), *J. Func. Anal.*, **252** (2007), 710-733

[R83] Extending Markov processes in weak duality by Poisson point processes of excursions, (with Z.-Q. Chen and J. Ying), in *The Abel Symposium 2005, Stochastic Analysis and Applications, A Symposium in Honor of Kiyosi Itô* (Oslo), F.E. Benth, G.Di Nunno, T. Lindstrom, B. Oksendal, T. Zhang (Eds), 153–196, Springer, 2007

[R84] One-point extensions of Markov processes by darning (with Z.-Q. Chen), *Probab. Th. Rel. Fields*, **141** (2008), 61-112

[R85] Flux and lateral conditions for symmetric Markov processes (with Z.-Q. Chen), *Potential Anal.*, **29** (2008), 241-269

[R86] On unique extension of time changed reflecting Brownian motions (with Z.-Q. Chen), *Ann. Inst. Henri Poincare, Probab. Statist.*, **45** (2009), 864-875

[R87] On extended Dirichlet spaces and the space of BL functions, *Potential Theory and Stochastics in Albac, Aurel Cornea Memorial Volume,* Eds. D. Bakry, L.Beznea, N. Boboc, G. Bucur, M. Roeckner, Theta Foundation, Bucharest, AMS distribution, 2009, 101-110

[R88] From one dimensional diffusions to symmetric Markov processes, *Stoch. Proc. Appl.*, **210** (2010), 590-604

[R89] A localization formula in Dirichlet form theory (with Z.-Q. Chen), *Proc. Amer. Math. Soc.*, **140** (2012), 1815-1822

[R90] Jump-type Hunt processes generated by lower bounded semi-Dirichlet forms (with T. Uemura), *Ann. Probab.*, **40** (2012), 858-889

[R91] On general boundary conditions for one-dimensional diffusions with symmetry, *J. Math. Soc. Japan*, **66** (2014), 289-316

[R92] Chordal Komatu-Loewner equation and Brownian motion with darning in multiply connected domains (with Z.-Q. Chen and S. Rohde), *Trans. Amer. Math. Soc.*, (to appear)

[R93] On Villat's kernels and BMD Schwarz kernels in Komatu-Loewner equations (with H. Kaneko), *Stochastic Analysis and Applications 2014, Springer Proceedings in Mathematics*, (to appear)

[R94] One-point reflection (with Z.-Q. Chen). *Stochastic Processes and their Applications*, (to appear)

[R95] Stochastic Komatu-Loewner evolution and BMD domain constant (with Z.-Q. Chen), arXiv:1410.8257 [math PR] (preprint)

Expository Writing

[E1] Boundary problems of Brownian motions and the Dirichlet spaces, Sūgaku, **20** (1968), 211–221 (in Japanese)

[E2] On the Theory of Markov Processes, *BUTSURI* (The Physical Society of Japan), **25** (1970), 37-41 (in Japanese)

[E3] On Random Spectra, *BUTSURI* (The Physical Society of Japan), **34** (1979), 153-159 (in Japanese)

[E4] On Spectral Analysis on Fractal, Manufacturing and Technology, **48** no. 1 (1996), 55-59 (in Japanese)

[E5] Fractal and Random walk –Toward the function and the shape of nature–, in *Shizennoshikumi to Ningennochie*, Osaka University Press, 1997, 211-222 (in Japanese)

[E6] Decompositions of symmetric diffusion processes and related topics in analysis, *Sugaku Expositions*, **14** (2001), 1–13. AMS

[E7] Stochastic Control Problem and Dynkin's Game Theory, Gien (Organization for Research and Development of Innovative Science and Technology (ORDIST)), Kansai University, **111** (2001), 35-39 (in Japanese)

[E8] Refined solutions of optimal stopping games for symmetric Markov processes, Technology Reports of Kansai University (with K. Menda), 48 (2006), 101-110

[E9] On the works of Kyosi Itô and stochastic analysis, *Japanese J. Math.*, **2** (2007), 45-53

[E10] A brief survey on stochastic calculus in Markov processes, *RIMS Kôkyûroku*, 1672 (2010), 191-197

[E11] On general boundary conditions for one-dimensional diffusions and symmetry, MinnHoKee Lecture at Seoul National University, 2012 June

[E12] Feller's contributions to the one - dimensional diffusion theory and beyond, In: *Selected Works of William Feller, 2 vols.*, Eds. R. Schilling, Z. Vondracek, W. Woyczynski, Springer Verlag, to appear

Seminar on Probability (in Japanese)

[S1] (with Kiyosi Itô and Shinzo Watanabe) On Diffusion processes, Seminar on Probability **3**, 1960

[S2] (with Ken-ichi Sato and Masao Nagasawa) Transformation of Markov processes and boundary problems, Seminar on Probability **16**, 1960

[S3] Dirichlet space and its representations, Seminar on Probability **31**, 1969

[S4] (with Hiroshi Kunita) Studies on Markov processes, Seminar on Probability **40**, 1969

[S5] (with Shintaro Nakao and Sin-ichi Kotani) On Random spectra, Seminar on Probability **45**, 1977

Monographs and Textbooks

[MT1] Dirichlet Forms and Markov Processes (Japanese), KINOKUNIYA Co. Ltd., 1975

[MT2] (with Kazunari Ishii) Natural Phenomena and Stochastic Processes (Japanese), Nippon-Hyoron-Sha Co., Ltd., 1980, (enlarged ed.) 1996

[MT3] Dirichlet Forms and Markov Processes, North-Holland mathematical library **23**, North-Holland, Amsterdam-New York/ Kodansha, Tokyo, 1980

[MT4] (with Yōichi Ōshima and Masayoshi Takeda) Dirichlet Forms and Symmetric Markov Processes, de Gruyter studies in mathematics **19**, de Gruyter, Berlin, 1994

[MT5] Probability Theory (Japanese), SHOKABO Pub. Co., Ltd., 1998

[MT6] (with Masayoshi Takeda) Markov Processes (Japanese), BAIFUKAN Pub. Co., Ltd, 2008

[MT7] (with Yōichi Ōshima and Masayoshi Takeda) Dirichlet Forms and Symmetric Markov Processes, Second revised and extended editions, de Gruyter, Berlin, 2011

[MT8] (with Z.-Q. Chen) Symmetric Markov Processes, Time Change, and Boundary Theory, Princeton University Press, Princeton NJ., 2012

Part 2

Contributions

Chapter 3

Quasi regular Dirichlet forms and the stochastic quantization problem

Sergio Albeverio*†, Zhi-Ming Ma‡ and Michael Röckner§

Dedicated to Masatoshi Fukushima for his 80th birthday

After recalling basic features of the theory of symmetric quasi regular Dirichlet forms we show how by applying it to the stochastic quantization equation, with Gaussian space-time noise, one obtains weak solutions in a large invariant set. Subsequently, we discuss non symmetric quasi regular Dirichlet forms and show in particular by two simple examples in infinite dimensions that infinitesimal invariance, does not imply global invariance. We also present a simple example of non-Markov uniqueness in infinite dimensions.

Contents

1. Introduction

The theory of symmetric Dirichlet forms is the natural extension to higher dimensional state spaces of the classical Feller theory of stochastic processes on the real line. Through ground breaking work by Beurling and Deny (1958-59), [BD58], [BD59], [Den70], [Sil74], [Sil76], and Fukushima (since 1971), [Fu71a], [Fu71b],

*Inst. Appl. Mathematics and HCM, University of Bonn; CERFIM (Locarno); albeverio@uni-bonn.de
†BiBoS (Bielefeld, Bonn)
‡Appl. Math., AMSS, CAS, Beijing
§Mathematics Faculty, University of Bielefeld

[Fu80], [FOT11], [CF12], it has developed into a powerful method of combining analytic and functional analytic, as well as potential theoretic and probabilistic methods to handle the basic construction of stochastic processes under natural conditions about the local characteristics (e.g., drift and diffusion coefficients), avoiding unnecessary regularity conditions.

Such processes arise in a number of applied problems where coefficients can be singular, locally or at infinity. For detailed expositions of the general theory, mainly concentrated on finite dimensional state spaces, see [Fu80], [F82], [FOT11], [Ma95] and, e.g., [Alb03], [BBCK09], [E99], [Fu10], [GriH08], [Hi10], [KaYa05], [KKVW09], [KT91]. For further new developments, e.g. concerning boundary conditions, or non symmetric processes, or processes with jumps, see, e.g., [CF12], [CMR94], [AFH11], [AMU98], [AU00], [Sta99a], [Sta99b], [BiT07], [Bou03], [KS03], [LW99], [Osh04], [Jac01], [JS00], [SU07]. An extension of the theory to processes with infinite dimensional state spaces of Lusin type has been first described indirectly, by a suitable map into a larger locally compact space and consequent reductions to the finite dimensional case, by Fukushima [Fu71a], [Fu71b] and then further developed in [Fu80], [Fu92], [FH01], [FOT11], [ST92]. This has been called "regularization method" (see [Fu80], [AMR93], [ST92], [Alb03]). Another approach on Wiener space has been developed by [BH91] and on certain Banach spaces by [FeyLa91]. Mainly motivated by applications to infinite dimensional processes connected with SPDE's, like those arising in quantum field theory, see, e.g., [Gro76], [AHK74], [AHK76], [AHK77a], [AHK77b], [AHKS77], [Alb97], [N73], [PaWu81], an extension of Dirichlet form theory to infinite dimensional state spaces of more general type, including spaces of distributions, has been developed in [AHK76], [AHK77a], [AHK77b], [AM91], [AM92], [AMR92a], [AMR92b], [AMR92c], [AMR93], [AFHMR92], [AR89a], [AR90b], [Ku92], [Kus82].

This theory is now known as the theory of quasi-regular Dirichlet forms and a systematic exposition of it is in [MR92]. For newer developments see also [AFHMR92], [Aid00], [Alb03], [ARü05], [ARW01], [Fu84], [DG97], [E99], [Ma95], [Kol06].

One main example of applications of the theory has been the construction of processes which arise in certain problems of quantum mechanics with singular potentials (for which we refer to [KaStr14]) and quantum field theory (stochastic quantization equations for invariant measures and Hamiltonians for $P(\varphi)_2$, $\exp(\varphi)_2$, $\sin(\varphi)_2$, φ_3^4 and other models). Recently there arose a renewed interest in SPDE's related to such problems, particularly in connection with Hairer's theory of regularity structures [Hai14] and related work by, a.a., Guminelli, Zambotti, see references in [Hai14].

Since the strength of results obtained by the theory of (quasi-regular) Dirichlet forms in connection with (infinite dimensional) processes has often not been fully realized in the literature, one of the aims of the present paper is to both recall main results and clarify both solved and not yet solved problems.

The structure of the paper is as follows.

In Section 2 we shall recall the basic setting of the theory of symmetric quasi regular Dirichlet forms and associated processes, including invariant measures and ergodicity questions.

Section 3 recalls the setting for classical infinite dimensional Dirichlet forms given by a probability measure, and its relation to infinite dimensional diffusion processes, their generators, associated Kolmogorov equations, and invariant probability measures.

Section 4 is devoted to the study of the stochastic quantization equation for the $P(\varphi)_2$ model, with remarks on related models. We distinguish clearly between the problem in a bounded domain and the global problem. We also recall our result on the ergodicity of the solution.

Section 5 is devoted to the discussion of further developments concerning non symmetric Dirichlet forms. It is pointed out by two examples in infinite dimension that local invariance is indeed weaker than invariance of measures. On the way we also give a new presentation of an example of non-Markov uniqueness (in fact in the language of SPDE), first constructed in [E99].

2. Symmetric quasi regular Dirichlet forms

We first recall some basic notations of the theory of symmetric quasi regular Dirichlet forms, for later use.

Let E be a Hausdorff topological space, m a σ-finite measure on E, and let $\mathcal{B}$ the smallest σ-algebra of subsets of E with respect to which all continuous functions on E are measurable.

Let $\mathcal{E}$ be a symmetric Dirichlet form acting in the real $L^2(m)$-space i.e. $\mathcal{E}$ is a positive, symmetric, bilinear, closed form with domain $D(\mathcal{E})$ dense in $L^2(m)$, and such that $\mathcal{E}(\Phi(u), \Phi(u)) \leq \mathcal{E}(u, u)$, for any $u \in D(\mathcal{E})$, where $\Phi(t) = (0 \vee t) \wedge 1$, $t \in \mathbb{R}$. The latter condition is known to be equivalent with the condition that the associated C_0-contraction semigroup T_t, $t \geq 0$, is submarkovian (i.e. $0 \leq u \leq 1$ m-a.e. implies $0 \leq T_t u \leq 1$ m-a.e., for all $u \in L^2(m)$; association means that $\lim_{t \downarrow 0} \frac{1}{t}(u - T_t u, v)_{L^2(m)} = \mathcal{E}(u, v)$, $\forall u, v \in D(\mathcal{E})$.

These conditions are also equivalent to other conditions expressed either in terms of $\mathcal{E}$ or the associated infinitesimal generator L of T_t, $t \geq 0$, resp. the resolvent, see, e.g., [Alb03] (Theorem 9). One observation which is important for an analytic construction is that for having the above contraction property in terms of Φ it is enough to verify it on a domain where the form is still closable and with Φ replaced by a regularized version Φ_ε of it (see, e.g., Definition 18 and Theorem 9 in [Alb03]). A symmetric Dirichlet form is called quasi-regular if the following holds:

(1) There exists a sequence $(F_k)_{k \in \mathbb{N}}$ of compact subsets of E such that $\bigcup_k D(\mathcal{E})_{F_k}$ is $\mathcal{E}_1^{\frac{1}{2}}$-dense in $D(\mathcal{E})$ (where $D(\mathcal{E})_{F_k} := \{u \in D(\mathcal{E}) | u = 0$ m-a.e. on $E - F_k\}$; $\mathcal{E}_1^{\frac{1}{2}}$

is the norm given by the scalar product in $L^2(m)$ defined by $\mathcal{E}_1$, where

$$\mathcal{E}_1(u,v) := \mathcal{E}(u,v) + (u,v)_{L^2(m)},$$

$(\ ,\ )_{L^2(m)}$ being the scalar product in $L^2(m)$. Such a sequence $(F_k)_{k\in\mathbb{N}}$ is called an $\mathcal{E}$-nest;

(2) there exists an $\mathcal{E}_1^{\frac{1}{2}}$-dense subset of $D(\mathcal{E})$ whose elements have $\mathcal{E}$-quasi-everywhere m-versions (where "quasi" is relative to the potential theory associated with $\mathcal{E}$, i.e. quasi-everywhere (noted q.e.) means with the possible exception of some $\mathcal{E}$-exceptional subset of E, i.e. a subset $N \subset \bigcap_k (E - F_k)$ for some sequence F_k of the above type; this is equivalent with N having $\mathcal{E}_1$-capacity 0, see [Alb03], [MR92]);

(3) there exists $u_n \in D(\mathcal{E})$, $n \in \mathbb{N}$, with $\mathcal{E}$-quasi continuous m-versions $\tilde{u}_n$ and there exists an $\mathcal{E}$-exceptional subset N of E s.t. $\{\tilde{u}_n\}_{n\in\mathbb{N}}$ separates the points of $E - N$ (a real function u on E is called quasi continuous when there exists an $\mathcal{E}$-nest s.t. u is continuous on each F_k, and is defined on a domain in E containing $\bigcup_k F_k$; for this it is sufficient that given $\varepsilon > 0$ there exists an U_ε open in E, s.t. cap $U_\varepsilon < \varepsilon$ and u is continuous on $E - U_\varepsilon$ ([MR92])).

If E is a locally compact separable metric space then $\mathcal{E}$ regular implies $\mathcal{E}$ quasi-regular (but not viceversa, in general).
The relation between these analytic notions and the probabilistic notions goes as follows. A submarkovian semigroup p_t acting in $\mathcal{B}_b(E) \cap L^2(m)$ is associated with a symmetric Dirichlet form $(\mathcal{E}, D(\mathcal{E}))$ on $L^2(m)$ if $p_t u$ is an m-version of $T_t u$, where T_t, $t \geq 0$ is the C_0-contraction semigroup associated with $(\mathcal{E}, D(\mathcal{E}))$.

A stochastic sub-Markov process $\boldsymbol{M} = (\Omega, \mathcal{F}, (\mathcal{F}_t)_{t\geq 0}, (X_t)_{t\geq 0}, (P_z)_{z\in E})$ with state space E and transition semigroup p_t, in the sense that $(p_t u)(z) = E^z u(X_t)$, $\forall z \in E$, $t \geq 0$, $u \in \mathcal{B}_b(E) \cap L^2(m)$, is said to be associated with $\mathcal{E}$ or $(T_t, t \geq 0)$ if p_t is associated with $\mathcal{E}$. p_t is m-symmetric (i.e. a symmetric semigroup in $L^2(m)$) and, in the Markov case, i.e. when $(P_t 1)(z) = 1$, has m has an invariant measure. If $p_t u$ is $\mathcal{E}$-quasi-continuous for all $t > 0$ then $\boldsymbol{M}$ is said to be properly associated with $\mathcal{E}$ (this is a substitute of the strong Feller property in the present case, where E is non necessarily locally compact). It turns out that

(1) $\mathcal{E}$ is quasi-regular iff $\boldsymbol{M}$ is an m-tight special standard process (in the sense of, e.g., [MR92]);

(2) $\mathcal{E}$ is a local (in the sense that $\mathcal{E}(u,v) = 0$ for all $u, v \in D(\mathcal{E})$ with $\mathrm{supp}(|u| \cdot m) \cap \mathrm{supp}(|v| \cdot m) = \varnothing$ and $\mathrm{supp}(|u| \cdot m), \mathrm{supp}(|v| \cdot m)$ compact)[a] and quasi regular iff $\boldsymbol{M}$ is an m-tight special standard process and it is a diffusion i.e. $P_z\{t \to X_t$ continuous on $[0, \zeta)\} = 1$ for all $z \in E$ (for some random variable ζ, with values in $[0, +\infty]$, the life time of X).

[a]These supports are well defined since E can be assumed, without loss of generality, to be a Lusin space, see [MR92] (p. 101).

A rough description of m-tight special standard processes is that they are right processes "concentrated on compacts", with almost-sure left limits and are almost-surely quasi left continuous. Right processes are adapted, strong Markov, normal (i.e. $P_z(X_0 = z) = 1$, for all $z = E \cup \{\Delta\}$, $\{\Delta\}$ being the 1-point set describing the "cemetery" for the process), and have right continuous paths (see [MR92], [Alb03] for details).
The connection of quasi regular Dirichlet forms with stochastic analysis goes beyond potential theory inasmuch as it permits an extension to the infinite dimensional case of Fukushima's decomposition.

Theorem 2.1. *Let $\boldsymbol{M}$ be a right process associated with a quasi regular symmetric Dirichlet form $(\mathcal{E}, D(\mathcal{E}))$ on a Hausdorff topological space E. If $u \in D(\mathcal{E})$ then there exists a martingale additive functional of finite energy $M^{[u]}$ and a continuous additive functional of zero energy N s.t. for any quasi-continuous version $\tilde{u}$ of u we have*

$$\tilde{u}(X_t) = \tilde{u}(X_0) + M_t^{[u]} + N_t^{[u]}.$$

For the proof of this theorem see [Fu80], [FOT11] in the locally compact case and [AMR92c], [MR92] (Chapter VI, Theorem 2.5) in the general case.
For the notions of martingale resp. continuous zero-energy additive functional see [MR92], [Alb03] (roughly $E^z\left(\left(M_t^{[u]}\right)^2\right) < \infty$, $E^z\left(M_t^{[u]}\right) = 0$, $\mathcal{E}$-q.e. $z \in E$, $\forall t \geq 0$; $M_t^{[u]}$ is a martingale under P_x for any $x \in E - N$, with N a properly exceptional set; $\lim_{t \downarrow 0} \frac{1}{2t} E^m\left(\left(N_t^{[u]}\right)^2\right) = 0$, $E^z\left(\left|N_t^{[u]}\right|\right) < \infty$ q.e. $z \in E$).
Let us conclude this section with a short discussion of problems of stochastic dynamics.
Given a probability measure μ on some space E one can ask the question whether there exists a Markov process $\boldsymbol{M}$ with Markov transition semigroup p_t, $t \geq 0$, μ-symmetric (in the sense that the adjoint p_t^* of p_t in $L^2(\mu)$ coincides with p_t, for all $t \geq 0$). In particular, then μ is p_t invariant i.e.

$$\int p_t u \, \mathrm{d}\mu = \int u \, \mathrm{d}\mu, \quad \forall u \in L^2(\mu). \tag{2.1}$$

One calls then p_t the "stochastic dynamics" associated with μ.
A probability measure μ is said to be infinitesimal invariant under a C_0-semigroup T_t, $t \geq 0$, if $\int Lu \, \mathrm{d}\mu = 0$ for all u in a subset of the domain $D(L)$ of L, which is dense in $L^2(E; \mu)$, where L is the infinitesimal generator of T_t.

In general, however, infinitesimal invariance of μ does not imply its invariance for the corresponding semigroup, i.e. does not imply (2.1). There are many counterexamples known (see e.g. [BRSt00] or [BKRS14]). In Section 5 we shall give an explicit (Gaussian) counterexample with state space being a Hilbert space, in which even the underlying martingale problem is well-posed.
For references on invariant and infinitesimally invariant measures see, e.g., [ABR99],

[AFe04], [AFe08], [AKR97a], [AKR97b], [AKR98], [AKR02], [ARW01], [ARü02], [MR10], [MZ10], [ARZ93b], [BoRöZh00], [DPZ92], [E99], [BKRS14].
To the above "inverse problem" there exists a direct problem: given a (Markov) stochastic process $\boldsymbol{M}$, find (if possible) a probability measure μ s.t. μ is is an invariant measure for μ in the sense that (2.1) holds.

We shall see in Section 4 how these problems are solved in the case of the processes associated with a special class of Dirichlet forms, namely classical Dirichlet forms (in the sense of [AR90a], [MR92], [Alb03], [AKR02]).

A further general question related with stochastic dynamics concerns the uniqueness of the invariant measure for a given stochastic process $\boldsymbol{M}$, see, e.g., [DPZ92], [PeZ07], [Eche82] for results on this problem.
Let us now briefly recall the relation of Dirichlet forms with martingale problems, related uniqueness problems, and large time asymptotic behaviour of associated semigroups.
Let us consider a topological space E, a set $\mathcal{A}$ of real-valued functions on E and a linear operator L defined on $\mathcal{A}$. A probability measure P on a probability space Ω consisting of continuous paths on E with possible finite life time is said to be a solution of the martingale problem for $(L, \mathcal{A})$ relative to a coordinate stochastic process X_t, $t \geq 0$ on $E \cup \{\Delta\}$ (Δ being the terminal point and $X_t(\omega) = \omega(t)$, $\omega \in \Omega$, with some initial distribution μ) if for all $f \in \mathcal{A}$ and $t \geq 0$ we have that $f(X_t)$ and $Lf(X_t)$ are in $L^1(P)$, the function $(\omega, s) \to (Lf)(X_s(\omega))$ is in $L^1(P \otimes \mathrm{d}s)$ for all $0 \leq s \leq t$, and

$$M_t^{[f]} := f(X_t) - \int_0^t (Lf)(X_s)\mathrm{d}s$$

is a martingale with the given initial probability measure μ with respect to the filtration generated by X_s, for all $0 \leq t < \infty$.
For a general discussion of martingale problems see [StV06], [EK86], [MR92], [AR89a], [AR90b], [ARZ93b], [E99], [BhK93].

The Markov uniqueness problem concerns the question whether in the class of Markov processes X_t, $t \geq 0$ the solution of the martingale problem with generator L acting in $L^2(\mu)$ is already determined by restricting the generator to a subset $\mathcal{A}$ strictly contained in the $L^2(\mu)$-domain of L, but still dense in $L^2(\mu)$. See [AHK82], [E99], [AR90b], [BK95], [AuR02], [SS03], [Tak92]. An example that Markov uniqueness can fail to hold is given in Section 5, which is originally from [E99], but which is presented here in an updated form.
The strong uniqueness problem concerns the question whether L is already essentially self-adjoint on $\mathcal{A}$ in $L^2(\mu)$, see [AKR02], [AKR92], [AKR93], [LR98].
The large time asymptotics of processes associated with Dirichlet forms can be discussed in terms of properties of the associated Dirichlet forms.
We recall some basic results in this direction.
Let (T_t), $t \geq 0$, be a submarkovian C_0-contraction semigroup in $L^2(m)$. T_t is called irreducible if $T_t(uf) = uT_tf$, for all $f \in L^\infty(m)$ and all $t > 0$ implies $u =$ constant

m-a.e.. A submarkovian C_0-contraction semigroup T_t in $L^2(m)$ is called $L^2(m)$-ergodic if $T_t u \to \int u\,dm$ as $t \to \infty$ in $L^2(m)$, for all $u \in L^2(m)$.

Let $\mathcal{E}$ be a symmetric Dirichlet form associated with an m-symmetric submarkovian C_0-contraction semigroup T_t in $L^2(m)$. Then the statement that T_t is $L^2(m)$-ergodic is equivalent with T_t being irreducible and is also equivalent with $\mathcal{E}$ irreducible. These properties are also equivalent with the statement that $u \in D(L)$ and $Lu = 0$ imply $u = \text{const}$ m-a.e., where L is the infinitesimal generator of T_t.
For the proof of these results see [AKR97a] (and also [AHK76], [AKR97b]).
For the connection of the above properties with properties of the right process associated to T_t resp. $\mathcal{E}$ via Dirichlet forms theory see Section 4.3 (and, for related problems, in the case of non local Dirichlet forms, e.g., [ARü02], [ARü05]).

3. Classical Dirichlet forms on Banach spaces and weak solutions to SDE

Let E be a separable real Banach space with dual E' and dualization ${}_{E'}\langle\ ,\ \rangle_E$. Let $\mathcal{B}(E)$ denote its Borel σ-algebra and let μ be a finite positive measure on $(E, \mathcal{B}(E))$ with $\text{supp}[\mu] = E$. Define for $K \subset E'$ the linear spaces

$$\mathcal{F}C_b^\infty(K) := \{f(l_1, \ldots, l_m) | m \in \mathbb{N}, f \in C_b^\infty(\mathbb{R}^m), l_1, \ldots, l_m \in K\}. \tag{3.1}$$

Set $\mathcal{F}C_b^\infty := \mathcal{F}C_b^\infty(E')$. Compared with the finite dimensional case we have that E, μ, $\mathcal{F}C_b^\infty(K)$ replace $\mathbb{R}^d$, $\mathrm{d}x$, $C_0^\infty(\mathbb{R}^d)$ respectively, where $\mathrm{d}x$ = Lebesgue measure. We want to define a gradient ∇. To this end fix $u = f(l_1, \ldots, l_m) \in \mathcal{F}C_b^\infty$, $z \in E$ and define for $k \in E$, $s \in \mathbb{R}$:

$$\frac{\partial u}{\partial k}(z) := \frac{\mathrm{d}}{\mathrm{d}s} u(z + sk)_{s=0} = \sum_{i=0}^{m} \frac{\partial f}{\partial x_i}(l_1(z), \ldots, l_m(z))\ {}_{E'}\langle l_i, k\rangle_E\,. \tag{3.2}$$

Furthermore, we assume that we are given a "tangent space" H to E at each point, in the sense that H is a separable real Hilbert space such that $H \subset E$ continuously and densely. Thus, identifying H with its dual H' we have

$$E' \subset H \subset E \text{ continuously and densely.} \tag{3.3}$$

We define $\nabla u(z)$ to be the unique element in H such that

$$\langle \nabla u(z), h\rangle_H = \frac{\partial u}{\partial h}(z) \text{ for all } h \in H (\subset E).$$

Now it is possible to define a positive definite symmetric bilinear form (henceforth briefly called *form*) on (real) $L^2(E;\mu)$ by

$$\mathcal{E}_\mu^0(u, v) := \frac{1}{2}\int_E \langle \nabla u, \nabla v\rangle_H \mathrm{d}\mu; \quad u, v \in \mathcal{F}C_b^\infty, \tag{3.4}$$

which is densely defined, since $\mathcal{F}C_b^\infty$ is dense in $L^2(E;\mu)$ (cf. [MR92, Chap. 4, Sect.b)]). An element $k \in E$ is called *well-μ-admissible* if there exists $\beta_k^\mu \in L^2(E;\mu)$

such that

$$\int_E u \, d\mu = -\int_E u \, \beta_k^\mu \, d\mu \text{ for all } u \in \mathcal{F}C_b^\infty. \tag{3.5}$$

Assume:

$$\begin{aligned}&\text{There exists a linear subspace } K \text{ of } E'(\subset H \subset E), \\ &\text{point separating on } E \text{ and consisting of well-}\mu\text{-admissible elements in } E,\end{aligned} \tag{3.6}$$

which we fix from now on in this section. Then it is easy to see that the form $(\mathcal{E}_\mu^0, \mathcal{F}C_b^\infty)$ defined in (3.4) is closable on $L^2(E;\mu)$ and that its closure $(\mathcal{E}_\mu, D(\mathcal{E}_\mu))$ is a Dirichlet form on $L^2(E;\mu)$ (cf. [AR90a], [MR92, Chap.II, Subsection 3b]). We also denote the closure of ∇ with domain $D(\mathcal{E}_\mu)$ by ∇, hence

$$\mathcal{E}_\mu(u,v) = \frac{1}{2}\int_E \langle \nabla u, \nabla v\rangle_H \, d\mu; \quad u, v \in D(\mathcal{E}_\mu). \tag{3.7}$$

The Dirichlet form in (3.7) is called *classical (gradient) Dirichlet form* given by μ (see [AR90a], [MR92]). Let L_μ with domain $D(L_\mu)$ be its generator (i.e. $\mathcal{E}_\mu(u,v) = (u, (-L_\mu)v)$, with $(\ ,\ )$ the scalar product in $L^2(E;\mu)$, $u \in D(\mathcal{E}_\mu)$, $v \in D(L_\mu)$) (cf. [MR92, Chap. I]). $(L_\mu, D(L_\mu))$ is a *Dirichlet operator*, i.e., $(e^{tL_\mu})_{t\geq 0}$ is sub-Markovian. It is immediate that if $u = f(l_1, \ldots, l_m) \in \mathcal{F}C_b^\infty(K)$ and $K_0 \subset K$ is an orthonormal basis of H having $l_1, \ldots, l_m$ in its linear span, then $u \in D(L_\mu)$ and

$$L_\mu u = \frac{1}{2}\sum_{k\in K_0}\left(\frac{\partial}{\partial k}\left(\frac{\partial u}{\partial k}\right) + \beta_k^\mu \frac{\partial u}{\partial k}\right). \tag{3.8}$$

Theorem 3.1. *The Dirichlet form* $(\mathcal{E}_\mu, D(\mathcal{E}_\mu))$ *defined in* (3.7) *is local and quasi-regular.*

Proof. See [MR92, Chap. 5, Example 1.12 (ii)] for the locality and for the quasi-regularity see [RS92] and also [MR92, Chap. 4, Sect. 4b]. □

Hence by Section 2 there exists a (Markov) diffusion process $\mathbf{M}_\mu = (\Omega, \mathcal{F}, (\mathcal{F}_t), (X_t), (P_z)_{z\in E})$ *properly associated with* the Dirichlet form $(\mathcal{E}_\mu, D(\mathcal{E}_\mu))$, i.e., for all $u \in L^2(E;\mu)$, $t > 0$,

$$\int u(X_t) \, dP_z = T_t u(z) \text{ for } \mu - \text{a.e. } z \in E \tag{3.9}$$

where $T_t u := \exp(tL_\mu)u$, and the function on the left hand side of (3.9) has an $\mathcal{E}_\mu$-quasi continuous version. $\mathbf{M}_\mu$ is easily seen to be conservative (i.e., has infinite lifetime) and to have μ as an invariant measure. By [MR92, Chap. 4, Sect. 4b] it follows that there is a point separating countable $\mathbb{Q}$-vector space $D \subset \mathcal{F}C_b^\infty(K)$. It is easy to see that there exists an ONB $K_0 \subset K$ of H such that the linear span of K_0 separates the points in E. Let us make for what follows a fix choice of such sets K_0 and D. Then as an immediate consequence of (a special case of) the Fukushima decomposition (see Sect.2) (in particular we use the version of [MR92], Chapter VI,

Theorem 2.5), we obtain the following result concerning the martingale problem for (L_μ, D):

Theorem 3.2. *There exists a set $S \in \mathcal{B}(E)$ such that $E\backslash S$ is* ***properly*** $\mathcal{E}_\mu$***-exceptional****, i.e.* $\mu(E\backslash S) = 0$ *and* $P_z[X_t \in S\ \forall t \geq 0] = 1$ for all $z \in S$, *and for all $u \in D$ (see above)*

$$u(X_t) - \int_0^t L_\mu u(X_s)\,\mathrm{d}s, \quad t \geq 0,$$

is an $(\mathcal{F}_t)$-martingale under P_z for all $z \in S$.

As a consequence of Theorem 3.2 and the results in [AR91] we obtain that $\mathbf{M}_\mu$ yields weak solutions to the corresponding stochastic differential equation (SDE) on E. More precisely, we have:

Theorem 3.3. *Let S be as in Theorem 3.2 and assume that there exists a $\mathcal{B}(E)/\mathcal{B}(E)$-measurable map $\beta^\mu : E \longrightarrow E$ such that*

(i) ${}_{E'}\langle k, \beta^\mu \rangle_E = \beta^\mu_k$ *μ-a.e for each $k \in K$;*
(ii) $\int_E ||\beta^\mu||_E^2\,\mathrm{d}\mu < \infty$.

Then there exists an E-valued $(\mathcal{F}_t)_{t\geq 0}$-Wiener process W_t, $t \geq 0$, on $(\Omega, \mathcal{F}, P)$ *such that $({}_{E'}\langle k, W_t \rangle_E)_{k\in K_0}$ is a cylindrical Wiener process in H and for every $z \in S$*

$$X_t = z + \frac{1}{2}\int_0^t \beta^\mu(X_s)\,\mathrm{d}s + W_t, \quad t \geq 0,$$

P_z-a.s. (where the integral is an E-valued Bochner integral).

Proof. [AR91, Theorem 6.10]. □

4. Applications to stochastic quantization in finite and infinite volume

Quantum field theory has its origin in physics (by work by Born, Dirac, Heisenberg, Jordan, Pauli, a.a., see, e.g., [Jo95]) as an attempt to quantize the classical theory of relativistic fields in a similar way as non relativistic quantum mechanics is the quantization of classical (non relativistic) field theory, see, e.g., [ASe14].
In contrast to non relativistic quantum mechanics, a mathematical sound construction of the dynamics encountered difficulties which are still not overcome. A systematic attempt to perform such a construction was initiated in the late 60's and culminated in the early 70's with the construction of models describing interactions, in the case of an idealized d-dimensional space-time world (for $d = 1, 2, 3$), see [Si74], [GJ81], [AFHKL86]. For the physically most interesting case $d = 4$ only partial results have been obtained, see [AGW97], [AGY05].

In these approaches a non Gaussian probability measure μ on a distributional space (e.g. $\mathcal{S}'(\mathbb{R}^d)$) is constructed having the heuristic expression:

$$\mu^*(\mathrm{d}\varphi) = Z^{-1} e^{-\int_{\mathbb{R}^d} v(\varphi(x))\mathrm{d}x} \mu_0^*(\mathrm{d}\varphi), \quad \text{with } Z = \int_{\mathcal{S}'(\mathbb{R}^d)} e^{-\int_{\mathbb{R}^d} v(\varphi(x))\mathrm{d}x} \mu_0^*(\mathrm{d}\varphi), \tag{4.1}$$

$\varphi \in \mathcal{S}'(\mathbb{R}^d)$ being a symbol for a space-time process connected with the physical quantum field in d-space-time dimensions. v is a real-valued function, describing the interaction. μ_0^* is a Gaussian random field, with mean 0 and covariance $E(\varphi(x)\varphi(y)) = (-\Delta + m^2)^{-1}(x,y)$, the fundamental solution or Green function of $-\Delta + m^2$, m being a positive constant (in physical terms m is the mass of particles described by the free field, below we shall always take for simplicity $m = 1$).
For $v \equiv 0$ we have heuristically $\mu^* = \mu_0^*$. A particularly well studied case for $v \not\equiv 0$ is where $v(y) = P(y)$, $y \in \mathbb{R}^d$, with P a polynomial with a positive even highest order term. μ^* is then called "Euclidean measure for the $P(\varphi)_d$-field" (for more general v μ^* yields the "Euclidean measure for the $v(\varphi)_d$-field").
The rigorous construction of μ^* uses tools of probability theory and statistical mechanics, see [Si74], [GRS75], [GJ81], [AFHKL86].
The analogy of the heuristic formula (4.1) with the (canonical, Gibbs) equilibrium measure in statistical mechanics makes it natural to ask, both from physics and mathematics, whether it is possible to find a Markov diffusion process $(X_t,\ t \geq 0)$ such that (4.1) appears as equilibrium measure (invariant measure) for this process. A formal computation can be performed to see that a solution X_t of the following heuristic stochastic quantization equation

$$\mathrm{d}X_t = (\Delta - 1)X_t\,\mathrm{d}t - v'(X_t)\mathrm{d}t + \mathrm{d}W_t,\ t \geq 0, \tag{4.2}$$

would have μ^* as an (heuristic) invariant measure, where $\mathrm{d}W_t$ stands for Gaussian white noise in all variables (t,x), $x \in \mathbb{R}^d$. In the physical interpretation, t here is a "computer time", whereas the Euclidean space-time-variables $x \in \mathbb{R}^d$ appear in $X_t = X_t(x)$. Heuristically, $X_0(x) = \varphi(x)$, with φ as in (4.1). $\Delta = \Delta_x$ is the Laplacian in $\mathbb{R}^d$. The physical reason for asking the above question is to exploit the dynamics of solutions of such an equation to perform Monte-Carlo simulations of physically relevant averages with respect to the equilibrium measure μ^* (see [PaWu81], and, e.g., [Mi89]). The mathematical interest arises because of the "typicality" of an equation of the form (4.2) for handling SPDE's with singular noise. Due to the singular character of the measure μ^* (whenever it exists!) one expects (for $d \geq 2$) troubles in giving a meaning to the term $v'(X_t)$ in (4.2). In this section we shall see that for $d = 2$ this programme can be rigorously achieved (the case $d = 1$ is simpler, since for $d = 1$, $x \to X_t(x)$, $x \in \mathbb{R}$, is continuous). For our discussion we shall separate 2 cases: the "finite volume case" (Section 4.1), where $\mathbb{R}^2$ is replaced by a rectangle Λ and the "infinite volume case" (Section 4.2), where the "space cut-off" Λ is eliminated and the problem is considered in the whole space $\mathbb{R}^2$.

4.1. *Finite volume*

Let Λ be an open rectangle in $\mathbb{R}^2$. Let $(-\Delta+1)_N$ be the generator of the following quadratic form on $L^2(\Lambda; \mathrm{d}x)$: $(u,v) \longrightarrow \int_\Lambda \langle \nabla u, \nabla v\rangle_{\mathbb{R}^d}\ \mathrm{d}x + \int_\Lambda u\ v\ \mathrm{d}x$ with $u,v \in \{g \in L^2(\Lambda;\mathrm{d}x) | \nabla g \in L^2(\Lambda;\mathrm{d}x)\}$ (where ∇ is in the sense of distributions, N reminds us to "Neumann boundary conditions"). Let $\{e_n | n \in \mathbb{N}\} \subset C^\infty(\bar{\Lambda})$ be the (orthonormal) eigenbasis of $(-\Delta+1)_N$ and $\{\lambda_n| \in \mathbb{N}\ \ (\subset]0,\infty[)$ the corresponding eigenvalues (cf. [RS75, p. 266]). Define for $\alpha \in \mathbb{R}$

$$H_\alpha := \{u \in L^2(\Lambda;\mathrm{d}x) | \sum_{n=1}^{\infty} \lambda_n^\alpha\ \langle u, e_n\rangle^2_{L^2(\Lambda;\mathrm{d}x)} < \infty\},$$

equipped with the inner product

$$\langle u,v\rangle_{H_\alpha} := \sum_{n=1}^{\infty} \lambda_n^\alpha\ \langle u, e_n\rangle_{L^2(\Lambda;\mathrm{d}x)} \langle v, e_n\rangle_{L^2(\Lambda;\mathrm{d}x)}.$$

Clearly, we have that

$$H_\alpha = \begin{cases} \text{completion of } C^\infty(\bar{\Lambda}) \text{ w.r.t. } \|\ \|_{H^\alpha} \text{ if } \alpha < 0 \\ \text{completion of } C_0^\infty(\Lambda) \text{ w.r.t. } \|\ \|_{H_\alpha} \text{ if } \alpha \geq 0 \end{cases} \tag{4.3}$$

(cf. [LM72, p. 79] for the latter).
To get into the framework of Section 3 we chose

$$H := L^2(\Lambda;\mathrm{d}x), \quad E := H_{-\delta},\ \delta > 0.$$

Then

$$E'(\ = H_\delta) \quad \subset H \quad \subset E. \tag{4.4}$$

Remark 4.1. In (4.4) we have realized the dual of $H_{-\delta}$ as H_δ using as usual the chain

$$H_\alpha \quad \subset L^2(\Lambda;\mathrm{d}x) \quad \subset H_{-\alpha}, \qquad \alpha \geq 0.$$

Fix $\delta > 0$. Since $\sum_{n=1}^\infty \lambda_n^{-1-\delta} < \infty$ (cf. [RS75]), we have, applying [Y89, Theorem 3.2] (i.e., the Gross-Minlos-Sazonov theorem) with $H := L^2(\Lambda;\mathrm{d}x)$, $||\cdot|| := ||\cdot||_{H_{-\delta}}$, $A_1 := (-\Delta+1)_N^{-\frac{\delta}{2}}$, and $A_2 := (-\Delta+1)_N^{-\frac{1}{2}}$, that there exists a unique mean zero Gaussian probability measure μ on $E := H_{-\delta}$ (called *free field* on Λ with Neumann boundary conditions; see [N73]) such that

$$\int_E {}_{E'}\langle l, z\rangle^2{}_E\ \mu(\mathrm{d}z) = ||l||^2_{H_{-1}} \text{ for all } l \in E' = H_\delta. \tag{4.5}$$

Clearly, supp $\mu = E$. For $h \in H_{-1}$ we define $X_h \in L^2(E;\mu)$ by

$$X_h := \lim_{n\to\infty}\ {}_{E'}\langle k_n, \cdot\rangle_E \quad \text{in } L^2(E;\mu), \tag{4.6}$$

where $(k_n)_{n\in\mathbb{N}}$ is any sequence in E' such that $k_n \underset{n\to\infty}{\longrightarrow} h$ in H_{-1}.

Let $h \in \mathrm{Dom}((-\Delta+1)_N)$ $(\subset L^2(\Lambda;\mathrm{d}x) \subset E)$, then

$$_{E'}\langle k,h\rangle_E = \langle k,h\rangle_{L^2(\Lambda,\mathrm{d}x)} = \langle k,(-\Delta+1)_N h\rangle_{H_{-1}}$$

for each $k \in E'$. Hence by [AR90a, Proposition 5.5] (4.5), and (4.6), h is well-μ-admissible and $\beta_h^\mu = X_{(-\Delta+1)_N h}$ (see (3.5)). Let K be the linear span of $\{e_n | n \in \mathbb{N}\}$. Below we shall consider the gradient Dirichlet form $(\mathcal{E}_{\bar{\mu}}, D(\mathcal{E}_{\bar{\mu}}))$ on $L^2(E;\bar{\mu})$ as introduced in Section 3, where $\bar{\mu}$ is the $P(\varphi)_2$-quantum field measure in the finite volume Λ with P being a polynomial, see below just before Proposition 4.1, for its mathematical description.

To this end we first have to introduce the so-called Wick powers $: z^n :$, $n \in \mathbb{N}$, which are renormalized powers of the Schwartz-Sobolev distribution $z \in E = H_{-\delta}$. Let $h \in L^2(\Lambda;\mathrm{d}x)$, $n \in \mathbb{N}$, and define $: z^n : (h)$ as a limit in $L^p(E;\mu)$, $p \in [1,\infty[$, as follows (cf., e.g., [GJ81, Sect. 8.5]): fix $n \in \mathbb{N}$ and let $H_n(t)$, $t \in \mathbb{R}$, be the nth Hermite polinomial, i.e., $H_n(t) := \sum_{m=0}^{[n/2]}(-1)^m \alpha_{nm} t^{n-2m}$, with $\alpha_{nm} := n!/[(n-2m)!2^m m!]$. Let $d \in C_0^\infty(\mathbb{R}^2)$, $d \geq 0$, $\int d(x)\mathrm{d}x = 1$, and $d(x) = d(-x)$ for each $x \in \mathbb{R}^2$. Define for $k \in \mathbb{N}$, $d_{k,x}(y) := 2^{2k} d(2^k(x-y))$; $x, y \in \mathbb{R}^2$. Let $z_k(x) := {}_{E'}\langle d_{k,x}, z\rangle_E$, $z \in E, x \in \Lambda$, and set

$$: z_k^n : (x) := c_k(x)^{n/2} H_n(c_k(x)^{-1/2} z_k(x)), \tag{4.7}$$

where $c_k(x) := \int z_k(x)^2 \mu(\mathrm{d}z)$. Then it is known that

$$: z_k^n : (h) := \int : z_k^n : (x) h(x)\mathrm{d}x \underset{k\to\infty}{\longrightarrow} : z^n : (h)$$

both in every $L^p(E;\mu)$, $+\infty > p \geq 1$, and for μ-a.e. $z \in E$ (cf.,e.g., [R86, Sect. 3] for the latter and [Si74] for the former). The map $z \mapsto \limsup_{k\to\infty} : z_k^n : (h)$ is then a μ-version of $: z^n : (h)$. From now on $: z^n : (h)$ shall denote this particular version. Since $H_n(s+t) = \sum_{m=0}^n \binom{n}{m} H_m(s) t^{n-m}$, $s,t \in \mathbb{R}$, we also have that if

$$z \in M := \{z \in E | \limsup_{k\to\infty} : z_k^n (h) : = \lim_{k\to\infty} : z_k^n : (h) (\in \mathbb{R})\}$$

then $\mu(M) = 1$, $z + k \in M$ for all $k \in K$ and

$$: (z+k)^n : (h) = \sum_{m=0}^n \binom{n}{m} : z^m : (k^{n-m} h) \tag{4.8}$$

(cf. [R86, Sect. 3] for details).

Now fix $N \in \mathbb{N}, a_n \in \mathbb{R}$, $0 \leq n \leq 2N$ with $a_{2N} > 0$ and define

$$V(z) :=: P(z) : (1_\Lambda) := \sum_{n=0}^{2N} a_n : z^n : (1_\Lambda), \quad z \in E,$$

where 1_Λ denotes the indicator function of Λ. Let

$$\varphi := \exp(-\frac{1}{2}V). \tag{4.9}$$

Then $\varphi > 0$ μ-a.e. and $\varphi \in L^p(E;\mu)$ for all $p \in [1,\infty[$ (cf., e.g., [Si74, Sect. 5.2] or [GJ81, Sect. 8.6]). Set

$$\bar{\mu} := \varphi^2 \cdot \mu.$$

Proposition 4.1. *Each $k \in K$ is well-μ-admissible and*

$$\beta_k^{\bar{\mu}} = -X_{[(-\Delta+1)_N k]} - \sum_{n=1}^{2N} n a_n : z^{n-1} : (k) \tag{4.10}$$

(cf. (3.5) above for the definition of $\beta_k^{\bar{\mu}}$).

Proof. [RZ92, Proposition 7.2]. □

It now follows that Theorem 3.1 applies to the corresponding classical (gradient) Dirichlet form $(\mathcal{E}_{\bar{\mu}}, D(\mathcal{E}_{\bar{\mu}}))$ introduced in Section 3.

So, let $\mathbf{M}_{\bar{\mu}} = (\Omega, \mathcal{F}, (\mathcal{F}_t), (X_t)_{t\geq 0}, (P_z)_{z\in E})$ be the corresponding (Markov) diffusion process as in Section 3. Then Theorem 3.2 applies, i.e. we have solved the martingale problem (in the sense of Theorem 3.2) for the corresponding operator $L_{\bar{\mu}}$ given by (3.8) with $\beta_k^{\bar{\mu}}$ replacing β_k^{μ}. Finally, taking δ for $E = H_{-\delta}$ large enough, there exists $\beta^{\bar{\mu}} : E \to E$, $\mathcal{B}(E)/\mathcal{B}(E)$ measurable and satisfying conditions *(i)* and *(ii)* in Theorem 3.3 (see [AR91, Proposition 6.9]). By construction and (4.10) we have that

$$\beta^{\bar{\mu}}(z) = -(-\Delta+1)_N\, z - \, : P'(z) :, \ z \in E. \tag{4.11}$$

Hence Theorem 3.3 implies the existence of a set $S \in \mathcal{B}(E)$ such that $\bar{\mu}(E\backslash S) = 0$ and S is invariant under the (Markov) diffusion process $(X_t)_{t\geq 0}$, i.e. $P_z[X_t \in S\ \forall t \geq 0] = 1$ for all $z \in S$, and for every $z \in S$

$$X_t = z \ - \ \frac{1}{2}\int_0^t [(-\Delta+1)_N X_s + \ : P'(X_s) :]\mathrm{d}s \ + \ W_t, \quad t \geq 0, \tag{4.12}$$

P_z-a.e., for some $L^2(\Lambda^2; \mathrm{d}x)$-cylindrical $(\mathcal{F}_t)$-Wiener process on $(\Omega, \mathcal{F}, P_z)$, i.e. we have a Markov weak solution to the SDE (4.12) ("weak" in the probabilistic sense). The solution to (4.12) is usually called *stochastic quantization process* for the $P(\varphi)_2$-quantum field with Neumann boundary condition in the finite volume Λ (and (4.12) is a rigorous "finite volume" version of the heuristic stochastic quantization equation (4.2)).

4.2. *Infinite volume*

For $n \in \mathbb{Z}$ define the space $\mathcal{S}_n$ to be the completion of $C_0^\infty(\mathbb{R}^2)$(= all compactly supported smooth functions on $\mathbb{R}^2$) with respect to the norm

$$\|k\|_n := \Big[\sum_{|\underline{m}|\leq n} \int_{\mathbb{R}} (1+|x|^2)^n \Big| \Big(\frac{\partial^{m_1}}{\partial x_1^{m_1}} \frac{\partial^{m_2}}{\partial x_2^{m_2}}\Big) k(x)\Big|^2 \mathrm{d}x\Big],$$

where $\underline{m} := (m_1, m_2) \in (\mathbb{Z}_+)^2$. To get into the framework of Section 3 we choose

$$K := C_0^\infty(\mathbb{R}^2),\ H := L^2(\mathbb{R}^2; \mathrm{d}x),\ E := \mathcal{S}_{-n},$$

where $n \in \mathbb{N}$, which later we shall choose large enough. So, we have

$$E' \subset H \subset E.$$

Let μ_0^* be the space time free field of mass 1 on $\mathbb{R}^2$, i.e. μ_0^* is the unique centered Gaussian measure on E with covariance operator $(-\Delta+1)^{-1}$. For $h \in L^{1+\varepsilon}(\mathbb{R}^2; \mathrm{d}x)$, with $\varepsilon > 0$, and $n \in \mathbb{N}$, let $: z^n : (h)$ be defined analogously as in Section 3, but with μ_0^* taking the role of μ (= the free field of mass 1 on Λ with Neumann boundary condition). From now on we fix $N \in \mathbb{N}$, $a_n \in \mathbb{R}$, $0 \le n \le 2N$, and define for $h \in L^{1+\varepsilon}(\mathbb{R}^2; \mathrm{d}x)$

$$: P(z) : (h) := \sum_{n=0}^{2N} a_n : z^n : (h) \ \text{ with } \ a_{2N} > 0. \tag{4.13}$$

We have that $: P(z) : (h),\ \exp(- : P(z) : (h)) \in L^p(E; \mu_0^*)$ for all $p \in [1, \infty[$, if $h \ge 0$ (cf. e.g. [Si74, §V.2]), hence the following probability measures (called *space-time cut-off quantum fields*) are well-defined for $\Lambda \in \mathcal{B}(\mathbb{R}^2)$, Λ bounded,

$$\mu_\Lambda^* := \frac{\exp(- : P(z) : (1_\Lambda))}{\int \exp(- : P(z) : (1_\Lambda))\mathrm{d}\mu_0^*} \cdot \mu_0^*. \tag{4.14}$$

It has been proven that the weak limit

$$\lim_{\Lambda \nearrow \mathbb{R}^2} \mu_\Lambda^* =: \mu^* \tag{4.15}$$

exists as a probability measure on $(E, \mathcal{B}(E))$ (see [GJ81] and also [R86], [R88]) having moments of all orders. Furthermore, by [AR90b, Proposition 2.7] we have that $\mathrm{supp}\mu^* = E$ (i.e. $\mu^*(U) > 0$ for each open subset U of E). It is, however, well-known that μ^* is not absolutely continuous with respect to μ_0^* (since $\mu^* \ne \mu_0^*$ is invariant under translations in the underlying $\mathbb{R}^2$-space, which act ergodically, see [AHK74], [Si74], [GRS75], [Frö74], [AL08]). Now we have:

Prop 1. Each $k \in K$ is well μ^*-admissible with

$$\beta_k^{\mu^*}(z) := -\sum_{n=1}^{2N} n a_n : z^{n-1} : (k) \ - \ {}_{E'}\langle (-\Delta+1)k, z \rangle_E,\ z \in E. \tag{4.16}$$

Proof. [AR91, Theorem 7.11]. □

As a consequence of Proposition 4.2, according to Section 3 we obtain the corresponding classical (gradient) Dirichlet form $(\mathcal{E}_{\mu^*}, D(\mathcal{E}_{\mu^*}))$, which is quasi regular by Theorem 3.1. So, let $\mathbf{M}_{\mu^*} = (\Omega, \mathcal{F}, (\mathcal{F}_t), (X_t)_{t\ge0}, (P_z)_{z\in E})$ be the corresponding (Markov) diffusion process as in Section 3. Then Theorem 3.2 applies, i.e. we have solved the martingale problem (in the sense of Theorem 3.2) for the corresponding operator L_{μ^*} given by (3.8) with $\beta_k^{\mu^*}$, as given by (4.16), replacing β_k^{μ}.

Finally, taking n large enough, there exists $\beta^{\mu^*} : E \to E$, $\mathcal{B}(E)/\mathcal{B}(E)$-measurable and satisfying condition *(i)* and *(ii)* in Theorem 3.3 (see [AR91, Proposition 6.9]). By construction and (4.16) we have that

$$\beta^{\mu^*}(z) = (\Delta - 1)z - \,:P'(z):, \; z \in E. \tag{4.17}$$

Hence Theorem 3.3 implies the existence of a set $S \in \mathcal{B}(E)$ such that $\mu^*(E\backslash S) = 0$ and S is invariant under the (Markov) diffusion process $(X_t)_{t\geq 0}$, i.e. $P_z[X_t \in S \; \forall t \geq 0] = 1$ for all $z \in S$, and, for every $z \in S$, X_t solves the stochastic integral equation

$$X_t = z + \frac{1}{2}\int_0^t \Big[(\Delta - 1)X_s - \,:P'(X_s):\Big] \mathrm{d}s \; + \; W_t, \; t \geq 0, \tag{4.18}$$

P_z-a.s. for some $L^2(\mathbb{R}^2; \mathrm{d}x)$-cylindrical $(\mathcal{F}_t)$-Wiener process on $(\Omega, \mathcal{F}, P_z)$, i.e. we have a Markov weak solution to the SDE (4.18) ("weak" in the probabilistic sense). The solution to (4.18) is usually called *stochastic quantization process* for the $P(\varphi)_2$-quantum field in infinite volume (thus, with Λ in Section 4.1 replaced by $\mathbb{R}^2$).

Remark 4.2. All the above results also hold for the "time zero quantum fields" associated with the $P(\varphi)_2$-quantum field in infinite volume, first discussed in [AHK74]. For details on this we refer to [AR91, Section 7, II b)].

4.3. *Ergodicity*

The use of Dirichlet form techniques is not limited to settling existence questions for solutions of SDE's with very singular coefficients. Also important special properties of solutions can be deduced. As one instance we consider the situation of the previous subsection, and ask about ergodic properties of the solution.

To start, we first mention that the construction of the infinite volume $P(\varphi)_2$-quantum field μ^*, which we took as a reference measure in the previous section, is quite specific. There is, in fact, quite a large set of possible reference measures that could replace μ^*, namely all so-called *Guerra-Rosen-Simon $P(\varphi)_2$-quantum fields* (see [GRS75]), which are defined as the convex set $\mathcal{G}$ of all Gibbs measures for a certain specification, i.e. they are defined through the classical Dobrushin-Landford-Ruelle equations appropriate to $P(\varphi)_2$-quantum fields. We do not go into details here and do not give the precise definition of the relevant specification, but rather refer to [R86] (see also [AKR97b, Section 4.3]). We only recall from [R86] that each ν from the convex set $\mathcal{G}$ can be represented as an integral over the set $\mathcal{G}_{ex}$ of all extreme points of $\mathcal{G}$. Furthermore, we recall that by the main result in [AR89a] (see also [AKR97b]) for every $\nu \in \mathcal{G}$ the corresponding form (3.4), with ν replacing μ, is closable on $L^2(E;\nu)$, so its closure $(\mathcal{E}_\nu, D(\mathcal{E}_\nu))$ (see (3.7)) is a classical (gradient) Dirichlet form for which all results from Section 4.2 apply with ν replacing μ^*. Then the following is a special case of [AKR97b, Theorem 4.14].

Theorem 4.1. *Suppose $\nu \in \mathcal{G}_{ex}$. Then $(\mathcal{E}_\nu, D(\mathcal{E}_\nu))$ is irreducible and (equivalently)* *the corresponding (Markov) diffusion process* $\boldsymbol{M}_\nu$ *is time ergodic under*

$$P_\nu := \int_E P_z\, \nu(\mathrm{d}z).$$

Remark 4.3.

(i) In fact, if we replace $(\mathcal{E}_\nu, D(\mathcal{E}_\nu))$ by the corresponding "maximal Dirichlet form" $(\mathcal{E}_\nu^{max}, D(\mathcal{E}_\nu^{max}))$ (see [AKR97b] for the definition), then the irreducibility of the latter implies that $\nu \in \mathcal{G}$ must be in $\mathcal{G}_{ex}$.

(ii) The equivalence of the irreducibility of $(\mathcal{E}_\nu, D(\mathcal{E}_\nu))$ and the time ergodicity of $\mathbf{M}_\nu$ under P_ν follows by the general theory (see [F82]). In this case the irreducibility of $(\mathcal{E}_\nu, D(\mathcal{E}_\nu))$ is also equivalent to the L^2-ergodicity of the corresponding semigroup (T_t^ν), $t \geq 0$, i.e.

$$\lim_{t\to\infty} \|T_t^\nu f - f\|_{L^2(E;\nu)} = 0 \quad \textit{for all} \quad f \in L^2(E;\nu).$$

For details we refer to [AKR97a, Proposition 2.3]. The latter paper was dedicated to Professor Masatoshi Fukushima on the occasion of his 60th birthday. A comprehensive study of the above relations between irreducibility of a Dirichlet form, the time ergodicity of the corresponding Markov process and the L^2-ergodicity of the corresponding semigroup and other related properties is contained in the forthcoming paper [BCR14] in a much more general context, including non-symmetric coercive and generalized Dirichlet forms.

(iii) Theorem 4.1, stated above for Guerra-Rosen-Simon Gibbs states of $P(\phi)_2$-Euclidean quantum field theory, is valid for many other Gibbs states of both lattice and continuum systems from statistical mechanics. We refer to [AKR97a, Section 5], [AKKR09] for lattice systems and to [AKR98, Section 7] for continuum systems.

4.4. *Additional remarks*

(1) Above we discussed the stochastic quantization equation as it was first proposed in [PaWu81], where the driving noise is $\mathrm{d}W_t$ i.e. a Gaussian space-time white noise. The same invariant measure can also be obtained by considering an SPDE with a space-regularized noise, in which case the drift term has to be modified accordingly. E.g., in the φ_2^4-case the regularized stochastic quantization equation has been discussed in the finite volume case, with various types of regularizations, in [DPT00], [HK98], [JLM85], [LR98], [GaGo96], [DPD03].

In particular in this finite volume case a proof of Markov uniqueness in $L^p(\mu)$, $1 \leq p < \infty$ has been achieved in [RZ92]. Strong solutions have been constructed in suitable Besov spaces [DPD03], and essential self-adjointness of generators has been proven in [LR98] and [DPT00]. The invalidity of a Girsanov formula has been shown in [MiRo99]. The discussion of uniqueness and ergodicity questions we sketched in Section 4.3 carry over to this case too.

The only papers, where the infinite volume processes which solve the stochastic quantization equation, in their original or regularized version, have been discussed, seem to be [AR89b], [AR91], [BCM88]. The uniqueness results which hold in the finite volume case do not carry over to the infinite volume case, where the question of uniqueness of generators is still open. See, however, [ARZ93a], [AR89a] for some partial results.

(2) The regularized stochastic quantization equation has also been discussed for other 2 space-time dimensional models. The starting point for the construction of solutions by the Dirichlet form method, namely the measure described heuristically given by (4.1) resp. its analogue in a bounded volume, has been constructed rigorously for the case, where e.g. v is an exponential function $v(y) = \exp(\alpha y)$ or a superposition of such functions, of the form $v(y) = \int \exp(\alpha y)\nu_v(\mathrm{d}y)$, with ν_v a bounded variation measure with $|\alpha| < \sqrt{4\pi}$ resp. $\operatorname{supp}\nu_v \in (-\sqrt{4\pi}, \sqrt{4\pi})$, in [AHK74], (see also [HK71], [Si74], [FrP77], [AGHK79], [AFHKL86], and [AHPRS90a], [AHPRS90b], [HKPS93] for an alternative construction using methods of white noise analysis).
The corresponding regularized stochastic quantization equation has been discussed in a bounded domain in [Mi06] and [AKMR]. Let us point out that models of this type are presently under intensive investigation in regard to their importance in completely different areas of research, see, e.g., [RhVa13].
For models given in terms of other functions v, e.g., superposition of trigonometric functions, like the function v appearing in the $\sin\alpha\varphi_2$-model (the quantized version of the Sine-Gordon equation in 2 space-time dimensions) constructions of the measure heuristically given by (4.1) as well as the definition of the dynamics in terms of Dirichlet forms have also been discussed, see [AHR01]. In this work the necessity of renormalization has been shown and strong solutions have been constructed in a suitable distributional setting. See also [AHPRS90a], [AHPRS90b] for a white noise analysis approach to such v's.

(3) Recent work has concerned both the stochastic quantization equation in space-time dimension $d = 1$ and $d = 3$.
For $d = 1$ the dynamics has been constructed in the strong probabilistic sense and L^p-uniqueness of the generators has been proven, see [Iwa85], [Iwa87] resp. [KR07], for v of polynomial type, resp. [AKaRö12], for v exponential or trigonometric type, both in finite and infinite volume.
In the case $d = 3$ an integration by parts formula has been established for the φ_3^4-model, as well as the existence of a pre-Dirichlet form [ALZ06]. However, the (generalized) logarithmic derivative does not seem to have good enough integrability properties. Therefore, the closability of this pre-Dirichlet form is an open problem, as is the existence of a global (Markov) dynamics. This might be put in relation with a recent approach to the local stochastic dynamics developed for this model by M. Hairer [Hai14].

5. Further developments

In applying Dirichlet form techniques to SDEs the symmetry assumption is, of course, very restrictive. Unfortunately, also the sector condition imposed on the (in general non-symmetric) sectorial Dirichlet forms analyzed in [MR92] is still too restrictive to cover important classes of SDE, in particular on infinite dimensional state spaces, including stochastic partial differential equations. Therefore, a theory of "fully non-symmetric" Dirichlet forms was developed in [Sta99a]. The main feature that this "theory of generalized Dirichlet forms" has in common with the symmetric and sectorial case is that it still requires a "reference measure" μ to be given or constructed beforehand. Below, we want to briefly describe an important subclass thereof, where one is just given an operator L and a measure μ, intrinsically related to L (see Definition 5.1 below), that will serve as a "reference measure". The underlying idea has been first put forward systematically in [ABR99] and [R98], and was one motivation that has eventually led to the recent monograph [BKRS14].

Consider the situation described at the beginning of Section 3, so E is a (real) separable Banach space, H a (real) separable Hilbert space such that $H \subset E$ continuously and densely, hence

$$E' \subset \ (H' \equiv) \ H \subset \ E,$$

continuously and densely. Fix an algebra $K \subset E'$, containing a countable subset, separating the points of E, thus K generates $\mathcal{B}(E)$, i.e.

$$\sigma(K) = \sigma(E') = \mathcal{B}(E). \tag{5.1}$$

Let L be a linear operator, whose domain contains $\mathcal{F}C_b^\infty(K)$ (cf. (3.1)).

Definition 5.1. A probability measure μ on $\mathcal{B}(E)$ is called L-infinitesimally invariant if $Lu \in L^1(E;\mu)$ for all $u \in \mathcal{F}C_b^\infty(K)$ and

$$\int_E Lu \, \mathrm{d}\mu = 0 \text{ for all } u \in \mathcal{F}C_b^\infty(K),$$

in short: if

$$L^*\mu = 0. \tag{5.2}$$

We note that for μ as in Definition 5.1, because of (5.1), we have that $\mathcal{F}C_b^\infty(K)$ is dense in $L^p(E;\mu)$ for all $p \in [1,\infty)$. Let us assume that L is an operator of type (3.8) or more generally a Kolmogorov or a diffusion operator (in the sense of [E99]). In this case quite general theorems are known to ensure the existence of measures μ satisfying (5.2) (see [BKRS14] and the references therein). It then follows immediately, that L is dissipative on $L^1(E;\mu)$ (see [E99, Lemma 1.8]), hence closable on $L^1(E;\mu)$. Let us denote its closure by $(L_\mu, D(L_\mu))$. It is, however, not true in general, that L_μ generates a C_0-(contraction) semigroup $T_t^\mu = e^{tL_\mu}$, $t \geq 0$, on $L^1(E;\mu)$. If it does, then this semigroup is always sub-Markovian (see [E99]). In this case one can ask whether there exists a (right) continuous Markov process on E whose transition semigroup is related to T_t^μ, $t \geq 0$, as in (3.9). So, in summary, one can realize the "Dirichlet form approach" for such general Kolmogorov operators L to construct a corresponding (right) continuous Markov process in this "fully non-symmetric" case, if the following three problems can be solved:

(1) Does there exist a probability measure μ on $(E, \mathcal{B}(E))$ having the property (5.2) in Definition 5.1?
(2) Does $(L_\mu, D(L_\mu))$ then generate a C_0-(contraction) semigroup T_t^μ, $t \geq 0$, on $L^1(E;\mu)$?
(3) Does T_t^μ, $t \geq 0$, come from a transition function p_t, $t \geq 0$, of a (right) continuous Markov?

We note that if the answer to (2) is yes, then μ is also invariant for T_t^μ, $t \geq 0$, i.e.

$$\int_E T_t^\mu u \, d\mu = \int_E u \, d\mu \tag{5.3}$$

for all $u \in \mathcal{F}C_b^\infty(K)$, equivalently for all $u \in L^1(E;\mu)$. A lot of work has been done on the three problems above in the past decade. We have already mentioned [BKRS14] as a good reference for problem (1), but also problem (2) is discussed there and further references are given in [BKRS14] concerning both (1) and (2). We only mention here that the answer "yes" to problem (2) is well-known to be equivalent to the "range condition", i.e. that $(1 - L)(\mathcal{F}C_b^\infty(K))$ is dense in $L^1(E;\mu)$ (see e.g. [BRSt00, Proposition 2.6]). But in concrete cases, in particular in infinite dimensions, this is a very hard problem (see, however, [Sta99b] for a useful characterization of the range condition if $\dim E < \infty$). Concerning (3) considerable progress has been made in [BeBoR06], [BeBoR08].

In the remainder of this subsection we want to address two important points concerning the above discussion, namely:

Questions:

(A) Suppose that all problems (1)-(3) above can be solved for some L-infinitesimally invariant measure $\mu := \mu_1$, which is hence invariant for $(p_t)_{t\geq 0}$. Is it possible that there exists some other L-infinitesimally invariant measure μ_2 which is not invariant for $(p_t)_{t\geq 0}$?
(B) Does there exist an L-infinitesimally invariant measure μ, so that the closure of $(L, \mathcal{F}C_b^\infty(K))$ in $L^2(E;\mu)$ does not generate a C_0-semigroup, but there exist two different closed extentions generating two Markov C_0-semigroups on $L^2(E;\mu)$? In short: can "Markov uniqueness" fail to hold?

Indeed, the answer to (A) is yes. There are well-known counterexamples $p_t, t \geq 0$, if $E = \mathbb{R}^d$ (see [BKRS14] and the references therein, in particular,

[Sta99b]). Below we give two simple examples when E is an infinite dimensional Hilbert space, μ_2 is a Gaussian measure and $(p_t)_{t\geq 0}$ is even a strong Feller transition semigroup of an Ornstein-Uhlenbeck process, i.e. is given explicitely by a Mehler-type formula. Also the answer to Question (B) is "yes". In fact, μ can even be chosen Gaussian and so that $(L, \mathcal{F}C_b(K))$ is symmetric in $L^2(E;\mu)$ and both Markov C_0-semigroups consist of self-adjoint operators in $L^2(E;\mu)$. We shall, however, not give all details here, but refer instead to [E99] where these can be found (see Remark 5.1 below). We rather concentrate on details for the examples to answer Question (A) by "yes".

First example:

Consider the open interval $(0,1) \subset \mathbb{R}$ and choose

$$H = E := L^2\Big((0,1); \mathrm{d}x\Big), \tag{5.4}$$

with the usual inner product $\langle\ ,\ \rangle$ and where $\mathrm{d}x$ denotes Lebesgue measure. Define $A_1 := -\Delta_D$, where Δ_D denotes the Dirichlet Laplacian on $(0,1)$, and $A_2 := -\Delta_{D,N}$, where $\Delta_{D,N}$ denotes the Dirichlet-Neumann-Laplacian on $(0,1)$. More precisely, we take Dirichlet boundary condition at $\xi = 0$ and Neumann boundary condition at $\xi = 1$. Define for $t \geq 0$, $x, y \in E$,

$$p_t^{(i)}(x, \mathrm{d}y) = N(e^{-tA_i}x, Q_t^{(i)})(\mathrm{d}y), \qquad i = 1, 2, \tag{5.5}$$

where N denotes the Gaussian measure on E with mean $e^{-tA_i}x$ and covariance operator

$$Q_t^{(i)} := \int_0^t e^{-2sA_i}\mathrm{d}s\ . \tag{5.6}$$

We note that, because we are on the one-dimensional underlying domain $(0,1)$, each Q_t^i is indeed trace class. It is well-known that $p_t^{(i)}$, $t \geq 0$, $i = 1, 2$, is the transition semigroup of the Ornstein-Uhlenbeck process $(X_t^{(i)})_{t\geq 0}$ solving in the mild sense the following SDE for $i = 1, 2$ respectively

$$\begin{aligned} \mathrm{d}X_t^{(i)} &= -A_i X_t^{(i)}\mathrm{d}t + \mathrm{d}W_t, \qquad t \geq 0, \\ X_0 &= x, \end{aligned} \tag{5.7}$$

where W_t, $t \geq 0$, is an H-cylindrical Wiener process, and that

$$\mu_i := N(0, \frac{1}{2}A_i^{-1}), \quad i = 1, 2, \tag{5.8}$$

is its respective invariant measure. Furthermore, it is well-known (and easy to see) that, for each $i \in \{1, 2\}$, $p_t^{(i)}$, $t \geq 0$, gives rise to a Markov C_0-semigroup $T_t^{\mu_i}$, $t \geq 0$, of symmetric contractions on $L^p(E;\mu_i)$ for all $p \in [1, \infty)$. Furthermore, for every bounded $\mathcal{B}(E)$-measurable function $f : E \to \mathbb{R}$

$$p_t^{(i)}f(x) := \int_E f(z)p_t^{(i)}(x, \mathrm{d}z), \quad x \in E, \tag{5.9}$$

is continuous in $x \in E$, i.e. both $p_t^{(1)}$, $t \geq 0$, and $p_t^{(2)}$, $t \geq 0$, are strong Feller (see [DPZ92]).

Let us now consider $K := C_0^2((0,1))$ and recall that as seen before, $\mathcal{F}C_b^\infty(K)$ is dense in $L^p(E,\mu_i)$ for all $p \in [1,\infty)$, $i = 1,2$. Then it is easy to check that for $i = 1,2$, $p_t^{(i)}, t \geq 0$, extends to a Markov C_0-semigroup $T_t^{\mu_i}, t \geq 0$, on $L^2(E;\mu_i)$ and for the generator $(L_{\mu_i}, D(L_{\mu_i}))$ of $T_t^{\mu_i}$, $t \geq 0$, on $L^2(E;\mu_i)$(but also poinwise, see [BoR95, Theorem 5.3]), we have that $\mathcal{F}C_b^\infty(K) \subset D(L_{\mu_i})$ and for all $u \in \mathcal{F}C_b^\infty(K)$, $u = F(\langle k_1, \cdot\rangle, \ldots, \langle k_N, \cdot\rangle)$, $F \in C_b^\infty(\mathbb{R}^N)$, $i = 1,2$,

$$\begin{aligned} L_{\mu_i}u(z) = &\sum_{j,j'=1}^{N} \langle k_j, k_{j'}\rangle \partial_{jj'} F(\langle k_1, z\rangle, \ldots, \langle k_N, z\rangle) \\ &+ \sum_{j=1}^{N} \langle A_i k_j, z\rangle \partial_j F(\langle k_1, z\rangle, \ldots, \langle k_N, z\rangle), \end{aligned} \tag{5.10}$$

where $\partial_j, \partial_{j,j'}$ mean partial derivatives in the j-th or in the j-th and j'-th variable respectively. Since $A_1 k = A_2 k$ for all $k \in K$, it follows that

$$L_{\mu_1}u = L_{\mu_2}u \tag{5.11}$$

for all $u \in \mathcal{F}C_b^\infty(K)$. Because of (5.11) we may define

$$Lu := L_{\mu_1}u \ (= L_{\mu_2}u), \ u \in \mathcal{F}C_b^\infty(K). \tag{5.12}$$

Then because $T_t^{\mu_i}$, $t \geq 0$, are symmetric in $L^2(E_i;\mu_i)$, $i = 1,2$, it follows that for $i = 1,2$

$$\int Lu\, v \, d\mu_i = \int u\, Lv \, d\mu_i \ \ \forall u,v \in \mathcal{F}C_b^\infty(K), \tag{5.13}$$

hence choosing $v \equiv 1$

$$\int Lu \, d\mu_i = 0 \ \ \forall u \in \mathcal{F}C_b^\infty(K), \tag{5.14}$$

i.e. μ_1 and μ_2 are L-infinitesimally invariant (for L with domain $\mathcal{F}C_b^\infty(K)$).

Remark 5.1.

(i) We have just seen an infinite dimensional example where "Markov uniqueness" fails, i.e. two generators of two different Markov C_0-(contraction) semigroups which coincide on a common domain $\mathcal{F}C_b^\infty(K)$ which is dense in $L^2(E;\mu_i)$, $i = 1,2$. And both semigroups are even strong Feller in this case. We note, however, that in contrast to the "classical" Markov uniqueness problem, our two C_0-semigroups live on different L^2-spaces, namely $L^2(E;\mu_1)$ and $L^2(E;\mu_2)$. However, one can show that if one considers the Friedrichs extension $(L_{\mu_2,F}, D(L_{\mu_2,F}))$ of the (by (5.13)) symmetric operator $(L, \mathcal{F}C_b^\infty(K))$ on $L^2(E;\mu_2)$, then the corresponding Dirichlet form does not coincide with the

Dirichlet form corresponding to the symmetric Markov C_0-semigroup $(T_t^{\mu_2})_{t\geq 0}$ introduced above (see [E99, Chap. 5b)1)]). Hence

$$(T_t^{\mu_2})_{t\geq 0} \neq \left(e^{tL_{\mu_2,F}}\right)_{t\geq 0},$$

and both are symmetric Markov C_0-semigroups on $L^2(E;\mu_2)$ with generators coinciding on $\mathcal{F}C_b^\infty(K)$, with L defined in (5.12). This is a "true" counterexample to Markov-uniqueness (first discorered in [E99]) even for symmetric C_0-semigroups and (at least) one of the two semigroups is even strong Feller. We stress that $\mathcal{F}C_b^\infty(K)$ is, of course, not an operator core for any of the generators of $(T_t^{\mu_2})_{t\geq 0}$ and $(e^{tL_{\mu_2,F}})_{t\geq 0}$ on $L^2(E;\mu)$.

(ii) Clearly, the above only occurs, because $\mathcal{F}C_b^\infty(K)$ is too small to determine L_{μ_2}, since it does not capture the boundary behaviour of A_2, since $K\subset C_0^2\Big((0,1)\Big)$, though $\mathcal{F}C_b^\infty(K)$ is dense in $L^2(E;\mu_2)$. For examples of non-Markov uniqueness avoiding this, we refer to [E99, Chap. 5b)2)].

It is easy to see that $\mathcal{F}C_b^\infty(K)$ is an operator core for $(L_{\mu_1}, D(L_{\mu_1}))$ on $L^2(E;\mu_1)$, i.e. $\mathcal{F}C_b^\infty(K)$ is dense in $D(L_{\mu_1})$ with respect to the graph norm $\|\cdot\|_1 := \|\cdot\|_{L^2(E;\mu_1)} + \|L_{\mu_1}\cdot\|_{L^2(E;\mu_1)}$. Indeed, consider the Sobolev space $H_0^1 := H_0^1\Big((0,1);\mathrm{d}x\Big)$ of order 1 in $L^2((0,1);\mathrm{d}x)$ with Dirichlet boundary conditions. Then it is obvious from (5.10) that $\mathcal{F}C_b^\infty(H_0^1)$ is in the closure of $\mathcal{F}C_b^\infty(K)$ with respect to the graph norm $\|\cdot\|_1$. Furthermore, it is also obvious from the definition of $p_t^{(1)}$, $t\geq 0$, in (5.5), that $p_t^{(1)}(\mathcal{F}C_b^\infty(H_0^1)) \subset \mathcal{F}C_b^\infty(H_0^1)$ for all $t\geq 0$, since $e^{-tA_1}(H_0^1)\subset H_0^1$ for all $t\geq 0$. Hence by a theorem of Nelson (see [RS75, Theorem X.49]) if follows that $\mathcal{F}C_b^\infty(H_0^1)$, hence $\mathcal{F}C_b^\infty(K)$ is dense in $D(L_{\mu_1})$ with respect to $\|\cdot\|_1$.

We recall that μ_2 is L-infinitesimally invariant (for the domain $\mathcal{F}C_b^\infty(K)$). Now we show that μ_2 is, however, not invariant for $p_t^{(1)}$, $t\geq 0$, hence giving the desired example for Question (A). This can be proved as follows. Fix $y\in E\backslash\{0\}$ and consider the function

$$f(z) := e^{i\langle y,z\rangle},\ z\in E. \tag{5.15}$$

Then

$$\int_E f(z)\,\mu_2(\mathrm{d}z) = e^{-\frac{1}{2}\langle\frac{1}{2}A_2^{-1}y,\,y\rangle}$$

and

$$p_t^{(1)}f(z) = e^{i\langle e^{-tA_1}y,z\rangle}e^{-\frac{1}{2}\langle Q_t^{(1)}y,\,y\rangle}, \tag{5.16}$$

hence

$$\int_E p_t^{(1)}f(z)\,\mu_2(\mathrm{d}z) = e^{-\frac{1}{2}\langle\frac{1}{2}A_2^{-1}e^{-2tA_1}y,\,y\rangle}\cdot e^{-\frac{1}{2}\langle Q_t^{(1)}y,\,y\rangle}.$$

But $Q_t^{(1)} = \frac{1}{2}A_1^{-1}(1-e^{-2tA_1})$, so in general (just choose y in (5.15) to be an eigenvector of A_1)

$$\int_E f(z)\,\mu_2(\mathrm{d}z) \neq \int_E p_t^{(1)} f(z)\,\mu_2(\mathrm{d}z) \tag{5.17}$$

So, we have proved that μ_2 is not invariant for $p_t^{(1)}$, $t \geq 0$.

Remark 5.2. Clearly, the martingale problem for $(L, \mathcal{F}C_b^\infty(K))$ is not well-posed, because the laws of the solutions $X^{(1)}$ and $X^{(2)}$ of (5.7) are solutions to this martingale problem, but do not coincide. The reason is explained in Remark 5.1 (ii). It is, however, interesting to note that since, as explained above, $\mathcal{F}C_b^\infty(K)$ is an operator core for $(L_{\mu_1}, D(L_{\mu_1}))$ in $L^2(E;\mu_1)$, it can be easily shown, that there exists at most one Markov selection for the martingale problem for $(L, \mathcal{F}C_b^\infty(K))$ such that the corresponding transition semigroup $(p_t)_{t\geq 0}$ extends to a C_0-semigroup on $L^2(E;\mu_1)$.

Second Example:

Define the measure μ by

$$\mu := N(1, \frac{1}{2}A_1^{-1}), \tag{5.18}$$

i.e. the image of μ_1, defined in (5.8), under the translation

$$E \ni z \mapsto z + 1 \in E,$$

with 1 being the constant function equal to one on $(0,1)$, we obtain a Gaussian (not centered) measure on E such that

$$L^*\mu = 0 \tag{5.19}$$

and μ is not invariant for $p_t^{(1)}, t \geq 0$. Indeed, (5.19) follows immediately from the definition of L (by (5.12), (5.10)) since

$$\langle A_1 k, 1\rangle = \int_0^1 -\Delta k \,\mathrm{d}x = 0 \text{ for all } k \in K.$$

On the other hand, we have for f as in (5.15)

$$\int_E f(z)\,\mu(\mathrm{d}z) = e^{i\langle y,1\rangle} e^{-\frac{1}{2}\langle \frac{1}{2}A_1^{-1}y,y\rangle}\,,$$

but by (5.16)

$$\begin{aligned}\int_E p_t^{(1)} f(z)\,\mu(\mathrm{d}z) &= \int_E p_t^{(1)} f(1+z)\,\mu_1(\mathrm{d}z)\\ &= e^{i\langle e^{-tA_1}y,1\rangle}\int_E p_t^{(1)} f(z)\,\mu_1(\mathrm{d}z)\\ &= e^{i\langle e^{-tA_1}y,1\rangle}\int_E f(z)\,\mu_1(\mathrm{d}z)\end{aligned}$$

$$= e^{i\langle y, e^{-tA_1}1-1\rangle} \int_E f(z)\, \mu(\mathrm{d}z). \tag{5.20}$$

But, since $A_1 := -\Delta_D$, we know that

$$e^{-tA_1}1 \neq 1.$$

Hence μ is not invariant for $p_t^{(1)}, t \geq 0$.

Remark 5.3. It is very easy to check that in the above example $p_t^{(1)}$, $t \geq 0$, is symmetric with respect to μ_1 and that $(L, \mathcal{F}C_b^\infty(K))$ is symmetric on $L^2(E, \mu)$. So, we even have

$$\int Lu\, v\, \mathrm{d}\mu = \int u\, Lv\, \mathrm{d}\mu \text{ for all } u, v \in \mathcal{F}C_b^\infty(K),$$

which is stronger than (5.19).

6. Acknowledgements

It is a special pleasure for us to contribute to this volume in Honour of Professor Fukushima. He has been a great mentor for us, providing us much inspiration over many years. We take the opportunity to express to him our sincere gratitude, our friendship and great admiration. We also would like to thank Zdzislaw Brzezniak for pointing out an error in an earlier version of this paper. Financial support of the DFG through SFB 701, IGK 1132, and the HCM, as well as of the NCMIS, 973 project (2011CB808000), NSFC (11021161) and NSERC (Grant No. 311945-2013) is gratefully acknowledged.

References

[Aid00] S. Aida, *On the irreducibility of Dirichlet forms on domains in infinite-dimensional spaces*, Osaka J. Math. **37** (2000), no.4, 953-966.

[Alb97] S. Albeverio, *Wiener and Feynman-Path integrals and their Applications*, AMS, Proceedings of Symposia in Applied Mathematics **52** (1997), 163-194.

[Alb03] S. Albeverio, *Theory of Dirichlet forms and applications*, In: P. Bernard (ed.) Lectures on probability theory and statistics (Saint-Flour, 2000), In: Lecture Notes in Math. **1816** (2003), 1-106, Springer, Berlin.

[ABR99] S. Albeverio, V. I. Bogachev and M. Röckner, *On uniqueness of invariant measures for finite and infinite dimensional diffusions*, Comm. Pure Appl. Math. **52** (1999), 325-362.

[AFHKL86] S. Albeverio, J.E. Fenstad, R. Høegh-Krohn and T. Lindstrøm, *Nonstandard methods in stochastic analysis and mathematical physics*, Pure and Applied Mathematics **122** (1986), Academic Press Inc., Orlando, FL. Reprint: Dover Publications, Mineola, NY (2009).

[AFe04] S. Albeverio and B. Ferrario, *Uniqueness of solutions of the stochastic Navier-Stokes equation with invariant measure given by the enstrophy*, Ann. Probab. **32**(2) (2004), 1632-1649.

[AFe08] S. Albeverio and B. Ferrario, *Some methods of infinite dimensional analysis in hydrodynamics: an introduction*, SPDE in hydrodynamic: recent progress and prospects, Lecture Notes in Math. **1942** (2008), 1-50.

[AFH11] S. Albeverio, R. Fan and F.S. Herzberg, *Hyperfinite Dirichlet Forms and Stochastic Processes*, Lecture Notes of the Unione Matematica Italiana **10** (2011), 1-63.

[AFHMR92] S. Albeverio, M. Fukushima, W. Hansen, Z.M. Ma and M. Röckner, *An invariance result for capacities on Wiener space*, J. Funct. Anal **106** (1992), 35-49.

[AGHK79] S. Albeverio, G. Gallavotti and R. Høegh-Krohn, *Some results for the exponential interaction in two or more dimensions*, Comm. Math. Phys. **70** (1979), 187-192.

[AGW97] S. Albeverio, H. Gottschalk and J.-L. Wu, *Models of local relativistic quantum fields with indefinite metric (in all dimensions)*, Comm. Math. Phys. **184**(3) (1997), 509-531.

[AGY05] S. Albeverio, H. Gottschalk and M.W. Yoshida, *Systems of classical particles in the grand canonical ensemble, scaling limits and quantum field theory*, Rev. Math. Phys. **17**(2) (2005), 175-226.

[AHR01] S. Albeverio, Z. Haba and F. Russo, *A two-space dimensional semilinear heat equation perturbed by (Gaussian) white space*, PTRF **121** (2001), 319-366.

[AHK74] S. Albeverio and R. Høegh-Krohn, *The Wightman axioms and the mass gap for strong interactions of exponential type in two-dimensional space-time*, J. Funct. Anal. **16** (1974), 39-82.

[AHK76] S. Albeverio and R. Høegh-Krohn, *Quasi invariant measures, symmetric diffusion processes and quantum fields*, Editions du CNRS, Proceedings of the International Colloquium on Mathematical Methods of Quantum Field Theory, Colloques Internationaux du Centre National de la Recherche Scientifique, No. **248** (1976), 11-59.

[AHK77a] S. Albeverio and R. Høegh-Krohn, *Dirichlet forms and diffusion processes on rigged Hilbert spaces*, Zeitschrift für Wahrscheinlichkeitstheorie und verwandte Gebiete **40** (1977), 1-57.

[AHK77b] S. Albeverio and R. Høegh-Krohn, *Hunt processes and analytic potential theory on rigged Hilbert spaces*, Ann. Inst. H. Poincaré (Probability Theory) **13** (1977), 269-291.

[AHK82] S. Albeverio and R. Høegh-Krohn, *Diffusions, quantum fields and groups of mappings*, Functional analysis in Markov processes, Lecture Notes in Math. **923** (1982), 133-145.

[AHKS77] S. Albeverio, R. Høegh-Krohn and L. Streit, *Energy forms, Hamiltonians, and distorted Brownian paths*, J. Math. Phys. **18** (1977), 907-917.

[AHPRS90a] S. Albeverio, T. Hida, J. Potthoff, M. Röckner and L. Streit, *Dirichlet forms in terms of white noise analysis I - Construction and QFT examples*, Rev. Math. Phys **1** (1990), 291-312.

[AHPRS90b] S. Albeverio, T. Hida, J. Potthoff, M. Röckner and L. Streit, *Dirichlet forms in terms of white noise analysis II - Closability and Diffusion Processes*, Rev. Math. Phys **1** (1990), 313-323.

[AKaRö12] S. Albeverio, H. Kawabi, M. Röckner, *Strong uniqueness for both Dirichlet operators and stochastic dynamics for Gibbs measures on a path space with exponential interactions*, JFA **262** (2012), no. 2, 602-638.

[AKKR09] S. Albeverio, Y. Kondratiev, Y. Kozitsky and M. Röckner, *Statistical Mechanics of Quantum Lattice Systems: A Path Integral Approach*, EMS

Tracts Math. **8** (2009), European Mathematical Society.

[AKMR] S. Albeverio, H. Kawabi, S. Mihalache and M. Röckner, *Dirichlet form approach to stochastic quantization under exponential interaction in finite volume*, in preparation.

[AKR92] S. Albeverio, Y. Kondratiev and M. Röckner, *An approximative criterium of essential selfadjointness of Dirichlet operators*, Potential Anal. **1**(3) (1992), 307-317.

[AKR93] S. Albeverio, Y. Kondratiev and M. Röckner, *Addendum to the paper 'An approximative criterium of essential selfadjointness of Dirichlet operators'*, Potential Anal. **2** (1993), 195-198.

[AKR97a] S. Albeverio, Y.G. Kondratiev and M. Röckner, *Ergodicity of L^2-semigroups and extremality of Gibbs states*, J. Funct. Anal. **144** (1997), 3394-423.

[AKR97b] S. Albeverio, Y.G. Kondratiev and M. Röckner, *Ergodicity for the stochastic dynamics of quasi-invariant measures with applications to Gibbs states*, J. Funct. Anal. **149** (1997), 415-469.

[AKR98] S. Albeverio, Y.G. Kondratiev and M. Röckner, *Analysis and geometry on configuration space. The Gibbsian case*, J. Funct. Anal. **157** (1998), 242-291.

[AKR02] S. Albeverio, Y.G. Kondratiev and M. Röckner, *Symmetryzing Measures for Infinite Dimensional Diffusions: An Analytic Approach*, in Lect. Notes Math., "Geometric Analysis and nonlinear partial differential equations", S. Hildebrandt et. al., eds, Springer, Berlin, (2003), 475-486.

[AL08] S. Albeverio, S. Liang, *A remark on the nonequivalence of the time-zero ϕ_3^4-mesure with the free field measure*, Markov Processes Rel. Fields **14** (2008), 159-164.

[ALZ06] S. Albeverio, S. Liang and B. Zegarliński, *Remark on the integration by parts formula for the ϕ_3^4-quantum field model*, Infin. Dimens. Anal. Quantum Probab. Relat. Top. **9** (2006), 149-154.

[AM91] S. Albeverio, Z.M. Ma, *Necessary and sufficient conditions for the existence of m- processes associeted with Dirichlet form*, Sém. de Prob. XXV, LNM 148 5, 374-406 (1991)

[AM92] S. Albeverio, Z.M. Ma, *A general correspondence between Dirichlet form and right processes*, Bull AMS **26** (1992), 245-252.

[AMR92a] S. Albeverio, Z.M. Ma and M. Röckner, *Non-symmetric Dirichlet forms and Markov processes on general state space*, CRA Sci., Paris **314** (1992), 77-82.

[AMR92b] S. Albeverio, Z.M. Ma and M. Röckner, *Regularization of Dirichlet spaces and applications*, CRAS, Ser. I, Paris **314**(11) (1992), 859-864.

[AMR92c] S. Albeverio, Z.M. Ma and M. Röckner, *A Beurling-Deny structure theorem for Dirichlet forms on general state space*, Cambridge University Press, R. Høegh-Krohn's Memorial Volume, Eds. S. Albeverio et al., Vol **1** (1992), 115-123.

[AMR93] S. Albeverio, Z.M. Ma and M. Röckner, *Quasi-regular Dirichlet forms and Markov processes*, J. Funct. Anal. **111** (1993), 118-154.

[AMU98] S. Albeverio, L.M. Morato and S. Ugolini, *Non-Symmetric Diffusions and Related Hamiltonians*, Potential Analysis **8** (1998), 195-204.

[AR89a] S. Albeverio and M. Röckner, *Dirichlet forms, quantum fields and stochastic quantization*, Stochastic analysis, path integration and dynamics (K.D. Elworthy and J.C. Zambrini, eds.) Research Notes in Mathematics, vol. 200, Longman, Harlow, 1989, pp 1-21.

[AR89b] S. Albeverio and M. Röckner, *Classical Dirichlet forms on topological vector spaces - the construction of the associated diffusion process*, Prob. Theory and Rel. Fields **83**(3) (1989), 405-434.

[AR90a] S. Albeverio and M. Röckner, *Classical Dirichlet forms on topological vector spaces-Closability and Cameron-Martin formula*, J. Funct. Anal. **88**(2) (1990), 395-436.

[AR90b] S. Albeverio and M. Röckner, *New developments in theory and applications od Dirichlet forms* "Stochastic Processes, Physics and Geometry, Ascona/Locarno, Switzerland, July 4-9, 1988" pp.27-76, World Scientific, Singapore, 1990

[AR91] S. Albeverio and M. Röckner, *Stochastic differential equations in infinite dimensions: solution via Dirichlet forms*, Prob. Th. Rel. Fields **89**(3) (1991), 347-386.

[ARZ93a] S. Albeverio, M. Röckner and T.S. Zhang, *Girsanov transform of symmetric diffusion processes in infinite dimensional space*, Ann. of Prob. **21** (1993), 961-978.

[ARZ93b] S. Albeverio, M. Röckner and T.S. Zhang, *Markov uniqueness and its applications to martingale problems, stochastic differential equation and stochastic quantization*, C.R. Math. Rep. Acad. Sci. Canada XV, no. 1, 1-6.

[ARW01] S. Albeverio, B. Rüdiger and J.-L. Wu, *Analytic and probabilistic aspects of Lévy processes and fields in quantum theory*, Lévy processes (2001), 187-224, Birkhäuser Boston, Boston.

[ARü02] S. Albeverio and B. Rüdiger, *Infinite dimensional Stochastic Differential Equations obtained by subordination and related Dirichlet forms*, J. Funct. Anal. **204**(1) (2003), 122-156.

[ARü05] S. Albeverio and B. Rüdiger, *Subordination of symmetric quasi-regular Dirichlet forms*, Random Oper. Stochastic Equations **13**(1) (2005), 17-38.

[ASe14] S. Albeverio and A. Sengupta, *From classical to quantum fields - A mathematical approach*, book in preparation.

[AU00] S. Albeverio and S. Ugolini, *Complex Dirichlet Forms: Non Symmetric Diffusion Processes and Schrödinger Operators*, Potential Analysis **12** (2000), 403-417.

[BBCK09] M.T. Barlow, R.F. Bass, Z.Q. Chen and Kassmann, *Non-local Dirichlet forms and symmetric jump processes*, TAMS **361** (2009), 1963-1999.

[BK95] Y.M. Berezansky and Y.G. Kondratiev, *Spectral Methods in Infinite-Dimensional Analysis*, Kluwer, Vol **1,2** (1995), first issued in URSS 1988.

[BD58] A. Beurling and J. Deny, *Espaces de Dirichlet*, Acta Math. **99** (1958), 203-224.

[BD59] A. Beurling and J. Deny, *Dirichlet spaces*, Proc. Nat. Acad. Sci. U.S.A. **45** (1959), 208-215.

[BeBoR06] L. Beznea, N. Boboc and M. Röckner, *Markov processes associated with L^p-resolvents and applications to stochastic differential equations on Hilbert space*, J. Evol. Equ. **6** (2006), no.4, 754-772.

[BeBoR08] L. Beznea, N. Boboc and M. Röckner, *Markov processes associated with L^p-resolvents,applications to quasi-regular Dirichlet forms and stochastic differential equations*, C. R. Math. Acad. Sci. Paris **346**(5-6) (2008), 323-328.

[BCR14] L. Beznea, J. Cimpian and M. Röckner, *Irreducible recurrence, ergodicity and extremality of invariant measures for L^p-resolvents*, pp.25, Preprint 2014.

[BhK93] A.G. Bhatt and R.L. Karandikar, *Invariant measures and evolution equations for Markov processes characterized via martingale problems*, Ann. Prob.

21 (1993), 2246-2268.

[BiT07] M. Biroli and N. Tchou, *Γ-convergence for strongly local Dirichlet forms in perforated domain with homogeneous Neumann boundary conditions*, Adv. Math. Sci. Appl. **17** (2007), 149-179.

[BKRS14] V.I. Bogachev, N.V. Krylov, M. Röckner and S. Shaposhnikov, *Fokker-Planck Kolmogorov equations*, Monograph, pp. 490, publication in preparation.

[BoR95] V.I. Bogachev and M. Röckner, *Mehler formula and capacities for infinite dimensional OrnsteinUhlenbeck processes with general linear drift*, Osaka J. Math. **32** (1995), 237-274.

[BoRöZh00] V. Bogachev, M. Röckner and T.S. Zhang, *Existence and uniqueness of invariant measures: an approach via sectorial forms*, Appl. Math. Optim. **41**(1) (2000), 87-109.

[BRSt00] V.I. Bogachev, M. Röckner and W. Stannat, *Uniqueness of invariant measures and maximal dissipativity of diffusion operators on L^1*, Infinite dimensional Stochastic Analysis (P. Clément et al., eds.), Royal Netherlands Academy of Arts and Sciences, Amsterdam, 2000, pp. 39-54.

[BCM88] V.S. Borkar, R.T. Chari and S.K. Mitter, *Stochastic quantization of field theory in finite and infinite volume*, J. Funct. Anal. **81**(1) (1988), 184-206.

[Bou03] N. Bouleau, *Error calculus for finance and physics: the language of Dirichlet form, de Gruyter*, Berlin (2003).

[BH91] N. Bouleau and F. Hirsch, *Dirichlet Forms and Analysis on Wiener Space*, deGruyter, Studies in Mathematics **14** (1991).

[CF12] Z.Q. Chen and M. Fukushima, *Symmetric Markov Processes, Time change, and Boundary Theory*, Princeton University Press, 2012.

[CMR94] Z.Q. Chen, Z.M. Ma and M.Röckner, *Quasi-homeomorphisms of Dirichlet forms*, Nagoja J.Math. **136** (1994), 1-15.

[DPD03] G. Da Prato and A. Debussche, *Strong solutions to the stochastic quantization equations*, Ann. Probab. **31**(4) (2003), 1900-1916.

[DPZ92] G. Da Prato and J. Zabczyk, *Stochastic equations in infinite dimensions*, Encyclopedia of Mathematics and its Applications **44**, Cambridge University Press, Cambridge (1992).

[DPT00] G. Da Prato and L. Tubaro, *Self-adjointness of some infinite dimensional elliptic operators and applications to stochastic quantization*, Prob. Theory Related Fields **118**(1) (2000), 131-145.

[Den70] J. Deny, *Méthodes hilbertiennes en théorie du potentiel*, In: Potential Theory (C.I.M.E., I Ciclo, Stresa, 1969) (1970), 121-201, Edizioni Cremonese, Rome.

[DG97] Z. Dong and F.Z. Gong, *A necessary and sufficient condition for quasiregularity of Dirichlet forms corresponding to resolvent kernels*, Acta Math. Appl. Sinica **20**(3) (1997), 378-385.

[E99] A. Eberle, *Uniqueness and Non-Uniqueness of Semigroups generated by Singular Diffusion Operators*, Lecture Notes in Math. Vol. 1718, Springer, Berlin 1999.

[Eche82] P.E. Echeverria, *A criterium for invariant measures of Markov processes*, W. Th. Vens. Geb. **61** (1982), 1-16.

[EK86] S.N. Ethier and T.G. Kurtz, *Markov Processes. Characterization and Convergence*, Wiley Series in Probability and Mathematical Statistics (1986).

[FeyLa91] D. Feyel and A. de la Pradelle, *Capacités gaussiennes*, Ann. Inst. Fourier **41** (1991), 49-76.

[Frö74] J. Fröhlich, *Verification of axioms for Euklidean and relativistic fields and*

Haag's theorem in a class of $P(\varphi)_2$ models, Ann. I. H. Poincaré **21** (1974), 271-317.

[FrP77] J. FRÖHLICH AND T.M. PARK, *Remarks on exponential interactions and the quantum sine-Gordon equation in two space-time dimensions*, HPA **50** (1977), 315-329.

[Fu71a] M. FUKUSHIMA, *Regular representations of Dirichlet spaces*, TAMS **155** (1971), 455-473.

[Fu71b] M. FUKUSHIMA, *Dirichlet spaces and strong Markov processes*, TAMS **162** (1971), 185-224.

[Fu80] M. FUKUSHIMA, *Dirichlet Forms and Markov processes*, North Holland (1980), Amsterdam.

[F82] M. FUKUSHIMA, *A note on irreducibility and ergodicity of symmetric Markov processes*, in "Proceedings Marseille", Lecture Notes in Physics, Vol. 173, pp. 200-207, Springer, Berlin, 1982.

[Fu84] M. FUKUSHIMA, *A Dirichlet form on the Wiener space and properties of Brownian motion*, Lecture Notes in Math. Vol. 1096, 290-300, Springer, 1984.

[Fu92] M. FUKUSHIMA, *(r,p) capacities and Hunt processes in infinite dimensions*, in A.N. Shiryaev et al., eds, 96-103, Probability, Theory and Mathematical Statistics (1992), World Scientific, Singapore.

[Fu10] M. FUKUSHIMA, *From one dimensional diffusions to symmetric Markov processes*, Stoch. Proc. Appl. **120** (2010), 590-604.

[FH01] M. FUKUSHIMA AND M. HINO, *On the space of BV functions and a related stochastic calculus in infinite dimensions*, J. Funct. Anal. **183** (2001), 245-268.

[FOT11] M. FUKUSHIMA, Y. OSHIMA AND M. TAKEDA, *Dirichlet forms and symmetric Markov processes*, 2nd revised and extended ed., de Gruyter Studies in Mathematics **19** (2011), Walter de Gruyter & Co., Berlin.

[GaGo96] D. GATAREK AND B. GOLDYS, *Existence, uniqueness and ergodicity for the stochastic quantization equation*, Studia Math. **119** (1996), 179-193.

[GJ81] J. GLIMM AND A. JAFFE, *Quantum Physics*, Springer, New York, 1981; 2nd ed., 1986.

[GriH08] A. GRIGORÝAN AND J. HU, *Off-diagonal upper estimates for the heat normal of the Dirichlet forms on metric spaces*, Inv. Math. **174** (2008), 81-126.

[Gro76] L. GROSS, *Logarithmic Sobolev inequalities*, Amer. J. Math. **97** (1976), 1061-1083.

[GRS75] F. GUERRA, J. ROSEN AND B. SIMON, *The $P(\Phi)_2$-Euclidean quantum field theory as classical statistical mechanics*, Ann. Math. **101** (1975), 111-259.

[Hai14] M. HAIRER, *A theory of regularity structures*, ArXive-prints (2013), arXiv: 1303.51113.

[HKPS93] T. HIDA, H.-H. KUO, J. POTTHOFF AND L. STREIT, *White Noise: An Infinite Dimensional Calculus*, Math. Appl. **253** (1993), Kluwer Academic Publishers Group, Dordrecht.

[Hi10] M. HINO, *Energy measures and indices of Dirichlet forms, with applications to derivatives on some fractals*, Proc. Lond. Math. foc. **100** (2010), 269-302.

[HK71] R. HØEGH-KROHN, *A general class of quantum fields without cut-offs in two space-time dimensions*, Comm. Math. Phys. **21** (1971), 244-255.

[HK98] Y. HU AND G. KALLIANPUR, *Exponential Integrability and Application to Stochastic Quantization*, Appl. Math. Optim. **37** (1998), 295-353.

[Iwa85] K. IWATA, *Reversible measures of a $P(\phi)_1$-time evolution*, Probabilistic methods in mathematical physics (1985), 195-209.

[Iwa87] K. IWATA, *An infinite dimensional stochastic differential equation with state *space $C(\mathbb{R})$*, Probab. Theory Related Fields **74** (1987), 141-159.

[Jac01] N. Jacob, *Pseudo-Differential Operators and Markov Processes*, Vol. I. Fourier analysis and semigroups (2001), Imperial College Press, London.

[JS00] N. Jacob and R.L. Schilling, *Fractional derivatives, non-symmetric and time-dependent Dirichlet forms and the drift form*, Z. Anal. Anwendungen **19**(3) (2000), 801-830.

[JLM85] G. Jona-Lasinio and P.K. Mitter, *On the stochastic quantization of field theory*, Comm. Math. Phys. **101** (1985), 409-436.

[Jo95] R. Jost, *Foundations of Quantum Field Theory*, pp. 153-169 in R. Jost, Das Märchen vom Elfenbeinernen Turm, Hrsg. K. Hepp et al., Springer, Berlin (1995).

[KaYa05] H. Kaneko, K. Yasuda, *Capacities associated with Dirichlet space on an infinite extension of a local field*, Forum Math. **17** (2005), 1011-1032.

[KKVW09] V. Kant, T. Klauss, J. Voigt, M. Weber, *Dirichlet forms for singular one-dimensional operators and on graphs*, J. Evol. Equ. **9** (2009), 637-659.

[KaStr14] W. Karwowski and L. Streit, *contribution to this volume.*

[KR07] H. Kawabi and M. Röckner, *Essential self-adjointness of Dirichlet operators on a path space with Gibbs measures via an SPDE approach*, J. Funct. Anal **242** (2007), 486-518.

[Kol06] A.V. Kolesnikov, *Mosco convergence of Dirichlet forms in infinite dimensions with changing reference measures*, JFA **230** (2006), 382-418.

[KT91] Y. Kondratiev and T.V. Tsycalenko, *Dirichlet operators and associated differential equations*, Selecta Math. Sovietica **10**(4) (1991), 345-397.

[Ku92] S. Kusuoka, *Høegh-Krohn's model of quantum fields and the absolute continuity of measures*, pp. 405-424 in S. Albeverio, J.E. Fenstad, H. Holden, T. Lindstrøm, eds., Ideas and Methods in Mathematical Analysis, stochastics, and Applications, Cambridge Univ. Press (1992).

[Kus82] S. Kusuoka, *Dirichlet forms and diffusion processes on Banach spaces*, J. Fac. Sc. Univ. Tokyo **29** (1982), 79-95.

[KS03] K. Kuwae and T. Shioya, *Sobolev and Dirichlet spaces over maps between metric spaces*, J. Reine Angew. Math. **555** (2003), 39-75.

[LW99] A. Laptev and T. Weidl, *Hardy inequalities for magnetic Dirichlet forms*, Oper. Theory Adv. Appl. **108** (1999), 299-305.

[LM72] J.L. Lions and E. Magenes, *Non-homogeneous boundary value problems and applications*, Grundlehren Math. Wiss., Springer, Berlin, 1972.

[LR98] V. Liskevich and M. Röckner, *Strong Uniqueness for Certain Infinite Dimensional Dirichlet Operators and Applications to Stochastic Quantization*, Ann. Scuola Norm. Pisa **27**(1) (1998), 69-91.

[Ma95] Z.M. Ma, *Quasiregular Dirichlet forms and applications*, Proc. ICM 1994, Zürich, Switzerland, 1994, Birkhäuser, Basel (1995).

[MR92] Z.M. Ma and M. Röckner, *Introduction to the theory of (non-symmetric) Dirichlet forms*, Springer, Berlin, 1992

[MR10] C. Marinelli and M. Röckner, *On uniqueness of mild solutions for dissipative stochastic evolution equations*, IDAQP **13** (2010), no. 3, 363-376.

[MZ10] C. Marinelli and G. Ziglio, *Ergodicity for nonlinear stochastic evolution equations with multiplicative Poisson noise*, Dyn. PDE **7** (2010), 1-23.

[Mi06] S. Mihalache, *Stochastische Quantisierung bei exponentieller Wechselwirkung*, Diploma Thesis, Bonn (2006).

[MiRo99] R. Mikulevicius and B. Rozovskii, *Martingale problems for stochastic PDE's*, pp. 243-325 in Stochastic Partial Differential Equations: Six Perspectives, R. Carmona and B. Rozovskii, eds., AMS (1999).

[Mi89] S. MITTER, *Markov random fields, stochastic quantization and image analysis*, Math. Appl. **56** (1989), 101-109.

[N73] E. NELSON, *The free Markov field*, J. Funct. Anal. **12** (1973), 211-227

[Osh04] Y. OSHIMA, *Time-dependent Dirichlet forms and related stochastic calculus*, IDAQP **7** (2004), 281-316.

[PaWu81] G. PARISI AND Y.S. WU, *Perturbation theory without gauge fixing*, Sci. Sinica **24** (1981), 483-496.

[PeZ07] S. PESZAT AND J. ZABCZYK, *Stochastic partial differential equations with Lévy noise. An evolution equation approach*, Encyclopedia of Mathematics and its Applications **113**, Cambridge University Press, Cambridge (2007).

[RS75] M. REED AND B. SIMON, *Methods of Modern Mathematical Physics. II: Fourier Analysis, Self-Adjointness*, Academic Press, Orlando, FL, 1975

[RhVa13] R. RHODES, V. VARGAS, *Gaussian multiplicative chaos and applications*, a seriens arXiv (2013)

[R86] M. RÖCKNER, *Specifications and Martin boundaries for $P(\Phi)_2$-random fields*, Commun. Math. Phys. **106** (1986), 105-135.

[R88] M. RÖCKNER, *Traces of harmonic functions and a new path space for the free quantum field*, J. Funct. Anal. **79** (1988), 211-249.

[R98] M. RÖCKNER, *L^p-analysis of finite and infinite dimensional diffusion operators*, Stochastic PDE's and Kolmogorov's equations in infinite dimensions (Giuseppe Da Prato, ed.), Lect. Notes Math., vol. 1715, Springer, Berlin, 1999, pp. 401-408.

[RS92] M. RÖCKNER AND B. SCHMULAND, *Tightness of general $C_{1,p}$-capacities on Banach space*, J. Funct. Anal. **108** (1992), 1-12

[RZ92] M. RÖCKNER AND T.S. ZHANG, *Uniqueness of generalized Schrödinger operators and applications*, J. Funct. Anal. **105** (1992), 187-231

[SU07] R.L. SCHILLING AND T. UEMURA, *On the Feller property of Dirichlet forms generated by pseudo differential operators*, Tohoku Math. J. (2) **59**(3) (2007), 401-422.

[SS03] B. SCHMULAND AND W. SUN, *The maximum Markovian self-adjoint extensions of Dirichlet operators for interacting particle systems*, Forum Math. **15**(4) (2003), 615-638.

[ST92] I. SHIGEKAWA AND S. TANIGUCHI, *Dirichlet forms on separable metric spaces*, Ed. A.N. Shiryaev et al., World Scientific, Singapore, Probability Theory and Mathematical Statistics (1992), 324-353.

[Sil74] M.L. SILVERSTEIN, *Symmetric Markov Processes*, Lect. Notes in Maths. **426** (1974), Springer, Berlin.

[Sil76] M.L. SILVERSTEIN, *Boundary Theory for Symmetric Markov Processes*, Lect. Notes in Maths. **516** (1976), Springer, Berlin.

[Si74] B. SIMON, *The $P(\Phi)_2$-Euclidean (Quantum) Field Theory*, Princeton Univ. Press, Princeton, NJ, 1974

[Sta99a] W. STANNAT, *Theory of generalized Dirichlet forms and its applications in analysis and stochastics*, Am. Math. Soc. **142**(678) (1999), 1-101.

[Sta99b] W. STANNAT, *(Nonsymmetric) Dirichlet operators on L^1: existence, uniqueness and associated Markov processes*, Annali. Scuola Norm. Sup. di Pisa Cl. Sci (4) **28** (1999), 99-140.

[StV06] D.W. STROOCK AND S.R.S. VARADHAN, *Multidimensional diffusion processes*, Reprint of the 1997 edition, Classics in Mathematics, Springer, Berlin (2006).

[Tak92] M. TAKEDA, *The maximum Markovian self-adjoint extensions of generalized*

Schrödinger operators, J. Math. Soc. Jap. **44** (1992), 113-130.

[Y89] J. A. YAN *Generalizations of Gross' and Minlos' theorems*, In: Azema, J., Meyer, P.A., Yor, M. (eds.) Séminaire de Probabilités. XXII (Lect. Notes Math., vol. 1372, pp. 395-404) Berlin Heidelberg New York: Springer 1989

Chapter 4

Comparison of quenched and annealed invariance principles for random conductance model: Part II

Martin Barlow*, Krzysztof Burdzy† and Adám Timár‡

We show that there exists an ergodic conductance environment such that the weak (annealed) invariance principle holds for the corresponding continuous time random walk but the quenched invariance principle does not hold. In the present paper we give a proof of the full scaling limit for the weak invariance principle, improving the result in an earlier paper where we obtained a subsequential limit.

Contents

1. Introduction

This article contains the completion of the project started in a previous paper [4], where we proved that there exists an ergodic conductance environment such that the weak (annealed) invariance principle holds for the corresponding continuous time random walk along a subsequence but the quenched invariance principle does not hold. In the present paper we give a proof of the full scaling limit for the weak invariance principle, improving the result in [4]. The improved result is, in a sense, a quantitative form of the invariance principle. The proof consists of several lemmas. Some of them are specific to our model but some of them have the more general character and may serve as technical elements for related projects. Since this paper is a continuation of [4], we start by presenting basic notation and definitions from that paper.

Let $d \geq 2$ and let E_d be the set of all non oriented edges in the d-dimensional integer lattice, that is, $E_d = \{e = \{x, y\} : x, y \in \mathbb{Z}^d, |x - y| = 1\}$. Let $\{\mu_e\}_{e \in E_d}$ be a random process with non-negative values, defined on some probability space $(\Omega, \mathcal{F}, \mathbb{P})$. The process $\{\mu_e\}_{e \in E_d}$ represents random conductances. We write $\mu_{xy} =$

*Department of Mathematics, University of British Columbia, Vancouver, B.C., Canada V6T 1Z2
†Department of Mathematics, Box 354350, University of Washington, Seattle, WA 98195, USA
‡Bolyai Institute, University of Szeged, Aradi v. tere 1, 6720 Szeged, Hungary

$\mu_{yx} = \mu_{\{x,y\}}$ and set $\mu_{xy} = 0$ if $\{x,y\} \notin E_d$. Set

$$\mu_x = \sum_y \mu_{xy}, \qquad P(x,y) = \frac{\mu_{xy}}{\mu_x},$$

with the convention that $0/0 = 0$ and $P(x,y) = 0$ if $\{x,y\} \notin E_d$. For a fixed $\omega \in \Omega$, let $X = \{X_t, t \geq 0, P^x_\omega, x \in \mathbb{Z}^d\}$ be the continuous time random walk on $\mathbb{Z}^d$, with transition probabilities $P(x,y) = P_\omega(x,y)$, and exponential waiting times with mean $1/\mu_x$. The corresponding expectation will be denoted E^x_ω. For a fixed $\omega \in \Omega$, the generator $\mathcal{L}$ of X is given by

$$\mathcal{L}f(x) = \sum_y \mu_{xy}(f(y) - f(x)). \tag{1.1}$$

In [3] this is called the *variable speed random walk* (VSRW) among the conductances μ_e. This model, of a reversible (or symmetric) random walk in a random environment, is often called the Random Conductance Model.

We are interested in functional Central Limit Theorems (FCLTs) for the process X. Given any process X, for $\varepsilon > 0$, set $X^\varepsilon_t = \varepsilon X_{t/\varepsilon^2}$, $t \geq 0$. Let $\mathcal{D}_T = D([0,T], \mathbb{R}^d)$ denote the Skorokhod space, and let $\mathcal{D}_\infty = D([0,\infty), \mathbb{R}^d)$. Write d_S for the Skorokhod metric and $\mathcal{B}(\mathcal{D}_T)$ for the σ-field of Borel sets in the corresponding topology. Let X be the canonical process on $\mathcal{D}_\infty$ or $\mathcal{D}_T$, P_{BM} be Wiener measure on $(\mathcal{D}_\infty, \mathcal{B}(\mathcal{D}_\infty))$ and let E_{BM} be the corresponding expectation. We will write W for a standard Brownian motion. It will be convenient to assume that $\{\mu_e\}_{e \in E_d}$ are defined on a probability space $(\Omega, \mathcal{F}, \mathbb{P})$, and that X is defined on $(\Omega, \mathcal{F}) \times (\mathcal{D}_\infty, \mathcal{B}(\mathcal{D}_\infty))$ or $(\Omega, \mathcal{F}) \times (\mathcal{D}_T, \mathcal{B}(\mathcal{D}_T))$. We also define the averaged or annealed measure $\mathbf{P}$ on $(\mathcal{D}_\infty, \mathcal{B}(\mathcal{D}_\infty))$ or $(\mathcal{D}_T, \mathcal{B}(\mathcal{D}_T))$ by

$$\mathbf{P}(G) = \mathbb{E} P^0_\omega(G). \tag{1.2}$$

Definition 1.1. For a bounded function F on $\mathcal{D}_T$ and a constant matrix Σ, let $\Psi^F_\varepsilon = E^0_\omega F(X^\varepsilon)$ and $\Psi^F_\Sigma = E_{\text{BM}} F(\Sigma W)$. We will use I to denote the identity matrix.

(i) We say that the *Quenched Functional CLT* (QFCLT) holds for X with limit ΣW if for every $T > 0$ and every bounded continuous function F on $\mathcal{D}_T$ we have $\Psi^F_\varepsilon \to \Psi^F_\Sigma$ as $\varepsilon \to 0$, with $\mathbb{P}$-probability 1.

(ii) We say that the *Weak Functional CLT* (WFCLT) holds for X with limit ΣW if for every $T > 0$ and every bounded continuous function F on $\mathcal{D}_T$ we have $\Psi^F_\varepsilon \to \Psi^F_\Sigma$ as $\varepsilon \to 0$, in $\mathbb{P}$-probability.

(iii) We say that the *Averaged (or Annealed) Functional CLT* (AFCLT) holds for X with limit ΣW if for every $T > 0$ and every bounded continuous function F on $\mathcal{D}_T$ we have $\mathbb{E}\Psi^F_\varepsilon \to \Psi^F_\Sigma$. This is the same as standard weak convergence with respect to the probability measure $\mathbf{P}$.

If we take Σ to be non-random then, since F is bounded, it is immediate that QFCLT $\Rightarrow$ WFCLT. In general for the QFCLT the matrix Σ might depend on the environment $\mu_{\cdot}(\omega)$. However, if the environment is stationary and ergodic, then Σ is a shift invariant function of the environment, so must be $\mathbb{P}$–a.s. constant. In [9] it is proved that if μ_e is a stationary ergodic environment with $\mathbb{E}\mu_e < \infty$ then the WFCLT holds. In [4, Theorem 1.3] it is proved that for the random conductance model the AFCLT and WFCLT are equivalent.

Definition 1.2. We say an environment (μ_e) on $\mathbb{Z}^d$ is *symmetric* if the law of (μ_e) is invariant under symmetries of $\mathbb{Z}^d$.

If (μ_e) is stationary, ergodic and symmetric, and the WFCLT holds with limit ΣW then the limiting covariance matrix $\Sigma^T\Sigma$ must also be invariant under symmetries of $\mathbb{Z}^d$, so must be a constant times the identity.

In a previous paper [4] we proved the following theorem:

Theorem 1.1. *Let $d = 2$ and $p < 1$. There exists a symmetric stationary ergodic environment $\{\mu_e\}_{e\in E_2}$ with $\mathbb{E}(\mu_e^p \vee \mu_e^{-p}) < \infty$ and a sequence $\varepsilon_n \to 0$ such that*
(a) the WFCLT holds for X^{ε_n} with limit W, i.e., for every $T > 0$ and every bounded continuous function F on $\mathcal{D}_T$ we have $\Psi^F_{\varepsilon_n} \to \Psi^F_I$ as $n \to \infty$, in $\mathbb{P}$-probability, but
(b) the QFCLT does not hold for X^{ε_n} with limit ΣW for any Σ.

In this paper we prove that for an environment similar to that in Theorem 1.1 the WFCLT holds for X^ε as $\varepsilon \to 0$, and not just along a subsequence.

Theorem 1.2. *Let $d = 2$ and $p < 1$. There exists a symmetric stationary ergodic environment $\{\mu_e\}_{e\in E_2}$ with $\mathbb{E}(\mu_e^p \vee \mu_e^{-p}) < \infty$ such that*
(a) the WFCLT holds for X^ε with limit W, i.e., for every $T > 0$ and every bounded continuous function F on $\mathcal{D}_T$ we have $\Psi^F_\varepsilon \to \Psi^F_I$ as $\varepsilon \to 0$, in $\mathbb{P}$-probability, but
(b) the QFCLT does not hold for X^ε with limit ΣW for any Σ.

For more remarks on this problem see [4].

Acknowledgment. We are grateful to Emmanuel Rio, Pierre Mathieu, Jean-Dominique Deuschel and Marek Biskup for some very useful discussions.

2. Description of the environment

Here we recall the environment given in [4]. We refer the reader to that paper for proofs of some basic properties.

Let $\Omega = (0,\infty)^{E_2}$, and $\mathcal{F}$ be the Borel σ-algebra defined using the usual product topology. Then every $t \in \mathbb{Z}^2$ defines a transformation $T_t(\omega) = \omega + t$ of Ω. Stationarity and ergodicity of the measures defined below will be understood with respect to these transformations.

All constants (often denoted c_1, c_2, etc.) are assumed to be strictly positive and finite. For a set $A \subset \mathbb{Z}^2$ let $E(A) \subset E_2$ be the set of all edges with both endpoints in A. Let $E_h(A)$ and $E_v(A)$ respectively be the set of horizontal and vertical edges in $E(A)$. Write $x \sim y$ if $\{x, y\}$ is an edge in $\mathbb{Z}^2$. Define the exterior boundary of A by

$$\partial A = \{y \in \mathbb{Z}^2 - A : y \sim x \text{ for some } x \in A\}.$$

Let also

$$\partial_i A = \partial(\mathbb{Z}^2 - A).$$

Define balls in the ℓ^∞ norm by $\mathcal{B}(x, r) = \{y : ||x - y||_\infty \leq r\}$; of course this is just the square with center x and side $2r$.

Let $\{a_n\}_{n\geq 0}$, $\{\beta_n\}_{n\geq 1}$ and $\{b_n\}_{n\geq 1}$ be strictly increasing sequences of positive integers growing to infinity with n, with

$$1 = a_0 < b_1 < \beta_1 < a_1 \ll b_2 < \beta_2 < a_2 \ll b_3 \ldots$$

We will impose a number of conditions on these sequences in the course of the paper. We collect the main ones here. There is some redundancy in the conditions, for easy reference.

(1) a_n is even for all n.
(2) For each $n \geq 1$, a_{n-1} divides b_n, and b_n divides β_n and a_n.
(3) $b_1 \geq 10^{10}$.
(4) $a_n/\sqrt{2n} \leq b_n \leq a_n/\sqrt{n}$ for all n, and $b_n \sim a_n/\sqrt{n}$.
(5) $b_{n+1} \geq 2^n b_n$ for all n.
(6) $b_n > 40 a_{n-1}$ for all n.
(7) b_n is large enough so that the estimates (5.1) and (6.1) of [4] hold.
(8) $100 b_n < \beta_n \leq b_n n^{1/4} < 2\beta_n < a_n/10$ for n large enough.

In addition, at various points in the proof, we will assume that a_n is sufficiently much larger than b_{n-1} so that a process $X^{(n-1)}$ defined below is such that for $a \geq a_n$ the rescaled process

$$(a^{-1} X^{(n-1)}_{a^2 t}, t \geq 0)$$

is sufficiently close to Brownian motion. We will mark the places in the proof where we impose these extra conditions by (♣).

We begin our construction by defining a collection of squares in $\mathbb{Z}^2$. Let

$$\begin{aligned} B_n &= [0, a_n]^2, \\ B'_n &= [0, a_n - 1]^2 \cap \mathbb{Z}^2, \\ \mathcal{S}_n(x) &= \{x + a_n y + B'_n : y \in \mathbb{Z}^2\}. \end{aligned}$$

Thus $\mathcal{S}_n(x)$ gives a tiling of $\mathbb{Z}^2$ by disjoint squares of side $a_n - 1$ and period a_n. We say that the tiling $\mathcal{S}_{n-1}(x_{n-1})$ is a refinement of $\mathcal{S}_n(x_n)$ if every square $Q \in \mathcal{S}_n(x_n)$ is a finite union of squares in $\mathcal{S}_{n-1}(x_{n-1})$. It is clear that $\mathcal{S}_{n-1}(x_{n-1})$ is a refinement of $\mathcal{S}_n(x_n)$ if and only if $x_n = x_{n-1} + a_{n-1} y$ for some $y \in \mathbb{Z}^2$.

Take $\mathcal{O}_1$ uniform in B_1', and for $n \geq 2$ take $\mathcal{O}_n$, conditional on $(\mathcal{O}_1, \ldots, \mathcal{O}_{n-1})$, to be uniform in $B_n' \cap (\mathcal{O}_{n-1} + a_{n-1}\mathbb{Z}^2)$. We now define random tilings by letting

$$\mathcal{S}_n = \mathcal{S}_n(\mathcal{O}_n),\ n \geq 1.$$

Let η_n, K_n be positive constants; we will have $\eta_n \ll 1 \ll K_n$. We define conductances on E_2 as follows. Recall that a_n is even, and let $a_n' = \frac{1}{2}a_n$. Let

$$C_n = \{(x, y) \in B_n \cap \mathbb{Z}^2 : y \geq x, x + y \leq a_n\}.$$

We first define conductances $\nu_e^{n,0}$ for $e \in E(C_n)$. Let

$$\begin{aligned} D_n^{00} &= \big\{(a_n' - \beta_n, y), a_n' - 10b_n \leq y \leq a_n' + 10b_n\big\}, \\ D_n^{01} &= \big\{(x, a_n' + 10b_n), (x, a_n' + 10b_n + 1), (x, a_n' - 10b_n), (x, a_n' - 10b_n - 1), \\ &\qquad a_n' - \beta_n - b_n \leq x \leq a_n' - \beta_n + b_n\big\}. \end{aligned}$$

Thus the set $D_n^{00} \cup D_n^{01}$ resembles the letter I (see Fig. 2).

For an edge $e \in E(C_n)$ we set

$$\begin{aligned} \nu_e^{n,0} &= \eta_n \quad \text{if } e \in E_v(D_n^{01}), \\ \nu_e^{n,0} &= K_n \quad \text{if } e \in E(D_n^{00}), \\ \nu_e^{n,0} &= 1 \quad \text{otherwise.} \end{aligned}$$

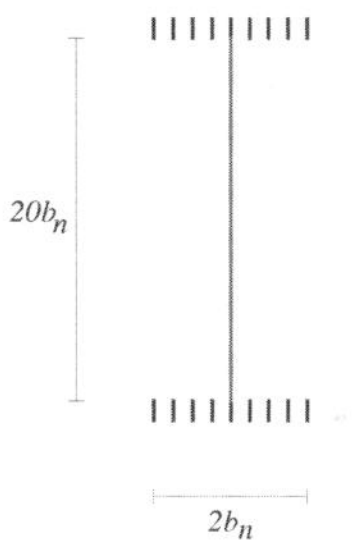

The set $D_n^{00} \cup D_n^{01}$ resembles the letter I. Blue edges have very low conductance. The red line represents edges with very high conductance. Drawing not to scale.

We then extend $\nu^{n,0}$ by symmetry to $E(B_n)$. More precisely, for $z = (x, y) \in B_n$, let $R_1 z = (y, x)$ and $R_2 z = (a_n - y, a_n - x)$, so that R_1 and R_2 are reflections in the lines $y = x$ and $x + y = a_n$. We define R_i on edges by $R_i(\{x, y\}) = \{R_i x, R_i y\}$ for $x, y \in B_n$. We then extend $\nu^{0,n}$ to $E(B_n)$ so that $\nu_e^{0,n} = \nu_{R_1 e}^{0,n} = \nu_{R_2 e}^{0,n}$ for $e \in E(B_n)$. We define the *obstacle* set D_n^0 by setting

$$D_n^0 = \bigcup_{i=0}^{1} \big(D_n^{0,i} \cup R_1(D_n^{0,i}) \cup R_2(D_n^{0,i}) \cup R_1 R_2(D_n^{0,i})\big).$$

Note that $\nu_e^{n,0} = 1$ for every edge adjacent to the boundary of B_n, or indeed within a distance $a_n/4$ of this boundary. If $e = (x, y)$, we will write $e - z = (x - z, y - z)$. Next we extend $\nu^{n,0}$ to E_2 by periodicity, i.e., $\nu_e^{n,0} = \nu_{e+a_n x}^{n,0}$ for all $x \in \mathbb{Z}^2$. We define the conductances ν^n by translation by $\mathcal{O}_n$, so that

$$\nu_e^n = \nu_{e-\mathcal{O}_n}^{n,0}, \; e \in E_2.$$

We also define the obstacle set at scale n by

$$D_n = \bigcup_{x \in \mathbb{Z}^2} (a_n x + \mathcal{O}_n + D_n^0). \tag{2.1}$$

We will sometimes call the set D_n the set of nth level obstacles.

We define the environment μ_e^n inductively by

$$\mu_e^n = \nu_e^n \quad \text{if } \nu_e^n \neq 1,$$
$$\mu_e^n = \mu_e^{n-1} \quad \text{if } \nu_e^n = 1.$$

Once we have proved the limit exists, we will set

$$\mu_e = \lim_n \mu_e^n. \tag{2.2}$$

Lemma 2.1. *(See [4, Theorem 3.1]).*
(a) The environments $(\nu_e^n, e \in E_2)$, $(\mu_e^n, e \in E_2)$ are stationary, symmetric and ergodic.
(b) The limit (2.2) exists $\mathbb{P}$–a.s.
(c) The environment $(\mu_e, e \in E_2)$ is stationary, symmetric and ergodic.

Now let

$$\mathcal{L}_n f(x) = \sum_y \mu_{xy}^n (f(y) - f(x)), \tag{2.3}$$

and $X^{(n)}$ be the associated Markov process. Set

$$\eta_n = b_n^{-(1+1/n)}, \; n \geq 1. \tag{2.4}$$

From Section 4 of [4] we have:

Theorem 2.1. *For each n there exists a constant K_n, depending on $\eta_1, K_1, \ldots \eta_{n-1}, K_{n-1}$, such that the QFCLT holds for $X^{(n)}$ with limit W.*

For each n the process $X^{(n)}$ has invariant measure which is counting measure on $\mathbb{Z}^2$. For $x \in \mathbb{R}^2$ and $a > 0$ write $[xa]$ for the point in $\mathbb{Z}^2$ closest to xa. (We use some procedure to break ties.) We have the following bounds on the transition probabilities of $X^{(n)}$ from [5]. We remark that the constant M_n below is not effective – i.e. the proof does not give any control on its value. Write $k_t(x, y) = (2\pi t)^{-1} \exp(-|x - y|^2/2t)$ for the transition density of Brownian motion in $\mathbb{R}^2$, and

$$p_t^{\omega,n}(x, y) = P_\omega^x(X_t^{(n)} = y)$$

for the transition probabilities for $X^{(n)}$.

Lemma 2.2. *For each $0 < \delta < T$ there exists $M_n = M_n(\delta, T)$ such that for $a \geq M_n$*

$$\frac{1}{2}k_t(x,y) \leq a^2 p^{\omega,n}_{a^2 t}([xa],[ya]) \leq 2k_t(x,y) \quad \text{for all } \delta \leq t \leq T, |x|, |y| \leq T^2. \tag{2.5}$$

3. Preliminary results

Since a proof of Theorem 1.1(b) was given in [4], all we need to prove is part (a) of Theorem 1.2. The argument consists of several lemmas. We start with some preliminary results on weak convergence of probability measures on the space of càdlàg functions. Recall the definitions of the measures $\mathbb{P}$ and P^0_ω.

Recall that $\mathcal{D} := \mathcal{D}_1 = D([0,1], \mathbb{R}^2)$ denotes the space of càdlàg functions equipped with the Skorokhod metric d_S defined as follows (see [6, p. 111]). Let Λ be the family of continuous strictly increasing functions λ mapping $[0,1]$ onto itself. In particular, $\lambda(0) = 0$ and $\lambda(1) = 1$. If $x(t), y(t) \in \mathcal{D}$ then

$$d_S(x,y) = \inf_{\lambda \in \Lambda} \max\Big(\sup_{t\in[0,1]} |\lambda(t) - t|, \sup_{t\in[0,1]} |y(\lambda(t)) - x(t)|\Big).$$

For $x(t) \in \mathcal{D}$, let $\mathrm{Osc}(x,\delta) = \sup\{|x(t) - x(s)| : s,t \in [0,1], |s-t| \leq \delta\}$.

Lemma 3.1. *Suppose that $\sigma : [0,1] \to [0,1]$ is continuous, non-decreasing and $\sigma(0) = 0$ (we do not require that $\sigma(1) = 1$). Suppose that $|\sigma(t) - t| \leq \delta$ for all $t \in [0,1]$. Let $\varepsilon \geq 0$, $\delta_1 > 0$, $x, y \in \mathcal{D}$ with $d_S(x(\cdot), y(\cdot)) \leq \varepsilon$, and $\mathrm{Osc}(x,\delta) \vee \mathrm{Osc}(y,\delta) \leq \delta_1$. Then $d_S(x(\sigma(\cdot)), y(\sigma(\cdot))) \leq \varepsilon + 2\delta_1$.*

Proof. For any $\varepsilon_1 > \varepsilon$ there exists $\lambda \in \Lambda$ such that,

$$\max\Big(\sup_{t\in[0,1]} |\lambda(t) - t|, \sup_{t\in[0,1]} |y(\lambda(t)) - x(t)|\Big) \leq \varepsilon_1.$$

We have for λ satisfying the above condition,

$$\begin{aligned}
&\sup_{t\in[0,1]} |y(\sigma(\lambda(t))) - x(\sigma(t))| \\
&\qquad \leq \sup_{t\in[0,1]} \big(|y(\sigma(\lambda(t))) - y(\lambda(t))| + |y(\lambda(t)) - x(t)| + |x(t) - x(\sigma(t))|\big) \\
&\qquad \leq \mathrm{Osc}(y,\delta) + \varepsilon_1 + \mathrm{Osc}(x,\delta) \leq \varepsilon_1 + 2\delta_1.
\end{aligned}$$

Hence,

$$\max\Big(\sup_{t\in[0,1]} |\lambda(t) - t|, \sup_{t\in[0,1]} |y(\sigma(\lambda(t))) - x(\sigma(t))|\Big) \leq \varepsilon_1 + 2\delta_1.$$

Taking infimum over all $\varepsilon_1 > \varepsilon$ we obtain $d_S(x(\sigma(\cdot)), y(\sigma(\cdot))) \leq \varepsilon + 2\delta_1$. □

Let $\mathbf{d}$ denote the Prokhorov distance between probability measures on a probability space defined as follows (see [6, p. 238]). Recall that $\Omega = (0,\infty)^{E_2}$ and $\mathcal{F}$ is the Borel σ-algebra defined using the usual product topology. We will use measurable spaces $(\mathcal{D}_T, \mathcal{B}(\mathcal{D}_T))$ and $(\Omega, \mathcal{F}) \times (\mathcal{D}_T, \mathcal{B}(\mathcal{D}_T))$, for a fixed T (often $T = 1$). Note that $\mathcal{D}_T$ and $\Omega \times \mathcal{D}_T$ are metrizable, with the metrics generating the usual topologies. A ball around a set A with radius ε will be denoted $\mathcal{B}(A, \varepsilon)$ in either space. For probability measures P and Q, $\mathbf{d}(P, Q)$ is the infimum of $\varepsilon > 0$ such that $P(A) \leq Q(\mathcal{B}(A, \varepsilon)) + \varepsilon$ and $Q(A) \leq P(\mathcal{B}(A, \varepsilon)) + \varepsilon$ for all Borel sets A. Convergence in the metric $\mathbf{d}$ is equivalent to the weak convergence of measures. By abuse of notation we will sometimes write arguments of the function $\mathbf{d}(\,\cdot\,,\,\cdot\,)$ as processes rather than their distributions: for example we will write $\mathbf{d}(\{(1/a)X^{(n)}_{ta^2}, t \in [0,1]\}, \mathrm{P}_{BM})$. We will use $\mathbf{d}$ for the Prokhorov distance between probability measures on $(\Omega, \mathcal{F}) \times (\mathcal{D}_T, \mathcal{B}(\mathcal{D}_T))$. We will write $\mathbf{d}_\omega$ for the metric on the space $(\mathcal{D}_T, \mathcal{B}(\mathcal{D}_T))$. It is straightforward to verify that if, for some processes Y and Z, $\mathbf{d}_\omega(Y, Z) \leq \varepsilon$ for $\mathbb{P}$–a.a. ω, then $\mathbf{d}(Y, Z) \leq \varepsilon$.

We will sometimes write $W(t) = W_t$ and similarly for other processes.

Lemma 3.2. *There exists a function $\rho : (0, \infty) \to (0, \infty)$ such that $\lim_{\delta \downarrow 0} \rho(\delta) = 0$ and the following holds. Suppose that $\delta, \delta' \in (0, 1)$ and $\sigma : [0, 1] \to [0, 1]$ is a non-decreasing stochastic process such that $t - \sigma_t \in [0, \delta]$ for all t, with probability greater than $1 - \delta'$. Suppose that $\{W_t, t \geq 0\}$ has the distribution P_{BM} and $W^*_t = W(\sigma_t)$ for $t \in [0, 1]$. Then $\mathbf{d}(\{W^*_t, t \in [0, 1]\}, P_{\mathrm{BM}}) \leq \rho(\delta) + \delta'$.*

Proof. Suppose that W, W^* and σ are defined on the sample space with a probability measure P. It is easy to see that we can choose $\rho(\delta)$ so that $\lim_{\delta \downarrow 0} \rho(\delta) = 0$ and $P(\mathrm{Osc}(W, \delta) \geq \rho(\delta)) < \rho(\delta)$. Suppose that the event $F := \{\mathrm{Osc}(W, \delta) < \rho(\delta)\} \cap \{\forall t \in [0, 1] : t - \sigma_t \in [0, \delta]\}$ holds. Then taking $\lambda(t) = t$,

$$
\begin{aligned}
\mathrm{d_S}(W, W^*) &\leq \max\Big(\sup_{t \in [0,1]} |\lambda(t) - t|, \sup_{t \in [0,1]} |W(\lambda(t)) - W^*(t)| \Big) \\
&= \sup_{t \in [0,1]} |W(t) - W(\sigma(t))| \leq \mathrm{Osc}(W, \delta) < \rho(\delta).
\end{aligned}
$$

We see that if F holds and $W \in A \subset \mathcal{D}$ then $W^*(\,\cdot\,) \in \mathcal{B}(A, \rho(\delta))$. Since $P(F^c) \leq \rho(\delta) + \delta'$, we obtain

$$
\begin{aligned}
&P(W \in A) \\
&\quad \leq P(\{W \in A\} \cap F) + P(F^c) \leq P(\{W^* \in \mathcal{B}(A, \rho(\delta))\} \cap F) + \rho(\delta) + \delta' \\
&\quad \leq P(W^* \in \mathcal{B}(A, \rho(\delta))) + \rho(\delta) + \delta'.
\end{aligned}
$$

Similarly we have $P(W^* \in A) \leq P(W \in \mathcal{B}(A, \rho(\delta))) + \rho(\delta) + \delta'$, and the lemma follows. □

Lemma 3.3. *Suppose that for some processes X, Y and Z on the interval $[0, 1]$ we have $Z = X + Y$ and $P(\sup_{0 \leq t \leq 1} |X_t| \leq \delta) \geq 1 - \delta$. Then $\mathbf{d}(\{Z_t, t \in [0, 1]\}, \{Y_t, t \in [0, 1]\}) \leq \delta$.*

Proof. Suppose that the event $F := \{\sup_{0\le t\le 1}|X_t| \le \delta\}$ holds. Then taking $\lambda(t) = t$,

$$\begin{aligned} d_S(Z,Y) &\le \max\Big(\sup_{t\in[0,1]}|\lambda(t)-t|,\ \sup_{t\in[0,1]}|Z(\lambda(t))-Y(t)|\Big) \\ &= \sup_{t\in[0,1]}|Z(t)-Y(t)| \le \delta. \end{aligned}$$

We see that if F holds and $Z \in A \subset \mathcal{D}$ then $Y(\cdot) \in \mathcal{B}(A,\delta)$. Since $P(F^c) \le \delta$, we obtain

$$\begin{aligned} P(Z\in A) &\le P(\{Z\in A\}\cap F) + P(F^c) \le P(\{Y\in\mathcal{B}(A,\delta)\}\cap F) + \delta \\ &\le P(Y\in\mathcal{B}(A,\delta)) + \delta. \end{aligned}$$

Similarly we have $P(Y\in A) \le P(Z\in\mathcal{B}(A,\delta)) + \delta$, and the lemma follows. □

Recall that the function $e \to \mu_e^n$ is periodic with period a_n. Hence the random field $\{\mu_e^n\}_{e\in E_2}$ takes only finitely many values – this is a much stronger statement than the fact that μ_e^n takes only finitely many values.

By Theorem 2.1 for each $n \ge 1$,

$$\lim_{a\to\infty} \mathbf{d}(\{(1/a)X^{(n)}_{ta^2}, t\in[0,1]\}, P_{\mathrm{BM}}) = 0.$$

Thus (♣) we can take a_{n+1} so large that for every ω, $n \ge 1$ and $a \ge a_{n+1}$,

$$\mathbf{d}_\omega(\{(1/a)X^{(n)}_{ta^2}, t\in[0,1]\}, P_{\mathrm{BM}}) \le 2^{-n}. \tag{3.1}$$

Let θ denote the usual shift operator for Markov processes, that is, $X_t^{(n)}\circ\theta_s = X^{(n)}_{t+s}$ for all $s,t\ge 0$ (we can and do assume that $X^{(n)}$ is the canonical process on an appropriate probability space). Recall that $\mathcal{B}(x,r) = \{y : ||x-y||_\infty \le r\}$ denote balls in the ℓ^∞ norm in $\mathbb{Z}^2$ (i.e. squares), $a'_n = a_n/2$, $B_n = [0,a_n]^2$ and $u_n = (a'_n, a'_n)$. Note that u_n is the center of B_n. We choose β_n so that

$$b_n n^{1/8} < \beta_n \le \lfloor b_n n^{1/4}\rfloor < 2\beta_n < a_n/10, \tag{3.2}$$

and we assume that n is large enough so that the above inequalities hold. Let $\mathcal{C}_n = \{u_n + \mathcal{O}_n + a_n\mathbb{Z}^2\}$ be the set of centers of the squares in $\mathcal{S}_n$, and let

$$\mathcal{K}(r) = \bigcup_{z\in\mathcal{C}_n} \mathcal{B}(z,r). \tag{3.3}$$

Now let

$$\begin{aligned} \Gamma_n^1 &= \mathcal{K}(2\beta_n), \\ \Gamma_n^2 &= \mathbb{Z}^2 \setminus \mathcal{K}(4\beta_n). \end{aligned}$$

Now define stopping times as follows.

$$\begin{aligned} &S_0^n = T_0^n = 0, \\ &U_k^n = \inf\{t \ge S_{k-1}^n : X_t^{(n)} \in \Gamma_n^2\}, \qquad k \ge 1, \end{aligned}$$

$$\begin{aligned}
S_k^n &= \inf\{t \geq U_k^n : X_t^{(n)} \in \Gamma_n^1\}, \qquad k \geq 1, \\
V_1^n &= \inf\Big\{t \in \bigcup_{k\geq 1} [U_k^n, S_k^n] : X_t^{(n)} \in X^{(n)}(T_0^n) + a_{n-1}\mathbb{Z}^2\Big\}, \\
T_k^n &= \inf\{t \geq V_k^n : X_t^{(n)} \in \Gamma_n^1\}, \qquad k \geq 1, \\
V_k^n &= V_1^n \circ \theta_{T_{k-1}^n}, \qquad k \geq 2.
\end{aligned}$$

Let

$$J = \bigcup_{k=1}^{\infty} [V_k^n, T_k^n];$$

for $t \in J$ the process $X^{(n)}$ is a distance at least β_n away from any nth level obstacle. Now set for $t \geq 0$,

$$\begin{aligned}
\sigma_t^{n,1} &= \int_0^t \mathbf{1}_J(s)ds = \sum_{k=1}^{\infty} (T_k^n \wedge t - V_k^n \wedge t)\,, \\
\sigma_t^{n,2} &= t - \sigma_t^{n,1} = \sum_{k=0}^{\infty} \left(V_{k+1}^n \wedge t - T_k^n \wedge t\right).
\end{aligned}$$

Let $\widehat{\sigma}^{n,j}$ denote the right continuous inverses of these processes, given by

$$\widehat{\sigma}_t^{n,j} = \inf\{s \geq 0 : \sigma_s^{n,j} \geq t\},\ j = 1, 2.$$

Finally let

$$\begin{aligned}
X_t^{n,1} &= X_0^{(n)} + \int_0^t \mathbf{1}_J(s)dX_s^{(n)} \\
&= X_0^{(n)} + \sum_{k=0}^{\infty} \left(X^{(n)}(T_k^n \wedge t) - X^{(n)}(V_k^n \wedge t)\right), \\
\widehat{X}_t^{n,1} &= X_0^{(n)} + X^{n,1}(\widehat{\sigma}_t^{n,1}), \\
X_t^{n,2} &= X_0^{(n)} + \int_0^t \mathbf{1}_{J^c}(s)dX_s^{(n)} \\
&= X_0^{(n)} + \sum_{k=0}^{\infty} \left(X^{(n)}(V_{k+1}^n \wedge t) - X^{(n)}(T_k^n \wedge t)\right), \\
\widehat{X}_t^{n,2} &= X_0^{(n)} + X^{n,2}(\widehat{\sigma}_t^{n,2}).
\end{aligned}$$

The point of this construction is the following. For every fixed ω, the function $e \to \mu_e^{n-1}$ is invariant under the shift by xa_{n-1} for any $x \in \mathbb{Z}^2$, and $X^{(n)}(V_{k+1}^n) = X^{(n)}(T_k^n) + xa_{n-1}$ for some $x \in \mathbb{Z}^2$. It follows that for each $\omega \in \Omega$, we have the following equality of distributions:

$$\{\widehat{X}_t^{n,1}, t \geq 0\} \stackrel{(d)}{=} \{X_t^{(n-1)}, t \geq 0\}. \tag{3.4}$$

The basic idea of the argument which follows is to write $X^{(n)} = X^{n,1} + X^{n.2}$. By Theorem 2.1, or more precisely by (3.1), the process $X^{n,1}$ is close to Brownian motion, so to prove Theorem 1.2 we need to prove that $X^{n,2}$ is small.

We state the next lemma at a level of generality greater than what we need in this article. A variant of our lemma is in the book [1] but we could not find a statement that would match perfectly our needs. Consider a finite graph $G = (\mathcal{V}, E)$ and suppose that for any edge $\overline{xy}$, μ_{xy} is a non-negative real number. Assume that $\sum_{y\sim x} \mu_{xy} > 0$ for all x. For $f : \mathcal{V} \to \mathbb{R}$ set

$$\mathcal{E}(f, f) = \sum_{\{x,y\}\in E} \mu_{xy}(f(y) - f(x))^2.$$

Suppose that $A_1, A_2 \subset \mathcal{V}$, $A_1 \cap A_2 = \emptyset$, and let

$$\begin{aligned} \mathcal{H} &= \{f : \mathcal{V} \to \mathbb{R} \text{ such that } f(x) = 0 \text{ for } x \in A_1, f(y) = 1 \text{ for } y \in A_2\}, \\ \mathbf{r}^{-1} &= \inf\{\mathcal{E}(f, f) : f \in \mathcal{H}\}. \end{aligned}$$

Thus $\mathbf{r}$ is the effective resistance between A_1 and A_2. Let Z be the continuous time Markov process on $\mathcal{V}$ with the generator $\mathcal{L}$ given by

$$\mathcal{L}f(x) = \sum_{y} \mu_{xy}(f(y) - f(x)). \tag{3.5}$$

Let $T_i = \inf\{t \geq 0 : Z_t \in A_i\}$ for $i = 1, 2$, and let $Z^{(i)}$ be Z killed at time T_i.

Lemma 3.4. *There exist probability measures ν_1 on A_1 and ν_2 on A_2 such that*

$$E^{\nu_2} T_1 + E^{\nu_1} T_2 = \mathbf{r}|\mathcal{V}|.$$

Moreover, for $i = 1, 2$, ν_i is the capacitary measure of A_i for the process $Z^{(3-i)}$.

Proof. Let $h_{12}(x) = P^x(T_1 < T_2)$. Set $D = \mathcal{V} - A_1$ and recall that $Z^{(i)}$ is Z killed at time T_i. Let G_2 be the Green operator for $Z^{(2)}$, and $g_2(x, y)$ be the density of G_2 with respect to counting measure, so that

$$E^x T_2 = \sum_{y\in\mathcal{V}} g_2(x, y).$$

Note that $g_2(x, y) = g_2(y, x)$. Let e_{12} be the capacitary measure of A_1 for the process $Z^{(2)}$. Then $\mathbf{r}^{-1} = \sum_{z\in A_1} e_{12}(z)$, and

$$h_{12}(x) = \sum_{z\in A_1} e_{12}(z) g_2(z, x).$$

So, if $\nu_1 = \mathbf{r}e_{12}$, then

$$\begin{aligned} \sum_{y\in\mathcal{V}} h_{12}(y) &= \sum_{y\in\mathcal{V}} \sum_{x\in A_1} e_{12}(x) g_2(x, y) \\ &= \mathbf{r}^{-1} \sum_{x\in A_1} \nu_1(x) \sum_{y\in\mathcal{V}} g_2(x, y) \end{aligned}$$

$$= \mathbf{r}^{-1} \sum_{x \in A_1} \nu_1(x) E^x T_1 = \mathbf{r}^{-1} E^{\nu_1} T_2.$$

Similarly if $h_{21}(x) = \mathbb{P}^x(T_2 < T_1)$ we obtain $\mathbf{r}^{-1} E^{\nu_2} T_1 = \sum_{y \in \mathcal{V}} h_{21}(y)$, and since $h_{12} + h_{21} = 1$, adding these equalities proves the lemma. □

4. Estimates on the process $X^{n,2}$

In this section we will prove

Proposition 4.1. *For every $\delta > 0$ there exists n_1 such that for all $n \ge n_1$, $u \ge a_n^2$, and ω such that $0 \notin \Gamma_n^1 \setminus \partial_i \Gamma_n^1$,*

$$P_\omega^0 \left(\sigma_u^{n,2}/u \le \delta, \sup_{0 \le s \le u} u^{-1/2} |X_s^{n,2}| \le \delta \right) \ge 1 - \delta. \tag{4.1}$$

The proof requires a number of steps. We begin with a Harnack inequality.

Lemma 4.1. *Let $1 \le \lambda \le 10$. There exist $p_1 > 0$ and $n_1 \ge 1$ with the following properties.*
(a) Let $x \in \mathbb{Z}^2$, let $B_1 = \mathcal{B}(x, \lambda\beta_n)$ and $B_2 = \mathcal{B}(x, (2/3)\lambda\beta_n)$. Let F be the event that $X^{(n)}$ makes a closed loop around B_2 inside $B_1 - B_2$ before its first exit from B_1. If $n \ge n_1$ and $D_n \cap B_1 = \emptyset$ then $P_\omega^y(F) \ge p_1$ for all $y \in B_2$.
(b) Let h be harmonic in B_1. Then

$$\max_{B_2} h \le p_1^{-1} \min_{B_2} h. \tag{4.2}$$

Proof. (a) Using (♣) and (3.1) we can make a Brownian approximation to $\beta_n^{-1} X^{(n)}$ which is good enough so that this estimate holds.
(b) Let $y \in B_1$ be such that $h(y) = \max_{z \in B_2} h(z)$. Then by the maximum principle there exists a connected path γ from y to $\partial_i B_1$ with $h(w) \ge h(y)$ for all $w \in \gamma$. Now let $y' \in B_2$. On the event F the process $X^{(n)}$ must hit γ, and so we have

$$h(y') \ge P_\omega^{y'}(F) \min_\gamma h \ge p_1 h(y),$$

proving (4.2). □

Lemma 4.2. *For some n_1 and c_1, for all $n \ge n_1$, $k \ge 1$, and ω such that $0 \notin \Gamma_n^1 \setminus \partial_i \Gamma_n^1$,*

$$E_\omega^0(U_k^n - S_{k-1}^n \mid \mathcal{F}_{S_{k-1}^n}) \le c_1 \beta_n^2. \tag{4.3}$$

Proof. Assume that ω is such that $0 \notin \Gamma_n^1 \setminus \partial_i \Gamma_n^1$. By the strong Markov property applied at S_{k-1}^n for $k > 1$, it is enough to prove the Lemma for $k = 1$, that is that $E_\omega^x(U_1^n) \le c_1 \beta_n^2$ for all $x \notin \Gamma_n^1 \setminus \partial_i \Gamma_n^1$. Let

$$\mathcal{V} = \mathcal{B}(u_n + \mathcal{O}_n, 4\beta_n + 1),$$
$$A_1 = \partial_i \mathcal{B}(u_n + \mathcal{O}_n, (3/2)\beta_n),$$

$$A_2 = \partial_i \mathcal{V},$$
$$A_3 = \partial_i \mathcal{B}(u_n + \mathcal{O}_n, 2\beta_n)$$
$$T_i = \inf\{t \geq 0 : X_t^{(n)} \in A_i\}, \qquad i = 1, 2, 3.$$

Let Z be the continuous time Markov chain defined on $\mathcal{V}$ by (3.5), relative to the environment μ^n. Note that the transition probabilities from x to one of its neighbors are the same for Z and $X^{(n)}$ if x is in the interior of $\mathcal{V}$, i.e., $x \notin \partial_i \mathcal{V} \cup (\mathbb{Z}^2 \setminus \mathcal{V})$. Note also that Z and $X^{(n-1)}$ have the same transition probabilities in the region between A_1 and A_3. The expectations and probabilities in this proof will refer to Z. By Lemma 3.4, there exists a probability measure ν_1 on A_1 such that $E^{\nu_1} T_2 \leq \mathbf{r}|\mathcal{V}|$. We have $|\mathcal{V}| \leq c_2 \beta_n^2$.

To estimate $\mathbf{r}$ note that by the choice of the constants η_{n-1} and K_{n-1} in Theorem 2.1, the resistance (with respect to μ_e^{n-1}) between two opposite sides of any square in $\mathcal{S}_{n-1}$ will be 1. It follows that the resistance between two opposite sides of any square side β_n which is a union of squares in $\mathcal{S}_{n-1}$ will also be 1. So, using Thompson's principle as in [2] we deduce that $\mathbf{r} \leq c_3$.

So, by Lemma 3.4 we have

$$E^{\nu_1} T_2 \leq c_4 \beta_n^2. \tag{4.4}$$

We have for some c_5, $p_1 > 0$ all n and $x \in \mathcal{V} \setminus \mathcal{B}(u_n + \mathcal{O}_n, (3/2)\beta_n)$,

$$P_\omega^x(T_1 \wedge T_2 \leq c_5 \beta_n^2) > p_1,$$

because an analogous estimate holds for Brownian motion and (♣) we have (3.1). This and a standard argument based on the strong Markov property imply that for $x \in A_3$,

$$E_\omega^x(T_1 \wedge T_2) \leq c_6 \beta_n^2.$$

Now for $y \in A_1$ and $x \in \mathcal{V}$ set

$$\nu_3^x(y) = P_\omega^x(X^{(n)}(T_1 \wedge T_2) = y).$$

(Note that there exist x with $\sum_{y \in A_1} \nu_3^x(y) < 1$.) We obtain for $n \geq n_2$ and $x \in A_3$,

$$\begin{aligned} E_\omega^x(T_2) &= E_\omega^x(T_1 \wedge T_2) + E_\omega^x((T_2 - T_1)\mathbf{1}_{T_1 < T_2}) \\ &= E_\omega^x(T_1 \wedge T_2) + E^{\nu_3^x} T_2 \leq c_6 \beta_n^2 + E_\omega^{\nu_3^x} T_2. \end{aligned} \tag{4.5}$$

For $y \in A_1$ the function $x \to \nu_3^x(y)$ is harmonic in $\mathcal{V} \setminus A_1$. So we can apply the Harnack inequality Lemma 4.1 to deduce that there exists c_7 such that

$$\nu_3^x(y) \leq c_7 \nu_3^{x'}(y) \text{ for all } x, x' \in A_3, y \in A_1. \tag{4.6}$$

The measure ν_1 is the hitting distribution on A_1 for the process Z starting with ν_2 (see [1, Chap. 3, p. 45]). So for any $x' \in A_3$,

$$\nu_1(y) = P_0^{\nu_2}(Z_{T_1} = y) = \sum_{x \in A_3} P_0^{\nu_2}(Z_{T_1} = x) P_\omega^x(Z_{T_1} = y)$$

$$\geq \sum_{x\in A_3} P_0^{\nu_2}(Z_{T_1}=x)P_\omega^x(Z_{T_1\wedge T_2}=y) \geq \min_{x\in A_3} \nu_3^x(y) \geq c_7^{-1}\nu_3^{x'}(y).$$

Hence for any $x \in A_3$,

$$E_\omega^{\nu_3^x} T_2 \leq c_7 E_\omega^{\nu_1} T_2 \leq c_8\beta_n^2,$$

and combining this with (4.5) completes the proof. □

Let

$$R_n^y = \inf\left\{t\geq 0 : X_t^{(n)} \in (y+a_{n-1}\mathbb{Z}^2)\cup\Gamma_n^1\right\}.$$

Lemma 4.3. *There exist $c_1>0$ and $p_1<1$ such that for all $x,y\in\mathbb{Z}^2$,*

$$P_\omega^x\left(R_n^y \geq c_1 b_n^2\right) \leq p_1, \tag{4.7}$$

$$P_\omega^x\left(\sup_{0\leq t\leq R_n^y} |x-X_t^{(n)}| \geq c_1 b_n\right) \leq p_1. \tag{4.8}$$

Proof. Recall that the family $\{\mu_{x+\cdot}^{n-1}\}_{x\in\mathbb{Z}^2}$ of translates of the environment $\mu_\cdot^{n-1}$ contains only a finite number of distinct elements. Since each square in $\mathcal{S}_{n-1}$ contains one point in $(y+a_{n-1}\mathbb{Z}^2)$, if b_n/a_{n-1} is sufficiently large (♣) then using the transition density estimates (2.5) as well as (3.1), we obtain (4.7) and (4.8). □

Lemma 4.4. *For some n_1 and c_1, for all $n\geq n_1$, $k\geq 1$, and ω such that $0\notin \Gamma_n^1\setminus\partial_i\Gamma_n^1$,*

$$E_\omega^0(V_k^n - T_{k-1}^n \mid \mathcal{F}_{T_{k-1}^n}) \leq c_1 b_n^2 n^{1/2}. \tag{4.9}$$

Proof. Assume that ω is such that $0\notin\Gamma_n^1\setminus\partial_i\Gamma_n^1$. Let

$$\widehat{R}_k^n = \inf\left\{t\geq U_k^n : X_t^{(n)} \in (X^{(n)}(T_0^n)+a_{n-1}\mathbb{Z}^2)\cup\Gamma_n^1\right\}.$$

Let $F_k=\{\widehat{R}_k^n < S_k^n\}$ and $G_k=\bigcap_{j=1}^k F_j^c$. Since $b_n n^{1/8}<\beta_n$ for large n, we obtain from (4.8) and definitions of $\Gamma_n^1,\Gamma_n^2,U_k^n$ and S_k^n that there exists $p_2>0$ such that for $x\in\Gamma_n^2$,

$$P_\omega^x(F_k\mid\mathcal{F}_{U_k^n}) > p_2.$$

Hence,

$$P_\omega^x(G_k) < (1-p_2)^k. \tag{4.10}$$

Note that if F_k occurs then $V_1^n\leq\widehat{R}_k^n$. We have, using (4.3), (4.7) and (4.10),

$$\begin{aligned}E_\omega^0(V_1^n-T_0^n) &\leq \sum_{k=1}^\infty E_\omega^0((U_k^n-S_{k-1}^n)\mathbf{1}_{G_{k-1}}) + \sum_{k=1}^\infty E_\omega^0((\widehat{R}_k^n-U_k^n)\mathbf{1}_{G_{k-1}})\\ &\leq \sum_{k=1}^\infty c_2\beta_n^2(1-p_2)^{k-1} + \sum_{k=1}^\infty c_3 b_n^2(1-p_2)^{k-1}\end{aligned}$$

$$\leq c_4 \beta_n^2 \leq c_5 b_n^2 n^{1/2}.$$

This proves the lemma for $k = 1$. The general case is obtained by applying this estimate to the process shifted by T_{k-1}^n; in other words, by using the strong Markov property. □

Lemma 4.5. *For every $\delta > 0$ there exists n_1 such that for all $n \geq n_1$, $u \geq a_n^2$, and ω such that $0 \notin \Gamma_n^1 \setminus \partial_i \Gamma_n^1$,*

$$P_\omega^0 \left(\sigma_u^{n,2} / u \leq \delta \right) \geq 1 - \delta/2. \tag{4.11}$$

Proof. Assume that ω is such that $0 \notin \Gamma_n^1 \setminus \partial_i \Gamma_n^1$. Fix an arbitrarily small $\delta > 0$, consider $u \geq a_n^2$ and let $j_* = \lceil u/(b_n^2 n^{5/8}) \rceil$. Then (4.9) implies that for some c_1 and n_2, all $n \geq n_2$, $u \geq a_n^2$,

$$E_\omega^0 \left(\frac{1}{j_*} \sum_{j=1}^{j_*} V_j^n - T_{j-1}^n \right) \leq c_1 b_n^2 n^{1/2}.$$

Hence, for some n_3, all $n \geq n_3$, $u \geq a_n^2$,

$$P_\omega^0 \left(\frac{1}{j_*} \sum_{j=1}^{j_*} V_j^n - T_{j-1}^n \geq \delta b_n^2 n^{9/16} \right) \leq \delta/8,$$

and, since $j_* \delta b_n^2 n^{9/16} \leq \delta u$,

$$P_\omega^0 \left(\sum_{j=1}^{j_*} V_j^n - T_{j-1}^n \geq \delta u \right) \leq \delta/8. \tag{4.12}$$

Recall $\mathcal{K}(r)$ from (3.3). Let

$$\widehat{V}_k^n = \inf\{t \geq V_k^n : X_t^{(n)} \in \mathbb{Z}^2 \setminus \mathcal{K}(b_n n^{3/8})\} \wedge T_k^n, \qquad k \geq 1,$$
$$\widetilde{V}_k^n = \inf\{t \geq \widehat{V}_k^n : |X_t^{(n)} - X^{(n)}(\widehat{V}_k^n)| \geq (1/2) b_n n^{3/8}\}, \qquad k \geq 1.$$

We can use estimates for Brownian hitting probabilities (♣) to see that for some c_2, c_3 and n_4, all $n \geq n_4$, k,

$$P_\omega^0(\widehat{V}_k^n < T_k^n \mid \mathcal{F}_{V_k^n}) \geq c_2 \frac{\log(4\beta_n) - \log(2\beta_n)}{\log(2b_n n^{3/8}) - \log(2\beta_n)} \geq c_3/\log n. \tag{4.13}$$

There exist (♣) c_4 and n_5, such that for all $n \geq n_5$, $k \geq 2$,

$$P_\omega^0(T_k^n - V_k^n \geq c_4 b_n^2 n^{3/4} \mid \widehat{V}_k^n < T_k^n, \mathcal{F}_{\widehat{V}_k^n})$$
$$\geq P_\omega^0(\widetilde{V}_k^n - \widehat{V}_k^n \geq c_4 b_n^2 n^{3/4} \mid \widehat{V}_k^n < T_k^n, \mathcal{F}_{\widehat{V}_k^n}) \geq 3/4.$$

This and (4.13) imply that the sequence $\{T_k^n - V_k^n\}_{k \geq 2}$ is stochastically minorized by a sequence of i.i.d. random variables which take value $c_4 b_n^2 n^{3/4}$ with probability

$c_3/\log n$ and they take value 0 otherwise. This implies that for some n_6, all $n \geq n_6$, $u \geq a_n^2$,

$$P_\omega^0\left(\frac{1}{j_*}\sum_{j=2}^{j_*} T_j^n - V_j^n \leq b_n^2 n^{3/4}/\log^2 n\right) \leq \delta/4$$

and, because $j_* b_n^2 n^{3/4}/\log^2 n \geq u$ assuming n_6 is large enough,

$$P_\omega^0\left(\sum_{j=2}^{j_*} T_j^n - V_j^n \leq u\right) \leq \delta/4.$$

We combine this with (4.12) and the definition of $\sigma_u^{n,2}$ to obtain for some n_7, all $n \geq n_7$, $u \geq a_n^2$,

$$P_\omega^0(\sigma_u^{n,2}/u \leq \delta) \geq 1 - 3\delta/8. \tag{4.14}$$

This completes the proof of the lemma. □

Let $Y_k^n = (Y_{k,1}^n, Y_{k,2}^n) = X^{(n)}(V_{k+1}^n) - X^{(n)}(T_k^n)$. Set $\bar{Y}_k^n = \sup_{T_k^n \leq t \leq V_{k+1}^n} |X^{(n)}(t) - X^{(n)}(T_k^n)|$. For $x \in \mathbb{Z}^2$, let $\Pi_n(x) \in B_n' - u_n + \mathcal{O}_n$ be the unique point with the property that $x - \Pi_n(x) = a_n y$ for some $y \in \mathbb{Z}^2$.

We next estimate the variance of $X^{n,2}(V_{m+1}^n) = \sum_{k=0}^m Y_k^n$.

Lemma 4.6. *There exist c_1, c_2 and n_1 such that for all $n \geq n_1$, $k \geq 0$, $j = 1, 2$, and ω,*

$$E_\omega^0|Y_{k,j}^n| \leq E_\omega^0|Y_k^n| \leq E_\omega^0|\bar{Y}_k^n| \leq c_1\beta_n, \tag{4.15}$$

$$\mathrm{Var} Y_{k,j}^n \leq \mathrm{Var}\bar{Y}_k^n \leq c_2\beta_n^2, \qquad \textit{under } P_\omega^x. \tag{4.16}$$

Proof. Let

$$\mathcal{X}_k^{(n)}(t) = X_t^{(n)} + \Pi_n(X^{(n)}(T_k^n)) - X^{(n)}(T_k^n), \qquad t \in [T_k^n, V_{k+1}^n], \tag{4.17}$$

and note that

$$Y_k^n = (Y_{k,1}^n, Y_{k,2}^n) = \mathcal{X}_k^{(n)}(V_{k+1}^n) - \mathcal{X}_k^{(n)}(T_k^n).$$

It follows from the definition that we have $\sup_{S_{k-1}^n \leq t \leq U_k^n} |X^{(n)}(t) - X^{(n)}(S_{k-1}^n)| \leq 16\beta_n$, a.s. This, (4.8) and the definition of V_{k+1}^n imply that $|\bar{Y}_k^n|$ is stochastically majorized by an exponential random variable with mean $c_3\beta_n$. This easily implies the lemma. □

Next we will estimate the covariance of $Y_{k,1}^n$ and $Y_{j,1}^n$ for $j \neq k$.

Lemma 4.7. *There exist c_1, c_2 and n_1 such that for all $n \geq n_1$, $j < k-1$ and ω such that $0 \notin \Gamma_n^1 \setminus \partial_i \Gamma_n^1$, under P_ω^0,*

$$\mathrm{Cov}(Y_{j,1}^n, Y_{k,1}^n) \leq c_1 e^{-c_2(k-j)}\beta_n^2. \tag{4.18}$$

Proof. Assume that ω is such that $0 \notin \Gamma_n^1 \setminus \partial_i \Gamma_n^1$. Let

$$\begin{aligned}\Gamma_n^3 &= \Gamma_n^1 \cap \mathcal{B}(u_n + \mathcal{O}_n, a_n/2) = \mathcal{B}(u_n + \mathcal{O}_n, 2\beta_n),\\ \Gamma_n^4 &= \partial_i \mathcal{B}(u_n + \mathcal{O}_n, 3\beta_n),\\ \tau(A) &= \inf\{t \geq 0 : \mathcal{X}_0^{(n)}(t) \in A\}.\end{aligned}$$

Suppose that $x, v \in \Gamma_n^3$ and $y \in \Gamma_n^4$. By the Harnack inequality proved in Lemma 4.1,

$$\frac{P_\omega^x(\mathcal{X}_0^{(n)}(\tau(\Gamma_n^4)) = y)}{P_\omega^v(\mathcal{X}_0^{(n)}(\tau(\Gamma_n^4)) = y)} \geq c_3. \tag{4.19}$$

Let $\mathcal{T}_k^n$ have the same meaning as T_k^n but relative to the process $\mathcal{X}_k^{(n)}$ rather than $X^{(n)}$. We obtain from (4.19) and the strong Markov property applied at $\tau(\Gamma_n^4)$ that, for any $x, v, y \in \Gamma_n^3$ we have

$$\frac{P_\omega^x(\mathcal{X}_0^{(n)}(\mathcal{T}_1^n) = y)}{P_\omega^v(\mathcal{X}_0^{(n)}(\mathcal{T}_1^n) = y)} \geq c_3.$$

Recall that $T_0^n = 0$. The last estimate implies that, for $x, v, y \in \Gamma_n^3$,

$$\frac{P_\omega(\mathcal{X}_1^{(n)}(T_1^n) = y \mid \mathcal{X}_0^{(n)}(T_0^n) = x)}{P_\omega(\mathcal{X}_1^{(n)}(T_1^n) = y \mid \mathcal{X}_0^{(n)}(T_0^n) = v)} \geq c_3.$$

Since the process $X^{(n)}$ is time-homogeneous, this shows that for $x, v, y \in \Gamma_n^3$ and all k,

$$\frac{P_\omega(\mathcal{X}_{k+1}^{(n)}(T_{k+1}^n) = y \mid \mathcal{X}_k^{(n)}(T_k^n) = x)}{P_\omega(\mathcal{X}_{k+1}^{(n)}(T_{k+1}^n) = y \mid \mathcal{X}_k^{(n)}(T_k^n) = v)} \geq c_3. \tag{4.20}$$

We now apply Lemma 6.1 of [8] (see Lemma 1 of [7] for a better presentation of the same estimate) to see that (4.20) implies that there exist constants C_k, $k \geq 1$, such that for every k and all $x, v, y \in \Gamma_n^3$,

$$\frac{P_\omega^x(\mathcal{X}_k^{(n)}(T_k^n) = y)}{P_\omega^v(\mathcal{X}_k^{(n)}(T_k^n) = y)} \geq C_k.$$

Moreover, $C_k \in (0, 1)$, C_k's depend only on c_3, and $1 - C_k \leq e^{-c_4 k}$ for some $c_4 > 0$ and all k. By time homogeneity of $X^{(n)}$, for $m \leq j < k$ and all $x, v, y, z \in \Gamma_n^3$,

$$\frac{P_\omega^z(\mathcal{X}_k^{(n)}(T_k^n) = y \mid \mathcal{X}_j^{(n)}(T_j^n) = x)}{P_\omega^z(\mathcal{X}_k^{(n)}(T_k^n) = y \mid \mathcal{X}_j^{(n)}(T_j^n) = v)} \geq C_{k-j},$$

and, by the strong Markov property applied at T_j^n,

$$\frac{P_\omega^z(\mathcal{X}_k^{(n)}(T_k^n) = y \mid \mathcal{X}_j^{(n)}(T_j^n) = x)}{P_\omega^z(\mathcal{X}_k^{(n)}(T_k^n) = y \mid \mathcal{X}_m^{(n)}(T_m^n) = v)} \geq C_{k-j}.$$

This and (4.15) imply that for $j < k-1$ and $x \in \mathbb{Z}^2$,

$$\begin{aligned}
|E^x_\omega(Y^n_{k,1} - E^x_\omega Y^n_{k,1} \mid \mathcal{F}_{T^n_{j+1}})| &= |E^x_\omega(Y^n_{k,1} \mid \mathcal{F}_{T^n_{j+1}}) - E^x_\omega Y^n_{k,1}| \\
&\le (1 - C_{k-j-1}) \sup_{y\in\mathbb{Z}^2} E^y_\omega |Y^n_{k,1}| \\
&\le e^{-c_4(k-j-1)} c_5 \beta_n \le c_6 e^{-c_4(k-j)} \beta_n. \qquad (4.21)
\end{aligned}$$

Hence for $j < k-1$,

$$\begin{aligned}
\mathrm{Cov}(Y^n_{j,1}, Y^n_{k,1}) &= E^x_\omega((Y^n_{j,1} - E^x_\omega Y^n_{j,1})(Y^n_{k,1} - E^x_\omega Y^n_{k,1})) \\
&= E^x_\omega(E^x_\omega((Y^n_{j,1} - E^x_\omega Y^n_{j,1})(Y^n_{k,1} - E^x_\omega Y^n_{k,1}) \mid \mathcal{F}_{T^n_{j+1}})) \\
&= E^x_\omega((Y^n_{j,1} - E^x_\omega Y^n_{j,1}) E^x_\omega(Y^n_{k,1} - E^x_\omega Y^n_{k,1} \mid \mathcal{F}_{T^n_{j+1}})) \\
&\le E^x_\omega(|Y^n_{j,1} - E^x_\omega Y^n_{j,1}| \cdot |E^x_\omega(Y^n_{k,1} - E^x_\omega Y^n_{k,1} \mid \mathcal{F}_{T^n_{j+1}})|) \\
&\le 2E^x_\omega |Y^n_{j,1}| c_6 e^{-c_4(k-j)} \beta_n \\
&\le c_7 e^{-c_4(k-j)} \beta_n^2.
\end{aligned}$$

□

Proof of Proposition 4.1. Assume that ω is such that $0 \notin \Gamma^1_n \backslash \partial_i \Gamma^1_n$. We combine (4.18) and (4.16) to see that for some c_1 and c_2 and all $m \ge 1$, we have under P^0_ω,

$$\begin{aligned}
\mathrm{Var}\left(\sum_{k=0}^m Y^n_{k,1}\right) &= \sum_{j=0}^m \sum_{k=0}^m \mathrm{Cov}(Y^n_{j,1}, Y^n_{k,1}) \qquad (4.22) \\
&\le \sum_{j=0}^m \sum_{k=0}^m c_1 e^{-c_3(k-j)} \beta_n^2 \le c_2 m \beta_n^2.
\end{aligned}$$

For fixed n and ω, the process $\{\mathcal{X}^{(n)}_k(T^n_k), k \ge 1\}$ is Markov with a finite state space and one communicating class, so it has a unique stationary distribution. We will call it $\mathbf{p}(n)$. We will argue that $E^{\mathbf{p}(n)}_\omega Y^n_{k,1} = 0$. Since $X^{(n)}$ and $X^{(n-1)}$ satisfy the quenched invariance principle and they are random walks among symmetric (in distribution) conductances, they have zero means. Recall that $X^{(n)} = X^{n,1} + X^{n,2}$ and $\widehat{X}^{n,1}$ has the same distribution as $X^{(n-1)}$. It follows that for some $c_4 > 0$ and $c_5 < 1/4$ and all large t, we have

$$P^{\mathbf{p}(n)}_\omega\left(\sup_{1\le s\le t} |\widehat{X}^{n,1}_s| \ge c_4\sqrt{t}\right) = P^{\mathbf{p}(n)}_\omega\left(\sup_{1\le s\le t} |X^{(n-1)}_s| \ge c_4\sqrt{t}\right) < c_5.$$

Since $\widehat{X}^{n,1}_t = X^{n,1}(\widehat{\sigma}^{n,1}_t)$ and $\widehat{\sigma}^{n,1}_t \ge t$, the last estimate implies that

$$P^{\mathbf{p}(n)}_\omega\left(\sup_{1\le s\le t} |X^{n,1}_s| \ge c_4\sqrt{t}\right) < c_5.$$

We also have for some $c_6 > 0$ and $c_7 < 1/4$, and all large t,

$$P^{\mathbf{p}(n)}_\omega\left(\sup_{1\le s\le t} |X^{(n)}_s| \ge c_6\sqrt{t}\right) < c_7.$$

Since $X^{n,2} = X^{(n)} - X^{n,1}$, we obtain for some $c_8 > 0$ and $c_9 < 1/2$ and all large t,

$$P_\omega^{\mathbf{p}(n)}\left(\sup_{1\le s\le t} |X_s^{n,2}| \ge c_8\sqrt{t}\right) < c_9.$$

This shows that $X^{n,2}$ does not have a linear drift. It is clear from the law of large numbers that $\liminf_{t\to\infty} \sigma_t^{n,2}/t > 0$, so $\widehat{X}^{n,2}$ does not have a linear drift either. We conclude that $E_\omega^{\mathbf{p}(n)} Y_{k,1}^n = 0$.

Now suppose that $X_0^{(n)}$ does not necessarily have the distribution $\mathbf{p}(n)$. The fact that $E_\omega^{\mathbf{p}(n)} Y_{k,1}^n = 0$ and a calculation similar to that in (4.21) imply that,

$$|E_\omega^0 Y_{k,1}^n| \le c_{10} e^{-c_{11}k}\beta_n.$$

Let c_{12} be the constant denoted c_1 in (4.15). The last estimate and (4.15) imply that for some c_{13} and all $m \ge 1$,

$$\begin{aligned}\left| E_\omega^0 \sum_{k=0}^m Y_{k,1}^n \right| &\le \sum_{k\ge 0} |E_\omega^0 Y_{k,1}^n| + \sup_{k\ge 1} E_\omega^0 |\bar{Y}_k^n| \\ &\le \sum_{k\ge 0} c_{10} e^{-c_{11}k}\beta_n + c_{12}\beta_n \le c_{13}\beta_n. \end{aligned} \tag{4.23}$$

All estimates that we derived for $Y_{k,1}^n$'s apply to $Y_{k,2}^n$'s as well, by symmetry.

Note that $|X^{(n)}(U_{k+1}^n) - X^{(n)}(T_k^n)| \ge \beta_n/2$. We have $V_{k+1}^n - T_k^n \ge U_{k+1}^n - T_k^n$ so we can assume (♣) that b_n/a_{n-1} is so large that for some $p_1 > 0$ and n_2, for all $n \ge n_2$ and $k \ge 1$,

$$P_\omega^x(V_{k+1}^n - T_k^n \ge \beta_n^2 \mid \mathcal{F}_{T_k^n}) \ge p_1.$$

Let $\mathcal{V}_m$ be a binomial random variable with parameters m and p_1. We see that $\sigma^{n,2}(V_m^n) = \sum_{k=0}^m V_{k+1}^n - T_k^n$ is stochastically minorized by $\beta_n^2\mathcal{V}_m$.

Recall that $u \ge a_n^2$. Let m_1 be the smallest integer such that

$$P_\omega^0(V_{m_1}^n \le u) < \delta/4. \tag{4.24}$$

Then

$$P_\omega^0(V_{m_1-1}^n \le u) \ge \delta/4. \tag{4.25}$$

Since δ in (4.14) can be arbitrarily small, we have for for some n_3 and all $n \ge n_3$,

$$P_\omega^0(\sigma_u^{n,2}/u \le \delta^4) \ge 1 - \delta/8. \tag{4.26}$$

The following estimate follows from the fact that $\sigma^{n,2}(V_{m_1-1}^n)$ is stochastically minorized by $\beta_n^2\mathcal{V}_{m_1-1}$, and from (4.25)-(4.26),

$$\begin{aligned} P_\omega^0(\beta_n^2\mathcal{V}_{m_1-1} \le \delta^4 u) &\ge P_\omega^0(\sigma^{n,2}(V_{m_1-1}^n) \le \delta^4 u) \\ &\ge P_\omega^0(\sigma_u^{n,2} \le \delta^4 u, V_{m_1-1}^n \le u) \ge \delta/8. \end{aligned}$$

This implies that for some c_{14}, we have $m_1 \le c_{14}\delta^3 u/\beta_n^2$. In other words, $u \ge m_1\beta_n^2/(c_{14}\delta^3)$. Note that for a fixed δ, we have for large n, (♣ $u^{1/2}\delta/4 - c_{13}\beta_n \ge$

$u^{1/2}\delta/8$. These observations, (4.22), (4.23) and the Chebyshev inequality imply that for $m \le m_1$,

$$P^0_\omega\left(u^{-1/2}\left(\left|\sum_{k=0}^m Y^n_{k,1}\right| + \left|\sum_{k=0}^m Y^n_{k,2}\right|\right) \ge \delta/2\right) \tag{4.27}$$
$$\le P^0_\omega\left(\left|\sum_{k=0}^m Y^n_{k,1}\right| \ge u^{1/2}\delta/4\right) + P^0_\omega\left(\left|\sum_{k=0}^m Y^n_{k,2}\right| \ge u^{1/2}\delta/4\right)$$
$$\le P^0_\omega\left(\left|\sum_{k=0}^m Y^n_{k,1} - E^0_\omega \sum_{k=0}^m Y^n_{k,1}\right| \ge u^{1/2}\delta/4 - c_{13}\beta_n\right)$$
$$+ P^0_\omega\left(\left|\sum_{k=0}^m Y^n_{k,2} - E^0_\omega \sum_{k=0}^m Y^n_{k,2}\right| \ge u^{1/2}\delta/4 - c_{13}\beta_n\right)$$
$$\le \frac{\mathrm{Var}\left(\sum_{k=0}^m Y^n_{k,1}\right)}{u\delta^2/64} + \frac{\mathrm{Var}\left(\sum_{k=0}^m Y^n_{k,2}\right)}{u\delta^2/64}$$
$$\le \frac{2c_2 m_1 \beta_n^2}{(c_{14}^{-1}\delta^{-3} m_1 \beta_n^2)\delta^2/64} \le c_{15}\delta.$$

Let $M = \min\{m \ge 1 : u^{-1/2}\left(\left|\sum_{k=0}^m Y^n_{k,1}\right| + \left|\sum_{k=0}^m Y^n_{k,2}\right|\right) \ge \delta\}$. By the strong Markov property applied at M and (4.27),

$$P^0_\omega\Bigg(\sup_{1\le m\le m_1} u^{-1/2}\left(\left|\sum_{k=0}^m Y^n_{k,1}\right| + \left|\sum_{k=0}^m Y^n_{k,2}\right|\right) \ge \delta, \tag{4.28}$$
$$u^{-1/2}\left(\left|\sum_{k=0}^{m_1} Y^n_{k,1}\right| + \left|\sum_{k=0}^{m_1} Y^n_{k,2}\right|\right) \le \delta/2\Bigg)$$
$$\le P^0_\omega\left(u^{-1/2}\left(\left|\sum_{k=0}^{m_1-M} Y^n_{k,1}\right| + \left|\sum_{k=0}^{m_1-M} Y^n_{k,2}\right|\right) \ge \delta/2 \mid M < m_1\right) \le c_{15}\delta.$$

Recall that $u \ge m_1\beta_n^2/(c_{14}\delta^3)$. For a fixed δ and large n, (♣) $u^{1/2}\delta - 2c_{12}\beta_n \ge u^{1/2}\delta/2$. It follows from this, (4.15) and (4.16) that

$$P^0_\omega\left(\exists k \le m_1 : |\bar{Y}^n_k| \ge u^{1/2}\delta\right) \tag{4.29}$$
$$\le m_1 \sup_{k\le m_1} P^0_\omega\left(|\bar{Y}^n_k| \ge u^{1/2}\delta\right)$$
$$\le m_1 \sup_{k\le m_1} P^0_\omega\left(|\bar{Y}^n_k| - E^0_\omega|\bar{Y}^n_k| \ge u^{1/2}\delta - c_{12}\beta_n\right)$$
$$\le m_1 \frac{c_{11}\beta_n^2}{u\delta^2/4} \le m_1 \frac{c_{11}\beta_n^2}{(c_{14}^{-1}\delta^{-3}m_1\beta_n^2)\delta^2} \le c_{16}\delta.$$

We use (4.24), (4.27), (4.28) and (4.29) to obtain

$$P^0_\omega\left(\sup_{0\le s\le u} u^{-1/2}|X^{n,2}_s| \ge 2\delta\right)$$

$$\leq P_\omega^0(V_{m_1}^n \leq u) + P_\omega^0\left(u^{-1/2}\left(\left|\sum_{k=0}^{m_1} Y_{k,1}^n\right| + \left|\sum_{k=0}^{m_1} Y_{k,2}^n\right|\right) \geq \delta/2\right)$$

$$+ P_\omega^0\left(\sup_{1\leq m\leq m_1} u^{-1/2}\left(\left|\sum_{k=0}^{m} Y_{k,1}^n\right| + \left|\sum_{k=0}^{m} Y_{k,2}^n\right|\right) \geq \delta, \right.$$

$$\left. u^{-1/2}\left(\left|\sum_{k=0}^{m_1} Y_{k,1}^n\right| + \left|\sum_{k=0}^{m_1} Y_{k,2}^n\right|\right) \leq \delta/2\right)$$

$$+ P_\omega^0\left(\exists k \leq m_1 : |\bar{Y}_k^n| \geq u^{1/2}\delta\right)$$

$$\leq \delta/4 + c_{15}\delta + c_{15}\delta + c_{16}\delta.$$

Since $\delta > 0$ is arbitrarily small, this implies that for every $\delta > 0$, some n_3 and all $n \geq n_3$,

$$P_\omega^0\left(\sup_{0\leq s\leq u} u^{-1/2}|X_s^{n,2}| \geq \delta\right) \leq \delta/2.$$

This and (4.11) yield the proposition. □

Recall from (1.2) the definition of the averaged measure $\mathbf{P}$.

Lemma 4.8. *For every $\delta > 0$ there exists n_1 such that for all $n \geq n_1$ and $u \geq a_n^2$,*

$$\mathbf{P}\left(\sigma_u^{n,2}/u \leq \delta, \sup_{0\leq s\leq u} u^{-1/2}|X_s^{n,2}| \leq \delta\right) \geq 1-\delta. \tag{4.30}$$

Proof. By Proposition 4.1 applied to $\delta/2$ in place of δ, for every $\delta > 0$ there exists n_2 such that for all $n \geq n_2$, $u \geq a_n^2$, and ω such that $0 \notin \Gamma_n^1 \setminus \partial_i\Gamma_n^1$,

$$P_\omega^0\left(\sigma_u^{n,2}/u \leq \delta, \sup_{0\leq s\leq u} u^{-1/2}|X_s^{n,2}| \leq \delta\right) \geq 1-\delta/2. \tag{4.31}$$

Let $|A|$ denote the cardinality of $A \subset \mathbb{Z}^2$. Since $|\Gamma_n^1| \leq 25\beta_n^2 \leq 25a_n^2 n^{-1/2} = 25n^{-1/2}|B_n'|$, the definitions of $\mathcal{O}_n$ and Γ_n^1 imply that $\mathbf{P}(0 \in \Gamma_n^1 \setminus \partial_i\Gamma_n^1) < \delta/2$ for some $n_3 \geq n_2$ and all $n \geq n_3$. This and (4.31) imply (4.30). □

In the following lemma and its proof, when we write the Prokhorov distance between processes such as $\{(1/a)X_{ta^2}^{(n-1)}, t \in [0,1]\}$, we always assume that they are distributed according to $\mathbf{P}$.

Lemma 4.9. *There exists a function $\rho^* : (0,\infty) \to (0,\infty)$ with $\lim_{\delta\downarrow 0}\rho^*(\delta) = 0$ and a sequence $\{a_n\}$ with the following properties,*

$$\mathbf{d}(\{(1/a)X_{ta^2}^{(n-1)}, t \in [0,1]\}, P_{\mathrm{BM}}) \leq 2^{-n}, \qquad a \geq a_n. \tag{4.32}$$

Moreover, suppose that for $\delta < 1/2$ and all $u \geq a_n^2$,

$$\mathbf{P}\left(\sigma_u^{n,2}/u \leq \delta, \sup_{0\leq s\leq u} u^{-1/2}|X_s^{n,2}| \leq \delta\right) \geq 1-\delta. \tag{4.33}$$

Then $\mathbf{d}(\{(1/a)X^{(n)}_{ta^2}, t \in [0,1]\}, P_{\mathrm{BM}}) \le 2^{-n} + \rho^*(\delta)$, *for all* $a \ge a_n$.

Proof. Formula (4.32) is special case of (3.1).

Fix some $a \ge a_n$. We will apply (4.33) with $u = a^2$. Note that on the event in (4.33) we have

$$1 - \sigma^{n,1}_{a^2}/a^2 = u/u - \sigma^{n,1}_u/u = \sigma^{n,2}_u/u \le \delta. \tag{4.34}$$

The function $t \to \sigma^{n,1}_{ta^2}/a^2$ is Lipschitz with the constant 1 and $\sigma^{n,1}_{ta^2}/a^2 \le t$ so (4.34) implies for $t \in [0,1]$,

$$t - \sigma^{n,1}_{ta^2}/a^2 \le 1 - \sigma^{n,1}_{a^2}/a^2 \le \delta. \tag{4.35}$$

Recall the function $\rho(\delta)$ from the proof of Lemma 3.2, such that $P_{\mathrm{BM}}(\mathrm{Osc}(W,\delta) \ge \rho(\delta)) < \rho(\delta)$ and $\lim_{\delta\downarrow 0}\rho(\delta) = 0$. By (4.35), we can apply Lemma 3.2 with $\sigma_t = \sigma^{n,1}_{ta^2}/a^2$. Recall that $W^*(t) = W(\sigma_t)$. By the definition of $\widehat{X}^{n,1}$,

$$\begin{aligned}
&\mathbf{d}(\{(1/a)X^{n,1}_{ta^2}, t \in [0,1]\}, P_{\mathrm{BM}}) \\
&\le \mathbf{d}(\{(1/a)X^{n,1}_{t/a^2}, t \in [0,1]\}, \{W^*_t, t \in [0,1]\}) + \mathbf{d}(\{W^*_t, t \in [0,1]\}, P_{\mathrm{BM}}) \\
&\le \mathbf{d}(\{(1/a)X^{n,1}_{ta^2}, t \in [0,1]\}, \{W^*_t, t \in [0,1]\}) + \rho(\delta) + \delta \\
&= \mathbf{d}(\{(1/a)\widehat{X}^{n,1}(\sigma^{n,1}_{ta^2}), t \in [0,1]\}, \{W(\sigma^{n,1}_{ta^2}/a^2), t \in [0,1]\}) + \rho(\delta) + \delta.
\end{aligned} \tag{4.36}$$

Recall from (3.4) that for a fixed $\omega \in \Omega$, the distribution of $\{\widehat{X}^{n,1}_t, t \ge 0\}$ is the same as that of $\{X^{n-1}_t, t \ge 0\}$. In view of Theorem 2.1, we can make a_n so large (♣) that $\mathbb{P}(\mathrm{Osc}(\widehat{X}^{n,1}, \delta) \ge 2\rho(\delta)) < 2\rho(\delta)$. This, Lemma 3.1 and the definition of the Prokhorov distance imply that

$$\begin{aligned}
\mathbf{d}(\{(1/a)&\widehat{X}^{n,1}(\sigma^{n,1}_{ta^2}),\ t \in [0,1]\}, \{W(\sigma^{n,1}_{ta^2}/a^2), t \in [0,1]\}) \\
&\le \mathbf{d}(\{(1/a)\widehat{X}^{n,1}_{ta^2}, t \in [0,1]\}, \{W_t, t \in [0,1]\}) + 4\rho(\delta) \\
&= \mathbf{d}(\{(1/a)X^{(n-1)}_{ta^2}, t \in [0,1]\}, \{W_t, t \in [0,1]\}) + 4\rho(\delta) \\
&\le 2^{-n} + 4\rho(\delta).
\end{aligned}$$

In the final two lines we used (3.4) and (4.32).

Combining the estimates above, since $P^0_\omega]\left(\sup_{0\le s\le u} u^{-1/2}|X^{n,2}_s| \le \delta\right) \ge 1 - \delta$ and $X^{(n)} = X^{n,1} + X^{n,2}$, Lemma 3.3 shows that

$$\begin{aligned}
&\mathbf{d}(\{(1/a)X^{(n)}_{ta^2}, t \in [0,1]\}, P_{\mathrm{BM}}) \\
&\le \mathbf{d}(\{(1/a)X^{(n)}_{ta^2}, t \in [0,1]\}, \{(1/a)X^{n,1}_{ta^2}, t \in [0,1]\}) \\
&\qquad + \mathbf{d}(\{(1/a)X^{n,1}_{ta^2}, t \in [0,1]\}, P_{\mathrm{BM}}) \\
&\le \delta + 2^{-n} + 5\rho(\delta) + \delta.
\end{aligned}$$

We conclude that the lemma holds if we take $\rho^*(\delta) = 5\rho(\delta) + 2\delta$. □

Proof of Theorem 1.2. Choose an arbitrarily small $\varepsilon > 0$. We will show that there exists a_* such that for every $a \geq a_*$,

$$\mathbf{d}(\{(1/a)X_{ta^2}, t \in [0,1]\}, P_{\text{BM}}) \leq \varepsilon. \tag{4.37}$$

Recall ρ^* from Lemma 4.9. Let n_1 be such that $2^{-n_1} \leq \varepsilon/4$ and let $\delta > 0$ be so small that $2^{-n_1} + \rho^*(\delta) < \varepsilon/2$. Let n_2 be defined as n_1 in Lemma 4.8, relative to this δ. Then, according to Lemma 4.9,

$$\mathbf{d}(\{(1/a)X^n_{ta^2}, t \in [0,1]\}, P_{\text{BM}}) \leq 2^{-n} + \rho^*(\delta) < \varepsilon/2, \tag{4.38}$$

for all $n \geq n_3 := n_1 \vee n_2$ and $a \geq a_n$.

For a set K let $\mathcal{B}(K,r) = \{z : \operatorname{dist}(z,K) < r\}$ and recall the definition of D_n given in (2.1). Let

$$\begin{aligned}
F_1 &= \{0 \in \mathcal{B}(D_{n+1}, a_{n+1}/\log(n+1))\},\\
F_2 &= \{0 \notin \mathcal{B}(D_{n+1}, a_{n+1}/\log(n+1))\} \cap \{\exists t \in [0, a^2_{n+1}] : X^{(n)}_t \in D_{n+1}\},\\
G^k_1 &= \{0 \in \mathcal{B}(D_k, b_k/k)\}, \qquad k > n+1,\\
G^k_2 &= \{0 \notin \mathcal{B}(D_k, b_k/k)\} \cap \{\exists t \in [0, a^2_{n+1}] : X^{(n)}_t \in D_k\}, \qquad k > n+1.
\end{aligned}$$

The area of $\mathcal{B}(D_{n+1}, a_{n+1}/\log(n+1))$ is bounded by $c_1(a_{n+1}/\log(n+1))^2$ so

$$\mathbb{P}(F_1) \leq c_1(a_{n+1}/\log(n+1))^2/a^2_{n+1} = c_1/\log^2(n+1). \tag{4.39}$$

We choose $n_4 > n_3$ such that $c_1/\log^2(n+1) < \varepsilon/8$ for $n \geq n_4$.

Note that D_{n+1} is a subset of a square with side $4\beta_{n+1} \leq 4a_{n+1}n^{-1/4}$. This easily implies that there exists $n_5 \geq n_4$ such that for $n \geq n_5$,

$$P_{\text{BM}}\left(\exists t \in [0, a^2_{n+1}] : W(t) \in D_{n+1} \mid 0 \notin \mathcal{B}(D_{n+1}, \frac{a_{n+1}}{\log(n+1)})\right) \leq \varepsilon/16.$$

We can assume (♣) that a_{n+1}/a_n is so large that for some $n_6 \geq n_5$ and all $n \geq n_6$,

$$\begin{aligned}
\mathbb{P}(F_2) &\leq \mathbb{P}\left(\exists t \in [0, a^2_{n+1}] : X^{(n)}_t \in D_{n+1} \mid 0 \notin \mathcal{B}(D_{n+1}, \frac{a_{n+1}}{\log(n+1)})\right) &(4.40)\\
&\leq \varepsilon/8. &(4.41)
\end{aligned}$$

The area of $\mathcal{B}(D_k, b_k/k)$ is bounded by $c_2b^2_k/k$ so

$$\mathbb{P}(G^k_1) \leq (c_2b^2_k/k)/a^2_k \leq c_3(b^2_k/k)/(kb^2_k) = c_3/k^2. \tag{4.42}$$

We let $n_7 > n_6$ be so large that $\sum_{k \geq n_7} c_3/k^2 < \varepsilon/8$. For all $k > n+1 \geq n_7+1$, we make b_k/k so large (♣) that

$$\mathbb{P}(G^k_2) \leq \mathbb{P}\left(\sup_{t \in [0, a^2_{n+1}]} |X^n_t| \geq b_k/k\right) \leq c_3/k^2. \tag{4.43}$$

We combine (4.39), (4.40), (4.42) and (4.43) to see that for $n \geq n_7$,

$$\mathbb{P}(\exists t \in [0, a^2_{n+1}]\ \exists k \geq n+1 : X^{(n)}_t \in D_k) \tag{4.44}$$

$$\leq \mathbb{P}(F_1) + \mathbb{P}(F_2) + \sum_{k>n+1} \mathbb{P}(G_1^k) + \sum_{k>n+1} \mathbb{P}(G_2^k)$$
$$\leq \varepsilon/8 + \varepsilon/8 + \varepsilon/8 + \varepsilon/8 = \varepsilon/2.$$

Let $R_{n+1} = \inf\{t \geq 0 : X_t \in \bigcup_{k \geq n+1} \mathcal{D}_k\}$. It is standard to construct X and $X^{(n)}$ on a common probability space so that $X_t = X_t^n$ for all $t \in [0, R_{n+1})$. This and (4.44) imply that for $n \geq n_7$ and all $a \in [a_n, a_{n+1}]$ we have

$$P(\exists t \in [0,1] : (1/a)X_{ta^2} \neq (1/a)X_{ta^2}^{(n)}) \leq \varepsilon/2.$$

We combine this with (4.38) to see that for all $a \geq a_{n_6}$,

$$\mathbf{d}(\{(1/a)X_{ta^2}, t \in [0,1]\}, P_{\mathrm{BM}}) \leq \varepsilon/2 + \varepsilon/2 = \varepsilon.$$

We conclude that (4.37) holds with $a_* = a_{n_7}$.

This completes the proof of AFCLT. The WFCLT then follows from Theorem 2.13 of [4]. □

5. Acknowledgements

Research supported in part by NSF Grant DMS-1206276, by NSERC, Canada, and Trinity College, Cambridge, and by MTA Rényi "Lendulet" Groups and Graphs Research Group. The third author's research was realized in the frames of TÁMOP 4.2.4. A/1-11-1-2012-0001 "National Excellence Program Elaborating and operating an inland student and researcher personal support system, which project was subsidized by the European Union and co-financed by the European Social Fund.

References

[1] D. Aldous and J. Fill, *Reversible Markov Chains and Random Walks on Graphs* (book in preparation, available online) `http://www.stat.berkeley.edu/~aldous/RWG/book.html`

[2] M. T. Barlow and R. F. Bass. On the resistance of the Sierpinski carpet. *Proc. R. Soc. London A.* **431** (1990) 345-360.

[3] M.T. Barlow and J.-D. Deuschel. Invariance principle for the random conductance model with unbounded conductances. *Ann. Probab.* **38** (2010), 234-276

[4] M.T. Barlow, K. Burdzy, Á. Timár. Comparison of quenched and annealed invariance principles for random conductance model. Preprint 2013. `Math arXiv 1304.3498`.

[5] M.T. Barlow, X. Zheng. The random conductance model with Cauchy tails. *Ann. Applied Probab.* **20** (2010), 869–889.

[6] P. Billingsley, *Convergence of probability measures.* Second edition. Wiley Series in Probability and Statistics: Probability and Statistics. A Wiley-Interscience Publication. John Wiley & Sons, Inc., New York, 1999.

[7] K. Burdzy and D. Khoshnevisan, Brownian motion in a Brownian crack *Ann. Appl. Probab.* **8** (1998), 708–748.

[8] K. Burdzy, E. Toby and R.J. Williams, On Brownian excursions in Lipschitz domains. Part II. Local asymptotic distributions, in *Seminar on Stochastic Processes 1988* (E. Cinlar, K.L. Chung, R. Getoor, J. Glover, editors), 1989, 55–85, Birkhäuser, Boston.

[9] A. De Masi, P.A. Ferrari, S. Goldstein, W.D. Wick. An invariance principle for reversible Markov processes. Applications to random motions in random environments. *J. Statist. Phys.* **55** (1989), 787–855.

Chapter 5

Some historical aspects of error calculus by Dirichlet forms

Nicolas Bouleau

We discuss the main stages of development of the error calculation since the beginning of XIX-th century by insisting on what prefigures the use of Dirichlet forms and emphasizing the mathematical properties that make the use of Dirichlet forms more relevant and efficient. The purpose of the paper is mainly to clarify the concepts. We also indicate some possible future research.

Contents

1. Introduction

There are several kinds of error calculations which have not followed the same historical development. The error calculus by Dirichlet forms that we will explain and trace the origins has to be distinguished from the following calculations:

a) The calculus of roundoff errors in numerical computations which appeared far before the representation of numbers in floating point be implemented on computers, and which possesses its specific difficulties. It has been much studied during the development of the numerical analysis for matrix discretization methods (cf. Hotelling [54], Von Neumann [57], Turing [58], Wilkinson [78], etc.);

b) The global evaluation of deterministic errors such as the interval calculus (cf. Moore [74], etc.). Some works of Laplace are related to this approach and also the paper of Cauchy [8];

c) The calculus of finite probabilistic errors where the errors are represented by random variables, which has been used by a very large number of authors to begin an argument and then, often, modified by supposing the errors to be small or gaussian in order to be able to pursue the calculation further (cf. Bienaymé [24], Birge [48], Bertrand [35], etc.) because the computation of image probability distributions is concretely inextricable what, in the second half of the XX-th century, justified the development of simulation methods (Monte-Carlo and quasi-Monte-Carlo).

The error calculus by Dirichlet forms assumes the errors to be both small, actually infinitesimal, and probabilistic. These two characteristics imply a peculiar differential calculus for the propagation of errors through models. As we will see the part of the calculation related to what is called today *the squared field operator* or more often *the carré du champ operator*, is ancient and dates back to the turn of the XVIII-th and XIX-th centuries in connection with the birth of the least squares method. Let us note, however, that our purpose is not to make a history of the method of least squares, broad topic that would lead to decline all the benefits of optimization in L^2 and its developments in statistics and analysis. I refer in this regard to the historical work of Kolmogorov and Yushkevich [84], also to the book of Pearson [85], and to the article of Sheynin [88] not always clear from a mathematical point of view probably because of an intrinsic ambiguity of the thought of the authors of the turn of the XVII-th and XIX-th centuries, but extremely well documented. Unfortunately it does not address at all the propagation of errors.

Let us note also in this introduction the very important phenomenon of *dichotomy of small errors* which allows to clarify the validity domains of the probabilistic and deterministic approaches. When we are concern with *small* errors that means mathematically that we are in an approximation procedure where – in principle – we are able to make the errors vanish. In such a situation the description of the asymptotic mechanism of the propagation of infinitely small errors is different according to the respective size of the variance and the bias of the error.

Three cases appear in the limited expansions. 1) If the variance is negligible with respect to the bias, then this property will persist by deterministic computations and it is enough for the asymptotic calculus to perform a first order differential calculus, i.e. a classical sensitivity calculation. 2) If the variance is of the same order of magnitude as the bias, then the calculus has to be a first order differential calculus for the variances which does not involve the bias, and the calculus for the bias is a second order differential calculus involving both biases and variances. 3) If the bias is negligible with respect to the variance then from the first calculation we return to the case 2). In these two last cases the propagation formulae are the following for a scalar erroneous quantity X:

$$\left.\begin{aligned} &\text{bias of error on } f(X)\\ &\qquad = (\text{bias of error on } X)f'(X) + \tfrac{1}{2}(\text{var of error on } X)f''(X)\\ &\text{var of error on } f(X)\\ &\qquad = (\text{var of error on } X)f'^2(X) \end{aligned}\right\} \tag{1.1}$$

We refer to [152] for more details and typical examples. Situations like case 1) are called *weakly stochastic*, and situations like 2) or 3) are called *strongly stochastic*. Let us mention that often in practice, in engineering for instance, we are not able to control the nature of the errors. Errors on the data in modelling are *exogenous*, we know few where they come from. It is wise to think according to the case of strongly stochastic errors, especially to take in account the randomness of the errors through the non-linearities of the model.

This is important because, by extending the ideas of Poincaré on the errors [37], it is possible to see that measurements done with graduated instruments are always *strongly stochastic* (because of the error on the choice of the nearest graduation cf. [143], [133], [149]). This is related to the theory of *arbitrary functions* (cf. von Kries [34], Fréchet [39], Borel [40], Hostinský [43], Hopf [49], [50], [51]).

In summary we can distinguish

<table>
<tr><th></th><th colspan="2">Infinitesimal errors</th><th>Finite errors</th></tr>
<tr><td>Deterministic approaches</td><td colspan="2">Deterministic sensitivity analysis: derivation with respect to the parameters of the model</td><td>Interval calculus, sup norm</td></tr>
<tr><td rowspan="2">Probabilistic approaches</td><td colspan="2">Error calculus using Dirichlet forms</td><td rowspan="2">Probability theory: images of probability distributions</td></tr>
<tr><td>first order calculus only dealing with variances</td><td>second order calculus with variance and biases</td></tr>
</table>

2. Gauss inventor of the carré du champ operator ?

Referring to the turn of the XVIII-th and the XIX-th centuries we see that two wakes were clearly drawn among researchers in matter of error: that of Laplace and that of Gauss.

Let us begin by Gauss whose works interess us particularly here. The current he initiates gives a fundamental place to his *law of errors*: assuming that, after several independent measures x_i the arithmetic average $\frac{1}{n}\sum_{i=1}^{n} x_i$ is the best value to take in account, he showed, with some additional assumptions, that the errors follow necessarly a normal law and the arithmetic average is both the most likely value and the one given by the least squares method.

Gauss tackled this question in the following way. First he admits – and this idea will be kept in the error theory with Dirichlet forms – that the quantity to be measured is random. It can vary in the scope of the measurement device following an *a priori* law. In modern language, let X be this random variable, μ its law. The results of the measurement operations are other random variables $X_1, \dots, X_n$ and Gauss assumes

a) the conditional law of X_1 given X to be of the form $\mathbb{P}\{X_i \in E | X = x\} = \int_E \varphi(x_1 - x)\, dx_1$,

b) the variables $X_1, \ldots, X_n$ to be conditionally independent given X.

Then he is easily able to compute the conditional law of X given the results of measure, it has a density with respect to μ and writing this density is maximal at the arithmetic average, he gets the relation

$$\frac{\varphi'(t-x)}{\varphi(t-x)} = a(t-x) + b$$

hence

$$\varphi(t-x) = \frac{1}{\sqrt{2\pi\sigma^2}} \exp(-\frac{(t-x)^2}{2\sigma^2}).$$

It is probably in the course *Calcul des Probabilités* of Poincaré [37] at the end of the XIXth century that Gauss' argument is the most clearly explained because Poincaré tries both to explicit all hypotheses and to generalize the proof[a]. He studies the case where the conditional law of X_1 given X is no more $\varphi(y - x)\, dy$ but of the more general form $\varphi(y, x)\, dy$. That gives him a way to explain the seeming paradox of *the error permanency*: "with a meter divided in millimeters, he writes, as often the measures be repeated, never will a length be determined up to a millionth of millimeter". This phenomenon is well known by physicists who noted of course that during the whole history of experimental sciences never a quantity has been precisely measured with rough instruments, cf [81]. The discussion leads Poincaré to suggest that measurements could be independent whereas the errors be not, when done by the same instrument. He doesn't develop a new mathematical formalism for this idea but emphasizes on the advantage of assuming small errors because then the argument of Gauss giving the normal law becomes compatible with non linear changes of variables and can be performed through differential calculus. It is the question of the *error propagation*.

It is in his *Theoria Combinationis Observatonum Erroribus Minimis Obnoxiae* published in 1823 that Gauss details his ideas about the errors propagation.

In the introduction, he cites Laplace and discusses the merits of reasonning with repeated observations or with observations immediately erroneous. Behind this discussion is the fact that Laplace gave the first analytical proof of the central limit theorem, and that Gauss intends to assert the interests of his demonstration that if the arithmetic mean is taken as the correct value then the law is necessarily normal, that he replaces in a more general approach of some kind of error calculus in an extended meaning. This dicussion is deepened in his section 17.

Gauss states the main problem in the following way: *Given a quantity* $U = F(V_1, V_2, V_3, \ldots)$ *function of the erroneous quantities* $V_1, V_2, V_3, \ldots$, *compute the*

[a]It is about this 'law of errors' that Poincaré writes "everybody believes in it because experimenters imagine it is a theorem of mathematics and mathematicians it is an experimental fact".

potential quadratic error to expect on U with the quadratic errors $\sigma_1^2, \sigma_2^2, \sigma_3^2, \ldots$ on $V_1, V_2, V_3, \ldots$ being known and assumed small and independent.

His answer is the following formula:

$$\sigma_U^2 = (\frac{\partial F}{\partial V_1})^2 \sigma_1^2 + (\frac{\partial F}{\partial V_2})^2 \sigma_3^2 + (\frac{\partial F}{\partial V_1})^2 \sigma_3^2 + \cdots \tag{2.1}$$

He also provides the covariance between an error on U and an error on another function of the V_i's. Formula (2.1) displays a property which makes it much to be preferred to other formulae encountered in textbooks throughout the XIXth and XXth centuries. It features a coherence property. With a formula such as

$$\sigma_U = |\frac{\partial F}{\partial V_1}|\sigma_1 + |\frac{\partial F}{\partial V_2}|\sigma_3 + |\frac{\partial F}{\partial V_1}|\sigma_3 + \cdots \tag{2.2}$$

errors may depend on the way in which the function F is written. These "ugly" formulae remain for instance in [53], [59]. Today we can understand that this difficulty does not arise with Gauss' calculus thanks to its connection with the theory of Dirichlet forms. Introducing the differential operator

$$L = \frac{1}{2}\sigma_1^2 \frac{\partial^2}{\partial V_1^2} + \frac{1}{2}\sigma_2^2 \frac{\partial^2}{\partial V_2^2} + \frac{1}{2}\sigma_3^2 \frac{\partial^2}{\partial V_3^2} \cdots \tag{2.3}$$

and supposing the functions to be smooth, we remark that formula (2.1) can be written as

$$\sigma_U^2 = L(F^2) - 2FL(F) \tag{2.4}$$

and coherence follows from the transport of a differential operator by an application. Today these intrinsic properties are understood also by the link with the stochastic calculus on manifolds and particularly the differential calculus of second order [124], [91], [144].

The errors on $V_1, V_2, V_3, \ldots$ are not necessarily supposed to be independent nor constant and may depend on $V_1, V_2, V_3, \ldots$, that gives formulas obtained by, what we call today, polarisation.

For the proof of (2.1) Gauss performs – as will do most of the applied textbooks thereafter – a computation supposing errors are *infinitely small* quantities and may replace differentials dV_1, dV_2, etc. But even Gaussian random variables have non compact support. There exists therefore a small probability that the errors be large. This is a difficulty that can be only treated with more precise notions of convergence or with the theory of Dirichlet forms, the proof of Gauss will then appear as a computation of what we call a *gradient in the sense of Dirichlet forms* often denoted by the term *sharp* (cf [133]), we will return to this point for clarification below in part V.

If we read the sections 19 and 20 of his treatise, we may ask whether the question posed by Gauss is not – in germ – the idea of a quadratic form which would be the ancestor of a Dirichlet form.

In some sense Gauss is on a product error structure (he assumes the errors of observation to be independent – in the sake of simplicity he says, because he is able to treat the general case). He obtains that the error on a quantity is given by a random quadratic function that he gets by local linearization with his change of variable formula, and whose expectation is the mean quadratic error: this random function is the "carré du champ". I do not say that Gauss had the idea of the carré du champ, but I say that his direction of research has something to do with Dirichlet forms what is not philosophically surprising since the landscape of potential theory was for him familiar.

Let us say now some words of the wake left by Laplace. By studying (cf [1], [4], [7] and Cauchy [8]) the mean of the positive errors and that of negative errors then taking the mean anew, Laplace, actually studies the first absolute moment of the error

$$\mathbb{E}[\,|\text{error}|\,]$$

instead of its variance. It is what seems for him the most natural. He has among available tools to make calculation the method of characteristic functions that he perfectly masters, and for this he must choose hypotheses on the law of this random variable |error|. But for the propagation he faces the usual difficulty of intractable calculations. He is then led to assume small and Gaussian errors following the argument provided by the central limit theorem when observations are repeated.

But when the errors are Gaussian the moments of order 1 and 2 are linked and we have, if V is $N(0,\sigma)$ distributed,

$$\mathbb{E}[\,|V|\,] = \sqrt{\frac{2}{\pi}}\,\sigma$$

so that, if the errors are small, the relation between variances is *equivalent* to the following relation for the first absolute moment:

$$\mathbb{E}[\,|e_F|\,] = \sqrt{F_1'^2(\mathbb{E}|e_1|)^2 + \cdots + F_n'^2(\mathbb{E}|e_n|)^2}.$$

We see that for first order moment a formula is obtained very similar to the one of Gauss (2.1) (instead of "ugly" formulae like (2.2)).

In the wake of the works of Gauss on errors we could cite almost all treatises on probability of the XIX-th century, let us quote particularly Faa di Bruno [32] who follows narrowly his approach and cites him explicitly (p38 and 53), also Liagre [33].

In the wake of Laplace we can cite Bravi [17], Bravais [20] who extends the arguments of Laplace to the multivariate case and introduces the error ellipsoids (cf [133] Chap.I), A. Meyer (1857) who in a treatise of very good mathematical level [30] opts for the approach of Laplace and argues this position, also the physicist Airy (1861) [31] who takes the whole theory of probability starting only from the treatise of Laplace.

Belonging clearly to both wakes are Hagen [16], Bienaymé [24] and also Biver [25] who, in 1853 introduces a general *cost functional* to be minimized to manage errors.

The propagation formula is encountered also in statistics if one is interested in the propagation of the Fisher information [41], [42]. If $\mathbb{P}_\theta$, $\theta \in \mathbb{R}^d$, is an indexed family of probability measures satisfying the conditions of regular models, having a density $f(.,\theta)$ with respect a measure $\mathbb{Q}$, the Fisher information related to the parameter θ is the matrix

$$J = 4\int (\frac{d\sqrt{f(.,\theta)}}{d\theta})(\frac{d\sqrt{f(.,\theta)}}{d\theta})^t \, d\mathbb{Q}.$$

If instead of the parameter θ we consider the parameter $\varphi = g(\theta)$ where g is smooth and injective, we find that the inverse matrix $\Gamma(\theta) = J^{-1}(\theta)$ that represents *a precision* on θ, is transported following Gauss formula.

$$\Gamma_{ij}(\varphi) = \sum_{k\ell} \frac{\partial g_{ij}}{\partial \theta_k}\frac{\partial g_{ij}}{\partial \theta_\ell}\Gamma_{k\ell}(\theta)$$

see on this topic [137] and Chorro [140]. This link with statistics gives a new interpretation of the carré du champ operator in terms of accuracy of a statistical estimate, there are no small errors more, only information and its inverse the precision. In this interpretation the Fréchet-Cramer-Rao inequality says that efficient estimates give the best precision.

We understand then that obtaining bounds for the propagation of the Fisher information is a means to express the regularity of a carré du champ matrix cf. Villani [128] p824 *et seq.*

3. Why should we ask the quadratic form to be closed?

The Dirichlet form strictly speaking appeared in potential theory in the classical case of the Laplacian operator far before the thought structures of the Hilbert spaces be available, it was simply a quantity whose minimum value was looked for. Dirichlet uses it [21] as soon as 1846, Thomson [22] and Riemann [23] near after, we refer to the historical sudies on the potential theory itself of Brelot [79], Temple [92], Ancona [94] among others on these developments.

After the works of Henri Cartan [55] and the thesis of Jacques Deny [60], [61] it is during the collaboration between Arne Beurling and Jacques Deny [68], [69], [77] that Hilbertian methods have taken a new importance especially by this discovery that the fact that contractions operate on a closed positive quadratic form is a necessary and sufficient condition for the associated contraction semi-group be positive on positive functions, i.e. be the transition probability semi-group of a Markov process.

The study of these questions under the angle of symmetric Markov processes developed quickly during the 1970s, Silvestein [80], Le Jan [86], (see also the references of the contributions to the Colloque en l'honneur de J. Deny, Orsay 1981), what allows Masatoshi Fukushima to discover these famous continuous additive functionals with zero quadratic variation but not finite variation [87].

The book of Fukushima [89], has been a reference for researchers and the starting point of several works. The notion of Dirichlet form appeared there so clearly – a primary notion somehow – that it came immediately to the mind that this notion could accompany any probability space from where the links with Malliavin calculus and error calculus.

From the point of view of error theory, the idea of imposing *a priori* that the quadratic form (the expectation of the carré du champ) be closed is a major step. The error calculus of Gauss contains the limitation of supposing that both the function F and the random variables $V_1, V_2, V_3, \ldots$ are analytically known. In probabilistic modelling however, we are often confronted to situations in which all the random variables, functions and covariances matrices are *given by limits*. For such situations, a means of extension thereby becomes essential.

Let us suppose that the quantities be defined on the probability space $(\Omega, \mathcal{A}, \mathbb{P})$. The extension tool lies in the following: we assume that if $X_n \to X$ in $L^2(\Omega, \mathcal{A}, \mathbb{P})$ and if the variance of the error $\Gamma(X_n - X_n)$ on $X_n - X_n$ can be made as small as we wish in $L^1(\Omega, \mathcal{A}, \mathbb{P})$ for m, n large enough, then the variance of the error $\Gamma(X_n - X)$ on $X_n - X$ goes to zero in L^1. This idea can be interpreted as a reinforced coherence principle, it means that the error on X is attached to X and furthermore, if the sequence of pairs (X_n, error on X_n) converges suitably, it converges necessarily to a pair (X, error on X).

The main benefit of the extension tool is that error theory based on Dirichlet forms extends to the infinite dimension, which allows error calculus on stochastic processes (Brownian motion, Poisson random measures, diffusions defined by stochastic differential equations). But also the calculus itself becomes more flexible: it allows now Lipschitzian changes of variables.

From a historical perspective, we can say that the main characteristic of the XX-th century with respect to XIX-th century, already so extraordinarily developed in mathematics, is the use of functional spaces (where points represent functions) and especially *complete* functional spaces like *Hilbert spaces, Banach spaces*, that permits to handle objects defined by limits. This idea has been the origin of gigantic progress in all domain of analysis. The closedness of Dirichlet forms allows to install an error calculus in most situations of classical analysis and stochastic analysis.

We arrive then to a kind of enriched error calculus, where any development – in particular asymptotic theorems – may be accompanied by errors. For instance the famous Donsker theorem on the approximation of the Brownian motion by a random walk extends in term of erroneous random walk which yields naturally the Ornstein-Uhlenbeck structure on the Wiener space cf [141].

4. Dirichlet form generated by an approximation

If we want to perform very precise error calculations – this is not an oxymoron! – we must be concern by the biases. Among the engineers who ask for a great precision from the observations, there are the geodesists and it seems that it was in this domain that has been used for the first time the fact that the random nature of errors imply a bias. No doubt everyone has long known that the image of the expectation for a non-linear mapping is not the expectation of the image, but geodesists write down differential error calculi with biases. This is explicit in Elkins (1950) [62], [64], [76], Jeudy (1988) [104], Hangos-Leisztner (1989) [109], Teunissen [105] [106], Coleman-Steele [107].

In Section 17 of the *Theoria Combinationis* Gauss writes "It is necessary to warn here that in the following research, it will issue only random errors reduced of their constant part. It is the observer's responsibility to remove the causes of constant error carefully. We reserve for another opportunity to examine the case where the observations are affected to an unknown error, and we will discuss this issue in another memory" maybe he would have encountered the question of the bias...

As we will see in the framework of Dirichlet forms the bias is represented by the generator of the semi-group associated with the form. The exact formula for the propagation of the bias is the following (see [111] p42 Exercise 6.2 and the precise hypotheses required for this formula)

$$A(F(f)) = \sum_{i=1}^{n} F_i'(f)Af_i + \frac{1}{2}\sum_{i,j} F_{i,j}''\Gamma(f_i, f_j).$$

A similar expression is used by the geodesists.

But the bias is a delicate notion because of some latitude in its definition. Let us resume what the theory says. If we consider two random variables Y and Y_n close together, the asymptotic behaviours of

$$\mathbb{E}[(\phi(Y_n) - \phi(Y))\chi(Y)]$$

and of

$$\mathbb{E}[(\phi(Y_n) - \phi(Y))\chi(Y_n)]$$

where ϕ and χ are test functions, are generally different. As a consequence several bias operators have to be distinguished (cf. [145]):

The asymptotics of $\mathbb{E}[(\varphi(Y_n) - \varphi(Y))\chi(Y)]$ yields the theoretical bias operator $\overline{A}$, the asymptotics of $\mathbb{E}[(\varphi(Y) - \varphi(Y_n))\chi(Y_n)]$ yields the practical bias operator noted $\underline{A}$ that of $\mathbb{E}[(\varphi(Y_n) - \varphi(Y))(\chi(Y_n) - \chi(Y))]$ gives the symmetric bias operator $\widetilde{A}$, and eventually $\mathbb{E}[(\varphi(Y_n)-\varphi(Y))(\chi(Y_n)+\chi(Y))]$ provides the singular bias operator $\not{A}$.

These operators are related thanks to the relations

$$\widetilde{A} = \frac{\overline{A}+\underline{A}}{2} \qquad \not{A} = \frac{\overline{A}-\underline{A}}{2}.$$

The symmetric bias operator satisfies

$$< \widetilde{A}[\varphi], \chi >_{L^2(\mathbb{P}_Y)} = < \varphi, \widetilde{A}[\chi] >_{L^2(\mathbb{P}_Y)}$$

under natural hypotheses it is indeed the generator of a Dirichlet form $\mathcal{E}$ which possesses a carré du champ operator Γ. It may be shown that this Dirichlet form is local iff the asymptotics (with the same weight) of $\mathbb{E}[(\varphi(Y_n)-\varphi(Y))^4]$ vanishes.

If the asymptotics of both variances $\mathbb{E}[(\varphi(Y_n)-\varphi(Y))^2\psi(Y)]$ and $\mathbb{E}[(\varphi(Y_n)-\varphi(Y))^2\psi(Y_n)]$ coincide – what is usually the case – the singular bias operator is a first order operator in the sense that it satisfies on the test functions

$$\not{A}[\varphi\chi] = \not{A}[\varphi]\chi + \varphi\not{A}[\chi].$$

Let us take an example. A very frequent situation in probability theory involve a triplet of real random variables (Y, Z, T), a real random variable G independent of (Y, Z, T) centered with variance one, and the approximation Y_ε of Y given by

$$Y_\varepsilon = Y + \varepsilon Z + \sqrt{\varepsilon}TG. \tag{4.1}$$

In that case the operator $\overline{A}$ may be shown to be given by

$$\overline{A}[\varphi](y) = \mathbb{E}[Z|Y=y]\ \varphi'(y) + \frac{1}{2}\mathbb{E}[T^2|Y=y]\ \varphi''(y)$$

and the Dirichlet form is given by

$$\mathcal{E}[\varphi, \chi] = \mathbb{E}[T^2\varphi'(Y)\chi'(Y)]$$

The operator $\widetilde{A}$ depends only on T, not on Z, $\underline{A}$ is obtained by difference.

A *weakly stochastic* approximation (see Part I above) may now be defined more precisely by the condition $\widetilde{A} = 0$.

5. "Small errors" what does it mean?

We can now clarify the question of small errors that has been historically a kind of conceptual blockage. We have seen that the theory of local Dirichlet forms with a carré du champ operator should be seen as a more accurate and rigorous form of the error calculation developed in the wake of Laplace and Gauss, allowing an extension of the sensitivity calculation to broader and more difficult situations involving stochastic calculus and Brownian motion.

In the setting of Dirichlet forms we know that the carré du champ operator represents the variance of the error and that the bias of the error can be represented by the symmetric generator (depending on which bias we speak as seen above). But the error itself, what is it? The error theory by Dirichlet forms considers that the error is $X_t - X_0$ where X_t is the symmetric Markov process associated to the Dirichlet form and which is taken as a *tangent process to the studied approximation procedure*. The error is not the "sharp". The "sharp" is mathematically a *gradient*. It is a linear operator which restores the carré du champ by taking the square. Actually it is a randomized gradient, for example in the classical case on $\mathbb{R}^2$ it writes $f^\sharp = f_1'\xi_1 + f_2'\xi_2$ where ξ_1 and ξ_2 are auxiliary orthogonal reduced random variables. This looks like an error by the fact that it is random but it is a tangent vector (a first order differential operator) which acts proportionally to itself.

On the contrary X_t evolves with a transition probability kernel. The crucial point is that for small t, $X_t - X_0$ is always a *sum of infinitely many quantities with independent sources of randomness*, it is not a quantity which decreases homothetically to itself, it does not fall under the Taylor formula

$$f(x+h) = (e^{hD}f)(x) = f(x) + hDf(x) + \cdots + \frac{h^n}{n!}D^n f(x) + \cdots$$

what is relevant for it is the *theory of semi-groups of operators*.

The errors may receive a rigorous mathematical treatment only if they are thought inside an approximation procedure. We can say that *throughout the nineteenth century* there was confusion between an error of the type hY with h a scalar number tending to zero and an error with accumulation of independence as a Markov process in small time. In this respect Louis Bachelier [36] with his players of infinitesimal games appears well marking the turning of the two centuries. He will be cited by Kolmogorov [46].

We find, more in Laplace's writings than in Gauss' ones (he keeps his distance) the concern that errors are sums of many independent terms. Chapters III and V of the *Théorie Analytique* insistently express this idea. But beyond the central limit theorem, Laplace is obviously not in position to make a theory where the Taylor formula be replaced by exponentials of operators, i.e. the theory of operators semi-groups.

In Section 17 of the *Theoria combinationis* Gauss writes "Laplace, considering the question under a different point of view, showed that this principle [of supposing that the errors follow a normal law] is better than any other choice, for any probability distribution of the errors of observation as soon as the number of observations is very large. But when this number is restricted, the question remains untouched; so that if our hypothetical law [the normal law] is rejected, the least squares method rests better than the others [e.g. than the use of absolute first moment], by the simple reason that it leads to simpler calculations."

This remark is relevant, but it shows also that Gauss somehow underestimates the mathematical importance of the accumulation of small errors.

Today we know that the infinitesimal reduction of random quantities is well described by the infinitesimal generators of Markov semi-groups which are second order differential operators (Kolmogorov [46], Ventsel [66], Hunt's complete maximum principle [67] Section 15, Meyer-Dellacherie [100] Theorem XIII 22-24) and have a fractional part (integro-differential operator) only in the presence of jumps (case of non local Dirichlet forms, cf below VI §e). Similarly the stochastic differential geometry uses second order tangent vectors (cf [144], Meyer [91]).

It emerges from this discussion that the "small errors" in Gauss' calculation have to be read as a computation of gradient, what reinforces the importance of the propagation formula under its analytical form (2.1) which is that of the carré du champ operator.

6. Trails of research

a) Obtaining numerical results is often difficult in probability theory because the spaces are high dimensional, often infinite dimensional. Therefore the simulation methods, also called Monte Carlo methods are very useful thanks also to their generality, cf for instance [117]. Intuitively the computation of the value of the carré du champ operator on a random variable in a parametrized model, say $X(\omega, \lambda)$, when λ is erroneous and the model is not, is very simple. It is enough to take a *cluster* of points in the neighborhood of the value of λ centered on λ_0 and with quadratic dispersion σ^2 around λ_0 and then to collect the dispersion of the corresponding values of X, with ω being fixed. The dispersion matrix will give the matrix $\Gamma(X)$ and the discrepancy between $X(\omega, \lambda_0)$ and the mean of the cluster of the values of X will yield the bias.

If the model is itself erroneous, the method has to be extended with a cluster around ω_0. This *clusters method* for the error calculus has already been used (Scotti [150]) but has not been theoretically studied up to now and many questions arise: optimal number of points in the cluster with respect to its concentration toward λ_0, obtention of the mean error (which is the square root of the Dirichlet form taken on X), use of acceleration by quasi-Monte-Carlo, etc.

At present theoretical results are in the opposite way: the methods of calculation of densities of probability distributions (cf for example [65], [71], [73], [82], [76], [121], [127], etc.) may be accelerated if an error structure is available (or a Malliavin calculus) see Caballero-Fernandez-Nualart [126], Kohatsu-Higa-Pettersson [132], Bouleau [139].

b) A purely mathematical enigma remains at present. It is the conjecture of the so-called *Energy Image Density property* or EID. This property is true for any local Dirichlet structure with carré du champ for real valued fonctions (cf Bouleau [93] Theorem 5 and Corollary 6), (see also the very close method of Davydov-Lifshitz [96], [97]). It has been conjectured in 1986 (Bouleau-Hirsch [99] p251) that (EID) be true for any local Dirichlet structure with carré du champ. This has been proved for the Wiener space equipped with the Ornstein-Uhlenbeck form and for some other Dirichlet structures by Bouleau-Hirsch (cf [111] Chap. II §5 and Chap. V example 2.2.4), it has been proved by Coquio [116] for random Poisson measures on $\mathbb{R}^d$ and by Bouleau-Denis as soon as the "bottom" space satisfies itself EID. But this conjecture is at present neither proved nor refuted in all generality, it has to be established in each particular framework. (EID) on the Wiener space is now a very frequently used tool to prove existence of density.

c) The error calculus by Dirichlet forms and the mathematically rigourous framework for the carré du champ operator may be used not only for the computation of measurement errors propagation, but to study the effect of *fluctuations* on physical systems. Very often the physicists handle fluctuations as small errors denoted ΔX and conduct calculations in the spirit of Gauss calculus. The program is then to write down anew the fluctuation theory and the deviation that it yields for (non linear) measurements thanks to Dirichlet forms. Theoretical advances have been already obtained in the direction cf for example Albeverio-Grothaus-Kondratiev-Röckner [129], Dembo-Deutschel [156].

In order to be concrete, let us look how L. Landau and E. Lifchitz in their famous textbook [75] compute the deviation of a stretched string due to the thermal fluctuations (Chapter XII exercise 8) :

Let ℓ be the length of the string, F the tension force. Let be a point at the distance x from an end of the string, y its transverse deplacement. To determine $\overline{y^2}$ we have to determine the equilibrium shape of the string for a given deplacement y of the point x; there are two segments of straight line betwen the ends and the point (x, y). The work spent for such a deformation of the string is equal to

$$R_{min} = F.(\sqrt{x^2 + y^2}) - x) + F.[\sqrt{x^2 + y^2} - (\ell - x)] \cong \frac{Fy^2}{2}(\frac{1}{x} + \frac{1}{\ell - x})$$

The mean square is therefore $\overline{y^2} = \frac{T}{F\ell}x(\ell - x)$.

These results are in perfect accordance with the approach that consists in taking the Ornstein-Uhlenbeck form on the Brownian bridge, because if

$$X_t = B_t - tB_1 = \int_0^1 (1_{[0,t]} - t)dB_s$$

then

$$\Gamma[X_t] = \sum_0^1 (1_{[0,t]} - t)^2 ds = t - t^2 = t(1 - t)$$
$$\Gamma[X_s, X_t] = \sum_0^1 (1_{[0,s]} - s)(1_{[0,t]} - t)du = s \wedge t - st = s(1 - t)$$

for $s < t$.

d) If we have an error structure, we have also a capacity theory associated with the Dirichlet form. This allows a refinement of the almost everywhere computations (see Fukushima [95] and works in the wake). This may be related with the error theory under the following regard: this helps to understand that during an approximation procedure some things *are not seen.*

The fact that there are errors, hence randomness, in the approximation procedure *erases* some features of reality. In our framework, the procedure is replaced by a Markov process X_t which is in a sense osculating when $t \to 0$. But the paths of X_t do not see very fines things.

An approximation with strongly stochastic errors (cf above and [152]), particularly when the object to be approximated is a point of a functional space or a path of an erroneous stochastic process, may make visible only some properties of this path.

e) Up to now we have dealt with local Dirichlet forms admitting a carré du champ operator, of course non local Dirichlet forms may possess also such an operator. It would mean to consider non local errors. This is not at all a crazy idea and such a concern appeared already in physics for example in Brillouin [70] Chap XV. About frequency rays of emission or absorption Brillouin distinguishes four possible cases of errors, one of which describes a very narrow ray with companion rays rather far from it with low probability. It seems that non local Dirichlet forms may be relevant in such cases for describing errors.

Of course no differential calculus is available under such hypotheses for the propagation through calculations.

Concluding remarks

With respect to the theory of Dirichlet forms, the error calculus à la Gauss looks quite like what the simply additive probability theory is with respect to the σ-additive theory.

The error calculus by Dirichlet forms allows to perform computations on complex objects defined by limits – as typically solutions to stochastic differential equations. But what yields it to engineers? The question here is still very similar to the one we could ask about the probability calculus axiomatised by Kolmogorov in the framework of measure theory. This can bring much to engineers dealing with stochastic processes.

Not only finance interested in processes! Especially all signal processings like Wiener and Kalman filtering, image improving, recognition, information transmission though channels à la Shannon, and particularly non linear treatments raise the question of an error calculus. We must get used to consider that any input process is accompanied with some accuracy defined by an error structure conditioning – depending on the stochasticity and the intrinsic accuracy of the treatment – the precision on the output expressed also by an error structure, so that a new treatment may be applied.

References

[1] Laplace P. S. "Mémoire sur les probabilités" *Mémoirs de l'Académie royale des Sciences de Paris*, 1780 Œuvres complètes 9, pp. 227-332

[2] Legendre A. M. *Nouvelles méthodes pour la détermination des orbites des planètes* Paris 1806.

[3] ADRAIN R. "Research concerning the probabilities of the errors which happens in making observations" Analyst math. museum, vol 1, n4, philadelphia 1808, 93-109.

[4] LAPLACE P. S. *Théorie analytique des probabilités* Paris 1812.

[5] GAUSS C. F. *Theoria motus corporum coelestium in sectionibus conicis solem ambientium* 1809.

[6] GAUSS C. F. *Disquisitio de elementis ellipticis Palladis* 1811

[7] LAPLACE P. S. *Essai philosophique sur les probabilités* Paris, 1814.

[8] CAUCHY A. "Mémoire [...] pour que la plus grande de toutes les erreurs, abstraction faite du signe, devienne un maximum" 1ère classe de l'Institut 18 fév.1814.

[9] GAUSS C. F. "Bestimmung der Genauigkeit der Beobachtungen" *Lindenau und Bohnenberger Zeitschrift* 1816

[10] GAUSS C. F. *Theoria combinationis observationum erroribus minimis obnoxiae* 1821.

[11] LACROIX S. F. *Traité élémentaire du calcul des probabilités*, 1822

[12] GAUSS C. F. *Supplementum Theoriae Combinationis Observationum* 1828.

[13] FOURIER J. "Second mémoire sur les résultats moyens et sur les erreurs de mesure" Paris 1829.

[14] BESSEL F. W. "Auszug aus einem Briefe des Herrn Prof. Bessel an den Herausgeber" *Astronomische Nachrichten*, vol 2, p.133, 1834

[15] POISSON S. D. *Recherches sur la probabilité des jugements en matière criminelle et en matière civile, précédées des règles générales du calcul des probabilités* 1837.

[16] HAGEN G. *Grûndzüge der Wahrscheinlichkeitsrechnung* Berlin, 1837.

[17] BRAVI G. *Teorica e pratica del probabile* Bergamo, 1840.

[18] COURNOT A. A. *Exposition de la théorie des chances et des probabilités* Paris, 1843.

[19] GERLING CH. L. *Die Angleichungsrechnungen der praktischen geometrie oder der Methode der kleinsten Quadrate* Hambourg, 1843.

[20] BRAVAIS A. *Analyse mathématique sur les probabilités des erreurs de situation d'un point.* Paris 1844.

[21] LEJEUNE-DIRICHLET G. "Sur un moyen général de vérifier l'expression du potentiel relatif à une masse quelconque, homogène ou hétérogène" *Crelle Journal für die reine und angewandte Mathematik*, Bd. 32, S. 80-94. (1846).

[22] THOMSON W. "Note sur une équation aux différences partielles qui se présente dans plusieurs questions de Physique mathématique." *Journal de mathématiques pures et appliquées* 1re série, tome 12 (1847), p. 493-496.

[23] RIEMANN B. "Grundlagen für eine allgemeine Theorie der Funktionen einer veränderlischen komplexen Grösse (Inaugural Dissertation) 1851.

[24] BIENAYMÉ I. J. "Sur la probabilité des erreurs" *Journal de mathématiques pures et appliquées* 1re série, tome 17 (1852), p. 33-78.

[25] BIVER P. E. "Théorie analytique des moindres carrés" *Journal de mathématiques pures et appliquées* 1re série, tome 18 (1853), p. 169-200.

[26] CAUCHY A. "Sur la probabilité des erreurs qui affectent des résultats moyens d'observations de même nature" *C. R. Acad. Sci.*, t. XXXVII, p. 264 (1853)

[27] CAUCHY A. "Sur la plus grande erreur à craindre dans un résultat moyen, et sur le système de facteurs qui rend cette plus grande erreur un minimum" *C. R. Acad. Sci.*, t. XXXVII, p. 326 (1853)

[28] GAUSS C. F. *Méthode des moindres carrés, mémoires sur la combinaison des observations* trad. J. Bertrand, Paris 1855.

[29] RIEMANN B. "Theorie der Abelschen Functionen" *Crelle Journal* 54, 115-155, 1857.

[30] MEYER A. *Essai sur une exposition nouvelle de la théorie analytique des probabilités a posteriori.* Liège, 1857.

[31] AIRY G. B. *On the algebraical and numerical theory of errors of observations and*

the combination of observations Cambridge, 1861.

[32] FAA DI BRUNO F. *Traité élémentaire du calcul des erreurs, avec des tables stéréotypées, ouvrage utile à ceux qui cultivent les sciences d'observation* (1869).

[33] LIAGRE J. B. J. *Calcul des Porbabilités et théories des erreurs* Paris, Bruxelles 1879.

[34] VON KRIES, J. *Die Prinzipien der Wahrscheinlichkeitsrechnung*, Freiburg 1886.

[35] BERTRAND J. *Calcul des probabilités* 1888.

XXth century

[36] BACHELIER L. "Théorie des probabilités continues", *J. de mathématiques pures et appliquées*, vol. 6, no 2, 1906, p. 259?327

[37] POINCARÉ, H. *Calcul des Probabilités* Gauthier-Villars, 1912.

[38] A DE FOREST PALMER *The Theory of Measurements* McGraw-Hill 1912.

[39] FRÉCHET M. "Remarque sur les probabilités continues" *Bull. Sci. Math.* 2e série, 45, (1921), 87-88.

[40] BOREL E. *Calcul des probabilités* Paris, 1924.

[41] FISHER R. A. *Statistical methods for research workers.* Genesis Publishing. (1925).

[42] FISHER R. A. *Theory of Statistical Information*, Proc. Cambridge Philo. Soc. Vol XXII, Pt5, 1925.

[43] HOSTINSKÝ B. "Sur la méthode des fonctions arbitraires dans le calcul des probabilités" *Acta Math.* 49, (1926), 95-113.

[44] SCARBOROUGH J. B. "The invalidity of commonly used method for computing a certain probable error" *Proc. Nat. Acad. Sci* 15, 665 (1929).

[45] SCHULTZ H. "Discussion" *J. of the American Statistical Association*, Vol. 24, No. 165, Supplement: Proceedings of the American Statistical Association (1929), pp. 86-89

[46] KOLMOGOROV A. N. "Ueber die analytischen Methoden in der Wahrscheinlichkeitsrecgnung", *Math. Annalen* 1931,

[47] HOSTINSKÝ B. *Méthodes générales de Calcul des Probabilités*, Gauthier-Villars 1931.

[48] BIRGE R. T. "The calculation of errors by the method of least squares" *Physical Review* V40, 1932.

[49] HOPF E. "On causality, statistics and probability" *J. of Math. and Physics* 18 (1934) 51-102.

[50] HOPF E. "Über die Bedeutung der willkürlichen Funktionen für die Wahrscheinlichkeitstheorie" *Jahresbericht der Deutschen Math. Vereinigung* XLVI, I, 9/12, 179-194, (1936).

[51] HOPF E. "Ein Verteilungsproblem bei dissipativen dynamischen Systemen" *Math. Ann.* 114, (1937), 161-186.

[52] BIRGE R. T. "The propagation of errors" *The Amer. Physics Teacher* V7, n6, 1939.

[53] BARET J. *Calcul des erreurs* Paris 1939.

[54] HOTELLING H. "Some new methods in matrix calculation" *Ann. Math. Statist.*, 14 (1943), pp. 1-34.

[55] CARTAN H. "Théorie du potentiel newtonien" *Bull. Soc. Math. de France* 73 (1945) 74-106.

[56] CRAMÉR,H. *Mathematical methods of statistic* Princeton Univ. Press 1946.

[57] VON NEUMANN J., GOLDSTINE H. H. "Numerical inverting of matrices of high order", *Bull. Amer. Math. Soc.*, 53 (1947), pp. 1021-1099.

[58] TURING A. M. "Rounding-off errors in matrix processes " *Quart. J. Mech.*, (1948), pp.287-308.

[59] VESSIOT E., MONTEL P. *Cours de Mathématiques Générales* Eyrolles 1947.

[60] DENY J. "Les potentiels d'énergie finie" *Acta Mathematica* 82, 107-183 (1950).

[61] DENY J. "Sur la définition de l'énergie en théorie du potentiel" *Ann. Inst. Fourier*

2, 83-99, 1950.

[62] ELKINS TH. "The second derivative method of gravity interpretation" *Society of Exploration Geophysicists* June 1950.

[63] ANGOT A. *Compléments de mathématiques à l'usage des ingénieurs de l'électrotechnique et des télécommunications* Paris 1952.

[64] ELKINS TH. "The effect of random errors in gravity data on second derivative values" *Geophysics*, January 1952, Vol. 17, No. 1 : pp. 70-88.

[65] PARZEN E. "On estimation of a probability density function and mode" *Ann. Inst. Statist. Math.* 6, 127-132, (1954).

[66] WENTZEL A. D. "Semi-groups of operators corresponding to a generalized differential operator of second order" *DAN SSSR*, Vol.111,2,1956,pp. 269-272.

[67] HUNT G. A. "Markov processes and potentials II" *Illinois J. of Math.* 1957, n1, 316-369.

[68] BEURLING A., DENY J. "Espaces de Dirichlet I. Le cas élémentaire", *Acta Math.* 99, (1958) 203-224.

[69] BEURLING A., DENY J."Dirichlet spaces" *Proc. Nat. Acad. Sci. U.S.A.* 45 (1959) 208-215.

[70] BRILLOUIN L. *La Science et la Théorie de l'Information* Paris, Masson 1959.

[71] BARTLETT,M.S. "Statistical estimation of density function" Sankhyä, sA, 25, 245-254 1963.

[72] BRELOT M. *Théorie Classique du Potentiel* cours polycopié CDU 3ème édition 1965.

[73] MURTHY,V. K. "Estimation of probability density" *The Annals of Mathematical Statistics* ,6, 1027-103, 1965.

[74] MOORE R. E. *Interval Analysis* Prentice Hall 1966.

[75] LANDAU L. LIFCHITZ E. *Physique Statistique* MIR 1967.

[76] ELKINS TH. "Cubical and Spherical Estimation of Multivariate Probability Density" *Journal of the American Statistical Association*, Vol. 63, No. 324 (Dec., 1968), pp. 1495-1513.

[77] DENY J. "Méthodes hilbertiennes en théorie du potentiel" Potential Theory, Centro Internazionale Matematico Estivo, Roma 1970.

[78] WILKINSON J. H. "Modern Error Analysis" *SIAM Review* Vol. 13, No. 4, October 1971.

[79] BRELOT M. "Les étapes et les aspects multiples de la théorie du potentiel" *L'Enseignement Mathématique* XVIII, fasc.1, 1972.

[80] SILVERSTEIN M. L. *Symmetric Markov processes*, L. N. in Math. vol 426, Springer, 1974.

[81] KLEIN H. A. *The science of measurement, a historical survey*, Dover, 1974.

[82] DEHEUVELS P. "Estimation non paramétrique de la densité par histogrammes généralisés" *R. Statist. Appl.* Vol 25, f3, 1-24, (1977).

[83] DELLACHERIE C. and MEYER P.-A. *Probability and Potential* North-Holland mathematics studies 29, 1978.

[84] KOLMOGOROV A. N., YUSHKEVICH A. P. *Mathematics of the 19th century (1978)*, Birkhaüser 1992.

[85] PEARSON K. *The History of Statistics in the 17th & 18th Centuries* Charles Griffin & Co 1978.

[86] LE JAN Y. "Mesures associées à une forme de Dirichlet, applications" *Bull. Soc. Math. France* 106, 61-112, 1978.

[87] FUKUSHIMA M. "A decomposition of additive functionals of finite energy" *Nagoya Math. J.* 74, 1979, 137-168.

[88] SHETNIN O. B. "C. F. Gauss and the theory of errors" *Archive for History of Exact*

Sciences 1979 Springer.

[89] FUKUSHIMA, M. *Dirichlet Forms and Symmetric Markov Processes* North-Holland 1980.

[90] IKEDA N., WATANABE S. *Stochastic Differential Equation and Diffusion Processes*, North-Holland, Koshanda 1981.

[91] MEYER P.-A. "Géométrie stochastique sans larmes" *Sém. de Probabilités XV* Lect. Notes in M. 850, Springer 1981.

[92] TEMPLE G. *100 Years of Mathematics* Duckworth 1981.

[93] BOULEAU N. "Décomposition de l'énergie par niveau de potentiel" *Colloque en l'honneur de Jacques Deny, Lect. Notes in M. 1096*, Springer(1984), http://hal.archives-ouvertes.fr/hal-00449195/fr/

[94] ANCONA A. "L'énergie et la théorie du potentiel dans l'œuvre de Jacques Deny" *Colloque en l'honneur de Jacques Deny, Lect. Notes in M. 1096*, Springer (1984).

[95] FUKUSHIMA M. "A Dirichlet form on the Wiener space and properties on Brownian motion ", *Colloque en l'honneur de Jacques Deny, Lect. Notes in M. 1096*, Springer (1984).

[96] DAVYDOV Y. A., LIFSHITS M. A. "The stratification method in some probability problems" *Prob. Theory, Math. Stat., Theor. Cybernetics*, 22,61-137, (1984).

[97] LIFSHITZ M.A. "An application of the stratification method to the study of functionals of Lévy processes and examples" *Theor. Probab. Appl.* 29, 753-764, (1984).

[98] RÖCKNER M. and WIELENS N. "Dirichlet forms — Closability and change of speed measure", *Infinite Dimensional Analysis and Stochastic Processes*, Research Notes in M., Albeverio ed. Pitman 124, 119-144, (1985).

[99] BOULEAU N. and HIRSCH F."Formes de Dirichlet générales et densité des variables aléatoires réelles sur l'espace de Wiener" *J. Funct. Analysis* 69, 2, 229-259, (1986).

[100] DELLACHERIE C. and MEYER P.-A. *Probabilités et Potentiel* Chap XII à XVI, Hermann 1987.

[101] BICHTELER K., GRAVEREAUX J.-B., JACOD J. *Malliavin Calculus for Processes with Jumps* (1987).

[102] BOULEAU N. "Autour de la variance comme forme de Dirichlet" in *Séminaire de Théorie du Potentiel Paris No.8*, Lect. Notes in Math. 1235, Springer 1987.

[103] WU L. "Construction de l'opérateur de Malliavin sur l'espace de Poisson" *Sém. Probabilité XXI* Lect. Notes in M. 1247, Springer (1987).

[104] JEUDY L. "Generalyzed variance-covariance propagation law formulae and application to explicit least-squares adjustments" *Bulletin géodésique* 1988, Volume 62, Issue 2, pp 113-124.

[105] TEUNISSEN P. J. G. "Nonlinearity and Least Squares" *CISM journ. ACSGC*, V42, n4, 1988.

[106] TEUNISSEN P. J. G. "First and Second Moments of non-Linear Least-Squares Estimators" *Bull. Géod.* 63 (1989) pp. 253-262.

[107] COLEMAN H. W., STEELE W. G. *Experimentation and uncertainty analysis for engineers* Wiley 1989.

[108] BOULEAU N. and LAMBERTON D. "Residual risks and hedging strategies in Markovian markets", *Stochastic processes and their applications* 33,131-150, 1989.

[109] HANGOS K. M., LEISZTNER L. "Towards properly controlled analytical measurement methods" *J. of Automatic Chemistry* Vol. 11, No. 2 (March-April 1989), pp. 76-79.

[110] ALBEVERIO S. and RÖCKNER M. "Classical Dirichlet forms on topological vector spaces — closability and a Cameron-Martin formula" *J. Funct. Analysis*, 88, 395-436, (1990).

[111] BOULEAU N. and HIRSCH F. *Dirichlet Forms and Analysis on Wiener Space* De Gruyter (1991).
[112] DELLACHERIE C., MAISONNEUVE B. and MEYER P.-A. *Probabilités et Potentiel* Chap XVII à XXIV, Hermann 1992.
[113] ENGEL E. "A road to randomness in physical systems" Lect. notes in Stat. 71, Springer 1992.
[114] MA and RÖCKNER M. *Introduction to the Theory of (Non-Symmetric) Dirichlet forms* Springer 1992.
[115] ALBEVERIO S., MA Z.-M., RÖCKNER M. "Quasi-regular Dirichlet forms and Markov processes", *J. Funct. Anal.* Ill (1993), 118-154.
[116] COQUIO A. "Formes de Dirichlet sur l'espace canonique de Poisson et application aux équations différentielles stochastiques" *Ann. Inst. Henri Poincaré* vol 19, n1, 1-36, (1993).
[117] BOULEAU N. and LÉPINGLE D. *Numerical Methods for Stochastic Processes* Wiley (1994).
[118] FUKUSHIMA M., OSHIMA Y. and TAKEDA M. *Dirichlet Forms and Symmetric Markov Processes* De Gruyter (1994).
[119] NUALART, D. *Malliavin Calculus and Related Topics* Springer 1995.
[120] BALLY V., TALAY D. "The law of the Euler scheme for stochastic differential equations : I. Convergence rate of the distribution function", *Prob. Th. and Rel. Fields* vol 2 No2, 93-128 (1996).
[121] BALLY V., TALAY D. "The law of the Euler scheme for stochastic differential equations : II. Convergence rate of the density", *Monte Carlo Methods and Appl.* vol 104, No1, 43-80 (1996).
[122] JACOB N. *Pseudo-Differential Operators and Markov Processes*, Akademic Verlag (1996).
[123] BOGACHEV V., ROECKNER M., SCHMULAND B. "Generalized Mehler semigroups and applications" *Probability Theory and R.F.* V 105, N2, 193-225, (1996).
[124] MALLIAVIN P., *Sochastic Analysis*, Springer 1997.
[125] BRELOT M. *Théorie Classique du Potentiel* Ass. Laplace-Gauss Paris (1997).
[126] CABALLERO M. E., B. FERNANDEZ, D. NUALART "Estimating densities and applications" *J. of Theoretical Probability* Vol 11, Nr3, (1998).
[127] SILVERMAN B. W. *Density Estimation for Statistics and Data Analysis* Chapman and Hall, 1998.
[128] VILLANI C. "Fisher Information Estimates for Boltzmann's Collision operator" *J. Math. Pures Appl.*, 77, 1998, p. 821-837.

XXIth century

[129] ALBEVERIO S., GROTHAUS M., KONDRATIEV Y. G., RÖCKNER M. "Fluctuations in classical continuous systems" Univ. Bielefeld 2000.
[130] MA and RÖCKNER M. "Construction of diffusion on configuration spaces" *Osaka J. Math.* 37, 273-314, (2000).
[131] PRIVAULT N. "Extended covariance identities and inequalities" *Statistics & Prob. Letters* 55, 247-255 (2001).
[132] KOHATSU-HIGA A., PETTERSSON R. "Variance reduction methods for simulation of densities on Wiener space" SIAM J. Numer. Anal., 40(2), 431-450, 2002.
[133] BOULEAU N. *Error Calculus for Finance and Physics, the Language of Dirichlet Forms*, De Gruyter (2003).
[134] BOULEAU N. "Error calculus and path sensitivity in financial models" *Mathematical Finance* vol 13 n^0 1, 115-134, (Jan. 2003).

[135] MALLIAVIN P., THALMAIER A. "Numerical error for SDE: Asymptotic expansion and hyperdistributions", *C. R. Acad. Sci. Paris* ser. I 336 (2003) 851-856.

[136] B. BOUCHARD, I. EKELAND, N. TOUZI "On the Malliavin approach to Monte Carlo approximation of conditional expectations" *Finance Stochast.* 8, 45-71, (2004).

[137] BOULEAU N., CHORRO CH., "Error structures and parameter estimation" *C. R. Acad. Sci. Paris*, Ser. I 338, 305-310 (2004).

[138] FITZSIMMONS P. "Superposition operators on Dirichlet spaces" Tohoku mathematical journal, 2004.

[139] BOULEAU N. "Dirichlet forms in simulation" *Monte Carlo Methods and Appl.* Vol 11, n 4, 385-396, (2005).

[140] CHORRO C. "Calculs d'erreur par formes de Dirichlet : liens avec l'information de Fisher et les théorèmes limites" Thèse, Univ. Paris I, 2005.

[141] BOULEAU N. "Th?or?me de Donsker et formes de Dirichlet" *Bulletin des Sc. Mathématiques*, 129, (2005), 369-380.

[142] CHORRO C. "Convergence in Dirichlet Law of Certain Stochastic Integrals" *Electr. J. of Probability* Vol 10, 1005-1025, (2005).

[143] BOULEAU N."On some errors related to the graduation of measuring instruments" Oct. 2006 http://hal.archives-ouvertes.fr/hal-00105452/fr/

[144] BOULEAU N. "Bringing errors into focus" arXiv : 0705.0519

[145] BOULEAU N. "When and how an error yields a Dirichlet form" *Journal of Functional Analysis* Vol 240, Issue 2 , (2006) 445-494.

[146] BALLY V., BAVOUZET M.-P., MESSAOUD M."Intergration by parts formula for locally smooth laws and application to sensitivity computation" *The Annals of Appl. Prob.* Vol. 17, No. 1, 33?66, 2007.

[147] PRIVAULT N., WEI X. "Integration by parts for point processes and Monte Carlo simulation" *J. of Appl. Prob.* 44, n3, 806-823, (2007).

[148] SCOTTI S."Errors Theory using Dirichlet Forms, Linear Partial Differential Equations and Wavelets" arXiv:0708.1073, 2007.

[149] BOULEAU N. "On error operators related to the arbitrary functions principle" Jour. Functional Analysis 251, (2007) 325-345.

[150] SCOTTI S. *Applications de la Théorie des Erreurs par Formes de Dirichlet*, Thesis Univ. Paris-Est, Scuola Normale Pisa, 2008. (http://pastel.paristech.org/4501/)

[151] SCOTTI S."Perturbative Approach on Financial Markets" arXiv:0806.0287, 2008.

[152] BOULEAU N. "How to specify an approximate numerical result" *Proceedings of RIMS 2006 Workshop on Stochastic Analysis and Applications*, Kyoto 2007, *RIMS Kôkyûroku Bessatsu* B6, p39-53, 2008. http://hal.archives-ouvertes.fr/hal-00781414.

[153] SCOTTI S., LY VATH V. "Bid-ask Spread Modelling, a perturbation appoach" finance-innovation.org (2008).

[154] BOULEAU N. and DENIS L. "Energy image density property and the lent particle method for Poisson measures" *Jour. of Functional Analysis* 257 (2009) 1144-1174. http//dx.doi.org/10.1016/j.jfa.2009.03.004

[155] PRIVAULT N. *Stochastic Analysis in Discrete and Continuous Settings with Normal Martingales* Springer 2009.

[156] DEMBO A., DEUSCHEL J.-D."Markovian perturbation, response and fluctuation dissipation theorem" arXiv:0710.4394v2 (2010).

[157] LIM TH.-S. "Quelques applications du contrôle stochastique aux risques de défaut et de liquidité" Thesis, Univ. Paris 7, 2010.

[158] BOULEAU N. and DENIS L. "Application of the lent particle method to Poisson driven SDE's", *Probability Theory and Related Fields* Vol 151, Issue 3-4, pp 403-433, December 2011.

[159] MALICET D. and POLY G. "Propriétés de convergence dans les structures d'erreur" 2012 http://hal.archives-ouvertes.fr/hal-00608007

[160] BOULEAU N., DENIS L. *Dirichlet Forms Methods for Poisson Point Measures and Lévy Processes, with emphasis on creation-annihilation techniques* 265p, March 2013 to appear.

Chapter 6

Stein's method, Malliavin calculus, Dirichlet forms and the fourth moment theorem

Louis H.Y. Chen* and Guillaume Poly†

The fourth moment theorem provides error bounds of the order $\sqrt{\mathbb{E}(F^4) - 3}$ in the central limit theorem for elements F of Wiener chaos of any order such that $\mathbb{E}(F^2) = 1$. It was proved by Nourdin and Peccati [31] using Stein's method and the Malliavin calculus. It was also proved by Azmoodeh, Campese and Poly [2] using Stein's method and Dirichlet forms. This paper is an exposition on the connections between Stein's method and the Malliavin calculus and between Stein's method and Dirichlet forms, and on how these connections are exploited in proving the fourth moment theorem.

Contents

*Department of Mathematics, National University of Singapore, 10 Lower Kent Ridge Road, Singapore 119076, Republic of Singapore, matchyl@nus.edu.sg
†Mathematical research unit, University of Luxembourg, 6 rue Richard Coudenhove-Kalergi, L-1359 Luxembourg, Grand Duchy of Luxembourg, poly@crans.org

1. Introduction

In 2005, Nualart and Peccati [36], discovered the remarkable fact that a sequence of multiple Wiener-Itô integrals, that is, members of a Wiener chaos, converges in distribution to a Gaussian random variable if and only if their second and fourth moments converge to the corresponding moments of the limiting random variable. The proof in [36] relies on a theorem in stochastic calculus, namely the Dubins-Schwarz Theorem. Although the proof is elegant, it does not provide good bounds on the distances (for instance, the Kolmogorov distance) between the sequence and its limit.

A few years later, the result of Nualart and Peccati [36] was generalized and developed further by Peccati and Tudor [38] and also by Nourdin and Peccati [31]. Peccati and Tudor [38] considered the multivariate central limit theorem for vectors of multiple Wiener-Itô integrals and established that if the covariance matrices of the vectors of multiple Wiener-Itô integrals converge to that of a Gaussian vector, then joint convergence in distribution to Gaussian is equivalent to coordinate-wise convergence in distribution to Gaussian. As an application of this result, the central limit theorem for any sequence of random variables can be established by proving a central limit theorem for each term in the chaos expansion.

Another significant development from the work of Nualart and Peccati [36] is that of Nourdin and Peccati [31]. Developing considerably a remarkable intuition of Nualart and Ortiz-Latorre in [35], Nourdin and Peccati [31] established a fundamental connection between Stein's method and the Malliavin calculus via the use of the Malliavin operators $(D, \delta, \mathbf{L})$. This connection resulted in the derivation of errors bounds, often optimal, in the central limit theorems for random variables in the Wiener space. Of special interest in this paper is the proof of error bounds in the central limit theorem for multiple Wiener-Itô integrals, which we call the fourth moment theorem.

The success of such a connection relies on the fact that both Stein's method and the Malliavin calculus are built on some integration by parts techniques. In addition, the operators of the Malliavin calculus, $D, \delta, \mathbf{L}$, satisfy several nice integration by parts formulae which fit in perfectly with the so-called Stein equation. For a good overview of these techniques, we refer to the following website.

`https://sites.google.com/site/malliavinstein/home`

The work of Nourdin and Peccati [31] has added a new dimension to Stein's method. Their approach of combining Stein's method with the Malliavin calculus has led to improvements and refinements of many results in probability theory, such as the Breuer-Major theorem [10]. More recently, this approach has been successfully used to obtain central limit theorems in stochastic geometry, stochastic calculus, statistical physics, and for zeros of random polynomials. It has also been extended to different settings as in non-commutative probability and Poisson chaos. Of particular interest is the connection between the Nourdin-Peccati analysis and information theory, which was recently revealed in [27, 34]. An overview of these new developments can also be found in the above website.

The approach of Nourdin and Peccati [31] entails the use of the so-called product formula for Wiener integrals. The use of this formula makes the proofs rather involved since it relies on subtle combinatorial arguments. Very recently, starting with the work of Ledoux [26], a new approach to the fourth moment theorem was developed by Azmoodeh, Campese and Poly [2] by combining Stein's method with the Dirichlet form calculus. An advantage of this new approach is that it provides a simpler proof of the theorem by avoiding completely the use of the product formula. Moreover, since a Dirichlet space is a more general concept than the Wiener space, the former contains examples of fourth moment theorems that cannot be realized on the latter.

A more algebraic flavor of this approach has enabled Azmoodeh, Malicet and Poly [3] to prove that convergence of pairs of moments other than the 2nd and 4th (for example, the 6th and 68th) also implies the central limit theorem. This new approach seems to open up new possibilities and perhaps also central limit theorems on manifolds.

This paper is an exposition on the connections between Stein's method and the Malliavin calculus and between Stein's method and Dirichlet forms, and on how these connections are exploited in proving the fourth moment theorem.

2. Stein's method

2.1. *How it began*

Stein's method began with Charles Stein using his own approach in the 1960's to prove the combinatorial central limit theorems of Wald and Wolfowitz [44] and of Hoeffding [24]. Motivated by permutation tests in nonparametric statistics, Wald and Wolfowitz [44] proved that under certain conditions, $\sum_{i=1}^n a_i b_{\pi(i)}$ converges in distribution to the standard normal distribution $\mathcal{N}(0,1)$, where $\{a_i, b_j : i, j = 1, \ldots, n\}$ are real numbers and π a random permutation of $\{1, \ldots, n\}$. Hoeffding [24] generalized the result of Wald and Wolfowitz [44] to $\sum_{i=1}^n c_{i\pi(i)}$, where $\{c_{ij} : i, j = 1, \ldots, n\}$ is a square array of real numbers and π a random permutation of $\{1, \ldots, n\}$.

Let $W = \sum_{i=1}^n c_{i\pi(i)}$ and let ϕ be the characteristic function of W. Assume $c_{i\cdot} = c_{\cdot j} = 0$ where $c_{i\cdot} = \sum_{j=1}^n c_{ij}/n$ and $c_{\cdot j} = \sum_{i=1}^n c_{ij}/n$ (which implies $\mathbb{E}[W] = 0$), and also assume $\mathrm{Var}(W) = 1$. While using exchangeable pairs to show that

$$\phi'(t) \simeq -t\phi(t) \tag{2.1}$$

Stein realized that there was nothing special about the complex exponentials. Since

$\phi'(t) = i\mathbb{E}\big[We^{itW}\big]$ and $-t\phi(t) = -t\mathbb{E}\big[e^{itW}\big] = -\frac{1}{i}\mathbb{E}\left[\frac{\partial}{\partial W}e^{itW}\right]$, replacing the complex exponential by an arbitrary function f, (2.1) becomes

$$\mathbb{E}\big[Wf(W)\big] \simeq \mathbb{E}\big[f'(W)\big].$$

By letting f be a solution, say f_h, of the differential equation

$$f'(w) - wf(w) = h(w) - \mathbb{E}\big[h(Z)\big],$$

where h is a bounded function and $Z \sim \mathcal{N}(0,1)$, one obtains

$$\mathbb{E}\big[h(W)\big] - \mathbb{E}\big[h(Z)\big] = \mathbb{E}\big[f_h'(W) - Wf_h(W)\big].$$

How close $\mathscr{L}(W)$ is to $\mathcal{N}(0,1)$ can then be determined by bounding $\mathbb{E}\big[f_h'(W) - Wf_h(W)\big]$. There is no inversion of the characteristic function.

This story of how Stein's method began is based on personal communications with Charles Stein and Persi Diaconis and also on an interview of Stein in Leong [28].

2.2. *A general framework*

Stein's method for normal approximation was published in his seminal 1972 paper in the Proceedings of the Sixth Berkeley Symposium. Although the method was developed for normal approximation, Stein's ideas were very general and the method was modified by Chen [13] for Poisson approximation. Since then the method has been constantly developed and applied to many approximations beyond normal and Poisson and in finite as well as infinite dimensional spaces. As Stein's method works well for dependent random variables, it has been applied, and continues to be applied, to a large number of problems in many different fields. The method, together with its applications, continues to grow and remains a very active research area. See, for example, Stein [43], Arratia, Goldstein and Gordon [1], Barbour, Holst and Janson [8], Diaconis and Holmes [21], Barbour and Chen [6, 7], Chatterjee, Diaconis and Meckes [12], Chen, Goldstein and Shao [15], Ross [40], Shih [41], Nourdin and Peccati [32], and Chen and Röllin [16, 17].

In a nutshell, Stein's method can be described as follows. Let W and Z be random elements taking values in a space $\mathcal{S}$ and let $\mathcal{X}$ and $\mathcal{Y}$ be some classes of real-valued functions defined on $\mathcal{S}$. In approximating the distribution $\mathscr{L}(W)$ of W by the distribution $\mathscr{L}(Z)$ of Z, we write $\mathbb{E}\big[h(W)\big] - \mathbb{E}\big[h(Z)\big] = \mathbb{E}\big[Lf_h(W)\big]$ for a test function $h \in \mathcal{Y}$, where L is a linear operator (Stein operator) from $\mathcal{X}$ into $\mathcal{Y}$ and $f_h \in \mathcal{X}$ a solution of the equation

$$Lf = h - \mathbb{E}\big[h(Z)\big] \qquad \text{(Stein equation)}. \tag{2.2}$$

The error $\mathbb{E}\big[Lf_h(W)\big]$ can then be bounded by studying the solution f_h and exploiting the probabilistic properties of W. The operator L characterizes $\mathscr{L}(Z)$ in the sense that $\mathscr{L}(W) = \mathscr{L}(Z)$ if and only if for a sufficiently large class of functions f we have

$$\mathbb{E}\big[Lf(W)\big] = 0 \qquad \text{(Stein identity)}. \tag{2.3}$$

In normal approximation, where $\mathscr{L}(Z)$ is the standard normal distribution, the operator used by Stein [42] is given by $Lf(w) = f'(w) - wf(w)$ for $w \in \mathbb{R}$, and in Poisson approximation, where $\mathscr{L}(Z)$ is the Poisson distribution with mean $\lambda > 0$, the operator L used by Chen [13] is given by $Lf(w) = \lambda f(w+1) - wf(w)$ for $w \in \mathbb{Z}_+$. However the operator L is not unique even for the same approximating distribution but depends on the problem at hand. For example, for normal approximation L can also be taken to be the generator of the Ornstein-Uhlenbeck process, that is, $Lf(w) = f''(w) - wf'(w)$, and for Poisson approximation, L taken to be the generator of an immigration-death process, that is, $Lf(w) = \lambda[f(w+1) - f(w)] + w[f(w-1) - f(w)]$. This generator approach, which is due to Barbour [4], allows extensions to multivariate and process settings. Indeed, for multivariate normal approximation, $Lf(w) = \Delta f(w) - w \cdot \nabla f(w)$, where f is defined on the Euclidean space; see Barbour [5] and Götze [23], and also Reinert and Röllin [39] and Meckes [30].

2.3. *Normal approximation*

In many problems of normal approximation, the random variable W whose distribution is to be approximated satisfies this equation

$$\mathbb{E}\big[Wf(W)\big] = \mathbb{E}\big[T_1 f'(W + T_2)\big]. \tag{2.4}$$

where T_1 and T_2 are some random variables defined on the same probability space as W, and f is an absolutely continuous function for which the expectations in (2.4) exist. Examples of W satisfying this equation include sums of locally dependent random variables as considered in Chen and Shao [19] and exchangeable pairs as defined in Stein [43]. More generally, a random variable W satisfies (2.4) if there is a Stein coupling (W, W', G) where W, W', G are defined on a common probability space such that $\mathbb{E}\big[Wf(W)\big] = \mathbb{E}\big[Gf(W') - Gf(W)\big]$ for absolutely continuous functions f for which the expectations exist (see Chen and Röllin [17]). In all cases it is assumed that $\mathbb{E}\big[W\big] = 0$ and $\mathrm{Var}(W) = 1$. Letting $f(w) = w$, we have $1 = \mathbb{E}\big[W^2\big] = \mathbb{E}\big[T_1\big]$.

As an illustration, let (W, W') be an exchangeable pair of random variables with $\mathbb{E}(W) = 0$ and $\mathrm{Var}(W) = 1$ such that $\mathbb{E}\big[W' - W|W\big] = -\lambda W$ for some $\lambda > 0$. Since the function $(w, w') \longmapsto (w' - w)(f(w') + f(w))$ is anti-symmetric, the exchangeability of (W, W') implies

$$E\big[(W' - W)(f(W') + f(W))\big] = 0.$$

From this we obtain

$$\mathbb{E}\big[Wf(W)\big] = \frac{1}{2\lambda}\mathbb{E}\Big[(W' - W)(f(W') - f(W))\Big]$$

$$= \frac{1}{2\lambda}\mathbb{E}\Big[(W' - W)^2 \int_0^1 f'(W + (W' - W)t)dt\Big] = \mathbb{E}\big[T_1 f'(W + T_2)\big]$$

where $T_1 = \dfrac{1}{2\lambda}(W' - W)^2$, $T_2 = (W' - W)U$, and U uniformly distributed on $[0,1]$ and independent of W, W', T_1 and T_2.

Let f_h be the unique bounded solution of the Stein equation

$$f'(w) - wf(w) = h(w) - \mathbb{E}\big[h(Z)\big], \tag{2.5}$$

where $w \in \mathbb{R}$, $Z \sim \mathcal{N}(0,1)$, and h a bounded test function.

The following boundedness properties of f_h are useful for bounding the errors in the approximation.

If h is bounded, then

$$\|f_h\|_\infty \le \sqrt{2\pi}\|h\|_\infty, \quad \|f_h'\|_\infty \le 4\|h\|_\infty.$$

If h is absolutely continuous, then

$$\|f_h\|_\infty \le 2\|h'\|_\infty, \quad \|f_h'\|_\infty \le \sqrt{2/\pi}\|h'\|_\infty, \quad \|f_h''\|_\infty \le 2\|h'\|_\infty.$$

If $h = \mathbf{1}_{(-\infty,x]}$, then for all $w, v \in \mathbb{R}$,

$$0 \le f_h(w) \le \sqrt{2\pi}/4, \quad |wf_h(w)| \le 1, \quad |f_h'(w)| \le 1, \quad |f_h'(w) - f_h'(v)| \le 1.$$

These can be found in Lemmas 2.4 and 2.5 of Chen, Goldstein and Shao [15] and in Lemma 2.2 of Chen and Shao [20]

Assume that $\mathbb{E}\big[W\big] = 0$ and $\mathrm{Var}(W) = 1$. From (2.4) and (2.5),

$$\begin{aligned}\mathbb{E}\big[h(W)\big] - \mathbb{E}\big[h(Z)\big] &= \mathbb{E}\big[f_h'(W) - T_1 f_h'(W + T_2)\big] \\ &= \mathbb{E}\big[T_1(f_h'(W) - f_h'(W + T_2))\big] + \mathbb{E}\big[(1 - T_1)f_h'(W)\big]\end{aligned} \tag{2.6}$$

Different techniques have been developed for bounding the error term on the right side of (2.6). We will consider two special cases.

Case 1. Assume that $T_1 = 1$. This is the case of zero-bias coupling. See Goldstein and Reinert [22], and also Chen, Goldstein and Shao [15]. Let h be absolutely continuous such that h' is bounded. From (2.6),

$$|\mathbb{E}\big[h(W)\big] - \mathbb{E}\big[h(Z)\big]| = |\mathbb{E}\Big[\int_0^{T_2} f_h''(W + t)dt\Big]| \le \|f_h''\|_\infty \mathbb{E}\big[|T_2|\big] \le 2\|h'\|_\infty \mathbb{E}\big[|T_2|\big].$$

From this we obtain the following bound on the Wasserstein distance between $\mathscr{L}(W)$ and $\mathcal{N}(0,1)$.

$$d_{\mathrm{W}}(\mathscr{L}(W), \mathcal{N}(0,1)) := \sup_{|h(x)-h(y)| \le |x-y|} |\mathbb{E}\big[h(W)\big] - \mathbb{E}\big[h(Z)\big]| \le 2\mathbb{E}\big[|T_2|\big].$$

Note that $d_{\mathrm{W}}(\mathscr{L}(W), \mathcal{N}(0,1)) = \|F - \Phi\|_1$ where $F(x) = P(W \le x)$ and $\Phi(x) = P(Z \le x)$.

If $W = X_1 + \cdots + X_n$ where $X_1, \ldots, X_n$ are independent random variables with $\mathbb{E}[X_i] = 0$, $\mathrm{Var}(X_i) = \sigma^2$ and $\mathbb{E}\big[|X_i|^3\big] = \gamma_i < \infty$, then $T_2 = \xi_I - X_I$ where the X_i, the ξ_i and I are independent, $\mathbb{E}\big[|\xi_i|\big] = \gamma_i/2\sigma_i^2$, and $P(I = i) = \sigma_i^2$. Note that

$\sum \sigma_i^2 = \mathrm{Var}(W) = 1$. Then the bound on the Wasserstein distance between $\mathscr{L}(W)$ and $\mathcal{N}(0,1)$ is $2\mathbb{E}\big[|\xi_I - X_I|\big] \leq 3\sum \gamma_i$.

It is more difficult to obtain a bound on the Kolmogorov distance between $\mathscr{L}(W)$ and $\mathcal{N}(0,1)$, namely $\sup_{x\in\mathbb{R}} |P(W \leq x) - \Phi(x)|$. Such a bound can be obtained by induction or the use of a concentration inequality. For induction, see [9]. For the use of a concentration inequality, see Chen [14] and Chen and Shao [18] for sums of independent random variables, and Chen and Shao [19] for sums of locally dependent random variables. See also Chen, Goldstein and Shao [15]. For sums of independent random variables, Chen and Shao [18] obtained a bound of $4.1\sum \gamma_i$ on the Kolmogorov distance.

Case 2. Assume that $T_2 = 0$. This is the case if W is a functional of independent Gaussian random variables as considered by Chatterjee [11] or a functional of Gaussian random fields as considered by Nourdin and Peccati [31]. In this case, (2.6) becomes

$$\mathbb{E}\big[h(W)\big] - \mathbb{E}\big[h(Z)\big] = \mathbb{E}\big[(1-T_1)f_h'(W)\big] = \mathbb{E}\big[(1-\mathbb{E}[T_1|W])f_h'(W)\big].$$

Let h be such that $|h| \leq 1$. Then we obtain the following bound on the total variation distance between $\mathscr{L}(W)$ and $\mathcal{N}(0,1)$.

$$\begin{aligned} d_{\mathrm{TV}}(\mathscr{L}(W), \mathcal{N}(0,1)) &:= \frac{1}{2} \sup_{|h|\leq 1} \big|\mathbb{E}\big[h(W)\big] - \mathbb{E}\big[h(Z)\big]\big| \\ &\leq \frac{1}{2}\|f_h'\|_\infty \mathbb{E}\big[|1-\mathbb{E}[T_1|W]|\big] \\ &\leq 2\sqrt{\mathrm{Var}(\mathbb{E}[T_1|W])} \end{aligned}$$

While Chatterjee [11] used a Poincaré inequality of second order to bound $2\sqrt{\mathrm{Var}(\mathbb{E}[T_1|W])}$, Nourdin and Peccati [31] deployed the Malliavin calculus. In the next two sections, we will discuss how the Malliavin calculus is used to bound $2\sqrt{\mathrm{Var}(\mathbb{E}[T_1|W])}$.

3. Malliavin calculus

3.1. *A brief history*

The Malliavin calculus was born in 1976 in a symposium in Kyoto. Paul Malliavin presented a remarkable theory which extended the powerful calculus of variations to the framework of stochastic calculus. The initial goal of the theory was to provide a probabilistic proof of the Hörmander criterion (Hörmander [25]) of hypoellipticity by relating the smoothness of the solutions of a second order partial differential equation with the smoothness of the law of the solution of a stochastic differential equation. In order to prove that a random variable X has a smooth density, Paul Malliavin introduced the following very efficient criterion.

Lemma 3.1. *Assume that for each $k \geq 1$, there is $C_k > 0$ such that for any $\phi \in \mathcal{C}_C^\infty$, the class of $\mathcal{C}^\infty$ functions with compact support, we have*

$$\left|\mathbb{E}\left[\phi^{(k)}(X)\right]\right| \leq C_k \|\phi\|_\infty, \tag{3.1}$$

then the distribution of X has a $\mathcal{C}^\infty$ density.

In order to prove the inequality (3.1) for X, Malliavin showed, by integration by parts, that for some suitable weight H_k,

$$\mathbb{E}\big[\phi^{(k)}(X)\big] = \mathbb{E}\big[\phi(X)H_k\big]. \tag{3.2}$$

Hence, (3.1) holds with $C_k = \mathbb{E}[|H_k|]$. In many situations of interest, $X = F(G_1, \cdots, G_n)$ where the $\{G_i\}_{i\geq 1}$ are i.i.d. Gaussian and F is smooth. The equation (3.2) can be proved by integration by parts through the "Gaussian structure" of X together with some "non-degeneracy" assumption on F. Since its introduction, the Malliavin calculus has been extended and used in many different areas of probability theory. However, regardless of the application, the central role of the Malliavin calculus always consists in proving that some integration by parts formula holds. In the present exposition, we use integration by parts not for proving the smoothness of a density, but for establishing Stein's bounds.

3.2. *Malliavin derivatives*

Let $\mathscr{H}$ be a real separable Hilbert space, typically $\mathscr{H} = L^2(\mathbb{R}_+)$ but the particular choice of $\mathscr{H}$ does not matter. We denote by $X = \{X(h), h \in \mathscr{H}\}$ an *isonormal Gaussian process* over $\mathscr{H}$. That means, X is a centered Gaussian family of random variables defined in some probability space $(\Omega, \mathcal{F}, P)$, with covariance given by

$$\mathbb{E}[X(h)X(g)] = \langle h, g\rangle_{\mathscr{H}},$$

for any $h, g \in \mathscr{H}$. We also assume that $\mathcal{F}$ is generated by X.

Let $\mathcal{S}$ be the set of all cylindrical random variables of the form:

$$F = g\left(X(\phi_1), \ldots, X(\phi_n)\right), \tag{3.3}$$

where $n \geq 1$, $g : \mathbb{R}^n \to \mathbb{R}$ is an infinitely differentiable function such that its partial derivatives have polynomial growth, and $\phi_i \in \mathscr{H}$, $i = 1, \ldots, n$. The *Malliavin derivative* of F with respect to X is the element of $L^2(\Omega, \mathscr{H})$ defined as

$$DF = \sum_{i=1}^n \frac{\partial g}{\partial x_i}\left(X(\phi_1), \ldots, X(\phi_n)\right)\phi_i. \tag{3.4}$$

In particular, $DX(h) = h$ for every $h \in \mathscr{H}$. By iteration, one can define the m-th derivative $D^m F$, which is an element of $L^2(\Omega, \mathscr{H}^{\odot m})$ for every $m \geq 2$, where $\mathscr{H}^{\odot m}$ commonly stands for the m-th symmetric tensor product of $\mathscr{H}$. Indeed, we set

$$D^m F = \sum_{i_1,\cdots,i_m=1}^{n} \frac{\partial^m g}{\partial x_{i_1}\cdots\partial x_{i_m}}\Big[X(\phi_1),\cdots,X(\phi_n)\Big]\phi_{i_1}\otimes\cdots\otimes\phi_{i_m}. \tag{3.5}$$

We stress that we rather use a symmetric tensor product instead of the usual one because of the celebrated Schwarz rule $\partial_x\partial y = \partial_y\partial_x$, which forces $D^m F$ to be a *symmetric* element of $\mathscr{H}^{\odot m}$. For $m \geq 1$ and $p \geq 1$, $\mathbb{D}^{m,p}$ denotes the closure of $\mathcal{S}$ with respect to the norm $\|\cdot\|_{m,p}$, defined by the relation

$$\|F\|_{m,p}^p = \mathbb{E}\left[|F|^p\right] + \sum_{i=1}^{m}\mathbb{E}\left[\|D^i F\|_{\mathcal{H}^{\otimes i}}^p\right].$$

We often use the notation $\mathbb{D}^\infty := \bigcap_{m\geq 1}\bigcap_{p\geq 1}\mathbb{D}^{m,p}$. To justify properly the validity of the closure procedure of $\mathcal{S}$ with respect to the norm $\|\cdot\|_{m,p}$ once needs to prove that the Malliavin derivatives are *closable.* Indeed, the closability is required to ensure that the limit of DF_n does not depend on the choice of the approximating sequence F_n.

Another operator closely related to D is the Ornstein-Uhlenbeck operator. For $F = \phi(X(h_1),\cdots,X(h_n))$, we set

$$\mathbf{L}[F] = \Delta\phi(X(h_1),\cdots,X(h_n)) - \sum_{i=1}^{n} X(h_i)\frac{\partial g}{\partial x_i}(X(h_1),\cdots,X(h_n)). \tag{3.6}$$

3.3. *Wiener chaos and multiple integrals*

For every $k \geq 1$, we denote by $\mathcal{H}_k$ the k-th *Wiener chaos* of X defined as the closed linear subspace of $L^2(\Omega)$ generated by the family of random variables $\{H_k(X(h)), h \in \mathscr{H}, \|h\|_{\mathscr{H}} = 1\}$, where H_k is the k-th Hermite polynomial given by

$$H_k(x) = (-1)^k e^{\frac{x^2}{2}}\frac{d^k}{dx^k}\left(e^{-\frac{x^2}{2}}\right).$$

For any $k \geq 1$, we denote by $\mathscr{H}^{\otimes k}$ the k-th tensor product of $\mathscr{H}$. Set, for any $h \in \mathscr{H}$ such that $\|h\|_{\mathscr{H}} = 1$,

$$I_k(h^{\otimes k}) = H_k(X(h)). \tag{3.7}$$

When $\phi = h_1^{\otimes k_1}\otimes\cdots\otimes h_p^{\otimes k_p}$ with $(h_i)_{1\leq i\leq p}$ an orthonormal system and $k_1+\cdots+k_p = k$, we extend (3.7) by

$$I_k(\phi) = \prod_{j=1}^{p} H_{k_j}(X(h_j)). \tag{3.8}$$

Then I_k is a linear isometry between the symmetric tensor product $\mathscr{H}^{\odot k}$ (equipped with the modified norm $\sqrt{k!}\|\cdot\|_{\mathscr{H}^{\otimes k}}$) and the k-th Wiener chaos $\mathcal{H}_k$. In the particular case $\mathscr{H} = L^2(A,\mathcal{A},\mu)$, where μ is a σ-finite measure without atoms, then $\mathscr{H}^{\odot k}$ coincides with the space $L^2_s(\mu^k)$ of symmetric functions which are square integrable with respect to the product measure μ^k, and for any $f \in \mathscr{H}^{\odot k}$ the random variable $I_k(f)$ is commonly denoted as the multiple stochastic integral of f with respect to the centered Gaussian measure generated by X.

The following fact is fundamental in the theory of Gaussian spaces.

Theorem 3.1. *Any random variable $F \in L^2(\Omega)$ admits an orthogonal decomposition of the form*

$$F = \mathbb{E}[F] + \sum_{k=1}^{\infty} I_k(f_k),$$

where the kernels $f_k \in \mathscr{H}^{\odot k}$ are uniquely determined by F. In the sequel, we shall also denote $J_k(F) = I_k(f_k)$.

The random variables $I_k(f_k)$ inherit some properties from the algebraic structure of the Hermite polynomials, such that the product formula (3.9) below. To state it, let us introduce a definition. Let $\{e_i, i \geq 1\}$ be a complete orthonormal system in $\mathscr{H}$.

Definition 3.1. Given $f \in \mathscr{H}^{\odot k}$ and $g \in \mathscr{H}^{\odot j}$, for every $r = 0, \dots, k \wedge j$, the *contraction* of f and g of order r is the element of $\mathscr{H}^{\otimes(k+j-2r)}$ defined by

$$f \otimes_r g = \sum_{i_1,\dots,i_r=1}^{\infty} \langle f, e_{i_1} \otimes \cdots \otimes e_{i_r} \rangle_{\mathscr{H}^{\otimes r}} \otimes \langle g, e_{i_1} \otimes \cdots \otimes e_{i_r} \rangle_{\mathscr{H}^{\otimes r}}.$$

When $\mathscr{H} = L^2(\mathbb{R}_+)$, the latter formula simply becomes

$$\begin{aligned} &f \otimes_r g(\mathrm{x}, \mathrm{y}) \\ &= \int_{\mathbb{R}_+^r} f(x_1, x_2, \cdots, x_{k-r}, t_1, \cdots, t_r) g(y_1, y_2, \cdots, y_{k-r}, t_1, \cdots, t_r) dt_1 \cdots dt_r. \end{aligned}$$

The contraction $f \otimes_r g$ is not necessarily symmetric, and we denote by $f \widetilde{\otimes}_r g$ its symmetrization. We end this section by summarizing four important properties of *Wiener chaos* which will be used in the sequel. Throughout the sequel, $\mathbf{K}er$ denotes the kernel of a linear operator and $\mathbf{I}d$ stands for the identity operator.

(1) $\mathcal{H}_k = \mathbf{K}er(\mathbf{L} + k\mathbf{I}d)$,
(2) $\mathcal{H}_k \subset \mathbb{D}^\infty$,
(3) For every $(p, q) \in \mathbb{N}^2$, we have the following product formula

$$I_p(f) I_q(g) = \sum_{r=0}^{p \wedge q} r! \binom{p}{r} \binom{q}{r} I_{p+q-2r}(f \widetilde{\otimes}_r g). \tag{3.9}$$

(4) For any $f \in \mathscr{H}^{\odot k}$, we have the following isometry:

$$\mathbb{E}[I_k(f)^2] = k! \|f\|^2_{\mathscr{H}^{\otimes k}}.$$

3.4. *Main properties of Malliavin operators*

The Malliavin derivative D, defined in (3.4), obeys the following *chain rule*. If $\varphi : \mathbb{R}^n \to \mathbb{R}$ is continuously differentiable with bounded partial derivatives and if $F = (F_1, \dots, F_n)$ is a vector of elements of $\mathbb{D}^{1,2}$, then $\varphi(F) \in \mathbb{D}^{1,2}$ and

$$D\,\varphi(F) = \sum_{i=1}^{n} \frac{\partial \varphi}{\partial x_i}(F) DF_i. \tag{3.10}$$

The domain $\mathbb{D}^{1,2}$ can be precisely related to the Wiener chaos decomposition. Indeed,

$$\mathbb{D}^{1,2} = \Big\{ F \in L^2(\Omega) \,:\, \sum_{k=1}^{\infty} k \|J_k F\|^2_{L^2(\Omega)} < \infty \Big\}. \tag{3.11}$$

In the particular case where $\mathscr{H} = L^2(A, \mathcal{A}, \mu)$ (with μ non-atomic), then the derivative of a random variable F in $L^2(\Omega)$ whose chaotic expansion is

$$F = \mathbb{E}[F] + \sum_{k=1}^{\infty} I_k(f_k), \text{ with } f_k \in \mathscr{H}^{\odot^k},$$

can be identified with the element of $L^2(A \times \Omega)$ given by

$$D_x F = \sum_{k=1}^{\infty} k I_{k-1}\left(f_k(\cdot, x)\right), \quad x \in A. \tag{3.12}$$

The notation $I_{k-1}(f_k(\cdot, x))$ means that we freeze one coordinate and take the Wiener-Itô integral of order $k-1$ with respect to the $k-1$ remaining coordinates. One should notice that, since the f_k are taken symmetric, then the above notation do not depend on the choice of the frozen coordinate. As a matter of fact, $D_x F$ ($x \in A$) is an element of $L^2(A \times \Omega)$. We denote by δ the adjoint of the operator D, also called the *divergence operator*. We recall that δ exists since the operator D is closed. A random element $u \in L^2(\Omega, \mathcal{H})$ belongs to the domain of δ, noted $\operatorname{Dom}\delta$, if and only if it verifies $|\mathbb{E}\left[\langle DF, u\rangle_{\mathcal{H}}\right]| \le c_u \|F\|_{L^2(\Omega)}$ for any $F \in \mathbb{D}^{1,2}$, where c_u is a constant depending only on u. If $u \in \operatorname{Dom}\delta$, then the random variable $\delta(u)$ is defined by the duality relationship (customarily called *integration by parts formula*)

$$\mathbb{E}[F\delta(u)] = \mathbb{E}[\langle DF, u\rangle_{\mathcal{H}}], \tag{3.13}$$

which holds for every $F \in \mathbb{D}^{1,2}$. More generally, if $F \in \mathbb{D}^{1,2}$ and $u \in \operatorname{Dom}\delta$ are such that the three expectations $\mathbb{E}\big[F^2\|u\|^2_{\mathscr{H}}\big]$, $\mathbb{E}\big[F^2\delta(u)^2\big]$ and $\mathbb{E}\big[\langle DF, u\rangle^2_{\mathscr{H}}\big]$ are finite, then $Fu \in \operatorname{Dom}\delta$ and

$$\delta(Fu) = F\delta(u) - \langle DF, u\rangle_{\mathscr{H}}. \tag{3.14}$$

The *Ornstein-Uhlenbeck* operator $\mathbf{L}$, defined in (3.6) satisfies the following relation

$$\mathbf{L} = \sum_{k=0}^{\infty} -kJ_k, \tag{3.15}$$

expressing that $\mathbf{L}$ is diagonalizable with spectrum $-\mathbb{N}$ with the Wiener chaos being its eigenspaces. Besides, the domain of $\mathbf{L}$ is

$$\mathrm{Dom}(\mathbf{L}) = \{F \in L^2(\Omega) : \sum_{k=1}^{\infty} k^2 \left\| J_k F \right\|_{L^2(\Omega)}^2 < \infty\} = \mathbb{D}^{2,2}. \tag{3.16}$$

There is an important relation between the operators D, δ and $\mathbf{L}$. A random variable F belongs to $\mathbb{D}^{2,2}$ if and only if $F \in \mathrm{Dom}\,(\delta D)$ (i.e. $F \in \mathbb{D}^{1,2}$ and $DF \in \mathrm{Dom}\,\delta$) and, in this case,

$$\delta DF = -\mathbf{L}F. \tag{3.17}$$

In particular, if $F \in \mathbb{D}^{2,2}$ and $H, G \in \mathbb{D}^{1,2}$ are such that $HG \in \mathbb{D}^{1,2}$, then

$$\begin{aligned} -\mathbb{E}[HG\,\mathbf{L}F] &= \mathbb{E}[HG\,\delta DF] \\ &= \mathbb{E}[\langle D(HG), DF\rangle_{\mathscr{H}}] \\ &= \mathbb{E}[H\langle DG, DF\rangle_{\mathscr{H}}] + \mathbb{E}[G\langle DH, DF\rangle_{\mathscr{H}}]. \end{aligned}$$

We end this section by introducing the operator $\mathbf{L}^{-1}$which is central in the next section in concern with applications of the Malliavin calculus to Stein's method. For any $F \in L^2(\Omega)$, we know that F can be expanded over the Wiener chaos, namely one has $F = \mathbb{E}[F] + \sum_{k=1}^{\infty} J_k F$.

Thus we set

$$\mathbf{L}^{-1}F = \mathbf{L}^{-1}\Big[F - \mathbb{E}[F]\Big] = \sum_{k=1}^{\infty} \frac{-1}{k} J_k F.$$

One can show by using (3.16) that $\mathbf{L}^{-1}F \in \mathrm{Dom}(\mathbf{L})$. Thus, relying on (3.15) we have $\mathbf{L}\mathbf{L}^{-1}F = F - \mathbb{E}[F]$. The operator $\mathbf{L}^{-1}$ is called the *pseudo-inverse* of $\mathbf{L}$. Notice that, since $\mathrm{Dom}(\mathbf{L}) \subset \mathbb{D}^{1,2}$ $\Big[$compare (3.11) and (3.16)$\Big]$, for any $F \in \mathbb{D}^{1,2}$ the quantity $< DF, -D\mathbf{L}^{-1}F >_{\mathscr{H}}$ is well defined.

4. Connecting Stein's method with Malliavin calculus

As is discussed in Section 2, the Stein operator L for normal approximation is given by $Lf(w) = f'(w) - wf(w)$ and the equation

$$\mathbb{E}\big[f'(Z) - Zf(Z)\big] = 0 \tag{4.1}$$

holds for absolutely continuous functions f for which the expectations exist if and only if $Z \sim \mathcal{N}(0,1)$. Interestingly, this equation is nothing but a simple consequence of integration by parts. Since there is the integration by parts formula of the Malliavin calculus for functionals of general Gaussian processes, there is a natural connection between Stein's method and the Malliavin calculus. Indeed, integration by parts has been used in less general situations to construct the equation

$$\mathbb{E}[Wf(W)] = E[Tf'(W)] \tag{4.2}$$

which is Case 2 of (2.4) discussed in Subsection 2.3. Let us provide two examples below.

Example 4.1. Assume $\mathbb{E}[W] = 0$ and $\mathrm{Var}(W) = 1$. Then we have $\mathbb{E}[T] = 1$. If W has a density $\rho > 0$ with respect to the Lebesgue measure, then by integration by parts, W satisfies (4.2) with $T = h(W)$, where

$$h(x) = \frac{\int_x^\infty y\rho(y)dy}{\rho(x)}.$$

If ρ is the density of $\mathcal{N}(0,1)$, then $h(w) = 1$ and (4.2) reduces to (4.1).

Example 4.2. Let $X = (X_1, \ldots, X_d)$ be a vector of independent Gaussian random variables and let $g : \mathbb{R}^d \to \mathbb{R}$ be an absolutely continuous function. Let $W = g(X)$. Chatterjee in [11] used Gaussian interpolation and integration by parts to show that W satisfies (4.2) with $T = h(X)$ where

$$h(x) = \int_0^1 \frac{1}{2\sqrt{t}} \mathbb{E}\Big[\sum_{i=1}^d \frac{\partial g}{\partial x_i}(x) \frac{\partial g}{\partial x_i}(\sqrt{t}x + \sqrt{1-t}X)\Big] dt.$$

If $d = 1$ and g the identity function, then $W \sim \mathcal{N}(0,1)$, $h(x) = 1$, and again (4.2) reduces to (4.1).

As the previous example shows (see Chatterjee [11] for details), it is possible to construct the function h when one deals with sufficiently smooth functionals of a Gaussian vector. This is part of a general phenomenon discovered by Nourdin and Peccati in [31]. Indeed, consider a functional F of an isonormal Gaussian process $X = \{X(h), h \in \mathscr{H}\}$ over a real Hilbert space $\mathscr{H}$. Assume $F \in \mathbb{D}^{1,2}$, $\mathbb{E}[F] = 0$ and $\mathrm{Var}(F) = 1$. Let $f : \mathbb{R} \to \mathbb{R}$ be a bounded $\mathcal{C}^1$ function having a bounded derivative. Since $\mathbf{L}^{-1}F \in \mathrm{Dom}(\mathbf{L})$, $\mathbf{L}^{-1}F \in \mathbb{D}^{2,2}$ and $D\mathbf{L}^{-1}F \in \mathrm{Dom}\,\delta$. By (3.17) and $\mathbb{E}[F] = 0$, we have

$$F = \mathbf{L}\mathbf{L}^{-1}F = \delta(-D\mathbf{L}^{-1}F).$$

Therefore

$$\mathbb{E}[Ff(F)] = \mathbb{E}[\mathbf{L}\mathbf{L}^{-1}F \times f(F)] = \mathbb{E}[\delta(-D\mathbf{L}^{-1}F)f(F)].$$

By the integration by parts formula (3.13),

$$\mathbb{E}[\delta(-D\mathbf{L}^{-1}F)f(F)] = \mathbb{E}[\langle Df(F), -D\mathbf{L}^{-1}F\rangle_{\mathscr{H}}]$$

and by the chain rule,

$$\mathbb{E}[\langle Df(F), -D\mathbf{L}^{-1}F\rangle_{\mathscr{H}}] = \mathbb{E}[f'(F)\langle DF, -D\mathbf{L}^{-1}F\rangle_{\mathscr{H}}].$$

Hence

$$\mathbb{E}[Ff(F)] = \mathbb{E}[f'(F)\langle DF, -D\mathbf{L}^{-1}F\rangle_{\mathscr{H}}]$$

and F satisfies (4.2) with $T = \langle DF, -D\mathbf{L}^{-1}F\rangle_{\mathscr{H}}$.

Now let f_h be the unique bounded solution of the Stein equation (2.5) where $h : \mathbb{R} \to \mathbb{R}$ is continuous and $|h| \leq 1$. Then $f_h \in \mathcal{C}^1$ and $\|f_h'\|_\infty \leq 4\|h\|_\infty \leq 4$, and we have

$$\begin{aligned}\mathbb{E}[h(F)] - \mathbb{E}[h(Z)] &= \mathbb{E}\{f_h'(F)[1 - \langle DF, -D\mathbf{L}^{-1}F\rangle_{\mathscr{H}}]\}\\ &= \mathbb{E}\{f_h'(F)[1 - \mathbb{E}(\langle DF, -D\mathbf{L}^{-1}F\rangle_{\mathscr{H}}|F)]\}.\end{aligned}$$

Therefore

$$\begin{aligned}\sup_{h\in\mathcal{C},|h|\leq 1} |\mathbb{E}[h(F)] - \mathbb{E}[h(Z)]| &\leq \|f_h'\|_\infty \mathbb{E}\Big[|1 - \mathbb{E}(\langle DF, -D\mathbf{L}^{-1}F\rangle_{\mathscr{H}}|F)|\Big]\\ &\leq 4\mathbb{E}\Big[|1 - \mathbb{E}(\langle DF, -D\mathbf{L}^{-1}F\rangle_{\mathscr{H}}|F)|\Big].\end{aligned}$$

If F has a density with respect to the Lebesgue measure, then

$$\begin{aligned}d_{\mathrm{TV}}(\mathscr{L}(F), \mathcal{N}(0,1)) &:= \frac{1}{2}\sup_{|h|\leq 1} |\mathbb{E}[h(F)] - \mathbb{E}[h(Z)]|\\ &= \frac{1}{2}\sup_{h\in\mathcal{C},|h|\leq 1} |\mathbb{E}[h(F)] - \mathbb{E}[h(Z)]|\\ &\leq 2\mathbb{E}\Big[|1 - \mathbb{E}(\langle DF, -D\mathbf{L}^{-1}F\rangle_{\mathscr{H}}|F)|\Big].\end{aligned}$$

If, in addition, $F \in \mathbb{D}^{1,4}$, then $\langle DF, -D\mathbf{L}^{-1}F\rangle_{\mathscr{H}}$ is square-integrable and

$$\mathbb{E}\Big[|1 - \mathbb{E}(\langle DF, -D\mathbf{L}^{-1}F\rangle_{\mathscr{H}}|F)|\Big] \leq \sqrt{\mathrm{Var}[\mathbb{E}(\langle DF, -D\mathbf{L}^{-1}F\rangle_{\mathscr{H}}|F)]}.$$

Thus we have the following theorem of Nourdin and Peccati (2011).

Theorem 4.1. *Let $F \in \mathbb{D}^{1,2}$ such that $\mathbb{E}[F] = 0$ and $\mathrm{Var}(F) = 1$. If F has a density with respect to the Lebesgue measure, then*

$$d_{\mathrm{TV}}(\mathscr{L}(F), \mathcal{N}(0,1)) \leq 2\mathbb{E}\Big[|1 - \mathbb{E}(\langle DF, -D\boldsymbol{L}^{-1}F\rangle_{\mathscr{H}}|F)|\Big]. \tag{4.3}$$

If, in addition, $F \in \mathbb{D}^{1,4}$, then

$$d_{\mathrm{TV}}(\mathscr{L}(F), \mathcal{N}(0,1)) \leq 2\sqrt{\mathrm{Var}[\mathbb{E}(\langle DF, -D\boldsymbol{L}^{-1}F\rangle_{\mathscr{H}}|F)]}. \tag{4.4}$$

The bound (4.4) is optimal for normal approximation for functionals of Gaussian processes. Many examples can be found in the literature, and the reader is referred to this website

`https://sites.google.com/site/malliavinstein/home`

for a complete overview. One can also consult the good survey Peccati [37] with an emphasis on recent developments. For the sake of completeness, we shall illustrate the optimality of the bound (4.4) using the example of the quadratic variation of a fractional Brownian motion. To do this, let $H \in (0, 1)$ be the Hurst parameter of a fractional Brownian motion $\{B_t^H\}_{t>0}$. It is well known from the Breuer-Major Theorem that if $0 < H < \frac{3}{4}$, then for some suitable $\sigma_H > 0$,

$$F_{n,H} := \frac{1}{\sigma_H \sqrt{n}} \sum_{k=1}^{n} \left((B_{k+1}^H - B_k^H)^2 - 1 \right) \xrightarrow[n\to\infty]{\mathscr{L}} \mathcal{N}(0,1).$$

Similarly, if $H = \frac{3}{4}$, then one can prove that for some $\sigma_{\frac{3}{4}} > 0$, we have

$$F_{n,\frac{3}{4}} := \frac{1}{\sigma_{\frac{3}{4}} \sqrt{n \log n}} \sum_{k=1}^{n} \left((B_{k+1}^{\frac{3}{4}} - B_k^{\frac{3}{4}})^2 - 1 \right) \xrightarrow[n\to\infty]{\mathscr{L}} \mathcal{N}(0,1).$$

Using the fact that $(B_{k+1}^H - B_k^H)^2 - 1$ is an element of the second Wiener chaos, and the equation (3.4), one can obtain explicit bounds on $\mathrm{Var}[\mathbb{E}(\langle DF_{n,H}, -D\mathbf{L}^{-1}F_{n,H}\rangle_{\mathscr{H}}]$ (see Nourdin and Peccati [33]). Since

$$\mathrm{Var}[\mathbb{E}(\langle DF, -D\mathbf{L}^{-1}F\rangle_{\mathscr{H}}|F)] \le \mathrm{Var}[\mathbb{E}(\langle DF, -D\mathbf{L}^{-1}F\rangle_{\mathscr{H}}],$$

it follows from (4.4) that

$$d_{TV}(F_{n,H}, G) \le c_H \begin{cases} \frac{1}{\sqrt{n}} & \text{if } H \in (0, \frac{5}{8}) \\ \frac{(\log n)^{\frac{3}{2}}}{\sqrt{n}} & \text{if } H = \frac{5}{8} \\ n^{4H-3} & \text{if } H \in (\frac{5}{8}, \frac{3}{4}) \\ \frac{1}{\log n} & \text{if } H = \frac{3}{4}. \end{cases}$$

These bounds are shown to be optimal in Nourdin and Peccati [33]. We wish to mention that for $H > \frac{3}{4}$, $F_{n,H}$ does not converge to a Gaussian distribution. Instead, it converges to the so-called Hermite distribution, which belongs to the second Wiener chaos and is therefore not Gaussian.

5. The Nualart-Peccati criterion of the fourth moment and Ledoux's idea

5.1. *Some history*

During the year 2005, in the seminal article [36], David Nualart and Giovanni Peccati discovered the following remarkable fact. Take $F_n = I_k(f_n)$ a sequence of elements of the k-th Wiener chaos. Then, F_n converges in law towards the Gaussian measure $\mathcal{N}(0,1)$ *if and only if* $\mathbb{E}[F_n^2] \to 1$ and $\mathbb{E}[F_n^4] \to 3$. This result can be seen as a drastic simplification of the so-called *method of moments* which consists in proving that $\mathbb{E}[F^p] \to \int_{\mathbb{R}} x^p e^{-\frac{x^2}{2}} \frac{dx}{\sqrt{2\pi}}$, *for each positive integer* p. A bit later, in Peccati and Tudor [38], the strength of this theorem was considerably reinforced by its multivariate counterpart. Indeed, Giovanni Peccati and Ciprian Tudor proved that a random vector $F_n = (F_{1,n}, \cdots, F_{d,n})$ with chaotic components converges in distribution towards a Gaussian vector with covariance C *if and only if* the covariance of F_n converges to C and for each $i \in \{1, \cdots, d\}$, $F_{i,n}$ converges in distribution to $\mathcal{N}\big[0, C(i,i)\big]$. That is to say, for chaotic random variables, the *componentwise* converges implies the *joint* convergence. This observation yielded to a very efficient strategy for proving central convergence in the Wiener space, by somehow *decomposing* the convergence on each Wiener chaos. This approach, which is more and more used by practitioners as an alternative to the semi-martingale approach, is at the heart of a very active line of research. See this website for details and an exhaustive overview.

`https://sites.google.com/site/malliavinstein/home`

5.2. *Overview of the proof of Nourdin and Peccati*

Whereas the original proof of the fourth moment Theorem relied on some tools of stochastic analysis, Ivan Nourdin and Giovanni Peccati produced a new proof in [31] based on a suitable combination of Malliavin calculus and Stein's method. As noticed by the Nourdin and Peccati in [31] and in conclusion of previous sections 3 and 4, one is left to show that

$$\mathrm{Var}(\mathbb{E}[T_1|W]) \le \mathrm{Var}\Big[\langle DF_n, -D\mathbf{L}^{-1}F_n\rangle_{\mathscr{H}}\Big] \to 0,$$

under the assumptions that $\mathbb{E}[F_n^4] \to 3$ and $\mathbb{E}[F_n^2] \to 1$, if $\{F_n\}$ is a sequence of elements in the k-th Wiener chaos. In fact, following their strategy we will prove that

$$\mathrm{Var}\Big[\langle DF_n, -D\mathbf{L}^{-1}F_n\rangle_{\mathscr{H}}\Big] \le \frac{k-1}{3k}\Big[\mathbb{E}[F_n^4] - 3\mathbb{E}[F_n^2]\Big]. \tag{5.1}$$

<u>Step 1: computing $\mathrm{Var}\big[\langle DF_n, -D\mathbf{L}^{-1}F_n\rangle_{\mathscr{H}}\big]$.</u>

By equation (3.12), one has $DF_n(t) = DI_k(f_n)(t) = kI_{k-1}(f_n(\cdot,t))$. By applying the product formula for multiple integrals (3.9), we get:

$$\begin{aligned}
\frac{1}{k}\langle DF_n, DF_n\rangle_{\mathscr{H}} &= k\int_0^\infty I_{k-1}(f_n(\cdot,t))^2 dt \\
&= k\int_0^\infty \sum_{r=0}^{k-1} r!\binom{k-1}{r}^2 I_{2k-2-2r}\big[f_n(\cdot,t)\widetilde{\otimes}_r f_n(\cdot,t)\big]dt \\
&= k\sum_{r=0}^{k-1} r!\binom{k-1}{r}^2 I_{2k-2-2r}\Big[\int_0^\infty f_n(\cdot,t)\widetilde{\otimes}_r f_n(\cdot,t)dt\Big] \\
&= k\sum_{r=1}^{k}(r-1)!\binom{k-1}{r-1}^2 I_{2k-2r}\Big[f_n\widetilde{\otimes}_r f_n\Big]. \\
&= k\sum_{r=1}^{k-1}(r-1)!\binom{k-1}{r-1}^2 I_{2k-2r}\Big[f_n\widetilde{\otimes}_r f_n\Big] + k!\|f\|_{\mathscr{H}}^2.
\end{aligned}$$

Taking into account that $\mathbb{E}[\langle DF_n, DF_n\rangle_{\mathscr{H}}] = k\mathbb{E}[F_n^2] = kk!\|f\|^2_{\mathscr{H}}$, the orthogonality of the Wiener chaos entails that:

$$\mathrm{Var}\left[\langle DF_n, -D\mathbf{L}^{-1}F_n\rangle_{\mathscr{H}}\right] = \sum_{r=1}^{k-1} \frac{r^2}{k^2}(r!)^2 \binom{k}{r}^4 (2k-2r)!\|f_n\widetilde{\otimes}_r f_n\|^2_{\mathscr{H}^{\odot(2k-2r)}}. \quad (5.2)$$

Step 2: computing $\mathbb{E}[F_n^4]$.

By product formula again, we have

$$F_n^2 = \sum_{r=0}^{k} r!\binom{k}{r}^2 I_{2k-2r}(f_n\widetilde{\otimes}_r f_n).$$

This yields to

$$\mathbb{E}[F_n^4] = \sum_{r=0}^{k} (r!)^2 \binom{k}{r}^4 (2k-2r)!\|f_n\widetilde{\otimes}_r f_n\|^2_{\mathscr{H}^{\odot(2k-2r)}}.$$

Unfortunately the latter expression is not immediately comparable with (5.2) because the "zero contractions" $\|f_n\widetilde{\otimes}_0 f_n\|^2_{\mathscr{H}^{\odot(2k-2r)}}$ do not appear in the decomposition of $\langle DF_n, DF_n\rangle_{\mathscr{H}}$. To face this problem, one has to take into account an intermediary expression, namely $\mathbb{E}[F_n^2\langle DF_n, DF_n\rangle_{\mathscr{H}}] = \frac{k}{3}\mathbb{E}[F_n^4]$. After few combinatorial arguments respective to the symmetrizations of the contractions of the kernels f_n appearing in the scalar product, and by a suitable comparison of the terms $\mathbb{E}[F_n^4]$, $\mathrm{Var}(\langle DF_n, DF_n\rangle_{\mathscr{H}})$ and $\mathbb{E}[F_n^2\langle DF_n, DF_n\rangle_{\mathscr{H}}]$ one may show that:

$$\mathbb{E}[F_n^4] - 3\mathbb{E}[F_n^2]^2 = \frac{3}{k}\sum_{r=1}^{k-1} r(r!)^2 \binom{k}{r}^4 (2k-2r)!\|f_n\widetilde{\otimes}_{2k-2r} f_n\|^2_{\mathscr{H}^{\odot(2k-2r)}}. \quad (5.3)$$

Since this part is rather technical and irrelevant for our purpose, we refer the reader to the book Nourdin Peccati [32] for more precise arguments. By comparing the equations (5.2) and (5.3), one may recover (5.1).

5.3. *About Ledoux's generalization*

As we just showed, the usual way of proving the inequality (5.1) relies heavily on the product formula for multiple Wiener integrals. In particular, this gives the impression that the various combinatorics coefficients appearing in the formulae (5.1), (5.2) and (5.3) are playing a major role in the phenomenon. In 2012, in the very insightful article [26], Michel Ledoux tried to tackle this problem by using a more "algebraic" approach based on spectral theory and Γ-calculus. In particular, he was able to prove the inequality (5.1) without using the product formula, under some suitable spectral conditions. Unfortunately, the provided conditions seemed rather difficult to check in practice, and no new structure (in addition to the Wiener space one) with a fourth moment phenomenon was given.

6. The general fourth moment Theorem for Dirichlet forms

In this section, we keep the remarkable intuition of Ledoux in [26] of exploiting the algebraic and spectral properties of the chaotic random variables rather than product formulae techniques. However, at the very beginning, we will take a different path. As we will show, the fourth moment Theorem turns out to be a direct consequence of a very simple spectral assumption. This simplification will allow us to produce various examples of new structures where the phenomenon holds. We will make a crucial use of the very powerful formalism of Dirichlet forms to achieve our goals.

6.1. *The Dirichlet structures*

Originated from potential theory and physics, the Dirichlet form theory has become a central object in analysis. Under the impulsion of Beurling, Deny, Fukushima, Meyer or Mokobodzki, just to name a few, the Dirichlet forms theory has unveiled beautiful ramifications in many area of mathematics such as geometric measure theory, partial differential equations, Markov processes, Malliavin calculus... Here, we will adopt a modern formalism, namely the notion of Dirichlet structure.

We will say that a Dirichlet structure is a set $(E, \mathcal{F}, \mu, \mathbb{D}, \Gamma)$ such that:

(1) $(E, \mathcal{F}, \mu)$ is a probability space,
(2) $\mathbb{D}$ is a vector space *dense* in $L^2(\mu)$,
(3) $\Gamma : \mathbb{D} \times \mathbb{D} \to L^1(\mu)$ is a *bilinear, symmetric* and *non-negative* operator,
(4) $\mathbb{D}$ endowed with the norm $\|X\|_{\mathbb{D}} = \sqrt{\mathbb{E}\big[X^2\big] + \mathbb{E}\big[\Gamma[X,X]\big]}$ is *complete.*
(5) For any vectors $X = (X_1, \cdots, X_p) \in \mathbb{D}^p$ and $Y = (Y_1, \cdots, Y_q) \in \mathbb{D}^q$, for any functions $F \in \mathcal{C}^1(\mathbb{R}^p, \mathbb{R}) \cap \mathrm{Lip}$ and $G \in \mathcal{C}^1(\mathbb{R}^q, \mathbb{R}) \cap \mathrm{Lip}$ we have $F(X) \in \mathbb{D}$ and $G(Y) \in \mathbb{D}$. Besides, one has the following *functional calculus*

$$\Gamma[F(X), G(Y)] = \sum_{i=1}^{p} \sum_{j=1}^{q} \partial_i F(X) \partial_j G(Y) \Gamma[X_i, Y_j]. \tag{6.1}$$

The mapping $(X,Y) \to \mathcal{E}\big[X,Y\big] = \mathbb{E}\big[\Gamma[X,Y]\big]$ is customarily called the *Dirichlet form* associated with the *carré-du-champ* operator Γ . For any of theses structures, one may associate another operator $\mathbf{L}$ defined on some domain dom($\mathbf{L}$) dense in $\mathbb{D}$ such that the next integration by parts holds for any $(X,Y) \in \mathrm{dom}(\mathbf{L}) \times \mathrm{dom}(\mathbf{L})$.

$$\mathbb{E}\big[\Gamma[X,Y]\big] = -\mathbb{E}\big[X\mathbf{L}[Y]\big] = -\mathbb{E}\big[Y\mathbf{L}[X]\big]. \tag{6.2}$$

Besides, one may derive from (1-5) the next relation between Γ and $\mathbf{L}$.

$$2\Gamma[X,Y] = \mathbf{L}[XY] - Y\mathbf{L}[X] - X\mathbf{L}[Y] \tag{6.3}$$

In addition to assumptions (1-5) which characterize a general Dirichlet structure, we assume further that

(H1) $-\mathbf{L}$ is *diagonalizable* with *spectrum* $\{0 = \lambda_0 < \lambda_1 < \lambda_2 < \cdots < \lambda_p < \cdots\}$,
(H2) For each $X \in \mathbf{K}er(\mathbf{L} + \lambda_k \mathbf{I}d)$,

$$X^2 \in \bigoplus_{\alpha \leq 2\lambda_k} \mathbf{K}er(\mathbf{L} + \alpha \mathbf{I}d).$$

Before stating and proving our result, we stress that the previous framework covers the case of the Wiener structure. Indeed, for the Wiener structure, $\Gamma[F,G] = \langle DF, DG \rangle_{\mathscr{H}}$ defined on $\mathbb{D}^{1,2}$ and the corresponding $\mathbf{L}$ is the Ornstein-Uhlenbeck operator. Properties (1-3) are straightforward, (4) proceeds from the closability of $\mathbb{D}^{1,2}$ see subsection (3.2), and (5) is a consequence of the chain rule (3.10). (H1) comes from (3.15) and (H2) is a trivial consequence of the product formula (3.9).

6.2. *Fourth moment theorem for Dirichlet structures with (H1) and (H2)*

Let X be an eigenfunction of $-\mathbf{L}$ associated with eigenvalue λ_k with $\mathbb{E}[X^2] = 1$. We will show that

$$\mathrm{Var}(\Gamma[X,X]) \leq \frac{\lambda_k^2}{3}(\mathbb{E}[X^4] - 3) \tag{6.4}$$

Suppose that the inequality (6.4) is true and take ϕ any test function. On the one hand we have by the chain rule (5) that $\Gamma[\phi(X), X] = \phi'(X)\Gamma[X,X]$. On the other hand, by integrations by parts (6.2) we also have $\mathbb{E}[\phi'(X)\Gamma[X,X]] = \lambda_k \mathbb{E}[X\phi(X)]$. Thus we are in the setting of the case 2 of the section 2.3 and one has the inequality

$$d_{\mathrm{TV}}(\mathscr{L}(X), \mathcal{N}(0,1)) \leq \mathbb{E}\left|\frac{\Gamma[X,X]}{\lambda_k} - 1\right| \leq \frac{1}{\lambda_k}\sqrt{\mathrm{Var}(\Gamma[X,X])}. \tag{6.5}$$

We used above the fact that $\mathbb{E}\big[\Gamma[X,X]\big] = -\mathbb{E}\big[X\mathbf{L}[X]\big] = \lambda_k$. Now, using (6.3) and taking into account that X is actually an *eigenfuntion* of $\mathbf{L}$, we derive $\Gamma[X,X] = \frac{1}{2}\big[\mathbf{L} + 2\lambda_k \mathbf{I}d\big][X^2]$. Hence, we get $\big[\Gamma[X,X] - \lambda_k\big] = \frac{1}{2}\big[\mathbf{L} + 2\lambda_k \mathbf{I}d\big][X^2 - 1]$. The rest of the proof is as follows. We use (H2) to say that X^2 and hence $X^2 - 1$ are expanded in finitely many eigenspaces of $-\mathbf{L}$ with maximum eigenvalue being $2\lambda_k$.

$$\begin{aligned}
\mathrm{Var}(\Gamma[X,X]) &= \mathbb{E}\Big[\frac{1}{2}\big[\mathbf{L} + 2\lambda_k \mathbf{I}d\big][X^2 - 1] \times \frac{1}{2}\big[\mathbf{L} + 2\lambda_k \mathbf{I}d\big][X^2 - 1]\Big] \\
&= \frac{1}{2}\mathbb{E}\Big[\mathbf{L}[X^2 - 1] \times \frac{1}{2}\big[\mathbf{L} + 2\lambda_k \mathbf{I}d\big][X^2 - 1]\Big] \\
&+ \lambda_k \mathbb{E}\Big[(X^2 - 1) \times \frac{1}{2}\big[\mathbf{L} + 2\lambda_k \mathbf{I}d\big][X^2 - 1]\Big] \\
&\underset{(H2)}{=} \frac{1}{4}\sum_{\lambda_i \leq 2\lambda_k} (-\lambda_i)(2\lambda_k - \lambda_i)\mathbb{E}\big[J_k[X^2 - 1]^2\big] \\
&+ \lambda_k \mathbb{E}\Big[(X^2 - 1)\big[\Gamma[X,X] - \lambda_k\big]\Big] \\
&\underset{(H2)}{\leq} \lambda_k \mathbb{E}\Big[(X^2 - 1)\Gamma[X,X]\Big] - \lambda_k^2 \mathbb{E}[X^2 - 1]
\end{aligned}$$

$$\underset{(5)}{=} \lambda_k \mathbb{E}\Big[\Gamma[\frac{X^3}{3} - X, X]\Big]$$

$$\underset{(6.2)}{=} \lambda_k^2 \Big[\frac{\mathbb{E}[X^4]}{3} - 1\Big].$$

So, the proof of (6.4) is done and one has shown (by dividing by λ_k) that

$$\mathrm{Var}(\Gamma[X, \frac{X}{\lambda_k}]) \leq \Big[\frac{\mathbb{E}[X^4]}{3} - 1\Big]. \tag{6.6}$$

Relying on the equations (6.1) and (6.2) it is easy to see that for any function f bounded with bounded derivatives

$$\mathbb{E}\left[f'(X)\Gamma[X, \frac{X}{\lambda_k}]\right] = \mathbb{E}\left[f(X)X\right].$$

Since $\Gamma[X, \frac{X}{\lambda_k}]$ plays the same role as the term "$\langle DF, -DL^{-1}F\rangle_{\mathscr{H}}$" of Section 4, Theorem 4.1 applies and can be combined with (6.6). Now take X_n as a sequence of eigenfunctions of $\mathbf{L}$ with same eigenvalue, such that $\mathbb{E}[X_n^2] = 1$. We deduce that

$$d_{\mathrm{TV}}(\mathscr{L}(X_n), \mathcal{N}(0,1)) \leq \sqrt{\frac{\mathbb{E}[X_n^4]}{3} - 1}.$$

In particular, if $\mathbb{E}[X_n^4] \to 3$, then we have convergence to $\mathcal{N}(0,1)$ in total variation.

6.3. *Dirichlet structures with (H1) and (H2)*

Here we give two examples of Dirichlet structures which satisfy (H1) and (H2). The first and most important example is provided by the Wiener space. First from equation 3.15, one can see that the so-called Ornstein-Uhlenbeck operator is diagonalizable with spectrum $\mathbb{N}$ and the Wiener chaos are its eigenspaces. This fact shows that the assumption (H1) is fulfilled. One the other hand, take $k \in \mathbb{N}$ and $X = I_k(f)$ some element of the k-th Wiener chaos. By using the product formula (3.9),

$$X^2 = \sum_{r=0}^{k} k! \binom{k}{r}^2 I_{2k-2r}(f \tilde{\otimes}_r f).$$

From which one can deduce that

$$X^2 \in \bigoplus_{r \leq 2k} \mathbf{K}er(\mathbf{L} + r\mathbf{I}d).$$

Hence (H1) and (H2) hold true for the Wiener structure. One should notice, that the precise combinatorial coefficients arising in the product formula play no role in the assumption (H2). In some sense, the approach by Dirichlet forms being less combinatoric is more direct than the usual proof in Nualart Peccati [36].

Not only the Dirichlet forms approach simplifies the proof of the Nualart-Peccati criterion, but also enables one to give new examples of Dirichlet structures where a fourth moment phenomenon holds. In order to avoid technicalities, we restrict the exposition to the finite dimensional case. So, let us introduce the Laguerre Dirichlet structure. Let $\nu \geq -1$, and $\pi_{1,\nu}(dx) = x^{\nu-1}\frac{e^{-x}}{\Gamma(\nu)}\mathbf{1}_{(0,\infty)}(x)dx$ be the Gamma distribution with parameter ν on $\mathbb{R}_+$. The associated Laguerre generator is defined for any test function ϕ (in dimension one) by:

$$\mathbf{L}_{1,\nu}\phi(x) = x\phi''(x) + (\nu + 1 - x)\phi'(x). \tag{6.7}$$

By a classical tensorization procedure, we obtain the Laguerre generator in dimension d associated to the measure $\pi_{d,\nu}(dx) = \pi_{1,\nu}(dx_1)\pi_{1,\nu}(dx_2)\cdots\pi_{1,\nu}(dx_d)$, where $x = (x_1, x_2, \cdots, x_d)$.

$$\mathbf{L}_{d,\nu}\phi(x) = \sum_{i=1}^{d}\Big(x_i\partial_{i,i}\phi + (\nu + 1 - x_i)\partial_i\phi\Big) \tag{6.8}$$

It is well known that (see for example Ledoux [26]) that the spectrum of $\mathbf{L}_{d,\nu}$ is given by $-\mathbb{N}$ and moreover that

$$\mathbf{Ker}(\mathbf{L}_{d,\nu} + p\mathbf{I}d) = \left\{ \sum_{i_1+i_2+\cdots+i_d=p} \alpha(i_1, \cdots, i_d) \prod_{j=1}^{d} L_{i_j}^{(\nu)}(x_j) \right\}, \tag{6.9}$$

where $L_n^{(\nu)}$ stands for the Laguerre polynomial of order n with parameter ν which is defined by

$$L_n^{(\nu)}(x) = \frac{x^{-\nu}e^x}{n!}\frac{d^n}{dx^n}\left(e^{-x}x^{n+\nu}\right).$$

Again, we have the following decomposition:

$$L^2(\mathbb{R}^d, \pi_{d,\nu}) = \bigoplus_{p=0}^{\infty}\mathbf{Ker}(\mathbf{L}_{d,\nu} + p\mathbf{I}d) \tag{6.10}$$

As a matter of fact, assumption (H1) follows. Let us check assumption (H2). Assume now that X is an eigenfunction of $\mathbf{L}_{d,\nu}$ with eigenvalue $-\lambda_p = -p$. In particular, X is a multivariate polynomial of degree p. Therefore, X^2 is a multivariate polynomial of degree $2p$. Note that by expanding X^2 over the basis of multivariate Laguerre polynomials $\prod_{j=1}^{d} L_{i_j}^{(\nu)}(x_j), i_j \geq 0$, we get that X^2 has a finite expansion over the first eigenspaces of the generator $\mathbf{L}_{d,\nu}$, i.e.

$$X^2 \in \bigoplus_{p=0}^{M}\mathbf{K}er(\mathbf{L}_{d,\nu} + p\mathbf{I}d).$$

Finally, by taking care of degree reasons we infer that $M = 2p$ and thus the assumption (H2) is valid in this structure.

7. Acknowledgments

We would like to thank Giovanni Peccati for his valuable comments which have helped to improve the exposition of this paper. This work is partially supported by AFR grant 4897114 at the University of Luxembourg and by Grant C-146-000-034-001 and Grant R-146-000-182-112 at the National University of Singapore.

References

[1] Arratia, R., Goldstein, L and Gordon, L. (1990). Poisson approximation and the Chen-Stein method. *Statist. Sci.* **5**, 403–434. With comments and a rejoinder by the authors.

[2] Azmoodeh, E., Campese, S. and Poly, G. (2014). Fourth moment theorems for Markov diffusion generators. *J. Functional Anal.* **266**, no 4, 2341-2359.

[3] Azmoodeh, E., Malicet, D. and Poly, G. (2013). Generalization of the Nualart Peccati criterion. *Preprint*

[4] Barbour, A. D. (1988). Stein's method and Poisson process convergence. *J. Appl. Probab.* **25A**, 175-184.

[5] Barbour, A. D. (1990). Stein's method for diffusion approximations. *Probab. Theory Related Fields* **84**, 297-322.

[6] Barbour, A. D. and Chen, L. H. Y. editors (2005a). *An Introduction to Stein's Method*, Lecture Notes Series **No. 4**, Institute for Mathematical Sciences, National University of Singapore, Singapore University Press and World Scientific Publishing.

[7] Barbour, A. D. and Chen, L. H. Y. editors (2005b). *Stein's Method and Applications*, Lecture Notes Series **No. 5**, Institute for Mathematical Sciences, National University of Singapore, Singapore University Press and World Scientific Publishing.

[8] Barbour, A. D., Holst, L. and Janson, S. (1992). *Poisson Approximation*, Oxford Studies in Probability **No. 2**, Oxford University Press.

[9] Bolthausen, E. (1984). An estimate of the remainder in a combinatorial central limit theorem. *Z. Wahrsch. Verw. Gebiete.* **66**. 379-386.

[10] Breuer, P., Major, P. (1983). Central limit theorems for nonlinear functionals of Gaussian fields. *J. Multivariate Anal.*, **13**, no. 3, 425-441.

[11] Chatterjee, S (2009). Fluctuations of eigenvalues and second order Poincaré inequalities. *Probab. Theory Related Fields* **143**, 1-40.

[12] Chatterjee, S., Diaconis, P. and Meckes, E. (2005). Exchangeable pairs and Poisson approximation. *Probab. Surv.* **2**, 64–106.

[13] Chen, L. H. Y. (1975). Poisson approximation for dependent trials. *Ann. Probab.* **3**, 534–545.

[14] Chen, L. H. Y. (1998). Stein's method: some perspectives with applications. *Probability Towards 2000* (L. Accardi and C. C. Heyde, eds.), Lecture Notes in Statistics **No. 128**, Springer Verlag, 97-122.

[15] Chen, L. H. Y., Goldstein, L. and Shao, Q. M. (2011). *Normal Approximation by Stein's Method.* Probability and its Applications, Springer.

[16] Chen, L. H. Y. and Röllin, A. (2013). Approximating dependent rare events. *Bernoulli* **19**, 1243-1267.

[17] Chen, L. H. Y. and Röllin, A. (2013). Stein couplings for normal approximation. *Preprint.*

[18] Chen, L. H. Y. and Shao, Q.M. (2001). A non-uniform Berry-Esseen bound via Stein's method. *Prob. Theo. Rel. Fields* **120**, no 3, 236-254.

[19] Chen, L. H. Y. and Shao, Q.M. (2004). Normal approximation under local dependence. *Ann. Prob.* **32**, no 3, 1727-2303.
[20] Chen, L. H. Y. and Shao, Q. M. (2005). Stein's method for normal approximation. *An Introduction to Stein's Method* (A.D. Barbour and L. H. Y. Chen, eds), Lecture Notes Series **No. 4**, Institute for Mathematical Sciences, National University of Singapore, Singapore University Press and World Scientific, 1-59.
[21] Diaconis, P. and Holmes, S. editors (2004). *Stein's Method: Expository Lectures and Applications*, IMS Lecture Notes Monogr. Ser. **46**, Inst. Math. Statist., Beachwood, OH.
[22] Goldsetin, L. and Reinert, G. (1997). Stein's method and the zero bias transformation with application to simple random sampling. *Ann. Appl. Probab.* **7**, no 4, 837-1139.
[23] Götze, F. (1991). On the rate of convergence in the multivariate CLT. *Ann. Probab.* **19**, 724–739.
[24] Hoeffding, W. (1951). A combinatorial central limit theorem. *Ann. Math. Statist.* **22**, 558-566.
[25] Hörmander, L. (1967). Hypoelliptic second order differential equations. *Acta Math.* **119**, 147–171.
[26] Ledoux, M. (2012). Chaos of a Markov operator and the fourth moment condition. *Ann. Probab.*. **40**, no. 6, 2439-2459.
[27] Ledoux, M., Nourdin, I. and Peccati, G. (2014). Stein's method, logarithmic Sobolev and transport inequalities. *arXiv:1403.5855.*
[28] Leong, Y. K. (2010). Charles Stein: The Invariant, the Direct and the "Pretentious". In *Creative Minds, Charmed Lives: Interviews at Institute for Mathematical Sciences, National University of Singapore*, World Scientific Publishing, Singapore.
[29] Malliavin, P. (1978). Stochastic calculus of variations and hypoelliptic operators. *Proc. Int. Symp. on Stoch. Diff. Equations*, Kyoto 1976, Wiley, 195–263.
[30] Meckes, E. S. (2009). On Stein's method for multivariate normal approximation. *High dimensional probability V: the Luminy volume*, 153-178, Inst. Math. Stat. Collect., 5, Inst. Math. Statist., Beachwood, OH.
[31] Nourdin, I. and Peccati, G. (2009). Stein's method on Wiener chaos. *Probab. Theory and Related Fields.* **145**, no. 1-2, 75-118.
[32] Nourdin, I. and Peccati, G. (2012). *Normal Approximation with Malliavin Calculus: From Stein's Method to Universality*, Volume 192 of Cambridge Tracts in Mathematics, Cambridge University Press.
[33] Nourdin, I. and Peccati, G. (2013). Optimal fourth moment Theorem. *Proc. Amer. Math. Soc.*, to appear.
[34] Nourdin, I., Peccati, G. and Swan, Y. (2013). Entropy and the fourth moment phenomenon. *J. Funct. Anal.* **266**, 3170-3207.
[35] Nualart, D. and Ortiz-Latorre, S. (2008). Central limit theorems for multiple stochastic integrals and Malliavin calculus. *Stochastic Process. Appl.* **118**, no 4, 614-628.
[36] Nualart, D. and Peccati, G. (2005). Central limit theorems for sequences of multiple stochastic integrals. *Ann. Probab.*, **33**, 177-193.
[37] Peccati, G. (2014). Quantitative clts on a gaussian space: a survey of recent developments. *ESAIM Proc. and Surv.* **44**, 61-78.
[38] Peccati, G. and Tudor, C.A. (2004). Gaussian limits for vector-valued multiple stochastic integrals. *Séminaire de Probabilités* XXXVIII, 247-262.
[39] Reinert, G. and Röllin, A. (2009). Multivariate normal approximation with Stein's method of exchangeable pairs under a general linearity condition. *Ann. Probab.* **37**, 2150–2173.
[40] Ross, N. (2011). Fundamentals of Stein's method. *Probab. Surv.* **8**, 210–293.

[41] Shih, H. H. (2011). On Stein's method for infinite-dimensional Gaussian approximation in abstract Wiener spaces. *J. Funct. Anal.* **261**, 1236–1283.

[42] Stein, C. (1972). A bound for the error in the normal approximation to the distribution of a sum of dependent random variables. In *Proceedings of the Sixth Berkeley Symposium on Mathematical Statistics and Probability (Univ. California, Berkeley, Calif., 1970/1971), Vol. II: Probability Theory*, 583–602, Univ. California Press, Berkeley, Calif.

[43] Stein, C. (1986). *Approximate Computation of Expectations*, IMS Lecture Notes Monogr. Ser. **7**, Inst. Math. Statist., Hayward, CA.

[44] Wald, A. and Wolfowitz, J. (1944). Statistical tests based on permutations of the observations. *Ann. Math. Statist.* **15**, 358-372.

Chapter 7

Progress on Hardy-type inequalities

Mu-Fa Chen*

This paper surveys some of our recent progress on Hardy-type inequalities which consist of a well-known topic in Harmonic Analysis. In the first section, we recall the original probabilistic motivation dealing with the stability speed in terms of the L^2-theory. A crucial application of a result by Fukushima and Uemura (2003) is included. In the second section, the non-linear case (a general Hardy-type inequality) is handled with a direct and analytic proof. In the last section, it is illustrated that the basic estimates presented in the first two sections can still be improved considerably.

Contents

This paper mainly concerns with the following Hardy-type inequality

$$\left(\int_{-M}^{N} |f - \pi(f)|^q \mathrm{d}\mu\right)^{1/q} \le A\left(\int_{-M}^{N} |f'|^p \mathrm{d}\nu\right)^{1/p}, \tag{1}$$

where $p, q \in (1, \infty)$, μ and ν are Borel measures on an interval $[M, N]$ $(M, N \le \infty)$. Here, we assume that $\mu[-M, N] < \infty$ and define a probability measure $\pi = (\mu[-M, N])^{-1}\mu$. Then $\pi(f) := \int f \mathrm{d}\pi$. The functions f are assumed to be absolutely continuous on $(-M, N)$ and belong to $L^q(\mu)$. For simplicity, we may also write the inequality as

$$\|f - \pi(f)\|_{L^q(\mu)} \le A \, \|f'\|_{L^p(\nu)}.$$

To save our notation, assume the constant A to be optimal. The linear case that $p = q = 2$ is discussed in the next section, where a result by Fukushima and Uemura [14] plays an important role. The general case is studied in Section 2. In the last section, we show the possibility for improving further the basic estimates of the optimal constant.

*Beijing Normal University, Beijing 100875, China

1. Linear case: $p = q = 2$.

Let us recall the original probabilistic problem. Throughout this section, we fix $p = q = 2$. Consider a second-order elliptic operator on $(-M, N)$:

$$L = a(x)\frac{\mathrm{d}^2}{\mathrm{d}x^2} + b(x)\frac{\mathrm{d}}{\mathrm{d}x}, \qquad a(x) > 0 \text{ on } (-M, N).$$

Then the two measures used in inequality (1) are as follows

$$\mu(\mathrm{d}x) = \frac{e^{C(x)}}{a(x)}\mathrm{d}x, \qquad \nu(\mathrm{d}x) = e^{C(x)}\mathrm{d}x, \qquad C(x) := \int_\theta^x \frac{b}{a}, \tag{2}$$

here in the last integral and in what follows, the Lebesgue measure $\mathrm{d}x$ is omitted, $\theta \in (-M, N)$ is a reference point. Denote by $\{P_t\}_{t\ge 0}$ the (maximal) semigroup generated by L on $L^2(\mu)$. Here "maximal" means the Dirichlet form having the maximal domain which we learnt earlier from Fukushima [12] (cf. [1] §6.7). We are interested in the stability speed, for instance, the L^2-exponential convergence rate ε:

$$\|P_t f - \pi(f)\|_{L^2(\mu)} \le \|f - \pi(f)\|_{L^2(\mu)} e^{-\varepsilon t}, \qquad t \ge 0.$$

Then, it turns out that the largest rate $\varepsilon_{\max}$ coincides with A^{-1} given in (1) (cf. [1] Theorem 9.1).

We can state one of our recent results as follows.

Theorem 1.1. (Chen, [4, Theorem 10.2]) Let $a > 0$, a and b be continuous on $[-M, N]$ (or $(-M, N]$ if $M = \infty$, for instance). Assume that $\mu(-M, N) < \infty$. Then for the optimal constant A, we have the basic estimates: $\kappa \le A \le 2\kappa$, where

$$\kappa^{-2} = \inf_{-M<x<y<N} \left[\mu(-M, x)^{-1} + \mu(y, N)^{-1}\right]\hat{\nu}(x, y)^{-1},$$

$\mu(x, y) = \int_x^y \mathrm{d}\mu$, and $\hat{\nu}(\mathrm{d}x) = e^{-C(x)}\mathrm{d}x$.

The continuity assumption on a and b is not essential and will be removed in the next section. To understand the proof of this theorem, assume that $M, N < \infty$. Then the general case can be done by an approximating procedure. The optimal constant actually describes an eigenvalue $\lambda_1 (= A^{-2})$ defined by

$$\lambda_1 = \inf\left\{\|f'\|_{L^2(\nu)}^2 : \pi(f) = 0,\ \|f\|_{L^2(\mu)} = 1\right\}.$$

Let g be the eigenfunction corresponding to λ_1:

$$Lg = -\lambda_1 g, \qquad g \ne 0.$$

To ignore a constant (say $\pi(g)$) from g, making derivative on both sides of the equation and replacing g' by f, we get

$$L_S f = -\lambda_1 f,$$

where

$$L_S = a(x)\frac{\mathrm{d}^2}{\mathrm{d}x^2} + \big(a'(x) + b(x)\big)\frac{\mathrm{d}}{\mathrm{d}x} + b'(x)$$

which is a Schrödinger operator. Because $g'(-M) = 0 = g'(N)$, the boundary condition for L_S becomes $f(-M) = 0 = f(N)$. This leads to the principal eigenvalue of L_S:

$$\lambda_S = \sup_{f\in\mathscr{F}} \inf_{x\in(-M,N)} \frac{-L_S f}{f}(x),$$

where

$$\mathscr{F} = \big\{f \in \mathscr{C}^1[-M,N] \cap \mathscr{C}^2(-M,N) : f(-M) = f(N) = 0,\ f|_{(-M,N)} > 0\big\}.$$

Note that the zero-point of g in the original eigenequation for L is located inside of the interval $(-M, N)$, not explicitly known; the zero-points for the eigenvalue λ_S are located at the boundaries $-M$ and N only. This is the advantage of L_S. However, there is an extra term b' in operator L_S which costs some trouble as usual. To avoid this, we rewrite L as

$$L = \frac{\mathrm{d}}{\mathrm{d}\mu}\frac{\mathrm{d}}{\mathrm{d}\hat{\nu}}.$$

Then we define a dual operator of L as follows.

$$L^* = \frac{\mathrm{d}}{\mathrm{d}\mu^*}\frac{\mathrm{d}}{\mathrm{d}\hat{\nu}^*} := \frac{\mathrm{d}}{\mathrm{d}\hat{\nu}}\frac{\mathrm{d}}{\mathrm{d}\mu},$$

i.e. an exchange of μ and $\hat{\nu}$. More explicitly,

$$L^* = a(x)\frac{\mathrm{d}^2}{\mathrm{d}x^2} + \big(a'(x) - b(x)\big)\frac{\mathrm{d}}{\mathrm{d}x}.$$

Next, define

$$\lambda_0^* = \sup_{f^*\in\mathscr{F}} \inf_{x\in(-M,N)} \frac{-L^* f^*}{f^*}(x).$$

In view of the next result which is crucial in proving Theorem 1.1, it is clear that we have thus removed the extra term b' in the operator L_S.

Proposition 1.1. We have $A^{-2} = \lambda_1 = \lambda_S = \lambda_0^*$.

Here, in proving $\lambda_1 = \lambda_S$, we have used a mathematical tool — the coupling technique (cf. [8] or [2]). We have also used another tool — dual technique in proving $\lambda_S = \lambda_0^*$. It is interesting that they are the main tools used in the study on interacting particle systems (cf. [1] and references therein). To obtain the basic estimates listed in Theorem 1.1, we need one more mathematical tool — the capacitary method. The next result is taken from Fukushima & Uemura [14] and [2, 3], see also [13].

Theorem 1.2. For a regular transient Dirichlet form $(D, \mathscr{D}(D))$ with locally compact state space $(E, \mathscr{E})$, the optimal constant $A_{\mathbb{B}}$ in the Poincaré-type inequality

$$\|f^2\|_{\mathbb{B}} \le A_{\mathbb{B}}^2\, D(f), \qquad f \in \mathscr{C}_K^\infty(E)$$

satisfies $B_{\mathbb{B}} \le A_{\mathbb{B}} \le 2B_{\mathbb{B}}$, where $\|\cdot\|_{\mathbb{B}}$ is the norm in a normed linear space $\mathbb{B}$ and

$$B_{\mathbb{B}}^2 = \sup_{\text{compact } K} \operatorname{Cap}(K)^{-1} \|\mathbb{1}_K\|_{\mathbb{B}}.$$

The space $\mathbb{B}$ can be very general, for instance $L^p(\mu)$ $(p \ge 1)$ or the Orlicz spaces. In the present context, $D(f) = \int_{-M}^N f'^2 e^C = \|f'\|_{L^2(\nu)}^2$, $\mathscr{D}(D)$ is the closure of $\mathscr{C}_K^\infty(-M, N)$ with respect to the norm $\|\cdot\|_D$: $\|f\|_D^2 = \|f\|^2 + D(f)$, and

$$\operatorname{Cap}(K) = \inf\big\{D(f) : f \in \mathscr{C}_K^\infty(-M, N),\ f|_K \ge 1\big\}.$$

Note that we have the universal factor 2 here and the isoperimetric constant $B_{\mathbb{B}}$ has a very compact form. We now need to compute the capacity only. The problem is that the capacity is usually not explicitly computable. For instance, at the moment, we do not know how to compute it for Schrödinger operators even for the elliptic operators having killings. Very lucky, we are able to compute the capacity for the one-dimensional elliptic operators. The result has a simple expression:

$$B_{\mathbb{B}}^2 = \sup_{-M<x<y<N} \big[\hat\nu(-M, x)^{-1} + \hat\nu(y, N)^{-1}\big]^{-1} \|\mathbb{1}_{(x,\,y)}\|_{\mathbb{B}}.$$

It looks strange to have double inverse here. So, making inverse in both sides, we get

$$B_{\mathbb{B}}^{-2} = \inf_{-M<x<y<N} \big[\hat\nu(-M, x)^{-1} + \hat\nu(y, N)^{-1}\big]\, \|\mathbb{1}_{(x,y)}\|_{\mathbb{B}}^{-1}.$$

Applying this result to $\mathbb{B} = L^1(\mu)$, we obtain the solution to the case having double Dirichlet boundaries: $\lambda_0 = A_{L^1(\mu)}^{-2}$ and

$$\kappa_0^{-2} = B_{L^1(\mu)}^{-2} = \inf_{-M<x<y<N} \big[\hat\nu(-M, x)^{-1} + \hat\nu(y, N)^{-1}\big]\, \mu(x, y)^{-1}.$$

Applying the last result to the dual process, we have not only

$$\big(\kappa_0^*\big)^2 \le \lambda_1 = \lambda_S = \lambda_0^* \le 4\big(\kappa_0^*\big)^2$$

but also

$$\begin{aligned}\big(\kappa_0^*\big)^{-2} &= \inf_{x<y} \big[\hat\nu^*(-M, x)^{-1} + \hat\nu^*(y, N)^{-1}\big]\, \mu^*(x, y)^{-1}\\ &= \inf_{x<y} \big[\mu(-M, x)^{-1} + \mu(y, N)^{-1}\big]\, \hat\nu(x, y)^{-1}\\ &= \kappa^{-2}.\end{aligned}$$

We have thus arrived at the assertion of Theorem 1.1. Refer to [4] §10 and [5] for more details.

To conclude this section, we remark that the use of the capacity is natural in the higher dimensions, since in which the boundary may be very complicated. However, it seems unnecessary to use it in the present one-dimensional situation. This leads to a direct proof of Theorem 1.1, given in the next section, without using the three mathematical tools just mentioned above.

2. Non-linear case

We now return to our general inequality (1). First, we need a measure $\hat{\nu}$, as in the last section, deduced from ν. Let $\nu^{\#}$ be the absolutely continuous part of ν with respect to the Lebesgue measure. Then, define

$$\hat{\nu}(\mathrm{d}x) = \hat{\nu}_p(\mathrm{d}x) = \left(\frac{\mathrm{d}\nu^{\#}}{\mathrm{d}x}\right)^{-1/(p-1)} \mathrm{d}x, \qquad p > 1.$$

Next, we need a universal factor

$$k_{q,p} = \left[\frac{q-p}{p\mathrm{B}\left(\frac{p}{q-p}, \frac{p(q-1)}{q-p}\right)}\right]^{1/p-1/q} \leq 2 \qquad \text{if } q \geq p,$$

where $\mathrm{B}(\alpha, \beta)$ is the Beta function. In particular (as the limit of $q \downarrow p$),

$$k_{p,p} = p^{1/p}\left(p^*\right)^{1/p^*},$$

where p^* is the conjugate of $p \in (1, \infty)$: $1/p + 1/p^* = 1$.

Theorem 2.1. (Chen, [6, Theorem 2.6]) Let $\mu(-M, N) < \infty$. Then the optimal constant A in the Hardy-type inequality (1) satisfies

(1) the upper estimate $A \leq k_{2,p} B^*$ for $1 < p \leq 2 \leq q < \infty$ once the pure point part of μ (denoted by μ_{pp}) vanishes, and
(2) the lower estimate $A \geq B_*$ for $1 < p, q < \infty$, where

$$B^* = \sup_{x \leq y} \frac{\hat{\nu}(x, y)^{(p-1)/p}}{\left\{\mu(-M, x)^{\frac{p}{q(1-p)}} + \mu(y, N)^{\frac{p}{q(1-p)}}\right\}^{(p-1)/p}},$$

$$B_* = \sup_{x \leq y} \frac{\hat{\nu}(x, y)^{(p-1)/p}}{\left\{\mu(-M, x)^{\frac{1}{1-q}} + \mu(y, N)^{\frac{1}{1-q}}\right\}^{(q-1)/q}}.$$

Moreover, $B_* \leq B^* \leq 2^{1/p-1/q} B_*$ once $q \geq p$.

Here are some remarks on Theorem 2.1.

(a) The isoperimetric constants B^* and B_* are expressed explicitly in measures μ and $\hat{\nu}$.
(b) The boundaries $-M$ and N symmetric in the formulas of B^* and B_*.
(c) Even though $B^* \geq B_*$ in general, but the rough ratio $k_{q,p} 2^{1/p-1/q}$ of the upper and lower bounds is still ≤ 2.

(d) When $q = p$, we have

$$B^* = B_* = \sup_{x \le y} \frac{\hat{\nu}(x,y)^{(p-1)/p}}{\left\{\mu(-M,x)^{\frac{1}{1-p}} + \mu(y,N)^{\frac{1}{1-p}}\right\}^{(p-1)/p}}.$$

(e) Ignoring the $\mu(-M,x)$-term in the expression of B^* or B_*, we obtain

$$B^+ = \sup_y \hat{\nu}(-M,y]^{1/p^*}\,\mu(y,N)^{1/q}.$$

Similarly, ignoring the $\mu(y,N)$-term in the expression of B^* or B_*, we obtain

$$B^- = \sup_x \hat{\nu}(x,N)^{1/p^*}\,\mu(-M,x)^{1/q}.$$

We have thus returned to one of the main results in the study of Hardy-type inequalities.

Theorem 2.2. (1920–1992) For the Hardy-type inequalities

$$\|f\|_{L^q(\mu)} \le A^+ \|f'\|_{L^p(\nu)}, \qquad f(-M) = 0$$

and

$$\|f\|_{L^q(\mu)} \le A^- \|f'\|_{L^p(\nu)}, \qquad f(N) = 0,$$

we have the basic estimates $B^\pm \le A^\pm \le k_{q,p} B^\pm$. Moreover, the factor $k_{q,p}$ is sharp.

There is a long history about the development of Theorem 2.2. The reader is urged to refer to [15] and [6] for a long list of references including five books.

Having the experience in proving Theorem 1.1 and known the history of Theorem 2.2, it is hardly imaginable how to find a direct proof of Theorem 1.1, or even much more general Theorem 2.2, without using capacity. To have a test, let us introduce the proof of a hard part — the upper estimate of Theorem 2.2.

The idea is starting from Theorem 2.2. For this, we split the interval $(-M,N)$ into two parts: $(-M,\theta)$ and (θ,N),

Denote by A_θ^- the optimal constant on the left subinterval $(-M,\theta)$ and by A_θ^+ the one on the right subinterval (θ,N) with the same boundary condition $f(\theta) = 0$. Then, we can rewrite Theorem 2.2 as follows.

Known Theorem Let $q \ge p$. Then we have $B_\theta^\pm \le A_\theta^\pm \le k_{q,p} B_\theta^\pm$, where

$$B_\theta^- = \sup_{x<\theta} \hat{\nu}(x,\theta)^{1/p^*}\,\mu(-M,x)^{1/q}, \qquad B_\theta^+ = \sup_{y>\theta} \hat{\nu}(\theta,y)^{1/p^*}\,\mu(y,N)^{1/q}.$$

Proof of the upper estimate: $k_{2,p} B^* \ge A$.

Rewrite B^* as

$$\begin{aligned} B^* &= \sup_{x\le y} \frac{\hat\nu(x,y)^{(p-1)/p}}{\left\{\mu(-M,x)^{\frac{p}{q(1-p)}} + \mu(y,N)^{\frac{p}{q(1-p)}}\right\}^{(p-1)/p}} \\ &= \left\{\sup_{x\le y} \frac{\hat\nu(x,y)}{\mu(-M,x)^{\frac{p}{q(1-p)}} + \mu(y,N)^{\frac{p}{q(1-p)}}}\right\}^{(p-1)/p} \\ &=: \left\{\sup_{x\le y} \frac{\hat\nu(x,y)}{\varphi(x)+\psi(y)}\right\}^{1/p^*}. \end{aligned}$$

By proportional property, we have

$$\frac{\hat\nu(x,y)}{\varphi(x)+\psi(y)} = \frac{\hat\nu(x,\theta)+\hat\nu(\theta,y)}{\varphi(x)+\psi(y)} \ge \left\{\frac{\hat\nu(x,\theta)}{\varphi(x)} \bigwedge \frac{\hat\nu(\theta,y)}{\psi(y)}\right\}, \qquad \theta\in(x,y).$$

Here $a\wedge b=\min\{a,b\}$ and similarly $a\vee b=\max\{a,b\}$. Hence (omit what in $\{\cdots\}$)

$$\frac{\hat\nu(x,y)}{\varphi(x)+\psi(y)} \ge \sup_{\theta\in(x,y)} \{\cdots\}.$$

Then

$$\begin{aligned} \sup_{x\le y}\frac{\hat\nu(x,y)}{\varphi(x)+\psi(y)} &\ge \sup_{x\le y}\sup_{\theta\in(x,y)}\{\cdots\} \\ &= \sup_{\theta}\sup_{(x,y)\ni\theta}\{\cdots\} \\ &= \left\{\left[\sup_{x\le\theta}\frac{\hat\nu(x,\theta)}{\varphi(x)}\right] \bigwedge \left[\sup_{y\ge\theta}\frac{\hat\nu(\theta,y)}{\psi(y)}\right]\right\}. \end{aligned}$$

Making power $1/p^*$ on both sides, by definition of $B_\theta^\pm$, we obtain

$$B^* \ge \sup_\theta \left(B_\theta^- \wedge B_\theta^+\right).$$

Here a problem appears: we need $\vee$ rather than $\wedge$ on the right-hand side. To overcome this, we assume that $\mu_{pp}=0$. Then, there exists $\bar\theta\in(-M,N)$ such that $B_{\bar\theta}^- = B_{\bar\theta}^+$. Therefore

$$B^* \ge \sup_\theta \left(B_\theta^- \wedge B_\theta^+\right) \ge B_{\bar\theta}^- = B_{\bar\theta}^- \vee B_{\bar\theta}^+$$

and then

$$\begin{aligned} k_{q,p}B^* &\ge \left(k_{q,p}B_{\bar\theta}^-\right) \vee \left(k_{q,p}B_{\bar\theta}^+\right). \\ &\ge A_{\bar\theta}^- \vee A_{\bar\theta}^+ \quad \text{(Known Theorem)} \\ &\ge \inf_\theta \left(A_\theta^- \vee A_\theta^+\right) \\ &\ge A \quad \text{(splitting technique)}. \end{aligned}$$

Here each step holds for all $q\ge p$ except the last one. In which, some additional work is required, due to the appearance of $f-\pi(f)$ rather than f only. We prove the conclusion first for $q = 2 \geq p$ and then extend it to $q \geq 2$, even to a large class of normed linear space $\mathbb{B}$, as used in Theorem 1.2, using a known lifting procedure (cf. [2] §6.3). Note that in the proof above, we use a bridge θ to combine the known results on two subintervals together. But then remove it, otherwise, $\bar{\theta}$ for instance, may not be computable. Nevertheless, it should be understandable that the present analytic proof does not use the coupling, duality, or capacitary techniques.

Actually, much more topics are studied in [6]: the bilateral Dirichlet boundaries, logarithmic Sobolev inequalities, Nash inequalities, and so on.

3. Improvements of the basic estimates

Note that for two given numbers having smaller ratio, their difference can be still quite big. Hence, it is meaningful to improve the basic estimates introduced in the last two sections. In this section, we show the possibility in doing so by a simplest example: $\mu = \mathrm{d}x$ and $\nu = \mathrm{d}x$ on $(0, 1)$. We need to consider the following Hardy-type inequality

$$\|f\|_{L^q(\mu)} \leq A \|f'\|_{L^p(\nu)}, \qquad f(0) = 0 \tag{3}$$

only since the other cases (the ergodic case in particular) can be reduced to this one by symmetry. The basic estimates for the optimal constant A in (3) are given in Theorem 2.2.

Example 3.1. Let $\mu = \mathrm{d}x$ and $\nu = \mathrm{d}x$ on $(0, 1)$. Then the optimal constant A in (3) is given as follows.

(1) When $p = q = 2$, we have $A = 2/\pi$.
(2) When $p = q \in (1, \infty)$, we have

$$A = \frac{p}{\pi (p-1)^{1/p}} \sin \frac{\pi}{p}.$$

(3) For general $p, q \in (1, \infty)$, we have

$$A = \frac{p^{\frac{1}{q}} q^{1-\frac{1}{p}} (pq + p - q)^{\frac{1}{p} - \frac{1}{q}}}{(p-1)^{\frac{1}{p}} \, \mathrm{B}\left(\frac{1}{q}, 1 - \frac{1}{p}\right)}.$$

This simplest example already shows that it is nontrivial from the special case $p = q = 2$ to the general one.

Proposition 3.1. (Chen, [7]) For the optimal constant A in the last example, we have the following improved estimates:

$$B \leq \bar{\delta}_1 \leq A \leq A^* \leq \delta_1 \leq k_{q,p} B,$$

where

$$B = \frac{p^{1/q}((p-1)q)^{1-1/p}}{(pq+p-q)^{1-1/p+1/q}},$$

$$\bar{\delta}_1 = \frac{p^{1/q}((p-1)(q+1))^{1-1/p}}{(pq+p-q)^{1-1/p+1/q}},$$

$$A^* = \left[\frac{p^*}{q}\right]^{\frac{1}{q}} \left[\frac{p^*+q}{\pi p^*} \sin \frac{\pi p^*}{p^*+q}\right]^{\frac{1}{p^*}+\frac{1}{q}} \qquad \boxed{= A \text{ if } q = p}$$

$$\delta_1 = \frac{1}{(q\gamma^*/p^*+1)^{\frac{1}{q}}} \left[\sup_{x\in(0,1)} \frac{1}{x^{\gamma^*}} \int_0^x \left(1 - y^{q/\gamma^* p^*+1}\right)^{\frac{p^*}{q}} \mathrm{d}y\right]^{\frac{1}{p^*}}, \quad \gamma^* := \frac{q}{p^*+q}$$

The results in Proposition 3.1 are shown by Figures 1–4.

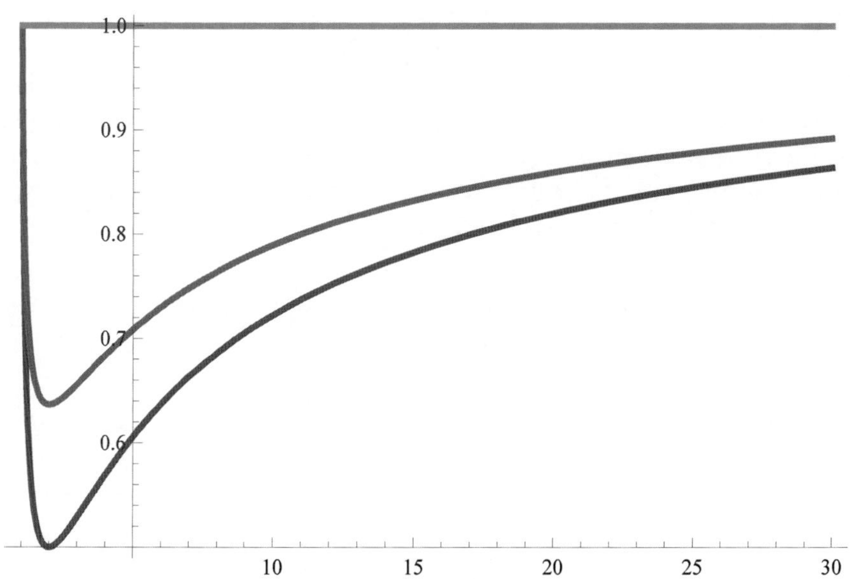

Figure 1 The basic estimates of A: $p = q \in (1, 30)$.

Note that in the case that $p = q$, we have $A^* = A$. So there are five curves only in Figure 2.

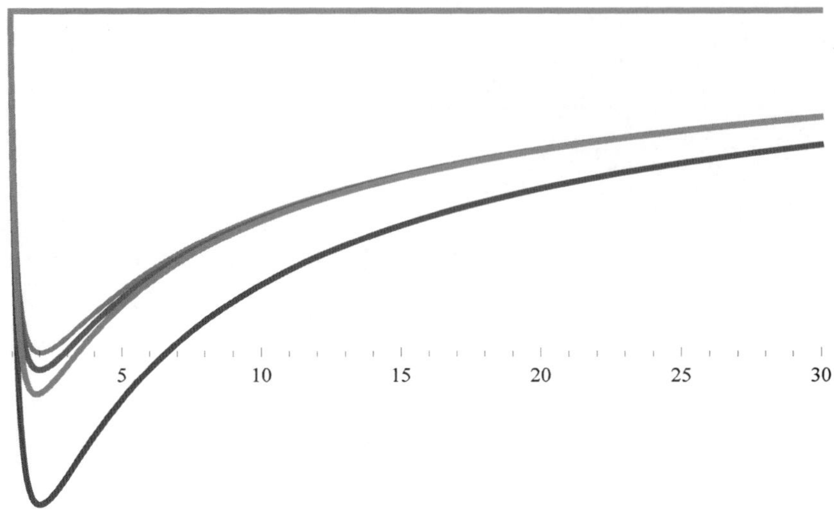

Figure 2 The basic estimates of A and their improvements: $p = q \in (1, 30)$.

The improvements are surprisingly effective. Note that a suitable convex mean of the new upper and lower bounds should provides a quite precise approximation of A. However, the convex means of the basic estimates do not have this property. When $p = q$, much more refined results can be found from [9–11].

Next, since $q \geq p$, we may write $q = p + r$ for some $r \geq 0$. In Figures 3 and 4, there are six curves, three of them are upper estimates and two of them are lower ones. The third curve from the bottom is the exact one; the top and the bottom curves consist of the basic estimates of the exact one. The other three curves are the improvements of the basic estimates.

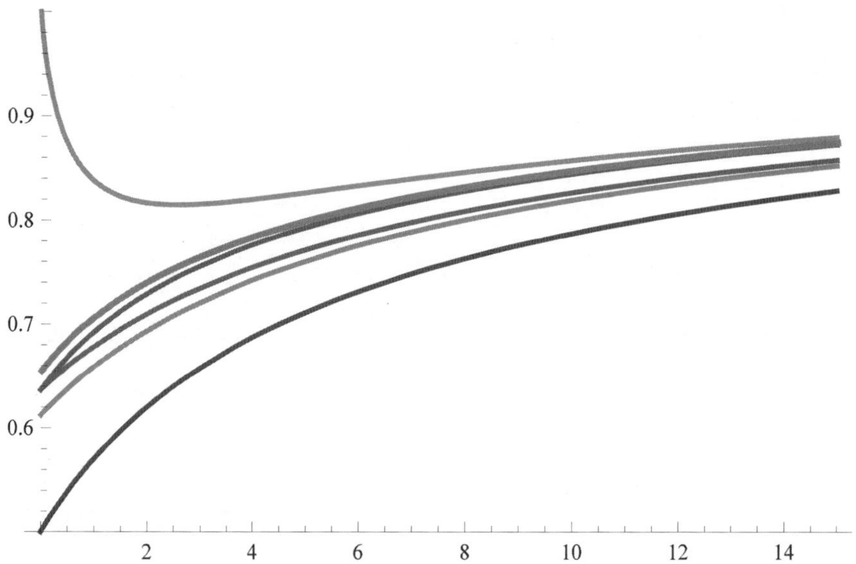

Figure 3 The basic estimates and their improvements: $p = 2$, $r \in (0, 15)$.

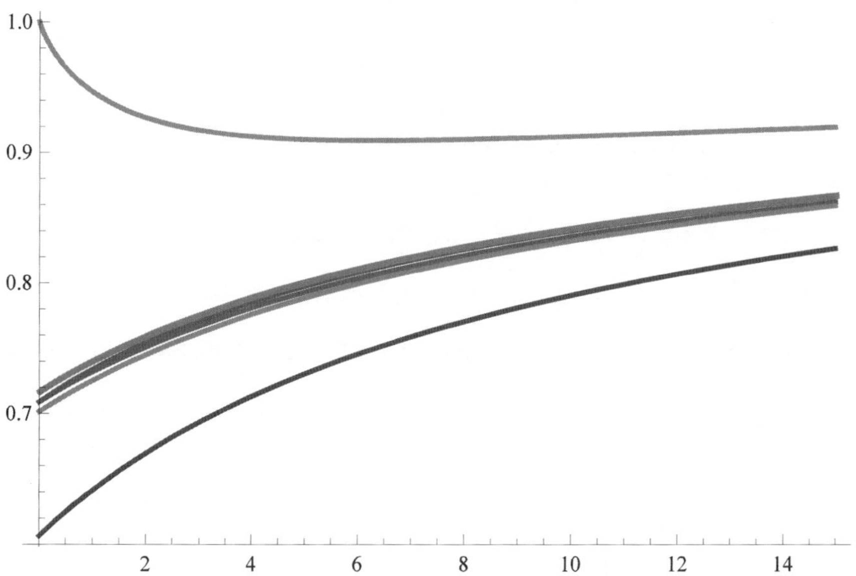

Figure 4 The basic estimates and their improvements: $p = 5$, $r \in (0, 15)$.

Note that in Figure 4, the new upper bounds and lower bounds are almost overlapped with A. In general, they are closer when p or $q \geq p$ is larger.

Finally, we mention that the main results in this note: Theorems 1.1 and 2.1, and Proposition 3.1 are new addition to the context of Hardy-type inequalities.

Acknowledgments. This short note introduces an example to illustrate the great help we have obtained from Professor Masatoshi Fukushima in the past decades. It is a nice chance to express our deepest thanks to him. Research supported in part by the National Natural Science Foundation of China (No. 11131003), and by the "985" project from the Ministry of Education in China.

References

[1] Chen, M.F. (2004). *From Markov Chains To Non-Equilibrium Particle Systems*, 2nd ed. World Sci., Singapore.

[2] Chen, M.F. (2005a). *Eigenvalues, Inequalities, and Ergodic Theory.* Springer, London.

[3] Chen, M.F. (2005b). *Capacitary criteria for Poincaré-type inequalities.* Potential Analysis 23(4), 303-322.

[4] Chen, M.F. (2010). *Speed of stability for birth–death processes.* Front. Math. China 5(3), 379–515.

[5] Chen, M.F. (2012). *Basic estimates of stability rate for one-dimensional diffusions.* Chapter 6 in "Probability Approximations and Beyond", 75–99, Lecture Notes in Statistics 205, eds. A.D. Barbour, H.P. Chan and D. Siegmund.

[6] Chen, M.F. (2013a). *Bilateral Hardy-type inequalities.* Acta Math. Sin. Eng. Ser. 2013, 29:1, 1–32.

[7] Chen, M.F. (2013b). *The optimal constant in Hardy-type inequalities*, in preparation.

[8] Chen, M.F. and Wang, F.Y. (1997). Estimation of spectral gap for elliptic operators. Trans. Amer. Math. Soc. 349(3), 1239–1267.

[9] Chen M.F., Wang L.D., Zhang Y.H. (2013a) *Mixed principal eigenvalues in dimension one.* Front. Math. China, 2013, 8(2): 317-343.

[10] Chen, M.F., Wang, L.D., Zhang, Y.H. (2013b) *Mixed eigenvalues of discrete p-Laplacian*, preprint.

[11] Chen, M.F., Wang, L.D., Zhang, Y.H. (2013c) *Mixed Eigenvalues of p-Laplacian*, preprint.

[12] Fukushima, M. (1980). *Dirichlet Forms and Markov Processes.* North Holland Math. 23.

[13] Fukushima, M., Oshima, Y. and Takeda, M. (2011). *Dirichlet Forms and Symmetric Markov Processes,* 2nd ed. Walter de Gruyter.

[14] Fukushima, M. and Uemura, T. (2003). *Capacitary bounds of measures and ultracontractivity of time changed processes.* J. Math. Pure et Appliquées 82(5), 553–572.

[15] Kufner, A., Maligranda, L. and Persson, L.E. (2007). *The Hardy Inequality: About its History and Some Related Results.* Vydavatelsky Servis.

Chapter 8

Functional inequalities for pure-jump Dirichlet forms

Xin Chen*, Feng-Yu Wang† and Jian Wang‡

In this survey paper we summarize some recent progress on functional inequalities for Dirichlet forms of type

$$\mathcal{E}(f,g) := \int_{\mathbb{R}^d\times\mathbb{R}^d} (f(x)-f(y))(g(x)-g(y))q(x,y)\mu(\mathrm{d}y)\mu(\mathrm{d}x),$$

where μ is a probability measure on $\mathbb{R}^d$ and q is a nonnegative measurable function on $\mathbb{R}^d \times \mathbb{R}^d$ such that

$$\sup_{x\in\mathbb{R}^d}\int_{\mathbb{R}^d}(1\wedge|x-y|^2)q(x,y)\mu(\mathrm{d}y) < \infty.$$

Criteria and perturbation results of the weak Poincaré, Poincaré and super Poincaré inequalities are presented. Concrete examples are presented to illustrate the main results.

Contents

1. Introduction

Functional inequalities of Dirichlet forms are powerful tools in characterizing the properties of Markov semigroups and their generators, see [2, 9, 13, 17, 25] for accounts on functional inequalities and applications. To establish functional inequalities, many explicit criteria have been presented for diffusion processes and

*Grupo de Fisica Matematica, Universidade de Lisboa, Av Prof Gama Pinto 2, Lisbon 1649-003, Portugal
†School of Mathematical Sciences, Beijing Normal University, Beijing 100875, China and Department of Mathematics, Swansea University, Singleton Park, SA2 8PP, United Kingdom
‡School of Mathematics and Computer Science, Fujian Normal University, Fuzhou 350007, China

Markov chains, but less is known for Lévy-type jump processes. Of course, using subordination techniques, functional inequalities for a class of jump processes can be deduced from known ones of diffusion processes, see [4, 15, 21, 22] for details. However, a Lévy-type process is not necessarily subordinate to a diffusion processes and, in case it is, the subordinator is in general not easy to describe. So, it is necessary to provide general criteria on functional inequalities for Lévy-type jump processes. Recently, using harmonic analysis technique, a sufficient condition for the weighted Poincaré inequality has been presented in [18] for the α-stable like Dirichlet forms, which is then extended in [16] to an abstract situation for L^p weighted Poincaré inequalities. The purpose of this paper is to summarize some recent progress on functional inequalities for Lévy-type Dirichlet forms we made in [10–12, 27, 28]. For simplicity we only consider Dirichlet forms on $\mathbb{R}^d$, although an abstract state space has been taken in [12].

Let μ be a probability measure on $\mathbb{R}^d$. Let q be a nonnegative measurable function on $\mathbb{R}^d \times \mathbb{R}^d$ with $q(x,x) = 0$ for $x \in \mathbb{R}^d$ such that

$$\lambda := \sup_{x\in\mathbb{R}^d} \int_{\mathbb{R}^d} \big(1 \wedge |x-y|^2\big) q(x,y)\mu(\mathrm{d}y) < \infty. \tag{1.1}$$

Then

$$\begin{aligned} &\Gamma(f,g)(x) := \int_{\mathbb{R}^d} (f(x)-f(y))(g(x)-g(y))q(x,y)\mu(\mathrm{d}y), \quad \forall x \in \mathbb{R}^d, \\ &f,g \in \mathcal{A} := \big\{f \in \mathcal{B}_b(\mathbb{R}^d) : \Gamma(f,f) \in \mathcal{B}_b(\mathbb{R}^d)\big\} \end{aligned} \tag{1.2}$$

gives rise to a non-negative definite bilinear map from $\mathcal{A} \times \mathcal{A}$ to $\mathcal{B}_b(\mathbb{R}^d)$. Since $C_b^2(\mathbb{R}^d) \subset \mathcal{A}$ is dense in $L^2(\mu)$, it is standard that the form

$$\begin{aligned} \mathcal{E}(f,g) :&= \mu(\Gamma(f,g)) \\ &= \iint_{\mathbb{R}^d\times\mathbb{R}^d} (f(x)-f(y))(g(x)-g(y))q(x,y)\mu(\mathrm{d}y)\mu(\mathrm{d}x), \quad f,g \in \mathcal{A} \end{aligned} \tag{1.3}$$

is closable in $L^2(\mu)$ and its closure $(\mathcal{E},\mathcal{D}(\mathcal{E}))$ is a symmetric conservative Dirichlet form, see for instance [14, Example 1.2.6]. A typical example of the framework is the stable-like Dirichlet form, where

$$q(x,y) = \frac{j(x,y)}{u(y)}, \quad \mu(\mathrm{d}x) = u(x)\mathrm{d}x$$

for some nonnegative $j \in \mathcal{B}(\mathbb{R}^d \times \mathbb{R}^d)$ such that $j(x,x) = 0$ for all $x \in \mathbb{R}^d$ and $\sup_{x\in\mathbb{R}^d} \int_{\mathbb{R}^d} \big(1 \wedge |x-y|^2\big) j(x,y)\mathrm{d}y < \infty$, and positive $u \in \mathcal{B}(\mathbb{R}^d)$ such that $\mu(\mathrm{d}x)$ is a probability measure. In particular, when $j(x,y) = \frac{\tilde q(x,y)}{|x-y|^{d+\alpha}}$ for some $\alpha \in (0,2)$ and nonnegative $\tilde q \in \mathcal{B}_b(\mathbb{R}^d \times \mathbb{R}^d)$, $(\mathcal{E},\mathcal{D}(\mathcal{E}))$ is a perturbed α-stable Dirichlet form.

The remainder of this paper is arranged as follows. We will present criteria of functional inequalities for the α-stable-like Dirichlet forms in Section 2, and criteria of functional inequalities for the truncated α-stable-like Dirichlet forms in Section 3. Section 4 is devoted to weighted functional inequalities, and in the final section we investigate perturbations of the functional inequalities. The results in each section will be illustrated by concrete examples.

2. Functional inequalities for stable-like Dirichlet forms

Let us start with a simple example, i.e. the Ornstein-Uhlenbeck process driven by an α-stable process. Let Δ be the Laplacian on $\mathbb{R}^d$. Consider the Ornstein-Uhlenbeck operator

$$L_\alpha f(x) := -(-\Delta)^{\alpha/2} f(x) - \langle x, \nabla f(x)\rangle, \quad f \in C_0^\infty(\mathbb{R}^d)$$

for some $\alpha \in (0,2)$, where $C_0^\infty(\mathbb{R}^d)$ is the set of all smooth functions on $\mathbb{R}^d$ with compact support. Then the associated Markov semigroup has a unique invariant (but not reversible, see [1]) probability measure μ_α, which is identified by the Fourier transformation

$$\hat{\mu}_\alpha(\xi) := \int_{\mathbb{R}^d} \mathrm{e}^{i\langle x,\xi\rangle}\, \mu_\alpha(\mathrm{d}x) = \mathrm{e}^{-\frac{1}{\alpha}|\xi|^\alpha}, \quad \xi \in \mathbb{R}^d.$$

For any $f \in C_0^\infty(\mathbb{R}^d)$, we have (see [20, (1.9)])

$$\mathcal{E}_\alpha(f,f) := -\int_{\mathbb{R}^d} f L_\alpha f \mathrm{d}\mu_\alpha = \frac{1}{2}\iint_{\mathbb{R}^d\times\mathbb{R}^d} \frac{|f(x)-f(y)|^2}{|x-y|^{d+\alpha}}\, \mathrm{d}y\mu_\alpha(\mathrm{d}x). \tag{2.1}$$

Let $\mathcal{D}(\mathcal{E}_\alpha) = \{f \in L^2(\mu_\alpha) : \mathcal{E}_\alpha(f,f) < \infty\}$. According to [20, Example 3.2(2)], the semigroup P_t^α generated by L_α is not hyperbounded, i.e. $\|P_t^\alpha\|_{L^p(\mu_\alpha)\to L^q(\mu_\alpha)} = \infty$ for any $t \geq 0$ and $q > p \geq 1$. Therefore, $\mathcal{E}_\alpha$ does not satisfy the log-Sobolev inequality, i.e. for any constant $C > 0$ the log-Sobolev inequality

$$\mu_\alpha(f^2 \log f^2) \leq C\mathcal{E}_\alpha(f,f), \quad f \in \mathcal{D}(\mathcal{E}_\alpha), \mu_\alpha(f^2) = 1 \tag{2.2}$$

fails, where $\mu(f) := \int_{\mathbb{R}^d} f \mathrm{d}\mu$ for a measure μ on $\mathbb{R}^d$ and a function $f \in L^1(\mu)$. In fact, since

$$\frac{1}{c(1+|x|^2)^{(d+\alpha)/2}}\,\mathrm{d}x \leq \mu_\alpha(\mathrm{d}x) \leq \frac{c}{(1+|x|^2)^{(d+\alpha)/2}}\,\mathrm{d}x \tag{2.3}$$

holds for some constant $c > 1$, see e.g. [5, Theorem 2.1] or [7, (1.5)], Corollary 2.1(2) below provides a stronger statement, i.e. the super Poincaré inequality is not available either. Recall that the log-Sobolev inequality (2.2) holds for some constant $C > 0$ if and only if the super Poincaré inequality

$$\mu_\alpha(f^2) \leq r\mathcal{E}_\alpha(f,f) + \exp\Big(c\big(1+r^{-1}\big)\Big)\mu_\alpha(|f|)^2, \quad r > 0, f \in \mathcal{D}(\mathcal{E}_\alpha)$$

holds for some constant $c > 0$, see e.g. [24, Corollary 1.2]. On the other hand, Corollary 2.1(1) below implies that the Poincaré inequality

$$\mu_\alpha(f^2) \leq C\mathcal{E}_\alpha(f,f), \quad f \in \mathcal{D}(\mathcal{E}_\alpha), \mu_\alpha(f) = 0$$

holds for some constant $C > 0$. Therefore, for this typical example, the best possibility among functional inequalities mentioned above is the Poincaré inequality.

Now, as a generalization of (2.1), we consider

$$\mathcal{E}_{\alpha,V}(f,g) := \iint_{\mathbb{R}^d\times\mathbb{R}^d} \frac{(f(x)-f(y))(g(x)-g(y))}{|x-y|^{d+\alpha}}\,\mathrm{d}y\mu_V(\mathrm{d}x),$$
$$\mathcal{D}(\mathcal{E}_{\alpha,V}) := \Big\{f\in L^2(\mu_V):\ \mathcal{E}_{\alpha,V}(f,f)<\infty\Big\},$$

where V is a measurable function on $\mathbb{R}^d$ such that $\mu_V(\mathrm{d}x) := C_V\,\mathrm{e}^{-V(x)}\mathrm{d}x$ is a probability measure. Then $(\mathcal{E}_{\alpha,V},\mathcal{D}(\mathcal{E}_{\alpha,V}))$ is a symmetric Dirichlet form on $L^2(\mu_V)$. Let

$$h(r)=\inf_{|x|\le r}\mathrm{e}^{V(x)},\quad H(r)=\sup_{|x|\le r}\mathrm{e}^{V(x)},$$

$$\Phi(r)=\inf_{|x|\ge r}\frac{\mathrm{e}^{V(x)}}{(1+|x|)^{d+\alpha}},\quad \Phi^{-1}(r)=\inf\big\{s\ge 0:\ \Phi(s)\ge r\big\},\quad r>0,$$

where we set $\inf\emptyset=\infty$ by convention. Moreover, let

$$\Psi_1(r)=\left(\sup_{|x|\le r}\frac{(1+|x|)^{d+\alpha}}{\mathrm{e}^{V(x)}}\right)\sup_{x\in\mathbb{R}^d}\frac{\mathrm{e}^{V(x)}}{(1+|x|)^{d+\alpha}},$$

$$\Psi_2(r)=\frac{1}{\mu_V(B(0,r))^2}\sup_{x\in B(0,r)}\int_{B(0,r)}|y-x|^{d+\alpha}\mathrm{e}^{-2V(y)}\mathrm{d}y,\quad r>0.$$

The main result of this section is the following

Theorem 2.1. (Wang et al. [27]) *Let* $\int_{\mathbb{R}^d}\mathrm{e}^{-V(x)}\mathrm{d}x<\infty$ *such that* μ_V *is a well defined probability measure.*

(1) *If* $\mathrm{e}^{-V}\in\mathcal{C}:=\{f\in C^2(\mathbb{R}^d):\ \|\nabla f\|_\infty,\|\nabla^2 f\|_\infty<\infty\}$ *such that*

$$\limsup_{r\to\infty}\left\{r^{d+\alpha-1}\sup_{|x|\ge r-1}|\nabla \mathrm{e}^{-V}(x)|+r^{d+\alpha-2}\sup_{|x|\ge r-1}\mathrm{e}^{-V(x)}\right\}=0 \tag{2.4}$$

and $\Phi(0)>0$, *then the Poincaré inequality*

$$\mu_V(f^2)\le C\mathcal{E}_{\alpha,V}(f,f),\quad f\in\mathcal{D}(\mathcal{E}_{\alpha,V}),\mu_V(f)=0 \tag{2.5}$$

holds for some constant $C>0$.

(2) *If* $\mathrm{e}^{-V}\in\mathcal{C}$ *such that* (2.4) *holds and* $\Phi(r)\uparrow\infty$ *as* $r\uparrow\infty$, *then there exist constants* $c_1,c_2>0$ *such that the super Poincaré inequality*

$$\mu_V(f^2)\le r\mathcal{E}_{\alpha,V}(f,f)+\beta(r)\mu_V(|f|)^2,\quad r>0, f\in\mathcal{D}(\mathcal{E}_{\alpha,V}) \tag{2.6}$$

holds with

$$\beta(r)=c_1\Big(1+r^{-d/\alpha}\{h\circ\Phi^{-1}(c_2r^{-1})\}^{-1-d/\alpha}\{H\circ\Phi^{-1}(c_2r^{-1})\}^{2+d/\alpha}\Big)$$

for $r>0$.

(3) *There exists a universal constant* $c>0$ *such that the weak Poincaré inequality*

$$\mu_V(f^2)\le\tilde\beta(r)\mathcal{E}_{\alpha,V}(f,f)+r\|f\|_\infty^2,\quad r>0, f\in\mathcal{D}(\mathcal{E}_{\alpha,V}),\mu_V(f)=0 \tag{2.7}$$

holds for

$$\tilde{\beta}(r) := \inf\left\{\big(c\Psi_1(R)\big)\wedge\Psi_2(R):\ \mu_V(B(0,R)^c)\leq\frac{r}{1+r}\right\}<\infty,\quad r>0.$$

Although we assume in Theorem 2.1(1)-(2) that e^{-V} is at least C^2-smooth, the assertions work also for the singular case by using perturbation results of functional inequalities, see Section 5 for details. To illustrate this result, below we consider some typical families of V with different growth rates: for faster growth rate of V one derives stronger functional inequality. When we apply Theorem 2.1(3) to derive weak Poincaré inequalities for these families of V, the function Ψ_1 in the definition of $\tilde{\beta}$ is better than Ψ_2. On the other hand, however, Ψ_2 is always finite but in some cases Ψ_1 is infinite. So, in general these two functions are incomparable.

According to (2.3), in the following result μ_V is a natural extension to μ_α, i.e. when $\varepsilon=\alpha$ a functional inequality for $\mathcal{E}_{\alpha,V}$ in $L^2(\mu_V)$ is equivalent to that for $\mathcal{E}_\alpha$ in $L^2(\mu_\alpha)$. In particular, as mentioned above, this result implies that $\mathcal{E}_\alpha$ satisfies the Poincaré inequality but not the super Poincaré inequality.

Corollary 2.1. (Wang et al. [27]) *Let* $V(x)=\frac{1}{2}(d+\varepsilon)\log(1+|x|^2)$ *for some constant* $\varepsilon>0$.

(1) *The Poincaré inequality* (2.5) *holds for some constant* $C>0$ *if and only if* $\varepsilon\geq\alpha$.
(2) *The super Poincaré inequality* (2.6) *holds for some function* $\beta:(0,\infty)\to(0,\infty)$ *if and only if* $\varepsilon>\alpha$, *and in this case there exists a constant* $c>0$ *such that the inequality holds with*

$$\beta(r)=c\Big(1+r^{-\frac{d}{\alpha}-\frac{(d+\varepsilon)(2\alpha+d)}{\alpha(\varepsilon-\alpha)}}\Big),\qquad r>0.$$

(3) *If* $\varepsilon\in(0,\alpha)$, *then there exists a constant* $c>0$ *such that the weak Poincaré inequality* (2.7) *holds for*

$$\tilde{\beta}(r)=c\Big(1+r^{-(\alpha-\varepsilon)/\varepsilon}\Big),\quad r>0.$$

This $\tilde{\beta}$ *is sharp in the sense that* (2.7) *does not hold if* $\lim_{r\to0}r^{(\alpha-\varepsilon)/\varepsilon}\tilde{\beta}(r)=0$.

Since $\varepsilon=\alpha$ in Corollary 2.1 is the critical situation for the Poincaré inequality, we consider below lower order perturbations of the corresponding V.

Corollary 2.2. (Wang et al. [27]) *Let* $V(x)=\frac{1}{2}(d+\alpha)\log(1+|x|^2)+\varepsilon\log\log(\mathrm{e}+|x|^2)$, $\varepsilon\in\mathbb{R}$.

(1) *The super Poincaré inequality* (2.6) *holds for some* β *if and only if* $\varepsilon>0$, *and in this case it holds with*

$$\beta(r)=\exp\Big[c\Big(1+r^{-1/\varepsilon}\Big)\Big]$$

for some constant $c>0$.

(2) *The super Poincaré inequality in* (1) *is sharp in the sense that* (2.6) *does not hold if*

$$\lim_{r\to 0} r^{1/\varepsilon} \log \beta(r) = 0.$$

(3) *The log-Sobolev inequality*

$$\mu_V(f^2 \log f^2) \le C\mathcal{E}_{\alpha,V}(f,f), \quad f \in \mathcal{D}(\mathcal{E}_{\alpha,V}), \mu_V(f^2) = 1$$

holds for some constant $C > 0$ *if and only if* $\varepsilon \ge 1$.

(4) *The Poincaré inequality* (2.5) *holds for some constant* $C > 0$ *if and only if* $\varepsilon \ge 0$, *and there exists a universal constant* $c > 0$ *such that for* $\varepsilon < 0$ *the weak Poincaré inequality* (2.7) *holds with*

$$\widetilde{\beta}(r) = c\Big(1 + \log^{-\varepsilon}\left(1 + r^{-1}\right)\Big), \quad r > 0.$$

This $\tilde{\beta}$ *is sharp in the sense that for* $\varepsilon < 0$ *the weak Poincaré inequality* (2.7) *does not hold if* $\lim_{r\to 0} \tilde{\beta}(r) \log^{\varepsilon}(1 + r^{-1}) = 0$.

Below we consider a family of functions V with slower growth rate such that μ_V is a probability measure, for which merely the weak Poincaré inequality is available.

Corollary 2.3. (Wang et al. [27]) *Let* $V(x) = \frac{d}{2}\log(1+|x|^2)+\varepsilon \log\log(\mathrm{e}+|x|^2)$, $\varepsilon > 1$. *Then there exist some constants* $c_1, c_2 > 0$ *such that the weak Poincaré inequality* (2.7) *holds with*

$$\widetilde{\beta}(r) = c_1 \exp\left[c_2 r^{-1/(\varepsilon-1)}\right].$$

This $\tilde{\beta}$ *is sharp in the sense that the weak Poincaré inequality* (2.7) *does not hold if* $\lim_{r\to 0} r^{1/(\varepsilon-1)} \log \tilde{\beta}(r) = 0$.

Finally, we consider two families of V with faster growth rates than all those presented above, so that the much stronger super Poincaré inequality is available.

Corollary 2.4. (Wang et al. [27]) (1) *Let* $V(x) = \log^{1+\varepsilon}(1 + |x|^2)$, $\varepsilon > 0$. *Then there exists a constant* $c > 0$ *such that the super Poincaré inequality* (2.6) *holds for*

$$\beta(r) = c + cr^{-2(\alpha+d)/\alpha} \exp\left[c \log^{1/(1+\varepsilon)}(1 + r^{-1})\right], \quad r > 0.$$

(2) *Let* $V(x) = (1 + |x|^2)^{\varepsilon}$, $\varepsilon > 0$. *Then there exists a constant* $c > 0$ *such that the super Poincaré inequality* (2.6) *holds for*

$$\beta(r) = c\Big(1 + r^{-2(\alpha+d)/\alpha} \log^{(2\alpha+d)(d+\alpha)/(2\varepsilon\alpha)}(1 + r^{-1})\Big), \quad r > 0.$$

We remark that a sufficient condition for $\mathcal{E}_{\alpha,V}$ to satisfy the Poincaré inequality has been presented in [18, Theorem 1.2]. More precisely, let $V \in C^2(\mathbb{R}^d)$ satisfy

$$\lim_{|x|\to\infty} \{\varepsilon|\nabla V|^2 - \Delta V\} = \infty \ \text{ for some constant } \varepsilon \in (0, 1/2). \tag{2.8}$$

Then there exist constants $\delta, C > 0$ such that the inequality

$$\int_{\mathbb{R}^d} (f(x) - \mu_V(f))^2 (1 + |\nabla V(x)|^\alpha) \mu_V(\mathrm{d}x) \le C \mathcal{E}_{\alpha,\delta,V}(f, f) \tag{2.9}$$

holds for all $f \in C_b^2(\mathbb{R}^d)$, where

$$\mathcal{E}_{\alpha,\delta,V}(f, f) := \iint_{\mathbb{R}^d \times \mathbb{R}^d} \frac{(f(x) - f(y))^2}{|x - y|^{d+\alpha}} \mathrm{e}^{-\delta|x-y|} \mathrm{d}y \mu_V(\mathrm{d}x). \tag{2.10}$$

Obviously, condition (2.8) does not hold for V in Corollaries 2.1-2.4(1), so that this result does not give us the Poincaré inequality (2.4). So, for this typical model our results are more efficient for the Poincaré inequality. On the other hand, however, our above results do not imply the stronger weighted Poincaré inequality (2.9). Thus, to cover and improve this result, we will consider in Section 4 the weighted functional inequalities, see Example 4.7 below for details.

We would like to mention that, besides the Poincaré-type inequalities stated in Theorem 2.1, entropy inequalities and Beckner-type inequalities for non-local Dirichlet forms $\mathcal{E}_{\alpha,\delta,V}$ are also established in [28] by using straightforward and elementary arguments.

3. Functional inequalities for truncated stable-like Dirichlet forms

This section is concerned with the following truncated stable-like Dirichlet form with small jumps only:

$$\mathcal{E}_{\alpha,V,\le 1}(f, g) := \iint_{\{|x-y|\le 1\}} \frac{(f(x) - f(y))(g(x) - g(y))}{|x - y|^{d+\alpha}} \mathrm{d}y\, \mu_V(\mathrm{d}x),$$

$$\mathcal{D}(\mathcal{E}_{\alpha,V,\le 1}) := \Big\{ f \in L^2(\mu_V) : \ \mathcal{E}_{\alpha,V,\le 1}(f, f) < \infty \Big\},$$

where μ_V is the probability measure as in Section 2, and $\alpha \in (0, 2)$. Comparing with the form $\mathcal{E}_{\alpha,V}$ studied in Section 2, this Dirichlet form is of finite range of jump and is thus more closer to the local ones. To see that conditions ensuring a functional inequality for these two Dirichlet forms are essentially different, we look at the corresponding concentration of the reference measure μ. According to Corollary 2.1, $\mathcal{E}_{\alpha,V}$ satisfies the super Poincaré inequality for a class of μ_V with $\mu_V(|\cdot| > r) = \mathrm{O}(r^{-\varepsilon})$ for some $\varepsilon > \alpha$ and r large enough (i.e. μ_V has an algebraic concentration). But as shown in Example 3.1 below that to ensure the super Poincaré inequality for $\mathcal{E}_{\alpha,V,\le 1}$ the concentration property of μ_V should be stronger than exponential as in the local setting addressed in [23, Section 6].

For any $r > 0$, define

$$k(r) = \inf_{|x|\leqslant r+1} \mathrm{e}^{-V(x)}, \quad K(r) = \sup_{|x|\leqslant r} \mathrm{e}^{-V(x)}.$$

Theorem 3.1. (Chen et al. [11])

(1) *If*

$$\liminf_{|x|\to\infty} \frac{\inf_{|x|-1\le|z|\leqslant|x|-1/2} e^{-V(z)}}{\sup_{|x|\le|z|\le|x|+1} e^{-V(z)}} > \frac{1}{\alpha} 2^{2d+1}(e+e^{1/2})(2^\alpha - 1),$$

then the Poincaré inequality

$$\mu_V(f^2) \leqslant C_1\mathcal{E}_{\alpha,V,\le 1}(f,f), \quad f\in\mathcal{D}(\mathcal{E}_{\alpha,V,\le 1}),\ \mu_V(f)=0 \tag{3.1}$$

holds for some constant $C_1 > 0$.

(2) *If*

$$\liminf_{|x|\to\infty} \frac{\inf_{|x|-1\le|z|\leqslant|x|-1/2} e^{-V(z)}}{\sup_{|x|\le|z|\le|x|+1} e^{-V(z)}} = \infty,$$

then there exist constants C_2, $C_3 > 0$ *such that the super Poincaré inequality*

$$\begin{aligned}\mu_V(f^2) \leqslant s\mathcal{E}_{\alpha,V,\le 1}(f,f) + \beta(s)\mu_V(|f|)^2,\\ s>0, f\in\mathcal{D}(\mathcal{E}_{\alpha,V,\le 1})\end{aligned} \tag{3.2}$$

holds for

$$\begin{aligned}\beta(s) :=& C_2\Big((1+s^{-d/\alpha})[\Lambda^{-1}(C_3s^{-1})]^{d+d^2/\alpha}\\ &\times [K(\Lambda^{-1}(C_3s^{-1}))]^{1+d/\alpha}[k(\Lambda^{-1}(C_3s^{-1}))]^{-2-d/\alpha}\Big),\end{aligned}$$

where $\Lambda(r) := \inf_{|x|\geqslant r}\left(e^{V(x)}\inf_{|x|-1\le|z|\leqslant|x|-1/2} e^{-V(z)}\right)$, $r>0$.

(3) *There is a constant* $C_4 > 0$ *such that the weak Poincaré inequality*

$$\begin{aligned}\mu_V(f^2) \leqslant C_4\alpha(s)\mathcal{E}_{\alpha,V,\le 1}(f,f) + s\|f\|_\infty^2,\\ s>0, f\in\mathcal{D}(\mathcal{E}_{\alpha,V,\le 1}), \mu_V(f)=0\end{aligned}$$

holds for

$$\alpha(s) := \inf\left\{\frac{r^{3d}K(r)}{k(r)} :\ \mu_V(B(0,r)^c) \leqslant \frac{s}{1+s}\right\}.$$

The following example shows that the criteria in Theorem 3.1 are qualitatively sharp.

Example 3.1. (Chen et al. [11]) (1) Consider $\mu_V(dx) = C_\lambda e^{-\lambda|x|}\,dx$ for $\lambda > 0$, where $C_\ll$ is the normalization constant. If

$$\lambda > \lambda_0 := 2\log\left[\frac{1}{\alpha}2^{2d+1}(e+e^{1/2})(2^\alpha-1)\right],$$

then the Poincaré inequality (3.1) holds.

(2) Let $\mu_{V_\delta}(dx) = C_\delta e^{-(1+|x|^\delta)}\,dx$ with $\delta > 0$. Then the super Poincaré inequality (3.2) holds if and only if $\delta > 1$, and in this case, it holds with

$$\beta(s) = c_1\exp\left(c_2\big(1+\log^{\frac{\delta}{(\delta-1)}}(1+1/s)\big)\right), \quad s>0 \tag{3.3}$$

for some constants $c_1, c_2 > 0$. Moreover, β in (3.3) is sharp for the super Poincaré inequality (3.2) to hold since the inequality does not hold if $\beta(s)$ satisfies

$$\lim_{s\to 0} \frac{\log \beta(s)}{\log^{\frac{\delta}{\delta-1}}(1+s^{-1})} = 0.$$

(3) For the probability measure $\mu_{V_\theta}(\mathrm{d}x) = C_\theta \mathrm{e}^{-|x|\log^\theta(1+|x|)}\,\mathrm{d}x$ with $\theta \in \mathbb{R}$, the super Poincaré inequality (3.2) holds if and only if $\theta > 0$, and in this case, it holds with

$$\beta(s) = c_3 \exp\left(1 + \mathrm{e}^{c_4 \log^{\frac{1}{\theta}}(1+1/s)}\right), \quad s > 0 \tag{3.4}$$

for some positive constants $c_3, c_4 > 0$. Moreover, β in (3.4) is sharp for the super Poincaré inequality (3.2) to hold since this inequality does not hold if $\beta(s)$ satisfies

$$\lim_{s\to 0} \frac{\log\log \beta(s)}{\log^{\frac{1}{\theta}}(1+s^{-1})} = 0.$$

Example 3.1 shows that the property of the probability measure μ_V fulfilling Poincaré inequalities and super Poincaré inequalities for $\mathcal{E}_{\alpha,V,\le 1}(f,f)$ is similar to that for the corresponding local Dirichlet form

$$\mathcal{D}_V(f,f) := \int_{\mathbb{R}^d} |\nabla f(x)|^2\,\mu_V(\mathrm{d}x),$$

e.g. see [25, Chapters 1 and 3]. On the other hand, from Example 3.1 we see that functional inequalities for $\mathcal{E}_{\alpha,V,\le 1}(f,f)$ requires weaker concentration of the reference measure than those for $\mathcal{D}_V(f,f)$. For instance, given the probability measure $\mu_{V_\delta}(\mathrm{d}x) = C_\delta \mathrm{e}^{-(1+|x|^\delta)}\,\mathrm{d}x$ with $\delta > 0$, Example 3.1(2) indicates that the measure μ_{V_δ} satisfies log-Sobolev inequalities for $\mathcal{E}_{\alpha,V_\delta,\le 1}(f,f)$ if $\delta > 1$; however, μ_{V_δ} satisfies log-Sobolev inequalities for $\mathcal{D}_{V_\delta}(f,f)$ only if $\delta \ge 2$, see e.g. [25, Chapters 3 and 5].

To illustrate Theorem 3.1, we consider the following example where V does not satisfy (2.8) as required in reference [18].

Example 3.2. (Chen et al. [11]) (1) Let $\mu_V(\mathrm{d}x) = C_\lambda \mathrm{e}^{-\lambda|x|}\,\mathrm{d}x$ with $\lambda > 0$. Then for any $\delta \ge 0$, there is a constant $c_5 := c_5(\delta,\lambda)$ such that the Poincaré inequality

$$\mu_V\big(f - \mu_V(f)\big)^2 \leqslant c_5 \mathcal{E}_{\alpha,\delta,V}(f,f), \quad f \in C_b^2(\mathbb{R}^d)$$

holds.

(2) Let $\mu_V(\mathrm{d}x) = C_\lambda \mathrm{e}^{-\lambda|x|\log(1+|x|)}\,\mathrm{d}x$ with $\lambda > 0$. Then for any $\delta \ge 0$, there is a constant $c_6 := c_6(\delta,\lambda)$ such that the following log-Sobolev inequality holds

$$\mu_V(f^2\log f^2) - \mu_V(f^2)\log\mu_V(f^2) \le c_6 \mathcal{E}_{\alpha,\delta,V}(f,f), \quad f \in C_b^2(\mathbb{R}^d).$$

4. Weighted functional inequalities for Lévy-type Dirichlet forms

As a continuation of Sections 2 and 3, we consider in this section the truncated stable Dirichlet form with exclusively large jumps of the type

$$\mathcal{E}_{\alpha,V,\geq 1}(f,g) := \iint_{\{|x-y|\geq 1\}} \frac{(f(x)-f(y))(g(x)-g(y))}{|x-y|^{d+\alpha}}\,\mathrm{d}y\,\mu_V(\mathrm{d}x),$$

$$\mathcal{D}(\mathcal{E}_{\alpha,V,\geq 1}) := \Big\{f\in L^2(\mu_V):\ \mathcal{E}_{\alpha,V,\geq 1}(f,f)<\infty\Big\},$$

where μ_V is the probability measure as in Section 2, and $\alpha\in(0,2)$.

Theorem 4.1. (Chen et al. [11]) (1) *If*

$$\liminf_{|x|\to\infty}\frac{\mathrm{e}^{V(x)}}{|x|^{d+\alpha}}>0, \tag{4.1}$$

then the weighted Poincaré inequality

$$\int (f(x)-\mu_V(f))^2\frac{\mathrm{e}^{V(x)}}{1+|x|^{d+\alpha}}\,\mu_V(\mathrm{d}x)\leqslant C_1\mathcal{E}_{\alpha,V,\geq 1}(f,f) \tag{4.2}$$

holds for some constant $C_1>0$ and all $f\in\mathcal{D}(\mathcal{E}_{\alpha,V,\geq 1})$. In particular, the Poincaré inequality

$$\mu_V(f^2)\leqslant C_2\mathcal{E}_{\alpha,V,\geq 1}(f,f),\quad f\in\mathcal{D}(\mathcal{E}_{\alpha,V,\geq 1}),\ \mu_V(f)=0 \tag{4.3}$$

holds for some constant $C_2>0$.

(2) *For any locally bounded function V, the following super Poincaré inequality*

$$\mu_V(f^2)\leq s\mathcal{E}_{\alpha,V,\geq 1}(f,f)+\beta(s)\mu_V(|f|)^2,\quad s>0,\ f\in\mathcal{D}(\mathcal{E}_{\alpha,V,\geq 1}) \tag{4.4}$$

does not hold for any rate function $\beta:(0,\infty)\to(0,\infty)$.

To see that condition (4.1) is sharp for the Poincaré inequality (4.3) to hold, let $\mu_V(\mathrm{d}x)=C_\varepsilon(1+|x|)^{-d-\varepsilon}\,\mathrm{d}x$ with $\varepsilon>0$. According to Corollary 2.1, the following Poincaré inequality

$$\mu_V(f^2)\leqslant c_1\mathcal{E}_{\alpha,V}(f,f),\quad f\in\mathcal{D}(\mathcal{E}_{\alpha,V}),\mu_V(f)=0$$

holds if and only if $\varepsilon\geq\alpha$. Note that $\mathcal{E}_{\alpha,V,\geq 1}(f,f)\leq\mathcal{E}_{\alpha,V}(f,f)$, which along with (4.1) indicates that for the above probability measure μ_V, the Poincaré inequality (4.3) holds if and only if $\varepsilon\geq\alpha$. On the other hand, according to the proof of Theorem 4.2 below and the fact that $\mathcal{E}_{\alpha,V,\geq 1}(f,f)\leq\mathcal{E}_{\alpha,V}(f,f)$, one can verify that the weight function $w(x)=\frac{\mathrm{e}^{V(x)}}{1+|x|^{d+\alpha}}$ in (4.2) is optimal.

According to Theorem 4.1, for locally bounded V the Dirichlet form $\mathcal{E}_{\alpha,V,\geq 1}$ never satisfies the super Poincaré inequality (4.4). Indeed, the local super Poincaré inequalities do not hold for $\mathcal{E}_{\alpha,V,\geq 1}$. Thus, to derive super Poincaré inequalities for non-local Dirichlet forms, small jumps play an essential role. This leads us to consider the weighted Poincaré inequality for $\mathcal{E}_{\alpha,V}$ and links to the super Poincaré inequality.

Theorem 4.2. (Chen et al. [10]) *If for some constant* $\alpha_0 \in (0, \alpha/2)$,

$$\liminf_{|x|\to\infty} \frac{e^{V(x)}}{|x|^{d+\alpha-\alpha_0}} = \infty, \tag{4.5}$$

then there exists a constant $C_3 > 0$ *such that the following weighted Poincaré inequality*

$$\int \big(f(x) - \mu_V(f)\big)^2 \frac{e^{V(x)}}{(1+|x|)^{d+\alpha}}\, \mu_V(dx) \le C_3 \mathcal{E}_{\alpha,V}(f,f) \tag{4.6}$$

holds for $f \in \mathcal{D}(\mathcal{E}_{\alpha,V})$.

The weight function $\omega(x) := \frac{e^{V(x)}}{(1+|x|)^{(d+\alpha)}}$ in (4.6) is optimal in the sense that this inequality fails if $\omega(x)$ is replaced by a positive function $\omega^*(x)$ such that $\liminf_{|x|\to\infty} \frac{\omega^*(x)}{\omega(x)} = \infty$. The following example shows that in some case, the weight function ω may satisfy $\lim_{|x|\to\infty} \omega(x) = 0$ so that the inequality (4.6) is weaker than the Poincaré inequality.

Example 4.1. (Chen et al. [10]) Let $\mu_V(dx) = C_\varepsilon (1+|x|)^{-d-\varepsilon}\, dx$ be a probability measure, where $\varepsilon, C_\varepsilon > 0$ are constants. According to Theorem 4.2, if $\varepsilon > \alpha/2$ then the weighted Poincaré inequality

$$\int \big(f(x) - \mu_V(f)\big)^2 \frac{1}{(1+|x|)^{\alpha-\varepsilon}}\, \mu_V(dx) \le c\mathcal{E}_{\alpha,V}(f,f), \quad f \in \mathcal{D}(\mathcal{E}_{\alpha,V})$$

holds for some constant $c > 0$. However, by Corollary 2.1(1), if $\varepsilon \in (\alpha/2, \alpha)$ then for any constant $C > 0$ the Poincaré inequality

$$\mu_V(f - \mu_V(f))^2 \le C\mathcal{E}_{\alpha,V}(f,f), \quad f \in \mathcal{D}(\mathcal{E}_{\alpha,V})$$

does not hold.

The next result shows that, if the weight function $\omega(x) \to \infty$ as $|x| \to \infty$, the weighted Poincaré inequality for $\mathcal{E}_{\alpha,V}$ implies the super Poincaré inequality. For any $r > 0$, define

$$h(r) = \inf_{|x|\le r} e^{V(x)}, \quad H(r) = \sup_{|x|\le r} e^{V(x)}.$$

Theorem 4.3. (Chen et al. [10]) *Let* $\omega : \mathbb{R}^d \to (0,\infty)$ *be a positive continuous function such that* $\lim_{|x|\to\infty} \omega(x) = \infty$. *Suppose that there is a constant* $C_4 > 0$ *such that the weighted Poincaré inequality*

$$\int_{\mathbb{R}^d} \big(f(x) - \mu_V(f)\big)^2 \omega(x)\, \mu_V(dx) \leqslant C_4 \mathcal{E}_{\alpha,V}(f,f), \quad f \in \mathcal{D}(\mathcal{E}_{\alpha,V}) \tag{4.7}$$

holds. Then the super Poincaré inequality

$$\mu_V(f^2) \le r\mathcal{E}_{\alpha,V}(f,f) + \beta(r)\mu_V(|f|)^2, \quad r > 0, \ f \in \mathcal{D}(\mathcal{E}_{\alpha,V}) \tag{4.8}$$

holds with

$$\beta(r)=\inf\bigg\{C_5H(t)^{2+d/\alpha}h(t)^{-1-d/\alpha}\big(1+s^{-d/\alpha}\big):\\ \frac{2C_4}{\inf\limits_{|x|\ge t}\omega(x)}+s\le r, t>1, s>0\bigg\}.$$

Let μ_V be the probability measure given in Example 4.1 with $\varepsilon>\alpha$. According to Theorem 4.3 and the weighted Poincaré inequality obtained in Example 4.1, we know that the super Poincaré inequality (4.8) holds for such μ_V and $\mathcal{E}_{\alpha,V}$ with

$$\beta(r)=c\Big(1+r^{-\frac{d}{\alpha}-\frac{(d+\varepsilon)(2\alpha+d)}{\alpha(\varepsilon-\alpha)}}\Big),$$

which coincides with the sharp one presented in Corollary 2.1(2).

Below we present a sufficient condition for the weighted Poincaré inequality to hold with arbitrary positive weight functions in the Dirichlet form.

Theorem 4.4. (Chen et al. [10]) *Let the function V satisfy*

$$\liminf_{|x|\to\infty}\frac{e^{V(x)}}{|x|^{d+\alpha}}>0, \tag{4.9}$$

and let $\omega:\mathbb{R}^d\to\mathbb{R}_+$ be a continuous and positive function. Then there exists a constant $C_6(\omega)>0$ such that the weighted Poincaré inequality

$$\begin{aligned}\int_{\mathbb{R}^d}\big(f(x)-\mu_V(f)\big)^2\frac{e^{V(x)}}{(1+|x|)^{d+\alpha}}\mu_V(dx)\\ \le C_6(\omega)\int_{\mathbb{R}^d}\omega(x)\,\mu_V(dx)\int_{\mathbb{R}^d}\frac{(f(y)-f(x))^2}{|y-x|^{d+\alpha}}\,dy\end{aligned} \tag{4.10}$$

holds for all $f\in C_0^\infty(\mathbb{R}^d)$.

Obviously, (4.9) implies (4.5) and thus also (4.6) due to Theorem 4.2. But (4.10) is much stronger than (4.6) since the weight function $\omega(x)>0$ is arbitrary. Such a result is impossible in the diffusion setting, since the Poincaré inequality for the weighted local Dirichlet form

$$\int_{\mathbb{R}^d}\omega(x)|\nabla f|^2(x)\mu_V(dx)$$

implies the exponential concentration of μ_V w.r.t. the intrinsic distance

$$\rho(x,0):=\sup\{|f(x)-f(0)|:\omega|\nabla f|^2\le 1\},$$

which is not true when $\omega(x)\to 0$ fast enough as $|x|\to\infty$.

Finally, to improve the main result in [18], we consider the weighted Poincaré inequalities for the Dirichlet form $\mathcal{E}_{\alpha,\delta,V}$.

Theorem 4.5. (Chen et al. [10]) *Suppose that for some constants $\delta > 0$ and $\alpha_0 \in (0,1)$,*

$$\limsup_{|x|\to\infty} \frac{e^{V(x)-\delta|x|}}{|x|^{d+\alpha-\alpha_0}} = \infty. \tag{4.11}$$

Then there exists a constant $C > 0$ such that the weighted Poincaré inequality

$$\int_{\mathbb{R}^d} \big(f(x) - \mu_V(f)\big)^2 \frac{e^{V(x)-\delta|x|}}{(1+|x|)^{d+\alpha}} \, \mu_V(dx) \leqslant C\mathcal{E}_{\alpha,\delta,V}(f,f) \tag{4.12}$$

holds for all $f \in C_0^\infty(\mathbb{R}^d)$.

The following example shows that Theorem 4.5 can be much better than [18, Theorem 1.2].

Example 4.2. (Chen et al. [10]) (1) Let $\delta > 0$ and $V(x) = \varepsilon(1+|x|)$ for some $\varepsilon > \delta$. Then (4.11) is fulfilled. By Theorem 4.5, the weighted Poincaré inequality (4.12) holds. But [18, Theorem 1.2] does not apply since (2.8) fails.

(2) Let $V(x) = 1 + |x|^2$. By Theorem 4.5, for any $\delta > 0$ there exists a constant $C > 0$ such that for all $f \in C_0^\infty(\mathbb{R}^d)$,

$$\int_{\mathbb{R}^d} \big(f(x) - \mu_V(f)\big)^2 \exp\Big(\frac{1}{2}(1+|x|^2)\Big)\mu_V(dx) \leq C\mathcal{E}_{\alpha,\delta,V}(f,f).$$

However, (2.9) only implies that there exist two constants δ, $C > 0$ such that for all $f \in C_0^\infty(\mathbb{R}^d)$,

$$\int_{\mathbb{R}^d} \big(f(x) - \mu_V(f)\big)^2 (1+|x|^\alpha)\mu_V(dx) \leq C\mathcal{E}_{\alpha,\delta,V}(f,f).$$

5. Perturbations of functional inequalities for non-local Dirichlet forms

To extend functional inequalities derived in previous sections to more general reference probability measures, we introduce in this section some perturbation results.

In the diffusion setting, sharp growth conditions have been presented in [3] for perturbations of super Poincaré and weak Poincaré inequalities. The aim of this section is to extend perturbation results derived in [3] to the non-local setting, so that combining with the existing sufficient conditions in previous sections we are able to establish functional inequalities for more general non-local Dirichlet forms.

To investigate perturbations of functional inequalities for the non-local Dirichlet form given in (1.3) using growth conditions, we let U be a locally bounded measurable function on $\mathbb{R}^d$ such that $\mu(e^U) = 1$, and denote $\mu_U(dx) = e^{U(x)}\mu(dx)$. Since for every $f \in \mathcal{A}$, $\Gamma(f,f)$ in (1.2) is a bounded measurable function, we have

$$\mathcal{E}_U(f,f) := \int_{\mathbb{R}^d} \Gamma(f,f)(x)\,\mu_U(dx) < \infty.$$

By the argument in [14, Example 1.2.6], the form

$$\mathcal{E}_U(f,g) := \iint_{\mathcal{R}^d\times\mathcal{R}^d} (f(x)-f(y))(g(x)-g(y))q(x,y)\mu(\mathrm{d}y)\mu_U(\mathrm{d}x)$$

defined for $f,g \in \mathcal{A}$ is closable in $L^2(\mu_U)$ and its closure $(\mathcal{E}_U, \mathcal{D}(\mathcal{E}_U))$ is a symmetric conservative Dirichlet form.

We shall assume that $\mathcal{E}$ satisfies a functional inequality and then search for conditions on U such that $\mathcal{E}_U$ satisfies the same type of functional inequality. In the following two subsections, we study perturbations of the super Poincaré inequality and the weak Poincaré inequality respectively.

5.1. *Perturbations of the super Poincaré inequality*

We consider the super Poincaré inequality

$$\mu(f^2) \le r\mathcal{E}(f,f) + \beta(r)\mu(|f|)^2, \quad r>0, f\in\mathcal{D}(\mathcal{E}), \tag{5.1}$$

where, by taking $\tilde\beta(r) := \inf_{s\in(0,r]}\beta(s)$ in place of $\beta(r)$, $\beta : (0,\infty)\to(0,\infty)$ is without loss of generality a decreasing function.

To establish a super Poincaré inequality for $\mathcal{E}_U$, we need the following quantities. For any $n\ge 1$ and $k\ge 1$, let

$$\begin{aligned}
K_{n,k}(U) &= \sup_{|x|\le n+2} U(x) - \inf_{|x|\le n+k+2} U(x),\\
J_{n,k}(U) &= \sup_{|x|\le n+1} U(x) - 2\inf_{|x|\le n+k+2} U(x),\\
\varepsilon_{n,k}(U) &= \sup_{m\ge n}\Big\{\beta^{-1}\big(1/[2\mu(|\cdot|>m-1)]\big)\mathrm{e}^{K_{m,k}(U)}\Big\},
\end{aligned}$$

where $\beta^{-1}(s) := \inf\{r>0 : \beta(r)\le s\}$ for $s>0$, with $\inf\emptyset := \infty$ by convention.

When the jump is of finite range, i.e. there exists $k_0>0$ such that $q(x,y)=0$ for $|x-y|>k_0$, we have the following result similar to [3, Theorem 3.1] for local Dirichlet forms.

Theorem 5.1. (Chen et al. [12]) *Assume that* (5.1) *holds and there exists* $k_0\ge 1$ *such that* $q(x,y)=0$ *for* $|x-y|>k_0$.

(1) *If* $\inf_{n\ge 1, k\ge k_0} \varepsilon_{n,k}(U)=0$, *then the super Poincaré inequality*

$$\mu_U(f^2)\le r\mathcal{E}_U(f,f)+\beta_U(r)\mu_U(|f|)^2, \quad r>0, \quad f\in\mathcal{D}(\mathcal{E}_U) \tag{5.2}$$

holds with

$$\beta_U(r) := \inf\Big\{(1+8\lambda r')\mathrm{e}^{J_{n,k}(U)}\beta(s) : s>0, r'\in(0,r], n\ge 1, k\ge k_0 \\ \text{such that } 8\varepsilon_{n,k}(U)+s\mathrm{e}^{K_{n,k}(U)}\le \frac{r'}{2+16\lambda r'}\Big\}.$$

(2) *If* $\inf_{n\geq 1,k\geq k_0} \varepsilon_{n,k}(U) < \infty$, *then the defective Poincaré inequality*

$$\mu_U(f^2) \leq C_1\mathcal{E}_U(f,f) + C_2\mu_U(|f|)^2, \quad f \in \mathcal{D}(\mathcal{E}_U) \tag{5.3}$$

holds for some constants $C_1, C_2 > 0$.

We note that according to [26, Corollary 1.2], if $\mathcal{E}_U$ is irreducible, i.e. $\mathcal{E}_U(f) = 0$ implies that f is constant, then the defective Poincaré inequality (5.3) implies the Poincaré inequality

$$\mu_U(f^2) \leq C\mathcal{E}_U(f,f), \quad f \in \mathcal{D}(\mathcal{E}_U), \mu_U(f) = 0$$

for some constant $C > 0$.

When the jump is of infinite range, we will need additional notation and assumptions to control the uniform norm appearing in the perturbed functional inequalities, which is an essentially different feature from the diffusion setting. For any $n, k \geq 1$ and $\delta > 1$, let

$$\begin{aligned}
Z_n(U) &= \sup_{|x|\leq n+1} U(x),\\
\zeta_n(U) &= \sup_{m\geq n} \Big\{\beta^{-1}\big(1/[2\mu(|\cdot| > m-1)]\big)\mathrm{e}^{Z_{m+1}(U)}\Big\},\\
t_{i,n,k}(\delta, U) &:= \beta^{-1}\Big(\frac{1}{4}\delta^i \mathrm{e}^{-J_{n,k}(U)}\Big).
\end{aligned}$$

Moreover, let

$$\begin{aligned}
\gamma_{n,k} &= \iint_{\{|x-y|>k,|y|\geq n-1\}} q(x,y)\mu(\mathrm{d}y)\mu(\mathrm{d}x),\\
\eta_{n,k} &= \iint_{\{|x|>n+k+2,|y|\leq n+1\}} q(x,y)\mu(\mathrm{d}y)\mu(\mathrm{d}x).
\end{aligned}$$

By (1.1), we see that $\gamma_{n,k} + \eta_{n,k} \downarrow 0$ as $n \uparrow \infty$ holds for any $k \geq 1$.

We assume

(A) There exist $\delta > 1$ and sequences $\{(n_i, k_i)\}_{i\geq 1} \subset \mathbb{N}^2$ such that $n_i \uparrow \infty$ and

A1 $\lim_{i\to\infty}(\varepsilon_{n_i,k_i}(U) + t_{i,n_i,k_i}(\delta, U)\mathrm{e}^{K_{n_i,k_i}(U)}) = 0$;

A2 $\sum_{i=1}^{\infty} \big\{\zeta_{n_i}(U)\gamma_{n_i,k_i} + t_{i,n_i,k_i}(\delta, U)\mathrm{e}^{Z_{n_i}(U)}\eta_{n_i,k_i}\big\} < \infty$.

We shall let I_δ denote the set of all sequences $\{(n_i, k_i)\} \subset \mathbb{N}^2$ such that $n_i \uparrow \infty$ and A1-A2 hold. Moreover, for any $r > 0$ and $\{n_i, k_i\} \in I_\delta$, let $D(r, \{(n_i, k_i)\})$ be the set of $j \in \mathbb{N}$ such that

$$\sup_{i\geq j}\big(8\varepsilon_{n_i,k_i}(U) + t_{i,n_i,k_i}(\delta, U)\mathrm{e}^{K_{n_i,k_i}(U)}\big) \leq \frac{1}{64\lambda} \wedge \frac{c(\delta)r}{16}, \quad i \geq j,$$

where $c(\delta) := \big(\frac{\sqrt{\delta}-1}{\delta}\big)^2$, and such that

$$\sum_{i=j}^{\infty}\Big(6\zeta_{n_i}(U)\gamma_{n_i,k_i} + t_{i,n_i,k_i}(\delta, U)\mathrm{e}^{Z_{n_i}(U)}\eta_{n_i,k_i}\Big)\delta^{i+2} \leq \frac{1}{256}.$$

By **(A)**, we see that for any $r > 0$ and $\{(n_i, k_i)\} \in I_\delta$, the set $D(r, \{(n_i, k_i)\})$ is non-empty.

Theorem 5.2. (Chen et al. [12]) *Assume that* (5.1) *holds.*

(1) *If* **(A)** *is satisfied, then the super Poincaré inequality* (5.2) *holds with*

$$\beta_U(r) := \inf\left\{2\delta^j : \{(n_i, k_i)\} \in I_\delta, \ j \in D(r, \{(n_i, k_i)\})\right\} < \infty, \quad r > 0.$$

(2) *If* A2 *is satisfied and* A1 *is replaced by the following weaker assumption*

$$\limsup_{i\to\infty}(\varepsilon_{n_i,k_i}(U) + t_{i,n_i,k_i}(\delta, U)) < \infty,$$

then the defective Poincaré inequality (5.3) *holds for some* $C_1, C_2 > 0$.

The following example shows that Theorem 5.2 is sharp in some specific situations.

Example 5.1. Let

$$q(x,y) = \frac{(1+|y|)^{d+\alpha}\log(1+|y|)}{|x-y|^{d+\alpha}}, \quad \mu(\mathrm{d}x) = \frac{c_{d,\alpha}\mathrm{d}x}{(1+|x|)^{d+\alpha}\log(1+|x|)},$$

where $\alpha \in (0, 2)$ and $c_{d,\alpha} > 0$ is the normalizing constant such that μ is a probability measure. Let U satisfy

$$-s\varepsilon \log\log(\mathrm{e} + |x|) - K \le U(x) \le (1-s)\varepsilon\log\log(\mathrm{e}+|x|) + K, \quad x \in \mathbb{R}^m \tag{5.4}$$

for some constants $\varepsilon \in (0, 1]$, $s \in [0, 1]$ and $K \in \mathbb{R}$ such that $\mu(\mathrm{e}^U) = 1$.

(1) If $\varepsilon < 1$ then (5.2) holds with

$$\beta_U(r) = \exp\left(C_1(1 + r^{-1/(1-\varepsilon)})\right), \tag{5.5}$$

for some constant $C_1 > 0$.

(2) β_U in (1) can not be replaced by any essentially smaller functions, i.e. when $U(x) = \varepsilon\log\log(\mathrm{e}+|x|) + K_0$ for some constant $K_0 \in \mathbb{R}$ such that $\mu(\mathrm{e}^U) = 1$, the estimate (5.5) is sharp in the sense that the super Poincaré inequality (5.2) does not hold if

$$\lim_{r\to 0} r^{1/(1-\varepsilon)}\log\beta_U(r) = 0.$$

(3) In (5.4) the constant K can not be replaced by any unbounded positive function, i.e. for any increasing function $\phi : [0,\infty) \to [0,\infty)$ with $\phi(r) \uparrow \infty$ as $r \uparrow \infty$, there exists U such that

$$\varepsilon\log\log(\mathrm{e}+|x|) - \phi(|x|) \le U(x) \le \varepsilon\log\log(\mathrm{e}+|x|) + \phi(|x|)$$

with $\mu(\mathrm{e}^U) < \infty$, but the super Poincaré inequality (5.2) with β_U given by (5.5) does not hold for $\mu_U(\mathrm{d}x) := \frac{\mathrm{e}^{U(x)}\mathrm{d}x}{\mu(\mathrm{e}^U)}$.

(4) If $\varepsilon = 1$, then $(\mathcal{E}_U, \mathcal{D}(\mathcal{E}_U))$ satisfies the Poincaré inequality

$$\mu_U(f^2) \le C\mathcal{E}_U(f,f) + \mu_U(f)^2, \quad f \in \mathcal{D}(\mathcal{E}_U) \tag{5.6}$$

with some constant $C > 0$.

Finally, we consider the following example to illustrate Theorem 5.1.

Example 5.2. Let

$$q(x,y) = \frac{e^{|y|^\kappa}}{|x-y|^{d+\alpha}} 1_{\{|x-y|\le 1\}}, \quad \mu(dx) = c_{d,\kappa} e^{-|x|^\kappa} dx$$

for some constants $0 < \alpha < 2$, $\kappa > 1$ and $c_{d,\kappa} \ge 1$ such that μ is a probability measure. Let U satisfy

$$-C_1(1 + |x|^{\theta-1}) - K \le U(x) \le C_1(1 + |x|^{\theta-1}) + K, \quad x \in \mathbb{R}^d \tag{5.7}$$

for some constants $\theta \in (1, \kappa]$, $C_1 > 0$ and $K \in \mathbb{R}$ such that $\mu(e^U) = 1$.

(1) If $\theta < \kappa$, then the super Poincaré inequality holds for $(\mathcal{E}_U, \mathcal{D}(\mathcal{E}_U))$ with

$$\beta_U(r) = C_2 \exp\left[C_2 \log^{\frac{\kappa}{\kappa-1}}(1 + r^{-1})\right]$$

for some positive constants $C_2 > 0$.

(2) Let $\theta = \kappa$. Then if $C_1 > 0$ is small enough, then the Poincaré inequality (5.6) holds for some constants $C > 0$.

To show that in Example 5.2(2) it is essential to assume that $C_1 > 0$ is small, we present below a counterexample inspired by [3, Proposition 5.1].

Proposition 5.1. (Chen et al. [12]) *In the situation of Example* 5.2, *let* $\theta = \kappa, \alpha \in (0,1)$ *and* $d = 1$. *Let*

$$U(x) = K_0 + L \sum_{n=1}^{\infty} 1_{[nH,(n+1)H)}(x)(n+1)^{\kappa-1}\Big(2n + 1 - \frac{2x}{H}\Big),$$

where $H > 4, L > \frac{\kappa H^\kappa}{H-2}$ *and* $K_0 \in \mathbb{R}$ *are constants such that* $\mu(e^U) = 1$. *Then* (5.7) *holds for some constant* $C_1 > 0$ *and* $K \in \mathbb{R}$; *however, for any* $C > 0$, *the Poincaré inequality* (5.6) *does not hold.*

Finally, we study perturbations for the super Poincaré inequality of non-local Dirichlet forms under a variation condition. It is well known that in the diffusion case the super Poincaré inequality is stable under Lipschitz perturbations (see [23, Proposition 2.6]). The following statement is to extend this result to the non-local setting using a variation condition on $\operatorname{supp} q := \{(x,y) : q(x,y) > 0\}$.

Theorem 5.3. (Chen et al. [12]) *Assume that* (5.1) *holds. If there exists a constant* $\kappa_1 > 0$ *such that* $\kappa_2 := \mu(e^{-2U}) < \infty$ *and*

$$|U(x) - U(y)| \le \kappa_1(1 \wedge d(x,y)), \quad (x,y) \in \operatorname{supp} q, \tag{5.8}$$

then (5.2) *holds for*

$$\beta_U(s) := \inf\Big\{16\kappa_2\beta(r)^3(4+\lambda\kappa_1^2 s') : \ s' \in (0,s], 0 < r \le \frac{s'\mathrm{e}^{-\kappa_1}}{4+\lambda\kappa_1^2 s'}\Big\}.$$

If the jump is of finite range, i.e. there is a constant $k \ge 1$ such that $q(x,y) = 0$ for $|x-y| > k$, then (5.8) holds for any Lipschitz function U. Therefore, the above theorem implies that the super Poincaré inequality is stable for all Lipschitiz perturbations as is known in the diffusion case. In particular, since the defective log-Sobolev inequality

$$\mu(f^2\log f^2) \le C_1\mathcal{E}(f) + C_2, \quad f \in \mathcal{D}(\mathcal{E}), \mu(f^2) = 1$$

holds for some $C_1, C_2 > 0$ if and only if the super Poincaré inequality (5.1) holds for $\beta(r) = \mathrm{e}^{c(1+r^{-1})}$ for some $c > 0$, see [24, Corllary 1.1] for $\delta = 1$, we conclude from Theorem 5.3 that the defective log-Sobolev inequality is stable under perturbations of Lipschitz functions U. See Example 3.1(3) with $\theta = 1$ for examples of μ and q having finite range of jumps such that the log-Sobolev inequality holds.

5.2. *Perturbations for the weak Poincaré inequality*

Suppose that the weak Poincaré inequality

$$\mu(f^2) \le \beta(r)\mathcal{E}(f,f) + r\|f\|_\infty^2, \quad r > 0, f \in \mathcal{D}(\mathcal{E}), \mu(f) = 0, \tag{5.9}$$

holds for some decreasing function $\beta : (0,\infty) \to (0,\infty)$. To derive the weak Poincaré inequality for $\mathcal{E}_U$ using growth conditions on U, for any $n, k \ge 1$ let

$$\begin{aligned}
\tilde{K}_{n,k}(U) &= \sup_{\rho\le n} U - \inf_{\rho\le n+k+1} U, \qquad \tilde{Z}_n(U) = \sup_{\rho\le n} U,\\
\tilde{\eta}_{n,k} &= \iint_{\{\rho(x)>n+k+1,\rho(y)\le n+1\}} q(x,y)\mu(\mathrm{d}y)\mu(\mathrm{d}x),\\
\tilde{\gamma}_k &= \iint_{\{d(x,y)>k\}} q(x,y)\mu(\mathrm{d}y)\mu(\mathrm{d}x).
\end{aligned}$$

It is clear that $\tilde{\eta}_{n,k} \le \tilde{\gamma}_k$. By (1.1) we have $\tilde{\eta}_{n,k} \downarrow 0$ as $n \uparrow \infty$ or $k \uparrow \infty$.

Theorem 5.4. (Chen et al. [12]) *Assume that the weak Poincaré inequality* (5.9) *holds. If for any* $\varepsilon > 0$

$$\inf_{n,k\ge 1} \mathrm{e}^{\tilde{Z}_n(U)}\beta\big(\varepsilon\mathrm{e}^{-\tilde{Z}_n(U)}\big)\big(\tilde{\eta}_{n,k} + \tilde{\gamma}_k + \mu(\rho > n-k)\big) = 0,$$

then

$$\mu_U(f^2) \le \beta_U(r)\mathcal{E}_U(f,f) + r\|f\|_\infty^2, \quad r > 0, f \in \mathcal{D}(\mathcal{E}_U), \mu_U(f) = 0 \tag{5.10}$$

holds with

$$\begin{aligned}
\beta_U(r) := \inf\Big\{&2\beta\Big(\frac{r}{8}\mathrm{e}^{-\tilde{Z}_n(U)}\Big)\mathrm{e}^{\tilde{K}_{n,k}(U)} : \ 6\mu_U(\rho > n)\\
&+ 2\mathrm{e}^{\tilde{Z}_n(U)}\beta\Big(\frac{r}{8}\mathrm{e}^{-\tilde{Z}_n(U)}\Big)\big(4\tilde{\eta}_{n,k} + \tilde{\gamma}_k + 4\lambda\mu(\rho > n-k)\big) \le \frac{r}{2}\Big\}\\
<&\infty, \qquad r > 0.
\end{aligned}$$

In the following example, $(\mathcal{E}, \mathcal{D}(\mathcal{E}))$ satisfies the Poincaré inequality, i.e. the weak Poincaré inequality (5.9) holds for a constant function β.

Example 5.3. Let

$$q(x,y) = \frac{(1+|y|)^{d+\alpha}}{|x-y|^{-(d+\alpha)}}, \quad \mu(\mathrm{d}x) = \frac{c_{d,\alpha}\mathrm{d}x}{(1+|x|)^{d+\alpha}}$$

for some constant $0 < \alpha < 2$, where $c_{d,\alpha}$ is a normalizing constant such that μ is a probability measure. Then $\mathcal{A} \supset C_0^2(\mathbb{R}^d)$, and according to Corollary 1.2(1), (5.9) holds for a constant rate function $\beta(r) \equiv \beta > 0$.

Now, let U satisfy

$$-s\varepsilon \log(1+|x|) - K \le U(x) \le (1-s)\varepsilon \log(1+|x|) + K, \quad x \in \mathbb{R}^d$$

for some constants $\varepsilon \in [0, \alpha)$, $s \in [0,1]$ and $K \in \mathbb{R}$ such that $\mu(\mathrm{e}^U) = 1$. Then the weak Poincaré inequality (5.10) holds with

$$\beta_U(r) = C\big(1 + r^{-\varepsilon/(\alpha-(1-s)\varepsilon)}\big) \tag{5.11}$$

for some constant $C > 0$.

Moreover, the assertion is sharp in the following two cases with $s = 0$.

(i) β_U in (5.11) is sharp, i.e. β_U can not be replaced by any essentially smaller functions: if

$$\lim_{r\to 0} r^{\varepsilon/(\alpha-\varepsilon)}\beta_U(r) = 0,$$

then the weak Poincaré inequality (5.10) does not hold.

(ii) The constant K can not be replaced by any unbounded functions: for

$$U(x) = \varepsilon \log(1+|x|) + \phi(|x|) + K_0,$$

where $\varepsilon \in [0, \alpha)$, $\phi : [0, +\infty) \to [0, +\infty)$ is an increasing function with $\phi(r) \uparrow \infty$ as $r \uparrow \infty$ such that $\mu(\mathrm{e}^{\varepsilon \log(1+|\cdot|)+\phi(|\cdot|)}) < \infty$, and $K_0 \in \mathbb{R}^d$ is such that $\mu(\mathrm{e}^U) = 1$, the weak Poincaré inequality (5.10) with the rate function β_U given by (5.11) does not hold.

Acknowledgements We would like to thank the referee for very useful comments and corrections.

References

[1] S. Albeverio, B. Rudiger, J.-L. Wu, *Invariant measures and symmetry property of Lévy type operator,* Pot. Anal. 13(2000), 147–168.

[2] D. Bakry, *L'hypercontractivité et son utilisation en théorie des semigroups,* Lecture Notes in Math. 1581, pp. 1–114, Springer, Berlin, 1994.

[3] D. Bakry, M. Ledoux, F.-Y. Wang, *Perturbations of functional inequalities using growth conditions,* J. Math. Pures Appl. 87(2007), 394–407.

[4] A. Bendikov, P. Maheux, *Nash type inequalities for fractional powers of non-negative self-adjoint operators*, Trans. Amer. Math. Soc. 359(2007), 3085–3097.

[5] R.M. Blumenthal, R.K. Getoor, *Some theorems on stable processes*, Trans. Amer. Math. Soc. 95(1960), 263–273.

[6] S.G. Bobkov, M. Ledoux, *Weighted Poincaré-type inequalities for Cauchy and other convex measures*, Ann. Probab. 37(2009), 403–427.

[7] Z.-Q. Chen, *Symmetric jump processes and their heat kernel estimates,* Sci. China Ser. A 52(2009), 1423–1445.

[8] P. Cattiaux, A. Guillin, L. Wu, *Some remarks on weighted Logarithmic Sobolev inequality,* Indiana Univ. Math. J. 60(2011), 1885–1904.

[9] M.-F. Chen, *Eigenvalues, Inequalities and Ergodic Theory,* Springer, Berlin, 2004.

[10] X. Chen, J. Wang, *Weighted Poincaré inequalities for non-local Dirichlet forms,* arXiv:1207.7140.

[11] X. Chen, J. Wang, *Functional inequalities for nonlocal Dirichlet forms with finite range jumps or large jumps,* Stoch. Proc. Appl. 124(2014), 123–153.

[12] X. Chen, F.-Y. Wang, J. Wang, *Perturbations of functional inequalities for Lévy type Dirichlet forms,* to appear in Forum Math.

[13] E. B. Davies, *Heat Kernels and Spectral Theory,* Cambridge University Press, Cambridge, 1989.

[14] M. Fukushima, Y. Oshima, M. Takeda, *Dirichlet Forms and Symmetric Markov Processes*, de Gruyter, Berlin, 2011, 2nd edn.

[15] I. Gentil, P. Maheux, *Nash-type inequalities, super-Poincaré inequalities for subordinated semigroups,* arXiv:1105.3095

[16] P. T. Gressman, *Fractional Poincaré and logarithmic Sobolev inequalities for measure spaces,* J. Funct. Anal. 265(2013), 867–889.

[17] L. Gross, *Logarithmic Sobolev inequalities and contractivity properties of semigroups,* Lecture Notes in Math. 1563, pp. 54–88, Springer, Berlin, 1993.

[18] C. Mouhot, E. Russ, Y. Sire, *Fractional Poincaré inequalities for general measures,* J. Math. Pures Appl. 95(2011), 72–84.

[19] M. Röckner, F.-Y. Wang, *Weak Poincaré inequalities and L^2-convergence rates of Markov semigroups,* J. Funct. Anal. 185(2001), 564–603.

[20] M. Röckner, F.-Y. Wang, *Harnack and functional inequalities for generalized Mehler semigroups,* J. Funct. Anal. 203(2003), 237–261.

[21] R. Schilling, R. Song, Z. Vondraček, *Bernstein Function, Theory and Applications,* de Gruyter, Berlin, 2012, 2nd edn.

[22] R. L. Schilling, J. Wang, *Functional inequalities and subordination: stability of Nash and Poincaré inequalities,* Math. Zeit. 272(2012), 921–936.

[23] F.-Y. Wang, *Functional inequalities for empty essential spectrum,* J. Funct. Anal. 170(2000), 219–245.

[24] F.-Y. Wang, *Functional inequalities, semigroup properties and spectrum estimates,* Infin. Dimens. Anal. Quant. Probab. Relat. Topics 3(2000), 263–295.

[25] F.-Y. Wang, *Functional Inequalities, Markov Processes and Spectral Theory,* Science Press, Beijing, 2005.

[26] F.-Y. Wang, *Criteria of spectral gap for Markov operators,* J. Funct. Anal. 266(2014), 2137–2152.

[27] F.-Y. Wang, J. Wang, *Functional inequalities for stable-like Dirichlet forms*, to appear in J. Theoret. Probab. arXiv:1205.4508

[28] J. Wang, *A simple approach to functional inequalities for non-local Dirichlet forms,* to appear in ESAIM: Probab. Stat. arXiv: 1306.2854.

Chapter 9

Additive functionals and push forward measures under Veretennikov's flow

Shizan Fang* and Andrey Pilipenko†

Dedicated to Professor Masatoshi Fukushima with admiration

In this work, we will be interested in the push forward measure $(\varphi_t)_*\gamma$, where φ_t is defined by the stochastic differential equation

$$d\varphi_t(x) = dW_t + \mathbf{a}(\varphi_t(x))dt, \quad \varphi_0(x) = x \in \mathbb{R}^m,$$

and γ is the standard Gaussian measure. We will prove the existence of density under the hypothesis that the divergence $\mathrm{div}(\mathbf{a})$ is not a function, but a signed measure belonging to a Kato class; the density will be expressed with help of the additive functional associated to $\mathrm{div}(\mathbf{a})$.

Contents

1. Introduction

Let $(X_t)_{t\geq 0}$ be a Brownian flow on $\mathbb{R}^m$, that is, $W_t = X_t - X_0$ is a standard Brownian motion; then for a function $u \in C^2(\mathbb{R}^m)$, Itô formula says that

$$u(X_t) - u(X_0) = \int_0^t \nabla u(X_s) \cdot dW_s + \int_0^t \frac{1}{2}\Delta u(X_s)\, ds. \tag{1.1}$$

In a celebrated paper [13] M. Fukushima extended a C^2 function u in (1.1) to a function u in the Sobolev space $H^1(\mathbb{R}^m)$; in order to reach this end, he used an additive functional $N_t^{[u]}$ of $X.$ to express the last term in (1.1), moreover he showed that $N_t^{[u]}/t$ tends to $\frac{1}{2}\Delta u$ in distribution sense.

*Institut de Mathematiques, Universite de Bourgogne, Dijon, France
†Institute of Mathematics NAS, Kiev, Ukraine

In this work, we will be concerned with the stochastic differential equation (SDE) on $\mathbb{R}^m$

$$d\varphi_t(x) = dW_t + \mathbf{a}(\varphi_t(x))dt, \quad \varphi_0(x) = x \in \mathbb{R}^m, \tag{1.2}$$

where $\mathbf{a} : \mathbb{R}^m \to \mathbb{R}^m$ is a measurable map.

The SDE (1.2), due to the non-degenerated noise W_t, makes illuminating difference with ordinary differential equations (ODE). In the context of ODE, the existence of a flow of quasi-invariant measurable maps associated to a vector field $\mathbf{a}$ on $\mathbb{R}^m$ belonging to Sobolev space, having a bounded divergence div($\mathbf{a}$), was established in a seminal paper by Di Perna and Lions in Ref. [8]; their result was extended later in Ref. [1] by L. Ambrosio to a vector field having only bounded variation regularity and bounded divergence (see also Ref. [7]).

There are various considerations to SDE (1.2). When $\mathbf{a}$ is bounded, it was proved by Veretennikov in Ref. [25] that there exists a unique strong solution $\varphi_t(x)$ to SDE (1.2). Moreover if $\mathbf{a}$ is Hölderian, it was proved in Ref. [12] as well as in Ref. [27] that $x \to \varphi_t(x)$ is a flow of diffeomorphisms. Recently, it was proved in Ref. [3] that if $\mathbf{a}$ is of bounded variation, and $\mu_{k,j} = \frac{\partial a_k}{\partial x_j}$ are signed measures satisfying (2.11) for all $k, j,$, then the solution φ_t to SDE (1.2) is in Sobolev space:

$$\varphi_t(\cdot) \in \cap_{p\geq 1} W^1_{p,loc}(\mathbb{R}^m, \mathbb{R}^m), t \geq 0.$$

Moreover, the Sobolev derivative $\nabla\varphi_t$ is a solution to the equation

$$\nabla\varphi_t = I + \int_0^t \bar{A}^\varphi(ds)\nabla\varphi_s(x), t \geq 0,$$

where $\bar{A}^\varphi$ is the additive functional associated to $\nabla\mathbf{a}$.

In Ref. [26], X. Zhang allowed $\mathbf{a}$ to be time-dependent, and established the existence of strong solutions under integrability conditions on the drift $\mathbf{a}$, while in Ref. [19], Krylov and Röckner considered such a SDE on a domain of $\mathbb{R}^m$ and established the existence of strong solutions. In another direction, in Ref. [4] Bass and Chen took the point of view of additive functionals

$$A^i_t = \int_0^t \mathbf{a}^i(\varphi_s(x))\, ds,$$

where $\mathbf{a}^i$ denotes the ith-component of $\mathbf{a}$, to generalize the drift $\mathbf{a}$; $\mathbf{a}^i(x)dx$ seen as the Revuz measure associated to A^i_t, was extended to the Kato class K_α for some $\alpha > 0$, where K_α is the class of signed measures on $\mathbb{R}^m$ defined by

$$K_\alpha = \Big\{\pi(dx);\ \overline{\lim_{\varepsilon\to 0}}\ \sup_{x\in\mathbb{R}^m} \int_{B(x,\varepsilon)} |x-y|^\alpha\ |\pi|(dy) = 0\Big\} \tag{1.3}$$

where $|\pi|$ denotes the total variation of π. More precisely, they proved that if the Revuz measures $\pi^1, \ldots, \pi^m$ are in K_{m-1} with $m \geq 3$, then with the help of associated additive functionals $A_t = (A_t^1, \ldots, A_t^m)$, the SDE

$$X_t = x + W_t + A_t$$

admits a unique weak solution. The interest of considering $\pi^1, \ldots, \pi^m$ in Kato class is that they are not necessarily absolutely continuous with respect to the Lebesgue measures. In the case where $\mathbf{a} = \nabla \log \rho$, by considering

$$\mathcal{E}(u,v) = \int_{\mathbb{R}^m} \nabla u \cdot \nabla v \; \rho(x) dx,$$

the theory of Dirichlet forms provides a powerful tool, which allows ρ to be only locally integrable (see Refs. [14, 15]).

In this work, we are interested in push-forward measures under the map $x \to \varphi_t(x)$ defined by SDE (1.2). It is well-known that if $\mathbf{a}$ is smooth and bounded, then $x \to \varphi_t(x)$ is a diffeomorphism of $\mathbb{R}^m$ and the inverse flow φ_t^{-1} can be expressed by SDE with reversed Brownian motion. More precisely, for $t > 0$ given, let $W_s^t = W(t-s) - W(t)$, and ψ_s^t solve the SDE

$$d\psi_s^t(x) = -\mathbf{a}(\psi_s^t(x))dt + dW_s^t, \; s \in [0,t], \; \psi_0^t(x) = x; \tag{1.4}$$

then $\varphi_t^{-1} = \psi_t^t$. Let γ be the standard Gaussian measure on $\mathbb{R}^m$. By Kunita [21] the push forward measure $(\varphi_t^{-1})_*\gamma$ admits the density $\tilde{K}_t$ with respect to γ given by

$$\tilde{K}_t(x) = \exp\Big(-\int_0^t \langle \varphi_s(x), \circ dW_s\rangle - \int_0^t \delta(\mathbf{a})(\varphi_s(x))\, ds\Big) \tag{1.5}$$

where $\circ dW_s$ means the stochastic integral in Stratanovich's sense, and $\delta(\mathbf{a})$ is the divergence with respect to γ, that is,

$$\int_{\mathbb{R}^m} \langle \nabla f, \mathbf{a}\rangle \, d\gamma = \int_{\mathbb{R}} f \delta(\mathbf{a})\, d\gamma \quad \text{for all } f \in C_0^1(\mathbb{R}^m).$$

We have $\delta(\mathbf{a})(x) = \langle \mathbf{a}, x\rangle - \mathrm{div}\,(\mathbf{a})$ so that

$$\int_0^t \delta(\mathbf{a})(\varphi_s(x))\, ds = \int_0^t \langle \mathbf{a}(\varphi_s(x)), \varphi_s(x)\rangle\, ds - \int_0^t \mathrm{div}\,(\mathbf{a})(\varphi_s(x))\, ds. \tag{1.6}$$

Here is the main result of this paper

Theorem 1.1. *Let* $\mathbf{a} : \mathbb{R}^m \to \mathbb{R}^m$ *be a bounded measurable map. Assume that the divergence* $\mathrm{div}\,(\mathbf{a})$ *in generalized sense is a signed measure* μ *satisfying the condition*

$$\lim_{t\to 0} \sup_{x\in\mathbb{R}^m} \int_{\mathbb{R}^m} \Big(\int_0^t s^{-m/2} e^{-|x-y|^2/(2s)}\, ds\Big)\, |\mu|(dy) = 0, \tag{1.7}$$

where $|\mu|$ *denotes the total variation of* μ. *Let* φ_t *be given by SDE* (1.2); *then almost surely the push forward measure* $(\varphi_t)_*\gamma$ *is equivalent to* γ, *and the density* $\tilde{K}_t$ *of the push forward measure* $(\varphi_t^{-1})_*\gamma$ *with respect to* γ *has the expression*

$$\tilde{K}_t(x) = \exp\Big(-A_t + \int_0^t \langle \mathbf{a}(\varphi_s(x)), \varphi_s(x)\rangle\, ds - \int_0^t \langle \varphi_s(x), dW_s\rangle - \frac{mt}{2}\Big), \tag{1.8}$$

where A_t *is the additive functional associated to* div $(\mathbf{a})$.

Notice that if f is a positive function in the Kato class K_{m-2}, then (see Ref. [4]),

$$\lim_{t\to 0} \sup_{x\in\mathbb{R}^m} \mathbb{E}\Big(\int_0^t f(W_s + x)\, ds\Big) = 0,$$

that is nothing but (1.7) for $\mu(dy) = f(y)dy$.

The organization of the paper is as follows. In section 2, we will recall and collect some facts concerning continuous additive functionals. Section 3 is devoted to the proof of Theorem 1.1. In section 4, we will consider some examples of vector fields $\mathbf{a}$ whose divergence div $(\mathbf{a})$ is a signed measure, but is not absolutely to Lebesgue measure. In section 5, we will discuss briefly generalizations of Theorem 1.1.

2. Continuous additive functionals

In this section, we recall some definitions and facts about continuous additive functionals of Markov processes. There are a lot of publications in the litterature on this topic, see for example Refs. [9, 14, 16, 17, 23, 24]. Here we will follow Chapters 6 - 8 in Ref. [9] Chapter II, section 6 in Ref. [17]. We don't need the theory on the whole generality; so some assumptions, statements or definitions are simplified in our exposition.

Let $\{X_t, t \geq 0\}$ be a continuous $\mathbb{R}^m$-valued homogeneous Markov processes adapted to a filtration $\{\mathcal{F}_t, t \geq 0\}$ with infinite life-time, P_x be the distribution of X given $X_0 = x$. Denote $\mathcal{N}_t = \sigma(X_s, s \in [0, t])$.

Definition 2.1. A non negative additive functional of X is a $\mathbb{R}_+$-valued, $\mathcal{N}_t$-adapted process $A = \{A_t(X), t \geq 0\}$ such that

1) it is almost surely continuous in t and $A_0(X) = 0$;

2) it is additive, i.e. $\forall t \geq 0\ \forall s \geq 0\ \forall x \in \mathbb{R}^m$:

$$A_{t+s}(X) = A_s(X) + A_t(\theta_s X), \quad P_x\text{-a.s.},$$

where θ_s is the shift operator.

Following the terminology of Dynkin [9] , we introduce the notion of W-functional.

Definition 2.2. A non negative continuous additive functional $A_t(X)$ is called W-functional if

$$\forall t \geq 0 : \ \sup_x \mathbb{E}_x(A_t(X)) < \infty. \tag{2.1}$$

The function $f_t(x) = \mathbb{E}_x(A_t(X))$ is called the characteristics of $A_t(X)$.

Here is an obvious example

Example 2.1. Let $b : \mathbb{R}^m \to [0, \infty)$ be a bounded measurable function. Then

$$A_t(X) := \int_0^t b(X_s)ds \tag{2.2}$$

is a W-functional of X.

Assume that for any $t > 0$, X_t has a transition density $p(t, x, y)$. Then the characteristics of $A_t(X)$ defined in (2.2) is equal to

$$\begin{aligned} f_t(x) = \mathbb{E}_x \int_0^t b(X_s)ds &= \int_0^t \int_{\mathbb{R}^m} b(y)p(s, x, y)dyds \\ &= \int_{\mathbb{R}^m} \Big(\int_0^t p(s, x, y)ds \Big) b(y)dy. \end{aligned} \tag{2.3}$$

There are close relations between convergence of W-functionals and their characteristics. The first one is the following

Proposition 2.1. (see Ref. Dynkin, [9, Theorem 6.3]) *A W-functional is defined by its characteristics uniquely up to the equivalence.*

The second one concerns the convergence, that is,

Theorem 2.1. (Dynkin, [9, Theorem 6.4, Lemma 6.1′]) *Let $\{A_t^{(n)}(X)\}$ be a sequence of W-functionals of X and $f_t^{(n)}(x) = \mathbb{E}_x\big(A_t^{(n)}(X)\big)$ be their characteristics. Assume that a function $f_t(x)$ is such that for each $t > 0$*

$$\lim_{n\to\infty} \sup_{0\le s\le t} \sup_{x\in \mathbb{R}^m} |f_s^{(n)}(x) - f_s(x)| = 0. \tag{2.4}$$

Then $f_t(x)$ is the characteristics of a W-functional $A_t(X)$. Moreover, for each $t > 0$,

$$\lim_{n\to+\infty} \mathbb{E}_x(|A_t(X) - A_t^{(n)}(X)|^2) = 0,$$

and in probability,

$$\lim_{n\to+\infty} \sup_{s\in[0,t]} |A_s^{(n)}(X) - A_s(X)| = 0.$$

Example 2.2. Let $\{X_t = B_t,\ t \ge 0\}$ be a one-dimensional Brownian motion; set

$$A_t^{(n)} := \int_0^t 2n\mathbf{1}_{B(s)\in\left[-\frac{1}{n},\frac{1}{n}\right]} ds.$$

Then the function b in expression (2.2) is equal to $2n\mathbf{1}_{\{|x|\le 1/n\}}$ and converges to the Dirac mass δ_0 at 0. It is easy to verify that (2.4) holds with

$$f_t(x) = \int_{\mathbb{R}} \int_0^t p(s, x, y)ds\delta_0(dy) = \int_0^t p(s, x, 0)ds,$$

where

$$p(s, x, y) = \frac{1}{\sqrt{2\pi s}} \exp\Big\{ \frac{|x - y|^2}{2s} \Big\}$$

is the transition density of a Brownian motion. The limiting additive functional is the local time of a Brownian motion at 0. Now let us write (2.3) as

$$f_t(x) = \int_{\mathbb{R}^d} \Big(\int_0^t p(s,x,y)ds\Big)\mu(dy), \tag{2.5}$$

with $\mu(dy) = a(y)dy$. Note that representation (2.5) makes a sense even μ is not absolutely continuous with respect to Lebesgue measure.

Similarly to Example 2.2, sometimes it is possible to assign a W-functional to a measure. For example, assume that there exists a sequence of non-negative bounded, measurable functions $\{b_n, n \geq 1\}$ such that for $t > 0$

$$\lim_{n\to 0} \sup_x \Big| \int_{\mathbb{R}^m} \Big(\int_0^t p(s,x,y)ds\Big) (\mu_n(dy) - \mu(dy))\Big| = 0,$$

where $\mu_n(dy) = b_n(y)dy$. Then a function $f_t(x)$ defined in (2.5) is the characteristic of a W-functional. We will formally denote it by

$$A_t := \int_0^t \frac{d\mu}{dy}(X_s)ds.$$

If there are some *a priori* estimates on the transition density of X_t, then using the described approach it is possible to characterize a class of measures corresponding to its W-functionals. See for example Ref. [9], Ch.8 for W-functionals of a Brownian motion.

Let's come back to SDE (1.2). It is known in Ref. [2] that the transition density of $\varphi_t(x)$ exists and there are constants $c_1, c_2 > 0$ depending only on $\sup_x |\mathbf{a}(x)|$ such that $\forall t \in (0,T]$,

$$c_1^{-1} t^{-m/2} \exp\Big\{ - \frac{|x-y|^2}{c_2 t}\Big\} \leq p(t,x,y) \leq c_1 t^{-m/2} \exp\Big\{ - \frac{c_2|x-y|^2}{t}\Big\}. \tag{2.6}$$

Observe that (Ref. [9] Ch.8)

$$\int_0^t p(s,x,y)ds \asymp \omega(|x-y|),$$

where

$$\omega(r) = \begin{cases} 1, m = 1, \\ \ln(r-1) \vee 1, m = 2, \\ r^{2-m}, m \geq 3, \end{cases}$$

More precisely, for each $t > 0$, there exists a positive constant C such that for all $x \neq y \in \mathbb{R}^m$ with $m > 1$,

$$C^{-1}\omega(|x-y|) \leq \int_0^t p(s,x,y)ds \leq C\omega(|x-y|). \tag{2.7}$$

So, a function $f_t(x)$ defined in (2.5) is finite if and only if $\int_{\mathbb{R}^m} \omega(x-y)\mu(dy) < \infty$. Hence, assumption (2.1) is equivalent to

$$\sup_x \int_{\mathbb{R}^m} \omega(|x-y|)\mu(dy) < \infty. \tag{2.8}$$

Assume that (2.8) is satisfied. It follows from (Ref. [9] , Theorem 6.6) that condition

$$\lim_{t\to 0} \sup_x f_t(x) = 0 \tag{2.9}$$

ensures that $f_t(x) = \int_{\mathbb{R}^m} \left(\int_0^t p(s,x,y)ds\right)\mu(dy)$ is a characteristic of W-functional. It follows from (2.6) that (2.9) is equivalent to

$$\lim_{t\to 0} \sup_x \int_{\mathbb{R}^m} \Big(\int_0^t s^{-m/2} \exp\Big\{-\frac{|x-y|}{2s}\Big\}ds\Big)\mu(dy) = 0. \tag{2.10}$$

Remark 2.1. If μ satisfies (2.10), then μ satisfies (2.8).

Now we deal with signed additive functionals.

Definition 2.3. We say that $A_t(X)$ is a signed continuous additive functional if it has the decomposition $A_t(X) = A_t^+(X) - A_t^-(X)$, where $\{A_t^{\pm}(X), t \geq 0\}$ are continuous non negative additive functionals of X.

For a signed measure $\mu = \mu^+ - \mu^-$ such that

$$\lim_{t\to 0} \sup_x \int_{\mathbb{R}^m} \Big(\int_0^t s^{-m/2} \exp\Big\{-\frac{|x-y|^2}{2s}\Big\}ds\Big)|\mu|(dy) = 0, \tag{2.11}$$

where $|\mu|$ is the total variation of μ, we can construct a signed W-functional $A_t = A_t^+ - A_t^-$, where functionals A_t^+, A_t^- correspond to μ^+, μ^- respectively (see Ref. [9]).

3. Proof of Theorem 1.1

Let $\mathbf{a}$ be a bounded measurable vector field on $\mathbb{R}^m$.

Definition 3.1. We say that a signed measure μ on $\mathbb{R}^m$ is the divergence in a generalized sense of $\mathbf{a}$ if for any test function $g \in C_0^\infty(\mathbb{R}^m)$:

$$\int_{\mathbb{R}^m} \langle \mathbf{a}(x), \nabla g(x)\rangle dx = -\int_{\mathbb{R}^m} g(x)\mu(dx),$$

where dx on the left hand side denotes the Lebesgue measure; we denote $\mu = \operatorname{div}(\mathbf{a})$.

In what follows, we will assume that $\operatorname{div}(\mathbf{a})$ exists and satisfies condition (1.7).

Let $\{g_n, n \geq 1\} \subset C_0^\infty(\mathbb{R}^m)$ be a sequence of non-negative smooth functions with compact support such that

$$\int_{\mathbb{R}^m} g_n(x)dx = 1, \text{ and } g_n(x) = 0 \text{ for } |x| > \frac{1}{n}.$$

Put

$$\mathbf{a}_n(x) := \mathbf{a} * g_n(x) = \int_{\mathbb{R}^m} \mathbf{a}(x-y) g_n(y) dy. \tag{3.1}$$

Note that $\mathbf{a}_n \in C^\infty(\mathbb{R}^m, \mathbb{R}^m)$,

$$\|\mathbf{a}_n\|_\infty = \sup_x |\mathbf{a}_n(x)| \le \sup_x |\mathbf{a}(x)| = \|\mathbf{a}\|_\infty, \tag{3.2}$$

and $\mathbf{a}_n$ converges to $\mathbf{a}$ in all $L^p_{loc}(\mathbb{R}^m, \mathbb{R}^m)$. Without loss of generality we may assume that as $n \to +\infty$,

$$\mathbf{a}_n(x) \to \mathbf{a}(x) \text{ for almost everywhere } x \in \mathbb{R}^m. \tag{3.3}$$

Let $\varphi_t^n(x)$ be the stochastic flow of diffeomorphisms defined by

$$d\varphi_t^n(x) = dW_t + \mathbf{a}_n(\varphi_t^n(x))dt, \quad \varphi_0^n(x) = x \in \mathbb{R}^m. \tag{3.4}$$

Let γ be the standard Gaussian measure on $\mathbb{R}^m$. We set

$$K_t^n(x) = \frac{d(\varphi_t^n)_*\gamma}{d\gamma}, \quad \tilde{K}_t^n(x) = \frac{d(\varphi_t^n)_*^{-1}\gamma}{d\gamma}.$$

It is well-known (see Ref. [21]) that

$$K_t^n(\varphi_t^n(x)) = \frac{1}{\tilde{K}_t^n(x)}, \tag{3.5}$$

and

$$\begin{aligned}\tilde{K}_t^n(x) &= \exp\Big\{ -\int_0^t (\delta\mathbf{a}_n)(\varphi_s^n(x))ds - \int_0^t \langle \varphi_s^n(x), \circ dW_s\rangle \Big\} \\ &= \exp\Big\{ -\int_0^t (\delta\mathbf{a}_n)(\varphi_s^n(x))ds - \int_0^t \langle \varphi_s^n(x), dW_s\rangle - \frac{mt}{2}\Big\},\end{aligned} \tag{3.6}$$

where $\delta\mathbf{a}_n(x) = (\operatorname{div}\mathbf{a}_n)(x) - \langle \mathbf{a}_n(x), x\rangle$. In Ref. [10], the L^p estimates on densities were established and used to prove the absolute continuity for a limit of push-forward measures. Here we will use the following result of Gikhman and Skorokhod Ref. [18].

Theorem 3.1. (see Ref. [18]) *Let $(X_1, \mathcal{F}, \mu_1)$ be a probability space, X_2 be a complete separable metric space, μ_2 be a probability measure on the Borel σ-algebra $\mathcal{B}(X_2)$.*

Assume that a sequence of measurable mappings $\{F_n : X_1 \to X_2, n \ge 0\}$ is such that

1) as $n \to +\infty$, F_n converges to F_0 in measure μ_1;

2) for all $n \ge 1$, the push forward measure $(F_n)_*\mu_1$ is absolutely continuous with respect to μ_2;

3) the sequence of the densities $\left\{\rho_n := \frac{d(F_n)_*\mu_1}{d\mu_2}, n \geq 1\right\}$ *is uniformly integrable with respect to* μ_2.
Then the push forward measure $(F_0)_*\mu_1$ *is absolutely continuous with respect to* μ_2. *Moreover, if* ρ_n *converges to* ρ *in measure* μ_2*, then* $\rho = \frac{d(F_0)_*\mu_1}{d\mu_2}$.

Let us apply Theorem 3.1 to the sequence $\{\varphi_t^n, n \geq 1\}$. First of all, we remark that although in Ref. [22] D. Luo assumed that the drift admits the divergence as a function satisfying some integrability condition, in the proof of Theorem 3.4 in Ref. [22], he only used Krylov estimate for non-degenerated diffusions, without involving the divergence. Since $\mathbf{a}_n$ converges to $\mathbf{a}$ in all L^p_{loc}, we can use directly Theorem 3.4 in Ref. [22] to get that for each $x \in \mathbb{R}^m$, we have

$$\lim_{n\to+\infty} \mathbb{E}\Big(\sup_{t\in[0,T]} |\varphi_t^n(x) - \varphi_t(x)|\Big) = 0. \tag{3.7}$$

Applying Fubini's theorem and choosing a subsequence if necessary we get

$$P\Big(\big\{\omega;\ \text{for } \gamma \text{ almost surely } \ x,\ \lim_{n\to\infty} \sup_{t\in[0,T]} |\varphi_t^n(x) - \varphi_t(x)| = 0\big\}\Big) = 1. \tag{3.8}$$

It follows that for almost surely ω, for all $t \in [0,T]$, φ_t^n converges to φ_t in measure with respect to γ.

Next, we will establish the uniform integrability of $\{K_t^n;\ n \geq 1\}$.

Proposition 3.1. *We have*

$$\sup_{n\geq 1} \mathbb{E}\Big(\sup_{t\in[0,T]} \int_{\mathbb{R}^m} K_t^n(x)|\ln K_t^n(x)|\gamma(dx)\Big) < +\infty. \tag{3.9}$$

Proof. We have

$$\int_{\mathbb{R}^m} K_t^n(x)|\ln(K_t^n(x))|\, d\gamma(x) = \int_{\mathbb{R}^m} |\ln(K_t^n(\varphi_t^n(x)))|\, d\gamma(x).$$

But by (3.5) and (3.6), we have

$$\begin{aligned}\ln(K_t^n(\varphi_t^n(x))) &= \int_0^t (\mathrm{div}\,(\mathbf{a}_n)(\varphi_s^n(x))\, ds - \int_0^t \langle \mathbf{a}_n(\varphi_s^n(x)), \varphi_s^n(x)\rangle\, ds \\ &\quad + \int_0^t \langle \varphi_s^n(x), dW_s\rangle + \frac{mt}{2}.\end{aligned}$$

Let $T > 0$ be fixed; then

$$\mathbb{E}\Big(\sup_{t\in[0,T]}\int_{\mathbb{R}^m} K_t^n(x)|\ln K_t^n(x)|\gamma(dx)\Big)$$
$$\leq \int_{\mathbb{R}^m}\Big[\mathbb{E}\big(\sup_{t\in[0,T]}|\int_0^t (\operatorname{div}\mathbf{a}_n)(\varphi_s^n(x))ds|\big)+\mathbb{E}\big(\sup_{t\in[0,T]}|\int_0^t\langle\varphi_s^n(x),dW_s\rangle|\big) \tag{3.10}$$
$$+\mathbb{E}\big(\int_0^T|\mathbf{a}_n(\varphi_s^n(x))|\,|\varphi_s^n(x)|ds\big)+mT\Big]\,d\gamma(x).$$

By (3.2), it is well-known that there exists a constant $c_0>0$ independent of n such that

$$\sup_n \mathbb{E}\Big(\sup_{t\in[0,T]}|\varphi_t^n(x)|\Big)\leq c_0(1+|x|). \tag{3.11}$$

Using Burkholder's inequality and (3.11), we also have

$$\sup_n \mathbb{E}\Big(\sup_{t\in[0,T]}|\int_0^t\langle\varphi_s^n(x),dW_s\rangle|\Big)\leq c_0(1+|x|). \tag{3.12}$$

Let us estimate $\mathbb{E}\Big(\sup_{t\in[0,T]}|\int_0^t(\operatorname{div}\mathbf{a}_n)(\varphi_s^n(x))ds|\Big)$.
Denote

$$\mu(dy)=(\operatorname{div}\mathbf{a})(dy),\;\;\mu_n(dy)=\operatorname{div}\mathbf{a}_n(y)dy.$$

We have

$$\operatorname{div}(\mathbf{a}_n)=\operatorname{div}(\mathbf{a}*g_n)=\operatorname{div}\mathbf{a}*g_n=\mu*g_n.$$

Let

$$A_n(t)=\int_0^t(\operatorname{div}\mathbf{a}_n)(\varphi_s^n(x))ds;$$

then A_n is a signed additive functional of φ^n. Let $p_n(t,x,y)$ be the transition density of $\omega\to\varphi_t^n(x,\omega)$. By (2.6), there are two constants $c_1,c_2>0$ independent of n such that

$$\frac{e^{-|x-y|^2/c_2 2t}}{c_1 t^{m/2}}\leq p_n(t,x,y)\leq\frac{c_1 e^{-c_2|x-y|^2/2t}}{t^{m/2}}.$$

So there exists a constant $\beta>0$ independent of n such that

$$\int_0^t p_n(s,x,y)\,ds\leq\beta\int_0^{t/c_2}\frac{e^{-|x-y|^2/2s}}{s^{m/2}}\,ds.$$

Let

$$k_t(r)=\int_0^{t/c_2}\frac{e^{-r^2/2s}}{s^{m/2}}\,ds.$$

It follows that

$$\begin{aligned}\mathbb{E}(|A_n(t)|) &\leq \int_{\mathbb{R}^m}\int_0^t p_n(s,x,y)|\mathrm{div}\,\mathbf{a}_n(y)|\,dsdy \\ &\leq \beta\int_{\mathbb{R}^m} k_t(|x-y|)|\mu_n|(dy).\end{aligned} \tag{3.13}$$

We have

$$|\mathrm{div}\,(\mathbf{a}_n)(y)| \leq \int_{\mathbb{R}^m} g_n(y-z)\,|\mu|(dz),$$

so that

$$\begin{aligned}\int_{\mathbb{R}^m} k_t(|x-y|)|\mu_n(y)|\,dy &\leq \int_{\mathbb{R}^m\times\mathbb{R}^m} k_t(|x-y|)g_n(y-z)\,|\mu|(dz)dy \\ &= \int_{\mathbb{R}^m}\Big(\int_{\mathbb{R}^m} k_t(|x+z-y|)g_n(y)\,|\mu|(dz)\Big)\,dy \leq \sigma(t),\end{aligned}$$

where

$$\sigma(t) = \sup_{x,y}\int_{\mathbb{R}^m} k_t(|x+z-y|)\,|\mu|(dz).$$

Then by condition (1.7), $C_T := \sup_{t\in[0,T]}\sigma(t) < +\infty$. Now combining this with (3.11) and (3.12), and by (3.10), we finally obtain (3.9). □

Now by Fatou's lemma,

$$\begin{aligned}&\mathbb{E}\left(\underline{\lim}_{n\to\infty}\sup_{t\in[0,T]}\int_{\mathbb{R}^m} K_t^n(x)|\ln K_t^n(x)|\nu(dx)\right) \\ &\leq \underline{\lim}_{n\to\infty}\mathbb{E}\Big(\sup_{t\in[0,T]}\int_{\mathbb{R}^m} K_t^n(x)|\ln K_t^n(x)|\nu(dx)\Big) < \infty.\end{aligned}$$

So, for almost surely ω :

$$\underline{\lim}_{n\to\infty}\sup_{t\in[0,T]}\int_{\mathbb{R}^m} K_t^n(x)|\ln K_t^n(x)|\nu(dx) < \infty.$$

Hence for almost surely ω, there exists a random subsequence $\{n_k\}$ such that

$$\sup_k\sup_{t\in[0,T]}\int_{\mathbb{R}^m} K_t^{n_k}(x)|\ln K_t^{n_k}(x)|\nu(dx) < \infty. \tag{3.14}$$

Now we can apply Theorem 3.1 to conclude that for almost surely ω, and for all $t\in[0,T]$, the push-forward measure $(\varphi_t)_*\gamma$ is absolutely continuous with respect to γ. Actually it remains to prove that $(\varphi_t)_*\gamma$ is equivalent to γ.

Proposition 3.2. *The map $x\to\varphi_t(x)$ admits an inverse map $x\to\psi_t(x)$, which is given by the reserved SDE*

$$d\psi_s(x) = dW_s^t - \mathbf{a}(\psi_s(x))\,ds, \quad \psi_0(x) = x, \quad s\in[0,t]. \tag{3.15}$$

Proof. For each n, the inverse map of $x \to \varphi_t^n(x)$ is given by ψ_t^n where ψ_s^n solves

$$d\psi_s^n(x) = dW_s^t - \mathbf{a}(\psi_s^n(x))\,ds, \quad \psi_0^n(x) = x, \quad s \in [0,t].$$

In the same way, we have

$$\lim_{n\to+\infty} \mathbb{E}\Big(\sup_{s\in[0,t]} |\psi_s^n(x) - \psi_s(x)|\Big) = 0.$$

In order to prove that ψ_t is the inverse map of φ_t, we will use the following result

Lemma 3.1. *Let X, Y be complete, separable metric spaces, ν be a finite measure on X.*
Assume that a sequence of X-valued random elements $\{\xi_n, n \geq 0\}$ and a sequence of measurable functions $f_n : X \to Y$ are such that

1) $\xi_n \to \xi_0$ in probability, as $n \to \infty$;
2) $f_n \to f_0$ in measure with respect to ν, as $n \to \infty$;
3) the push forward measure $(\xi_n)_* P$ is absolutely continuous with respect to ν;
4) the sequence of densities $\Big\{\frac{d(\xi_n)_*P}{d\nu}, n \geq 1\Big\}$ is uniformly integrable with respect to ν.

Then as $n \to +\infty$

$$f_n(\xi_n) \to f_0(\xi_0) \quad \textit{in probability.}$$

We refer to Corollary 9.9.11 in Ref. [5], as well as to Lemma 2 in Ref. [20] for a proof. □

Proof (continued) of Proposition 3.2. For almost surely ω, up to a subsequence, the family of densities $\{K_t^n;\ n \geq 1\}$ is uniformly integrable. In Lemma 3.1, we take $X = Y = \mathbb{R}^m$, $\xi_n = \varphi_t^n$, $f_n = \psi_t^n$. Then $\varphi_t^n(\psi_t^n)$ converges to $\varphi_t(\psi_t)$ in probability. So that $\varphi_t \circ \psi_t = Id$. In the same way, we prove that $\psi_t \circ \varphi_t = Id$. □

End of the proof of Theorem 1.1. Let A_t be a signed additive functional of φ_t that corresponds to the measure $\mu = \operatorname{div} \mathbf{a}$. Then using Theorem 2.1, similarly to the proof of Lemma 3 in Ref. [3], we get

$$\int_0^t \operatorname{div} \mathbf{a}_n(\varphi_s^n(x))ds \to A_t \text{ in } L^2, \text{ as } n \to \infty.$$

Using again Lemma 3.1, we have for s fixed, $\mathbf{a}_n \circ \varphi_s$ converges to $\mathbf{a} \circ \varphi_s$ in measure. Therefore by expression

$$\tilde{K}_t^n(\omega, x) = \exp\Big\{ -\int_0^t (\operatorname{div} \mathbf{a}_n)(\varphi_s^n(x))\,ds + \int_0^t \langle \mathbf{a}_n(\varphi_s(x)), \varphi_s^n(x)\rangle\,ds - \int_0^t \langle \varphi_s^n(x), dW_s\rangle - \frac{mt}{2}\Big\},$$

when $n \to +\infty$, K_t^n converges in measure $P \otimes \gamma$ to

$$\exp\Big\{ - A_t + \int_0^t \langle \mathbf{a}(\varphi_s(x)), \varphi_s(x)\rangle ds - \int_0^t \langle \varphi_s(x), dW_s\rangle - \frac{mt}{2}\Big\}. \tag{3.16}$$

The proof of Theorem 1.1 is completed. □

4. Examples

In this section, we will construct examples of vector fields **a** satisfying the condition in Theorem 1.1.

a) Examples of W-functionals.
Let $\{X(t), t \geq 0\}$ be a Markov process in $\mathbb{R}^m$ with transition density satisfying condition (2.6). Let $D_1, \ldots, D_k$ be bounded domains of $\mathbb{R}^m$ with C^1 boundary, and $\sigma_{\partial D_j}$ be the surface measure on ∂D_j.
Let μ be a signed measure defined by

$$\mu(dx) = b_0(x)dx + \sum_{j=1}^k b_k(x)\sigma_{\partial D_j}(dx), \tag{4.1}$$

where $b_0, \ldots, b_k$ are bounded measurable functions. Then conditions (2.8) and (2.10) are satisfied. So the additive functional

$$A(t) = \int_0^t \frac{d\mu(X(s))}{dx} ds$$

is well-defined.
Remark that for $m = 1$, any finite measure μ satisfies condition (2.10). Indeed,

$$\sup_x \int_{\mathbb{R}^m} \int_0^t s^{-1/2} \exp\Big\{ - \frac{|x-y|^2}{s}\Big\} ds\mu(dy) \leq \mu(\mathbb{R}) \int_0^t s^{-1/2} ds \to 0, t \to 0+.$$

b) Functions of bounded variation.
Assume that derivatives $\frac{\partial \mathbf{a}_i}{\partial x_k}$ considered in a generalized sense are measures. Such function $\mathbf{a} = (\mathbf{a}_1, \ldots, \mathbf{a}_m)$ are called functions of bounded variation (BV). If this measures are of the form (4.1) with bounded b_j, then **a** satisfies condition (1.7). Let now $g \in C^1(\mathbb{R}^m, \mathbb{R}^m), D$ be a bounded domain with C^1 boundary. Then $\mathbf{a}(x) = g(x)\mathbf{1}_{\{x \in D\}}$ also satisfies condition (1.7) since the generalized divergence $\operatorname{div} \mathbf{a}$ equals to

$$(\operatorname{div} g(x))\mathbf{1}_{\{x\in D\}} dx + \langle g(x), n(x)\rangle \sigma_{\partial D}(dx),$$

where $n(x)$ is the normal vector at $x \in \partial D$ (see Ref. [11]).
Linear combinations of the form

$$b_0(x) + \sum_{j=1}^k g_j(x)\mathbf{1}_{x\in D_j}$$

also satisfy condition (1.7), if $b_0 \in \mathrm{Lip},\ g_j \in C^1, D_j$ are bounded with C^1 boundary.

It should be noted that if $\mathbf{a} = (a_1, \ldots, a_m)$ is a vector field of bounded variation and $\mu_{k,j} = \frac{\partial a_k}{\partial x_j}$ satisfies (2.11) for all k, j, it has been proved in Ref. [3]

$$P\big(\varphi_t(\cdot) \in \cap_{p\geq 1} W^1_{p,loc}(\mathbb{R}^m, \mathbb{R}^m), t \geq 0\big) = 1.$$

Moreover, the Sobolev derivative is a solution of the equation

$$\nabla\varphi_t = I + \int_0^t \bar{A}^\varphi(ds)\nabla\varphi_s(x), t \geq 0, \tag{4.2}$$

where

$$\bar{A}^\varphi(t) = \int_0^t \nabla\mathbf{a}(\varphi_s)ds, t \geq 0 \tag{4.3}$$

was defined in section 2. It follows from (4.2) that a.s.

$$\det \nabla\varphi_t(x) = \exp\{\mathrm{tr}(\bar{A}^\varphi(t))\} > 0.$$

Hence it follows from Ch. 9.2 in Ref. [6] the absolute continuity $(\varphi_t)_*\gamma$ with respect to γ.

c) Example of $\mathbf{a} \notin BV$ with $\mathrm{div}\,\mathbf{a} = 0$.
Functions of bounded variation is not unique example satisfying condition (1.7). For $m = 2$, let

$$\mathbf{a}(x_1, x_2) = (g(x_1 - x_2), g(x_2 - x_1)),$$

where g is only measurable, bounded function. Then $\mathrm{div}\,\mathbf{a} = 0$ in the generalized sense, but partial derivatives $\frac{\partial \mathbf{a}}{\partial x_k}$ may not be measures.

5. Generalizations and localization

In this section, we give briefly some generalization of Theorem 1.1.

Assume now that the vector field $\mathbf{a}$ is locally bounded and for any $x \in \mathbb{R}^m$, SDE (1.2) is conservative in the sense of Kunita [21], that is, if $\tau(x)$ is the life-time of $\varphi_t(x)$, then

$$P(\{\omega; \tau(x) = +\infty\}) = 1.$$

For example, this is the case if $\mathbf{a}$ has a linear growth.

Let $\{f_n;\ n \geq 1\}$ be a sequence of functions in $C_0^\infty(\mathbb{R}^m)$ such that

$$\sup_{n,x}(|f_n(x)| + |\nabla f_n(x)|) < \infty;\ f_n(x) = 1, \text{ for } |x| \leq n.$$

Denote

$$\mathbf{a}_n(x) = \mathbf{a}(x) f_n(x),$$

and

$$\tau_n(x) = \inf\{t \geq 0 : |\varphi_t(x)| \geq n\}.$$

Let φ_t^n be the solution to SDE (1.2) with $\mathbf{a}_n$ instead of $\mathbf{a}$. Observe that $\mathbf{a}_n$ is a bounded vector field on $\mathbb{R}^m$. By uniqueness of solutions, almost surely, for $t \le \tau_n(x)$, $\varphi_t(x) = \varphi_t^n(x)$. So for any bounded Borel function $h : \mathbb{R}^m \to \mathbb{R}$,

$$\int_{\{\tau_n(x)\ge t\}} h(\varphi_t(x))\, d\gamma(x) = \int_{\{\tau_n(x)\ge t\}} h(\varphi_t^n(x))\, d\gamma(x). \tag{5.1}$$

Observe that $(\varphi_t^n)_*(\mathbf{1}_{\{\tau_n(x)\ge t\}}\gamma)$ is absolutely continuous with respect to $(\varphi_t^n)_*\gamma$ and for almost surely ω, $(\varphi_t)_*(\mathbf{1}_{\{\tau_n(x)\ge t\}}\gamma)$ converges to $(\varphi_t)_*\gamma$ weakly as $n \to +\infty$ since $\tau_n(x) \to +\infty, n \to \infty$.

Assume that for each $n \ge 1$

$$\lim_{t\to 0} \sup_{|x|\le n} \int_{|y|\le n} \int_0^t s^{-m/2} \exp\Big\{ -\frac{|x-y|^2}{2s} \Big\} ds\, |\mu|(dy) = 0, \tag{5.2}$$

where $\mu = \operatorname{div} \mathbf{a}$. Then for any n, the vector field $\mathbf{a}_n$ satisfies condition (1.7) in Theorem 1.1; therefore the push forward measure $(\varphi_t^n)_*\gamma$ is absolutely continuous with respect to γ. Now let E be a Borel subset of $\mathbb{R}^m$ such that $\gamma(E) = 0$; then by (5.1), then

$$\int_{\{\tau_n(x)\ge t\}} \mathbf{1}_E(\varphi_t(x))\, d\gamma(x) = \int_{\{\tau_n(x)\ge t\}} \mathbf{1}_E(\varphi_t^n(x))\, d\gamma(x) \le [(\varphi_t^n)_*\gamma](E) = 0.$$

Letting $n \to +\infty$ yields $[(\varphi_t)_*\gamma](E) = 0$. in other words, $(\varphi_t)_*\gamma$ is absolutely continuous with respect to γ.
Note also that in this case

$$A_n^{\varphi^n}(t) = A_m^{\varphi^m}(t),\ t \in [0, \tau_n(x)] \text{ a.s.}$$

for all $m \ge n$. Therefore we can define $A^{\varphi}(t) = \lim_{n\to\infty} A_n^{\varphi^n}(t)$ and expression (1.8) also holds true if the reverse SDE (3.15) is conservative. □

Acknowledgment: The authors thank the referee for his careful reading.

References

[1] Ambrosio L., *Transport equation and Cauchy problem for BV vector fields.* Invent. Math. 158 (2004), 227-260.
[2] Aronson D. G.*Bounds for the fundamental solution of a parabolic equation.* Bull. Amer. Math. Soc., 73:890-896, 1967.
[3] Aryasova O.V., Pilipenko A.Yu. *On differentiability of stochastic flow for multidimensional SDE with discontinuous drift.* arXiv:1306.4816v1 [math.PR].
[4] Bass R., Chen Z.Q. *Brownian motion with singular drift*, Ann. Proba. 31 (2003), 791-817.
[5] Bogachev V. I. *Measure Theory*, volume 2. Springer, Berlin, 2007.
[6] Bogachev V.I. *Differentiable measures and the Malliavin calculus.* American Mathematical Society, Providence, Rhode Island, 2010.

[7] Cipriano F. and Cruzeiro A.B., *Flows associated with irregular $\mathbb{R}^d$-vector fields.* J. Diff. Equations 210 (2005), 183-201.

[8] Di Perna R.J. and Lions P.L., *Ordinary differential equations, transport theory and Sobolev spaces.* Invent. Math. 98 (1989), 511-547.

[9] Dynkin E. B. *Markov Processes.* Fizmatlit, Moscow, 1963. [Translated from the Russian. Academic Press, New York; Springer, Berlin, 1965. vol. 1, xii + 365 pp.; vol. 2, viii + 274 pp.].

[10] Fang S., Luo D., Thalmaier A. *Stochastic differential equations with coefficients in Sobolev spaces.* J. Funct. Analysis, 259 (2010), 1129-1168.

[11] Federer H. *Geometric measure theory.* Die Grundlehren der mathematischen Wissenschaften, 153, Springer-Verlag New York Inc., New York 1969.

[12] F. Flandoli, M. Gubinelli and E. Priola, *Well-posedness of the transport equation by stochastic perturbation.* Invent. Math. (2009).

[13] Fukushima M. *A decomposition of additive functionals of finite energy.* Nagoya Math. J. 74(1979), 137-168.

[14] Fukushima M. *Dirichlet forms and Markov processes*, vol. 23, North-Holland publishing compagny, 1980.

[15] Fukushima M. *On a stochastice calculus related to Dirichlet forms and distorted Brownian motion*, Physic Reports, 77-3(1981), 255-262.

[16] Gihman I. I., Skorohod A. V. *The theory of stochastic processes. I.* Translated from the Russian. Corrected reprint of the first English edition. Grundlehren der Mathematischen Wissenschaften, 210, Springer-Verlag, Berlin-New York, 1980.

[17] Gikhman I. I., Skorokhod A. V. *The theory of stochastic processes. II.* Nauka, Moscow, 1973. [Translated from the Russian. Corrected printing of the first edition. Berlin: Springer, 2004. viii, 441 p.].

[18] Gihman I. I., Skorohod A. V. *Densities of probability measures in function spaces.* (Russian) Uspehi Mat. Nauk 21, 1966, no. 6 (132), 83-152.

[19] Krylov N.V., Röckner M., *Strong solutions of stochastic equations with singular time dependent drift*, Prob. Th. Related Fields, 131 (2005), 154-196.

[20] Kulik A. M., Pilipenko A. Yu. *Nonlinear transformations of smooth measures on infinite-dimensional spaces.* (Russian) Ukrain. Mat. Zh. 52 (2000), no. 9, 1226-1250; translation in Ukrainian Math. J. 52 (2000), no. 9, 1403-1431.

[21] Kunita H. *Stochastic flows and stochastic differential equations.* Cambridge Studies in Advanced Mathematics, 24. Cambridge University Press, Cambridge, 1990.

[22] Luo D. *Absolute continuity under flows generated by SDE with measurable drift coefficients.* Stochastic Process. Appl. 121 (2011), no. 10, 2393-2415.

[23] Revuz D. and Yor M. *Continuous martingales and Brownian motion*, Grundlehren der Mathematischen Wissenschaften, 293, Springer-Verlag, 1991.

[24] Uemura H. *Positive continuous additive functionals of mutidimensional Brownian motion and the Brownian local time.* J. Math. Kyoto, 47-2 (2007), 371-390.

[25] Veretennikov A. Y. *On strong solutions and explicit formulas for solutions of stochastic integral equations.* Math. USSR Sborn, 39(3):387-403, 1981.

[26] Xicheng Zhang, *Strong solutions of SDES with singular drift and Sobolev diffusion coefficients.* Stochastic Process. Appl. 115 (2005), 1805-1818.

[27] Xicheng Zhang, *Stochastic flows of SDEs with irregular coefficients and stochastic transport equations.* Bull. Sci. Math. 134-4 (2009), 340-378.

Chapter 10

On a result of D.W. Stroock

Patrick J. Fitzsimmons*

We modify an argument of Stroock to show that an additive function from one Banach space to another, if universally Gaussian measurable, is necessarily continuous and linear.

Recently, D.W. Stroock gave a simple probabilistic proof of L. Schwartz' "Borel graph theorem", which states (in the context of Banach spaces) that if E and F are separable Banach spaces and $\psi : E \to F$ is a linear map with Borel measurable graph, then ψ is continuous. In fact, Stroock obtained the continuity of ψ under the weaker hypothesis that ψ is μ-measurable for every centered Gaussian measure μ on E. My aim here is to show that Stroock's argument, slightly amended, proves an infinite dimensional version of the familiar fact [4] that Lebesgue measurable solutions of Cauchy's functional equation must be continuous (and linear).

A map $\psi : E \to F$ between Banach spaces in *additive* provided

$$\psi(x+y) = \psi(x) + \psi(y) \qquad \forall x, y \in E.$$

An additive ψ is necessarily linear over the rationals:

$$\psi(rx+sy) = r\psi(x) + s\psi(y) \qquad \forall r, s \in \mathbb{Q},\ \forall x, y \in E, \tag{0.1}$$

Theorem 0.1. *Let E and F be Banach spaces and let $\psi : E \to F$ be additive. If ψ is μ-measurable for every centered Gaussian Radon measure μ on E, then ψ is continuous (and linear).*

(We note that in Stroock's *proof*, the Gaussian measures used are all Radon measures, hence our slight relaxation of his measurability assumption.)

Let us begin with a brief discussion of Gaussian measures on Banach spaces. A probability measure μ on the Borel σ-algebra $\mathcal{B}(E)$ of a Banach space E is a *Radon* measure provided it is inner regular.

Let $\mathcal{E} := \sigma\{x^* : x^* \in E^*\}$ denote the cylinder σ-algebra on E. A Radon probability measure μ on $\mathcal{B}(E)$ is uniquely determined by its restriction to $\mathcal{E}$, and $\mathcal{B}(E)$ is contained in the μ-completion $\mathcal{E}_\mu$ of $\mathcal{E}$; see [1, A.3.12].

A Radon probability measure μ on $\mathcal{B}(E)$ is a centered Gaussian measure if each $x^* \in E^*$, viewed as a random variable on the probability space $(E, \mathcal{E}_\mu, \mu)$,

*Department of Mathematics, U.C. San Diego, La Jolla, CA 92093–0112, pfitzsim@ucsd.edu

is normally distributed with mean 0 and variance $\sigma^2(x^*) \in [0,\infty)$. The following characterization of Gaussian Radon measures (due to X. Fernique) is crucial to Stroock's argument. Let μ be a Radon probability measure on $\mathcal{B}(E)$, and let X and Y be independent random elements of E with distribution μ (defined on some probability space $(\Omega, \mathcal{A}, \mathbb{P})$). If μ is centered Gaussian, then for each pair of real numbers (α,β) with $\alpha^2+\beta^2=1$, the random vector $(\alpha X+\beta Y,\ \beta X-\alpha Y)$ has the same distribution as the pair (X,Y), namely the product measure $\mu\otimes\mu$ (on $\mathcal{E}_\mu\otimes\mathcal{E}_\mu$). (Notice that the map $(x,y)\mapsto(\alpha x+\beta y,\ \beta x-\alpha y)$ is $\mathcal{E}_\mu\otimes\mathcal{E}_\mu/\mathcal{E}_\mu\otimes\mathcal{E}_\mu$-measurable.) Conversely, if this equality in distribution holds for $\alpha=\beta=1/\sqrt{2}$ alone, then μ is centered Gaussian.

Proof of Theorem 0.1. The proof in [5] needs to be supplemented at the two points where the full linearity of ψ is used: (i) in showing that the image $\psi_*\mu$ of a centered Gaussian measure μ on E is a centered Gaussian measure on F, and (ii) in the third display on page 6 of [5].

Let us take up point (ii) first. The display referred to makes use of the fact that

$$\langle\psi(tx),y^*\rangle = t\langle\psi(x),y^*\rangle \qquad \forall t\in\mathbb{R},\ x\in E,\ y^*\in F^*, \tag{0.2}$$

where F^* is the dual space of F. To see that this partial linearity follows from our hypotheses, fix $x\in E$ and consider the centered Gaussian Radon measure μ_x, the image of the standard normal distribution on $\mathbb{R}$ under the mapping $\mathbb{R}\ni t\mapsto tx\in E$. The assumed μ_x-measurability of ψ then implies that the additive function $f(t):=\langle\psi(tx),y^*\rangle$, $t\in\mathbb{R}$, is Lebesgue measurable. It is well known [4] that such an f is necessarily linear, and so (0.2) holds.

Turning to (i), we require the following simple fact.

Lemma 0.1. *There is a sequence* $\{(\alpha_n,\beta_n): n\ge 1\}$ *of pairs of rational numbers such that* $\alpha_n^2+\beta_n^2=1$ *for all* n, *and* $\lim_n\alpha_n=\lim_n\beta_n=1/\sqrt{2}$.

Proof. We produce the required pairs by an appeal to Euclid's construction of Pythagorean triples [2]. Abbreviate

$$\kappa := \sqrt{\frac{\sqrt{2}-1}{\sqrt{2}+1}},$$

choose a sequence of positive integers $m_n\in\{1,2,\dots,n-1\}$ such that

$$\lim_n \frac{m_n}{n} = \kappa,$$

and define rationals

$$\alpha_n := \frac{n^2-m_n^2}{n^2+m_n^2}, \qquad \beta_n := \frac{2m_n n}{n^2+m_n^2}.$$

Clearly $\alpha_n^2+\beta_n^2=1$ and $\lim_n\alpha_n=(1-\kappa^2)/(1+\kappa^2)=1/\sqrt{2}$, as desired. □

We now fix a centered Gaussian Radon measure μ on $\mathcal{B}(E)$ and proceed to show that $\psi_*\mu$ is a centered Gaussian measure on F. Let X and Y be independent random elements of E with distribution μ. Let (α_n, β_n), $n \geq 1$, be as in Lemma 4. Then, using (0.1) for the first equality below,

$$\begin{aligned}(\alpha_n\psi(X) + \beta_n\psi(Y),\ \beta_n\psi(X) - \alpha_n\psi(Y)) \\ &= (\psi(\alpha_n X + \beta_n Y),\ \psi(\beta_n X - \alpha_n Y)) \\ &\stackrel{d}{=} (\psi(X),\ \psi(Y)),\end{aligned}$$

the $\stackrel{d}{=}$ indicating equality in distribution. Sending $n \to \infty$ we obtain

$$\left(\frac{\psi(X)+\psi(Y)}{\sqrt{2}},\ \frac{\psi(X)-\psi(Y)}{\sqrt{2}}\right) \stackrel{d}{=} (\psi(X), \psi(Y)),$$

so $\psi_*\mu$, the distribution of $\psi(X)$, is a centered Gaussian Radon probability measure on $\mathcal{B}(F)$. □

Remark 0.1. Concerning the "universal Gaussian measurability" hypothesis in the Theorem, we follow [5] in noting that if $\psi : E \to F$ is additive and has a Borel measurable graph $G \subset E \times F$, and if E and F are separable, then $\psi^{-1}(B) = \pi_E(G \cap (E \times B))$ is an analytic subset of E (hence universally measurable) for each $B \in \mathcal{B}(F)$, so Theorem 2 applies. The separability condition on E is harmless, since the function ψ is continuous if and only if it is sequentially continuous.

References

[1] Bogachev, V.: *Gaussian Measures*, Mathematical Surveys and Monographs, **62**, American Mathematical Society, Providence, 1998.

[2] Euclid: *Elements*, Book X, Proposition 29, Lemma 1.

[3] Fernique, X.: Régularité des trajectoires des fonctions aléatoires gaussiennes, In *Écoles d' été de probabilités de Saint-Flour IV-1974*, Lecture Notes in Mathematics, **480**, Springer-Verlag, Berlin, 1975, pp. 1–96.

[4] Letac, G.: Cauchy functional equation again, *Amer. Math. Monthly* **85** (1978) 663–664.

[5] Stroock, D.W.: On a theorem of Laurent Schwartz, *C. R. Acad. Sci. Paris, Ser. I*, **349** (2011) 5–6.

Chapter 11

Consistent risk measures and a non-linear extension of backwards martingale convergence

Hans Föllmer* and Irina Penner†

We study the behavior of conditional risk measures along decreasing σ-fields. Under a condition of consistency, we prove a non-linear extension of backwards martingale convergence. In particular we show the existence of a limiting conditional risk measure with respect to the tail field, we describe its dual representation in terms of a limiting penalty function, and we show that consistency extends to the tail field. Moreover, we clarify the structure of global risk measures which are consistent with the given sequence of conditional risk measures.

Contents

1. Introduction

Consider a filtration $(\mathcal{F}_n)_{n\in\mathbb{Z}}$, indexed by the integers, on some measurable space $(\Omega, \mathcal{F})$. In the forward direction we define the asymptotic σ-field $\mathcal{F}_\infty := \sigma\left(\bigcup_n \mathcal{F}_n\right)$, in the backward direction the tail field $\mathcal{F}_{-\infty} := \bigcap_n \mathcal{F}_n$.

For a given probability measure P and for any bounded measurable function X on $(\Omega, \mathcal{F})$, let us denote by

$$\eta_n(X) := E_P\left[-X|\mathcal{F}_n\right], \qquad n \in \mathbb{Z} \tag{1.1}$$

the conditional expectation of $-X$ with respect to $\mathcal{F}_n$ under the measure P. Since we are using the minus sign, the functional η_n can be regarded as the special linear case of a conditional convex risk measure, as explained below.

Due to the projectivity of conditional expectations, the sequence $(\eta_n)_{n\in\mathbb{Z}}$ is

*Humboldt-Universität zu Berlin, Institut für Mathematik, Unter den Linden 6, 10099 Berlin, Germany, foellmer@math.hu-berlin.de
†Humboldt-Universität zu Berlin, Institut für Mathematik, Unter den Linden 6, 10099 Berlin, Germany, penner@math.hu-berlin.de

consistent in the sense that

$$\eta_n(-\eta_{n+1}(X)) = \eta_n(X), \qquad n \in \mathbb{Z}. \tag{1.2}$$

Martingale convergence forwards and backwards yields the existence of the limits

$$\eta_\infty(X) := \lim_{n\uparrow\infty} \eta_n(X), \quad \eta_{-\infty}(X) := \lim_{n\downarrow-\infty} \eta_n(X)$$

P-a.s. and in $L^1(P)$, and these limits are identified as conditional expectations

$$\eta_\infty(X) = E_P\left[-X|\mathcal{F}_\infty\right], \quad \eta_{-\infty}(X) = E_P\left[-X|\mathcal{F}_{-\infty}\right]$$

with respect to the limiting σ-fields $\mathcal{F}_\infty$ and $\mathcal{F}_{-\infty}$. Again by projectivity, we see that the consistency relation (1.2) extends to infinity in both directions, that is,

$$\eta_n(-\eta_\infty) = \eta_n \quad \text{and} \quad \eta_{-\infty}(-\eta_n) = \eta_{-\infty} \tag{1.3}$$

for any $n \in \mathbb{Z}$. Let us summarize these classical facts by saying that the sequence $(\eta_n)_{n\in\mathbb{Z}}$ is *asymptotically precise* in both directions.

In this paper, we study the question whether asymptotic precision extends from the linear case of conditional expectation to the non-linear case of conditional risk measures. For each $n \in \mathbb{Z}$, let ρ_n denote a conditional convex risk measure on $L^\infty(\Omega, \mathcal{F}, P)$ with respect to $\mathcal{F}_n$, and let

$$\mathcal{A}_n := \left\{ X \in L^\infty(\Omega, \mathcal{F}, P) \mid \rho_n(X) \le 0 \right\}$$

denote the corresponding acceptance set; see, e.g., [1, Chapter 11]. Under an additional continuity assumption, the conditional risk measure ρ_n admits the dual representation

$$\rho_n(X) = \operatorname*{ess\,sup}_{\substack{Q\ll P\\ Q\approx P \text{ on } \mathcal{F}_n}} \left(E_Q\left[-X|\mathcal{F}_n\right] - \alpha_n(Q)\right) \tag{1.4}$$

with penalty function

$$\alpha_n(Q) = \operatorname*{ess\,sup}_{X\in\mathcal{A}_n} E_Q\left[-X|\mathcal{F}_n\right].$$

In the special coherent case where ρ_n is also positively homogeneous, this reduces to the representation

$$\rho_n(X) = \operatorname*{ess\,sup}_{Q\in\mathcal{Q}_n} E_Q\left[-X|\mathcal{F}_n\right] \tag{1.5}$$

with a suitable class $\mathcal{Q}_n$ of probability measures Q. Under the additional condition of comonotonicity, the coherent risk measure in (1.5) can also be regarded as a conditional Choquet integral

$$\rho_n(X) = \int (-X)dC_n,$$

where $C_n(A) := \rho_n(-I_A)$ is a conditional Choquet capacity, in analogy to the discussion in [1, Section 4.7]. Clearly, we recover the conditional expectation $\eta_n(X)$

in (1.1) in the simple special case, where the set $\mathcal{Q}_n$ reduces to the single probability measure P.

Let us now assume that the sequence $(\rho_n)_{n\in\mathbb{Z}}$ of conditional risk measures is consistent in the sense of (1.2), i.e.,

$$\rho_n(-\rho_{n+1}t(X)) = \rho_n(X), \qquad n \in \mathbb{Z} \tag{1.6}$$

for any $X \in L^\infty(\Omega, \mathcal{F}, P)$. Consistency can be characterized in terms of the acceptance sets $(\mathcal{A}_n)_{n\in\mathbb{Z}}$, in terms of the penalty functions $(\alpha_n)_{n\in\mathbb{Z}}$, and also by supermartingale criteria for the joint behavior of (ρ_n) and (α_n); this is recalled in Section 2.2.

In the forward direction, the behavior of the consistent sequence (ρ_n) along the filtration $(\mathcal{F}_n)_{n\geq 0}$ has been studied in [2]. The supermartingale criteria for consistency yield existence of the limit

$$\rho_\infty(X) := \lim_n \rho_n(X).$$

The question is whether ρ_∞ has good properties as a conditional risk measure with respect to $\mathcal{F}_\infty$. In the case $\mathcal{F}_\infty = \mathcal{F}$, asymptotic precision in the forward direction amounts to the condition $\rho_\infty(X) = -X$. However, neither asymptotic precision nor the weaker condition $\rho_\infty(X) \geq -X$ of *asymptotic safety* may hold; see [2, Section 5] for criteria and for counterexamples.

In this paper, we focus on the backward direction, and so it is enough to consider the filtration $(\mathcal{F}_n)_{n\leq 0}$. Under a mild condition on the penalties for our reference measure P, we show in Section 3 that asymptotic precision is indeed satisfied along decreasing σ-fields. More precisely, an application of the supermartingale criteria for consistency yields the existence of the limit

$$\rho_{-\infty}(X) = \lim_{n\downarrow -\infty} \rho_n(X).$$

We then show that the functional $\rho_{-\infty}$ defines a conditional convex risk measure with respect to the tail field $\mathcal{F}_{-\infty}$, that this risk measure is continuous from above, and that its dual representation (1.4) for $n = -\infty$ is given by the limiting penalty function

$$\alpha_{-\infty}(Q) = \lim_n \alpha_n(Q).$$

Moreover, we show that the consistency condition (1.6) extends to $-\infty$, that is,

$$\rho_{-\infty}(-\rho_n) = \rho_{-\infty}$$

for any $n \leq 0$, in analogy to (1.3). In particular, these properties of asymptotic precision in the backward direction hold for a consistent sequence of conditional coherent risk measures, and also for the special case of conditional Choquet integrals.

In the final Section 4 we study the structure of the set $\mathcal{R}$ of all global (unconditional) risk measures ρ on $L^\infty(\Omega, \mathcal{F}, P)$, which are consistent with the given sequence $(\rho_n)_{n\leq 0}$. Under additional continuity conditions, we show that such risk measures are of the form

$$\rho = \hat{\rho}(-\rho_{-\infty}),$$

where $\hat{\rho}$ is a convex risk measure on the tail field; $\hat{L} := L^\infty(\Omega, \mathcal{F}_{-\infty}, P)$ with respect to the tail field, the precise formulation is given in Theorem 4.1 and Corollary 4.1.

Our discussion of the behavior of conditional convex risk measures along decreasing σ-fields is motivated by the problem of clarifying the structure of spatial risk measures consistent with a given local specification in a large network. Under a condition of local law-invariance, the local conditional risk measures must be entropic, and then the problem can be solved explicitly, as shown in [3]. Without this condition, the main problem consists in extending the local specification to the tail-field, and this can be done by using the general convergence results of the present paper. The application to spatial risk measures will be discussed in [4].

2. Preliminaries

Throughout this paper fix a probability space $(\Omega, \mathcal{F}, P)$. We write $L^\infty := L^\infty(\Omega, \mathcal{F}, P)$ and denote by $\mathcal{M}_1(P)$ the set of all probability measures absolutely continuous with respect to P.

In this section we recall some basic facts about conditional convex risk measures and about consistency that will be used later on. For further details see, for example, [2, 5–9], and [1, Chapter 11].

2.1. *Conditional convex risk measures*

Let $\mathcal{F}_0$ be a sub-σ-field of $\mathcal{F}$ and write $L_0^\infty := L^\infty(\Omega, \mathcal{F}_0, P)$.

Definition 2.1. A map

$$\rho_0 : L^\infty \to L_0^\infty$$

is called a *conditional convex risk measure with respect to* $\mathcal{F}_0$ if it satisfies the following properties for any $X, Y \in L^\infty$:

- Conditional cash invariance: For all $X_0 \in L_0^\infty$,

$$\rho_0(X + X_0) = \rho_0(X) - X_0$$

- Monotonicity: $X \leq Y \;\Rightarrow\; \rho_0(X) \geq \rho_0(Y)$
- Conditional convexity: For all $\lambda \in L_0^\infty$ such that $0 \leq \lambda \leq 1$,

$$\rho_0(\lambda X + (1-\lambda)Y) \leq \lambda \rho_0(X) + (1-\lambda)\rho_0(Y)$$

- Normalization: $\rho_0(0) = 0$.

A conditional convex risk measure ρ_0 is called a *conditional coherent risk measure* if it has in addition the following property:

- Conditional positive homogeneity: For all $\lambda \in L_0^\infty$ such that $\lambda \geq 0$,

$$\rho_0(\lambda X) = \lambda \rho_0(X).$$

Remark 2.1. A conditional convex risk measure ρ_0 is uniquely determined by the associated *acceptance set*

$$\mathcal{A}_0 := \left\{ X \in L^\infty \mid \rho_0(X) \leq 0 \right\},$$

since

$$\rho_0(X) = \operatorname{ess\,inf} \left\{ Y \in L_0^\infty \mid X + Y \in \mathcal{A}_0 \right\}. \tag{2.1}$$

Thus $\rho_0(X)$ has the financial interpretation of a capital requirement, namely the minimal amount which should be added to the position X to make it acceptable.

Note that $\mathcal{A}_0$ is conditionally convex and solid, and that $\rho_0(0) = 0$ implies $0 \in \mathcal{A}_0$ and $\operatorname{ess\,inf} \left\{ X \in L_0^\infty \mid X \in \mathcal{A}_0 \right\} = 0$. Conversely, any set $\mathcal{A}_0$ with these properties defines via (2.1) a conditional convex risk measure ρ_0.

Under an additional continuity condition, the conditional convex risk measure ρ_0 admits the following dual representation in terms of suitably penalized probability measures $Q \in \mathcal{M}_1(P)$; this is also called the *robust representation* of ρ_0.

For any $Q \in \mathcal{M}_1(P)$ we define

$$\alpha_0(Q) := \operatorname{ess\,sup}_{X \in \mathcal{A}_0} E_Q \left[-X | \mathcal{F}_0 \right]. \tag{2.2}$$

Q-almost surely, taking the essential supremum under Q. Clearly, $\alpha_0(Q)$ is well defined P-almost surely if Q is equivalent to P on $\mathcal{F}_0$, and in that case (2.2) can be read as well as an essential supremum under P.

Remark 2.2.

(1) Since $0 \in \mathcal{A}_0$, we have $\alpha_0(Q) \geq 0$ Q-a.s., and hence P-a.s. if $Q \approx P$ on $\mathcal{F}_0$.
(2) For any $X \in L^\infty$ we have $X + \rho_0(X) \in \mathcal{A}_0$, and so (2.2) implies

$$\rho_0(X) \geq E_Q \left[-X | \mathcal{F}_0 \right] - \alpha_0(Q) \quad Q\text{-a.s.} \tag{2.3}$$

for any $Q \in \mathcal{M}_1(P)$.

With this definition of the penalty function α_0 the following equivalence holds; see [2, 5, 6, 8–10], and [1].

Theorem 2.1. *For a conditional convex risk measure ρ_0 with respect to $\mathcal{F}_0$, the following are equivalent:*

(1) ρ_0 has the robust representation

$$\rho_0(X) = \operatorname*{ess\,sup}_{\substack{Q \in \mathcal{M}_1(P) \\ Q \approx P \text{ on } \mathcal{F}_0}} \left(E_Q \left[-X | \mathcal{F}_0 \right] - \alpha_0(Q) \right), \quad X \in L^\infty, \tag{2.4}$$

where the essential supremum is taken under P.

(2) ρ_0 *is continuous from above, i.e.,*

$$X_k \searrow X \quad P\text{-}a.s \quad \Longrightarrow \quad \rho_0(X_k) \nearrow \rho_0(X) \quad P\text{-}a.s$$

for $X \in L^\infty$ *and any sequence* $(X_k) \subseteq L^\infty$.

Remark 2.3. The penalty function α_0 is *minimal* in the following sense: If the representation (2.4) holds with some function $\tilde{\alpha}_0$, then

$$\tilde{\alpha}_0(Q) \geq \alpha_0(Q) \quad P\text{-a.s.} \tag{2.5}$$

for any $Q \in \mathcal{M}_1(P)$ such that $Q \approx P$ on $\mathcal{F}_0$. Indeed, (2.4) implies

$$\tilde{\alpha}_0(Q) \geq E_Q\left[-X|\mathcal{F}_0\right] - \rho_0(X) = E_Q\left[-(X + \rho_0(X))|\mathcal{F}_0\right] \quad P\text{-a.s.},$$

and hence (2.5) in view of (2.2), since $X + \rho_0(X) \in \mathcal{A}_0$.

Remark 2.4. Continuity from above is equivalent to the following condition, also called the *Fatou property*:

$$\rho_0(X) \leq \liminf_{k\to\infty} \rho_0(X_k)$$

for any uniformly bounded sequence $(X_k) \subset L^\infty$ which converges P-a.s. to some $X \in L^\infty$. We say that ρ_0 has the *Lebesgue property*, if the inequality in the preceding condition can be replaced by the equality

$$\rho_0(X) = \lim_{k\to\infty} \rho_0(X_k).$$

The Lebesgue property holds if and only if ρ_0 is not only continuous from above but also *continuous from below*, that is,

$$X_k \nearrow X \quad P\text{-a.s} \quad \Longrightarrow \quad \rho_0(X_k) \searrow \rho_0(X) \quad P\text{-a.s.}.$$

Moreover, it can be shown that the Lebesgue property is equivalent to the condition that the essential supremum in (2.4) is actually attained by some measure Q depending on X; for a proof in the unconditional case $\mathcal{F}_0 = \{\Omega, \emptyset\}$ see [11, Theorem 2].

The proof of Theorem 2.1 shows that the robust representation in (2.4) can actually be refined in the sense that we can use a smaller set of probability measures; see, e.g., [2] or [1, Chapter 11].

Corollary 2.1. *If* ρ_0 *is continuous from above then we have*

$$\rho_0(X) = \operatorname*{ess\,sup}_{Q\in\mathcal{Q}_0} \left(E_Q\left[-X|\mathcal{F}_0\right] - \alpha_0(Q)\right), \quad X \in L^\infty,$$

where

$$\mathcal{Q}_0 := \left\{ Q \in \mathcal{M}_1(P) \;\middle|\; Q = P \text{ on } \mathcal{F}_0,\ E_Q[\alpha_0(Q)] < \infty \right\}.$$

Remark 2.5. In the special case $\mathcal{F}_0 = \{\Omega, \emptyset\}$, the preceding discussion reduces to standard definitions and basic facts for (unconditional) *convex risk measures*

$$\rho : L^\infty \to \mathbb{R}$$

on the Banach space L^∞; see, [12–16], and also [1, Chapter 4].

2.2. *Consistency*

Let us now fix two sub-σ-fields $\mathcal{F}_0 \subseteq \mathcal{F}_1$ of $\mathcal{F}$. For $i = 0, 1$, we write $L_i^\infty := L^\infty(\Omega, \mathcal{F}_i, P)$, and we consider a conditional convex risk measure $\rho_i : L^\infty \to L_i^\infty$ with respect to $\mathcal{F}_i$.

Definition 2.2. We say that the conditional risk measures ρ_0 and ρ_1 are *consistent* if

$$\rho_0 = \rho_0(-\rho_1),$$

that is, if $\rho_0(-\rho_1(X)) = \rho_0(X)$ for all $X \in L^\infty$.

From now on we assume that both ρ_0 and ρ_1 are continuous from above. Let $\mathcal{A}_i$ and α_i denote the acceptance set and the minimal penalty function corresponding to ρ_i. Consistency of ρ_0 and ρ_1 can then be characterized in terms of the acceptance sets, in terms of the minimal penalty functions, and in terms of the joint behavior of (ρ_i) and (α_i). To this end, consider the restriction of ρ_0 to the subspace L_1^∞ of L^∞ and denote by

$$\mathcal{A}_{0,1} := \left\{ X \in L_1^\infty \mid \rho_0(X) \leq 0 \ P\text{-a.s.} \right\}$$

the acceptance set and by

$$\alpha_{0,1}(Q) := \operatorname*{ess\,sup}_{X \in \mathcal{A}_{0,1}} E_Q\left[-X | \mathcal{F}_0\right], \qquad Q \in \mathcal{M}_1(P)$$

the minimal penalty function associated to this restriction in analogy to (2.2). As shown in [2, 7, 8, 10, 17, 18] and [9, Theorem 1.20], consistency can now be characterized as follows

Theorem 2.2. *The following conditions are equivalent:*

(1) ρ_0 *and* ρ_1 *are consistent.*

(2) $\mathcal{A}_0 = \mathcal{A}_{0,1} + \mathcal{A}_1$.

(3) For any $Q \in \mathcal{M}_1(P)$,

$$\alpha_0(Q) = \alpha_{0,1}(Q) + E_Q[\,\alpha_1(Q) \,|\, \mathcal{F}_0\,] \quad Q\text{-a.s.}$$

(4) For $X \in L^\infty$ *and any* $Q \in \mathcal{M}_1(P)$,

$$E_Q[\,\rho_1(X) + \alpha_1(Q) \,|\, \mathcal{F}_0] \leq \rho_0(X) + \alpha_0(Q) \quad Q\text{-a.s.}$$

Remark 2.6. All our penalty functions are non-negative, since we have assumed that all our risk measures are normalized. Thus property (3) of Theorem 2.2 implies that

$$\alpha_0(Q) \geq E_Q[\alpha_1(Q)|\mathcal{F}_0] \quad Q\text{-a.s. for all } Q \in \mathcal{M}_1(P). \tag{2.6}$$

In particular, $(\alpha_i(Q))_{i=0,1}$ is a non-negative supermartingale with respect to Q for all $Q \in \mathcal{M}_1(P)$ such that $E_Q[\alpha_0(Q)] < \infty$. Note that the consistency criterion

(3) of Theorem 2.2 provides, in addition to the supermartingale inequality (2.6), a special form of the predictable increasing process in the Doob decomposition of $(\alpha_i)_{i=0,1}$.

Condition (2.6) is equivalent to *weak consistency* of $(\rho_i)_{i=0,1}$, that is, to the condition that

$$\rho_1(X) \le 0 \quad \Longrightarrow \quad \rho_0(X) \le 0$$

for any $X \in L^\infty$; cf. [9, Proposition 8]. Note that weak consistency amounts to the relaxation $\mathcal{A}_1 \subseteq \mathcal{A}_0$ of the consistency criterion (2) in Theorem 2.2. For other relaxations of the strong notion of consistency in Definition 2.2 see, for example, [9, 19–23], and in the law-invariant case [24].

In Section 4 we are going to use the Lebesgue property of conditional risk measures that was introduced in Remark 2.4, and we will apply the criterion of Proposition 2.1. This involves the following notion of strong sensitivity; see also [22].

Definition 2.3. We call a conditional convex risk measure ρ_0 *strongly sensitive* with respect to P if

$$P\left[\rho_0(X) < \rho_0(Y)\right] > 0$$

whenever $X, Y \in L^\infty$ satisfy $X \ge Y$ P-a.s. and $P[X > Y] > 0$.

Proposition 2.1. *Let ρ_0 and ρ_1 be consistent, and assume that ρ_0 has the Lebesgue property and is strongly sensitive. Then ρ_1 inherits the Lebesgue property and is strongly sensitive.*

Proof. For $X \in L^\infty$ and a uniformly bounded sequence (X_k) in L^∞ such that $X_k \to X$ P-a.s., the Fatou property of ρ_1 yields

$$\rho_1(X) \le \liminf_k \rho_1(X_k) \le \limsup_k \rho_1(X_k) \quad P\text{-a.e.} \tag{2.7}$$

To prove the Lebesgue property of ρ_1, we have to show that

$$\rho_1(X) = \limsup_k \rho_1(X_k) \quad P\text{-a.e.}.$$

In view of (2.7), this will follow from

$$\rho_0(-\rho_1(X)) = \rho_0(-\limsup_k \rho_1(X_k)),$$

due to the strong sensitivity of ρ_0. Indeed, using consistency, monotonicity of ρ_0 applied to (2.7), and first the Fatou property and then the Lebesgue property of ρ_0, we obtain

$$\begin{aligned}\rho_0(X) &= \rho_0(-\rho_1(X)) \le \rho_0(-\limsup_k \rho_1(X_k)) \\ &= \rho_0(\liminf_k \rho_1(-X_k)) \le \liminf_k \rho_0(-\rho_1(X_k)) \\ &= \liminf_k \rho_0(X_k) = \rho_0(X).\end{aligned}$$

To see that ρ_1 is strongly sensitive, take $X, Y \in L^\infty$ such that $X \geq Y$ and $P[X > Y] > 0$. Then we have $P[\rho_1(X) < \rho_1(Y)] > 0$, since $\rho_1(X) = \rho_1(Y)$ P-a.s. would imply

$$\rho_0(X) = \rho_0(-\rho_1(X)) = \rho_0(-\rho_1(Y)) = \rho_0(Y)$$

in contradiction to the strong sensitivity of ρ_0. □

3. Backwards convergence

From now on we fix a filtration $(\mathcal{F}_n)_{n\geq 0}$ on our probability space $(\Omega, \mathcal{F}, P)$. Thus, the σ-fields $\mathcal{F}_n \subseteq \mathcal{F}$ are decreasing as n decreases to $-\infty$. We denote by

$$\mathcal{F}_{-\infty} := \bigcap_{n\leq 0} \mathcal{F}_n$$

the corresponding *tail field* and write $L_n^\infty = L^\infty(\Omega, \mathcal{F}_n, P)$.

Let $(\rho_n)_{n\leq 0}$ be a sequence of conditional convex risk measures

$$\rho_n : L^\infty \to L_n^\infty.$$

We denote by $\mathcal{A}_n$ the acceptance set of ρ_n, and we assume that each ρ_n is continuous from above. Thus ρ_n admits a dual representation (2.4) in terms of its minimal penalty function α_n. We also assume that the sequence is *consistent* in the sense that

$$\rho_n(-\rho_{n+1}) = \rho_n \tag{3.1}$$

for all $n < 0$.

Example 3.1. For $\beta \geq 0$ consider the conditional *entropic* risk measures $(\rho_n)_{n\leq 0}$ defined by

$$\rho_n(X) := \frac{1}{\beta} \log E_P\left[e^{-\beta X} | \mathcal{F}_n\right]; \tag{3.2}$$

for $\beta = 0$ this is interpreted as the linear case (1.1), that is, as the limiting case of (3.2) as β decreases to 0. For $\beta > 0$, the corresponding penalty functions are given by

$$\alpha_n(Q) = \frac{1}{\beta} H_n(Q|P),$$

where $H_n(Q|P)$ denotes the conditional relative entropy with respect to $\mathcal{F}_n$; see [2] or [1, Chapter 11]. It is easy to check that the sequence $(\rho_n)_{n\leq 0}$ is consistent. Note that ρ_n is *law-invariant* in the sense that $\rho_n(X)$ only depends on the conditional distribution of X with respect to $\mathcal{F}_n$ under P. Conversely, law-invariance together with consistency implies that the risk measures ρ_n are entropic, if the parameter β is allowed to be tail-measurable with values in $[0, \infty)$; see [3] and also [25]. In this special entropic case, the sequence $(\rho_n)_{n\leq 0}$ admits an immediate extension

$$\rho_{-\infty}(X) = \frac{1}{\beta} \log E_P\left[e^{-\beta X} | \mathcal{F}_{-\infty}\right]$$

to the tail field, and the properties of asymptotic precision are clearly satisfied.

Remark 3.1. In general, let $(\tilde{\rho}_n)_{n\le 0}$ be any sequence of conditional convex risk measures, not necessarily consistent. Defining recursively

$$\rho_0 := \tilde{\rho}_0 \quad \text{and} \quad \rho_n := \tilde{\rho}_n(-\rho_{n+1}t) \quad \text{for } n < 0,$$

we obtain a sequence $(\rho_n)_{n\le 0}$ which is indeed consistent.

Our goal in this section is to extend the sequence $(\rho_n)_{n\le 0}$ to a conditional convex risk measure $\rho_{-\infty}$ with respect to the tail field, to show that this risk measure is continuous from above, and to identify its dual representation. To this end we will make use of the supermartingale properties implied by the consistency condition (3.1), as they are stated in Theorem 2.2 and Remark 2.6.

Theorem 3.1. *Let us assume*

$$\sup_{n\le 0} E_P[\alpha_n(P)] < \infty. \tag{3.3}$$

Then the limit

$$\rho_{-\infty}(X) := \lim_{n\downarrow -\infty} \rho_n(X)$$

exists P-a.s. and in $L^1(P)$ for all $X \in L^\infty$. Moreover, the resulting map

$$\rho_{-\infty} : L^\infty \to L^\infty(\Omega, \mathcal{F}_{-\infty}, P)$$

defines a conditional convex risk measure with respect to the tail-field $\mathcal{F}_{-\infty}$, and it satisfies the consistency condition

$$\rho_{-\infty} = \rho_{-\infty}(-\rho_n) \tag{3.4}$$

for all $n \le 0$.

Proof. Fix $X \in L^\infty$. Due to our assumption (3.3), Theorem 2.2 together with Remark 2.6 shows that $(\alpha_n(P))_{n\le 0}$ is a backwards supermartingale under P which is bounded in $L^1(\Omega, \mathcal{F}, P)$. In view of part (4) of Theorem 2.2, the same is true for the process

$$V_n(P, X) := \rho_n(X) + \alpha_n(P), \qquad n \le 0,$$

since it is bounded from below by $-\|X\|_\infty$ and satisfies

$$\sup_{n\le 0} E_P\left[V_n(P, X)\right] \le \|X\|_\infty + \sup_{n\le 0} tE_P\left[\alpha_n(P)\right] < \infty.$$

Applying supermartingale convergence backwards under P, we obtain the existence of finite limits

$$V_{-\infty}(P, X) := \lim_{n\downarrow -\infty} V_n(P, X) \tag{3.5}$$

and

$$\alpha_{-\infty}(P) := \lim_{n\downarrow-\infty} \alpha_n(P)$$

both P-a.s. and in $L^1(P)$; cf. [26]. This yields the existence of the limit

$$\rho_{-\infty}(X) := \lim_{n\downarrow-\infty} \rho_n(X) = V_{-\infty}(P, X) - \alpha_{-\infty}(P) \tag{3.6}$$

both P-a.s. and in $L^1(P)$. Moreover, we have $|\rho_{-\infty}(X)| \leq \|X\|_\infty$, and it is easy to check that the resulting map

$$\rho_{-\infty} : L^\infty \to L^\infty(\Omega, \mathcal{F}_{-\infty}, P)$$

has the properties of a conditional convex risk measure with respect to the tail field $\mathcal{F}_{-\infty}$, as stated in Definition 2.1. To prove the consistency property (3.4) of $\rho_{-\infty}$, note that property (3.1) of the sequence (ρ_n) implies

$$\rho_{-\infty}(-\rho_n(X)) = \lim_{m\downarrow-\infty} \rho_m(-\rho_n(X)) = \lim_{m\downarrow-\infty} \rho_m(X) = \rho_{-\infty}(X)$$

for any $X \in L^\infty$ and $n \leq 0$. □

In the preceding proof, we can replace the reference measure P by any measure Q belonging to the set

$$\mathcal{Q}_P := \left\{ Q \in \mathcal{M}_1(P) \mid Q = P \text{ on } \mathcal{F}_{-\infty}, \ \sup_{n\leq 0} E_Q[\alpha_n(Q)] < \infty \right\}.$$

This yields the following result.

Corollary 3.1. *For any $Q \in \mathcal{Q}_P$, the limit*

$$\alpha_{-\infty}(Q) := \lim_{n\downarrow-\infty} \alpha_n(Q) \tag{3.7}$$

exists Q-a.s and in $L^1(Q)$, and we have

$$E_Q\left[\alpha_{-\infty}(Q)\right] = \lim_{n\downarrow-\infty} E_Q[\alpha_n(Q)] < \infty.$$

Let us denote by

$$\alpha_{n,n+1}(Q) := \operatorname*{ess\,sup}_{X\in\mathcal{A}_n\cap L^\infty_{n+1}} E_Q\left[-X|\mathcal{F}_n\right] \geq 0$$

the one-step penalty function of $Q \in \mathcal{M}_1(P)$ for $n \leq 0$; we put $L_1^\infty := L^\infty$ so that $\alpha_{0,1}(Q) = \alpha_0(Q)$.

Lemma 3.1. *For any $Q \in \mathcal{Q}_P$ the limit $\alpha_{-\infty}(Q)$ in (3.7) is given by*

$$\alpha_{-\infty}(Q) = E_Q\left[\sum_{l=-\infty}^{0} \alpha_{l,l+1}(Q)|\mathcal{F}_{-\infty}\right], \tag{3.8}$$

and we have

$$\alpha_{-\infty}(Q) = \lim_{n\downarrow-\infty} E_Q\left[\alpha_n(Q)|\mathcal{F}_{-\infty}\right] \tag{3.9}$$

Q-a.s. and in $L^1(Q)$.

Proof. Iterating condition (3) of Theorem 2.2 for $l = n, \dots, -1$, we obtain

$$\alpha_n(Q) = \alpha_{n,n+1}(Q) + E_Q\left[\alpha_{n+1}(Q)|\mathcal{F}_n\right] = E_Q\left[\sum_{l=n}^{0} \alpha_{l,l+1}(Q)|\mathcal{F}_n\right] \tag{3.10}$$

for any $n \leq 0$. Combining monotone convergence with martingale convergence ("Hunt's lemma"), we obtain

$$\alpha_{-\infty}(Q) = \lim_{n\downarrow-\infty} \alpha_n(Q) = E_Q\left[\sum_{l=-\infty}^{0} \alpha_{l,l+1}(Q)|\mathcal{F}_{-\infty}\right]$$

Q-a.s. and in $L^1(Q)$. Moreover, (3.10) implies

$$E_Q\left[\alpha_n(Q)|\mathcal{F}_{-\infty}\right] = E_Q\left[\sum_{l=n}^{0} \alpha_{l,l+1}(Q)|\mathcal{F}_{-\infty}\right],$$

and so equation (3.9) follows by monotone convergence. □

Remark 3.2. For any $Q \in \mathcal{Q}_P$, equation (3.10) shows that the *Doob decomposition* of the backwards Q-supermartingale $(\alpha_n(Q))_{n\leq 0}$ is given by

$$\alpha_n(Q) = M_n(Q) - A_n(Q), \qquad n \leq 0,$$

where the predictable increasing process is given by

$$A_n(Q) := \sum_{l<n} \alpha_{l,l+1}(Q) \in L^1(Q)$$

and the backwards martingale by

$$M_n(Q) := E_Q\left[\alpha_0(Q) + A_0(Q)|\mathcal{F}_n\right] = E_Q\left[\sum_{l=-\infty}^{0} \alpha_{l,l+1}(Q)|\mathcal{F}_n\right].$$

Our next goal is to show that the conditional risk measure $\rho_{-\infty}$ has the Fatou property, and that the minimal penalty function in its dual representation is given by the limits $\alpha_{-\infty}(Q)$ for $Q \in \mathcal{Q}_P$. To this end we consider the functional $\rho_P : L^\infty \to \mathbb{R}$ defined by

$$\rho_P(X) := E_P\left[\rho_{-\infty}(X)\right]. \tag{3.11}$$

Lemma 3.2. *ρ_P is a convex risk measure, and it has the Fatou property.*

Proof. It is easy to see that ρ_P has the properties of a convex risk measure. It remains to prove the Fatou property. Take $X \in L^\infty$ and a uniformly bounded sequence $(X_k)_{k\in\mathbb{N}}$ such that $X_k \to X$ P-a.s.. For any $n \leq 0$, the Fatou property of ρ_n implies that the functional $V_n(P, \cdot)$ has the Fatou property as well. Thus we obtain

$$E_P\left[V_n(P,X)\right] \leq \liminf_k E_P\left[V_n(P,X_k)\right] \leq \liminf_k E_P\left[V_{-\infty}(P,X_k)\right]$$

for all n, where the last inequality follows from the supermartingale property of $(V_n(P, X_k))_{n\le 0}$. Using the supermartingale convergence in (3.5), this implies

$$E_P[V_{-\infty}(P, X)] \le \liminf_k E_P[V_{-\infty}(P, X_k)].$$

Subtracting $E_P[\alpha_{-\infty}(P)]$ from both sides and recalling (3.6), we obtain the Fatou property for ρ_P:

$$E_P[\rho_{-\infty}(X)] \le \liminf_k E_P[\rho_{-\infty}(X_k)].$$

□

Theorem 3.2. *Under our assumptions* (3.3) *and* (3.1), *the conditional convex risk measure* $\rho_{-\infty}$ *has the Fatou property, and it admits the representation*

$$\rho_{-\infty}(X) = \operatorname*{ess\,sup}_{Q\in\mathcal{Q}_P} \left(E_Q[-X|\mathcal{F}_{-\infty}] - \alpha_{-\infty}(Q)\right), \quad X \in L^\infty$$

in terms of the limiting penalty function $\alpha_{-\infty}$ *in* (3.7) *and* (3.8). *Moreover,* $\alpha_{-\infty}$ *coincides with the minimal penalty function of* $\rho_{-\infty}$, *i.e.,*

$$\alpha_{-\infty}(Q) = \operatorname*{ess\,sup}_{X\in\mathcal{A}_{-\infty}} E_Q[-X|\mathcal{F}_{-\infty}] \quad P\text{-a.s.} \tag{3.12}$$

for any $Q \in \mathcal{Q}_P$, *where*

$$\mathcal{A}_{-\infty} := \left\{ X \in L^\infty \;\middle|\; \rho_{-\infty}(X) \le 0 \right\}$$

denotes the acceptance set of $\rho_{-\infty}$.

Proof. To prove the Fatou property, we show that $\rho_{-\infty}$ is continuous from above. Take $X \in L^\infty$ and a decreasing sequence (X_k) in L^∞ with $X_k \searrow X$ P-a.s.. Monotonicity of $\rho_{-\infty}$ yields

$$\rho_{-\infty}(X) \ge \lim_k \rho_{-\infty}(X_k) \quad P\text{-a.s.}. \tag{3.13}$$

On the other hand, the unconditional convex risk measure ρ_P in (3.11) is continuous from above by Lemma 3.2 and Remark 2.4. This implies

$$\begin{aligned} E_P[\rho_{-\infty}(X)] &= \rho_P(X) = \lim_k \rho_P(X_k) = \lim_k E_P[\rho_{-\infty}(X_k)] \\ &= E_P\left[\lim_k \rho_{-\infty}(X_k)\right], \end{aligned}$$

using monotone convergence in the last step. Combined with (3.13) this yields

$$\rho_{-\infty}(X) = \lim_k \rho_{-\infty}(X_k) \quad P\text{-a.s.},$$

and hence the Fatou property of $\rho_{-\infty}$. By Corollary 2.1 it follows that $\rho_{-\infty}$ has the robust representation

$$\rho_{-\infty}(X) = \operatorname{ess\,sup}_{Q\in\mathcal{Q}_P} \left(E_Q[-X|\mathcal{F}_{-\infty}] - \tilde{\alpha}_{-\infty}(Q)\right), \quad X \in L^\infty,$$

where we denote by $\tilde{\alpha}_{-\infty}(Q)$ the right-hand side of (3.12).

For a given $Q \in \mathcal{Q}_P$, we now show that $\tilde{\alpha}_{-\infty}(Q) = \alpha_{-\infty}(Q)$ P-a.s.. Note first that, due to (3.4) and Theorem 2.2,

$$\begin{aligned}\tilde{\alpha}_{-\infty}(Q) &= \tilde{\alpha}_{-\infty,n}(Q) + E_Q\left[\alpha_n(Q)|\mathcal{F}_{-\infty}\right] \\ &\geq E_Q\left[\alpha_n(Q)|\mathcal{F}_{-\infty}\right] \qquad Q\text{-a.s.}\end{aligned}$$

for any $n \leq 0$, where $\tilde{\alpha}_{-\infty,n}(Q) \geq 0$ denotes the minimal penalty function of $\rho_{-\infty}$ restricted to $\mathcal{F}_n$. This implies

$$\tilde{\alpha}_{-\infty}(Q) \geq \alpha_{-\infty}(Q) \quad P\text{-a.s.}, \tag{3.14}$$

using equation (3.9) in Lemma 3.1 and the equality of Q and P on $\mathcal{F}_{-\infty}$. To obtain the converse inequality, take any $X \in L^\infty$. We have

$$\rho_n(X) \geq E_Q\left[-X|\mathcal{F}_n\right] - \alpha_n(Q) \quad Q\text{-a.s.}$$

for any $n \leq 0$ by Remark 2.2, and hence

$$\begin{aligned}\rho_{-\infty}(X) = \lim_n \rho_n(X) &\geq \lim_n \left(E_Q\left[-X|\mathcal{F}_n\right] - \alpha_n(Q)\right) \\ &= E_Q\left[-X|\mathcal{F}_{-\infty}\right] - \alpha_{-\infty}(Q) \quad Q\text{-a.s.}.\end{aligned}$$

Since $Q = P$ on $\mathcal{F}_{-\infty}$, we obtain

$$\rho_{-\infty}(X) \geq E_Q\left[-X|\mathcal{F}_{-\infty}\right] - \alpha_{-\infty}(Q) \quad P\text{-a.s.}. \tag{3.15}$$

This holds for all $Q \in \mathcal{Q}_P$, and so we get

$$\begin{aligned}\rho_{-\infty}(X) &= \operatorname{ess\,sup}_{Q\in\mathcal{Q}_P}\left(E_Q\left[-X|\mathcal{F}_{-\infty}\right] - \tilde{\alpha}(Q)_{-\infty}\right) \\ &\leq \operatorname{ess\,sup}_{Q\in\mathcal{Q}_P}\left(E_Q\left[-X|\mathcal{F}_{-\infty}\right] - \alpha_{-\infty}(Q)\right) \\ &\leq \rho_{-\infty}(X),\end{aligned}$$

where we have used (3.14) for the first and (3.15) for the second inequality. The resulting equality shows that $\rho_{-\infty}$ has a robust representation with penalty function $\alpha_{-\infty}$. Since $\tilde{\alpha}_{-\infty}$ is the minimal penalty function, we obtain $\tilde{\alpha}_{-\infty}(Q) \leq \alpha_{-\infty}(Q)$ P-a.s. for any $Q \in \mathcal{Q}_P$. Combined with (3.14), this yields equality (3.12). □

Thus we have shown backwards convergence of ρ_n as $n \to -\infty$ to a nice conditional risk measure $\rho_{-\infty}$ with respect to the tail field. This can be seen as a backward analogue to the properties of asymptotic safety and asymptotic precision in the forward direction for a consistent sequence $(\rho_n)_{n\geq 0}$ along a filtration $(\mathcal{F}_n)_{n\geq 0}$; see the discussion in [2, Section 5]. As shown in [2, Theorem 5.4], asymptotic safety can be characterized in terms of various asymptotic properties of acceptance sets and of penalty functions as n tends to ∞. The following corollary states backward analogues to those properties as n tends to $-\infty$.

For $n \leq 0$, we denote by

$$\mathcal{A}_{-\infty,n} := \mathcal{A}_{-\infty} \cap L_n^\infty$$

the acceptance set of $\rho_{-\infty}$ restricted to $\mathcal{F}_n$, and by

$$\alpha_{-\infty,n}(Q) := \operatorname{ess\,sup}_{X\in\mathcal{A}_{-\infty,n}} E_Q\left[-X|\mathcal{F}_n\right],$$

the corresponding minimal penalty function for $Q \in \mathcal{Q}_P$. [2, Theorem 5.6]:

Corollary 3.2.

(1) $\bigcap_n \mathcal{A}_{-\infty,n} = L^\infty_+(\mathcal{F}_{-\infty})$.

(2) $\lim_{n\downarrow-\infty} \alpha_{-\infty,n}(Q) = 0$ *Q-a.s. for all* $Q \in \mathcal{Q}_P$.

(3) A position $X \in L^\infty$ *belongs to* $\mathcal{A}_{-\infty}$ *if and only if there exists a uniformly bounded sequence* $X_n \in \mathcal{A}_n$, $n \leq 0$, *such that* $\exists \lim_{n\downarrow-\infty} X_n \leq X$.

Proof. Since each ρ_n and hence $\rho_{-\infty}$ is normalized, we obtain $L^\infty_+(\mathcal{F}_{-\infty}) \subseteq \mathcal{A}_{-\infty,n}$ for any $n \leq 0$, and this shows the inclusion "$\supseteq$" in (1). Conversely, if $X \in \mathcal{A}_{-\infty}$ is $\mathcal{F}_n$-measurable for all $n \leq 0$, then X is $\mathcal{F}_{-\infty}$-measurable, and conditional cash invariance of $\rho_{-\infty}$ yields $-X = \rho_{-\infty}(X) \leq 0$ P-a.s.. This proves (1).

By Theorem 2.2 combined with (3.12), the consistency relation

$$\rho_{-\infty}(-\rho_n) = \rho_{-\infty}$$

in Theorem 3.1 implies

$$\alpha_{-\infty}(Q) = \alpha_{-\infty,n}(Q) + E_Q\left[\alpha_n(Q)|\mathcal{F}_{-\infty}\right]$$

for any $Q \in \mathcal{Q}_P$, and so the convergence in (2) follows from the second equality in Lemma 3.1.

In order to prove (3), take $X \in \mathcal{A}_{-\infty}$. Note that $X_n := X + \rho_n(X) \in \mathcal{A}_n$ for all $n \leq 0$, that the sequence (X_n) is uniformly bounded by $2\|X\|_\infty$, and that $\lim_n X_n = X + \rho_{-\infty}(X) \leq X$. Conversely, let $X \in L^\infty$ satisfy the condition $X \geq \lim_n X_n$ for some uniformly bounded sequence (X_n) such that $X_n \in \mathcal{A}_n$ for each $n \leq 0$. Since $\rho_n(X_n) \leq 0$, monotonicity and the Fatou property of $\rho_{-\infty}$ together with the consistency condition $\rho_{-\infty}(-\rho_n) = \rho_{-\infty}$ yield

$$\begin{aligned}\rho_{-\infty}(X) &\leq \rho_{-\infty}(\lim_n X_n) \leq \liminf_n \rho_{-\infty}(X_n)\\ &= \liminf_n \rho_{-\infty}(-\rho_n(X_n)) \leq 0,\end{aligned}$$

and so X belongs to $\mathcal{A}_{-\infty}$. □

For the rest of this section we focus on the special case where each ρ_n is *coherent.* Let us denote by $\mathcal{M}_1^e(P)$ the class of all probability measures Q on $(\Omega, \mathcal{F})$ which are equivalent to P.

Definition 3.1. A class $\mathcal{Q} \subseteq \mathcal{M}_1^e(P)$ of probability measures on $(\Omega, \mathcal{F})$ is called *stable* with respect to the filtration $(\mathcal{F}_n)_{n\leq 0}$ if, for any $Q_1, Q_2 \in \mathcal{Q}$ and any $n \leq 0$, the probability measure Q defined by

$$E_Q[X] := E_{Q_1}\left[E_{Q_2}[X|\mathcal{F}_n]\right]$$

belongs again to $\mathcal{Q}$.

Corollary 3.3. *Under assumptions* (3.3) *and* (3.1), *the conditional risk measure* $\rho_{-\infty}$ *defined in Theorem 3.1 is coherent if and only if each* ρ_n, $n \leq 0$, *is coherent. In this case both* ρ_n *and* $\rho_{-\infty}$ *have robust representations in terms of the set*

$$\mathcal{Q}_P^e := \mathcal{Q}_P \cap \mathcal{M}_1^e(P) = \left\{ Q \approx P \;\middle|\; \alpha_{-\infty}(Q) = 0 \right\},$$

i.e.,

$$\rho_n(X) = \underset{Q \in \mathcal{Q}_P^e}{\operatorname{ess\,sup}} E_Q \left[-X | \mathcal{F}_n \right], \qquad X \in L^\infty$$

for $n \leq 0$ *and* $n = -\infty$. *Moreover, the set* $\mathcal{Q}_P^e$ *is stable, and the process* $(\rho_n(X))_{n \leq 0}$ *is a backward* Q*-supermartingale for any* $Q \in \mathcal{Q}_P^e$.

Proof. It is straightforward to see that the limiting risk measure $\rho_{-\infty}$ is coherent, if each ρ_n is coherent. The converse as well as all other statements of the corollary follow from Theorem 3.2 and [2, Corollary 4.12], due to the consistency condition (3.4). □

Remark 3.3. Under an additional condition of *comonotonicity*, a conditional version of the arguments in [1, Corollary 4.95] shows that the coherent risk measures ρ_n can be interpreted as conditional Choquet integrals

$$\rho_n(X) = \int (-X) dC_n,$$

that is, as Choquet integrals with respect to the conditional submodular capacity C_n defined by

$$C_n(A) := \rho_n(-I_A)$$

for any $A \in \mathcal{F}$; see also [27] and the references therein. If these conditional Choquet integrals are consistent in the sense of (3.1), then Theorem 3.2 shows that they converge along our decreasing σ-fields, that is,

$$\lim_n \int X dC_n = \int X dC_{-\infty} \qquad P\text{-a.s.}$$

for any $X \in L^\infty$.

4. The structure of global risk measures consistent with $(\rho_n)_{n \leq 0}$

Let us denote by $\mathcal{R}$ the class of all convex risk measures ρ on L^∞ which are consistent with the sequence $(\rho_n)_{n \leq 0}$, that is, ρ satisfies the condition

$$\rho = \rho(-\rho_n), \qquad n \leq 0. \tag{4.1}$$

Note that $\mathcal{R} \neq \emptyset$, since the risk measure ρ_P defined in (3.11) belongs to $\mathcal{R}$.

From now on we focus on risk measures $\rho \in \mathcal{R}$ which have the Lebesgue property. We denote be $\mathcal{R}_L$ the class of all those risk measures, and by $\mathcal{R}_{L,S}$ the subclass of all $\rho \in \mathcal{R}_L$ which are strongly sensitive in the sense of Definition 2.3. In this section, our aim is to clarify the structure of the sets $\mathcal{R}_L$ and $\mathcal{R}_{L,S}$.

Lemma 4.1. *For any $\rho \in \mathcal{R}_L$ the consistency condition* (4.1) *extends to the tail field $\mathcal{F}_{-\infty}$, that is,*

$$\rho(-\rho_{-\infty}) = \rho.$$

Proof. For any $X \in L^\infty$ the sequence $(\rho_n(X))_{n\geq 0}$ is uniformly bounded by $\|X\|_\infty$ and P-a.e. convergent to $\rho_{-\infty}(X)$. Combining (4.1) with the Lebesgue property of ρ, we obtain

$$\rho(-\rho_{-\infty}(X)) = \lim_n \rho(-\rho_n(X)) = \rho(X). \tag{4.2}$$

□

Proposition 4.1. *We have $\mathcal{R}_{L,S} \neq \emptyset$ if and only if the conditional risk measure $\rho_{-\infty}$ has the Lebesgue property and is strongly sensitive in the sense of Definition 2.3.*

Proof. Suppose that $\mathcal{R}_{L,S} \neq \emptyset$. For any $\rho \in \mathcal{R}_{L,S}$, we have $\rho(-\rho_{-\infty}) = \rho$ due to Lemma 4.1. Applying Proposition 2.1 to ρ and to $\mathcal{F}_1 := \mathcal{F}_{-\infty}$ we see that $\rho_{-\infty}$ has the Lebesgue property and is strongly sensitive.

Conversely, the Lebesgue property of $\rho_{-\infty}$ implies that the risk measure ρ_P defined in (3.11) belongs to $\mathcal{R}_L$. If, moreover, $\rho_{-\infty}$ is strongly sensitive in the sense of Definition 2.3, then $\rho := \rho_P$ is strongly sensitive as well. □

Our description of the risk measures in $\mathcal{R}_L$ and $\mathcal{R}_{L,S}$ will involve the conditional risk measure $\rho_{-\infty}$ with respect to the tail filed $\mathcal{F}_{-\infty}$ and an unconditional risk measure on the tail field. More precisely, let us denote by $\hat{\mathcal{R}}$ the class of convex risk measures $\hat{\rho}$ on $\hat{L}^\infty := L^\infty(\Omega, \mathcal{F}_{-\infty}, P)$ which have the Lebesgue property on $\hat{L}^\infty$, and by $\hat{\mathcal{R}}_{L,S}$ the subclass of all $\hat{\rho} \in \hat{\mathcal{R}}_L$ which are strongly sensitive on $\hat{L}^\infty$.

Theorem 4.1. *Suppose that $\rho_{-\infty}$ has the Lebesgue property. Then the class R_L has the following structure:*

$$\mathcal{R}_L = \left\{ \hat{\rho}(-\rho_{-\infty}) \mid \hat{\rho} \in \hat{\mathcal{R}}_L \right\}. \tag{4.3}$$

Proof. Take any $\rho \in \mathcal{R}_L$, and denote by $\hat{\rho}$ the restriction of ρ to $\hat{L}^\infty$. Clearly, $\hat{\rho}$ belongs to $\hat{\mathcal{R}}_L$, and Lemma 4.1 implies

$$\rho = \rho(-\rho_{-\infty}) = \hat{\rho}(-\rho_{-\infty}).$$

This shows the inclusion "$\subseteq$" in (4.3).

Conversely, take any $\hat{\rho} \in \hat{\mathcal{R}}_L$. Then $\rho := \hat{\rho}(-\rho_{-\infty})$ defines a convex risk measure on L^∞. We have $\rho \in \mathcal{R}$ since

$$\rho(-\rho_n(X)) = \hat{\rho}(-\rho_{-\infty}(-\rho_n(X))) = \hat{\rho}(-\rho_{-\infty}(X)) = \rho(X)$$

due to (3.4). Moreover, ρ has the Lebesgue property on L^∞. Indeed, for any uniformly bounded sequence (X_k) in L^∞ such that $X_k \to X$ P-a.e., the Lebesgue property of $\rho_{-\infty}$ implies

$$\rho_{-\infty}(X) = \lim_k \rho_{-\infty}(X_k) \quad P\text{-a.e.},$$

hence

$$\rho(X) = \hat{\rho}(-\rho_{-\infty}(X)) = \lim_k \hat{\rho}(-\rho_{-\infty}(X_k)) = \lim_k \rho(X_k)$$

due to the Lebesgue property of $\hat{\rho}$. □

Corollary 4.1. *If $\mathcal{R}_{L,S} \neq \emptyset$, then*

$$\mathcal{R}_{L,S} = \left\{ \hat{\rho}(-\rho_{-\infty}) \mid \hat{\rho} \in \hat{\mathcal{R}}_{L,S} \right\}.$$

Proof. By Proposition 4.1, the existence of some $\rho \in \mathcal{R}_{L,S}$ implies that $\rho_{-\infty}$ has the Lebesgue property and is strongly sensitive. For any $\rho \in \mathcal{R}_{L,S}$, the restriction $\hat{\rho}$ of ρ to $\hat{L}^\infty$ clearly belongs to $\hat{\mathcal{R}}_{L,S}$, and we have $\rho = \hat{\rho}(-\rho_{-\infty})$ due to (4.2).

Conversely, take $\hat{\rho} \in \hat{\mathcal{R}}_{L,S}$. Then Theorem 4.1 shows that $\rho := \hat{\rho}(-\rho_{-\infty})$ belongs to $\mathcal{R}_L$. Moreover, ρ is strongly sensitive. Indeed, for X and Y in n with $X \leq Y$ P-a.e. and $P[X < Y] > 0$, we obtain $\rho_{-\infty}(X) \geq \rho_{-\infty}(Y)$ P-a.e. and $P[\rho_{-\infty}(X) > \rho_{-\infty}(Y)] > 0$ due to Proposition 4.1, and so the strong sensitivity of $\hat{\rho}$ implies

$$\rho(X) = \hat{\rho}(-\rho_{-\infty}(X)) > \hat{\rho}(-\rho_{-\infty}(Y)) = \rho(Y).$$

□

Acknowledgments

Convex risk measures contain, as a special case, the class of Choquet integrals with respect to a submodular Choquet capacity. In this sense they are related to probabilistic potential theory; see, e.g., [28] for a discussion of Choquet capacities in this context. Masatoshi Fukushima and the first author were both working in this area at the time when they first met at the Sixth Berkeley Symposium on Mathematical Statistics and Probability in 1970. It is a great pleasure to dedicate this paper to Masatoshi Fukushima on the occasion of his 80th birthday.

References

[1] H. Föllmer and A. Schied, *Stochastic Finance: An introduction in discrete time*, 3rd revised and extended edition, Walter de Gruyter & Co., Berlin (2011)
[2] H. Föllmer and I. Penner, Convex risk measures and the dynamics of their penalty functions, *Statist. Decisions.*, **24** (2006), 61–96
[3] H. Föllmer, Spatial risk measures and their local specification: the locally law-invariant case, *Statistics & Risk Modeling.*, **31** (2014), 79–103
[4] H. Föllmer and C. Klüppelberg, Spatial risk measures: local specification and boundary risk, *forthcoming.*
[5] K. Detlefsen and G. Scandolo, Conditional and dynamic convex risk measures, *Finance Stoch.*, **9** (2005), 539–561

[6] J. Bion-Nadal, Conditional risk measure and robust representation of convex conditional risk measures, (2004). CMAP preprint 557, Ecole Polytechnique Palaiseau. http://www.cmap.polytechnique.fr/preprint/repository/557.pdf

[7] J. Bion-Nadal, Dynamic risk measures: time consistency and risk measures from BMO martingales, *Finance Stoch.*, **12** (2008), 219–244

[8] P. Cheridito, F. Delbaen, and M. Kupper, Dynamic monetary risk measures for bounded discrete-time processes, *Electron. J. Probab.*, **11** (2006), 57–106

[9] B. Acciaio and I. Penner, Dynamic risk measures, In G. Di Nunno and B. Øksendal (Eds.) Advanced Mathematical Methods for Finance, pp. 1–34, Springer, 2011

[10] J. Bion-Nadal, Time consistent dynamic risk processes, *Stochastic Processes and their Applications*, **119** (2008), 633–654

[11] F. Delbaen, Differentiability properties of utility functions, In *Optimality and risk – modern trends in mathematical finance*, pp. 39–48. Springer, (2009)

[12] P. Artzner, F. Delbaen, J.-M. Eber, and D. Heath, Thinking coherently, *RISK*, **10** (1997), 68–71

[13] P. Artzner, F. Delbaen, J.-M. Eber, and D. Heath, Coherent measures of risk, *Math. Finance.*, **9** (1999), 203–228

[14] F. Delbaen. Coherent risk measures on general probability spaces. In K. Sandmann and P.J. Schönbucher (Eds.) *Advances in Finance and Stochastics*, pp. 1–37, Springer, 2002

[15] H. Föllmer and A. Schied, Convex measures of risk and trading constraints, *Finance Stoch.*, **6** (2002), 429–447

[16] M. Frittelli and E. Rosazza Gianin, Putting order in risk measures, *Journal of Banking Finance*, **26** (2002), 1473–1486

[17] F. Delbaen, The structure of m-stable sets and in particular of the set of risk neutral measures, In *memoriam Paul-Andre Meyer: Seminaire de Probabilités XXXIX*, **1874**, Lecture Notes in Math., pp. 215–258, Springer, 2006

[18] P. Cheridito and M. Kupper, Composition of time-consistent dynamic monetary risk measures in discrete time, *International Journal of Theoretical and Applied Finance*, **14** (2011), 137–162

[19] S. Tutsch, Konsistente und konsequente dynamische Risikomaße und das Problem der Aktualisierung, PhD thesis, Humboldt-Universität zu Berlin (2006). *URL http://edoc.hu-berlin.de/dissertationen/ tutsch-sina-2006-11-06/PDF/tutsch.pdf*

[20] S. Tutsch, Update rules for convex risk measures, *Quant. Finance.*, **8** (2008), 833–843

[21] B. Roorda and J. M. Schumacher, Time consistency conditions for acceptability measures, with an application to Tail Value at Risk, *Insurance Math.Econom.*, **40** (2007), 209–230

[22] B. Roorda and J. M. Schumacher, Membership conditions for consistent families of monetary valuations, To appear in *Statistics & Risk Modelling*, 10.1524/strm.2013.1131.

[23] V. Fasen and A. Svejda, Time consistency of multi-period distortion measures, *Stat. Risk Model.*, **29** (2012), 133–153

[24] S. Weber, Distribution-invariant risk measures, information, and dynamic consistency, *Mathematical Finance.*, **16** (2006), 419–441

[25] M. Kupper and W. Schachermayer, Representation results for law invariant time consistent functions, *Mathematics and Financial Economics*, **2** (2009), 189–210

[26] J. Neveu, *Discrete-parameter martingales*, Translated by T. P. Speed. North-Holland Mathematical Library. Vol. 10, North- Holland Publishing Company; New York: American Elsevier Publishing Company, Inc. VIII, 236, 1975

[27] A. Chateauneuf, R. Kast, and A. Lapied, Conditioning capacities and Choquet integrals: the role of comonotony, *Theory and Decision*, **51** (2002) (2001), 367–386
[28] M. Fukushima, *Dirichlet forms and Markov processes*, vol. 23, North-Holland Mathematical Library, North-Holland Publishing Co., Amsterdam, 1980

Chapter 12

Unavoidable collections of balls for processes with isotropic unimodal Green function

Wolfhard Hansen*

Let us suppose that we have a right continuous Markov semigroup on $\mathbb{R}^d$, $d \geq 1$, such that its potential kernel is given by convolution with a function $G_0 = g(|\cdot|)$, where g is decreasing, has a mild lower decay property, and a very weak upper decay property at infinity. This captures not only the Brownian semigroup (classical potential theory) and isotropic α-stable semigroups (Riesz potentials), but also more general isotropic Lévy processes, where the characteristic function has a certain lower scaling property, and various geometric stable processes.

There always exists a corresponding Hunt process. A subset A of $\mathbb{R}^d$ is called unavoidable, if the process hits A with probability 1, wherever it starts. It is known that, for any locally finite union of pairwise disjoint balls $B(z, r_z)$, $z \in Z$, which is unavoidable, $\sum_{z \in Z} g(|z|)/g(r_z) = \infty$. The converse is proven assuming, in addition, that, for some $\varepsilon > 0$, $|z - z'| \geq \varepsilon |z| (g(|z|)/g(r_z))^{1/d}$, whenever $z, z' \in Z$, $z \neq z'$. It also holds, if the balls are regularly located, that is, if their centers keep some minimal mutual distance, each ball of a certain size intersects Z, and $r_z = g(\phi(|z|))$, where ϕ is a decreasing function.

The results generalize and, exploiting a zero-one law, simplify recent work by A. Mimica and Z. Vondraček.

Contents

1. Introduction and main results

Let $\mathbb{P} = (P_t)_{t>0}$ be a right continuous Markov semigroup on $\mathbb{R}^d$, $d \geq 1$, such that its potential kernel $V_0 := \int_0^\infty P_t \, dt$ is given by convolution with a $\mathbb{P}$-excessive function

$$G_0 = g(|\cdot|),$$

where g is a decreasing function on $[0, \infty)$ such that $0 < g < \infty$ on $(0, \infty)$, $\lim_{r \to 0} g(r) = g(0) = \infty$ and the following holds:

*Department of Mathematics, University of Bielefeld, 33501 Bielefeld, Germany, hansen@math.uni-bielefeld.de

(LD) *Lower decay property*: There are $R_0 \geq 0$ and $C_G \geq 1$ such that

$$d \int_0^r s^{d-1} g(s)\, ds \leq C_G\, r^d g(r), \qquad \text{for all } r > R_0. \tag{1.1}$$

(UD) *Upper decay property at infinity*: There are $R_1 \geq 0$, $\eta \in (0,1)$, and $K > 1$ such that

$$g(Kr) \leq \eta g(r), \qquad \text{for all } r > R_1. \tag{1.2}$$

Remark 1.1. The inequality (1.1) has a very intuitive meaning: If B is a ball with radius r and center 0, and λ_B denotes normed Lebesgue measure on B, then the potential $G\lambda_B := G_0 * \lambda_B$ of λ_B satisfies $G\lambda_B(0) \leq C_G g(r)$ (and hence $G\lambda_B \leq C_G g(r)$ on $\mathbb{R}^d$), where $g(r)$ is the value of G_0 at the boundary of B.

In Section 2 we shall see the following.

1. For the Brownian semigroup (classical potential theory) and isotropic α-stable semigroups (Riesz potentials) we have $g(r) = r^{\alpha-d}$, $\alpha \in (0,2]$, $\alpha < d$, and our assumptions are satisfied with $R_0 = R_1 = 0$. This holds as well for the more general isotropic unimodal Lévy semigroups considered in [11].

2. If (LD) is satisfied for some $R_0 > 0$, then, for *every* $R > 0$, there exists $C_G \geq 1$ such that (1.1) holds for all $r > R$ (and hence the restriction $r_z > R_0$, for all $z \in Z$, imposed below, reduces to the requirement that $\inf_{z\in Z} r_z > 0$). Analogously for (UD).

3. If $\int_0^1 s^{d-1} g(s)\, ds < \infty$ and $g(r) \approx r^{\beta-d}$, $0 < \beta < d$, as $r \to \infty$, then (LD) and (UD) hold with arbitrary $R_0, R_1 \in (0,\infty)$.

Subordinate Brownian semigroups with subordinators having Laplace exponents of the form

$$\phi(\lambda) = \ln^\delta(1 + \lambda^{\alpha/2}), \qquad 0 < \delta \leq 1,\ \alpha \in (0,2],\ \alpha < d,$$

provide examples (symmetric geometric stable processes, if $\delta = 1$), where (LD) does not hold with $R_0 = 0$. Here g satisfies

$$g(r) \approx r^{-d} \ln^{-(1+\delta)}(1/r) \text{ as } r \to 0 \quad \text{and} \quad g(r) \approx r^{\delta\alpha - d} \text{ as } r \to \infty.$$

Let $\mathfrak{X} = (\Omega, \mathfrak{M}, \mathfrak{M}_t, X_t, \theta_t, P^x)$ be an associated Hunt process on $\mathbb{R}^d$ (for its existence see Remark 3.1,1). A Borel measurable set A in $\mathbb{R}^d$ is called *unavoidable*, if

$$P^x[T_A < \infty] = 1 \qquad \text{for every } x \in \mathbb{R}^d,$$

where $T_A(\omega) := \inf\{t \geq 0\colon X_t(\omega) \in A\}$. Otherwise, it is called *avoidable*, that is, A is avoidable, if there exists $x \in \mathbb{R}^d$ such that $P^x[T_A < \infty] < 1$.

For all $x \in \mathbb{R}^d$ and $r > 0$, let $B(x,r)$ denote the open ball with center x and radius r, and let $\overline{B}(x,r)$ be its closure. Let us introduce two properties for families of balls which, in the the classical case, have already been considered in [3, 4] (and where it does not make a real difference, if we look at open or closed balls, since a union of open balls is unavoidable if and only if the union of the corresponding closed balls is unavoidable; see Remark 3.1,2).

Let Z be a countable set in $\mathbb{R}^d \setminus \{0\}$ and $r_z > R_0$, $z \in Z$, such that the balls $B(z, r_z)$ are pairwise disjoint. We say that the balls $B(z, r_z)$, $z \in Z$, satisfy the *separation condition*, if Z is locally finite and

$$\inf_{z,z' \in Z,\, z \neq z'} \frac{|z - z'|^d}{|z|^d} \frac{g(r_z)}{g(|z|)} > 0. \tag{1.3}$$

We say that they are *regularly located*, if the following holds:

(a) There exists $\varepsilon > 0$ such that $|z - z'| \geq \varepsilon$, for all $z, z' \in Z$, $z \neq z'$.
(b) There exists $R > 0$ such that $B(x, R) \cap Z \neq \emptyset$, for every $x \in \mathbb{R}^d$.
(c) There is a decreasing function $\phi\colon [0, \infty) \to (0, \infty)$ such that $r_z = \phi(|z|)$.

Our main results are the following (where we might bear in our mind that $1/g(r)$ is approximately the capacity of balls having radius r, that is, the total mass of their equilibrium measure; see Proposition 3.2).

Theorem 1.1. *If the balls $B(z, r_z)$, $z \in Z$, satisfy the separation condition, then their union A is unavoidable provided*

$$\sum_{z \in Z} \frac{g(|z|)}{g(r_z)} = \infty.$$

Corollary 1.1. *Suppose that the balls $B(z, r_z)$, $z \in Z$, are regularly located. Then their union A is unavoidable if and only if*

$$\int_1^\infty \frac{r^{d-1} g(r)}{g(\phi(r))}\, dr = \infty.$$

The converse in Theorem 1.1 is already known without any restriction on the balls and assuming only $\lim_{r \to \infty} g(r) = 0$ instead of (UD) (see [8, Theorem 6.8]; the inequality $R_1^{\overline{B}(z,r_z)} \leq g(|z|)/g(r_z)$, which is used in its proof, holds trivially, since g is decreasing).

Proposition 1.1. *Let A be an unavoidable union of balls $B(z, r_z)$, $z \in Z$. Then $\sum_{z \in Z} g(|z|)/g(r_z) = \infty$ and $\sum_{z \in Z} 1/g(r_z) = \infty$.*

Remark 1.2. 1. In the classical case, Theorem 1.1 is [4, Theorem 6] (for unavoidableness under a weaker separation property see [3]) and Corollary 1.1 is [3, Theorem 2].

2. In the more general case of isotropic unimodal Lévy processes, where the characteristic function satisfies a lower scaling condition (and (LD), (UD) hold with $R_0 = R_1 = 0$), both Theorem 1.1, its converse, and Corollary 1.1 are proven in [11]. We shall use the same method of considering finitely many countable unions of concentric shells, but have to overcome additional difficulties caused by having only a rather weak estimate for the exit distribution of balls (compare [11, Lemma 2.2],

going back to [5, Corollary 2], and Proposition 3.3). Nevertheless our proof for Theorem 1.1 can be simpler, since starting with an avoidable union A and an arbitrary $\delta > 0$, we may assume without loss of generality that $P^0[T_A < \infty] < \delta$ (using Proposition 3.1 and translation invariance).

3. If the balls $B(z, r_z)$, $z \in Z$, are regularly located, then

$$\sum\nolimits_{z\in Z} \frac{g(|z|)}{g(r_z)} = \infty \quad \text{if and only if} \quad \int_1^\infty \frac{r^{d-1} g(r)}{g(\phi(r))}\, dr = \infty.$$

This is fairly obvious (see [11, Lemma 4.1]) and allows us to reduce Corollary 1.1 to a consequence of Theorem 1.1 by first treating a simple case (see Proposition 5.1).

In view of the second statement in Proposition 1.1 let us mention the following part of [8, Theorem 6.8] (where only $\lim_{r\to\infty} g(r) = 0$ instead of (UD) is needed). See also [7] for the result in classical potential theory.

Theorem 1.2. *Suppose that* (LD) *holds with* $R_0 = 0$. *Let* $h\colon (0,1) \to (0,1)$ *with* $\lim_{t\to 0} h(t) = 0$, *let* $\varphi \in \mathcal{C}(\mathbb{R}^d)$, $\varphi > 0$, *and* $\delta > 0$. *Then there exist a locally finite set* Z *in* $\mathbb{R}^d$ *and* $0 < r_z < \varphi(z)$, $z \in Z$, *such that the balls* $\overline{B}(z, r_z)$ *are pairwise disjoint, the union of all* $\overline{B}(z, r_z)$ *is unavoidable, and*

$$\sum\nolimits_{z\in Z} h(r_z)/g(r_z) < \delta.$$

In Section 2, we shall first take a closer look at the properties (LD) and (UD) and then show that our assumptions cover the isotropic unimodal processes considered in [11] and geometric stable processes. In Section 3, we shall discuss some general potential theory of the semigroup $\mathbb{P}$, where, as in [8], at the beginning (LD) and (UD) are replaced by the weaker properties $\int_0^1 r^{d-1} g(r)\, dr < \infty$ and $\lim_{r\to\infty} g(r) = 0$. In Section 4, we prove Theorem 1.1, and the proof of Corollary 1.1 is given in Section 5.

2. Examples

Let us first consider an arbitrary positive decreasing function g on $(0, \infty)$ and write down a few elementary facts justifying, in particular, our statements in Remark 1.1.

Given $R_0 \geq 0$, we say that (LD) holds on (R_0, ∞), if there exists $C \geq 1$ such that

$$d \int_0^r s^{d-1} g(s)\, ds \leq C r^d g(r), \qquad \text{for every } r > R_0. \tag{2.1}$$

Similarly, given $0 \leq R_1 < \infty$, we say that (UD) holds on (R_1, ∞), if there exist $K > 1$ and $\eta \in (0,1)$ such that

$$g(Kr) \leq \eta g(r), \qquad \text{for every } r > R_1. \tag{2.2}$$

Lemma 2.1.

1. *If there is a function* $\varphi > 0$ *on* $(0,1)$ *with* $\int_0^1 \gamma^{d-1}\varphi(\gamma)\,d\gamma < \infty$ *and*
$$g(\gamma r) \le \varphi(\gamma) g(r), \qquad \text{for all } \gamma \in (0,1) \text{ and } r > 0,$$
then (LD) *holds on* $(0,\infty)$.
2. *Let* $f(r) := r^d g(r)$, $R_0 \ge 0$, $\kappa, C \in (0,\infty)$. *If* $\int_0^1 s^{-1} f(s)\,ds < \infty$, $f \ge \kappa$ *on* (R_0,∞), *and*
$$\int_{R_0}^r s^{-1} f(s)\,ds \le C f(r), \qquad \text{for every } r > R_0,$$
then (LD) *holds on* (R_0,∞).
3. *If* $0 < R < R_0$ *and* (LD) *holds on* (R_0,∞), *then* (LD) *holds on* (R,∞).

Proof. 1. For every $r > 0$,
$$\int_0^r s^{d-1} g(s)\,ds = r^d \int_0^1 \gamma^{d-1} g(\gamma r)\,d\gamma \le r^d g(r) \int_0^1 \gamma^{d-1}\varphi(\gamma)\,d\gamma.$$

2. Clearly, $c := \int_0^{R_0} s^{d-1} g(s)\,ds < \infty$. For every $r > R_0$,
$$\int_0^r s^{d-1} g(s)\,ds = c + \int_{R_0}^r s^{-1} f(s)\,ds \le (c\kappa^{-1} + C) f(r) = (c\kappa^{-1} + C) r^d g(r).$$

3. Let $0 < R < R_0 < R_1$ and assume that (2.1) holds. Defining $\tilde{C} := C(R_1/R)^d$ we obtain that, for every $r \in [R, R_0]$,
$$d\int_0^r s^{d-1} g(s)\,ds \le C R_1^d g(R_1) = \tilde{C} R^d g(R_1) \le \tilde{C} r^d g(r).$$
□

Lemma 2.2. *Let* $0 \le R_1 < \infty$.

1. *If there is a function* $\varphi > 0$ *on* (R,∞), $R > 0$, *with* $\lim_{\lambda\to\infty} \varphi(\lambda) = 0$ *and* $g(\lambda r) \le \varphi(\lambda) g(r)$, *for all* $\lambda \ge R$ *and* $r > R_1$, *then* (UD) *holds on* (R_1,∞).
2. *If* $0 < R < R_1$ *and* (UD) *holds on* (R_1,∞), *then* (UD) *holds on* (R,∞).
3. *If* (UD) *holds on* (R_1,∞), *then, for every* $\delta > 0$, *there exists* $K > 1$ *such that* $g(Kr) \le \delta g(r)$ *for every* $r > R_1$.

Proof. 1. We take $K \ge R$ such that $\varphi(K) < \eta$.

2. Let $c := R_1/R$, $r > R$. Then $cr > R_1$ and $g(Kcr) \le \eta g(cr) \le \eta g(r)$.

3. We choose $m \in \mathbb{N}$ such that $\eta^m < \delta$ and replace K by K^m. □

If $0 < \alpha < d$ and $g(r) = r^{\alpha-d}$, then, by Lemmas 2.1 and 2.2, (LD) and (UD) hold on $(0,\infty)$. So our assumptions are satisfied by Brownian motion and isotropic α-stable processes with $0 < \alpha \le 2$, $\alpha < d$.

Let us observe next that, more generally, our assumptions are satisfied by the isotropic unimodal Lévy processes $\mathfrak{X} = (X_t, P^x)$ studied in [11], where the characteristic function ψ for $\mathfrak{X}$ (characterized by $e^{-t\psi(|x|)} = E^0[e^{i\langle x, X_t\rangle}]$, $t > 0$) is supposed to satisfy the following *weak lower scaling condition*: There exist $\alpha > 0$ and $C_L > 0$ such that

$$\psi(\lambda r) \geq C_L \lambda^\alpha \psi(r), \qquad \text{for all } \lambda \geq 1 \text{ and } r > 0$$

(see [11, (1.4)] and the subsequent list of examples in [11]). Then, by [11, Lemma 2.1] (see also [5, Proposition 1, Theorem 3]), there exists a constant $C \geq 1$ such that, for all $r > 0$ and $0 < \gamma \leq 1$,

$$\frac{C^{-1}}{r^d \psi(1/r)} \leq g(r) \leq \frac{C}{r^d \psi(1/r)}, \tag{2.3}$$

$$C^{-1}\gamma^{2-d} g(r) \leq g(\gamma r) \leq C\gamma^{\alpha-d} g(r). \tag{2.4}$$

By Lemma 2.1,1 and the second inequality of (2.4), (LD) holds on $(0,\infty)$. Replacing, in the first inequality of (2.4), r by λr and γ by $1/\lambda$, we see that $g(\lambda r) \leq C\lambda^{2-d} g(r)$, for all $r > 0$ and $\lambda \geq 1$. Hence (UD) holds on $(0,\infty)$, by Lemma 2.2,1, provided $d \geq 3$. For the case $d \leq 2$, see [11, Section 6].

Further, since the transition kernels P_t are given by convolution with positive functions p_t (see, for example, [9]) satisfying $p_s * p_t = p_{s+t}$, $s, t > 0$, we have

$$G_0 = \int_0^\infty p_t \, dt \in E_{\mathbb{P}}.$$

Moreover, the separation condition (1.6) in [11, Theorem 1.1] is our separation condition (1.3).

Now let us look at a subordinate Brownian semigroup, where $0 < \delta \leq 1$, $\alpha \in (0,2]$, $\alpha < d$, and the Laplace exponent of the subordinator is

$$\phi(\lambda) = \ln^\delta(1 + \lambda^{\alpha/2}).$$

If $\delta = 1$, then, by [13, Theorem 3.2 and Remark 3.3]),

$$g(r) \approx r^{-d} \ln^{-2}(1/r) \text{ as } r \to 0 \quad \text{and} \quad g(r) \approx r^{\alpha-d} \text{ as } r \to \infty,$$

and (LD) certainly does not hold with $R_0 = 0$, since, for $r > 0$,

$$\int_0^r s^{-1} \ln^{-2}(1/s) \, ds = \ln^{-1}(1/r). \tag{2.5}$$

In the general case[a], we have $\phi'(\lambda)/\phi^2(\lambda) \approx \lambda^{-1} \ln^{-(1+\delta)}(\lambda)$ as $\lambda \to \infty$. By [10, Proposition 4.5], we obtain that

$$g(r) \approx r^{-d-2} \phi'(r^{-2})/\phi^2(r^{-2}) \approx r^{-d} \ln^{-(1+\delta)}(1/r) \qquad \text{as } r \to 0.$$

Further, by [12, Theorem 3.3], $g(r) \approx r^{\delta\alpha - d}$ as $r \to \infty$. Thus, by Lemmas 2.1 and 2.2, our assumptions in Section 1 are satisfied taking any $R_0, R_1 \in (0,\infty)$.

[a] The author is indebted to T. Grzywny for informations in this case.

3. Potential theory of $\mathbb{P}$

For the moment, let us assume that the right continuous semigroup $\mathbb{P}$ is only sub-Markov and that, instead of (LD) and (UD),

$$\int_0^1 r^{d-1} g(r)\,dr < \infty \quad \text{and} \quad \lim_{r\to\infty} g(r) = 0. \tag{3.1}$$

Let $\mathcal{B}(\mathbb{R}^d)$ ($\mathcal{C}(\mathbb{R}^d)$, respectively) denote the set of all Borel measurable numerical functions (continuous real functions, respectively) on $\mathbb{R}^d$. We recall that the potential kernel $V_0 = \int_0^\infty P_t\,dt$ is given by

$$V_0 f(x) := G_0 * f(x) = \int G_0(x-y) f(y)\,dy, \qquad f \in \mathcal{B}^+(\mathbb{R}^d),\ x \in \mathbb{R}^d.$$

Let $E_{\mathbb{P}}$ denote the set of all $\mathbb{P}$-excessive functions, that is, $E_{\mathbb{P}}$ is the set of all $v \in \mathcal{B}^+(X)$ such that $\sup_{t>0} P_t v = v$. We note that $V_0(\mathcal{B}^+(\mathbb{R}^d)) \subset E_{\mathbb{P}}$. If $f \in \mathcal{B}^+(\mathbb{R}^d)$ is bounded and has compact support, then $V_0 f \in \mathcal{C}(\mathbb{R}^d)$ and $V_0 f$ vanishes at infinity, by (3.1). This leads to the following results in [8, Section 6] (for the definition of balayage spaces and their connection with sub-Markov semigroups see [2], [6], or [8, Section 8]).

Theorem 3.1. *$(\mathbb{R}^d, E_{\mathbb{P}})$ is a balayage space such that every point in $\mathbb{R}^d$ is polar and Borel measurable finely open sets $U \neq \emptyset$ have strictly positive Lebesgue measure.*

Remark 3.1. 1. There exists a Hunt process $\mathfrak{X} = (\Omega, \mathfrak{M}, \mathfrak{M}_t, X_t, \theta_t, P^x)$ on $\mathbb{R}^d$ with transition semigroup $\mathbb{P}$ (see [2, IV.7.6]).

2. Every open ball $B(x,r)$, $x \in \mathbb{R}^d$, $r > 0$, is finely dense in the closed ball $\overline{B}(x,r)$ (see [8, Proposition 6.4]; the fine topology is the coarsest topology such that every function in $E_{\mathbb{P}}$ is continuous).

For every subset A of $\mathbb{R}^d$, we have a reduced function R_1^A:

$$R_1^A := \inf\{v \in E_{\mathbb{P}} \colon v \geq 1 \text{ on } A\}.$$

Obviously, $R_1^A \leq 1$, since $1 \in E_{\mathbb{P}}$. Hence $R_1^A = 1$ on A. If A is open, then $R_1^A \in E_{\mathbb{P}}$. For a general subset A, the greatest lower semicontinuous minorant $\hat{R}_1^A$ of R_1^A (which is also the greatest finely lower semicontinuous minorant of R_1^A) is contained in $E_{\mathbb{P}}$. It is known that $\hat{R}_1^A = R_1^A$ on A^c. If A is Borel measurable, then, for every $x \in \mathbb{R}^d$,

$$R_1^A(x) = P^x[T_A < \infty] \tag{3.2}$$

(see [2, VI.3.14]). The zero-one law (3.3) will be the key to our proofs of both Theorem 1.1 and Corollary 1.1.

Proposition 3.1. *Suppose that $\mathbb{P}$ is a Markov semigroup. Then the constant function 1 is harmonic and, for each set A in $\mathbb{R}^d$,*

$$R_1^A = 1 \quad \textit{or} \quad \inf_{x\in\mathbb{R}^d} R_1^A(x) = 0. \tag{3.3}$$

Proof. Having $P_t 1 = 1$, for every $t > 0$, we know, by [8, Theorem 6.3,4]), that 1 is harmonic. Moreover, by [8, Proposition 2.3,1]), (3.3) holds. □

To illustrate that (3.3) is almost trivial, let us suppose that $R_1^A \in E_{\mathbb{P}}$ (which is true in our applications) and let $\gamma := \inf_{x\in\mathbb{R}^d} R_1^A(x)$. Since $E_{\mathbb{P}}$ is a cone, we trivially have $R_{1-\gamma}^A = (1-\gamma)R_1^A \in E_{\mathbb{P}}$. So $v := \gamma + R_{1-\gamma}^A \in E_{\mathbb{P}}$ and $v = 1$ on A, hence $v \ge R_1^A$. Moreover, $w := R_1^A - \gamma \in E_{\mathbb{P}}$ and $w = 1-\gamma$ on A, hence $w \ge R_{1-\gamma}^A$. Therefore $R_1^A = v = \gamma + R_{1-\gamma}^A = \gamma + (1-\gamma)R_1^A$. Thus $\gamma R_1^A = \gamma$, that is, $\gamma = 0$ or $R_1^A = 1$.

For all $x, y \in \mathbb{R}^d$, let

$$G_y(x) := G(x,y) := G_0(x-y),$$

and let us recall that, by definition, a potential is a positive superharmonic function with greatest harmonic minorant 0. The next result is essentially [8, Theorem 6.6].

Theorem 3.2.

1. *The function G is symmetric and continuous.*
2. *For every $y \in \mathbb{R}^d$, G_y is a potential with superharmonic support $\{y\}$.*
3. *If μ is a measure on $\mathbb{R}^d$ with compact support, then $G\mu := \int G_y \, d\mu(y)$ is a potential, and the support of μ is the superharmonic support of $G\mu$.*
4. *For every potential p on $\mathbb{R}^d$, there exists a (unique) measure μ on $\mathbb{R}^d$ such that $p = G\mu$.*

For every ball B, let $|B|$ denote the Lebesgue measure of B and let λ_B denote normalized Lebesgue measure on B (the measure on B having density $1/|B|$ with respect to Lebesgue measure).

Let us now fix $R_0 \ge 0$ and assume that (1.1) holds, that is, for $r > R_0$,

$$G\lambda_{B(0,r)}(0) = \frac{1}{|B(0,r)|} \int_{B(0,r)} G_y(0)\, dy \le C_G\, g(r). \tag{3.4}$$

Then, in fact (see [8, (6.9)]),

$$G\lambda_{B(0,r)} \le C_G g(r), \qquad \text{for every } r > R_0. \tag{3.5}$$

Moreover, since $g(r/2) \le g$ on $(0, r/2)$ and $d\int_0^{r/2} s^{d-1}\, ds = (r/2)^d$, we see that there exists $1 \le C_D \le 2^d C_G$ such that, for all $r > R_0$,

$$g(r/2) \le C_D g(r) \qquad \text{(doubling property).} \tag{3.6}$$

To simplify our estimates, let us define, once and for all,

$$c := \max\{C_D, C_G\}.$$

If B is an open ball, then $R_1^B = R_1^{\overline{B}}$ is a continuous potential and, by Theorem 3.2, there exists a unique measure μ on $\overline{B}$, the *equilibrium measure* for B, such that $R_1^B = G\mu$. For measures ν on $\mathbb{R}^d$, let $\|\nu\|$ denote their total mass. The following holds (cf. [8, Proposition 6.7] in the case $r_0 = \infty$; assuming $r > R_0$ its simple proof carries over word by word).

Proposition 3.2. *Let $r > R_0$, $B := B(0,r)$, and let μ be the equilibrium measure for B. Then*

$$\frac{g(|\cdot|)}{g(r)} \geq R_1^B \geq c^{-1}\frac{g(|\cdot|+r)}{g(r)} \quad \text{and} \quad c^{-1}\frac{1}{g(r)} \leq \|\mu\| \leq c\,\frac{1}{g(r)}.$$

The following well known fact will be used in the proofs of Proposition 3.3 and Lemma 4.1. By (3.2), it is an immediate consequence of the strong Markov property. For the convenience of the reader we write down its short proof (a corresponding argument based on iterated balayage can be given using [2, VI.2.9]).

Lemma 3.1. *Let A be a Borel measurable set in an open set $U \subset \mathbb{R}^d$ and $\gamma > 0$ such that $R_1^A \leq \gamma$ on U^c. Then $P^x[T_A < T_{U^c}] \geq R_1^A(x) - \gamma$, for every $x \in U$.*

Proof. Let $\tau := T_{U^c}$ and $x \in U$. Obviously,

$$[T_A < \infty] \setminus [T_A < \tau] = [\tau \leq T_A < \infty] \subset [\tau < \infty] \cap \theta_\tau^{-1}([T_A < \infty]).$$

Since $X_\tau \in U^c$ on $[\tau < \infty]$, the strong Markov property yields that

$$P^x([\tau < \infty] \cap \theta_\tau^{-1}[T_A < \infty]) = \int_{[\tau<\infty]} P^{X_\tau}[T_A < \infty]\, dP^x \leq \gamma.$$

Therefore $P^x[T_A < \infty] - P^x[T_A < \tau] \leq \gamma$. □

For every $r > 0$, we introduce the (closed) shell

$$S(r) := \overline{B}(0,3r) \setminus B(0,r).$$

The following estimate of the probability for hitting a shell $S(r)$ before leaving a ball $B(0,Mr)$, M large, will be sufficient for us (see [11, Lemma 2.2], going back to [5, Corollary 2], for a much stronger estimate which is used [11]).

Proposition 3.3. *Let $r > R_0$, $\eta := c^{-3}/2$, $M > 3$, and $g((M-2)r) \leq \eta g(r)$. Then*

$$P^0[T_{S(r)} < T_{B(0,Mr)^c}] \geq \eta. \tag{3.7}$$

Proof. We choose $z \in \partial B(0,2r)$ and take $B := B(z,r)$. Then B is contained in $S(r)$. By Proposition 3.2,

$$R_1^B(0) \geq c^{-1}\frac{g(|z|+r)}{g(r)} = c^{-1}\frac{g(3r)}{g(r)} \geq c^{-3} = 2\eta, \tag{3.8}$$

whereas, for every $y \in B(0,Mr)^c$,

$$R_1^B(y) \leq \frac{g(|y-z|)}{g(r)} \leq \frac{g((M-2)r)}{g(r)} \leq \eta. \tag{3.9}$$

The proof is finished by Lemma 3.1. □

The next simple result on comparison of potentials will be sufficient for us (see the proof of [8, Theorem 5.3] for a much more delicate version; cf. also the proof of [1, Theorem 3]).

Lemma 3.2. *Let $Z \subset \mathbb{R}^d$ be finite and $r_z > R_0$, $z \in Z$, such that, for $z \neq z'$, $B(z, r_z) \cap B(z', 3r_{z'}) = \emptyset$. Let $w \in E_{\mathbb{P}}$ and, for every $z \in Z$, let μ_z, ν_z be measures on $\overline{B}(z, r_z)$ such that $G\mu_z \in \mathcal{C}(\mathbb{R}^d)$, $G\mu_z \leq w$, and $\|\mu_z\| \leq \|\nu_z\|$. Then $\mu := \sum_{z \in Z} \mu_z$ and $\nu := \sum_{z \in Z} \nu_z$ satisfy*

$$G\mu \leq w + cG\nu. \tag{3.10}$$

Proof. Let $z, z' \in Z$, $z' \neq z$, and $x \in \overline{B}(z, r_z)$. For all $y, y' \in \overline{B}(z', r_{z'})$, $|y - y'| \leq 2r_{z'} \leq |x - y'|$, hence $R_0 < r_{z'} < |x - y| \leq 2|x - y'|$ and $g(|x - y'|) \leq cg(|x - y|)$. By integration, $G\mu_{z'}(x) \leq cG\nu_{z'}(x)$. Therefore

$$G\mu(x) = G\mu_z(x) + \sum_{z' \in Z, z' \neq z} G\mu_{z'}(x) \leq w(x) + cG\nu(x). \tag{3.11}$$

Thus $G\mu \leq w + cG\nu$ on the union of the balls $\overline{B}(z, r_z)$, $z \in Z$. By the minimum principle [2, III.6.6], the proof is finished. □

Remark 3.2. If each $G\mu_z$ is only bounded by some potential in $\mathcal{C}(\mathbb{R}^d)$, but there exists $\gamma > 1$ such that $B(z, \gamma r_z) \cap B(z', 3r_{z'}) = \emptyset$, whenever $z \neq z'$, then (3.11) holds for all $x \in B(z, \gamma r_z)$, $z \in Z$, and (3.10) follows as well.

4. Proof of Theorem 1.1

From now on let us suppose that the assumptions introduced at the beginning of Section 1 are satisfied. We recall that in many cases (LD) and (UD) hold with $R_0 = 0$ and $R_1 = 0$. If not, we may assume without loss of generality that R_0 and R_1, respectively, while being strictly positive, are as small as we want.

We prepare the proof of Theorem 1.1 by a first application of Lemma 3.2.

Lemma 4.1. *Let $\rho > \max\{R_0, R_1\}$, $0 < \varepsilon \leq 1/4$. Let Z be a finite subset of $S(\rho)$ and $R_0 < r_z \leq |z|/4$, $z \in Z$, such that the balls $B(z, 4r_z)$ are pairwise disjoint and*

$$|z - z'| \geq 4\varepsilon|z|\big(g(|z|)/g(r_z)\big)^{1/d}, \qquad \text{whenever } z \neq z'.$$

Moreover, let $C := 1 + (4/\varepsilon)^d c^3$, $\delta := (2Cc^4)^{-1}$, $M > 4$, and suppose that $g((M-3)\rho) \leq \delta g(\rho)$.

Then the union A of the balls $B(z, r_z)$, $z \in Z$, satisfies

$$P^x[T_A < T_{B(0,M\rho)^c}] \geq \delta \sum_{z \in Z} g(|z|)/g(r_z), \qquad \text{for every } x \in B(0, 3\rho).$$

Proof. Let $B := B(0, 4\rho)$. By (3.5),

$$G\lambda_B \leq cg(4\rho). \tag{4.1}$$

For $z \in Z$, let

$$\tilde{r}_z := \max\{r_z, \varepsilon|z|\big(g(|z|)/g(r_z)\big)^{1/d}\}$$

so that $B(z, \tilde{r}_z) \cap B(z', 3\tilde{r}_{z'}) = \emptyset$, whenever $z \neq z'$.

For the moment, we fix $z \in Z$. Since $\max\{r_z, \varepsilon|z|\} \leq |z|/4 < \rho$ and $g(|z|)/g(r_z) \leq 1$, we know that $B(z, \tilde{r}_z) \subset B$. Moreover,

$$\tilde{r}_z^{-d} \leq \varepsilon^{-d} \frac{g(r_z)}{|z|^d g(|z|)} \leq \varepsilon^{-d} \frac{g(r_z)}{\rho^d g(4\rho)}. \tag{4.2}$$

Let μ_z be the equilibrium measure for $B(z, r_z)$, that is, $G\mu_z = R_1^{B(z,r_z)}$. Then $\|\mu_z\| \leq c g(r_z)^{-1}$, by Proposition 3.2. We define

$$\nu_z := \|\mu_z\| \lambda_{B(z,\tilde{r}_z)} = \beta_z 1_{B(z,\tilde{r}_z)} \lambda_B, \tag{4.3}$$

where, by (4.2),

$$\beta_z = \|\mu_z\| \frac{|B|}{|B(z, \tilde{r}_z)|} \leq c g(r_z)^{-1} (4\rho/\tilde{r}_z)^d \leq (4/\varepsilon)^d c g(4\rho)^{-1} =: \beta. \tag{4.4}$$

Let $\nu := \sum_{z \in Z} \nu_z$. Since the balls $B(z, \tilde{r}_z)$, $z \in Z$, are pairwise disjoint subsets of B, we conclude, by (4.3), (4.4), and (4.1), that

$$G\nu \leq \beta G\lambda_B \leq (4/\varepsilon)^d c^2. \tag{4.5}$$

Next let $\mu := \sum_{z \in Z} \mu_z$ so that

$$p := \sum_{z \in Z} R_1^{B(z,r_z)} = G\mu.$$

By Lemma 3.2, $G\mu \leq 1 + cG\nu$, and hence $p \leq C$, by (4.5) and our definition of C. Therefore, using the minimum principle [2, III.6.6], we obtain that $C^{-1}p \leq R_1^{\overline{A}} = R_1^A$. Trivially, $R_1^A \leq p$. Thus

$$C^{-1}p \leq R_1^A \leq p. \tag{4.6}$$

Let $U := B(0, M\rho)$ and $z \in Z$. By Proposition 3.2, for $y \in U^c$,

$$g(r_z) R_1^{B(z,r_z)}(y) \leq g(|y - z|) \leq g((M-3)\rho) \leq \delta g(\rho),$$

whereas, for every $x \in B(0, 3\rho)$,

$$g(r_z) R_1^{B(z,r_z)}(x) \geq c^{-1} g(|x - z| + \rho) \geq c^{-1} g(7\rho) \geq c^{-4} g(\rho).$$

Defining $\gamma := \sum_{z \in Z} g(\rho)/g(r_z)$ we hence see, by (4.6), that

$$R_1^A \leq \delta\gamma \quad \text{on } U^c, \qquad R_1^A \geq 2\delta\gamma \quad \text{on } B(0, 3\rho).$$

By Lemma 3.1, for every $x \in B(0, 3\rho)$,

$$P^x[T_A < T_{B(0,M\rho)^c}] \geq \delta\gamma.$$

Observing that $g(\rho) \geq g$ on $S(\rho)$ the proof is finished. □

Now let us fix a locally finite subset Z of $\mathbb{R}^d\setminus\{0\}$ and $r_z > 4R_0$, $z \in Z$, such that the balls $B(z, r_z)$ are pairwise disjoint and satisfy the separation condition (1.3). Let A denote the union of these balls. For a proof of Theorem 1.1 we show the following.

Proposition 4.1. *If A is avoidable, then $\sum_{z\in Z} g(|z|)/g(r_z) < \infty$.*

Proof. So let us suppose that A is avoidable. To prove that $\sum_{z\in Z} g(|z|)/g(r_z) < \infty$ we may assume that $|z| > 8R_0$, for every $z \in Z$ (we simply omit finitely many points from Z). Further, we may assume that the balls $B(z, 4r_z)$ are pairwise disjoint. Indeed, since $g(r) \le g(r/4) \le c^2 g(r)$, $r > R_0$, a replacement of r_z by $r_z/4$ does neither affect (1.3) nor the convergence of $\sum_{z\in Z} g(|z|)/g(r_z)$, and the new, smaller union is, of course, avoidable. Moreover, similarly as at the beginning of the proof of [11, Theorem 1.1]), we may assume without loss of generality that

$$r_z \le |z|/8, \qquad \text{for every } z \in Z. \tag{4.7}$$

Indeed, replacing r_z by $r'_z := \min\{r_z, |z|/8\}$ our assumptions are preserved as well. Suppose we have shown that $\sum_{z\in Z} g(|z|)/g(r'_z) < \infty$. Since $g(|z|)/g(|z|/8) \ge c^{-3}$, we see that the set Z' of all $z \in Z$ such that $r'_z = |z|/8$ is finite, and hence certainly $\sum_{z\in Z'} g(|z|)/g(r_z) < \infty$. So we may assume without loss of generality that $r'_z = r_z$, for all $z \in Z$, that is, (4.7) holds.

By (1.3), we may choose $0 < \varepsilon < 1/4$ such that, for $z, z' \in Z$, $z \ne z'$,

$$|z - z'| \ge 8c^{1/d}\varepsilon|z|\big(g(|z|)/g(r_z)\big)^{1/d}. \tag{4.8}$$

As in Lemma 4.1, we define

$$C := 1 + (4/\varepsilon)^d c^3, \qquad \delta := (2Cc^4)^{-1}.$$

By Lemma 2.2, there exists $M := 3^m$, $m \in \mathbb{N}t$, such that

$$g((M-3)\rho) \le \delta g(\rho), \qquad \text{for every } \rho > R_1.$$

Moreover, let us define

$$R := 1 + \max\{R_0, R_1\}.$$

By Proposition 3.1, there is a point x_0 in $\mathbb{R}^d$ such that

$$P^{x_0}[T_A < \infty] = R_1^A(x_0) < \delta/2. \tag{4.9}$$

Deleting finitely many points from Z, we obtain $Z \cap B(0, 2|x_0| + R) = \emptyset$. Then, for every $z \in Z$,

$$|z|/2 \le |z - x_0| \le 2|z|, \qquad c^{-1}g(|z|) \le g(|z - x_0|) \le cg(|z|). \tag{4.10}$$

Hence, by (4.7) and (4.8), $r_z < |z - x_0|/4$ and, for $z, z' \in Z$, $z \ne z'$,

$$|z - z'| \ge 4\varepsilon|z - x_0|\big(g(|z - x_0|)/g(r_z)\big)^d. \tag{4.11}$$

By translation invariance, we may therefore assume without loss of generality that $x_0 = 0$, $Z \cap B(0,R) = \emptyset$, and (4.11) holds instead of (4.8).

For every $0 \le j < m$, let

$$Z_j := \bigcup_{n=0}^{\infty} Z \cap S(3^{nm+j} R).$$

Then Z is the union of $Z_0, Z_1, \ldots, Z_{m-1}$. Therefore it suffices to show that

$$\sum_{z \in Z_j} g(|z|)/g(r_z) < \infty, \qquad \text{for every } 0 \le j < m. \tag{4.12}$$

So let us fix $0 \le j < m$. For the moment, we also fix $n \in \{0,1,2,\ldots\}$ and define $\rho := 3^{nm+j} R$,

$$S := T_{S(\rho)}, \quad \tau := T_{B(0,\rho)^c}, \quad \tau' := T_{B(0,M\rho)^c}, \quad T := \min\{T_A, \tau'\}.$$

By Lemma 4.1,

$$P^y[T_A < \tau'] \ge \delta \sum_{z \in Z \cap S(\rho)} g(|z|)/g(r_z), \qquad \text{for every } y \in S(\rho).$$

By Proposition 3.3, $P^0[S < \tau'] \ge \delta$, and hence, by (4.9),

$$P^0[S < T] \ge P^0[S < \tau'] - P^0[T_A < \infty] > \delta/2.$$

Clearly, $S + T_A \circ \theta_S = T_A$ and $S + \tau' \circ \theta_S = \tau'$ on $[S < T]$. Hence

$$[S < T_A < \tau'] = [S < T, T_A < \tau'] = [S < T] \cap \theta_S^{-1}([T_A < \tau']).$$

Since $X_S \in S(\rho)$ on $[S < \infty]$, the strong Markov property yields that

$$P^0[S < T_A < \tau'] = \int_{[S<T]} P^{X_S}[T_A < \tau'] \, dP^0 \ge (\delta^2/2) \sum_{z \in Z \cap S(\rho)} \frac{g(|z|)}{g(r_z)}.$$

Of course, $\tau \le S$. Hence the sets $[S < T_A < \tau']$, obtained for different n, are pairwise disjoint subsets of $[T_A < \infty]$ (recall that $M = 3^m$). Thus, by (4.9),

$$\sum_{z \in Z_j} g(|z|)/g(r_z) \le (2/\delta^2) P^0[T_A < \infty] \le 1/\delta.$$

□

Let us note that the preceding proof could also be presented in a purely analytic way using iterated balayage of measures.

5. Proof of Corollary 1.1

Again we suppose that the assumptions from the beginning of Section 1 are satisfied. Let Z be a countable set in $\mathbb{R}^d$ and $r_z > 0$, $z \in Z$, such that the balls $B(z, r_z)$ are pairwise disjoint and regularly located. So there exist $\varepsilon, R \in (0,\infty)$ such that the points in Z have a mutual distance which is at least ε and every open ball of radius R contains some point of Z. Moreover, $r_z = \phi(|z|)$, where the function ϕ is decreasing. If (LD) does not hold with $R_0 = 0$, we assume that $\kappa := \inf_{x \in \mathbb{R}^d} \phi(x) > 0$. By

Lemma 2.1, we then know that (LD) holds, if we define $R_0 := \kappa/8$. By Lemma 2.2, (UD) holds with, say, $R_1 := R_0 + 1$. Of course, we may assume that $R \geq 1 + \phi(1)$.

We already know (see Remark 1.2,3 and Proposition 1.1) that it suffices to show that the union A of all $B(z, r_z)$, $z \in Z$, is unavoidable provided

$$\sum_{z \in Z} g(|z|)/g(r_z) = \infty. \tag{5.1}$$

Let us assume for the moment that

$$\limsup_{\rho \to \infty} \rho^d g(\rho)/g(\phi(\rho)) < \infty. \tag{5.2}$$

Then $\beta := \inf_{z \in Z} g(r_z)(|z|^d g(|z|))^{-1} > 0$. Since $|z - z'| \geq \varepsilon > 0$, whenever $z \neq z'$, this implies that

$$\inf_{z,z' \in Z, z \neq z'} \frac{|z - z'|^d}{|z|^d} \frac{g(r_z)}{g(|z|)} \geq \varepsilon^d \beta.$$

So the balls $B(z, r_z)$, $z \in Z$, satisfy the separation condition (1.3), and hence, by Theorem 1.1, (5.1) implies that A is unavoidable. Thus already the following lemma would finish the proof of Corollary 1.1.

Lemma 5.1. *If* $\limsup_{\rho \to \infty} \rho^d g(\rho)/g(\phi(\rho)) = \infty$, *the set A is unavoidable.*

For Lévy processes considered in [11], this is [11, Lemma 4.2]. However, its proof (by contradiction) is almost as involved as the proof of [11, Theorem 1.1].

By Proposition 3.1, we only have to show that $\inf_{x \in \mathbb{R}^d} R_1^A(x) > 0$. Hence a second application of Lemma 3.2, which is yet another variation of the arguments for quasi-additivity of capacities in [1], will allow us even to prove the following.

Proposition 5.1. *Suppose that* $\limsup_{\rho \to \infty} \rho^d g(\rho)/g(\phi(\rho)) > \eta > 0$. *Then the union A of all $B(z, r_z)$, $z \in Z$, is unavoidable.*

Proof. We define

$$a := (2c)^{-1}(18R)^{-d}, \qquad b := c^2 R^{-d},$$

and fix $x \in \mathbb{R}^d$. There exists $\rho > 9R + 2|x| + 4R_1$ such that

$$\gamma := \rho^d g(\rho)/g(\phi(\rho)) > \eta. \tag{5.3}$$

Let

$$r := \phi(\rho), \qquad B := B(0, \rho) \quad \text{and} \quad S := \overline{B}(0, \rho/2) \setminus B(\rho/4).$$

There exist finitely many points $y_1, \ldots, y_m \in S$ such that the open balls $B(y_1, 3R), \ldots, B(y_m, 3R)$ are pairwise disjoint and S is covered by $B(y_1, 9R), \ldots, B(y_m, 9R)$. Obviously, $m \geq (1/2)(\rho/18R)^d$. There exist points $z_j \in Z \cap B(y_j, R)$, $1 \leq j \leq m$. Then, for all $i, j \in \{1, \ldots, m\}$ with $i \neq j$, $|z_i - z_j| \geq |y_i - y_j| - 2R \geq 4R$, and hence

$$B(z_i, R) \cap B(z_j, 3R) = \emptyset. \tag{5.4}$$

Let $1 \le j \le m$. Clearly, $\rho \ge \rho/2 + R \ge |z_j| \ge \rho/4 - R \ge R \ge 1$, and hence

$$r = \phi(\rho) \le \phi(|z_j|) = r_{z_j} \le \phi(1) \le R. \tag{5.5}$$

Moreover, $r + |x - z_j| \le R + |x| + \rho/2 + R \le \rho$, and hence $g(|x - z_j| + r) \ge g(\rho)$. So, by translation invariance and Proposition 3.2,

$$R_1^{B(z_j,r)}(x) \ge c^{-1} g(|x - z_j| + r)/g(r) \ge c^{-1} g(\rho)/g(r). \tag{5.6}$$

Let

$$A_x := B(z_1, r) \cup \cdots \cup B(z_m, r) \quad \text{and} \quad p := \sum\nolimits_{j=1}^m R_1^{B(z_j,r)}.$$

Then $A_x \subset A$, by (5.5). So, by (5.6) and our definitions of γ, a, and r,

$$R_1^{A_x} \le R_1^A \quad \text{and} \quad p(x) \ge mc^{-1} g(\rho)/g(r) \ge a\gamma. \tag{5.7}$$

Now let μ_0 denote the equilibrium measure for $B(0,r)$, $G\mu_0 = R_1^{B(0,r)}$. By Proposition 3.2, $\|\mu_0\| \le c g(r)^{-1}$. We define

$$\nu_j := \|\mu_0\| \lambda_{B(z_j,R)} = \|\mu_0\| (\rho/R)^d 1_{B(z_j,R)} \lambda_B, \qquad 1 \le j \le m,$$

and $\nu := \sum_{j=1}^m \nu_j$. Since $B(z_1, R), \dots, B(z_j, R)$ are pairwise disjoint subsets of B and $G\lambda_B \le c g(\rho)$, we see that

$$G\nu \le \|\mu_0\| (\rho/R)^d G\lambda_B \le c^2 R^{-d} g(r)^{-1} \rho^d g(\rho) = b\gamma.$$

For every $1 \le j \le m$, $R_1^{B(z_j,r)} = G\mu_j$, where μ_j is obtained from μ_0 translating by z_j. Let $\mu := \sum_{j=1}^m \mu_j$. By (5.4), (5.5), and Lemma 3.2,

$$p = G\mu \le 1 + cG\nu \le 1 + cb\gamma.$$

Since μ is supported by $\overline{A}_x$ and p is continuous, we get that

$$R_1^{A_x} = R_1^{\overline{A}_x} \ge (1 + cb\gamma)^{-1} p,$$

by the minimum principle [2, III.6.6]. In particular,

$$R_1^A(x) \ge R_1^{A_x}(x) \ge \frac{a\gamma}{1 + cb\gamma} = \frac{a}{\gamma^{-1} + cb} > \frac{a}{\eta^{-1} + cb}$$

by (5.7) and (5.3). Thus A is unavoidable, by Proposition 3.1. □

References

[1] H. Aikawa and A. A. Borichev, Quasiadditivity and measure property of capacity and the tangential boundary behavior of harmonic functions, *Trans. Amer. Math. Soc.* **348**, 1013–1030 (1996).

[2] J. Bliedtner and W. Hansen, Potential Theory – An Analytic and Probabilistic Approach to Balayage, Universitext, Springer, Berlin-Heidelberg-New York-Tokyo (1986).

[3] T. Carroll and J. Ortega-Cerdà, Configurations of balls in Euclidean space that Brownian motion cannot avoid, *Ann. Acad. Sci. Fenn. Math.* **32**(1), 223–234 (2007).

[4] S. J. Gardiner and M. Ghergu, Champagne subregions of the unit ball with unavoidable bubbles, *Ann. Acad. Sci. Fenn. Math.*, **35** (1), 321–329 (2010).
[5] T. Grzywny, On Harnack inequality and Hölder regularity for isotropic unimodal Lévy processes, *Potential Anal.*, DOI 10.1007/s11118-013-9360-y, arXiv:1306.5909v4.
[6] W. Hansen, *Three views on potential theory.* A course at Charles University (Prague), Spring 2008, http://www.karlin.mff.cuni.cz/ hansen/lecture/ course-07012009.pdf.
[7] W. Hansen and I. Netuka, Champagne subdomains with unavoidable bubbles, *Adv. Math.* **244**, 106–116 (2013).
[8] W. Hansen and I. Netuka, Unavoidable sets and harmonic measures living on small sets. Preprint 13047, CRC 701, University of Bielefeld, arXiv:1306.5433v2.
[9] J. Hawkes, Potential theory of Lévy processes, *Proc. London Math. Soc.* **38**, 335–352 (1979).
[10] P. Kim and A. Mimica, Harnack inequalities for subordinate Brownian motions, *Electron. J. Probab.* **17**(37), 1-23 (2012).
[11] A. Mimica and Z. Vondraček, Unavoidable collections of balls for isotropic Lévy processes, *Stochastic Process. Appl.* **124**(3), 1303–1334 (2014).
[12] M. Rao, R. Song, and Z. Vondraček, Green function estimates and Harnack inequality for subordinate Brownian motions, *Potential Analysis* **25**, 1–27 (2006).
[13] H. Šikić, R. Song, and Z. Vondraček, Potential theory of geometric stable processes, *Probab. Theory Relat. Fields* **135**, 547–575 (2004).

Chapter 13

Functions of locally bounded variation on Wiener spaces

Masanori Hino*

We introduce the concept of functions of locally bounded variation on abstract Wiener spaces and study their properties. Some nontrivial examples and applications to stochastic analysis are also discussed.

Contents

1. Introduction

Functions of bounded variation (BV functions) on abstract Wiener spaces were first studied in [5, 6] for their applications in stochastic analysis, but recently BV functions on infinite dimensional spaces have attracted attention for a variety of reasons. In this paper, we newly introduce the space $\dot{BV}_{\mathrm{loc}}$ of functions of locally bounded variation (local BV functions) on the abstract Wiener space (E, H, μ). Needless to say, there could be several ways of localizing the concept of bounded variation. In this article, we adopt the ideas in the theory of (quasi-)regular Dirichlet forms, which are suitable for application to stochastic analysis. Indeed, we show that a Dirichlet form of type

$$\mathcal{E}^{\rho}(f,g) = \frac{1}{2}\int_E \langle \nabla f, \nabla g\rangle_H \,\rho\, d\mu$$

associated with a nonnegative function ρ in $\dot{BV}_{\mathrm{loc}}$ with some extra assumptions provides a diffusion process that has the Skorokhod representation (Theorem 2.4). This result is regarded as a natural generalization of [6, Theorem 4.2], where ρ is assumed to be a BV function instead. We also consider the classical Wiener space on $\mathbb{R}^d$ as E and provide a sufficient condition for an open set O of $\mathbb{R}^d$ so that the indicator function of the set of all paths staying in the closure $\overline{O}$ is a local BV function. Accordingly, we can construct the (modified) reflecting Ornstein–Uhlenbeck process with the Skorokhod representation on the set of paths staying in $\overline{O}$ under a rather weak condition on O. This is a study related to another paper [9],

*Graduate School of Engineering Science, Osaka University, hino@sigmath.es.osaka-u.ac.jp

in which a sufficient condition was given for the above-mentioned indicator function to be a BV function on either pinned path spaces or one-sided pinned spaces.

The remainder of this paper is organized as follows. In Section 2, we introduce the concept of local BV functions on a Wiener space and study their properties. In Section 3, we provide a sufficient condition for a class of indicator functions to be local BV functions.

2. The space $\dot{BV}_{\mathrm{loc}}$ on Wiener spaces

Let (E, H, μ) be an abstract Wiener space. That is, E is a separable Banach space, H is a separable Hilbert space densely and continuously embedded in E, and μ is a Gaussian measure on E that satisfies

$$\int_E \exp\left(\sqrt{-1}l(z)\right)\mu(dz) = \exp\left(-|l|_H^2/2\right), \quad l \in E^*.$$

Here, we regard the topological dual of E, E^*, as a dense subspace of H by the natural identification $H^* \simeq H$. The inner product and the norm of H are denoted by $\langle \cdot, \cdot \rangle_H$ and $|\cdot|_H$, respectively. We set

$$\mathcal{F}C_b^1 = \left\{ u\colon E \to \mathbb{R} \;\middle|\; \begin{array}{l} u(z) = f(l_1(z), \dots, l_m(z)),\ l_1, \dots, l_m \in E^*, \\ f \in C_b^1(\mathbb{R}^m) \text{ for some } m \in \mathbb{N} \end{array} \right\}$$

and define $\mathcal{F}C_b^1(E^*)$ as the set of all linear combinations of H-valued functions $u(\cdot)l$ on E with $u \in \mathcal{F}C_b^1$ and $l \in E^* \subset H$. For $u \in \mathcal{F}C_b^1$, ∇u denotes the H-derivative of u, which is an H-valued function on E that is characterized by the identity

$$\langle \nabla u(z), l \rangle_H = \lim_{\varepsilon \to 0}(u(z + \varepsilon l) - u(z))/\varepsilon, \quad l \in E^* \subset H \subset E.$$

In the same way, ∇u is defined for $u \in \mathcal{F}C_b^1(E^*)$ as an $H \otimes H$-valued function on E.

Let $(X, \mathcal{X}, \nu)$ be a measure space and let Y be a separable Hilbert space with norm $|\cdot|_Y$. For $p \in [1, +\infty]$, we denote by $L^p(X \to Y; \nu)$ the space of all Y-valued L^p-functions on $(X, \mathcal{X}, \nu)$ with the norm defined as

$$\|f\|_{L^p(\nu)} = \begin{cases} \left(\displaystyle\int_X |f(x)|_Y^p \, \nu(dx) \right)^{1/p}, & p \in [1, +\infty), \\ \nu\text{-}\operatorname*{ess\,sup}_{x \in E} |f(x)|_Y, & p = +\infty. \end{cases}$$

As usual, two functions that are equal a.e. are identified. If $X = E$ or $\nu = \mu$ or $Y = \mathbb{R}$, we often omit these symbols from the notation. In particular, L^p and $\|\cdot\|_p$ represent $L^p(E; \mu)$ and $\|\cdot\|_{L^p(\mu)}$, respectively. For $p \in [1, +\infty)$, the closure of $\mathcal{F}C_b^1$ (resp. $\mathcal{F}C_b^1(E^*)$) with respect to the norm $(\|\cdot\|_p^p + \|\nabla \cdot\|_p^p)^{1/p}$ will be denoted by $\mathbb{D}^{1,p}$ (resp. $\mathbb{D}^{1,p}(H)$). The operator ∇ extends to a continuous map from $\mathbb{D}^{1,p}$ to $L^p(E \to H)$. Let ∇^* denote the adjoint operator of ∇. This is regarded as a bounded operator from $\mathbb{D}^{1,p}(H)$ to L^p for $p \in (1, \infty)$. We note that both ∇ and ∇^* have the local property in the sense that, if $f \in \mathbb{D}^{1,2}$ (resp. $G \in \mathbb{D}^{1,2}(H)$) satisfies

$f = 0$ (resp. $G = 0$) μ-a.e. on some measurable set A, then $\nabla f = 0$ (resp. $\nabla^* G = 0$) μ-a.e. on A. See, e.g., [11, Propositions 1.3.16 and 1.3.15] for the proof.

Let $L(\log L)^{1/2}$ denote the set of all real-valued μ-measurable functions f on E such that $f(0 \vee \log|f|)^{1/2} \in L^1$. For $\rho \in L(\log L)^{1/2}$, define

$$V(\rho) = \sup\left\{ \int_E (\nabla^* G)\rho \, d\mu \,\middle|\, G \in \mathcal{F}C_b^1(E^*),\ |G(z)|_H \le 1 \text{ for every } z \in E \right\}$$
$$(\le +\infty).$$

The function space BV on E is defined as

$$BV = \{\rho \in L(\log L)^{1/2} \mid V(\rho) < \infty\}.$$

A function in BV is called a BV function or a function of bounded variation. We remark that the following inequality holds for $p \in (1, +\infty]$ and $\rho \in BV \cap L^p$.

$$\int_E (\nabla^* G)\rho \, d\mu \le V(\rho)\|G\|_\infty, \quad G \in \mathbb{D}^{1,q}(H), \tag{2.1}$$

where q is the conjugate exponent of p.

For a function space $\mathcal{C}$ on E and a μ-measurable subset A of E, $\mathcal{C}_A$ denotes the space of all functions in $\mathcal{C}$ vanishing μ-a.e. on $E \setminus A$, and $\mathcal{C}_b$ denotes the set of all bounded functions in $\mathcal{C}$. Moreover, $\mathcal{C}_{A,b}$ denotes $\mathcal{C}_A \cap \mathcal{C}_b$.

For $\xi \in L^1$, F^ξ denotes the support of the measure $|\xi| \cdot \mu$. For $\xi \in L^1$ with $\xi \ge 0$ μ-a.e., we define a bilinear form $(\mathcal{E}, \mathcal{F}C_b^1)$ on $L^2(F^\xi, \xi \cdot \mu)$ by

$$\mathcal{E}^\xi(f,g) = \frac{1}{2}\int_E \langle \nabla f, \nabla g \rangle_H \, \xi \, d\mu, \quad f, g \in \mathcal{F}C_b^1.$$

The set of all ξ such that $(\mathcal{E}, \mathcal{F}C_b^1)$ is closable on $L^2(F^\xi; \xi \cdot \mu)$ is denoted by QR. For $\xi \in QR$, the closure of $(\mathcal{E}, \mathcal{F}C_b^1)$, denoted by $(\mathcal{E}^\xi, \mathcal{F}^\xi)$, is a quasi-regular Dirichlet form on $L^2(F^\xi; \xi \cdot \mu)$ (see, e.g., [12] and [5, Theorem 2.1] for the proof). The space $\mathcal{F}^\xi$ is regarded as a Hilbert space by using the inner product $(f,g) \mapsto \mathcal{E}^\xi(f,g) + \int_{F^\xi} fg\,\xi\,d\mu$. The associated capacity Cap^ξ is then defined as

$$\mathrm{Cap}^\xi(A) = \inf\left\{ \mathcal{E}^\xi(f,f) + \|f\|_{L^2(\xi\cdot\mu)}^2 \,\middle|\, \begin{array}{l} f \in \mathcal{F}^\xi \text{ and } f \ge 1 \ \xi\cdot\mu\text{-a.e. on} \\ \text{some open set including } A \end{array} \right\}$$

for $A \subset F^\xi$. The concept "$\mathcal{E}^\xi$-quasi-everywhere" ($\mathcal{E}^\xi$-q.e.) is based on this capacity. An increasing sequence $\{F_k\}_{k=1}^\infty$ of closed sets in F^ξ is called an $\mathcal{E}^\xi$-nest if $\lim_{k\to\infty} \mathrm{Cap}^\xi(F^\xi \setminus F_k) = 0$. It is known that $\{F_k\}_{k=1}^\infty$ is an $\mathcal{E}^\xi$-nest if and only if $\bigcup_{k=1}^\infty \mathcal{F}_{F_k}^\xi$ is dense in $\mathcal{F}^\xi$. In this situation, the set S^ξ of all smooth measures is described as the totality of all positive Borel measures ν on F^ξ such that ν charges no set of zero capacity Cap^ξ and there exists an $\mathcal{E}^\xi$-nest $\{F_k\}_{k=1}^\infty$ such that $\nu(F_k) < \infty$ for all k. A function f on E is called $\mathcal{E}^\xi$-quasi-continuous if there exists an $\mathcal{E}^\xi$-nest $\{F_k\}_{k=1}^\infty$ such that $f|_{F_k}$ is continuous on F_k for every k. Any function $f \in \mathcal{F}^\xi$ has an $\mathcal{E}^\xi$-quasi-continuous modification $\tilde{f}$. When $\xi \equiv 1$, we write $\mathcal{E}$, $\mathcal{F}$, and Cap instead of $\mathcal{E}^\xi$, $\mathcal{F}^\xi$, and Cap^ξ, respectively. We note that $\mathcal{F} = \mathbb{D}^{1,2}$. For a subset A

of E, e_A denotes the equilibrium potential of A with respect to $(\mathcal{E}, \mathcal{F})$: that is, e_A attains the infimum of $\{\mathcal{E}(f,f) + \|f\|_2^2 \mid f \in \mathbb{D}^{1,2}$ and $\tilde{f} \geq 1$ $\mathcal{E}$-q.e. on $A\}$, which is equal to $\mathrm{Cap}(A)$ (cf. [7, Theorem 2.1.5]). Henceforth, we will always assume that e_A is $\mathcal{E}$-quasi-continuous by itself. Any $G \in \mathbb{D}^{1,2}(H)$ also has an $\mathcal{E}$-quasi-continuous modification $\tilde{G}$ (cf. [2, Chapter VII, Theorem 1.3.1]).

One of the fundamental properties of BV functions is given by the following theorem.

Theorem 2.1 (cf. [6, Theorem 3.9]). *For $\rho \in BV$, there exists a positive finite measure ν on E and an H-valued Borel function σ on E such that $|\sigma|_H = 1$ ν-a.e., and for every $G \in \mathcal{F}C_b^1(E^*)$,*

$$\int_E (\nabla^* G)\rho \, d\mu = \int_E \langle G, \sigma \rangle_H \, d\nu. \tag{2.2}$$

The measure ν belongs to $S^{|\rho|+1}$. If $\rho \in QR$ in addition, then $\nu|_{E \setminus F^\rho} = 0$ and $\nu|_{F^\rho} \in S^\rho$. Also, ν and σ are uniquely determined in the following sense: if ν' and σ' are another pair satisfying (2.2) *for all $G \in \mathcal{F}C_b^1(E^*)$, then $\nu = \nu'$ and $\sigma = \sigma'$ ν-a.e.*

We now introduce some localized function spaces.

Definition 2.1. Let ρ be a real-valued μ-measurable function on E.

(i) For $p \in [1, +\infty]$, we say $\rho \in \dot{L}^p_{\mathrm{loc}}$ if there exists an $\mathcal{E}$-nest $\{F_k\}_{k=1}^\infty$ such that $\rho \cdot \mathbf{1}_{F_k} \in L^p$ for every $k \in \mathbb{N}$.
(ii) We say $\rho \in \dot{BV}_{\mathrm{loc}}$ if there exists an $\mathcal{E}$-nest $\{F_k\}_{k=1}^\infty$ and $\{\rho_k\}_{k=1}^\infty \subset BV$ such that $\rho = \rho_k$ μ-a.e. on F_k for every $k \in \mathbb{N}$.

The following are some properties of $\dot{BV}_{\mathrm{loc}}$.

Theorem 2.2. *For $\rho \in \dot{L}^2_{\mathrm{loc}}$, the implications* (a) $\Leftrightarrow$ (b) $\Leftarrow$ (c) *hold.*

(a) $\rho \in \dot{BV}_{\mathrm{loc}}$.
(b) *There exists an $\mathcal{E}$-nest $\{F_k\}_{k=1}^\infty$ and positive numbers $\{c_k\}_{k=1}^\infty$ such that for every $k \in \mathbb{N}$, $\rho \cdot \mathbf{1}_{F_k} \in L^2$ and*

$$\int_E \rho \nabla^* G \, d\mu \leq c_k \|G\|_\infty, \quad G \in \mathbb{D}^{1,2}(H)_{F_k,b}.$$

(c) *There exists an $\mathcal{E}$-nest $\{F_k\}_{k=1}^\infty$ and $\{f_m\}_{m=1}^\infty \subset \mathbb{D}^{1,1}$ such that for every $k \in \mathbb{N}$, $(f_m - \rho) \cdot \mathbf{1}_{F_k}$ converges to 0 in L^1 as $m \to \infty$ and $\|(\nabla f_m) \cdot \mathbf{1}_{F_k}\|_1$ is bounded in m.*

Unlike the case of finite-dimensional spaces, we must take care with the localized test functions on E. Indeed, except for the zero function there are no functions f in $\mathcal{F}C_b^1$ such that f vanishes outside a bounded set. This is why we introduce $\mathbb{D}^{1,2}(H)_{F_k,b}$ in the theorem above.

For the proof of Theorem 2.2, we introduce some lemmas.

Lemma 2.1. *Let $\{F_k\}_{k=1}^\infty$ be an $\mathcal{E}$-nest. Then, there exists another $\mathcal{E}$-nest $\{F'_k\}_{k=1}^\infty$ and functions $\{\varphi_k\}_{k=1}^\infty \subset \bigcup_{n=1}^\infty \mathbb{D}^{1,2}_{F_n}$ such that for any k, $F'_k \subset F_l$ for some $l \ge k$, φ_k is $\mathcal{E}$-quasi-continuous, $0 \le \varphi_k \le 1$ on E, $\varphi_k = 1$ on F'_k, and $\lim_{k\to\infty} \mathcal{E}(\varphi_k, \varphi_k) = 0$.*

Proof. We denote $E \setminus F_k$ by F_k^c, etc. Take an increasing sequence $\{n(k)\}_{k=1}^\infty$ of natural numbers such that $\mathrm{Cap}(F^c_{n(k)}) < 2^{-k}$ for every k. There exists an $\mathcal{E}$-nest $\{\hat{F}_l\}_{l=1}^\infty$ such that $e_{F^c_{n(k)}}$ is continuous on each $\hat{F}_l$ for every k (cf. [7, Theorem 2.1.2]). For $k \in \mathbb{N}$, let $\varphi_k = 1 \wedge 2(1 - e_{F^c_{n(k)}})$ and $\check{F}_k = \{\varphi_k = 1\}$. Since $\check{F}_k^c = \{e_{F^c_{n(k)}} > 1/2\}$,

$$\mathrm{Cap}(\check{F}_k^c) \le \mathcal{E}(2e_{F^c_{n(k)}}, 2e_{F^c_{n(k)}}) + \|2e_{F^c_{n(k)}}\|_2^2 = 4\mathrm{Cap}(F^c_{n(k)}) < 2^{-k+2}.$$

Define $F'_k = \bigcap_{l=k}^\infty (\check{F}_l \cap \hat{F}_l \cap F_{n(l)})$. Then, $\{F'_k\}_{k=1}^\infty$ is a nondecreasing sequence of closed sets and

$$\begin{aligned}
\mathrm{Cap}((F'_k)^c) &= \mathrm{Cap}\left(\left(\bigcup_{l=k}^\infty \check{F}_l^c\right) \cup \hat{F}_k^c \cup F^c_{n(k)}\right) \\
&\le \sum_{l=k}^\infty \mathrm{Cap}(\check{F}_l^c) + \mathrm{Cap}(\hat{F}_k^c) + \mathrm{Cap}(F^c_{n(k)}) \\
&\le 2^{-k+3} + \mathrm{Cap}(\hat{F}_k^c) + \mathrm{Cap}(F^c_{n(k)}),
\end{aligned}$$

which converges to 0 as $k \to \infty$. Furthermore, $\mathcal{E}(\varphi_k, \varphi_k) \le 4\mathcal{E}(e_{F^c_{n(k)}}, e_{F^c_{n(k)}}) < 2^{-k+2} \to 0$ as $k \to \infty$. □

Lemma 2.2. *Let $p \in (1, +\infty]$ and q denote the conjugate exponent of p.*

(i) *Let $f \in BV \cap L^p$ and $g \in \mathbb{D}_b^{1,q}$. Then, $fg \in BV$ and $V(fg) \le V(f)\|g\|_\infty + \|f\nabla g\|_1$.*
(ii) *Let $f \in BV$ and $g \in \mathbb{D}^{1,q}_{F,b}$ for some μ-measurable set $F \subset E$. If $f \cdot \mathbf{1}_F \in L^p$, then $fg \in BV \cap L^p$ and $V(fg) \le V(f)\|g\|_\infty + \|f\nabla g\|_1$.*

Proof. (i) Let $\{T_t\}_{t>0}$ be the Ornstein–Uhlenbeck semigroup on E. For $t > 0$, $T_t f \in \mathbb{D}^{1,p}$, thus $(T_t f)g \in \mathbb{D}^{1,1}$ and

$$\begin{aligned}
\|\nabla((T_t f)g)\|_1 &\le \|\nabla(T_t f)\|_1 \|g\|_\infty + \|(T_t f)\nabla g\|_1 \\
&\le e^{-t} V(f)\|g\|_\infty + \|(T_t f)\nabla g\|_1,
\end{aligned}$$

from, e.g., [6, Proposition 3.6]. Since $\lim_{t\to 0}(T_t f)g = fg$ in L^1 and $\lim_{t\to 0}(T_t f)\nabla g = f\nabla g$ in $L^1(H)$, Theorem 3.7 of [6] completes the proof.

(ii) For $n \in \mathbb{N}$, let $f_n = (-n) \vee (f \wedge n)$. From [6, Corollary 3.8], $f_n \in BV$ and $V(f_n) \le V(f)$. By the assumption and (i), $f_n g \in BV \cap L^p$, $\|f_n g\|_p \le \|fg\|_p$, and

$$V(f_n g) \le V(f_n)\|g\|_\infty + \|f_n \nabla g\|_1 \le V(f)\|g\|_\infty + \|f\nabla g\|_1,$$

which is bounded for each n. Since $\lim_{n\to\infty} f_n g = fg$ in L^1, the claim follows. □

Lemma 2.3. *For $\rho \in \dot{BV}_{\mathrm{loc}} \cap \dot{L}^2$, there exists an $\mathcal{E}$-nest $\{F_k\}_{k=1}^\infty$ and functions $\{\rho_k\}_{k=1}^\infty$ in $BV \cap L^2$ such that $\rho = \rho_k$ on F_k for every $k \in \mathbb{N}$.*

Proof. There exist an $\mathcal{E}$-nest $\{F_k\}_{k=1}^\infty$ and $\{\rho_k\}_{k=1}^\infty \subset BV$ such that for every k, $\rho \cdot \mathbf{1}_{F_k} \in L^2$ and $\rho = \rho_k$ on F_k. Take the $\mathcal{E}$-nest $\{F'_k\}_{k=1}^\infty$ and the functions $\{\varphi_k\}_{k=1}^\infty$ in Lemma 2.1. From Lemma 2.2, the assertion is true by taking $\{F'_k\}_{k=1}^\infty$ and $\{\rho\varphi_k\}_{k=1}^\infty$ as $\{F_k\}_{k=1}^\infty$ and $\{\rho_k\}_{k=1}^\infty$, respectively. □

Proof of Theorem 2.2. (a)⇒(b): Take the $\mathcal{E}$-nest $\{F_k\}_{k=1}^\infty$ and $\{\rho_k\}_{k=1}^\infty \subset BV \cap L^2$ as in Lemma 2.3. Let $k \in \mathbb{N}$ and take $G \in \mathbb{D}^{1,2}(H)_{F_k,b}$. Since $\nabla^* G = 0$ μ-a.e. on $E \setminus F_k$,

$$\int_E \rho \nabla^* G \, d\mu = \int_E \rho_k \nabla^* G \, d\mu \le V(\rho_k) \|G\|_\infty,$$

where the last inequality follows from (2.1). Thus, (b) holds.

(b)⇒(a): Take the $\mathcal{E}$-nest $\{F'_k\}_{k=1}^\infty$ and the functions $\{\varphi_k\}_{k=1}^\infty$ as in Lemma 2.1. For $k \in \mathbb{N}$, take $l \in \mathbb{N}$ such that $\varphi_k \in \mathbb{D}^{1,2}_{F_l}$ and let $\rho_k = \rho\varphi_k$. Then, $\rho_k \in L^2$ and $\rho_k = \rho$ on F'_k. Moreover, for any $G \in \mathcal{F}C_b^1(E^*)$,

$$\begin{aligned}\int_E \rho_k \nabla^* G \, d\mu &= \int_E \rho\varphi_k \nabla^* G \, d\mu \\ &= \int_E \rho\{\nabla^*(\varphi_k G) + \langle \nabla\varphi_k, G\rangle_H\} \, d\mu \\ &\le c_l \|\varphi_k G\|_\infty + \|\rho \cdot \mathbf{1}_{F_l}\|_2 \|\nabla\varphi_k\|_2 \|G\|_\infty \\ &\le (c_l + \|\rho \cdot \mathbf{1}_{F_l}\|_2 \|\nabla\varphi_k\|_2)\|G\|_\infty.\end{aligned}$$

Therefore, $\rho_k \in BV$.

(c)⇒(b): We may assume that $\rho \cdot \mathbf{1}_{F_k} \in L^2$ for every k. For $M > 0$, define $\Phi_M(t) = (-M) \vee (t \wedge M)$ for $t \in \mathbb{R}$. Then, for $k \in \mathbb{N}$ and $G \in \mathbb{D}^{1,2}(H)_{F_k,b}$,

$$\begin{aligned}\int_E \Phi_M(f_m)\nabla^* G \, d\mu &= \int_E \langle \nabla(\Phi_M(f_m)), G\rangle_H \, d\mu \\ &\le \|(\nabla f_m) \cdot \mathbf{1}_{F_k}\|_1 \|G\|_\infty \le c_k \|G\|_\infty,\end{aligned}$$

where $c_k := \sup_{m\in\mathbb{N}} \|(\nabla f_m) \cdot \mathbf{1}_{F_k}\|_1 < \infty$. Since $\nabla^* G \in L^2$,

$$\lim_{m\to\infty} \int_E \Phi_M(f_m)\nabla^* G \, d\mu = \int_E \Phi_M(\rho)\nabla^* G \, d\mu.$$

Since $\lim_{M\to\infty}(\Phi_M(\rho) - \rho) \cdot \mathbf{1}_{F_k} = 0$ in L^2,

$$\lim_{M\to\infty} \int_E \Phi_M(\rho)\nabla^* G \, d\mu = \int_E \rho \nabla^* G \, d\mu.$$

Therefore, we have $\int_E \rho\nabla^* G \, d\mu \le c_k \|G\|_\infty$. □

The following are some basic lemmas that are used later.

Lemma 2.4. *The following claims hold.*

(i) *For $\rho \in L^\infty$, $\mathcal{F}^{|\rho|+1} = \mathbb{D}^{1,2}$ and their norms are equivalent.*

(ii) *Let* $\rho \in QR \cap L^\infty$ *and* $\{F_k\}_{k=1}^\infty$ *be an* $\mathcal{E}$*-nest. Then,* $\mathbb{D}^{1,2}|_{F^\rho}$ *is continuously embedded in* $\mathcal{F}^\rho$ *and there exists* $c > 0$ *such that* $\mathrm{Cap}^\rho(A) \le c\mathrm{Cap}(A)$ *for any* $A \subset F^\rho$*. In particular,* $\{F_k \cap F^\rho\}_{k=1}^\infty$ *is an* $\mathcal{E}^\rho$*-nest. Moreover,* $\bigcup_{k=1}^\infty \mathbb{D}^{1,2}_{F_k,b}|_{F^\rho}$ *is dense in* $\mathcal{F}^\rho$*.*

Proof. We prove only the last claim of (ii). It suffices to prove that for any function $f \in \mathcal{F}C_b^1$, $f|_{F^\rho}$ can be approximated by elements of $\bigcup_{k=1}^\infty \mathbb{D}^{1,2}_{F_k,b}|_{F^\rho}$ in $\mathcal{F}^\rho$. Since $\bigcup_{k=1}^\infty \mathbb{D}^{1,2}_{F_k,b}$ is dense in $\mathbb{D}^{1,2}$, there exists $\{f_n\}_{n=1}^\infty$ in $\bigcup_{k=1}^\infty \mathbb{D}^{1,2}_{F_k,b}$ such that f_n converges to f in $\mathbb{D}^{1,2}$. Then, $f_n|_{F^\rho}$ converges to $f|_{F^\rho}$ in $\mathcal{F}^\rho$. □

Lemma 2.5. *Let* $\{F_k\}_{k=1}^\infty$ *be an* $\mathcal{E}$*-nest. Let* ν *and* ν' *be* $\mathcal{E}$*-smooth measures on* E *such that* $\nu(F_k) < \infty$ *and* $\nu'(F_k) < \infty$ *for every* $k \in \mathbb{N}$*. Let* σ *and* σ' *be* H*-valued Borel functions on* E *such that* $|\sigma|_H = 1$ ν*-a.e. and* $|\sigma'|_H = 1$ ν'*-a.e. If* $\int_E \langle \tilde{G}, \sigma\rangle_H \, d\nu = \int_E \langle \tilde{G}, \sigma'\rangle_H \, d\nu'$ *for every* $G \in \bigcup_{k=1}^\infty \mathbb{D}^{1,2}(H)_{F_k,b}$*, then* $\nu = \nu'$ *and* $\sigma = \sigma'$ ν*-a.e.*

Proof. Let $\xi = \nu + \nu'$ and $\gamma = \sigma \frac{d\nu}{d\xi} - \sigma' \frac{d\nu'}{d\xi}$. Then, $\int_E \langle \tilde{G}, \gamma\rangle_H \, d\xi = 0$ for every $G \in \bigcup_{k=1}^\infty \mathbb{D}^{1,2}_{F_k,b}$. Taking a uniformly bounded sequence $\{G_n\}_{n=1}^\infty$ from $\bigcup_{k=1}^\infty \mathbb{D}^{1,2}_{F_k,b}$ such that $\langle \tilde{G}_n, \gamma\rangle_H \to |\gamma|_H$ ξ-a.e. as $n \to \infty$, we obtain $\gamma = 0$ ξ-a.e. Therefore, $|\sigma|_H \frac{d\nu}{d\xi} = |\sigma'|_H \frac{d\nu'}{d\xi}$ ξ-a.e. Since $|\sigma|_H = 1$ ν-a.e., $|\sigma|_H \frac{d\nu}{d\xi} = \frac{d\nu}{d\xi}$ ξ-a.e. Similarly, $|\sigma'|_H \frac{d\nu'}{d\xi} = \frac{d\nu'}{d\xi}$ ξ-a.e. Then, $\frac{d\nu}{d\xi} = \frac{d\nu'}{d\xi}$ ξ-a.e., which implies $\nu = \nu'$. The identity $\sigma = \sigma'$ ν-a.e. follows from $\gamma = 0$ ξ-a.e. and $\nu = \nu'$. □

Theorem 2.3. *For* $\rho \in \dot{BV}_{\mathrm{loc}} \cap \dot{L}^2$*, there exist an* H*-valued Borel function* σ *on* E*, an* $\mathcal{E}$*-smooth measure* ν *on* E*, and an* $\mathcal{E}$*-nest* $\{F_k\}_{k=1}^\infty$ *such that* $\rho \cdot \mathbf{1}_{F_k} \in L^2$ *and* $\nu(F_k) < \infty$ *for every* $k \in \mathbb{N}$*,* $|\sigma|_H = 1$ ν*-a.e., and*

$$\int_E \rho \nabla^* G \, d\mu = \int_E \langle \tilde{G}, \sigma\rangle_H \, d\nu \quad \text{for every } G \in \bigcup_{k=1}^\infty \mathbb{D}^{1,2}(H)_{F_k,b}, \tag{2.3}$$

where $\tilde{G}$ *denotes an* $\mathcal{E}$*-quasi-continuous modification of* G*. The pair* ν *and* σ *is uniquely determined in the following sense: if another pair* ν' *and* σ' *with some* $\mathcal{E}$*-nest satisfies the conditions above, then* $\nu = \nu'$ *and* $\sigma = \sigma'$ ν*-a.e.*

If $\rho \in QR \cap L^\infty$ *in addition, then* $\nu|_{E \setminus F^\rho} = 0$ *and* $\nu|_{F^\rho} \in S^\rho$*.*

Proof. Take an $\mathcal{E}$-nest $\{F_k\}_{k=1}^\infty$ and $\{\rho_k\}_{k=1}^\infty \subset BV \cap L^2$ as in Lemma 2.3. From Theorem 2.1, for each k, there exist an H-valued Borel function σ_k on E and an $\mathcal{E}$-smooth finite measure ν_k on E such that $|\sigma_k|_H = 1$ ν_k-a.e. and

$$\int_E \rho_k \nabla^* G \, d\mu = \int_E \langle G, \sigma_k\rangle_H \, d\nu_k \tag{2.4}$$

for every $G \in \mathcal{F}C_b^1(E^*)$. By approximation, (2.4) holds for $G \in \mathbb{D}^{1,2}(H)_b$ with $\langle G, \sigma_k\rangle_H$ replaced by $\langle \tilde{G}, \sigma_k\rangle_H$.

Take an $\mathcal{E}$-nest $\{F'_k\}_{k=1}^\infty$ and functions $\{\varphi_k\}_{k=1}^\infty$ as in Lemma 2.1. For $k \in \mathbb{N}$, let $n(k)$ denote a number such that $\varphi_k \in \mathbb{D}^{1,2}_{F_{n(k)}}$. We may assume that $\{n(k)\}_{k=1}^\infty$ is an increasing sequence. Then, for $l \geq k$,

$$\varphi_k \cdot \nu_{n(k)} = \varphi_k \cdot \nu_{n(l)} \text{ and } \sigma_{n(k)} = \sigma_{n(l)} \ (\varphi_k \cdot \nu_{n(k)})\text{-a.e.} \tag{2.5}$$

Indeed, for any $G \in \mathbb{D}^{1,2}(H)_b$, we have $\varphi_k G \in \mathbb{D}^{1,2}(H)_{F_{n(k)},b}$ and

$$\begin{aligned}\int_E \varphi_k \langle \tilde{G}, \sigma_{n(k)} \rangle_H \, d\nu_{n(k)} &= \int_E \rho_{n(k)} \nabla^*(\varphi_k G) \, d\mu \\ &= \int_E \rho_{n(l)} \nabla^*(\varphi_k G) \, d\mu \\ &= \int_E \varphi_k \langle \tilde{G}, \sigma_{n(l)} \rangle_H \, d\nu_{n(l)}.\end{aligned}$$

Thus, (2.5) follows from Lemma 2.5. Therefore, we can define σ and ν so that $\sigma = \sigma_{n(k)}$ on F'_k and $\nu|_{F'_k} = \nu_{n(k)}|_{F'_k}$ for $k \in \mathbb{N}$, and $\nu(E \setminus \bigcup_{k=1}^\infty F'_k) = 0$. Then, the conditions described in the theorem are satisfied with $\{F_k\}_{k=1}^\infty$ replaced by $\{F'_k\}_{k=1}^\infty$.

Suppose that another ν', σ', and an $\mathcal{E}$-nest $\{F''_k\}_{k=1}^\infty$ satisfy the required conditions. Applying Lemma 2.5 to the $\mathcal{E}$-nest $\{F_k \cap F''_k\}_{k=1}^\infty$, we obtain that $\nu = \nu'$ and $\sigma = \sigma'$ ν-a.e.

Lastly, suppose $\rho \in QR \cap L^\infty$ in addition. Let $l \in E^*$ and $g \in \mathbb{D}^{1,2}_{F_k,b}$ for some $k \in \mathbb{N}$. Letting $G = g(\cdot)l$ in (2.3), we obtain

$$\int_{F^\rho} (\partial_l g + g \cdot l(\cdot))\rho \, d\mu = \int_E \tilde{g} \langle l, \sigma \rangle_H \, d\nu. \tag{2.6}$$

Denote the left-hand side of (2.6) by $I(g)$. Then, I provides a bounded functional on $\mathcal{F}^\rho$ as well as on $\mathbb{D}^{1,2}$. From the observation that $I(g)$ does not change if g is replaced by $(-M) \vee (g \wedge M)$ with $M = \mu\text{-ess sup}|g \cdot \mathbf{1}_{F_k}|$, the inequality $I(g) \leq \|g|_{F^\rho}\|_{L^\infty(\rho\cdot\mu)} |l|_H \nu(F_k)$ holds. Then, from Lemma 2.4 and [8, Theorem 2.18], there exists a unique $\nu_l \in S^\rho$ such that $I(g) = \int_{F^\rho} \tilde{g} \, d\nu_l$ for all $g \in \bigcup_{k=1}^\infty \mathbb{D}^{1,2}_{F_k,b}$. From the uniqueness of the integral representation of $I(g)$ as a functional on $\mathbb{D}^{1,2}$, we conclude that $(\langle l, \sigma \rangle_H \cdot \nu)|_{E \setminus F^\rho} = 0$ and $(\langle l, \sigma \rangle_H \cdot \nu)|_{F^\rho} = \nu_l \in S^\rho$. Because l is arbitrary, we deduce that $\nu|_{E \setminus F^\rho} = 0$ and $\nu|_{F^\rho} \in S^\rho$. □

We will denote ν and σ in the theorem above by $\|D\rho\|$ and σ_ρ, respectively.

Theorem 2.4 (Skorokhod representation). *Let $\rho \in BV_{\mathrm{loc}} \cap QR \cap L^\infty$. Then the sample path of the diffusion process $\mathbf{M}^\rho = (X_t, \mathcal{M}_t, P_z)$ associated with $(\mathcal{E}^\rho, \mathcal{F}^\rho)$ admits the following expression as a sum of three E-valued continuous additive functionals.*

$$\begin{aligned}&X_t(\omega) - X_0(\omega) \\ &= W_t(\omega) - \frac{1}{2}\int_0^t X_s(\omega) \, ds + \frac{1}{2}\int_0^t \sigma_\rho(X_s(\omega)) \, dA_s^{\|D\rho\|}(\omega), \quad t \geq 0.\end{aligned} \tag{2.7}$$

Here, $A^{\|D\rho\|}$ is a real-valued positive continuous additive functional associated with $\|D\rho\|$ via the Revuz correspondence. Moreover, for $\mathcal{E}^\rho$-q.e. $z \in F^\rho$, $\{W_t\}_{t\ge 0}$ is the $\{\mathcal{M}_t\}$-Brownian motion on E under P_z.

Proof. The proof is provided along the same lines as those of [6, Theorem 4.2] and [5, Theorem 3.2] with the use of [4, Theorem 6.1] instead of [4, Theorem 6.2] (or [5, Theorem 2.2]). We define $\{W_t\}_{t\ge 0}$ so that (2.7) holds. Then, $\{W_t\}_{t\ge 0}$ is an E-valued continuous additive functional with the same defining and exceptional sets as $A^{\|D\rho\|}$. Take the $\mathcal{E}$-nest $\{F_k\}_{k=1}^\infty$ in Theorem 2.3. Let $l \in E^*$ and consider the identity

$$l(X_t) - l(X_0) = l(W_t) - \frac{1}{2}\int_0^t l(X_s)\,ds + \frac{1}{2}\int_0^t \langle l, \sigma_\rho(X_s)\rangle_H\,dA_s^{\|D\rho\|}.$$

We note that $l(\cdot)|_{F^\rho} \in \mathcal{F}^\rho$. From (2.3), the identity

$$\mathcal{E}^\rho(l(\cdot), g) = \frac{1}{2}\int_{F^\rho} g l(\cdot)\rho\,d\mu - \frac{1}{2}\int_{F^\rho} \tilde{g}\langle l, \sigma_\rho\rangle_H\,d\|D\rho\|$$

holds for $g \in \bigcup_{k=1}^\infty \mathbb{D}^{1,2}_{F_k,b}$. Take a nest $\{F'_k\}_{k=1}^\infty$ and functions $\{\varphi_k\}_{k=1}^\infty$ as in Lemma 2.1. For any $g \in \mathcal{F}^\rho_{F'_k,b}$ with $k \in \mathbb{N}$, there exist $\{g_n\} \subset \bigcup_{m=1}^\infty \mathbb{D}^{1,2}_{F_m,b}$ such that $\{g_n\}$ are uniformly bounded and $g_n|_{F^\rho}$ converges to g in $\mathcal{F}^\rho$ as $n \to \infty$. Since $\mathcal{E}(g_n\varphi_k, g_n\varphi_k)^{1/2} \le \mathcal{E}(g_n, g_n)^{1/2}\|\varphi_k\|_\infty + \mathcal{E}(\varphi_k, \varphi_k)^{1/2}\|g_n\|_\infty$, $\|g_n\varphi_k\|_2 \le \|g_n\|_\infty\|\varphi_k\|_2$, and $(g_n\varphi_k)|_{F^\rho} \to g$ in $L^2(\rho\cdot\mu)$, the Cesàro means of a certain subsequence of $\{(g_n\varphi_k)|_{F^\rho}\}_{n=1}^\infty$, which are all elements of $\mathbb{D}^{1,2}_{F_l,b}|_{F^\rho}$ for some l, converges to g in $\mathcal{F}^\rho$. A further suitable subsequence of their $\mathcal{E}^\rho$-quasi-continuous modifications converges $\mathcal{E}^\rho$-q.e. from [7, Theorem 2.1.4]. Therefore, the above identity holds for $g \in \bigcup_{k=1}^\infty \mathcal{F}^\rho_{F'_k,b}$, where $\tilde{g}$ is interpreted as an $\mathcal{E}^\rho$-quasi-continuous modification of g. From [4, Theorem 6.1], the Fukushima decomposition of $l(X_t) - l(X_0)$ is given by the sum of $M_t^l := l(W_t)$ and $N_t^l := -\frac{1}{2}\int_0^t l(X_s)\,ds + \frac{1}{2}\int_0^t \langle l, \sigma_\rho(X_s)\rangle_H\,dA_s^{\|D\rho\|}$. Moreover, the quadratic variation of $\{M_t^l\}$ is equal to $\{t|l|_H^2\}$. From [6, Lemma 4.1], we conclude that $\{W_t\}_{t\ge 0}$ is the E-valued Brownian motion, which completes the proof. □

3. Indicator functions in $\dot{BV}_{\mathrm{loc}}$

In this section, we provide some nontrivial examples of functions in $\dot{BV}_{\mathrm{loc}}$. Let $d \in \mathbb{N}$ and $T > 0$. We consider the classical d-dimensional Wiener space as (E, H, μ): that is,

$$E = \{w \in C([0,T] \to \mathbb{R}^d) \mid w(0) = 0\},$$

$$H = \left\{ h \in E \,\middle|\, h \text{ is absolutely continuous and } \int_0^T |\dot{h}(s)|^2_{\mathbb{R}^d}\,ds < \infty \right\},$$

and μ is the Wiener measure on E. For a subset A of $\mathbb{R}^d$, $\overline{A}$ (resp. ∂A) denotes the closure (resp. boundary) of A, and A^c denotes $\mathbb{R}^d \setminus A$. Define some subsets of E as follows:

$$\Xi_A = \{w \in E \mid w(t) \in A \text{ for all } t \in [0,T]\},$$
$$\Theta_A = \{w \in E \mid w(t) \in A \text{ for some } t \in [0,T]\},$$
$$\partial\Xi_A = \left\{ w \in E \,\middle|\, \begin{array}{l} w(t) \in \overline{A} \text{ for all } t \in [0,T] \text{ and} \\ w(s) \in \partial A \text{ for some } s \in [0,T] \end{array} \right\},$$
$$\partial'\Xi_A = \left\{ w \in E \,\middle|\, \begin{array}{l} w(t) \in \overline{A} \text{ for all } t \in [0,T] \text{ and } w(s) \in \partial A \text{ for} \\ \text{some } unique\ s \in [0,T] \end{array} \right\}.$$

We note that $\partial\Xi_A$ is the topological boundary of Ξ_A in E with the uniform topology. A sufficient condition for $\mathbf{1}_{\Xi_A}$ to belong to BV was given in [9, 13]. Here, we provide a sufficient condition for $\mathbf{1}_{\Xi_A}$ to belong to $\dot{BV}_{\mathrm{loc}}$.

For $x \in \mathbb{R}^d$ and $r \geq 0$, we write $B(x,r)$ and $\overline{B}(x,r)$ for $\{z \in \mathbb{R}^d \mid |z-x|_{\mathbb{R}^d} < r\}$ and $\{z \in \mathbb{R}^d \mid |z-x|_{\mathbb{R}^d} \leq r\}$, respectively. Let O be a proper open subset of $\mathbb{R}^d$ such that $0 \in O$. For $y \in \partial O$, we define $\delta(y) \in [0, +\infty]$ as

$$\delta(y) = \sup\{r \geq 0 \mid \text{there exists } z \in O^c \text{ such that } \overline{O} \cap \overline{B}(z,r) = \{y\}\}.$$

From [9, Theorem 5.1], $\mathbf{1}_{\Xi_{\overline{O}}} \in BV$ if the uniform exterior ball condition

$$\inf_{y \in \partial O} \delta(y) > 0 \tag{3.1}$$

holds. In order to describe a weaker condition which ensures that $\mathbf{1}_{\Xi_{\overline{O}}} \in \dot{BV}_{\mathrm{loc}}$, we introduce the Riesz (and logarithmic) capacities on $\mathbb{R}^d$. For a Borel set A of $\mathbb{R}^d$, let $\mathscr{P}(A)$ denote the set of all Borel probability measures on A. For $\beta \geq 0$, the β-capacity $\mathrm{Cap}_\beta(A)$ of A is defined as

$$\mathrm{Cap}_\beta(A) = \left(\inf_{\lambda \in \mathscr{P}(A)} \iint_{A \times A} g_\beta(|x-y|_{\mathbb{R}^d})\, \lambda(dx)\lambda(dy) \right)^{-1},$$

where $g_\beta(t) = t^{-\beta}$ for $\beta > 0$ and $g_0(t) = \log(t^{-1} \vee e)$. For $\beta < 0$, we define $\mathrm{Cap}_\beta(A) = 1$ for $A \neq \emptyset$ and $\mathrm{Cap}_\beta(\emptyset) = 0$. These are Choquet capacities.

For $\beta \geq 0$, let $\mathscr{H}^\beta(A)$ denote the β-dimensional Hausdorff measure of $A \subset \mathbb{R}^d$. It is known that, if a Borel set A of $\mathbb{R}^d$ satisfies $\mathscr{H}^\beta(A) < \infty$, then $\mathrm{Cap}_\beta(A) = 0$ (see, e.g., [3, §1, Theorem 1]).

For $r > 0$, a closed subset Σ_r of $\mathbb{R}^d$ is defined as

$$\Sigma_r = \overline{\{y \in \partial O \mid \delta(y) < r\}}. \tag{3.2}$$

Denote $\bigcap_{r>0} \Sigma_r$ by Σ. The following is the main theorem of this section.

Theorem 3.1. *Suppose that*

$$\mathrm{Cap}_{d-4}(\Sigma) = 0. \tag{3.3}$$

Then, $\mathbf{1}_{\Xi_{\overline{O}}} \in \dot{BV}_{\mathrm{loc}} \cap QR$. *In particular, the conclusion of Theorem 2.4 holds with* $\rho = \mathbf{1}_{\Xi_{\overline{O}}}$. *Moreover,* $\mathrm{Cap}^{\mathbf{1}_{\Xi_{\overline{O}}}}(\partial\Xi_{\overline{O}} \setminus \partial'\Xi_{\overline{O}}) = 0$.

Remark 3.1.

(i) The set $\Xi_{\overline{O}}$ coincides with the closure of Ξ_O in E.
(ii) If $d \geq 4$ and there exists a sequence of Borel subsets $\{B_n\}_{n=1}^\infty$ of $\mathbb{R}^d$ such that $\bigcup_{n=1}^\infty B_n = \mathbb{R}^d$ and $\mathscr{H}^{d-4}(\Sigma \cap B_n) < \infty$ for every $n \in \mathbb{N}$, then (3.3) holds.
(iii) Since $\inf_{z\in\partial O} \delta(z) > 0$ implies $\Sigma = \emptyset$, the uniform exterior condition (3.1) implies condition (3.3).
(iv) The diffusion $\{X_t\}$ associated with $(\mathcal{E}^\rho, \mathcal{F}^\rho)$ for $\rho = \mathbf{1}_{\Xi_{\overline{O}}}$ is regarded as the modified reflecting Ornstein–Uhlenbeck process on $\Xi_{\overline{O}}$; the term "modified" is added since the domain $\mathcal{F}^\rho$ is defined as the closure of smooth functions and it is not clear whether $\mathcal{F}^\rho$ is the maximal domain.
(v) The fact that $\mathrm{Cap}^{\mathbf{1}_{\Xi_{\overline{O}}}}(\partial\Xi_{\overline{O}} \setminus \partial'\Xi_{\overline{O}}) = 0$ implies that the measure $\|D\mathbf{1}_{\Xi_{\overline{O}}}\|$ concentrates on $\partial'\Xi_{\overline{O}}$, which means that the process $\{X_t\}$ reflects only at $\partial'\Xi_{\overline{O}}$ in view of (2.7).

Example 3.1. Let $\varphi\colon [1,2) \to \mathbb{R}_+$ be a convex function such that $\varphi(1) = 0$ and $\varphi(t) > 0$ for $t \in (1,2)$. Suppose $d \geq 4$ and consider an open subset O of $\mathbb{R}^d$ that is defined by

$$O = B(0,2) \setminus \{(x_1,\ldots,x_d) \in \mathbb{R}^d \mid x_d \in [1,2) \text{ and } \varphi(x_d)^2 \geq x_1^2 + \cdots + x_{d-1}^2\}.$$

Then $\Sigma = \{(0,\ldots,0,1)\}$ and $\mathscr{H}^{d-4}(\Sigma) < \infty$. Therefore, $\mathbf{1}_{\Xi_{\overline{O}}} \in BV_{\mathrm{loc}}$ from Theorem 3.1 and Remark 3.1 (ii). It is likely that $\mathbf{1}_{\Xi_{\overline{O}}} \notin BV$ in general, but I have no proof to offer at the moment.

The capacity Cap_{d-4} is involved in Theorem 3.1 for the following reason. From Theorem 1.1 of [10] and the arguments in Section 7 in that paper, for $a \in (0,T]$ and $M > 0$, there exists a constant $C > 0$ depending only on a and M such that, for any closed set A in $\overline{B}(0,M) \subset \mathbb{R}^d$,

$$\begin{aligned} C^{-1}\mathrm{Cap}_{d-4}(A) &\leq \mathbf{P}(\{Z_s(t) \in A \text{ for some } (s,t) \in [a,T] \times [a,T]\}) \\ &\leq C\,\mathrm{Cap}_{d-4}(A), \end{aligned}$$

where $(\{Z_s\}_{s\geq 0}, \mathbf{P})$ is the Ornstein–Uhlenbeck process on E with initial distribution μ. This implies the following property.

If a closed subset A of $\mathbb{R}^d \setminus \{0\}$ satisfies $\mathrm{Cap}_{d-4}(A) = 0$, then $\mathrm{Cap}(\Theta_A) = 0$. (3.4)

In particular, since $0 \notin \Sigma$, condition (3.3) implies

$$\mathrm{Cap}(\Theta_\Sigma) = 0. \tag{3.5}$$

For the proof of Theorem 3.1, we provide some quantitative estimates as in [9, 13]. We define a Lipschitz continuous function q on $\mathbb{R}^d$ by

$$q(x) = \inf_{y\in O^c} |x-y|_{\mathbb{R}^d} - \inf_{y\in O} |x-y|_{\mathbb{R}^d}, \quad x \in \mathbb{R}^d.$$

For $r \geq 0$, set $O_r = \{x \in \mathbb{R}^d \mid q(x) > r\}$. Note that $O_0 = O$ and $\{q(x) \geq 0\} = \overline{O}$.

Let $W = C([0,\infty) \to \mathbb{R}^d)$ and $\mathcal{B}(W)$ be the Borel σ-field of W. Let $\{\hat{P}_x\}_{x\in\mathbb{R}^d}$ be the probability measures on W such that the coordinate process $\{\omega_t\}_{t\geq 0}$ is the

d-dimensional Brownian motion starting at x under $\hat{P}_x$ for each $x \in \mathbb{R}^d$. For $t \geq 0$, let $\hat{\mathcal{F}}_t$ denote the σ-field generated by $\{\omega_s \mid s \in [0,t]\}$. For an $\{\hat{\mathcal{F}}_t\}$-stopping time τ, define $\hat{\mathcal{F}}_\tau$ as $\{A \in \mathcal{B}(W) \mid A \cap \{\tau \leq t\} \in \hat{\mathcal{F}}_t \text{ for all } t \geq 0\}$. We denote the integral with respect to $\hat{P}_x$ by $\hat{E}_x$. For $s > 0$, the shift operator $\theta_s \colon W \to W$ is defined by $(\theta_s\omega)_t = \omega_{s+t}$, $t \geq 0$.

The following claims (Lemma 3.1–Proposition 3.2) are slight modifications of those in [9].

Lemma 3.1. *Let $x \in \overline{O}$ and choose $y \in \partial O$ such that $q(x) = |x-y|_{\mathbb{R}^d}$. Suppose that $\delta(y) > 0$. Let $\delta \in (0, \delta(y))$ and take $z \in O^c$ such that $B(z,\delta) \cap \overline{O} = \{y\}$. Let $C_\delta = (d-1)/(2\delta)$ and $R_t = |\omega_t - z|_{\mathbb{R}^d}$ for $\omega = \{\omega_t\} \in W$. Then, for each $u > 0$,*

$$\{R_t \geq \delta \text{ for all } t \in [0,u]\} \subset \{R_t \leq q(x) + \delta + C_\delta t + S_t \text{ for all } t \in [0,u]\}$$

up to a $\hat{P}_x$-null set. Here, S_t is the 1-dimensional Brownian motion under $\hat{P}_x$ starting at 0 that is defined by

$$S_t(\omega) = \sum_{i=1}^d \int_0^t \frac{\omega_s^{(i)} - z^{(i)}}{R_s}\, d\omega_s^{(i)}, \quad \omega_s = (\omega_s^{(1)}, \dots, \omega_s^{(d)}),\ z = (z^{(1)}, \dots, z^{(d)}),$$

up to the $\{\hat{\mathcal{F}}_t\}$-stopping time $\inf\{t \geq 0 \mid R_t = 0\}$.

Proof. Define $\sigma = \inf\{t \geq 0 \mid R_t = 0\}$. Note that $R_0 = |x-z|_{\mathbb{R}^d} = q(x) + \delta$ $\hat{P}_x$-a.e. By virtue of Itô's formula,

$$R_t = q(x) + \delta + \int_0^t \frac{d-1}{2R_s}\, ds + S_t \quad \text{on } \{t < \sigma\} \quad \hat{P}_x\text{-a.e.}$$

Therefore, the assertion holds. □

Proposition 3.1. *In the same situation as in Lemma 3.1, for every $u > 0$,*

$$\hat{P}_x\left[\inf_{t\in[0,u]} q(\omega_t) \geq 0\right] \leq \left(\frac{d-1}{\delta(y)} + u^{-1/2}\right) q(x).$$

Proof. Take $\delta \in (0, \delta(y))$. From Lemma 3.1,

$$\hat{P}_x\left[R_t \geq \delta \text{ for all } t \in [0,u]\right] \leq \hat{P}_x\left[q(x) + \delta + C_\delta t + S_t \geq \delta \text{ for all } t \in [0,u]\right].$$

Let $r > q(x)$ and define $\eta = \inf\{t \geq 0 \mid C_\delta t + S_t \leq -r\}$. The law of η under $\hat{P}_x$ is given by

$$\hat{P}_x[\eta \in dt] = \mathbf{1}_{(0,\infty)}(t) \frac{r}{\sqrt{2\pi t^3}} \exp\left(-\frac{(r + C_\delta t)^2}{2t}\right) dt + (1 - e^{-2C_\delta r})\delta_\infty(dt),$$

where δ_∞ is the delta measure at ∞ (see, e.g., [1, p. 295]). Then, we have

$$\begin{aligned}\hat{P}_x\left[\inf_{t\in[0,u]} q(\omega_t) \geq 0\right] &\leq \hat{P}_x\left[R_t \geq \delta \text{ for all } t \in [0,u]\right] \leq \hat{P}_x[\eta > u] \\ &= \int_u^\infty \frac{r}{\sqrt{2\pi t^3}} \exp\left(-\frac{(r + C_\delta t)^2}{2t}\right) dt + 1 - e^{-2C_\delta r}\end{aligned}$$

$$\leq \int_u^\infty \frac{r}{\sqrt{2\pi t^3}}\,dt + 2C_\delta r = \sqrt{\frac{2}{\pi}}\frac{r}{\sqrt{u}} + 2C_\delta r.$$

Letting $r \to q(x)$ and $\delta \to \delta(y)$, we obtain the claim. □

For $r > 0$, define an $\{\hat{\mathcal{F}}_t\}$-stopping time τ_r by $\tau_r = \inf\{t \geq 0 \mid \omega_t \notin O_r\}$. Let $\hat{P}_x^r$ be the law of τ_r under $\hat{P}_x$. The following lemma is the same as [9, Lemma 3.2], so the proof is omitted.

Lemma 3.2. *$\hat{P}_x^r([0,t])$ is differentiable in t on $(0,\infty)$ and there exists a constant $\hat{C}_1 > 0$ independent of x, r and t such that $\frac{d}{dt}\hat{P}_x^r([0,t]) \leq \hat{C}_1 t^{-1}$.*

For a closed subset A of $\mathbb{R}^d$, a stopping time σ_A is defined as $\sigma_A = \inf\{t \geq 0 \mid \omega_t \in A\}$. For $r > 0$, let

$$A_r = \{x \in \mathbb{R}^d \mid \overline{B}(x,r) \cap \Sigma_r \neq \emptyset\}, \tag{3.6}$$

where Σ_r is defined in (3.2).

Proposition 3.2. *For $\gamma > 0$, $u > 0$, $r \in (0,\gamma]$, and $x \in \overline{O}$, the inequality*

$$\hat{P}_x\left[0 \leq \inf_{t\in[0,u]} q(\omega_t) \leq r,\ \sigma_{A_\gamma} > u\right] \leq \left(\frac{d-1}{\gamma} + \hat{C}_2 u^{-1/2}\right) r \tag{3.7}$$

holds with $\hat{C}_2 = 4\hat{C}_1 + 2$.

Proof. Denote $(d-1)/\gamma$ by C'_γ. If $x \in A_\gamma$, (3.7) is trivial since the left-hand side is 0. Suppose $x \in \overline{O} \setminus (O_r \cup A_\gamma)$. Then, $q(x) < r \leq \gamma$. Take $y \in \partial O$ such that $|x-y|_{\mathbb{R}^d} = q(x)$. Because $x \notin A_\gamma$ and $q(x) < \gamma$, y does not belong to Σ_γ: that is, $\delta(y) \geq \gamma$. From Proposition 3.1,

$$\hat{P}_x\left[0 \leq \inf_{t\in[0,u]} q(\omega_t) \leq r\right] \leq \left(\frac{d-1}{\delta(y)} + u^{-1/2}\right) q(x) \leq (C'_\gamma + u^{-1/2})r.$$

Next, suppose $x \in O_r \setminus A_\gamma$. Then,

$$\left\{0 \leq \inf_{t\in[0,u]} q(\omega_t) \leq r\right\} = \left\{\tau_r \leq u,\ 0 \leq \inf_{t\in[0,u-\tau_r]} q((\theta_{\tau_r}\omega)_t)\right\}$$
$$\subset \bigcup_{k=1}^\infty \left\{2^{-k}u < u - \tau_r \leq 2^{-k+1}u,\ 0 \leq \inf_{t\in[0,2^{-k}u]} q((\theta_{\tau_r}\omega)_t)\right\} \cup \{\tau_r = u\}.$$

From Lemma 3.2, $\hat{P}_x[\tau_r = u] = 0$. From the strong Markov property and Proposition 3.1,

$$\hat{P}_x\left[2^{-k}u < u - \tau_r \leq 2^{-k+1}u,\ 0 \leq \inf_{t\in[0,2^{-k}u]} q((\theta_{\tau_r}\omega)_t),\ \sigma_{A_\gamma} > u \,\middle|\, \hat{\mathcal{F}}_{\tau_r}\right]$$
$$\leq \mathbf{1}_{\{2^{-k}u < u-\tau_r \leq 2^{-k+1}u\}} \cdot \hat{E}_x\left[\hat{P}_{\omega_{\tau_r}}\left[0 \leq \inf_{t\in[0,2^{-k}u]} q(\omega_t)\right]; \omega_{\tau_r} \in \partial O_r \setminus A_\gamma\right]$$
$$\leq \mathbf{1}_{\{2^{-k}u < u-\tau_r \leq 2^{-k+1}u\}} \cdot (C'_\gamma + (2^{-k}u)^{-1/2})r$$

$$\leq \mathbf{1}_{\{2^{-k}u<u-\tau_r\leq 2^{-k+1}u\}}\cdot (C'_\gamma+((u-\tau_r)/2)^{-1/2})r.$$

Therefore,

$$\begin{aligned}
&\hat{P}_x\left[0\leq \inf_{t\in[0,u]} q(\omega_t)\leq r,\ \sigma_{A_\gamma}>u\right]\\
&\leq r\hat{E}_x[C'_\gamma+((u-\tau_r)/2)^{-1/2};\ \tau_r\leq u]\\
&\leq rC'_\gamma+r\left(\frac{u}{4}\right)^{-1/2}\hat{P}_x[\tau_r\leq u/2]+r\int_{u/2}^{u}\left(\frac{u-s}{2}\right)^{-1/2}\hat{C}_1 s^{-1}\,ds\\
&\qquad\text{(from Lemma 3.2)}\\
&\leq rC'_\gamma+2ru^{-1/2}+\frac{2\hat{C}_1 r}{u}\int_{u/2}^{u}\left(\frac{u-s}{2}\right)^{-1/2}ds\\
&=rC'_\gamma+2ru^{-1/2}+4\hat{C}_1 ru^{-1/2}.
\end{aligned}$$

This completes the proof. □

Lemma 3.3. *Let $\{B_n\}_{n=1}^\infty$ be a decreasing sequence of closed subsets of $\mathbb{R}^d$. If* $\mathrm{Cap}_{d-4}\left(\bigcap_{n=1}^\infty B_n\right)=0$*, then* $\mathrm{Cap}(\Theta_{B_n})$ *converges to* 0 *as* $n\to\infty$.

Proof. From (3.4), $\mathrm{Cap}(\Theta_{\bigcap_{n=1}^\infty B_n})=0$. For $\varepsilon>0$, take a compact subset F of E such that $\mathrm{Cap}(E\setminus F)<\varepsilon$. Since $\{\Theta_{B_n}\cap F\}_{n=1}^\infty$ is a decreasing sequence of compact sets,

$$\lim_{n\to\infty}\mathrm{Cap}(\Theta_{B_n}\cap F)=\mathrm{Cap}(\Theta_{\bigcap_{n=1}^\infty B_n}\cap F)=0.$$

Therefore,

$$\lim_{n\to\infty}\mathrm{Cap}(\Theta_{B_n})\leq\lim_{n\to\infty}\mathrm{Cap}(\Theta_{B_n}\cap F)+\mathrm{Cap}(E\setminus F)\leq\varepsilon.$$

Since $\varepsilon>0$ is arbitrary, we obtain the claim. □

Proof of Theorem 3.1. Recall the definition of A_r in (3.6). We note that

$$\bigcap_{k=1}^\infty A_{1/k}=\Sigma. \tag{3.8}$$

From Proposition 3.2 and (3.5),

$$\begin{aligned}
\mu(\Xi_{\overline{O}}\setminus\Xi_O)&=\hat{P}_0\left[\inf_{t\in[0,T]}q(\omega_t)=0\right]\\
&\leq\lim_{n\to\infty}\hat{P}_0\left[\left\{\inf_{t\in[0,T]}q(\omega_t)=0\right\}\cap\{\sigma_{A_{1/n}}>T\}\right]+\mu(\Theta_\Sigma)\\
&=0.
\end{aligned}$$

That is, $\mathbf{1}_{\Xi_{\overline{O}}}=\mathbf{1}_{\Xi_O}$ μ-a.e. Since $\mathbf{1}_{\Xi_O}$ is a lower semicontinuous function, $\mathbf{1}_{\Xi_{\overline{O}}}\in QR$ (cf. [5, p. 230]).

For $k \in \mathbb{N}$, let B_k denote the closure of $A^c_{1/k}$, and let $F_k = \Xi_{B_k}$. Then, $\{F_k\}_{k=1}^\infty$ is an $\mathcal{E}$-nest from (3.5), (3.8), and Lemma 3.3. Since $A_{1/(k+1)} \subset B_k^c$, $\Theta_{A_{1/(k+1)}} \subset E\setminus F_k$ for each k. Let $h(w) = \inf_{t\in[0,T]} q(w(t))$ for $w \in E$ and $\chi(t) = (0 \vee t) \wedge 1$ for $t \in \mathbb{R}$. For $m \in \mathbb{N}$, let $f_m(w) = \chi(mh(w))$ for $w \in E$. Then, $f_m \in \mathbb{D}^{1,1}$ and f_m converges to $\mathbf{1}_{\Xi_O}$ in L^1 as $m \to \infty$. Moreover, for $k \in \mathbb{N}$ and $m \geq k+1$,

$$\begin{aligned}\|(\nabla f_m)\cdot \mathbf{1}_{F_k}\|_1 &\leq \int_{F_k} m\cdot \mathbf{1}_{\{0\leq h\leq 1/m\}}\,d\mu\\ &\leq m\mu\left(\{0\leq h\leq 1/m\}\setminus \Theta_{A_{1/(k+1)}}\right)\\ &\leq (k+1)(d-1)+\hat{C}_2T^{-1/2}\quad \text{(from Proposition 3.2)}.\end{aligned}$$

Therefore, $\sup_{m\in\mathbb{N}} \|(\nabla f_m)\cdot \mathbf{1}_{F_k}\|_1 < \infty$ for each $k \in \mathbb{N}$. From Theorem 2.2, we conclude that $\mathbf{1}_{\Xi_{\overline{O}}} \in \dot{BV}_{\mathrm{loc}}$.

The identity $\mathrm{Cap}^{\mathbf{1}_{\Xi_{\overline{O}}}}(\partial\Xi_{\overline{O}} \setminus \partial'\Xi_{\overline{O}}) = 0$ is proved in a similar way to the proof of [9, Theorem 2.4]. We provide details for the readers' convenience. Take an $\mathcal{E}$-nest $\{F'_k\}_{k=1}^\infty$ and functions $\{\varphi_k\}_{k=1}^\infty$ obtained by applying Lemma 2.1 to the $\mathcal{E}$-nest $\{F_k\}_{k=1}^\infty$ defined above. For $s \in (0,T)$, we define a closed set V_s by

$$V_s = \left\{ w\in E \;\middle|\; \inf_{t\in[0,s]} q(w(t)) = 0 \text{ and } \inf_{t\in[s,T]} q(w(t)) = 0\right\}.$$

Then, $\partial\Xi_{\overline{O}} \setminus \partial'\Xi_{\overline{O}} \subset \bigcup_{s\in(0,T)\cap\mathbb{Q}} V_s$. Fix $s \in (0,T)$ and define a map $f\colon E \to \mathbb{R}^2$ by

$$f(w) = \Bigl(\inf_{t\in[0,s]} q(w(t)),\ \inf_{t\in[s,T]} q(w(t))\Bigr).$$

Fix $k \in \mathbb{N}$ and take $l \in \mathbb{N}$ such that $\varphi_k \in \mathbb{D}^{1,2}_{F_l}$. Let $\varepsilon > 0$. Henceforth, C denotes an unimportant positive constant independent of ε that may vary line by line. Take a smooth function g on $[0,\infty)$ such that

$$g(t) = \begin{cases} 1 & t\in[0,e^{-2/\varepsilon}],\\ -3\varepsilon\log t - 4 & t\in[e^{-14/(9\varepsilon)}, e^{-13/(9\varepsilon)}],\\ 0 & t\in[e^{-1/\varepsilon},\infty),\end{cases}$$

and $-3\varepsilon/t \leq g'(t) \leq 0$ for all $t > 0$. We define a function $\zeta\colon \mathbb{R}^2 \to \mathbb{R}$ by $\zeta(x,y) = \sqrt{x^2+y^2}$ and set $\iota = g\circ\zeta$. Since $\iota\circ f$ is a bounded H-Lipschitz continuous function, it belongs to $\mathbb{D}^{1,2}_b$. Let $\eta = (\iota\circ f)\varphi_k + e_{E\setminus F'_k} \in \mathbb{D}^{1,2}$. Then, $\eta \geq 1$ μ-a.e. on some open set including V_s. Since $(\iota\circ f)\varphi_k \in \mathbb{D}^{1,2}_{F_l}$,

$$\begin{aligned}&\mathcal{E}^{\mathbf{1}_{\Xi_{\overline{O}}}}(\eta|_{\Xi_{\overline{O}}},\eta|_{\Xi_{\overline{O}}})\\ &\leq 2\mathcal{E}^{\mathbf{1}_{\Xi_{\overline{O}}}}(((\iota\circ f)\varphi_k)|_{\Xi_{\overline{O}}},((\iota\circ f)\varphi_k)|_{\Xi_{\overline{O}}}) + 2\mathcal{E}^{\mathbf{1}_{\Xi_{\overline{O}}}}(e_{E\setminus F'_k}|_{\Xi_{\overline{O}}}, e_{E\setminus F'_k}|_{\Xi_{\overline{O}}})\\ &\leq 2\int_{\Xi_{\overline{O}}\cap F_l} |\nabla(\iota\circ f)|_H^2\,d\mu + 4\mathcal{E}(\varphi_k,\varphi_k) + 2\mathcal{E}(e_{E\setminus F'_k}, e_{E\setminus F'_k}).\end{aligned}$$

Denoting the gradient operator on $\mathbb{R}^2$ by $\nabla_{\mathbb{R}^2}$, we have

$$\int_{\Xi_{\overline{O}}\cap F_l} |\nabla(\iota\circ f)|_H^2\,d\mu = \int_{\Xi_{\overline{O}}\cap B_l} |\langle(\nabla f)(w), (\nabla_{\mathbb{R}^2}\iota)(f(w))\rangle_{\mathbb{R}^2}|_H^2\,\mu(dw)$$

$$
\begin{aligned}
&\le C\int_{\Xi_{\overline{O}\cap B_l}} |(\nabla_{\mathbb{R}^2}\,\iota)(f(w))|^2_{\mathbb{R}^2}\,\mu(dw)\\
&= C\int_{\{(x,y)\in\mathbb{R}^2 \mid x\ge 0, y\ge 0\}} |\nabla_{\mathbb{R}^2}\,\iota|^2_{\mathbb{R}^2}\, d(f_*(\mu|_{\Xi_{\overline{O}\cap B_l}}))\\
&=: I_1.
\end{aligned}
$$

In the first line, $\langle\cdot,\cdot\rangle_{\mathbb{R}^2}$ denotes a pairing between elements in $H\otimes\mathbb{R}^2$ and in $\mathbb{R}^2$ that takes values in H. We note that

$$
\begin{aligned}
|\nabla_{\mathbb{R}^2}\,\iota|^2_{\mathbb{R}^2} &= (\partial\iota/\partial x)^2 + (\partial\iota/\partial y)^2\\
&= (g'\circ\zeta(x,y))^2\,\frac{x^2}{x^2+y^2} + (g'\circ\zeta(x,y))^2\,\frac{y^2}{x^2+y^2}\\
&= (g'\circ\zeta(x,y))^2 .
\end{aligned}
$$

By letting $\psi = (\zeta\circ f)_*(\mu|_{\Xi_{\overline{O}\cap B_l}})$, we obtain

$$
I_1 = C\int_0^\infty g'(r)^2\,\psi(dr) \le 9C\varepsilon^2\int_{e^{-2/\varepsilon}}^{e^{-1/\varepsilon}} r^{-2}\,\psi(dr) =: I_2.
$$

We now have

$$
\begin{aligned}
\Psi(r) &:= \psi([0,r]) = (f_*(\mu|_{\Xi_{\overline{O}\cap B_l}}))(\zeta^{-1}([0,r]))\\
&= \mu\left[\left\{ w\in\Xi_{\overline{O}\cap B_l} \,\middle|\, \inf_{t\in[0,s]} q(w(t))^2 + \inf_{t\in[s,T]} q(w(t))^2 \le r^2 \right\}\right]\\
&\le \mu\left[\left\{ w\in\Xi_{\overline{O}\setminus A_{1/(l+1)}} \,\middle|\, \begin{array}{l} 0\le \inf_{t\in[0,s]} q(w(t)) \le r, \\ 0\le \inf_{t\in[s,T]} q(w(t)) \le r \end{array} \right\}\right]\\
&\le \hat{E}_0\left[\hat{P}_{\omega_s}\left[0\le \inf_{t\in[0,T-s]} q(\omega_t)\le r,\ \sigma_{A_{1/(l+1)}} > T-s\right];\right.\\
&\qquad\qquad \left. 0\le \inf_{t\in[0,s]} q(\omega_t)\le r,\ \sigma_{A_{1/(l+1)}} > s\right].
\end{aligned}
$$

From Proposition 3.2, if $r\le 1/(l+1)$, this expectation is dominated by

$$
\begin{aligned}
&((d-1)(l+1) + C(T-s)^{-1/2})r\hat{P}_0\left[0\le \inf_{t\in[0,s]} q(\omega_t)\le r,\ \sigma_{A_{1/(l+1)}} > s\right]\\
&\le ((d-1)(l+1) + C(T-s)^{-1/2})((d-1)(l+1) + Cs^{-1/2})r^2\\
&= Cr^2.
\end{aligned}
$$

Thus, for $\varepsilon\le 1/\log(l+1)$,

$$
\begin{aligned}
I_2 &= 9C\varepsilon^2\int_{e^{-2/\varepsilon}}^{e^{-1/\varepsilon}} r^{-2}\,d\Psi(r)\\
&= C\varepsilon^2\left\{\left[\frac{\Psi(r)}{r^2}\right]_{e^{-2/\varepsilon}}^{e^{-1/\varepsilon}} + \int_{e^{-2/\varepsilon}}^{e^{-1/\varepsilon}} \frac{2\Psi(r)}{r^3}\,dr\right\}
\end{aligned}
$$

$$\leq C\varepsilon^2\Big(C + \int_{e^{-2/\varepsilon}}^{e^{-1/\varepsilon}} \frac{C}{r}\,dr\Big) = C\varepsilon^2(1+1/\varepsilon).$$

Therefore,

$$\begin{aligned}
\mathrm{Cap}^{\mathbf{1}_{\Xi_{\overline{O}}}}(V_s) &\leq \mathcal{E}^{\mathbf{1}_{\Xi_{\overline{O}}}}(\eta|_{\Xi_{\overline{O}}}, \eta|_{\Xi_{\overline{O}}}) + \|\eta|_{\Xi_{\overline{O}}}\|^2_{L^2(\mu|_{\Xi_{\overline{O}}})} \\
&\leq 2\int_{\Xi_{\overline{O}}\cap F_l} |\nabla(\iota\circ f)|_H^2\,d\mu + 4\mathcal{E}(\varphi_k,\varphi_k) + 2\mathcal{E}(e_{E\setminus F'_k}, e_{E\setminus F'_k}) \\
&\quad + 2\Psi(e^{-1/\varepsilon}) + 2\|e_{E\setminus F'_k}\|_2^2 \\
&\leq C\varepsilon^2(1+1/\varepsilon) + 4\mathcal{E}(\varphi_k,\varphi_k) + 2\mathrm{Cap}(E\setminus F'_k) + Ce^{-2/\varepsilon}.
\end{aligned}$$

By letting $\varepsilon \to 0$ and then $k \to \infty$, we obtain $\mathrm{Cap}^{\mathbf{1}_{\Xi_{\overline{O}}}}(V_s) = 0$. Therefore,

$$\mathrm{Cap}^{\mathbf{1}_{\Xi_{\overline{O}}}}(\partial\Xi_{\overline{O}} \setminus \partial'\Xi_{\overline{O}}) \leq \sum_{s\in(0,T)\cap\mathbb{Q}} \mathrm{Cap}^{\mathbf{1}_{\Xi_{\overline{O}}}}(V_s) = 0. \qquad \square$$

Remark 3.2. In this section, we considered only one-sided pinned path spaces as the underlying space for simplicity. However, the general idea of the argument is also valid for pinned path spaces, as discussed in [9].

Acknowledgements

This research was partially supported by JSPS KAKENHI Grant Number 24540170.

References

[1] Borodin, A. N. and Salminen, P., *Handbook of Brownian motion—facts and formulae, 2nd ed.*, Probability and its Applications, Birkhäuser Verlag, Basel, 2002.

[2] Bouleau, N. and Hirsch, F., *Dirichlet forms and analysis on Wiener space*, de Gruyter Studies in Mathematics **14**, Walter de Gruyter, Berlin, 1991.

[3] Carleson, L., *Selected problems on exceptional sets*, Van Nostrand Mathematical Studies **13**, D. Van Nostrand Co., Inc., Princeton, N.J.-Toronto, Ont.-London, 1967.

[4] Fukushima, M., On semimartingale characterizations of functionals of symmetric Markov processes, *Electron. J. Probab.* **4** (1999), no. 18, pp. 1–32.

[5] Fukushima, M., *BV* functions and distorted Ornstein Uhlenbeck processes over the abstract Wiener space, *J. Funct. Anal.* **174** (2000), 227–249.

[6] Fukushima, M. and Hino, M., On the space of BV functions and a related stochastic calculus in infinite dimensions, *J. Funct. Anal.* **183** (2001), 245–268.

[7] Fukushima, M., Oshima, Y., and Takeda, M., *Dirichlet forms and symmetric Markov processes*, 2nd ed., de Gruyter Studies in Mathematics **19**, Walter de Gruyter, Berlin, 2011.

[8] Hino, M., Integral representation of linear functionals on vector lattices and its application to BV functions on Wiener space, in: *Stochastic Analysis and Related Topics in Kyoto in honour of Kiyosi Itô*, 121–140, Advanced Studies in Pure Mathematics **41**, 2004.

[9] Hino, M. and Uchida, H., Reflecting Ornstein–Uhlenbeck processes on pinned path spaces, in: *Proceedings of RIMS Workshop on Stochastic Analysis and Applications*, 111–128, RIMS Kôkyûroku Bessatsu **B6**, Res. Inst. Math. Sci. (RIMS), Kyoto, 2008.

[10] Khoshnevisan, D. and Shi, Z., Brownian sheet and capacity, Ann. Probab. **27** (1999), 1135–1159.
[11] Nualart, D., *The Malliavin calculus and related topics*, 2nd ed., Probability and its Applications, Springer-Verlag, Berlin, 2006.
[12] Röckner, M. and Schmuland, B., Tightness of general $C_{1,p}$ capacities on Banach space, J. Funct. Anal. **108** (1992), 1–12.
[13] Uchida, H., Reflecting Ornstein–Uhlenbeck processes on path spaces (in Japanese), Master thesis in Kyoto University, 2006.

Chapter 14

A Dirichlet space on ends of tree and superposition of nodewise given Dirichlet forms with tier linkage

Hiroshi Kaneko*

Stochastic processes on the field of p-adic numbers have been studied for more than two decades. The theory of Dirichlet space is applied in important parts of the studies. In some recent observations, the hierarchical structure is handled by tree structure and the structure gives an overview for measure symmetric stochastic processes on the field of p-adic numbers and ends of a tree. The existing researches were undertaken by relying on structure of eigenfunctions determined by the structural design of branches at every node. The objective of the present article is focusing on the case where a Dirichlet form is given at every node and tier linkage is given by bilinear forms coming from pairs of nodes connected by branch or stemmed directly from their identical parental node. In other words, we will deal with the case where the structure of eigenfunctions differs from the one in existing studies.

Contents

1. Introduction

After Albeverio and Karwowski constructed a Hunt process on the field $\mathbb{Q}_p$ of p-adic numbers whose infinitesimal operator covers the Vladimirov operator in [Albeverio and Karwowski (1994)], many schemes were launched as generalized variants ruling Hunt processes. Exemplary attempts of them are undertaken by Albeverio, Kigami, Yasuda, Karwowski and Vilela-Mendes. Karwowski and Vilela-Mendes focused on the case where the infinitesimal generators are given as modified Vladimirov operator in [Karwowski and Vilela-Mendes (1994)]. The study undertaken by the present author addressed a class of Hunt processes which admits infinitesimal generators with no apparent symmetry in their description in [Kaneko (2000)]. Yasuda emphasized that rotational invariance is not essestial in constructing semi-stable processes by taking limit of suitably scaled sums in [Yasuda (2006)].

*Tokyo University of Science, 1-3 Kagurazaka, Shinjuku, Tokyo 162-8601, stochos@rs.kagu.tus.ac.jp

Subsequently to those schemes, a research in another direction was undertaken by Albeverio and Karwowski ([Albeverio and Karwowski (2008)]). They proposed a space consisting of ends of a tree as an even more generalized state space than $\mathbb{Q}_p$. They constructed jump processes on the space consisting of the ends which illustrates a random motion of particle driven by jump to an end from another, where the randomness of the motion is interpreted by the notion of conductance. They revealed that the whole figure of their class is captured within the scheme of the Dirichlet space theory, the infinitesimal generator is written in terms directly of eigenfuctions as in [Albeverio and Karwowski (1994)]. This approach can be viewed as the intuitive method where the conductance comes first and the families of eigenfunctions and eigenvalues emerges later. In this case, tangible description of the symmetry of the operator is not given similarly to the case in [Kaneko (2000)].

Recently, Kigami in [Kigami (2013)] proposed a class of Hunt process on the space Σ^+ consisting of all ends as a wider coverage to overview the existing frameworks on measure symmetric Markov processes on non-Archimedean spaces where the eigenvalues and the eigenfunctions are provided first so as to be the ones associated with the nodewise given Dirichlet forms. Since it turns out that eventually these are the ones of the infinitesimal generator of Hunt process in the scheme, this approach is grasped as the study where the families of eigenfunctions and eigenvalues comes first and the conductance emerges later.

A set of the ends which originated in an identical node is called branch in [Albeverio and Karwowski (2008)] and denoted by Σ_x^+ in [Kigami (2013)] when the node shared by those ends as their origin is given by x. The notion plays important roles in [Albeverio and Karwowski (2008)] and [Kigami (2013)] along with functions taking constant on each branch. Especially, functions taking constant on every branch Σ_y^+ growing directly from some identical node x and vanishing on any other branches are crucially important. As a matter of fact, the family of such functions whose integral over Σ_x^+ vanish is in the domain of Dirichlet space as eigenfunctions in typical existing studies. Hereafter, we will denote the family of functions with these properties for the node x by $\mathcal{C}_{x,0}$. In the present author's research in [Kaneko (2013)], the orthogonal property $\mathcal{E}(\varphi, \psi) = 0$ for $\varphi \in \mathcal{C}_{x,0}, \psi \in \mathcal{C}_{y,0}$ with distinct nodes x, y was addressed coherently in the existing frameworks based on Dirichlet space $(\mathcal{E}, \mathcal{F})$ on $L^2(\Sigma^+; \mu)$. A close attention to some characteristic features coming from the orthogonal property was paid and an attempt on a construction of a Hunt process was given based on the importance of the features.

In this article, we will explore construction of a Hunt process which admits less restrictive condition on the orthogonal property than the one given as $\mathcal{E}(\varphi, \psi) = 0$ for $\varphi \in \mathcal{C}_{x,0}, \psi \in \mathcal{C}_{y,0}$ with distinct nodes x, y in [Kaneko (2013)]. We will seek for a sufficient condition for the construction in such a situation with a less restrictive conditions by overlaying some effective existing methods.

In the second section, we will recall some fundamental notions and terminologies introduced in [Kigami (2013)] and propose a less restrictive orthogonal condition on Dirichlet spaces than [Kaneko (2013)] so that a more general framework on Hunt processes on the ends of tree is provided by taking a superposition of Dirichlet forms with a sort of tier linkage. Despite of the tier linkage between distinct nodes given in terms of bilinear forms, our scheme will be shown also within the scope where nodewise decomposition of Dirichlet forms is feasible and the absolute continuity of so called jump measure $J(d\eta, d\zeta)$ in the Beurling-Deny formula with respect to the product measure $\mu(d\eta)\mu(d\zeta)$ of the ambient measure μ is validated. However, we will see that the density may depend on the pair of branches which are not directly growing from an identical node due to the tier linkage.

In the third section, we will start with a family of nodewise given Dirichlet forms as in [Kaneko (2013)] and bilinear forms yielding the tier-linkage. We will then introduce a Dirichlet space on Σ^+ which is cohesively formulated to the less restrictive orthogonality involving the tier linkage. While keeping the decomposablity of the Dirichlet space along the method in the previous section, we will find a sufficient condition on the family of nodeswise given symmetric bilinear forms to meet a construction method of Dirichlet space on Σ^+ and to cope with a random motion driven by a Dirichlet form at primary influential position and marginally by bilinear forms at secondary influential positions. The observations in the section will give a method for wider class of Hunt processes under such less restrictive situation. This results a class of so-called pre-Dirichlet forms with the jump measure which meets the class described by the densities in the previous section.

The final section will establish a wider framework than the one in [Kaneko (2013)], however it largely still relies on the spectral method invented by Kigami in [Kigami (2013)] based on the complete orthogonal system of the family of square integrable functions on Σ^+. A flexibility in randomness of the moving particle is allowed by an attentive use of the method so that a superposition of Dirichlet forms and bilinear forms with our tier linkage is covered. Our result can be viewed as a wider coverage of Hunt processes on Σ^+ than other studies previously mentioned where the density depends simply on node or on the pair of branches directly jointed with their confluent node.

2. Decomposition into nodewise given Dirichlet forms

We take a set T consisting of countably infinite vertices and a map $\mathcal{A} : T \times T \to \{0, 1\}$ satisfying $\mathcal{A}(x, x) = 0$ for any $x \in T$ and $\mathcal{A}(x, y) = \mathcal{A}(y, x)$ for any $x, y \in T$. Each element in T is called a node as in Introduction and the set $\{x, y\}$ of two distinct nodes x, y satisfying $\mathcal{A}(x, y) = 1$ will be called an edge. A sequence $(a_0, a_1, \ldots, a_n)$ of nodes in T is called a path, if $\mathcal{A}(a_i, a_{i+1}) = 1$ is satisfied for any $i = 0, \ldots, n-1$. If any pair of distinct elements y, z in T admits a path $(a_0, a_1, \ldots, a_n)$ with $a_0 = y$ and $a_n = z$, the pair $(T, \mathcal{A})$ is called a non-directed tree. If a sequence $(a_0, a_1, \ldots, a_n)$ satisfies $a_i \neq a_j$ for any distinct integers i, j, the sequence said to be simple. The set $V(x)$ of nodes directly tied with $x \in T$ is given by $V(x) = \{y \mid \mathcal{A}(x,y) = 1\}$. Throughout the present article, we suppose that a non-directed tree $(T, \mathcal{A})$ satisfying the following properties is given as in [Kigami (2013)]:

(i) the tree does not admit any path $(a_0, a_1, \ldots, a_n)$ satisfying $a_0 = a_n$ with distinct edges $\{a_0, a_1\}, \{a_1, a_2\}, \ldots, \{a_{n-1}, a_n\}$,

(ii) $V(x)$ is a finite set and $\#(V(x)) \geq 3$ at any node x in T.

We introduce the notion of end of the tree in the next similarly to [Kigami (2013)]. An infinite sequence $(a_0, a_1, \ldots)$ of nodes is called a geodesic ray if any finite subsequence of $(a_0, a_1, \ldots, a_n, \ldots)$ is simple path. We denote the set of geodesic rays by $\mathcal{R}$ and introduce an equivalence relation "$\sim$" on $\mathcal{R}$ defined by

$$(a_0, a_1, \ldots) \sim (b_0, b_1, \ldots) \Leftrightarrow \text{ there exists an integer } k \text{ satisfying } a_{k+m} = b_m \text{ for any } m \geq 0.$$

The quotient space $\mathcal{R}/\sim$ is denoted by Σ and each element in Σ is called an end. For coping with such hierarchical structure as in the field of p-adic numbers, we fix an element Δ in Σ and denote $\Sigma \setminus \{\Delta\}$ by Σ^+. Let us take a representative sequence $(\delta_0, \delta_1, \ldots)$ for Δ. Then any tree satisfying these assumptions provides us with the situation where any node x outside $\{\delta_0, \delta_1, \ldots\}$ and any δ_i are connected by a unique simple path $(a_0, \ldots, a_n)$ as $a_0 = x$ and $a_n = \delta_i$. The length of simple path $(a_0, \ldots, a_n)$ is defined as n i.e., the number of the nodes added to the initial node a_0 in the path. Our situation enables us to take a unique path with minimal length in the set $\{(a_0, \ldots, a_n) \mid a_0 = x, a_n = \delta_i, \text{ with some positive integer } n \text{ for some } i\}$ of the simple paths. Consequently, we can focus only on the path with the minimal length and denote the path $(a_0, \ldots, a_n)$ by $(x, x', \ldots, \delta_{i_{(x)}})$ and its length by $\ell(x)$.

Then, the map $\pi : T \to T$ is defined by

$$\pi(x) = \begin{cases} x' & \text{if } x \notin \{\delta_0, \delta_1, \ldots\}, \\ \delta_{i+1} & \text{if } x = \delta_i \text{ for some } i \geq 0. \end{cases}$$

T is represented as the disjoint union of its subsets $\{T_m \mid m \in \mathbb{Z}\}$ defined by $T_m = \{x \in T \mid \ell(x) - i_{(x)} = m\}$ for $m \in \mathbb{Z}$. Then, it turns out that $\pi(T_m) = T_{m-1}$ for any integer m.

Let us define $S_x = \{y \in T \mid \pi^k(y) = x \text{ for some non-negative integer } k\}$ for any $x \in T$ and $\Sigma_x^+ = \{\eta \in \Sigma^+ \mid \eta \text{ admits a geodesic ray } (a_0, a_1, \cdots) \text{ as a representative sequence of } \eta \text{ satsfying } a_0, a_1, \cdots \in S_x\}$. We can introduce a topology on Σ^+. As a matter of fact, the family $\{\Sigma_S^+ \mid S \subset T\}$ of subsets $\Sigma_S^+ = \cup_{x \in S} \Sigma_x^+$ determined by $S \subset T$ satisfies the axioms for open sets on Σ^+. We will regard Σ^+ as a topological space equipped with the family of open sets.

Example 2.1. (A tree $T_{\mathbb{Q}_p}$ associated with the field $\mathbb{Q}_p$ of p-adic numbers). Let $T_{\mathbb{Q}_p}$ be the set consisting of all balls in $\mathbb{Q}_p$ and denote the radius of ball B by $\mathrm{r}(B)$.

Then we define $\mathcal{A}_{\mathbb{Q}_p}(B, B')$ for $B, B' \in T_{\mathbb{Q}_p}$ by

$$\mathcal{A}_{\mathbb{Q}_p}(B, B') = \begin{cases} 1 & \text{if either } B \subset B', \mathrm{pr}(B) = \mathrm{r}(B') \text{ or } B' \subset B, \mathrm{pr}(B') = \mathrm{r}(B) \\ 0 & \text{otherwise.} \end{cases}$$

Then it is not difficult to see the pair $(T_{\mathbb{Q}_p}, \mathcal{A}_{\mathbb{Q}_p})$ is a tree satisfying condition (i) and (ii). Take the ball Δ_0 centered at the origin 0 and with the radius 1 and a sequence $(\Delta_0, \Delta_1, \cdots)$ of elements in $T_{\mathbb{Q}_p}$ specified by $\Delta_i \subset \Delta_{i+1}$ and $\mathrm{r}(\Delta_i) = p^i$ for any $i = 0, 1, 2, \ldots$. In accordance with the choice of the end $\Delta_{\mathbb{Q}_p} \in \Sigma$ represented by the geodesic ray $(\Delta_0, \Delta_1, \cdots)$, the map π is defined by $\pi(B) = B'$ with the ball B' characterized by $B \subset B'$ and $pr(B) = \mathrm{r}(B')$ and in addition a homeomorphism between Σ^+ and $\mathbb{Q}_p$ is obtained. In fact, any end $\eta \in \Sigma^+$ admits a geodesic ray $(B_0, B_1, \cdots)$ represented by a sequence of balls satisfying $B_0 \supsetneqq B_1 \supsetneqq \ldots$, which determines a singleton $\{a\} \subset \mathbb{Q}_p$ by $\{a\} = \cap_i B_i$. The map $\eta \mapsto a$ gives a bijection from Σ^+ to $\mathbb{Q}_p$ which is viewed as a homeomorphism.

Afterwards, we will mention a Dirichlet form associated with the tree T. Especially, in [Kigami (2010)], important observations on the Cantor set in noncommutative Riemannian geometry are shown along with a homeomorphism between the Cantor set and ends of a tree.

The main assertions in [Kigami (2010)] show that a Dirichlet form on the Cantor set is constructed so as to be a natural counterpart of the classical Douglas integral on the unit circle, where functions on the unit circle are replaced with ones on the Cantor set and the standard Brownian motion on the unit disk is replaced with a random walk on S_{Δ_0} for accommodating the Dirichlet form to the generalization of the classical Douglas integral based on its probabilistic reinterpretation as in [Doob (1962)]. In [Kigami (2013)], this sort of generalization is addressed in the case the unit circle is replaced with $\Sigma^+_{\Delta_0}$. A relationship between random walks on tree and a capacity on the ends of tree is discussed in [Lyons (1992)].

A function taking constant on every Σ^+_y for some disjoint open cover $\{\Sigma^+_y\}_{y \in S}$ of Σ^+ determined by some $S \subset T$ is said to be locally constant. The family of locally constant functions taking constant on every Σ^+_y with $y \in T_{m+1}$ will be denoted by $\mathcal{C}^m(\Sigma^+)$. The family of locally constant functions vanishing outside Σ^+_x will be denoted by $\mathcal{C}(\Sigma^+_x)$ for every $x \in T$. The Stone-Weierstrass theorem shows that $\mathcal{C}(\Sigma^+_x)$ is contained densely in the family of continuous functions with support in Σ^+_x. In what follows, the intersection $\mathcal{C}(\Sigma^+_x) \cap \mathcal{C}^m(\Sigma^+)$ given by $x \in T_m$ will play an important role and will be denoted by $\mathcal{C}_x$. In the present article, a node $x \in T$ will be called the confluent node for $y, z \in T$, if there exist positive integers m, ℓ such that $\pi^\ell(y) = \pi^m(z) = x$ and $\pi^{\ell-1}(y) \neq \pi^{m-1}(z)$. The confluent node for $y, z \in T$ will be denoted x by $[y, z]$.

We denote $V(x) \setminus \{\pi(x)\}$ by $S^+(x)$ and suppose that $n(x) = \#(S^+(x)) - 1 = \#(V(x)) - 2$ is determined as a positive integer for every $x \in T$. Throughout this article, we restrict our attention to the case that Σ^+ admits a Radon measure μ

on Σ^+ with the support Σ^+ and a complete orthonormal system Φ of $L^2(\Sigma^+;\mu)$ are given so that it is divided into orthonormal systems $\{\Phi_x\}_{x\in T}$ each of which is assigned by $\Phi_x = \{\varphi \in \Phi \cap \mathcal{C}_x \mid (\varphi, 1_{\Sigma^+})_{L^2(\Sigma^+;\mu)} = 0\}$ and consists of $n(x)$ elements. The existence of such complete orthonormal system is explicitly assumed in [Kigami (2013)] and is substantially ensured in [Albeverio and Karwowski (2008)] by employing the system of the eigenfunctions.

We take a regular Dirichlet space $(\mathcal{E},\mathcal{F})$ on $L^2(\Sigma^+;\mu)$ equipped with the inner product $\mathcal{E}_1(u,v) = \mathcal{E}(u,v) + (u,v)_{L^2(\Sigma^+;\mu)}$ for $u,v \in \mathcal{F}$. In accordance with the context in [Albeverio and Karwowski (2008)] and [Kigami (2013)], we assume that $\Phi \subset \mathcal{F}$ and look into a relationship between the Dirichlet form $\mathcal{E}$ and the orthogonal projection P_x to the linear subspace $\mathcal{C}_{x,0}$ spanned by Φ_x at each node $x \in T$. In this section we assume that there exists some positive number $\Lambda < \dfrac{1}{2}$ such that $|\mathcal{E}(P_x 1_{\Sigma^+_w} P_{\pi(x)} 1_{\Sigma^+_{w'}})| \le \Lambda \mathcal{E}(P_{\pi(x)} 1_{\Sigma^+_w}, P_{\pi(x)} 1_{\Sigma^+_{w'}})$ for any $w, w' \in S^+(x)$ and $x \in T$. In what follows, for any $x \in T$ and $u \in L^2(\Sigma^+,\mu)$, we will denote $\int_{\Sigma^+_x} u(\zeta)\mu(d\zeta)$ by $(u)_{\mu,x}$.

Lemma 2.1. (i) If a function u in the closure of the linear subspace spanned by $\cup_{k=1}^{\infty} \mathcal{C}_{\pi^k(x),0}$ in $L^2(\Sigma^+;\mu)$ satisfies $\mathcal{E}(P_{\pi^k(x)}u, P_{\pi^\ell(x)}u) = 0$ for any $x \in T$ and non-negative integers $k,\ell \ge 0$ with $|k-\ell| \ge 2$ and if $\mathcal{E}(P_{\pi^k(x)}u, P_{\pi^k(x)}u) \le \lambda (P_{\pi^k(x)}u, P_{\pi^k(x)}u)_{L^2(\Sigma^+;\mu)}$ is satisfied for any $k = 0,1,\ldots,$ by taking some $\lambda > 0$, then $\lim_{m\to\infty} \mathcal{E}(u - \sum_{k=1}^m P_{\pi^k(x)}u, u - \sum_{k=1}^m P_{\pi^k(x)}u) + (u - \sum_{k=1}^m P_{\pi^k(x)}u, u - \sum_{k=1}^m P_{\pi^k(x)}u)_{L^2(\Sigma^+;\mu)} = 0$.

(ii) $|\mathcal{E}(P_x u, 1_{\Sigma_x})| \le \Lambda (u)_{\mu,x} \mathcal{E}(1_{\Sigma_x}, 1_{\Sigma_x})$ for any non-negative locally constant function on Σ_x.

Proof. (i) Since Φ is a complete orthonormal system in $L^2(\Sigma^+;\mu)$, one sees $\lim_{m\to\infty}(u - \sum_{k=1}^m P_{\pi^k(x)}u, u - \sum_{k=1}^m P_{\pi^k(x)}u)_{L^2(\Sigma^+;\mu)} = 0$ and $\sum_{k=1}^{\infty}(P_{\pi^k(x)}u, P_{\pi^k(x)}u)_{L^2(\Sigma^+;\mu)} = (u,u)_{L^2(\Sigma^+;\mu)}$.

Thanks to the closability of the symmetric bilinear form $\mathcal{E}$, it suffices to show $\lim_{m,m'\to\infty} \mathcal{E}(\sum_{k=1}^m P_{\pi^k(x)}u - \sum_{k=1}^{m'} P_{\pi^k(x)}u, \sum_{k=1}^m P_{\pi^k(x)}u - \sum_{k=1}^{m'} P_{\pi^k(x)}u) = 0$. Since $|\mathcal{E}(u,v)| \le \dfrac{1}{2}(\mathcal{E}(u,u) + \mathcal{E}(v,v))$ for any $u,v \in \mathcal{F}$, the assumption in the lemma enables us to derive the convergence from the inequality

$$\mathcal{E}(\sum_{k=1}^{m} P_{\pi^k(x)}u - \sum_{k=1}^{m'} P_{\pi^k(x)}u, \sum_{k=1}^{m} P_{\pi^k(x)}u - \sum_{k=1}^{m'} P_{\pi^k(x)}u)$$
$$\le 3\lambda \sum_{k=\min\{m,m'\}+1}^{\max\{m,m'\}} (P_{\pi^k(x)}u, P_{\pi^k(x)}u)_{L^2(\Sigma^+;\mu)}$$

and the convergence to zero of the right-hand side as $m, m' \to \infty$.

(ii) It suffices to show that

$$|\mathcal{E}(P_x u, (u)_{\mu,x} 1_{\Sigma_x})| \le \Lambda (u)^2_{\mu,\pi(x)} \mathcal{E}(1_{\Sigma_x}, 1_x)$$

for any function u with $\text{supp}[u] \subset \Sigma_x^+$, the inequality in which is equivalent to

$$|\mathcal{E}(P_x u, \sum_{k=1}^{\infty} P_{\pi^k(x)} u)| \leq \Lambda \mathcal{E}(\sum_{k=1}^{\infty} P_{\pi^k(x)} u, \sum_{k=1}^{\infty} P_{\pi^k(x)} u).$$

This follows from

$$\begin{aligned}|\mathcal{E}(P_x u, P_{\pi(x)} u)| \leq \Lambda \Big(& \mathcal{E}(P_{\pi(x)} u, P_{\pi(x)} u) \\ & + 2\mathcal{E}(P_{\pi(x)} u, P_{\pi^2(x)} u) + \mathcal{E}(P_{\pi^2(x)} u, P_{\pi^2(x)} u) \\ & + \qquad \cdots \\ & + 2\mathcal{E}(P_{\pi^{m-1}(x)} u, P_{\pi^m(x)} u) + \mathcal{E}(P_{\pi^m(x)} u, P_{\pi^m(x)} u) \Big) \\ \leq \Lambda \mathcal{E}(& \sum_{k=1}^{m} P_{\pi^k(x)} u, \sum_{k=1}^{m} P_{\pi^k(x)} u)\end{aligned}$$

which is valid for any locally constant function u with $\text{supp}[u] \subset \Sigma_x^+$. □

The results obtained in the final section of [Albeverio and Karwowski (2008)] give us the existence of Φ satisfying our assumptions and they showed that the conditions imposed on u in Lemma 2.1 is satisfied with $1_{\Sigma_x^+}$ in their framework. Accordingly, without losing the general settings as in [Kaneko (2013)], we may assume that, for every $x \in T$, $1_{\Sigma_x^+}$ is described as the limit of the sequence $\{P^{\pi(x)}_{\pi^{k+1}(x)} 1_{\Sigma_x^+}\}_{k=1}^{\infty}$ with respect to the norm $\sqrt{\mathcal{E}_1(u,u)}$ on $\mathcal{F}$, where $P^{\pi(x)}_{\pi^{k+1}(x)}$ stands for the orthogonal projection to the linear subspace $\mathcal{C}^{\pi(x)}_{\pi^{k+1}(x),0}$ spanned by $\cup_{m=1}^{k+1} \Phi_{\pi^m(x)}$ in $L^2(\Sigma^+;\mu)$. This assumption clarifies the regularity of the Dirichlet space in terms of the complete orthonormal basis Φ. In fact, the linear subspace spanned by Φ is contained densely in $\mathcal{F}$ with respect to the norm $\sqrt{\mathcal{E}_1(u,u)}$. Due to the Stone-Weierstrass theorem and the discussion as in Chapter VII Lemma 1 in [Vladimirov, Volovich and Zelenov (1995)], it turns out that the family of continuous functions on Σ^+ with compact support contains the linear subspace spanned by finite linear combination of elements in $\{1_{\Sigma_x^+} \mid x \in T\}$ as a dense subset. We will admit a conventional notation π^0 to represent the identity map on T.

We suppose that a finite subset S of T with the property $y, z \in S, y \neq z \Rightarrow \Sigma_y^+ \cap \Sigma_z^+ = \emptyset$ and a symmetric bilinear form $\mathcal{E}$ on the real linear space $\mathcal{C}$ spanned by $\{1_{\Sigma_x^+} \mid x \in S\}$ are given. $(\mathcal{E}, \mathcal{C})$ is called a Dirichlet space, if it satisfies

(i) $\mathcal{E}(u,u) \geq 0$ for any $u \in \mathcal{C}$,
(ii) $\mathcal{E}(1_{\Sigma_y^+}, 1_{\Sigma_z^+}) \leq 0$ for any distinct $y, z \in S$,
(iii) $v = 1$ on $\cup_{x \in S} \Sigma_x^+$ implies $\mathcal{E}(u,v) = 0$ for any $u \in \mathcal{C}$.

When a Dirichlet space $(\mathcal{E}, \mathcal{C})$ is given, it admits the adjacent matrix A which is viewed as a linear operator in the Euclidean space $\mathcal{C}$ and specified by $\mathcal{E}(u,v) =$

$-\sum_{y,z\in S} A_{y,z}u(y)v(z)$ for any $u, v \in \mathcal{C}$. It is easy to see the identities $A_{y,z} = A_{z,y}$ for any $y, z \in S$ and $\sum_{y\in S} A_{y,z} = 0$ for any $z \in S$. Another representation $\mathcal{E}(u, v) = \frac{1}{2}\sum_{y,z\in S} A_{y,z}(u(y) - u(z))(v(y) - v(z))$ for any $u, v \in \mathcal{C}$ of the bilinear form follows from these observations. If a symmetric bilinear form $\mathcal{E}$ with domain $\mathcal{C}$ satisfies (i), $\mathcal{E}$ said to be non-negative definite.

In proving the following proposition, the Beurling-Deny formula which provides us with the representation

$$\mathcal{E}(u, v) = \frac{1}{2}\int\int_{\Sigma^+\times\Sigma^+\cap\{\eta\neq\zeta\}} (u(\eta) - u(\zeta))(v(\eta) - v(\zeta))J(d\eta, d\zeta)$$

in the Dirichlet space theory (see [Fukushima, Oshima and Takeda (2010)] for the detail) will be crucially applied. In fact, it is not difficult to see that there is no Σ^+-valued continuous function defined on any intervals in the real line. Hereafter, we take the linear subspace $\mathcal{C}^{S^+(x)}_{\pi^k(x)} = \oplus_{\pi(y)=x}\mathcal{C}_{y,0} \oplus \mathcal{C}_{x,0} \oplus \cdots \oplus \mathcal{C}_{\pi^{k-1}(x),0} \oplus \mathcal{C}_{\pi^k(x)}$ of $L^2(\Sigma^+;\mu)$ and denote symmetric bilinear form $\sum_{\pi(y)=\pi(z)=x}\big(\mathcal{E}_{y,z}(u, v) + \mathcal{E}_{y,x}(u, v) + \mathcal{E}_{x,z}(u, v)\big) + \mathcal{E}_{x,x}(u, v) + \sum_{\ell=0}^{k-1}\big(\mathcal{E}_{\pi^\ell(x),\pi^{\ell+1}(x)}(u, v) + \mathcal{E}_{\pi^{\ell+1}(x),\pi^\ell(x)}(u, v) + \mathcal{E}_{\pi^{\ell+1}(x),\pi^{\ell+1}(x)}(u, v)\big)$ by $\mathcal{E}^{S^+(x)}_{\pi^k(x)}(u, v)$ defined for $u, v \in \mathcal{C}_{\pi^k(x)}$.

Proposition 2.1. If $\mathcal{E}(\varphi, \psi) = 0$ for any $\varphi \in \mathcal{C}_{y,0}, \psi \in \mathcal{C}_{z,0}$ with $y, z \in T$ except $y = z$, $\pi(y) = z$,$\pi(z) = y$ and $\pi(y) = \pi(z)$, then a bilinear form $\mathcal{E}_{y,z}$ is defined on $\mathcal{C}_y \times \mathcal{C}_z$ such that the symmetric bilinear form satisfies:

(i) $\mathcal{E}_{y,z}(u, v) = \mathcal{E}_{z,y}(v, u)$ for any $u \in \mathcal{C}_{y,0}, v \in \mathcal{C}_{z,0}$,

(ii) $\mathcal{E}_{y,z}(u, 1_{\Sigma^+_z}) = 0$ for any $u \in \mathcal{C}_{y,0}$ and $\mathcal{E}_{y,z}(1_{\Sigma^+_y}, v) = 0$ for any $v \in \mathcal{C}_{z,0}$,

(ii) the symmetric bilinear form $\mathcal{E}^{S^+(x)}_{\pi^k(x)}$ determined by

$$\begin{aligned}\mathcal{E}^{S^+(x)}_{\pi^k(x)}(u, v) = &\sum_{\pi(y)=\pi(z)=x} \big(\mathcal{E}_{y,z}(u, v) + \mathcal{E}_{y,x}(u, v) + \mathcal{E}_{x,z}(u, v)\big) \\ &\qquad + \mathcal{E}_{x,x}(u, v) \\ &+ \sum_{\ell=0}^{k-1} \big(\mathcal{E}_{\pi^\ell(x),\pi^{\ell+1}(x)}(u, v) + \mathcal{E}_{\pi^{\ell+1}(x),\pi^\ell(x)}(u, v) \\ &\qquad + \mathcal{E}_{\pi^{\ell+1}(x),\pi^{\ell+1}(x)}(u, v)\big)\end{aligned}$$

with domain $\oplus_{\pi(y)=x}\mathcal{C}_{y,0} \oplus \mathcal{C}_{x,0} \oplus \cdots \oplus \mathcal{C}_{\pi^{k-1}(x),0} \oplus \mathcal{C}_{\pi^k(x)}$ is a Dirichlet space for any $x \in T$ and non-negative integer k.

Proof. First we note that the projection $P_x : C_x \to C_{x,0}$ in $L^2(\Sigma^+;\mu)$ is explicitly given by $P_x u = u - (u)_{\mu,x} = u - \big(\frac{1}{\mu(\Sigma^+_x)}\int_{\Sigma^+_x} u(\zeta)\mu(d\zeta)\big)1_{\Sigma^+_x}$. Any element in $\oplus_{\pi(y)=x}\mathcal{C}_{y,0} \oplus \mathcal{C}_{x,0} \oplus \cdots \oplus \mathcal{C}_{\pi^{k-1}(x),0} \oplus \mathcal{C}_{\pi^k(x)}$ is decomposed into direct sum $\oplus_{\pi(y)=x}\mathcal{C}_{y,0} \oplus \mathcal{C}_{x,0} \oplus \cdots \oplus \mathcal{C}_{\pi^k(x),0} \oplus \{\alpha 1_{\Sigma^+_{\pi^k(x)}} \mid \alpha \in \mathbb{R}\}$. More specifically, $u \in \oplus_{\pi(y)=x}\mathcal{C}_{y,0} \oplus \mathcal{C}_{x,0} \oplus \cdots \oplus \mathcal{C}_{\pi^{k-1}(x),0} \oplus \mathcal{C}_{\pi^k(x)}$ is described as the sum of $u - \left(\frac{1}{\mu(\Sigma^+_{\pi^k(x)})}\int_{\Sigma^+_{\pi^k(x)}} u(\zeta)\mu(d\zeta)\right)1_{\Sigma^+_{\pi^k(x)}} \in \oplus_{\pi(y)=x}\mathcal{C}_{y,0} \oplus \mathcal{C}_{x,0} \oplus \cdots \oplus \mathcal{C}_{\pi^{k-1}(x),0} \oplus \mathcal{C}_{\pi^k(x),0}$ and $\int_{\Sigma^+_{\pi^k(x)}} u(\zeta)\mu(d\zeta)1_{\Sigma^+_{\pi^k(x)}} \in \mathcal{C}_{\pi^k(x)}$. We define the bilinear form

$$\mathcal{E}_{y,z}(u,v) = \mathcal{E}(P_y u, P_z v)$$

for $u \in \mathcal{C}_y, v \in \mathcal{C}_z$.

On the other hand, the Beurling-Deny formula and Lemma 2.1 (ii) implies that

$$\begin{aligned}
&\sum_{\pi(y)=\pi(z)=x} \left(\mathcal{E}_{y,z}(1_{\Sigma^+_w}, 1_{\Sigma^+_{w'}}) + \mathcal{E}_{y,x}(1_{\Sigma^+_w}, 1_{\Sigma^+_{w'}}) + \mathcal{E}_{x,z}(1_{\Sigma^+_w}, 1_{\Sigma^+_{w'}})\right) \\
&\qquad + \mathcal{E}_{x,x}(1_{\Sigma^+_w}, 1_{\Sigma^+_{w'}}) \\
&\quad + \sum_{\ell=0}^{k-1} \left(\mathcal{E}_{\pi^\ell(x),\pi^{\ell+1}(x)}(1_{\Sigma^+_w}, 1_{\Sigma^+_{w'}}) + \mathcal{E}_{\pi^{\ell+1}(x),\pi^\ell(x)}(1_{\Sigma^+_w}, 1_{\Sigma^+_{w'}})\right. \\
&\qquad \left. + \mathcal{E}_{\pi^{\ell+1}(x),\pi^{\ell+1}(x)}(1_{\Sigma^+_w}, 1_{\Sigma^+_{w'}})\right) \\
&\le \mathcal{E}(1_{\Sigma^+_w}, 1_{\Sigma^+_{w'}}) - \frac{\mu(\Sigma^+_w)}{\mu(\Sigma^+_{\pi^k(x)})}\frac{\mu(\Sigma^+_{w'})}{\mu(\Sigma^+_{\pi^k(x)})}\mathcal{E}(P_{\pi^k(x)}1_{\Sigma^+_w}, 1_{\Sigma^+_{\pi^k(x)}}) \\
&\quad - \frac{\mu(\Sigma^+_w)}{\mu(\Sigma^+_{\pi^k(x)})}\frac{\mu(\Sigma^+_{w'})}{\mu(\Sigma^+_{\pi^k(x)})}\mathcal{E}(1_{\Sigma^+_{\pi^k(x)}}, P_{\pi^k(x)}1_{\Sigma^+_{w'}}) \\
&\quad + \frac{\mu(\Sigma^+_w)}{\mu(\Sigma^+_{\pi^k(x)})}\frac{\mu(\Sigma^+_{w'})}{\mu(\Sigma^+_{\pi^k(x)})}\mathcal{E}(1_{\Sigma^+_{\pi^k(x)}}, 1_{\Sigma^+_{\pi^k(x)}}) \\
&\le \mathcal{E}(1_{\Sigma^+_w}, 1_{\Sigma^+_{w'}}) - (1-2\Lambda)\frac{\mu(\Sigma^+_w)}{\mu(\Sigma^+_{\pi^k(x)})}\frac{\mu(\Sigma^+_{w'})}{\mu(\Sigma^+_{\pi^k(x)})}\mathcal{E}(1_{\Sigma^+_{\pi^k(x)}}, 1_{\Sigma^+_{\pi^k(x)}}) \\
&\le 0
\end{aligned}$$

for any $1_{\Sigma^+_w}, 1_{\Sigma^+_{w'}} \in \oplus_{\pi(y)=x}\mathcal{C}_{y,0} \oplus \mathcal{C}_{x,0} \oplus \cdots \oplus \mathcal{C}_{\pi^{k-1}(x),0} \oplus \mathcal{C}_{\pi^k(x)}$ satisfying $\Sigma^+_w \cap \Sigma^+_{w'} = \emptyset$ and $\pi^\ell(w) = \pi^{\ell'}(w') = \pi^k(x)$ for some ℓ, ℓ' with $\ell, \ell' = k+2$. Since $P_w 1_{\Sigma^+_{\pi^k(x)}} = 0$ for any $w \in T$ satisfying $\Sigma_w \subsetneqq \Sigma_{\pi^k(x)}$,

$$\begin{aligned}
&\sum_{\pi(y)=\pi(z)=x} \left(\mathcal{E}_{y,z}(u, 1_{\Sigma^+_{\pi^k(x)}}) + \mathcal{E}_{y,x}(u, 1_{\Sigma^+_{\pi^k(x)}}) + \mathcal{E}_{x,z}(u, 1_{\Sigma^+_{\pi^k(x)}})\right) \\
&\qquad + \mathcal{E}_{x,x}(u, 1_{\Sigma^+_{\pi^k(x)}}) \\
&\quad + \sum_{\ell=0}^{k-1} \left(\mathcal{E}_{\pi^\ell(x),\pi^{\ell+1}(x)}(u, 1_{\Sigma^+_{\pi^k(x)}}) + \mathcal{E}_{\pi^{\ell+1}(x),\pi^\ell(x)}(u, 1_{\Sigma^+_{\pi^k(x)}})\right. \\
&\qquad \left. + \mathcal{E}_{\pi^{\ell+1}(x),\pi^{\ell+1}(x)}(u, 1_{\Sigma^+_{\pi^k(x)}})\right)
\end{aligned}$$

$$= 0$$

for any $u \in \oplus_{\pi(y)=x}\mathcal{C}_{y,0} \oplus \mathcal{C}_{x,0} \oplus \cdots \oplus \mathcal{C}_{\pi^{k-1}(x),0} \oplus \mathcal{C}_{\pi^k(x)}$. By combining these facts with the identity

$$\begin{aligned}
&\sum_{\pi(y)=\pi(z)=x} \big(\mathcal{E}_{y,z}(u,u) + \mathcal{E}_{y,x}(u,u) + \mathcal{E}_{x,z}(u,u)\big) + \mathcal{E}_{x,x}(u,v) \\
&+ \sum_{\ell=0}^{k-1} \big(\mathcal{E}_{\pi^\ell(x),\pi^{\ell+1}(x)}(u,u) + \mathcal{E}_{\pi^{\ell+1}(x),\pi^\ell(x)}(u,u) + \mathcal{E}_{\pi^{\ell+1}(x),\pi^{\ell+1}(x)}(u,u)\big) \\
&= \mathcal{E}\Big(\sum_{\pi(y)=x} P_y u + \sum_{\ell=0}^{k} P_{\pi^\ell(x)} u, \sum_{\pi(y)=x} P_y u + \sum_{\ell=0}^{k} P_{\pi^\ell(x)} u\Big) \geq 0,
\end{aligned}$$

it turns out that the symmetric bilinear form with the domain $\oplus_{\pi(y)=x}\mathcal{C}_{y,0} \oplus \mathcal{C}_{x,0} \oplus \cdots \oplus \mathcal{C}_{\pi^{k-1}(x),0} \oplus \mathcal{C}_{\pi^k(x)}$ is a Dirichlet space. □

The observation on the eigenfunctions in [Albeverio and Karwowski (2008)] clipped in our attention after Lemma 2.1 shows that $P_{\pi^k(x)}(P_x 1_{\Sigma_z^+}) = 0$ for any positive integer k and $x, z \in T$ with $\pi(z) = x$, which implies a less restrictive property $\mathcal{E}(P_x 1_{\Sigma_z^+}, u) = 0$ for any $u \in \cup_{k=2}^\infty \mathcal{C}_{\pi^k(x),0}$ on the bilinear form $\mathcal{E}$ than the scheme in [Kaneko (2013)]. By combining this identity with the fact that $\mathcal{C}_{x,0}$ is an $n(x)$-dimensional real vector space given by $\{P_x u \mid u \in L^2(\Sigma^+;\mu)\} = \{P_x 1_{\Sigma_z^+} \mid z \in S^+(x)\}$, we can derive the property $\mathcal{E}(P_x u, 1_{\Sigma^+_{\pi^\ell(x)}}) = 0$ for any $u \in L^2(\Sigma^+;\mu)$ and $\ell \geq 1$ at any $x \in T$ due to the approximation seen in the proof of Theorem 4.3 in [Albeverio and Karwowski (2008)]. In the rest of the section, we assume $\mathcal{E}(P_x u, 1_{\Sigma^+_{\pi^\ell(x)}}) = 0$ for any $u \in L^2(\Sigma^+;\mu)$ and $\ell \geq 1$ at any $x \in T$ for coverage on a more general setting

Theorem 2.1. $\mathcal{E}(\varphi, \psi) = 0$ for any $\varphi \in \mathcal{C}_{y,0}, \psi \in \mathcal{C}_{z,0}$ with $y, z \in T$ except $y = z$, $\pi(y) = z$, $\pi(z) = y$ and $\pi(y) = \pi(z)$, if and only if the measure $J(d\eta, d\zeta)$ is absolutely continuous with respect to the product measure $\mu(d\eta)\mu(d\zeta)$ and the density function takes constant on every $\Sigma_x^+ \times \Sigma_y^+$ determined by $x, y \in T$ with $\pi(x) \neq \pi(y)$ and $\pi^2(x) = \pi^2(y)$.

Proof. First we derive the latter properties in the assertion from the assumption that $\mathcal{E}(\varphi, \psi) = 0$ for any $\varphi \in \mathcal{C}_{y,0}, \psi \in \mathcal{C}_{z,0}$ with $y, z \in T$ except $y = z$, $\pi(y) = z$, $\pi(z) = y$ and $\pi(y) = \pi(z)$. We take $w, w', y, z \in T$ with distinct $\pi(y), \pi(z)$ satisfying $\pi(w) = y, \pi(w') = z$ and $\pi^2(y) = \pi^2(z) = x$. Since $P_y 1_{\Sigma_w^+}, P_{\pi(z)} 1_{\Sigma_z^+}, 1_{\Sigma^+_{\pi(z)}} \in \mathcal{F}$ and $1_{\Sigma^+_{\pi(z)}} \in \oplus_{k=0}^\infty \mathcal{C}_{\pi^k(x),0}$, we have $\mathcal{E}(P_y 1_{\Sigma_w^+}, P_{\pi(z)} 1_{\Sigma_z^+}) = \mathcal{E}(P_y 1_{\Sigma_w^+}, 1_{\Sigma^+_{\pi(z)}}) = 0$, which implies $\mathcal{E}(P_y 1_{\Sigma_w^+}, 1_{\Sigma_z^+}) = 0$. Consequently,

$$J(\Sigma_w^+, \Sigma_z^+) - \frac{\mu(\Sigma_w^+)}{\mu(\Sigma_y^+)} J(\Sigma_y^+, \Sigma_z^+) = -\mathcal{E}(1_{\Sigma_w^+} - (1_{\Sigma_w^+})_{\mu,y}, 1_{\Sigma_z^+})$$

$$= -\mathcal{E}(P_y 1_{\Sigma_w^+}, 1_{\Sigma_z^+}) = 0.$$

On the other hand, $1_{\Sigma_w^+} \in \oplus_{k=0}^{\infty} \mathcal{C}_{\pi^k(y),0}$ implies $\mathcal{E}(P_z 1_{\Sigma_{w'}^+}, 1_{\Sigma_w^+}) = 0$. Therefore, we obtain

$$J(\Sigma_w^+, \Sigma_{w'}^+) - \frac{\mu(\Sigma_{w'}^+)}{\mu(\Sigma_z^+)} J(\Sigma_w^+, \Sigma_z^+) = -\mathcal{E}(1_{\Sigma_w^+}, 1_{\Sigma_{w'}^+} - (1_{\Sigma_{w'}^+})_{\mu,z}) = 0.$$

As a result, from these identities, we can derive

$$J(\Sigma_w^+, \Sigma_{w'}^+) - \frac{\mu(\Sigma_w^+)}{\mu(\Sigma_y^+)} \frac{\mu(\Sigma_{w'}^+)}{\mu(\Sigma_z^+)} J(\Sigma_y^+, \Sigma_z^+) = 0.$$

By repeating successively the procedures taken so far, we obtain

$$J(\Sigma_w^+, \Sigma_{w'}^+) = \frac{\mu(\Sigma_w^+)}{\mu(\Sigma_y^+)} \frac{\mu(\Sigma_{w'}^+)}{\mu(\Sigma_z^+)} J(\Sigma_y^+, \Sigma_z^+)$$

as long as $\Sigma_w^+ \subset \Sigma_y^+$ and $\Sigma_{w'}^+ \subset \Sigma_z^+$ for any $y, z \in T$ with $\pi(y) \neq \pi(z)$ and $\pi^2(y) = \pi^2(z)$. The absolute continuity of $J(d\eta, d\zeta)$ with respect to $\mu(d\eta)\mu(d\zeta)$ has been proved.

Conversely, we assume that

$$J(\Sigma_w^+, \Sigma_{w'}^+) = \frac{\mu(\Sigma_w^+)}{\mu(\Sigma_y^+)} \frac{\mu(\Sigma_{w'}^+)}{\mu(\Sigma_z^+)} J(\Sigma_y^+, \Sigma_z^+)$$

as long as $\Sigma_w^+ \subset \Sigma_y^+$ and $\Sigma_{w'}^+ \subset \Sigma_z^+$ for any $y, z \in T$ with $\pi(y) \neq \pi(z)$ and $\pi^2(y) = \pi^2(z)$. From the identity, we can derive

$$\begin{aligned}
\mathcal{E}(1_{\Sigma_w^+}, 1_{\Sigma_{w'}^+} - (1_{\Sigma_{w'}^+})_{\mu,z}) &= -J(\Sigma_w^+, \Sigma_{w'}^+) + \frac{\mu(\Sigma_{w'}^+)}{\mu(\Sigma_z^+)} J(\Sigma_w^+, \Sigma_z^+) \\
&= -J(\Sigma_w^+, \Sigma_{w'}^+) + \frac{\mu(\Sigma_w^+)}{\mu(\Sigma_y^+)} \frac{\mu(\Sigma_{w'}^+)}{\mu(\Sigma_z^+)} J(\Sigma_y^+, \Sigma_z^+) \\
&= 0
\end{aligned}$$

and

$$\begin{aligned}
\mathcal{E}(1_{\Sigma_y^+}, 1_{\Sigma_{w'}^+} - (1_{\Sigma_{w'}^+})_{\mu,z}) &= -J(\Sigma_y^+, \Sigma_{w'}^+) + \frac{\mu(\Sigma_{w'}^+)}{\mu(\Sigma_z^+)} J(\Sigma_y^+, \Sigma_z^+) \\
&= 0
\end{aligned}$$

for any w, w' satisfying $\pi(w) = y$ and $\pi(w') = z$. Consequently, we have $\mathcal{E}(P_y 1_{\Sigma_w^+}, P_z 1_{\Sigma_{w'}^+}) = 0$, which means $\mathcal{E}(\varphi, \psi) = 0$ for any $\varphi \in \mathcal{C}_{y,0}, \psi \in \mathcal{C}_{z,0}$ with distinct $\pi(y), \pi(z)$ and $\pi^2(y) = \pi^2(z)$.

The identity originally assumed shows that

$$J(\Sigma_w^+, \Sigma_{w'}^+) = \frac{\mu(\Sigma_w^+)}{\mu(\Sigma_y^+)} \frac{\mu(\Sigma_{w'}^+)}{\mu(\Sigma_z^+)} J(\Sigma_y^+, \Sigma_z^+)$$

holds valid even for $\Sigma_w^+ \subset \Sigma_y^+$, $\Sigma_{w'}^+ \subset \Sigma_z^+$ and $\pi^\ell(y) = \pi^m(z) = [y, z]$ with $\ell, m \geq 2$. In fact, by the assumption we can drive

$$J(\Sigma_w^+, \Sigma_{w'}^+) = \frac{\mu(\Sigma_w^+)}{\mu(\Sigma_{\pi^{\ell-2}(y)}^+)} \frac{\mu(\Sigma_{w'}^+)}{\mu(\Sigma_{\pi^{m-2}(z)}^+)} J(\Sigma_{\pi^{\ell-2}(y)}^+, \Sigma_{\pi^{m-2}(z)}^+)$$

and

$$J(\Sigma_y^+, \Sigma_z^+) = \frac{\mu(\Sigma_y^+)}{\mu(\Sigma_{\pi^{\ell-2}(y)}^+)} \frac{\mu(\Sigma_z^+)}{\mu(\Sigma_{\pi^{m-2}(z)}^+)} J(\Sigma_{\pi^{\ell-2}(y)}^+, \Sigma_{\pi^{m-2}(z)}^+),$$

which show the validity of the identity in our focus. By using the identity and taking a similar procedure to the previous case, we can conclude $\mathcal{E}(\varphi, \psi) = 0$ for any $\varphi \in \mathcal{C}_{y,0}, \psi \in \mathcal{C}_{z,0}$ for $y, z \in T$ satisfying $\pi^\ell(y) = \pi^m(z) = [y, z]$ with some $\ell, m \geq 2$.

Since we already assumed that $\mathcal{E}(P_y 1_{\Sigma_w^+}, 1_{\Sigma_{\pi^i(y)}^+}) = 0$ for any $w \in S^+(y)$ and $i \geq 1$, we can similarly deduce $\mathcal{E}(P_y 1_{\Sigma_w^+}, P_z 1_{\Sigma_{w'}^+}) = \mathcal{E}(P_y 1_{\Sigma_w^+}, 1_{\Sigma_{w'}^+} - \frac{\mu(\Sigma_{w'}^+)}{\mu(\Sigma_z^+)} 1_{\Sigma_z^+}) = 0$ for any $z = \pi(w')$, $y = \pi(w)$ and $w' = \pi^i(y)$ with $i \geq 1$.

Finally, we verify that $\mathcal{E}(P_y 1_{\Sigma_w^+}, P_z 1_{\Sigma_{w'}}) = 0$ for any $w \in S^+(y), w' \in S^+(z)$ with $\pi(y) \neq w'$ and $\pi^2(y) = z$. Since

$$J(\Sigma_w^+, \Sigma_{w'''}^+) = \frac{\mu(\Sigma_w^+)}{\mu(\Sigma_y^+)} \frac{\mu(\Sigma_{w'''}^+)}{\mu(\Sigma_{w''}^+)} J(\Sigma_y^+, \Sigma_{w''}^+)$$

as long as $\Sigma_w^+ \subset \Sigma_y^+$ and $\Sigma_{w'''}^+ \subset \Sigma_{w''}^+$ for any y, w'' with $\pi(y) \neq \pi(w'') = w'$ and $\pi^2(y) = \pi^2(w'')$, by taking $w'' = w'''$, one sees that $\mathcal{E}(P_y 1_{\Sigma_w^+}, 1_{\Sigma_{w''}^+}) = 0$ for any $w, y \in T$ with $\pi(w) = y$ and $w'' \in T$ with $\pi(y) \neq \pi(w'')$ and $\pi^2(y) = \pi^2(w'')$. This implies $\mathcal{E}(P_y 1_{\Sigma_w^+}, 1_{\Sigma_{w'}^+}) = \sum_{\pi(w'')=w'} \mathcal{E}(P_y 1_{\Sigma_w^+}, 1_{\Sigma_{w''}^+}) = 0$. By combining this with $\mathcal{E}(P_y 1_{\Sigma_w^+}, 1_{\Sigma_z^+}) = 0$, we obtain the required identity. □

Let $\{(\mathcal{E}_k, \mathcal{F}_k)\}_{k=0}^\infty$ be a sequence of Dirichlet spaces on $L^2(\Sigma^+; \mu)$ such that each $\mathcal{E}_k(\varphi, \psi)$ vanishes for any $\varphi \in \mathcal{C}_{y,0}, \psi \in \mathcal{C}_{z,0}$ with $y, z \in T$ except $y = z$, $\pi(y) = z$, $\pi(z) = y$ and $\pi(y) = \pi(z)$. If $\mathcal{F}_k \subset \mathcal{F}_{k+1}$ and $\mathcal{E}_{k,\pi^\ell(x)}^{S^+(x)}(u, v) = \mathcal{E}_{k+1,\pi^\ell(x)}^{S^+(x)}(u, v)$ for any $u, v \in \mathcal{F}_k$ for any non-negative integer k, ℓ and $x \in T$, the sequence is said to be consistent in the tier-linked sense, where $\mathcal{E}_{k,\pi^\ell(x)}^{S^+(x)}(u, v)$ stands for the symmetric bilinear form

$\mathcal{E}_{k,\pi^\ell(x)}^{S^+(x)}(u, v)$

$$= \sum_{\pi(y)=\pi(z)=x} \big(\mathcal{E}_{k,y,z}(u,v) + \mathcal{E}_{k,y,x}(u,v) + \mathcal{E}_{k,x,z}(u,v)\big) + \mathcal{E}_{k,x,x}(u,v)$$

$$+ \sum_{j=0}^{\ell-1} \big(\mathcal{E}_{k,\pi^j(x),\pi^{j+1}(x)}(u,v) + \mathcal{E}_{k,\pi^{j+1}(x),\pi^j(x)}(u,v) + \mathcal{E}_{k,\pi^{j+1}(x),\pi^{j+1}(x)}(u,v)\big)$$

determined by $\mathcal{E}_{k,y,z}(u,v) = \mathcal{E}_k(P_y u, P_z v)$.

Lemma 2.2. If a consistent sequence $\{(\mathcal{E}_k, \mathcal{F}_k)\}_{k=0}^{\infty}$ of Dirichlet spaces in the tier-linked sense on $L^2(\Sigma^+;\mu)$ each of which satisfies $\mathcal{E}_k(\varphi,\psi) = 0$ for the case $\varphi \in \mathcal{C}_{y,0}, \psi \in \mathcal{C}_{z,0}$ with $y, z \in T$ satisfying $y \neq z$, $\pi(y) \neq \pi(z)$, $\pi(y) \neq z$ and $\pi(z) \neq y$ or the case either φ or ψ is in $\oplus_{\ell=k+1}^{\infty} \mathcal{C}_{\pi^\ell(o),0}$ with $o \in T_n$ for some fixed n, then $\sup\{\lambda \mid \mathcal{E}_k(v,v) = \lambda(v,v)_{L^2(\Sigma^+;\mu)}$ for some $v = \sum_{w\in N} P_w u$ given by some $N \subset S^+(x) \cup \{x, \pi(x), \pi^2(x), \dots\}$ and some $u \in \mathcal{C}(\Sigma_x^+)\} < \infty$ for any $x \in T$ implies $\lim_{k\to\infty} \mathcal{E}_k(u,u)$ is determined as a real value for any function $u \in C(\Sigma_x^+)$ with some $x \in T$.

Proof. We first note that $C(\Sigma_x^+) \subset L^2(\Sigma^+;\mu)$. We may assume $\pi^\ell(x) = o$ for some $o \in T_n$ with some non-negative integer ℓ and the fixed integer n. Therefore, u admits the decomposition $u = u_m + \sum_{j=k+m+1}^{k+m'} P_{\pi^j(x)} u + \sum_{j=k+m'+1}^{\infty} P_{\pi^j(x)} u$ with $u_m \in \oplus_{\pi(y)=x} \mathcal{C}_{y,0} \oplus \mathcal{C}_{x,0} \oplus \dots \oplus \mathcal{C}_{\pi^{\ell-1}(x),0} \oplus \mathcal{C}_{\pi^\ell(x),0} \cdots \oplus \mathcal{C}_{\pi^{k+m}(x),0}$. Since the consistency in the tier-linked sense implies

$$\begin{aligned}
&\mathcal{E}_{k+m'}(u,u)\\
&= \mathcal{E}^{S^+(x)}_{k+m',\pi^{\ell+k+m'}(x)}\Big(u_m + \sum_{j=k+m+1}^{k+m'} P_{\pi^j(x)}u, u_m + \sum_{j=k+m+1}^{k+m'} P_{\pi^j(x)}u\Big)\\
&= \mathcal{E}^{S^+(x)}_{k+m',\pi^{k+m'}(x)}(u_m,u_m) + \mathcal{E}_{k+m'}\Big(\sum_{j=k+m+1}^{k+m'} P_{\pi^j(x)}u, \sum_{j=k+m+1}^{k+m'} P_{\pi^j(x)}u\Big)\\
&\quad +2\mathcal{E}_{k+m'}(P_{\pi^{k+m}(x)}u, P_{\pi^{k+m+1}(x)}u)\\
&= \mathcal{E}^{S^+(x)}_{k+m,\pi^{k+m}(x)}(u_m,u_m) + \mathcal{E}_{k+m'}\Big(\sum_{j=k+m+1}^{k+m'} P_{\pi^j(x)}u, \sum_{j=k+m+1}^{k+m'} P_{\pi^j(x)}u\Big)\\
&\quad +2\mathcal{E}_{k+m'}(P_{\pi^{k+m}(x)}u, P_{\pi^{k+m+1}(x)}u)\\
&= \mathcal{E}_{k+m}(u,u) + \mathcal{E}_{k+m'}\Big(\sum_{j=k+m}^{k+m'} P_{\pi^j(x)}u, \sum_{j=k+m}^{k+m'} P_{\pi^j(x)}u\Big)\\
&\quad +2\mathcal{E}_{k+m'}(P_{\pi^{k+m}(x)}u, P_{\pi^{k+m+1}(x)}u),
\end{aligned}$$

we obtain

$$|\mathcal{E}_{k+m'}(u,u) - \mathcal{E}_{k+m}(u,u)| \le 2\mathcal{E}_{k+m'}\Big(\sum_{j=k+m}^{k+m'} P_{\pi^j(x)}u, \sum_{j=k+m}^{k+m'} P_{\pi^j(x)}u\Big)$$

$$+\mathcal{E}_{k+m'}(P_{\pi^{k+m}(x)}u, P_{\pi^{k+m}(x)}u)$$
$$+\mathcal{E}_{k+m'}(P_{\pi^{k+m+1}(x)}u, P_{\pi^{k+m+1}(x)}u).$$

Our assumption shows that the right-hand side tends to zero as $m, m' \to \infty$, which implies that $\lim_{k\to\infty} \mathcal{E}_k(u,u)$ is determined as a real number. □

In the rest of this section, we assume that $\lim_{m\to\infty} \sup\{\lambda \mid \mathcal{E}(u,u) = \lambda(u,u)_{L^2(\Sigma^+;\mu)}$ for some $u \in \oplus_{\ell=m}^{\infty}\mathcal{C}_{\pi^\ell(x),0} \setminus \{0\}\} = 0$ for any $x \in T$.

Proposition 2.2. Under the same assumption in Proposition 2.1, the symmetric bilinear form

$$\mathcal{E}^{S^+(x)}_{\pi^k(x)}(u,v) = \sum_{\pi(y)=\pi(z)=x} \big(\mathcal{E}_{y,z}(u,v) + \mathcal{E}_{y,x}(u,v) + \mathcal{E}_{x,z}(u,v)\big) + \mathcal{E}_{x,x}(u,v)$$
$$+\sum_{\ell=0}^{k-1} \big(\mathcal{E}_{\pi^\ell(x),\pi^{\ell+1}(x)}(u,v) + \mathcal{E}_{\pi^{\ell+1}(x),\pi^\ell(x)}(u,v) + \mathcal{E}_{\pi^{\ell+1}(x),\pi^{\ell+1}(x)}(u,v)\big)$$

with domain $\mathcal{C}^{S^+(x)}_{\pi^k(x)} = \oplus_{\pi(y)=x}\mathcal{C}_{y,0} \oplus \mathcal{C}_{x,0} \oplus \cdots \oplus \mathcal{C}_{\pi^{k-1}(x),0} \oplus \mathcal{C}_{\pi^k(x)}$, is a Dirichlet space for each $x \in T$. The sequence $\{(\mathcal{E}^{S^+(x)}_{\pi^k(x)}, \mathcal{C}^{S^+(x)}_{\pi^k(x)})\}_{k=0}^{\infty}$ of Dirichlet spaces is consistent in the tier-linked sense.

Proof. In Proposition 2.1, it is already shown that $(\mathcal{E}^{S^+(x)}_{\pi^k(x)}, \mathcal{C}^{S^+(x)}_{\pi^k(x)})$ is a Dirichlet space for each non-negative integer k. It is easy to see that $\mathcal{E}^{S^+(x)}_{\pi^k(x)}(\varphi,\psi)$ vanishes for any $\varphi \in \mathcal{C}_{y,0}, \psi \in \mathcal{C}_{z,0}$ with $y, z \in T$ except $y = z$, $\pi(y) = z$, $\pi(z) = y$ and $\pi(y) = \pi(z)$.

We see that $\mathcal{C}_{x',0}, \mathcal{C}_{x'',0} \subset \mathcal{C}^{S^+(x)}_{\pi^k(x)}$ imples $\mathcal{E}^{S^+(x)}_{\pi^{k+1}(x)}(P_{x'}u, P_{x''}v) = \mathcal{E}(P_{x'}u, P_{x''}v) = \mathcal{E}_{x',x''}(u,v)$ for any $u, v \in \mathcal{C}^{S^+(x)}_{\pi^k(x)}$. This shows the identity $\mathcal{E}^{S^+(x)}_{\pi^{k+1}(x)}(u,v) = \mathcal{E}^{S^+(x)}_{\pi^k(x)}(u,v)$ for any $u, v \in \mathcal{C}^{S^+(x)}_{\pi^k(x)}$. The consistency in the tier-linked sense of $\{(\mathcal{E}^{S^+(x)}_{\pi^k(x)}, \mathcal{C}^{S^+(x)}_{\pi^k(x)})\}_{k=0}^{\infty}$ has been proved. □

3. Admissible family of nodewise given Dirichlet forms

In this section, we start with a family $\{(\mathcal{E}_{y,z}, \mathcal{C}_y \times \mathcal{C}_z)\}_{y,z\in T}$ of the pairs of bilinear form and its domain satisfying $\mathcal{E}_{y,z}(\varphi,\psi) = 0$ for any $\varphi \in \mathcal{C}_{y,0}, \psi \in \mathcal{C}_{z,0}$ with $y, z \in T$ except $y = z$, $\pi(y) = z$, $\pi(z) = y$ and $\pi(y) = \pi(z)$. In what follows, the bilinear form $\mathcal{E}_{x,x} + \sum_{\pi(y)=\pi(z)=x}(\mathcal{E}_{y,x} + \mathcal{E}_{x,z}) + \sum_{\pi(y)=\pi(z)=x, y\neq z} \mathcal{E}_{y,z}$ will be denoted by $\hat{\mathcal{E}}_x$ and its domain $(\mathcal{C}_x \times \mathcal{C}_x) + \sum_{\pi(y)=\pi(z)=x}((\mathcal{C}_y \times \mathcal{C}_x) + (\mathcal{C}_x \times \mathcal{C}_z)) + \sum_{\pi(y)=\pi(z)=x, y\neq z}(\mathcal{C}_y \times \mathcal{C}_z)$ by $\hat{\mathcal{C}}_x$.

A family $\{(\mathcal{E}_{y,z}, \mathcal{C}_y \times \mathcal{C}_z)\}_{y,z\in T}$ of the bilinear forms is said to be admissible, if the symmetric bilinear form

$$\begin{aligned}
&\mathcal{E}^{S^+(x)}_{\pi^k(x)}(u,v)\\
&= \sum_{\pi(y)=\pi(z)=x} (\mathcal{E}_{y,z}(u,v)+\mathcal{E}_{y,x}(u,v)+\mathcal{E}_{x,z}(u,v)) + \mathcal{E}_{x,x}(u,v)\\
&+\sum_{\ell=0}^{k-1} (\mathcal{E}_{\pi^\ell(x),\pi^{\ell+1}(x)}(u,v)+\mathcal{E}_{\pi^{\ell+1}(x),\pi^\ell(x)}(u,v)+\mathcal{E}_{\pi^{\ell+1}(x),\pi^{\ell+1}(x)}(u,v))
\end{aligned}$$

with the domain $\oplus_{\pi(y)=x}\mathcal{C}_{y,0} \oplus \mathcal{C}_{x,0} \oplus \cdots \oplus \mathcal{C}_{\pi^{k-1}(x),0} \oplus \mathcal{C}_{\pi^k(x)}$ satisfies $\mathcal{E}^{S^+(x)}_{\pi^k(x)}(1_{\Sigma^+_w}, 1_{\Sigma^+_{w'}}) \leq 0$ for any $x \in T$ and non-negative integer k as long as disjoint $\Sigma^+_w, \Sigma^+_{w'}$ determined by $w, w' \in T$ yield $1_{\Sigma^+_w}, 1_{\Sigma^+_{w'}} \in \oplus_{\pi(y)=x}\mathcal{C}_{y,0} \oplus \mathcal{C}_{x,0} \oplus \cdots \oplus \mathcal{C}_{\pi^{k-1}(x),0} \oplus \mathcal{C}_{\pi^k(x)}$, here and in the sequel $\oplus_{\pi(y)=x}\mathcal{C}_{y,0} \oplus \mathcal{C}_{x,0} \oplus \cdots \oplus \mathcal{C}_{\pi^{k-1}(x),0} \oplus \mathcal{C}_{\pi^k(x)}$ is denoted by $\mathcal{C}^{S^+(x)}_{\pi^k(x)}$.

Remark 3.1. In this situation, $\mathcal{E}^{S^+(x)}_{\pi^k(x)}(\varphi, \psi) = 0$ is satisfied for any $\varphi \in \mathcal{C}_{y,0}, \psi \in \mathcal{C}_{z,0}$ except the cases determined by $y, z \in T$ with $y = z$, $\pi(y) = z$, $y = \pi(z)$ and $\pi(y) = \pi(z)$ as long as the left-hand side makes sense. Therefore, as in the claims described just before Theorem 2.1, $\mathcal{E}^{S^+(x)}_{\pi^k(x)}(P_w u, 1_{\Sigma^+_{\pi^\ell(w)}}) = 0$ for any $u \in L^2(\Sigma^+;\mu)$ and positive integer ℓ.

Lemma 3.1. If the family $\{(\mathcal{E}_{y,z}, \mathcal{C}_y \times \mathcal{C}_z)\}_{y,z\in T}$ of the bilinear forms is admissible, then

(i) $(\mathcal{E}^{S^+(x)}_{\pi^k(x)}, \mathcal{C}^{S^+(x)}_{\pi^k(x)})$ is a Dirichlet space for any non-negative integer k and $x \in T$,

(ii) the sequence $(\mathcal{E}^{S^+(x)}_{\pi^k(x)}, \mathcal{C}^{S^+(x)}_{\pi^k(x)})$ of Dirichlet spaces is consistent in the tier-linked sense.

Proof. (i) The property $\mathcal{E}^{S^+(x)}_{\pi^k(x)}(u, 1_{\Sigma^+_{\pi^k(x)}}) = 0$ for any $u \in \mathcal{C}^{S^+(x)}_{\pi^k(x)}$ is clear. The identity $\mathcal{E}^{S^+(x)}_{\pi^k(x)}(1_{\Sigma^+_w}, 1_{\Sigma^+_{w'}}) \leq 0$ for any $1_{\Sigma^+_w}, 1_{\Sigma^+_{w'}} \in \mathcal{C}^{S^+(x)}_{\pi^k(x)}$ determined by disjoint $\Sigma^+_w, \Sigma^+_{w'}$ follows from the admissibility of $(\mathcal{E}^{S^+(x)}_{\pi^k(x)}, \mathcal{C}^{S^+(x)}_{\pi^k(x)})$. These properties imply the non-negative definiteness of $\mathcal{E}^{S^+(x)}_{\pi^k(x)}$. Since the non-negativity of $\mathcal{E}^{S^+(x)}_{\pi^k(x)}(u,u)$ for any $u \in \mathcal{C}^{S^+(x)}_{\pi^k(x)}$ is clear, the assertion is proved.

(ii) Since $(\mathcal{E}^{S^+(x)}_{\pi^k(x)}, \mathcal{C}^{S^+(x)}_{\pi^k(x)})$ is a Dirichlet space, the procedure in Proposition 2.1 for defining $\mathcal{E}_{y,z}$ from $\mathcal{E}$ is applicable to determine $\mathcal{E}^{S^+(x)}_{\pi^{k+1}(x),y,z}$ by $\mathcal{E}^{S^+(x)}_{\pi^{k+1}(x)}$ for any $y, z \in T$ except $y = z$, $\pi(y) = z$, $\pi(z) = y$ and $\pi(y) = \pi(z)$. Due to the description of $\mathcal{E}^{S^+(x)}_{\pi^{k+1}(x),y,z}(u,v)$ involving the projection P_y, P_z in the proof of Proposition 2.1, we have

$$\mathcal{E}_{y,z}(P_y 1_{\Sigma^+_w}, P_z 1_{\Sigma^+_{w'}}) = \mathcal{E}^{S^+(x)}_{\pi^{k+1}(x),y,z}(P_y 1_{\Sigma^+_w}, P_z 1_{\Sigma^+_{w'}})$$

for $w \in S^+(y), w' \in S^+(z)$ with $1_{\Sigma_w^+}, 1_{\Sigma_{w'}^+} \in \mathcal{C}_{\pi^k(x)}^{S^+(x)}$. We can derive the consistency of $\{(\mathcal{E}_{\pi^k(x)}^{S^+(x)}, \mathcal{C}_{\pi^k(x)}^{S^+(x)})\}_{k=0}^{\infty}$ in the tier-linked sense from the identity by applying the same method as in the proof of Proposition 2.2. □

We temporarily fix a point o in T and an integer $m \geq n$ with $o \in T_n$. Let us define the symmetric bilinear form

$$\begin{aligned}\mathcal{E}_o^{m+1}(u,v) = \hat{\mathcal{E}}_o(u,v) \\ + \sum_{x \in T_{n+1}, \pi(x)=o} \Big(\hat{\mathcal{E}}_x(u,v) + \cdots \\ + \sum_{x'' \in T_{m-1}, \pi(x'')=x'} \Big(\hat{\mathcal{E}}_{x''}(u,v) \\ + \sum_{x''' \in T_m, \pi(x''')=x''} \hat{\mathcal{E}}_{x'''}(u,v) \\ + \sum_{x'''' \in T_{m+1}, \pi(x'''')=x'''} \mathcal{E}_{x'''',x''''}(u,v)\Big) \cdots \Big)\end{aligned} \tag{3.1}$$

for $u, v \in \mathcal{F}_o^{m+1}$. Here and in the sequal, $\mathcal{F}_o^{m+1}$ stands for $\mathcal{C}(\Sigma_o^+) \cap \mathcal{C}^{m+1}(\Sigma^+)$ and $\Pi_1(\eta,\zeta)$ for the pair of nodes (y', z') characterized by $\eta \in \Sigma_{y'}^+$, $\zeta \in \Sigma_{z'}^+$, $\pi(y') \neq \pi(z')$ and $\pi^2(y') = \pi^2(z')$.

Proposition 3.1. If the family $\{(\mathcal{E}_{y,z}, \mathcal{C}_y \times \mathcal{C}_z)\}_{y,z \in T}$ of the bilinear forms is admissible, then there exists a function $K_{1,o}(y', z')$ depending only on the pair (y', z') of distinct nodes with $\pi(y') \neq \pi(z')$ and $\pi^2(y') = \pi^2(z')$ such that

$$\begin{aligned}&\mathcal{E}_o^{m+1}(u,v) \\ &= \frac{1}{2} \int\!\!\int_{\Sigma_o^+ \times \Sigma_o^+ \setminus \bigcup_{\pi^{m+2-n}(w)=o} \Sigma_w^+ \times \Sigma_w^+} (u(\eta) - u(\zeta))(v(\eta) - v(\zeta)) \\ &\qquad \times K_{1,o}(\Pi_1(\eta,\zeta))\mu(d\eta)\mu(d\zeta)\end{aligned}$$

for any $u, v \in \mathcal{F}_o^{m+1}$. The sequence $\{(\mathcal{E}_{\pi^k(o)}^{m+1}, \mathcal{F}_{\pi^k(o)}^{m+1})\}_{k=0}^{\infty}$ of Dirichlet spaces obtained by replacing o with $\pi^k(o)$ in (3.1) is consistent in tier-linked sense.

Proof. First we will show that $(\mathcal{E}_o^{m+1}, \mathcal{F}_o^{m+1})$ is a Dirichlet space. The identity $\mathcal{E}_o^{m+1}(1_{\Sigma_o^+}, v) = 0$ for any $v \in \mathcal{F}_o^{m+1}$ is clear. Therefore, it suffices to show that, by assuming $\pi^\ell(x) = o$ with some positive integer ℓ,

$$\mathcal{E}_o^{m+1}(1_{\Sigma_w^+}, 1_{\Sigma_{w'}^+}) = \frac{\mu(\Sigma_w^+)\mu(\Sigma_{w'}^+)}{\mu(\Sigma_{y'}^+)\mu(\Sigma_{z'}^+)} \mathcal{E}_o^{S^+(x)}(1_{\Sigma_{y'}^+}, 1_{\Sigma_{z'}^+}) \tag{3.2}$$

for any disjoint Σ_w^+, $\Sigma_{w'}^+$, where y, z stands for the pair of distinct nodes characterized by $\pi^i(w) = y'$, $\pi^j(w') = z'$ with some positive integer i, j and $\pi(y') \neq \pi(z')$ and $\pi^2(y') = \pi^2(z') = x$. In fact, the right-hand side is non-positive due to the admissibility of the family $\{(\mathcal{E}_{y,z}, \mathcal{C}_{y,z})\}_{y,z\in T}$ which shows that a non-negative measurable function $K_{1,o}(y', z')$ is defined by'

$$K_{1,o}(y', z') = -\frac{1}{\mu(\Sigma_{y'}^+)\mu(\Sigma_{z'}^+)}\mathcal{E}_o^{S^+(x)}(1_{\Sigma_{y'}^+}, 1_{\Sigma_{z'}^+}), \tag{3.3}$$

for any $y', z' \in T$ satisfying $\pi(y') \neq \pi(z')$ and $\pi^2(y') = \pi^2(z') = x$ so that we can derive from (3.2) that

$$\mathcal{E}_o^{m+1}(1_{\Sigma_w^+}, 1_{\Sigma_{w'}^+}) = -\frac{1}{2}\int_{\Sigma_w^+}\int_{\Sigma_{w'}^+} K_{1,o}(\Pi_1(\eta, \zeta))\mu(d\eta)\mu(d\zeta)$$

for any $w, w' \in T$ with $\pi^i(w) = y$, $\pi^j(w') = z$ and $\pi(y) = \pi(z) = x$ with some positive integer i, j.

In showing (3.3), we need the following expressions for $1_{\Sigma_w^+}, 1_{\Sigma_{w'}^+}$:

$$\begin{aligned}1_{\Sigma_w^+} &= P_{\pi(w)}1_{\Sigma_w^+} + \cdots + P_{\pi^i(w)}\frac{\mu(\Sigma_w^+)}{\mu(\Sigma_{\pi^{i-1}(w)}^+)}1_{\Sigma_{\pi^{i-1}(w)}^+} \\ &\quad + \frac{\mu(\Sigma_w^+)}{\mu(\Sigma_{\pi^i(w)}^+)}1_{\Sigma_{\pi^i(w)}^+}, \\ 1_{\Sigma_{w'}^+} &= P_{\pi(w')}1_{\Sigma_w^+} + \cdots + P_{\pi^j(w')}\frac{\mu(\Sigma_{w'}^+)}{\mu(\Sigma_{\pi^{j-1}(w')}^+)}1_{\Sigma_{\pi^{j-1}(w')}^+} \\ &\quad + \frac{\mu(\Sigma_{w'}^+)}{\mu(\Sigma\pi^j(w')^+)}1_{\Sigma_{\pi^j(w')}^+},\end{aligned}$$

the final terms of which are equal to $\frac{\mu(\Sigma_w^+)}{\mu(\Sigma_{y'}^+)}1_{\Sigma_{y'}^+}$ and $\frac{\mu(\Sigma_{w'}^+)}{\mu(\Sigma_{z'}^+)}1_{\Sigma_{z'}^+}$ respectively. By using the right-hand sides to describe $1_{\Sigma_w^+}, 1_{\Sigma_{w'}^+}$, we see

$$\mathcal{E}_o^{m+1}(1_{\Sigma_w^+}, 1_{\Sigma_{w'}^+}) = \frac{\mu(\Sigma_w^+)\mu(\Sigma_{w'}^+)}{\mu(\Sigma_{y'}^+)\mu(\Sigma_{z'}^+)}\mathcal{E}_o^{S^+(x)}(1_{\Sigma_{y'}^+}, 1_{\Sigma_{z'}^+})$$

with the choice of nodes y', z' for (3.3). This is because $\mathcal{E}_{y,z}(P_y u, P_z v) = 0$ for any $u, v \in L^2(\Sigma^+; \mu)$, unless $y = z$, $\pi(y) = z$, $\pi(z) = y$ or $\pi(y) = \pi(z)$. The non-negative definiteness of $\mathcal{E}_o^{m+1}(u, v)$ follows from these two properties.

Finally, we note that the consistency of $\{(\mathcal{E}_{\pi^k(o)}^{m+1}, \mathcal{F}_{\pi^k(o)}^{m+1})\}_{k=0}^{\infty}$ in the tier-linked sense derived from $\mathcal{E}_{\pi^k(o),y,z}^{m+1} = \mathcal{E}_{y,z}$ and $\mathcal{E}_{y,z}(\varphi, \psi) = 0$ for any $\varphi \in \mathcal{C}_{y,0}, \psi \in \mathcal{C}_{z,0}$ with $y, z \in T$ except $y = z$, $\pi(y) = z$, $\pi(z) = y$ and $\pi(y) = \pi(z)$. □

We may start with $\pi^k(o)$ instead of o, then the identity (3.3) needs to be replace with

$$K_{1,\pi^k(o)}(y',z') = \frac{1}{\mu(\Sigma^+_{y'})\mu(\Sigma^+_{z'})}\mathcal{E}^{S^+(x)}_{\pi^k(o)}(1_{\Sigma^+_{y'}}, 1_{\Sigma^+_{z'}}).$$

for any $y', z' \in T$ satisfying $\pi(y') \neq \pi(z')$ and $\pi^2(y') = \pi^2(z') = x$.

In the rest of the section, we discuss a construction of a regular Dirichlet space under a similar assumption to the previous section given as

$$\begin{aligned}\sup\{\lambda \mid \mathcal{E}^{S^+(x)}_{\pi^m(x)}(v,v) = \lambda(v,v)_{L^2(\Sigma^+;\mu)} \text{ for some positive integer } m \\ \text{and } v = \sum_{w\in N} P_w u \text{ given by some } N \subset S^+(x) \cup \\ \{x, \pi(x), \pi^2(x), \dots\} \text{ and some } u \in \mathcal{C}(\Sigma^+_x)\} < \infty \end{aligned} \tag{3.4}$$

for any $x \in T$ satisfying $\pi^\ell(x) = \pi^k(o)$ with some non-negative integer k and ℓ.

By combining Lemma 2.2 and Lemma 3.1 with the following fact that for any $y' \in S^+(y), z' \in S^+(z)$ with $y \neq z$ and $\pi(y) = \pi(z)$, by taking $x = \pi(y) = \pi(z)$ one sees $1_{\Sigma^+_{y'}} + 1_{\Sigma^+_{z'}}, 1_{\Sigma^+_{y'}} - 1_{\Sigma^+_{z'}} \in \oplus_{\pi(y)=x}\mathcal{C}_{y,0} \oplus \mathcal{C}_{x,0} \oplus \cdots \oplus \mathcal{C}_{\pi^{\ell-1}(x),0} \oplus \mathcal{C}_{\pi^\ell(x),0} \oplus \cdots$ yielding

$$\begin{aligned}\mathcal{E}^{S^+(x)}_{\pi^k(o)}(1_{\Sigma^+_{y'}}, 1_{\Sigma^+_{z'}}) = \frac{1}{4}\Big(&\mathcal{E}^{S^+(x)}_{\pi^k(o)}(1_{\Sigma^+_{y'}} + 1_{\Sigma^+_{z'}}, 1_{\Sigma^+_{y'}} + 1_{\Sigma^+_{z'}}) \\ &+\mathcal{E}^{S^+(x)}_{\pi^k(o)}(1_{\Sigma^+_{y'}} - 1_{\Sigma^+_{z'}}, 1_{\Sigma^+_{y'}} - 1_{\Sigma^+_{z'}})\Big),\end{aligned}$$

it turns out that the sequence $\{\mathcal{E}^{S^+(x)}_{\pi^k(o)}(1_{\Sigma^+_{y'}}, 1_{\Sigma^+_{z'}})\}$ admits the limit as $k \to \infty$. In this situation, we can introduce the subset $\mathcal{D}_{1,T} = \{(y',z') \mid \pi^2(y') = \pi^2(z') \text{ and } \pi(y') \neq \pi(z')\}$ of $T \times T$ and define the function

$$K_1(y',z') = -\lim_{k\to\infty} \frac{1}{\mu(\Sigma^+_{y'})\mu(\Sigma^+_{z'})}\mathcal{E}^{S^+(x)}_{\pi^k(o)}(1_{\Sigma^+_{y'}}, 1_{\Sigma^+_{z'}})$$

on $\mathcal{D}_{1,T}$. We can define the bilinear from

$$\begin{aligned}\mathcal{E}^{m+1}_*(u,v) = \lim_{k\to\infty}\frac{1}{2}\int\int_{(\Sigma^+_{\pi^k(o)}\times\Sigma^+_{\pi^k(o)})\setminus\bigcup_{\pi^{m-n+k+1}(y)=\pi^k(o)}\Sigma^+_y\times\Sigma^+_y} \\ \times(u(\eta)-u(\zeta))(v(\eta)-v(\zeta))K_1(\Pi_1(\eta,\zeta))\mu(d\eta)\mu(d\zeta)\end{aligned}$$

defined for any any locally constant functions $u, v \in \mathcal{C}^{m+1}(\Sigma^+)$ with compact supports.

Let us denote the projection to $\mathcal{F}^{m+1}_{\pi^k(o)}$ in $L^2(\Sigma^+;\mu)$ by $P^{m+1}_{\pi^k(o)}$. The consistency of the family of the Dirichlet spaces in the tier-linked sense provides us with a symmetric bilinear form

$$\mathcal{E}_*^{m+1}(u,v) = \lim_{k\to\infty} \mathcal{E}_*^{m+1}(P_{\pi^k(o)}^{m+1}u, P_{\pi^k(o)}^{m+1}v)$$

defined for $u, v \in \mathcal{F}_*^{m+1} = \{u \in \mathcal{C}^{m+1}(\Sigma^+) \cap L^2(\Sigma^+;\mu) \mid \lim_{k\to\infty} \mathcal{E}_*^{m+1}(P_{\pi^k(o)}^{m+1}u, , P_{\pi^k(o)}^{m+1}u) < \infty\}$. This is because $\{\mathcal{E}_*^{m+1}(P_{\pi^k(o)}^{m+1}u, P_{\pi^k(o)}^{m+1}u)\}$ is non-decreasing in k for any $u \in \mathcal{C}^{m+1}(\Sigma^+) \cap L^2(\Sigma^+;\mu)$ due to the representation of the bilinear form by the integral. Since any locally constant function with compact support is represented as a finite linear combination of functions in $\{1_{\Sigma_x^+} \mid x \in T\}$ as recalled in Section 2, we can see that any locally constant function with compact support is an element of $\mathcal{F}_{\pi^k(o)}^{m+1}$ for sufficiently large m and k. In fact, for any $y' \in T_{m+2}$, one can choose ℓ' and k such that $\pi^{\ell'}(y') = \pi^k(o)$ and we can derive $\lim_{k\to\infty} \mathcal{E}_*^{m+1}(P_{\pi^k(o)}^{m+1}1_{\Sigma_{y'}^+}, P_{\pi^k(o)}^{m+1}1_{\Sigma_{y'}^+}) = \lim_{k\to\infty} \mathcal{E}_{\pi^k(o)}^{m+1}(P_{\pi^k(o)}^{m+1}1_{\Sigma_{y'}^+}, P_{\pi^k(o)}^{m+1}1_{\Sigma_{y'}^+}) < \infty$ from Lemma 3.1 and the proof of Lemma 2.2.

Proposition 3.2. *If a family of the bilinear forms $\{(\mathcal{E}_{y,z}, \mathcal{C}_{y,z})\}_{y,z\in T}$ is admissible, the family $\{(\mathcal{E}_*^{m+1}, \mathcal{F}_*^{m+1})\}$ of the symmetric bilinear forms with their domains satisfies*

$$\mathcal{E}_*^{m+2}(u,v) = \mathcal{E}_*^{m+1}(u,v)$$

for any $u, v \in \mathcal{F}_*^{m+1}$ and any integer $m > n$ with $o \in T_n$.

Proof. The assertion is directly derived from the property $\mathcal{E}_{y,z}(u, P_z v) = 0$ for any $u \in \mathcal{C}_{y,0}$ and $v \in \mathcal{F}_*^{m+1}$ whenever $y \in \cup_{k=-\infty}^{m+1} T_k$ and $z \in T_{m+2}$. □

In constructing a regular Dirichlet space naturally associated with the admissible family $\{(\mathcal{E}_{y,z}, \mathcal{C}_{y,z})\}_{y,z\in T}$, a symmetric bilinear form with domain in $L^2(\Sigma^+;\mu)$ viewed as a so-called pre-Dirichlet form is required. In our case with $\{(\mathcal{E}_*^{m+1}, \mathcal{F}_*^{m+1})\}$, it is given by

$$\mathcal{E}_*(u,v) = \mathcal{E}_*^{m+1}(u,v)$$

for $u, v \in \mathcal{F}_*$ by taking sufficiently large m, where $\mathcal{F}_* = \cup_{m=n}^{\infty} \mathcal{F}_*^{m+1}$.

Theorem 3.1. If a family $\{(\mathcal{E}_{y,z}, \mathcal{C}_{y,z})\}_{x,y\in T}$ of bilinear forms is admissible and if the condition (3.4) is satisfied for any $x \in T$, then there exists a regular Dirichlet space $(\mathcal{E}, \mathcal{F})$ on $L^2(\Sigma^+;\mu)$ characterized by

$$\mathcal{E}_{y,z}(u,v) = \mathcal{E}(P_y u, P_z v)$$

for any $u \in \mathcal{C}_y, v \in \mathcal{C}_z$ and at any node $y, z \in T$ except $y = z$, $\pi(y) = z$, $\pi(z) = y$ and $\pi(y) = \pi(z)$.

Proof. In Section 2, we already see that the family of locally constant functions on Σ^+ with compact support is densely contained in the family of continuous functions on Σ^+ with compact support, the finiteness condition in the assertion implies any function in $\mathcal{C}^{m+1}(\Sigma^+)$ with compact support is an element of $\mathcal{F}_*^{m+1}$ for sufficiently large integer m as discussed before Proposition 3.2. These observations show that $\cup_{k=0}^{\infty}\mathcal{F}_{\pi^k(o)}^{m+1} \subset \mathcal{F}_*^{m+1}$ and

$$\begin{aligned}&\mathcal{E}_*^{m+1}(u,v)\\&= \lim_{k\to\infty}\frac{1}{2}\int\int_{(\Sigma^+_{\pi^k(o)}\times\Sigma^+_{\pi^k(o)})\setminus\bigcup_{\pi^{m-n+k+1}(y)=\pi^k(o)}\Sigma^+_y\times\Sigma^+_y}\\&\qquad\qquad\times(u(\eta)-u(\zeta))(v(\eta)-v(\zeta)K_1(\Pi_1(\eta,\zeta))\mu(d\eta)\mu(\zeta)\end{aligned}$$

for any $u, v \in \mathcal{F}_*^{m+1}$ by the description on the definition of K_1 before Proposition 3.2. In other words, we have

$$\begin{aligned}\mathcal{E}_*(u,v) = \frac{1}{2}\int\int_{(\Sigma^+\times\Sigma^+)\setminus\{\eta\neq\zeta\}} &(u(\eta)-u(\zeta))(v(\eta)-v(\zeta))\\&\times K_1(\Pi_1(\eta,\zeta))\mu(d\eta)\mu(\zeta)\end{aligned}$$

for $u, v \in \mathcal{F}_*$.

Here, we recall again that the family of continuous functions on Σ^+ with compact support is densely contained in $\mathcal{F}_*$ as a subspace of $L^2(\Sigma^+;\mu)$. The closability of $(\mathcal{E}_*, \mathcal{F}_*)$ on $L^2(\Sigma^+;\mu)$ which can be routinely shown will be proved in the next lemma. We denote the Dirichlet space obtained as the smallest closed extension of $(\mathcal{E}_*, \mathcal{F}_*)$ by $(\mathcal{E}, \mathcal{F})$.

The original family $\{(\mathcal{E}_{x,y}, \mathcal{C}_x \times \mathcal{C}_y)\}_{x,y\in T}$ of bilinear forms is given by

$$\mathcal{E}_{y,z}(u,v) = \mathcal{E}(P_y u, P_z v)$$

for any $u \in \mathcal{C}_x, v \in \mathcal{C}_y$. □

Lemma 3.2. *The symmetric bilinear form $\mathcal{E}_*$ with domain $\mathcal{F}_*$ is closable on $L^2(\Sigma^+;\mu)$.*

Proof. We assume that a sequence $\{u_n\}$ in $\mathcal{F}_*$ satisfies $\lim_{n,m\to\infty}\mathcal{E}_*(u_n - u_m, u_n - u_m) = 0$ and $\lim_{n\to\infty}\|u_n\|_{L^2(\Sigma^+;\mu)} = 0$. These conditions imply that, for any $\varepsilon > 0$, there exists a subsequence $\{u_{n'}\}$ satisfying $\mathcal{E}_*(u_{n'} - u_{m'}, u_{n'} - u_{m'}) \leq \varepsilon$ and $\lim_{n'\to\infty} u_{n'}(\eta) = 0$ almost every x in the support of the measure μ. Then, one sees

$$\mathcal{E}_*(u_{n'}, u_{n'})$$

$$\leq \liminf_{m'\to\infty} \frac{1}{2} \int\int_{(\Sigma^+\times\Sigma^+)\cap\{\eta\neq\zeta\}} (u_{n'}(\eta) - u_{m'}(\eta) - u_{n'}(\zeta) - u_{m'}(\zeta))^2$$
$$\times K_1(\Pi_1(\eta,\zeta))d\mu(\eta)d\mu(\zeta)$$
$$\leq \varepsilon.$$

By repeating the procedure of taking subsequence, one can take some subsequence $\{u_{n''}\}$ of $\{u_n\}$ such that $\lim_{n''\to\infty} \mathcal{E}_*(u_{n''}, u_{n''}) = 0$. This implies $\lim_{n\to\infty} \mathcal{E}_*(u_n, u_n) = 0$. □

4. A sufficient condition for admissibility and Hunt process

We introduce a map $\alpha_{\iota,\kappa} : \overline{T} \times \overline{T} \to [0,\infty)$, where $\overline{T} = \cup_{x\in T} N(x)$ and $N(x) = \{x\} \times \{1,\dots,n(x)\}$. Here, we recall the orthonormal system Φ_x is represented also as $\Phi_x = \{\varphi_\nu \mid \nu \in N(x)\}$. We will denote the orthogonal projection to the one dimensional linear span E_ν of φ_ν in $L^2(\Sigma^+;\mu)$ by P_ν and will focus on the symmetric bilinear form

$$\mathcal{E}_{y,z}(u,v) = \sum_{\iota\in N(y),\kappa\in N(z)} \alpha_{\iota,\kappa}(\varphi_\iota, u)_{L^2(\Sigma^+;\mu)}(\varphi_\kappa, v)_{L^2(\Sigma^+;\mu)},$$

with domain $L^2(\Sigma^+;\mu) \times L^2(\Sigma^+;\mu)$ determined by the family of non-negative real numbers $\{\alpha_{\iota,\kappa} \mid \iota \in N(y), \kappa \in N(z)$ for $y, z \in T$ with $y = z$, $\pi(y) = z$, $\pi(z) = y$ or $\pi(y) = \pi(z)\}$ satisfying $\alpha_{\iota,\kappa} = \alpha_{\kappa,\iota}$. When the domain is replaced with $\mathcal{C}_y \times \mathcal{C}_z$, the symmetric bilinear form $\mathcal{E}_{y,z}(u,v)$ turns to admit adjacent matrix $H_{y,z}$ which is characterized by $\mathcal{E}_{y,z}(u,v) = -(u, H^\mu_{y,z}v)_{L^2(\Sigma^+;\mu)}$, i.e., the identity

$$H^\mu_{y,z}\begin{pmatrix}\vdots\\ \psi\\ \vdots\end{pmatrix} = -\sum_{\iota\in N(y),\kappa\in N(z)} \begin{pmatrix}\vdots\\ \varphi_\iota\\ \vdots\end{pmatrix} \alpha_{\iota,\kappa}(\varphi_\kappa, \psi)_{L^2(\Sigma^+;\mu)}$$

is valid for any $\psi \in \mathcal{C}_z$ by taking the matrix $(H^\mu_{y,z})_{w_i w_j} = ((H_{y,z})_{w_i,w_j}\mu(\Sigma^+_{w_j}))$ weighted by μ. Here, we note that

(i) $\mathcal{E}_{y,z}(u,v) + \mathcal{E}_{z,y}(u,v) = \mathcal{E}_{y,z}(v,u) + \mathcal{E}_{z,y}(v,u)$ for any $u, v \in \mathcal{C}_y \oplus \mathcal{C}_z$,
(ii) $\mathcal{E}_{y,z}(u, 1_{\Sigma^+_z}) = 0$ for any $u \in \mathcal{C}_y$ whenever $\Sigma^+_y \subset \Sigma^+_z$.

We focus on the case that $(\mathcal{E}_{x,x}, \mathcal{C}_{x,x})$ is a Dirichlet space for any $x \in T$, equivalently $H_{x,x}$ is a non-positive definite symmetric matrix satisfying $\sum_{z\in S^+(x)}(H_{x,x})_{y,z}\mu(\Sigma^+_z) = 0$ for any $y \in S^+(x)$ and $H_{x,x}$ admits negative values only at diagonal components for any $x \in T$.

In this section, we will make an attempt on constructing Hunt process based on the family $\{(\mathcal{E}^{S^+(x)}_{\pi^k(x)}, \mathcal{C}^{S^+(x)}_{\pi^k(x)})\}_{x\in T,k=0,1,\cdots}$ of bilinear forms. For the purpose, we will introduce the notations $\overline{\lambda}(x) = \max_{\|u\|_{L^2(\Sigma_x^+;\mu)} \leq 1} \mathcal{E}_{x,x}(u,u)$ and a positive real number $\Lambda < \frac{1}{2}$ such that

$$\iota \in N(y), \kappa \in N(x) \Rightarrow |\alpha_{\iota,\kappa}(\varphi_\iota, 1_{\Sigma_w^+})_{L^2(\Sigma^+;\mu)}(\varphi_\kappa, 1_{\Sigma_z^+})_{L^2(\Sigma^+;\mu)}| \\ \leq \Lambda \frac{1}{n(y)} \frac{1}{n(x)} \frac{\mu(\Sigma_w^+)}{\mu(\Sigma_y^+)} |\mathcal{E}_{x,x}(1_{\Sigma_y^+}, 1_{\Sigma_z^+})|, \tag{4.1}$$

so that we obtain

$$|\mathcal{E}_{y,x}(1_{\Sigma_{y'}^+}, 1_{\Sigma_z^+})| \leq \Lambda \frac{\mu(\Sigma_{y'}^+)}{\mu(\Sigma_y^+)} |\mathcal{E}_{x,x}(1_{\Sigma_y^+}, 1_{\Sigma_z^+})|$$

for any $y' \in S^+(y)$, $z \in S^+(x)$ with $y \in S^+(x)$.

Aiming at a sufficient condition for the admissibility of family $\{(\mathcal{E}_{y,z}, \mathcal{C}_{y,z})\}_{y,z \in T}$ in terms of $\{\alpha_{\kappa,\mu}\}$ and μ, we introduce function I defined by

$$\begin{aligned} I(y',z') = \tfrac{1}{2}\Big(&(H_{\pi(y'),\pi(z')})_{y',z'} \\ &+ (1-2\Lambda)(H_{[y',z']})_{\pi(y'),\pi(z')} \\ &- (1+2\Lambda)\textstyle\sum_{i=1}^\infty \big(\overline{\lambda}(\pi^i([y',z']))\big(\frac{1}{\mu(\Sigma^+_{\pi^{i-1}([y',z'])})} - \frac{1}{\mu(\Sigma^+_{\pi^{i}([y',z'])})}\big)\Big) \end{aligned}$$

on $\mathcal{D}_{1,T} = \{(y',z') \mid \pi^2(y') = \pi^2(z'), y' \neq z'\}$. We assume that the sum in the right-hand side converges and the function I is non-negative valued. Since $\int\int_{\Sigma_y^+ \times \Sigma_z^+} (H_{\pi(y'),\pi(z')})_{y',z'} \mu(dy')\mu(dz') = 0$ for any distinct $y, z \in S^+(x)$, the assumption shows the non-negativity

$$\begin{aligned} &(H_{[y',z']})_{\pi(y'),\pi(z')} \\ &\quad -(1+2\Lambda)\textstyle\sum_{i=1}^\infty \big(\overline{\lambda}(\pi^i([y',z']))\big(\frac{1}{\mu(\Sigma^+_{\pi^{i-1}([y',z'])})} - \frac{1}{\mu(\Sigma^+_{\pi^{i}([y',z'])})}\big) \\ &\geq (1-2\Lambda)(H_{[y',z']})_{\pi(y'),\pi(z')} \\ &\quad -(1+2\Lambda)\textstyle\sum_{i=1}^\infty \big(\overline{\lambda}(\pi^i([y',z']))\big(\frac{1}{\mu(\Sigma^+_{\pi^{i-1}([y',z'])})} - \frac{1}{\mu(\Sigma^+_{\pi^{i}([y',z'])})}\big) \geq 0 \end{aligned}$$

for any $y' \in S^+(y), z' \in S^+(z)$ with distinct $y, z \in S^+(x)$, which will be applied in the first case in showing the admissibility in the proof of the following theorem:

Theorem 4.1. *If $I(y',z')$ is non-negative for any $(y',z') \in \mathcal{D}_{T,1}$, $\sup_{\ell \in \{0,1,\dots\}} \overline{\lambda}(\pi^\ell(x)) < \infty$ for any $x \in T$ and the conditions (3.4) is satisfied, then there exists a regular Dirichlet space $(\mathcal{E}, \mathcal{F})$ on $L^2(\Sigma^+;\mu)$ characterized by*

$$\mathcal{E}_{y,z}(u,v) = \mathcal{E}(P_y u, P_z v)$$

for any $u \in \mathcal{C}_y, v \in \mathcal{C}_z$ and at any node $y, z \in T$ except $y = z$, $\pi(y) = z$, $\pi(z) = y$ and $\pi(y) = \pi(z)$.

Proof. In showing that the bilinear form

$$\begin{aligned}\mathcal{E}_{\pi^k(x)}^{S^+(x)}(u,v) = &\sum_{\pi(y)=\pi(z)=x} \big(\mathcal{E}_{y,z}(u,v)+\mathcal{E}_{y,x}(u,v)+\mathcal{E}_{x,z}(u,v)\big)+\mathcal{E}_{x,x}(u,v)\\ &+\sum_{\ell=0}^{k-1}\big(\mathcal{E}_{\pi^\ell(x),\pi^{\ell+1}(x)}(u,v)+\mathcal{E}_{\pi^{\ell+1}(x),\pi^\ell(x)}(u,v)\\ &\qquad+\mathcal{E}_{\pi^{\ell+1}(x),\pi^{\ell+1}(x)}(u,v)\big)\end{aligned}$$

satisfies $\mathcal{E}_{\pi^k(x)}^{S^+(x)}(1_{\Sigma_w^+},1_{\Sigma_{w'}^+})\le 0$ for any $x\in T$ and non-negative integer k as long as disjoint $\Sigma_w^+,\Sigma_{w'}^+$ with $w,w'\in T$ yield $1_{\Sigma_w^+},1_{\Sigma_{w'}^+}\in\mathcal{C}_{\pi^k(x)}^{S^+(x)}$, we may restrict our attention to the case that $w,w'\in S^+(y)\cup S^+(z)$ with $\pi(y)=\pi(z)=x$.

In the case that $w,w'\in S^+(y)$, we can derive the non-positivity of $\mathcal{E}_{\pi^k(x)}^{S^+(x)}(1_{\Sigma_w^+},1_{\Sigma_{w'}^+})$ from (4.1) and our assumption on the function I as

$$\begin{aligned}\mathcal{E}_{\pi^k(x)}^{S^+(x)}(1_{\Sigma_w^+},1_{\Sigma_{w'}^+}) &= \mathcal{E}_{y,y}(P_y1_{\Sigma_w^+},P_y1_{\Sigma_{w'}^+})+\mathcal{E}_{y,x}(P_y1_{\Sigma_w^+},P_x1_{\Sigma_{w'}^+})\\ &\quad+\mathcal{E}_{x,y}(P_x1_{\Sigma_w^+},P_y1_{\Sigma_{w'}^+})+\mathcal{E}_{x,x}(P_x1_{\Sigma_w^+},P_x1_{\Sigma_{w'}^+})\\ &\quad+\sum_{\ell=0}^{k-1}\big(\mathcal{E}_{\pi^\ell(x),\pi^{\ell+1}(x)}(P_{\pi^\ell(x)}1_{\Sigma_w^+},P_{\pi^{\ell+1}(x)}1_{\Sigma_{w'}^+})\\ &\qquad+\mathcal{E}_{\pi^{\ell+1}(x),\pi^\ell(x)}(P_{\pi^{\ell+1}(x)}1_{\Sigma_w^+},P_{\pi^\ell(x)}1_{\Sigma_{w'}^+})\\ &\qquad+\mathcal{E}_{\pi^{\ell+1}(x),\pi^{\ell+1}(x)}(P_{\pi^{\ell+1}(x)}1_{\Sigma_w^+},P_{\pi^{\ell+1}(x)}1_{\Sigma_{w'}^+})\big)\\ &\le \mathcal{E}_{y,y}(1_{\Sigma_w^+},1_{\Sigma_{w'}^+})+(1+2\Lambda)\mathcal{E}_{x,x}(P_x1_{\Sigma_w^+},P_x1_{\Sigma_{w'}^+})\\ &\quad+(1+2\Lambda)\sum_{\ell=0}^{k-1}\mathcal{E}_{\pi^{\ell+1}(x),\pi^{\ell+1}(x)}(P_{\pi^{\ell+1}(x)}1_{\Sigma_w^+},P_{\pi^{\ell+1}(x)}1_{\Sigma_{w'}^+})\\ &\le (1-2\Lambda)\mathcal{E}_{y,y}(1_{\Sigma_w^+},1_{\Sigma_{w'}^+})+(1+2\Lambda)\mathcal{E}_{x,x}(P_x1_{\Sigma_w^+},P_x1_{\Sigma_{w'}^+})\\ &\quad+(1+2\Lambda)\sum_{\ell=0}^{k-1}\mathcal{E}_{\pi^{\ell+1}(x),\pi^{\ell+1}(x)}(P_{\pi^{\ell+1}(x)}1_{\Sigma_w^+},P_{\pi^{\ell+1}(x)}1_{\Sigma_{w'}^+})\\ &\le -2\mu(\Sigma_w^+)\mu(\Sigma_{w'}^+)I(w,w').\end{aligned}$$

This is because we see that

$$\begin{aligned}\mathcal{E}_{x,x}(P_x1_{\Sigma_w^+},P_x1_{\Sigma_{w'}^+}) &= \frac{\mu(\Sigma_w^+)}{\mu(\Sigma_y^+)}\frac{\mu(\Sigma_{w'}^+)}{\mu(\Sigma_y^+)}\mathcal{E}_{x,x}(1_{\Sigma_y^+},1_{\Sigma_y^+})\\ &\le \frac{\mu(\Sigma_w^+)}{\mu(\Sigma_y^+)}\frac{\mu(\Sigma_{w'}^+)}{\mu(\Sigma_y^+)}\overline{\lambda(x)}\|P_x1_{\Sigma_y^+}\|_{L^2(\Sigma^+;\mu)}^2 \qquad (4.2)\\ &= \mu(\Sigma_w^+)\mu(\Sigma_{w'}^+)\overline{\lambda(x)}\frac{(\mu(\Sigma_x^+)-\mu(\Sigma_y^+))}{\mu(\Sigma_x^+)\mu(\Sigma_y^+)}\end{aligned}$$

$$= \mu(\Sigma_w^+)\mu(\Sigma_{w'}^+)\overline{\lambda(x)}\Big(\frac{1}{\mu(\Sigma_y^+)} - \frac{1}{\mu(\Sigma_x^+)}\Big)$$

and that, for any non-negative integer ℓ,

$$\begin{aligned}
&\mathcal{E}_{\pi^{\ell+1}(x),\pi^{\ell+1}(x)}(P_{\pi^{\ell+1}(x)}1_{\Sigma_w^+}, P_{\pi^{\ell+1}(x)}1_{\Sigma_{w'}^+}) \\
&= \frac{\mu(\Sigma_w^+)}{\mu(\Sigma_{\pi^\ell(x)}^+)}\frac{\mu(\Sigma_{w'}^+)}{\mu(\Sigma_{\pi^\ell(x)}^+)}\mathcal{E}_{\pi^{\ell+1}(x),\pi^\ell+1(x)}(1_{\Sigma_{\pi^\ell(x)}^+}, 1_{\Sigma_{\pi^\ell(x)}^+}) \\
&\leq \frac{\mu(\Sigma_w^+)}{\mu(\Sigma_{\pi^\ell(x)}^+)}\frac{\mu(\Sigma_{w'}^+)}{\mu(\Sigma_{\pi^\ell(x)}^+)}\overline{\lambda(\pi^{\ell+1}(x))}\|P_{\pi^{\ell+1}(x)}1_{\Sigma_{\pi^\ell(x)}^+}\|_{L^2(\Sigma^+;\mu)}^2 \\
&= \mu(\Sigma_w^+)\mu(\Sigma_{w'}^+)\overline{\lambda(\pi^{\ell+1}(x))}\Big(\frac{1}{\mu(\Sigma_{\pi^\ell(x)}^+)} - \frac{1}{\mu(\Sigma_{\pi^{\ell+1}(x)}^+)}\Big).
\end{aligned} \tag{4.3}$$

Otherwise if $w \in S^+(y), w' \in S^+(z)$ for some distinct y, z with $\pi(y) = \pi(z) = x$, then from (4.1) and our assumption on the function I we deduce the non-positivity of

$$\begin{aligned}
&\mathcal{E}_{\pi^k(x)}^{S^+(x)}(1_{\Sigma_w^+}, 1_{\Sigma_{w'}^+}) \\
&= \mathcal{E}_{y,z}(P_y 1_{\Sigma_w^+}, P_z 1_{\Sigma_{w'}^+}) + \mathcal{E}_{y,x}(P_y 1_{\Sigma_w^+}, P_x 1_{\Sigma_{w'}^+}) \\
&+\mathcal{E}_{x,z}(P_x 1_{\Sigma_w^+}, P_z 1_{\Sigma_{w'}^+}) + \mathcal{E}_{x,x}(P_x 1_{\Sigma_w^+}, P_x 1_{\Sigma_{w'}^+}) \\
&+\sum_{\ell=0}^{k-1}\big(\mathcal{E}_{\pi^\ell(x),\pi^{\ell+1}(x)}(P_{\pi^\ell(x)}1_{\Sigma_w^+}, P_{\pi^{\ell+1}(x)}1_{\Sigma_{w'}^+}) \\
&\qquad +\mathcal{E}_{\pi^{\ell+1}(x),\pi^\ell(x)}(P_{\pi^{\ell+1}(x)}1_{\Sigma_w^+}, P_{\pi^\ell(x)}1_{\Sigma_{w'}^+}) \\
&\qquad +\mathcal{E}_{\pi^{\ell+1}(x),\pi^{\ell+1}(x)}(P_{\pi^{\ell+1}(x)}1_{\Sigma_w^+}, P_{\pi^{\ell+1}(x)}1_{\Sigma_{w'}^+})\big) \\
&\leq \mathcal{E}_{y,z}(P_y 1_{\Sigma_w^+}, P_z 1_{\Sigma_{w'}^+}) + (1-2\Lambda)\mathcal{E}_{x,x}(P_x 1_{\Sigma_w^+}, P_x 1_{\Sigma_{w'}^+}) \\
&+\sum_{\ell=0}^{k-1}\big(\mathcal{E}_{\pi^\ell(x),\pi^{\ell+1}(x)}(P_{\pi^\ell(x)}1_{\Sigma_w^+}, P_{\pi^{\ell+1}(x)}1_{\Sigma_{w'}^+}) \\
&\qquad +\mathcal{E}_{\pi^{\ell+1}(x),\pi^\ell(x)}(P_{\pi^{\ell+1}(x)}1_{\Sigma_w^+}, P_{\pi^\ell(x)}1_{\Sigma_{w'}^+}) \\
&\qquad +\mathcal{E}_{\pi^{\ell+1}(x),\pi^{\ell+1}(x)}(P_{\pi^{\ell+1}(x)}1_{\Sigma_w^+}, P_{\pi^{\ell+1}(x)}1_{\Sigma_{w'}^+})\big) \\
&\leq \mathcal{E}_{y,z}(P_y 1_{\Sigma_w^+}, P_z 1_{\Sigma_{w'}^+}) + (1-2\Lambda)\mathcal{E}_{x,x}(1_{\Sigma_w^+}, 1_{\Sigma_{w'}^+}) \\
&+(1+2\Lambda)\sum_{\ell=0}^{k-1}\mathcal{E}_{\pi^{\ell+1}(x),\pi^{\ell+1}(x)}(P_{\pi^{\ell+1}(x)}1_{\Sigma_w^+}, P_{\pi^{\ell+1}(x)}1_{\Sigma_{w'}^+}) \\
&\leq -2\mu(\Sigma_w^+)\mu(\Sigma_{w'}^+)I(w, w').
\end{aligned}$$

This is because $(1_{\Sigma_w^+}, H_{[w,w']}^\mu 1_{\Sigma_{w'}^+})_{L^2(\Sigma^+,\mu)} = \frac{\mu(\Sigma_w^+)}{\mu(\Sigma_y^+)}\frac{\mu(\Sigma_{w'}^+)}{\mu(\Sigma_z^+)}\mathcal{E}_{x,x}(P_x 1_{\Sigma_y^+},$

$P_x 1_{\Sigma_z^+}) = \mathcal{E}_{x,x}(P_x 1_{\Sigma_w^+}, P_x 1_{\Sigma_{w'}^+})$ for any $w \in S^+(y), w' \in S^+(z)$ with $y, z \in S^+(x)$.

Therefore, the admissibility of the family $\{(\mathcal{E}_{y,z}, \mathcal{C}_{y,z})\}_{x,y \in T}$ of bilinear forms is shown.

On the other hand, under our assumptions in the assertion, we can see that the condition (3.4) in the previous section is satisfied. In fact, $\mathcal{E}_{\pi^k(o)}^{S^+(x)}$ does not involve P_w with $\pi^{\ell''}(w) = x$ ($\ell'' \geq 2$) and we may assume u in (3.4) takes a constant on every Σ_w^+ with $\pi^2(w) = x$. The finiteness condition given as (3.4) for such function is derived from the convergence of

$$\sum_{\ell=0}^{\infty} \mathcal{E}_{\pi^{\ell+1}(x),\pi^{\ell+1}(x)}(P_{\pi^{\ell+1}(x)} 1_{\Sigma_w^+}, P_{\pi^{\ell+1}(x)} 1_{\Sigma_w^+})$$

which follows from the estimates (4.2) and (4.3) with $w = w'$.

As a result, we can obtain the Dirichlet space in the assertion by applying Theorem 3.1. □

References

[Albeverio and Karwowski (1994)] Albeverio, S. and Karwowski, W. (1994). A random walk on *p*-adics - the generator and its spectrum, *Stochastic Proc. Appl.* **53**, pp. 1-22.

[Albeverio and Karwowski (2008)] Albeverio, S. and Karwowski, W. (2008). Jump processes on leaves of multibranching trees, *J. Math. Phys.* **49**, 093503, 20pp.

[Chen and Fukushima (2012] Chen, Z.-Q. and Fukushima, M. (2012). *Symmetric Markov processes, time changes, and boundary theory*, Princeton Univesirty Press.

[Doob (1962)] Doob, J. L. (1962). Boundary properties for functions with finite Dirichlet integrals, *Ann. Inst. Fourier (Grenoble)* **12**, pp. 573-621.

[Fukushima, Oshima and Takeda (2010)] Fukushima, M. Oshima, M. and Takeda, M. (2010). *Dirichlet forms and symmetric Markov processes*, 2nd edn.,Walter de Gruyter.

[Kaneko (2000)] Kaneko, H. (2000). A class of spatially inhomogeneous Dirichlet spaces on the p-adic number field, *Stochastic Proc. Appl.* **88**, pp. 161-174.

[Kaneko (2013)] Kaneko, H. (2013). A Dirichlet space on ends of tree and Dirichlet forms with a nodewise orthogonal property, to appear in *Potential Analysis.*

[Karwowski and Vilela-Mendes (1994)] Karwowski W. and Vilela-Mendes, R. (1994). Hierarchical structures and asymmetric processes on p-adics and adeles, *J. Math. Phys.* **35**, pp. 4637-4650.

[Kigami (2001)] Kigami, J. (2001). *Analysis on Fractals*, Cambridge University Press.

[Kigami (2010)] Kigami, J. (2010). Dirichlet forms and associated kernels on the Cantor set induced by random walks on trees, *Advances in Math.* **225**, pp. 2674-2730.

[Kigami (2013)] Kigami, J. (2013), Transitions on a noncompact Cantor set and random walks on its defining tree, *Ann. Inst. H. Poincare, Probab. et Statist.* **49**, pp. 1090-1129.

[Lyons (1992)] Lyons, R. (1992). Random walks, capacity, and percolation on trees, *Ann. Probab.* **20**, pp. 2043-2088.

[Robert (2000)] Robert, A. (2000). *A course in p-adic Analysis*, Springer-Verlag.

[Vladimirov, Volovich and Zelenov (1995)] Vladimirov, V. S. Volovich I. V. and Zelenov, E. I. 1995). *P-Adic Analysis and Mathematical Physics*, Series on Soviet and East European Mathematics, World Scientific.

[Yasuda (2006)] Yasuda, K. (2006). Semi-stable processes on local fields, *Tohoku Math. J.* **58**, pp. 419-431.

Chapter 15

Dirichlet forms in quantum theory

Witold Karwowski and Ludwig Streit

We intend to highlight the impact that the theory of local Dirichlet forms has had on the formulation of dynamics in Schroedinger theory, in stochastic models for large scale natural phenomena, and in the theory of relativistic quantum fields where perturbative methods are bound to fail.

Contents

1. Introduction

Quantum physics and mathematical physics more generally, have been a prime mover of mathematical progress in the 20th century. This is true for the fields of group theory, of operator algebras, and very much so for functional and stochastic analysis.

Quantum dynamics in particular is in terms of a unitary group with self-adjoint generator H, and a central concern was and still is, the correct formulation of interactions, in the simplest case as perturbations of the Laplace operator in $L^2(\mathbf{R}^n, dx)$ by a "potential" V:

$$H = H_0 + V \tag{1.1}$$
$$= -\Delta_x + V(x). \tag{1.2}$$

The problem here is that in most physically interesting cases, the perturbation V will not be bounded, often not even Kato bounded w.r. the Laplacian [52].

And then there are singular perturbations, such as the so-called pseudopotentials, informally in terms of a Dirac distribution

$$V(x) = g\delta(x) \tag{1.3}$$

which physics invokes to describe zero-range interactions.

The situation is significantly aggravated when we move from non-relativistic quantum mechanics to relativistic quantum field theory. Here, by Haag's theorem [46], [66], we cannot even expect a common representation space for the perturbed and the unperturbed H. Hence it is no surprise that the search for alternate formulations of dynamics originated in quantum field theory, in [37] and [39]. The common idea of both these attempts was to formulate the dynamics not in terms of a perturbative ansatz for H as in (1.1), but in terms of an eigenvector corresponding to the infimum of its spectrum, the "ground state" or "vacuum vector" in quantum field theory parlance. Araki's paper was mathematically rigorous but abstract, while Coester and Haag tried a constructive but heuristic approach.

It was only with the advent of our understanding of Dirichlet form methods, to a large extent owed to the efforts and insights of M. Fukushima, that a mathematically solid construction of quantum (field) dynamics became possible starting from suitable ground state measures. A milestone in this direction was the fundamental article by Albeverio and Hoegh-Krohn [17].

While these new powerful methods thus were first applied to quantum field theory and only later to non-relativistic quantum mechanics, we will reverse this historical order and begin our review with the relatively simpler case of (1.2) before we have a look at what can be done with Dirichlet form techniques in quantum field theory.

2. Quantum Mechanics in Terms of the Ground State

In this section we shall exploit the relations between positive quadratic forms and positive self-adjoint operators in Hilbert space. In the context of quantum mechanics the Hilbert space is a complex L^2 space and the quadratic forms are sesquilinear. The corresponding operators are meant to describe dynamics of the quantum particles. However our results can be directly reformulated for real L^2 spaces. The quadratic forms will then be bilinear and actually appear to be the local Dirichlet forms i.e. the energy forms in sense of reference [44]. The corresponding self-adjoint operators generate the Markovian semigroups. In view of the Markov processes generated by these forms, applications will go beyond quantum mechanics proper and include properties and applications of these processes in other fields of physics.

Let us begin with two informal observations: the first is that indeed "the ground state ψ knows everything". If ψ solves the equation

$$(-\Delta_x + V(x))\,\psi(x) = 0 \tag{2.1}$$

we can recuperate from this equation the potential V as

$$V(x) = \frac{\Delta_x \psi(x)}{\psi(x)}. \tag{2.2}$$

Secondly it is important to observe in view of this formula that, as an input datum, the function ψ is less singular than the potential V: for a discontinuous V, ψ will be once continuously differentiable. So we may hope to deal with manageable functions ψ and obtain a well-defined dynamics in cases like (1.3) where a rigorous description by a potential function fails. It is true that this approach is not (yet?) customary in the physics community; nevertheless it is worth noting that the deuteron wave function - ground state ψ of the two-nucleon system - is known experimentally with extreme precision while on the other hand realistic nucleon-nucleon potentials V are obtained only with considerable effort [60].

We begin with a rather regular example.

Example 2.1. Recall that $H_0 = -\Delta$ with the domain $D(H_0) = \{f \in L^2(\mathbf{R}^n, dx) \cap C^1(\mathbf{R}^n);\ \frac{\partial f}{\partial x_i}$ *is a.c. and* $\frac{\partial f}{\partial x_i}, \frac{\partial^2 f}{\partial x_i^2} \in L^2, i = 1, ...n\}$ is the Hamilton operator for a free quantum particle. The set $C_0^\infty(\mathbf{R}^n)$ is a core for H_0. Assume $V(x)$ to be bounded measurable function. Then [52]

$$H = -\Delta_x + V(x) \tag{2.3}$$

is a self-adjoint operator in $L^2(\mathbf{R}^n, dx)$ with $D(H) = D(H_0)$ and $C_0^\infty(\mathbf{R}^n)$ is a core also for H. Adding a constant to V if necessary we have $\inf \sigma(H) = 0$. Let ϕ_0 be the (possibly generalized) eigenfunction of H corresponding to 0 [9], then ϕ_0 has absolutely continuous partial derivatives and no zeros and solves the equation

$$H\phi_0 = 0. \tag{2.4}$$

Hence

$$V(x) = \frac{\Delta_x \phi_0(x)}{\phi_0(x)}. \tag{2.5}$$

We further observe that $\phi_0 C_0^\infty(\mathbf{R}^n) \subset D(H)$. Put

$$\mathcal{E}(\phi_1, \phi_2) = (\phi_1, H\phi_2), \tag{2.6}$$

where $\phi_i = f_i\phi_0,\ f_i \in C_0^\infty(\mathbf{R}^n),\ i = 1, 2$. Applying integration by parts and (2.5) one obtains

$$\mathcal{E}(\phi_1, \phi_2) = \int_{\mathbf{R}^n} f_1(x)\phi_0(x)(-\Delta_x + V(x))f_1(x)\phi_0(x)dx \tag{2.7}$$

$$= \int_{\mathbf{R}^n} \overline{\nabla f_1} \cdot \nabla f_2 \phi_0^2 dx \equiv \mathcal{E}_{\phi_0^2}(f_1, f_2) \tag{2.8}$$

The last integral can be looked upon as a quadratic form in $L^2(\mathbf{R}^n, \phi_0^2 dx)$ defined on $C_0^\infty(\mathbf{R}^n)$. It is known that $\mathcal{E}_{\phi_0}$ is closable iff [52] the operator $\nabla : C_0^\infty \subset L^2(\mathbf{R}^n, \phi_0^2 dx) \to \oplus_{j=1}^n L^2(\mathbf{R}^n, \phi_0^2 dx)$ has densely defined adjoint.
Here $\oplus_{j=1}^n L^2(\mathbf{R}^n, \phi_0^2 dx)$ stands for the direct sum of n copies of $L^2(\mathbf{R}^n, \phi_0^2 dx)$. Then with obvious notation $\oplus_{j=1}^n C_0^\infty(R^n) \subset \oplus_{j=1}^n L^2(\mathbf{R}^n, \phi_0^2 dx)$.
By direct computation the adjoint

$$\nabla^* = -\nabla - \beta(x), \tag{2.9}$$

where

$$\beta(x) = 2\phi_0^{-1}(x)\nabla\phi_0(x). \tag{2.10}$$

Clearly $\nabla^* \oplus_{j=1}^n C_0^\infty(\mathbf{R}^n) \subset L^2(\mathbf{R}^n, \phi_0^2 dx)$ and since $\oplus_{j=1}^n C_0^\infty(\mathbf{R}^n)$ is dense in $\oplus_{j=1}^n L^2(\mathbf{R}^n, \phi_0^2 dx)$ the form $\mathcal{E}_{\phi_0}$ is closable and its closure $\overline{\mathcal{E}_{\phi_0}}$ defines a positive self-adjoint operator H_{ϕ_0} in $L^2(\mathbf{R}^n, \phi_0^2 dx)$. Let $f_1, f_2 \in C_0^\infty$. Then

$$\int_{\mathbf{R}^n} \overline{\phi_1(x)}\phi_2(x)dx = \int_{\mathbf{R}^n} \overline{f_1(x)} f_2(x)\phi_0^2(x)dx.$$

The map $\phi_0(x)f(x) \to f(x)$ is a densely defined isometry from $\phi_0 C_0^\infty \subset L^2(\mathbf{R}^n, dx)$ to $C_0^\infty \subset L^2(\mathbf{R}^n, \phi_0^2 dx)$ and can be extended by continuity to an unitary operator from $L^2(\mathbf{R}^n, dx)$ onto $L^2(\mathbf{R}^n, \phi_0^2 dx)$.
Consequently $\phi_0 H_{\phi_0} \phi_0^{-1}$ is a self-adjoint operator in $L^2(\mathbf{R}^n, dx)$ and (2.8) implies $\phi_0 H_{\phi_0} \phi_0^{-1} = H$. Thus the formula

$$(\phi_1, H\phi_2) = \int_{\mathbf{R}^n} \overline{\nabla f_1} \cdot \nabla f_2 \phi_0^2 dx, \ \phi_i = f_i \phi_0, \ i = 1, 2 \tag{2.11}$$

describes quantum dynamics in terms of ground state.

In our discussion the Hamiltonian was defined as a sum of free one plus potential with rather strong restrictions imposed on V. To deal with wider class of dynamics including one known as "point interaction" (1.3) we reverse the procedure. First we define a function ϕ_0 and then investigate the dynamics for which ϕ_0 is the ground state.
Let Ω be an open nonempty subset of $\mathbf{R}^n$. The possible choice of ϕ_0 is restricted by the conditions that the measures $\phi_0^2(x)dx$ and dx are equivalent in Ω and $\mathcal{E}(f_1, f_2) = \int_\Omega \overline{\nabla f_1} \cdot \nabla f_2 \phi_0^2 dx$ as defined on $C_0^\infty(\Omega)$ must be closable in $L^2(\Omega, \phi_0^2 dx)$. Such functions will be called admissible. Although complete characterization of admissible functions is not available the following theorem covers most of interesting cases.

Theorem 2.1. (Albeverio et al. [24]) *If ϕ_0 satisfies*
a) ϕ_0 is absolutely continuous in $\Omega \backslash N$, where N is a closed set of vanishing Lebesgue

measure,
b) $\phi_0 \in L^2_{loc}(\Omega, dx)$*,*
c) the measure $\phi_0^2(x)dx$ *and* dx *are equivalent,*
d) $\frac{\partial}{\partial x_i}\phi_0(x) \in L^2_{loc}(\Omega \setminus N),\ i = 1, ..., n.$
Then

$$\mathcal{E}(f_1, f_2) = \int_\Omega \overline{\nabla f_1} \cdot \nabla f_2 \phi_0^2 dx,\ f_i \in C_0^\infty(\Omega) \tag{2.12}$$

is closable in $L^2(\Omega, \phi_0^2 dx)$*, and there is a unique positive self-adjoint operator* H *in* $L^2(\Omega, dx)$ *with*

$$D(H) \subset \{f\phi_0 : f \in D(\overline{\mathcal{E}})\} \tag{2.13}$$

and

$$\overline{\mathcal{E}}(f_1, f_2) = (\phi_1, H\phi_2) \tag{2.14}$$

for all

$$\phi_1 = f_1\phi_0,\ f_1 \in D(\overline{\mathcal{E}}),\ \textit{and}\ \phi_2 = f_2\phi_0 \in D(H)$$

where $\overline{\mathcal{E}}$ *is the closure of* $\mathcal{E}$ *in* $L^2(\Omega, \phi_0^2 dx)$*. Under the unitary map between* $L^2(\Omega, \phi_0^2 dx)$ *and* $L^2(\Omega, dx)$ *defined by* ϕ_0*,* H *can be written in the form*

$$H = \phi_0 \nabla^* \nabla \phi_0^{-1}. \tag{2.15}$$

It is remarkable that according to Theorem 2.1 ϕ_0 does not have to be square integrable. In other words our procedure works not only for Hamiltonians with eigenvector ground state but also for the generalized ground states. Another fact which we shall exploit later is that ϕ_0 with nodes are also admissible. Finally the question of formulating Hamiltonians for the dynamics with singular interaction will be illustrated by the example of "point interaction".

Example 2.2.
Consider a ground state of a particle given by

$$\phi_0(x) = \sqrt{\frac{\lambda}{2\pi}} \frac{\exp\{-\lambda|x|\}}{|x|},\ x \in \mathbf{R}^3. \tag{2.16}$$

Obviously $\phi_0 \in L^2(\mathbf{R}^3, dx)$ and $\nabla\phi_0 \in \bigoplus_{i=1}^3 L^2_{loc}(\mathbf{R}^3 \setminus N)$, where $N = \{0\}$, so that by Theorem 2.1 ϕ_0 is an admissible ground state. The associated Hamiltonian has zero as an eigenvalue:

$$H\phi_0 = 0.$$

Let $f_1 \in C_0^2(\mathbf{R}^3)$ and $f_2 \in C_0^2(\mathbf{R}^3 \setminus \{0\})$. Then $\phi_i = f_i\phi_0 \in D(H),\ i = 1, 2$ and

$$(\phi_1, H\phi_2) = \mathcal{E}(\phi_1, \phi_2) = \int_{\mathbf{R}^3} \overline{\nabla} f_1 \cdot \nabla f_2 \phi_0^2 dx.$$

Integration by parts yields

$$(\phi_1, H\phi_2) = \int_{\mathbf{R}^3} \phi_1(-\Delta + \lambda^2)\phi_2 dx$$

or

$$H\phi = (-\Delta + \lambda^2)\phi \;\; if \;\; \frac{\phi}{\phi_0} \in C_0^2(\mathbf{R}^3 \setminus \{0\}) \tag{2.17}$$

that is H restricted to the wave functions vanishing near the origin equals to the free hamiltonian shifted by the constant λ^2. Obviously H differs from the free Hamiltonian $-\Delta$ because it has the eigenvector ϕ_0 given by (2.16).

Remark 2.1. The problem of point interaction had opened enormous research activity going under the name of "singular perturbations of differential operators" employing different techniques notably the Krein theory of self adjoint extensions of symmetric operators see [6], [8], [55], [29] and references there in. However at the early stage the Dirichlet forms proved to be the natural and useful language for mathematically precise study of the underlying physical problem [24].

3. Diffusion Processes in $\mathbf{R}^n$

In previous section we presented some problems of quantum mechanics and their solutions by the quadratic forms method. In this section we shall be mainly interested in the relations between the Dirichlet forms and Markov processes and prepare techniques for physical applications. Thus whenever we refer to a L^2 we always mean a real space. Our discussion will be based on [44] and further developments [45] and references there in.
Let Ω be an open subset of $\mathbf{R}^n$ and ϕ_0 an admissible function. Consider the form

$$\mathcal{E}(f_1, f_2) = \int_\Omega \nabla f_1 \cdot \nabla f_2 \phi^2 dx, \quad D(\mathcal{E}) = C_0^\infty(\Omega).$$

It is evident that if $f_1, f_2 \in C_0^\infty(\Omega)$ and suppf_1, suppf_2 are disjoint then $\mathcal{E}(f_1, f_2) = 0$. It is known see [44] (Chapt.2) that $\overline{\mathcal{E}}$ the closure of $\mathcal{E}$ in $L^2(\Omega, \phi^2 dx)$ has the same property and moreover it is a regular Dirichlet form. It follows from theorem 6.2.2 of reference [44] that there is a diffusion process in Ω whose Dirichlet form is $\overline{\mathcal{E}}$. Given an admissible function ϕ it can be used either to define a quantum dynamics or a diffusion process.

Example 2.2 indicates that a Schrödinger operator H corresponding to an admissible function ϕ cannot always be defined as the Laplacian Δ perturbed by a potential V. For various purposes, such as computations or development of scattering theory it is useful to approximate the singular operator H by the operators

$$H_n = -\Delta + V_n, \tag{3.1}$$

where V_n are measurable functions. Let ϕ_n, ϕ be the ground states for H_n, H respectively. The question arises; What are the conditions for the convergence

$\phi_n \to_{n\to\infty} \phi$ so that $H_n \to_{n\to\infty} H$ in the strong resolvent sense. In parallel, given the Dirichlet form $\mathcal{E}_\phi$, the associated semigroup T_t and process X_t, can it be approximated by a sequence of Dirichlet forms $\mathcal{E}^n$, $n \in \mathbf{N}$ defined by more regular functions ϕ_n, so that the sequence of corresponding semigroups $T_t{}^n$ converge to T_t as $n \to \infty$. This turns out to be a very complex problem. It has been investigated in several papers. An example of approximation of point interaction by short range potentials can be found in [14]. The following theorem [25] gives the sufficient conditions for approximate ground state to generate approximate Hamiltonian.

Theorem 3.1. *Let $\{\phi_n\}_{n\geq 0}$ be a sequence of functions admissible in the sense of Theorem 2.1 and such that*
(1) $0 < \phi_1 \leq \phi_2 \leq ... \leq \phi_0$ almost everywhere,
(2) $\frac{\phi_n}{\phi_0} \to_{n\to\infty} 1$ in the L^∞ sense.
Then the resolvents of H_n associated to ϕ_n converge strongly to that of H, associated to ϕ_0.

An analogous theorem can be proven for monotone decreasing sequences of functions [25].
The monotonicity conditions were relaxed in [38] but condition (2) of Theorem 3.1 or

$$\phi \leq \phi_n, \ \frac{\phi_n}{\phi} \to 1, \ in \ L^\infty(\mathbf{R^d}) \tag{3.2}$$

were still required. The strongest results appeared in [30]. The authors were able to relax also the assumptions (2) of Theorem 3.1 and (3.2). Instead they made assumptions (formulated below) eliminating strong singularities of ϕ_n, ϕ_n^{-1} and too strong growth of $x \cdot \nabla\phi_n(x)/\phi_n(x)$. They consider a sequence ϕ_n of $L^2_{loc}(\mathbf{R}^d)$ functions such that the energy forms

$$\mathcal{E}^n(u,u) \equiv \int |\nabla u(x)|^2 \phi_n^2(x)dx, \ u \in C_0^\infty(\mathbf{R}^d) \tag{3.3}$$

are closable. (For sufficient conditions for closability of $\mathcal{E}^n$ see [45] and references therein). The closure $\overline{\mathcal{E}^n}$ of $\mathcal{E}^n$ with

$$D(\overline{\mathcal{E}^n}) = \{f \in L^2(\mathbf{R}^n, \phi_n^2 dx); \ \nabla f \in L^2(\mathbf{R}^n, \phi_n^2 dx)\}$$

is the minimal Markovian extension of $\mathcal{E}^n$. Put P_t^n for the symmetric Markovian semigroup in $L^2(\mathbf{R}^n, \phi_n^2 dx)$ associated with $\mathcal{E}^n$. Then

$$T_t^n \equiv \phi_n P_t^n \phi_n^{-1} \tag{3.4}$$

is a symmetric positivity preserving contraction semigroup in $L^2(\mathbf{R}^n, dx)$. It is also assumed that

$$\phi_n \to_{n\to\infty} \phi \ in \ L^2_{loc}(\mathbf{R^d}). \tag{3.5}$$

Their analysis goes in several steps. First they prove existence of the limit T_t of T_t^n, with T_t a positivity preserving contraction semigroup. Under additional assumptions it is proved that $P_t = \phi^{-1}T_t\phi$ is the symmetric Markov semigroup in $L^2(\mathbf{R}^d, \phi^2 dx)$. Moreover the explicit formulas for the generator $\mathcal{L}$ of P_t and corresponding Dirichlet forms (on certain subsets of their domains) are given. Finally under still stronger assumptions it is shown that P_t can be identified with the Markovian semigroup associated to the energy form given by ϕ. More precisely;

Put

$$B_r \equiv \{x \in \mathbf{R}^d;\ |x| \leq r\}.$$

Assumption A. There are $R_0 > 0$, and $C > 0$ such that

$$\phi_n|(\mathbf{R}^d \setminus B_{R_0}) \in C^2(\mathbf{R}^d \setminus B_{R_0})$$

and

$$x \cdot \frac{\nabla \phi_n}{\phi_n} \leq C|x|^2$$

for all $x \in \mathbf{R}^d \setminus B_{R_0}$, and $n \in \mathbf{N}$.

Let $A \subset \mathbf{R}^d$ and $\chi_A(x)$, $x \in \mathbf{R}^d$ stands for the characteristic function of A.

Assumption B. There exists an increasing sequence $\{G_m\}$, $m > 1$, of bounded domains in $\mathbf{R}^d$ with smooth boundary such that
(a) $\chi_{G_m}(x)\phi_n(x)^{-1} \to \chi_{G_m}\phi(x)^{-1}$ in $L^2(\mathbf{R}^d)$ as $n \to \infty$, and
(b) $\mathbf{R}^d \setminus \bigcup_{m=1}^{\infty} G_m$ is of null Lebesgue measure.

Theorem 3.2. *Assume* ***A*** *and* ***B****. Then there exists a subsequence n_k and a positivity preserving contraction semigroup $\{T_t,\ t \geq 0\}$ on $L^2(\mathbf{R}^d)$ such that*

$$\{T_t^{n_k} \to T_t\}$$

strongly as $k \to \infty$, for any $t \geq 0$.

Assumption C. There exists a domain G in $\mathbf{R}^d$ such that
(i) $\chi_G\phi_n^{-1} \to \chi_G\phi^{-1}$ in $L^2_{loc}(G)$;
(ii) $\phi|G \in C^1(G)$ and $\phi > 0$ for all $x \in G$;
(iii) the Lebesgue measure of $\mathbf{R}^d \setminus G$ is zero.

Theorem 3.3. *Assume* ***A, B, C****. Let T_t^n be as in Theorem 3.2 with ϕ_n satisfying $\phi_n \in L^2_{loc}(\mathbf{R}^d)$ and $\phi_n(x) > 0$ a.e. Put T_t for the strong limit of T_t^n restricted to subsequences if necessary. Let P_t be the symmetric Markov semigroup in $L^2(\mathbf{R^d}, \phi^2 dx)$ given by $P_t = \phi^{-1}T^t\phi$, $\mathcal{L}$ be the generator of $(P_t, t > 0)$ and let $\mathcal{E}$ be the Dirichlet form associated with P_t. Then $C_0^2(G) \subset D(\mathcal{E})$ and*

$\mathcal{L}g(x) = \Delta g + 2(\nabla\phi(x)/\phi(x))\nabla g(x)$ *for all* $g \in C_0^2(G)$. *Moreover,* $C_0^1(G) \subset D(\mathcal{E})$ *and*

$$\mathcal{E}(g,g) = \int_{\mathbf{R^d}} |\nabla g(x)|^2 \phi(x)^2 dx \text{ for all } g \in C_0^1(\mathbf{R^d}).$$

Assumption D

(1) $\phi_n^{-1} \to \phi^{-1}$ *in* $L^2_{loc}(\mathbf{R}^d)$,

(2) there exists a constant $C > 1$ such that $\phi_n(x) \le C\phi$ for any $n \ge 1$, $x \in \mathbf{R^d}$.

Theorem 3.4. *Assume* ***A, B, C, D****. Let* T_t^n, T_t *be as in Theorem 3.3. Let* $D(\widetilde{\mathcal{E}}_0) = \{u \in L^2(\mathbf{R^d}, \phi^2 dx); \nabla u$ *is a measurable function and* $\int |\nabla u(x)|^2 \phi(x)^2 dx < \infty\}$; $\widetilde{\mathcal{E}}_0(u,u) = \int |\nabla u(x)|^2\phi(x)^2 dx$. *Then* $\widetilde{\mathcal{E}}_0$ *has closed extensions and let* $\mathcal{E}_0$ *be the smallest closed extension of* $\widetilde{\mathcal{E}}_0$. *Then* $\mathcal{E}_0$ *is the Dirichlet form associated with* $P_t = \phi^{-1}T_t\phi$.

4. Unpenetrable Barriers

We have seen that the functions admissible in the sense of Theorem 2.1 may have nods. In this section we will discuss the consequences of the ground state nods for corresponding quantum dynamics and stochastic process.

We begin with a simple example. Consider the free dynamics Hamiltonian H_0 in $L^2(\mathbf{R}, dx)$. It has continuous spectrum $\sigma(H_0) = \{x \ge 0\}$ and its ground state is a generalized eigenfunction equal to a constant, say, $\phi_0 \equiv 1$. Clearly $-\frac{d^2\phi_0(x)}{dx^2} \equiv 0$ and $V(x) = \frac{d^2\phi_0(x)}{dx^2}/\phi_0(x) \equiv 0$. However the space of solutions of the equation

$$-\frac{d^2\phi(x)}{dx^2} = 0 \tag{4.1}$$

is two dimensional hence in particular $\phi(x) = x$ satisfies equation (4.1) and by Theorem 2.1 $\phi(x) = x$ is admissible. The quadratic form

$$\mathcal{E}_x(u,v) = \int_{\mathbf{R}} \frac{du(x)}{dx}\frac{dv(x)}{dx} x^2 dx, \quad u,v \in C_0^\infty(\mathbf{R}) \tag{4.2}$$

in $L^2(\mathbf{R}, x^2dx)$ is closable. It can be shown by direct computation that for any $u \in C_0^\infty(\mathbf{R})$ the function $\chi_{\mathbf{R}^-}u \in D(\overline{\mathcal{E}}_x)$ and consequently $\chi_{\mathbf{R}^-}f \in D(\overline{\mathcal{E}}_x)$ for any $f \in D(\overline{\mathcal{E}}_x)$. Hence the quadratic form $\overline{\mathcal{E}}_x$ decomposes into the sum

$$\overline{\mathcal{E}}_x(f,g) = \overline{\mathcal{E}}_{-,x}(f,g) + \overline{\mathcal{E}}_{+,x}(f,g) \; f,g \in D(\overline{\mathcal{E}}_x). \tag{4.3}$$

where $\overline{\mathcal{E}}_{\mp}(f,g) = \overline{\mathcal{E}}(\chi_{\mathbf{R}^\mp}f, g)$ are the closed quadratic forms in $\mathcal{H}_{-,x} = L^2(\mathbf{R}^-, x^2dx)$ and $\mathcal{H}_{+,x} = L^2(\mathbf{R}^+, x^2dx)$ respectively. Put

$$P_\mp f = \chi_{\mathbf{R}^\mp} f, \; f \in L^2(\mathbf{R}, x^2dx)$$

and

$$Q_\mp f = \chi_{\mathbf{R}^\mp} f, \; f \in L^2(\mathbf{R}, dx)$$

then $P_{\mp}$ ($Q_{\mp}$) are the projections onto

$$\mathcal{H}_{-,x}\ (\mathcal{H}_{-} \equiv L^2(\mathbf{R}^-, dx))$$

and

$$\mathcal{H}_{+,x}\ (\mathcal{H}_{+} \equiv L^2(\mathbf{R}^+, dx))$$

respectively. It follows that the spectral projections of the operator H_x defined by the form (4.3) commute with $P_{\mp}$. Put $H \equiv x^{-1}H_x x$. Since $xQ_{\mp} = P_{\mp}x$ every spectral projection of H commutes with $Q_{\mp}$. Hence the quantum mechanical evolution e^{iHt} is reduced by the decomposition

$$\mathcal{H} = \mathcal{H}_{-} \oplus \mathcal{H}_{+}. \tag{4.4}$$

Physically if a particle is observed in negative semi axis at time t then it could have never been and never will be observed in positive semi axis. Analogically if $L^2(R, dx)$ is real then the Dirichlet form

$$\mathcal{E}_x(u,v) = \int_{\mathbf{R}} \frac{du(x)}{dx}\frac{dv(x)}{dx}x^2 dx$$

defines the positive self adjoint operator H in $L^2(\mathbf{R}, dx)$ such that the transition semigroup e^{-Ht} is reduced by the decomposition (4.4). Hence the corresponding diffusion process never crosses the point 0.

To summarise, if $\phi_0(x) \equiv 1$ then the operator H in $L^2(\mathbf{R}, dx)$ defined by the quadratic form $\mathcal{E}_x(u,v) = \int_{\mathbf{R}} \frac{du(x)}{dx}\frac{dv(x)}{dx}\phi_0^2(x)dx$ is
$H = -\frac{d^2}{dx^2}$
$D(H) = \{f \in L^2(\mathbf{R}, dx),\ f' \text{ is a.c.},\ f', f'' \in L^2(\mathbf{R}, dx)\}$.
If $\phi_0(x) = x$ then the node of ϕ_0 enforces the additional boundary condition. Namely
$H = -\frac{d^2}{dx^2}$
$D(H) = \{f \in L^2(\mathbf{R}, dx), f' \text{is a.c.},\ f', f'' \in L^2(\mathbf{R}, dx), f(0) = 0\}$.
Our example shows that zero of a ground state may be responsible for unpenetrable barrier at location of the node.

Let $\mathcal{H} = L^2(\Omega, m)$, where Ω is a nonvoid open subset of $\mathbf{R}^n$ and m a positive Radon measure on Ω. Let $\Omega_1 \subset \Omega$ be m-measurable, and let U_t be a one-parameter strongly continuous group of unitary maps of $\mathcal{H}$. We say that there is nontunneling through the boundary $\delta\Omega_1$ of Ω_1 in Ω iff U_t is reduced by the decomposition $L^2(\Omega, m) = L^2(\Omega_1, m) \oplus L^2(\Omega \setminus \Omega_1, m)$. In other words nontunneling is the same as existence of unpenetrable barrier on $\delta\Omega$. In general situation of diffusion processes in $\mathbf{R}^n$ the relation of unpenetrable barriers and the nodes of the ground state is described by following theorems [7]:

Theorem 4.1. *If the Hamiltonian H given by the energy form $\mathcal{E}$ as in Theorem 2.1 has no tunneling through the boundary of an open set Ω, then there is a decreasing sequence O_ν of open sets such that*

(1) Cap $O_\nu \to 0$,
(2) $\delta\Omega \subset \overline{O_\nu}$.

Theorem 4.2. *Let* $\mathcal{E}$, H *be as in Theorem 2.1 and* Ω *an open subset of* $\mathbf{R}^n$. *If there is a decreasing sequence* O_ν *of open sets such that*
(1) Cap $O_\nu \to 0$,
(2) $\delta\Omega \subset \overline{O_\nu}$,
(3) Any continuous path from Ω *to* $\mathbf{R}^n \setminus \Omega$ *must cross* O_ν *for some* ν
Then there is no tunneling through $\delta\Omega$.

Theorem 4.3. *Let* H *be given by* $\mathcal{E}$ *in* $L^2(\mathbf{R}, dx)$ *as in Theorem 2.1 and*
(a) $|\phi_0(x)|^2 \leq const|x|$ *at least near one side of the origin, then there is no tunneling through* $x = 0$.
(b) $|\phi_0(x)|^2 \geq const|x|^\alpha$ *for some* $0 < \alpha < 1$ *in a neighbourhood of the origin then there is tunneling through* $x = 0$.

The above mathematical mechanism of unpenetrable barriers formation is the appropriate language for description of various natural phenomena. Consider a system of a large number of particles moving in an external field of potential forces and randomly colliding among themselves. Such a situation is believed to have occurred in an early stage of formation of the solar system when a cloud consisting of gas and particles of stellar and interstellar origin moved in the field of central gravitational forces. Also planetary atmosphere can be seen as a layer of particles moving under the gravitational field and random collisions. The motion of such a system can be described by the so called Newtonian diffusion processes.
Let M be a smooth oriented Riemann manifold of dimension d, and X_t, $t \in T \subset \mathbf{R}_+$ a diffusion process with values in M. The concept of Newtonian diffusion process is based on analytical description of X_t. The infinitesimal generator L_t of X_t is assumed to be

$$L_t = \frac{1}{2}\Delta + \beta \cdot D$$

where $\beta \cdot D \equiv \beta^i D_i$, $\beta^i(x,t)$, $i = 1, ..., d$ is a $\mathcal{C}^\infty$ random field (the drift), D is covariant derivative and Δ is the Laplace-Beltrami operator on M. Let dx be the Riemaniann volume element on M. Then the probability distribution law of X_t has smooth density $\rho(x,t)$ with respect to dx i.e. $dP(X_t \in dx) = \rho(x,t)dx$. Put $\hat{X}_t$, $t \in -T$ for the time reversed process to X_t. Then $\hat{X}_t$ is also a Markov process [27] with infinitesimal generator

$$\hat{L}_t \equiv \frac{1}{2}\Delta - \hat{\beta} \cdot D$$

where $\hat{\beta}$ is the backward drift. The functions $u = \frac{1}{2}(\beta - \hat{\beta})$ and $v = \frac{1}{2}(\beta + \hat{\beta})$ are called osmotic velocity and current velocity respectively. Further we need the concepts of mean forward and mean backward derivatives [40], [61], [58] denoted by

D_+, D_-. They can be given from [40], [59]

$$D_+ = \frac{\partial}{\partial t} + \beta \cdot \nabla + \frac{1}{2}(\Delta - R)$$

$$D_- = \frac{\partial}{\partial t} + \hat{\beta} \cdot \nabla - \frac{1}{2}(\Delta - R)$$

where R is the Ricci tensor acting on vectors. Then the mean acceleration is defined by

$$a(X_t, t) = \frac{1}{2}(D_+D_- + D_-D_+)X_t \tag{4.5}$$

Definition 4.1. A Markov process is said to satisfy Newton law in the mean if there is a positive constant m and a real-valued function V on $M \times T$ such that

$$ma(X_t, t) = -\nabla V(X_t, t) \tag{4.6}$$

and there is a real function $S(x,t)$ such that the corresponding current velocity $v(x,t) = \nabla S(x,t)$.

A Markov process satisfying Newton law in the mean is called a conservative Newtonian diffusion.
It can be shown that if X_t is a conservative Newtonian diffusion then its probability distribution $\rho(x,t) = |\psi(x,t)|^2$, where ψ is a solution of the Schrödinger equation

$$i\frac{\partial \psi}{\partial t} = -\frac{1}{2}\Delta\psi + V\psi \tag{4.7}$$

with the initial condition $\psi(x,0)$ such that $|\psi(x,0)|^2$ is the initial distribution of X_t. Conversely if ψ is a solution of (4.7) then there is a conservative Newtonian diffusion X_t, $t \in T$ with the distribution $\rho(x,t) = |\psi(x,t)|^2$. Assuming that the distribution is stationary i.e. $\rho(x.t) = \rho(x)$ we have $|\psi(x,t)|^2$ is independent of t. Solving (4.7) under this condition and setting $u = \frac{1}{2}\nabla \log \rho$, and v so that the continuity equation holds we can compute β and hence the process X_t. The function ρ is then invariant distribution for X_t and $\frac{\partial \rho}{\partial t} = 0$.
If $v = 0$ and $u = \frac{1}{2}\nabla \log \rho$ then $\frac{\partial \rho}{\partial t} = 0$ follows.
$v = 0$ is equivalent with the symmetry of distribution P_t of X_t in $L^2(M, \rho dx)$ [12], which is equivalent with time reversal invariance. In this situation we can simplify construction of the process starting from the Dirichlet forms and following [31] and [8]. Then we can utilize the theory of barriers formation described in [7]. We are prepared now to discuss the physical applications.

4.1. *Formation of the solar system*

Historically it has been speculated (Descartes (1644), Kant (1755), Laplace (1796)) that the solar system has formed from a cloud of gas and particles of stellar and interstellar origin. Most of the material of such protosolar nebula concentrated to form the sun. The residual matter kept by gravity became confined in nearly planar rings around the sun. In the next stage matter of the rings aggregated in larger bodies finally forming the planets. According to astronomical observations average distances of planets in solar system exhibit remarkable regularity. In 1776 J. D. Titius found a simple formula (presently known as Titius-Bode law) for distances of planets to sun.

$$r(n) = 0,4 + 0,3 \cdot 2^{n-1}, \tag{4.8}$$

the unit being the average distance of Earth from Sun. The values of n are: Mercury (-∞), Venus (1), Earth (2), Mars (3), Jupiter (5), Saturn (6). Discovery of Uranus (7) and Ceres (4) confirmed merit of the Titius-Bode law.
It has been observed later that similar formulas hold for some moons of Jupiter, Saturn and Uranus. Thus it is reasonable to require that a model of formation of solar system should yield the Titius-Bode formula. For the history of attempts to explain the Titius-Bode law see [62] and [50].
In Alfvens theory the rings called jet-streams were of electromagnetic origin while aggregations were caused by other effects like for instance nuclear processes in sun. [36], [62]
The motion of a cloud of particles in the field of central gravitational forces and randomly colliding can be reasonably approximated by stationary time reversal invariant conservative Newtonian diffusion. As said above such a process can be constructed in terms of Dirichlet forms. We shall demonstrate that the formation of jet-streams naturally follows from the structure of such processes and be able to obtain radiuses of the rings being in good agreement with Titus-Bode law [2], [3], [4]. To be more precise we obtain location of barriers preventing transport of matter between the regions.
Let $V(x) = -\frac{gmM}{|x|}$ where g is the gravitational constant and m, M are the masses of the particle and sun respectively. Then
$H = -\nabla + V$
$D(H) = \{f, Vf \in L^2(\mathbf{R}^3, dx); \frac{\partial f}{\partial x_i}$ *is a.c. and* $\frac{\partial f}{\partial x_i}, \frac{\partial^2 f}{\partial x_i^2} \in L^2, i = 1, ..., 3\}$
is a self-adjoint operator in $L^2(\mathbf{R}^3, dx)$. The spectrum of H consists of negative eigenvalues $E_n = \frac{\alpha}{n^2}$, $n = 1, 2, ...$ where α is a negative constant and the continuous part from zero to infinity. The eigenspace corresponding to E_n is spanned by the functions given in spherical coordinates by

$$\psi_{nlm}(r, \vartheta, \varphi) = R_{nl}(r) P_l^m(cos\vartheta) e^{im\varphi} \tag{4.9}$$

where $n = 1, 2, ...,\ l = 0, ..., n-1,\ -l \le m \le l,\ R_{nl}(r) = e^{-\frac{r}{r_0}} (\frac{r}{r_0})^l L_{n+1}^{(2l+1)}(2\frac{r}{r_0})$.
The functions $L_{n+1}^{(2l+1)}$ and $P_l^m(cos\vartheta)$ are the Laguerre and Legendre polynomials respectively. r_0 is a constant depending on physical data of the problem. The eigenspace for E_1 is one dimensional and since L_2^1 as well as P_0^0 are positive so is the eigenfunction ψ_{100}. Thus $H_{E_1} \equiv -\nabla + V - E_1$, $D(H_{E_1}) = D(H)$ is a non negative self adjoint operator and its ground state ψ_{100} defines the Dirichlet form

as the closure of

$$\mathcal{E}_{\psi_{100}}(f_1, f_2) = \int_{\mathbf{R}^3} \nabla f_1 \cdot \nabla f_2 \psi_{100}^2 dx, \ f_i \in C_0^\infty(\mathbf{R}^3) \tag{4.10}$$

in real $L^2(\mathbf{R}^3, \psi_{100}^2 dx)$.
Since ψ_{100} has no nodes there are no barriers for the corresponding stochastic process. However by Theorem 2.1 every eigenfunction of H is admissible and so any ψ_{nlm}, defines a Dirichlet form as the closure of

$$\mathcal{E}_{\psi_{nml}}(f_1, f_2) = \int_{\mathbf{R}^3} \nabla f_1 \cdot \nabla f_2 \psi_{nml}^2 dx, \ f_i \in C_0^\infty(\mathbf{R}^3) \tag{4.11}$$

in $L^2(\mathbf{R}^3, \psi_{nlm}^2 dx)$. If $n > 1$ then the ψ_{nlm} have nodes. It follows from (4.9) that the distribution $\rho_{nlm}(x) = |\psi_{nlm}|^2$ vanishes at concentric spheres with radii given by zeros of L_{n+1}^{2l+1} and cones given by zeros of $P_l^m(\cos\vartheta)$. It is known that L_{n+1}^{2l+1} has $n - l - 1$ zeros and $P_l^m(\cos\vartheta)$ has $(l - |m|)$ zeros. The maxima of ρ_{nlm} inside the regions of confinement are located on concentric circles $\vartheta = 0$ and $r'_i \approx \frac{1}{2}(r_{i+1} + r_i)$ where r_i are the nodes of ρ_{nlm}. Taking $n = 16$ and $l = m = 6$ we get 9 spherical barriers i.e. 10 regions of confinement intersecting the plane $\vartheta = 0$. The maxima of $\rho_{16,6,6}$ reasonably approximate observed planetary orbits. The same model gives approximation for the orbits of moons of Jovian planets and of Saturn rings. We are also able to explain spikes in Saturn rings. The distribution $\rho_{nlm} = |\psi_{nlm}|^2$ does not depend on the longitudinal variable ϕ. Thus in particular $\rho_{nlm} = \rho_{nl(-m)}$. Set $\psi_{nlm}^c = \frac{1}{2}(\psi_{nlm} + \psi_{nl(-m)})$ and $\psi_{nl(-m)}^s = \frac{1}{2}(\psi_{nlm} - \psi_{nl(-m)})$. Then $\rho_{nlm}^c \equiv |\psi_{nlm}^c|^2$ and $\rho_{nlm}^s \equiv |\psi_{nlm}^s|^2$ have nods at $\cos(m\phi) = 0$ and $\sin(m\phi) = 0$ respectively. The Newtonian diffusions with distributions ρ_{nlm}^c, and ρ_{nlm}^s describe both the circles and spike formation.

4.2. *Wind zones in planetary atmospheres*

A planetary atmosphere can be looked upon as a spherical layer of randomly colliding particles moving in spherically symmetric gravitational field. Thus it is accessible to description by the Newtonian diffusion distributed by $\rho_{nlm} = |\psi_{nlm}|^2$. Since thickness of the layer is small when compared with its radius the function $R_{nl}(r)$ can be assumed constant, and the barriers i.e. boundaries of atmospheric zones are expected to appear at nods of $P_l^m(\cos\vartheta)$. The existence of atmospheric zones on Earth is well known. We have:
$\vartheta_1 = 0$ ($\vartheta'_1 = 0$), $\vartheta_2 = 30$ ($\vartheta'_2 = 28$), $\vartheta_3 = 60$ ($\vartheta'_3 = 56$), $\vartheta_4 = 90$ ($\vartheta'_4 = 90$), where ϑ_i and ϑ'_i are the observed latitudes of zone boundaries in northern hemisphere and nods of $P_6^1(\vartheta)$ respectively.

Astronomical observations prove existence of atmospheric zones at other planets. In case of Jupiter northern latitude of the zone boundaries are ($\vartheta'_1 = 2.9$), ($\vartheta'_2 = 8.7$), $\vartheta_3 = 16$ ($\vartheta'_3 = 14.5$), $\vartheta_4 = 20$ ($\vartheta'_4 = 20.5$), $\vartheta_5 = 30$ ($\vartheta'_5 = 26$), $\vartheta_6 = 34$ ($\vartheta'_6 =$

32), $\vartheta_7 = 38$ ($\vartheta'_7 = 37.6$), $\vartheta_8 = 42$ ($\vartheta'_8 = 43.6$), $\vartheta_9 = 50$ ($\vartheta'_9 = 49.5$), $\vartheta_{10} = 55$ ($\vartheta'_{10} = 55$), $\vartheta_{11} = 59$ ($\vartheta'_{11} = 60.5$), ($\vartheta'_{12} = 67$), ($\vartheta'_{13} = 73.7$), ($\vartheta'_{14} = 78.5$), where ϑ_i and ϑ'_i are the observed latitudes and nods of P_3^1 respectively.

4.3. *Galaxies*

The techniques of Newtonian diffusion can also be used in constructing models for formation of the galaxies [5]. A star in a galaxy is attracted by the center of mass of the system but also in a significant way by the nearby stars. The interaction with the neighbours can be taken as random. The variety of structures of the galaxies can be obtained by non stationary processes. Here are the examples of galaxies and corresponding distributions:
1) E0 galaxy - elliptical (spherical), $\rho(r, \vartheta = \frac{\pi}{2}, \phi) = ||\alpha\psi_{100}||^2$,
2) E7 galaxy - elliptical, $\rho(r, \vartheta = \frac{\pi}{2}, \phi) = ||\alpha\psi_{100} + \beta\psi_{320} + \gamma\psi_{322}||^2$,
3) S0 galaxy - lenticular, $\rho(r, \vartheta = \frac{\pi}{2}, \phi) = ||\alpha\psi_{100} + \beta\psi_{300}||^2$,
4) Sa galaxy - normal spiral, $\rho(r, \vartheta = \frac{\pi}{2}, \phi) = ||\alpha\psi_{100} + \beta_{\pm}\psi_{21(\pm 1)} + \gamma_{\pm}\psi_{31(\pm 1)}||^2$.

5. Dirichlet Forms and Quantum Field Theory

5.1. *An Overview*

In quantum physics the dynamics of particles is usually given heuristically as a sum of two self-adjoint Hilbert space operators

$$H = H_0 + H_1$$

where H_0 is essentially a Laplacian operator and H_1 a multiplication operator which describes the interaction of particles. Physically desirable choices of H_1 tend to lead to ill-defined operator sums, self-adjointness of H is far from evident for unbounded H_1, and perturbation theory fails in important cases where H_1 is not Kato-small w.r. to H_0.

Evidently local Dirichlet forms or "energy forms" are an extremely powerful tool to deal with this situation; as we have seen one arrives at a well defined positive generator H for the time development even in cases where the operator H_1 itself defies a mathematically sound definition.

If this scenario is a frequent one in non-relativistic quantum mechanics it is ubiquitous in quantum field theory. This fact is epitomized by Haag's theorem [46].

We can easily define an infinite dimensional Gaussian measure μ_0 such that the operator H_0 has a proper representation in the Hilbert space $L^2(d\mu_0)$ but Haag's theorem essentially says that any relativistic interactions will require mutually singular measures μ for the representation of H.

This being the case it would be natural to in fact define the quantum field dynamics in terms of suitable measures or "ground states", as proposed by Coester and Haag [39] and by Araki [37]. The challenge that remains is to come up with measures such that the unitary time development gives rise to local relativistically invariant fields. Short of solving this problem one will in any case ask

(1) Is an infinite dimensional version of local Dirichlet forms in principle suitable?
(2) Can known qft models be cast into this framework?
(3) And if so, what is the nature of the underlying measures?
(4) How does "Euclidean quantum field theory" fit into this framework?

5.2. *Canonical Quantum Fields*

Quasi-invariant probability measures on co-nuclear spaces N', typically the Schwartz distribution space $S^*(R^s)$ had been known to produce Hilbert spaces $L^2(S^*(R^s), d\mu)$ which admit continuous unitary representations of the Weyl group

$$U(f)V(g) = \exp(i\varphi(f))\exp(i\pi(g)) = \exp(i(f,g))\,U(f)V(g) \qquad f,g \in S(R^s),$$

and hence of the canonical commutation relations (CCR) of quantum field theory:

$$[\varphi(f), \pi(g)] \subset i\,(f,g)_{L^2(R^s)}\,.$$

U is realized as a multiplication operator

$$U(f) = \exp(i(\bullet, f))$$

on $L^2(N', d\mu)$, V is a translation operator, and $\Omega = 1 \in L^2(N', d\mu)$ is called the "vacuum" or "ground state".

For a canonical Hamiltonian H one requires that it annihilates the ground state and

$$[\varphi(f), H] \subset i\pi(f)\,,$$

or in terms of Weyl operators

$$[U(f), [H, U(g)]] \subset -(f,g)U(f+g).$$

A consequence of this is the "Araki relation" [37]

$$(U(f)\Omega, HU(g)\Omega) = (f,g)\,(U(f)\Omega, U(g)\Omega)\,. \tag{5.1}$$

5.3. *Dirichlet Forms for Quantum Fields*

The 1975 paper by Albeverio and Hoegh-Krohn [20] was seminal. It extends the energy form theory to sufficiently regular quasi-invariant measures μ on duals of nuclear spaces. Introducing first a gradient, the authors show that an infinite dimensional diffusion operator can be associated with (the closure) of the corresponding local Dirichlet form. It generates a Markov semigroup - related to the "Euclidean" quantum fields - , and conditions for ergodicity can be given in terms of the defining measure μ.

A first example is the class of non-interacting Bose fields for which the Gaussian measure can be given explicitly, but non-Gaussian vacuum measures obtained in constructive quantum field theory are shown to fit into this framework as well. For further development along these lines see [22] [23] [28].

In conventional methods of constructive qft one obtains ground state measures μ starting from a suitably regularized version of the Hamiltonian operator H, with renormalizations of the latter such that regularizations can be removed by limit taking. Usually this construction, as an intermediate step, proceeds via the construction of a measure μ_E and commutative "Euclidean" (Markov) fields $\Phi = \Phi(x)$, $x = (x_0, \vec{x}) \in R^{s+1}$ depending on space $\vec{x} \in R^s$ and (imaginary) time x_0.

Time ordered expectations of relativistic fields are then obtained by analytic continuation in the time parameters $x_{k,0}$ of the "Schwinger functions"

$$S_n(x_1, \ldots, x_n) = E_\nu\left(\Phi(x_1) \ldots \Phi(x_n)\right).$$

Using the measures μ so obtained for the time zero fields φ to then construct energy forms, one obtains an alternate expression for the Hamiltonian via the quadratic form. While both will agree on dense domains their self-adjoint extensions might differ.

In what follows - limited by space and expertise -, we shall focus on one particular aspect: a concrete realization of the measures in question.

5.4. *Concepts from White Noise Analysis*

White noise analysis has by now proven its usefulness as a framework of infinite dimensional analysis [47] [49] [56] [63].

Starting from the Gaussian white noise measure μ_0 with characteristic function

$$C(f) = \exp\left(-\frac{1}{2}\|f\|_2^2\right) = \int_{S^*(R^s)} d\mu_0(\omega) \exp\left(i(\omega, f)\right) \qquad f \in S(R^s),$$

$(L^2) \equiv L^2\left(S^*(R^s), d\mu_0\right)$ is embedded between spaces of smooth and generalized functions ("Hida distributions") of white noise, respectively:

$$(S) \subset (L^2) \subset (S)^*.$$

The construction of the Gelfand triple follows that of Schwartz test functions and distributions. We refer to the literature for details. There is a useful characterization theorem for the space $(S)^*$ in terms of Fourier Gauss transforms. Using the fact that white noise functions given by

$$e_f(\omega) = \exp\left(i\langle\omega, f\rangle\right)$$

are in (S), we can define the "T-transform" of generalized white noise functions $\Psi \in (S)^*$ by

$$T\Psi(f) = \langle \Psi, e_f \rangle$$

Theorem 5.1. *A complex valued function u on Schwartz space is the T-transform of a unique $\Psi \in (S)^*$ iff*

(1) The mapping $x \to u(f + xg)$ has an entire extension for all $f, g \in S(R^s)$
(2)

$$|u(zf)| \leq a \exp\left(b|z|^2 \|f\|^2\right) \tag{5.2}$$

where $a, b > 0$ and $\|\cdot\|$ is a continuous norm on $S(R^s)$.

In the present context, we shall make use of the fact that any positive Hida distribution gives rise to a unique measure on Schwartz distribution space.

Definition 5.1. We call $\Psi \in (S)^*$ positive if $\langle \Psi, F \rangle \geq 0$ for all nonnegative $F \in (S)$.

Theorem 5.2. *(Kondratiev [54]) For any positive $\Psi \in (S)^*$ there is a unique finite measure ν on the Borel sets of $S^*(\mathbb{R}^s)$ such that all $F \in (S)$*

$$\langle \Psi, F \rangle = \int_{S^*(\mathbb{R}^s)} F d\nu.$$

Test functionals $F \in (S)$ admit directional ("Gateaux") derivatives:

$$D_h F(\omega) = \lim_{\varepsilon \to 0} \frac{F(\omega + \varepsilon h) - F(\omega)}{\varepsilon}$$

exists for any generalized function $h \in (S)^*$. In particular for $h = \delta_x$, we introduce the Hida derivative

$$D_h \equiv \partial_x.$$

∂_x gives rise to a natural gradient (Frechet derivative)

$$\nabla F = \{\partial_x F : x \in \mathbb{R}\}.$$

I.e.

$$\nabla : (L)^2 \to L^2(\mathbb{R}) \otimes (L)^2$$

and

$$\nabla : (S) \to S(\mathbb{R}) \otimes (S).$$

The corresponding carré du champ functional is a test functional [48], for all $F \in (S)$

$$|\nabla F|^2 = (\nabla F, \nabla F)_{L^2(\mathbb{R})} = \int_{\mathbb{R}} dx \, |\partial_x F|^2 \in (S).$$

More generally we can e.g. introduce an orthonormal base of functions e_k in $S(\mathbb{R}^s)$ and define

$$|\nabla F|^2 = \sum_k |D_{e_k} F|^2 \in (S).$$

5.5. *Energy Forms in White Noise Analysis [48]*

We have seen that all positive Hida distributions Ψ give rise to measures.

Definition 5.2. We shall call them admissible if the ensuing Dirichlet form ("energy form")

$$\varepsilon(F) = \left\langle \Psi, |\nabla F|^2 \right\rangle$$

is closable.

A simple criterion is given in

Theorem 5.3. (Hida et al. [48]) *A positive $\Psi \in (S)^*$ is admissible if $D_{e_k}\Psi = B_k \Psi$ with $B_k \in (S)$ for all k.*

For more on closability see [31] [32] [34] [35].

For admissible measures one then shows

Theorem 5.4. (Hida et al. [48]) *Let*

$$\varepsilon(F) = \left\langle \Psi, |\nabla F|^2 \right\rangle$$

and $\bar{\varepsilon}$ its closure. Then $\bar{\varepsilon}$ is Markovian in the sense of Fukushima [45].

Following [21] the contraction semigroup P_t generated by $H = \nabla^* \nabla$ gives rise to a time homogeneous Markov process $\{X(t) : t \geq 0\}$with initial measure ν and transition probabilities $\{P_t : t \geq 0\}$.

5.6. *Application to Quantum Field Theory [11] [12]*

In view of theorems 5.1 and 5.2, moment estimates of the form

$$|(\Omega, \varphi^n(f)\Omega)| \leq a\sqrt{n!}\, \|f\|^n \tag{5.3}$$

for some continuous norm $\|\cdot\|$ on $S(R^s)$ ensure the existence of a unique measure μ on $S^*(\mathbb{R}^s)$ such that

$$|(\Omega, \exp(i\varphi(f))\,\Omega)| = \int_{S^*(\mathbb{R}^s)} \exp i\,\langle\omega, f\rangle\rangle\, d\mu.$$

Before addressing the question of admissibility let us see for which models of quantum field theory we do have bounds of the form (5.3).

The $\sqrt{n!}$ bound is characteristic of Gaussian fields, so it is not hard to show this bound for zero time as well as for Euclidean space time free fields, centered Gaussians, with covariance $\left(f, \left(-\Delta + m^2\right)^{-1/2} f\right)_{L^2(R^s)}$ and $\left(f, \left(-\Delta + m^2\right)^{-1} f\right)_{L^2(R^{s+1})}$, respectively, ensuring the existence of the ground state measure μ and the Euclidean measure ν as positive Hida distributions.

A less trivial example, with non-linear dynamics, is the $s = 1$ Hoegh-Krohn model with an exponential interaction, informally given by an interaction Hamiltonian

$$H_1(\varphi) = \int_R dx \int dm(\alpha) :\exp(\alpha\varphi(x)):$$

(the $: \cdot :$ standing for "normal ordering", effectively a renormalization of the exponential [47]). For suitable measures m of bounded support the Schwinger functions and the corresponding real time Wightman fuctions could be constructed [17] [18] [19] [13]. Since both the Schwinger and the time zero Wightman functions are bounded by those of free fields, the moment inequalities (5.3) are trivially fulfilled, and hence both the Euclidean and the time zero measure can be represented as positive Hida distributions.

For the $P(\varphi)_2$ models, with, again informally,

$$H_1(\varphi) = \int_R dx : P(\varphi) :$$

there are the Froehlich bounds [11] [42] [43]

$$\left|S_n(f^{\otimes n})\right| \leq (n!)^m \;\|f\|^n,$$

where $2m$ is the degree of the interaction polynomial P (with leading term $a_{2m}x^{2m} > 0$), ensuring as in the case of (5.3) the existence of the Euclidean measure in Hida space.

Finally, a general, model independent criterion for the ground state measure as positive element of $(S)^*$was given in [65].

Theorem 5.5. *If a field theory (φ, H, Ω) admits the Araki relation (5.1) and a "φ-bound" of the form*

$$\pm\varphi(f) \leq \alpha H + \beta \|f\|^2 + \gamma$$

(for some $\alpha, \beta, \gamma > 0$, and a Sobolev norm $\|\|$), as quadratic forms on the domain $D_f = F(\varphi(f))\,\Omega$ for Schwartz test functions F, then a moment bound as in (5.3) holds.

For the above mentioned measures, admissibility, i.e. closability of the corresponding forms ε then follows by the citeria of [31] [32] [34] [35].

5.6.1. *Supplementary remarks*

The Hamiltonian defined from the ground state measure coincides with the one from constructive field theory on a dense set, but coincidence of self-adjoint extensions is still open.

A local Dirichlet form in the $L^2\,(dv)$-space will give rise to fields with an extra time parameter ("computer time") as in the stochastic quantization program [64] which is based on solving an infinite dimensional stochastic differential equation for a stochastic process with the Euclidean measure as stationary distribution [35].

As a side remark, - nothing seems to be known (except for free fields), whether possibly the characteristic bound (5.2) might in some cases hold not only for the Schwinger functions, but also, after analytic continuation for the time ordered expectation values of the relativistic fields.

Apart from the measures of canonical quantum field theories, consider the Edwards polymer model. For two-dimensional Brownian motion B, Varadhan [68] has established the existence of the probability measure

$$dv = \frac{1}{Z} \exp\left(-gL_c\right) d\mu(\omega)$$

where L is the centered self-intersection local time.

$$L_c = \iint\limits_{0<s<t<T} \delta_c\left(B(t) - B(s)\right),$$

see e.g. [41] for a precise definition in terms of white noise. A local Dirichlet form and corresponding diffusion process were constructed in [26].

References

[1] S. Albeverio, Ph. Blanchard, R. Høegh-Krohn Newtonian diffusions and planets, with a remark on non-standard Dirichlet forms and polymers. Lecture Notes in Mathematics Volume 1095, 1984, pp 1-24.

[2] S. Albeverio, Ph. Blanchard, R. Hoegh-Krohn: A stochastic model for the orbits of planets and satellites: an interpretation of the Titus-Bode law. Expositiones Mathematicae **1**, 365-373 (1983).

[3] S. Albeverio, Ph. Blanchard, R. Hoegh-Krohn: Processus de diffusion, confinement et formation de "jet-streams" dans la nébuleuse protosolaire. CERN Preprint TH 3536, Février 1983.

[4] S. Albeverio, Ph. Blanchard, R. Hoegh-Krohn: Diffusion sur une variété riemanienne: barriéres infranchissables et applications, Colloque en l'honneur de Laurent Schwartz, Astérique Société Mathématique de France.

[5] Albeverio, S., Blanchard, Ph., Hoegh-Krohn, R.: Strata and voids in galactic structures - A probabilistic approach. BiBoS - Preprint (1989).

[6] Albeverio, S., Fenstad, J., E., Hoegh-Krohn, R., Lindstrom.: Nonstandard Methods in Stochastic Analysis and Mathematical Physics, volume 122 of Pure and Applied Mathematics. Academic Press, Orlando, Fla., 1986.

[7] Albeverio, S., Fukushima, M., Karwowski, W., Streit, L.: Capacity and Quantum Mechanical Tunneling, Comm. Math. Phys., **81** (1981) 501-513.

[8] Albeverio, S., Gesztesy, F., Hoegh-Krohn, R., Solvable Models in Quantum Mehcanics. Texts and Monographs in Physics. Springer, New York 1988.

[9] Albeverio, S., Gesztesy, F., Karwowski, W., Streit, L., On the connection between Schrödinger and Dirichlet forms. J. Math. Phys. **26**(1985) 2546-2553.

[10] S. Albeverio, A. Hahn, A. Sengupta : Chern-Simons Theory, Hida Distributions, and State Models, *IDAQP* **6** (Special Issue on Probability and Geometry) 65-81 (2003).

[11] S. Albeverio, T. Hida, J. Potthoff, M. Röckner, L. Streit, Dirichlet Forms in Terms of White Noise Analysis I:. Construction and Qft. Examples. Rev. Math. Physics **1**, 291-312 (1989).

[12] S. Albeverio, T. Hida, J. Potthoff, M. Röckner, L. Streit: Dirichlet Forms in Terms of White Noise Analysis II - Closability and Diffusion Processes - *Rev. Math. Phys.* **1**, 313 (1990)

[13] S. Albeverio, T. Hida, J. Potthoff, L. Streit: "The vacuum of the Hoegh-Krohn model as a generalized white noise functional" *Phys. Lett.* **B 217**, 511 (1989).

[14] Albeverio, S., Hoegh-Krohn, R., Point interactions as limit of short range interactions. J. Operator Theory **6** (1981), 313-339.

[15] Albeverio, S., Hoegh-Krohn, R., "Some Remarks on Dirichlet Forms and their Applications to Quantum Mechanics and Statistical Mechanics" and "Diffusion, quantum fields, and groups of mappings," in Functional Analysis and Markov Processes, Lecture Notes in Mathamatics, Vol. 923, edited by M. Fukushima (Springer, Berlin, 1982) pp. 120 and 133 respectively.

[16] Albeverio, S., Hoegh-Krohn, R., Let us talk about the wheather - A model for large scale atmospheric phenomena. Proc. Les Embiez (1980) Conference "Disordered Systems".

[17] S. Albeverio and R. Hoegh-Krohn, Uniqueness of the physical vacuum and the Wightman functions in the infinite volume limit for some non-polynomial interactions, Comun. Math. Phys. **30**, 171-200 (1973).

[18] S. Albeverio and R. Hoegh-Krohn, The scattering matrix for some non-polynomial interactions I and II, Helv. Phys. Acta **46**, 504-534, 535-545 (1973).

[19] S. Albeverio and R. Høegh-Krohn, The Wightman Axioms and the Mass Gap for Strong Interactions of Exponential Type in Two-Dimensional Space-Time. J. Funct. L Analysis **16**, 39-82 (1974)

[20] S. Albeverio and R. Høegh-Krohn, Quasi-invariant measures, symmetric diffusion processes and quantum fields, Les méthodes mathématiques de la théorie quantique de champs, Colloques Internationaux du C.N.R.S., no. 248, Marseille, 23-27 juin 1975, C.N.R.S., 1976.

[21] S. Albeverio, R. Høegh-Krohn, Dirichlet forms and diffusion processes on rigged Hilbert spaces. Z. Wahrsch. **40**, 1 (1977)

[22] S. Albeverio, R. Høegh-Krohn, The method of Dirichlet forms. Springer Lecture Notes in Physics 93, 250-258 (1979).

[23] S. Albeverio, R. Høegh-Krohn, Diffusion Fields, Quantum Fields and Fields with Values in Groups, pp. 1–98 in "Stochastic Analysis and Applications", Adv. in Probability, Edt. M. Pinsky, M. Dekker, New York (1984)

[24] Albeverio, S., Hoegh-Krohn, R., Streit, L., Energy forms, Hamilonians, and distorted Brownian paths, J. Math. Phys. **18**(1977) 907-917.

[25] Albeverio, S., Hoegh-Krohn, R., Streit, L., Regularisations of Hamiltonians and processes, J. Math. Phys. **21**(1980) 1636-1642.

[26] S. Albeverio , Y. Hu , M. Röckner , X. Zhou, Stochastic Quantization of the Two-Dimensional Polymer Measure. Appl. Math. Optim. **40**, 341 - 354 (1999).

[27] Albeverio, S., Karwowski, W., Röckner, M., Streit, L., Capacity, Dirichlet forms, Green's functions and Schrödinger operators" Research Notes in Mathematics **124** Infinite Dimensional Analysis and Stochastic Processes, Ed. S. Albeverio, pp. 197 - 215 Pitman Advanced Publishing Program, Boston-London-Melbourne (1985)

[28] S. Albeverio, Yuri G. Kondratiev, Michael Röckner, Infinite Dimensional Diffusions, Markov Fields, Quantum Fields and Stochastic Quantization. In: Stochastic Analysis and Applications in Physics, NATO ASI Series Volume 449, 1994, pp 1-34.

[29] Albeverio, S., Kurasov, P.: Singular Perturbations of Diferential Operators. London Mathematical Society Lecture Note Series 271, Cambridge University Press 2000.

[30] Albeverio, S., Kusuoka, S., Streit, L., Convergence of Dirichlet forms and associated Schrödinger operators, J. Funct. Anal., **68** (1986) 130-148.

[31] S. Albeverio and M. Roeckner, "Classical Dirichlet forms on topological vector spaces - Closability and a Cameron-Martin formula", J. Funct. Anal. **88**, 233-465 (1990).

[32] S. Albeverio and M. Roeckner, "Classical Dirichlet forms on topological vector spaces - Construction of an associated diffusion process", Prob. Th. Rel. Fields **83**, 405-434 (1989).

[33] S. Albeverio and M. Roeckner, "Stochastic differential equations in infinite dimensions: solutions via Dirichlet forms" Prob. Th. Rel. Fields, **89**, 347-386 (1991).

[34] S. Albeverio and M. Roeckner, "New developments in theory and applications of Dirichlet forms", in Stochastic Processes, Physics and Geometry, Proc. 2nd Int. Conj. Ascona- Locarno-Como 1988, eds. S. Albeverio, G. Casali, U. Cattaneo, D. Merlini, and D. Moresi, World Scientific, Singapore (1990).

[35] S. Albeverio and M. Roeckner, "Dirichlet forms, quantum fields and stochastic quantization", in Stochastic Analysis, Path Integration and Dynamics, ed. K. D. Elworthy, and J. C. Zambrini, Wiley, New York (1989).

[36] H. Alfven, G. Arrhenius: Structure and Evolutionary History of the Solar System. D. Reidel 1975

[37] H. Araki, Hamiltonian formalism and canonical commutation relations in quantum field theory, J. Math. Phys. **1**, 492 (1960).

[38] J. Brasche: "Störungen for Schrödinger operatoren durch Maße," Staatsexamenarbeit, Bielefeld; Perturbation of Schödinger Hamiltonians by measures-self-adjointness and lower semiboundedness, J. Math. Phys. **26** 621-626 (1985).

[39] F. Coester and R. Haag, Phys. Rev. **117**, 1137 (1960).

[40] D. Dohrn, F. Guerra: Nelson's stochastic mechanics on Remanian Manifolds. Lett. all Nouvo Cimento **22**, 121-127, (1978).

[41] M. de Faria, T. Hida, L. Streit, and H. Watanabe. Intersection local times as generalized white noise functionals. *Acta Applicandae Mathematica*, **46**, 351–362 (1997).

[42] J. Froehlich, Schwinger functions and their generating functionals. I Helv. Phys. Acta **47**, 265-305 (1973).

[43] J. Froehlich, Schwinger functions and their generating functionals. II. Markovian and generalized path space measures. Advances in Mathematics 23, 119–180 (1977).

[44] Fukushima, M., Dirichlet Forms and Markov Processes, North Holland, Amsterdam and Kodansha, Tokyo 1980.

[45] M. Fukushima, Y. Oshima, and M. Takeda, *Dirichlet forms and symmetric Markov processes.* De Gruyter Studies in Math., vol. 19, Walter de Gruyter, Berlin

[46] R. Haag, On quantum field theories, Matematisk-fysiske Meddelelser, **29**, 12 (1955).

[47] T. Hida, H. H. Kuo, J. Potthoff, L. Streit: *White Noise. An infinite dimensional calculus.* Kluwer Academic, 1993
[48] T. Hida, J. Potthoff, and L. Streit: Dirichlet Forms and White Noise Analysis. Commun. Math. Phys. **116**, 235-245 (1988).
[49] Huang, Z.Y. and J. A. Yan, Introduction to infinite dimensional stochastic analysis, Kluwer Academic Publishers, 2000
[50] S. L. Jaki: Planets and Planetarians. Scottish Acad. Press (1978).
[51] Kakutani, S., Two-dimentional Brownian motion and harmonic functions, Proc. Imp. Acad. Tokyo, **20** (1944), 706-714.
[52] T. Kato: Perturbation Theory for Linear Operators. Corrected Printing of the Second Edition. Springer-Verlag. Berlin Heidelberg New York 1980.
[53] A. Kolmogorov: Zur Umkehrbarkeit der statistischen Naturgesetze, Math. Ann. **112**, 155-160, 1936
[54] Yu. G. Kondratiev: "Nuclear spaces of entire functions in problems of infinite-dimensional analysis." *Sov. Math. Doklady* **22**, 588-592 (1980).
[55] Koshmanenko, V., Singular Quadratic Forms in Perturbation Theory. Kluwer Dordrecht/Boston/London 1999.
[56] H.-H. Kuo. *White Noise Distribution Theory.* CRC Press, Boca Raton, New York, London, Tokyo, 1996.
[57] W. H. McCrea, I. P. Williams: Segregation of materials in cosmology. Proc. R. Soc. **A 287**, 143-164, 1975.
[58] P. A. Meyer: Géometrie différentielle stochastique. Lecture Notes in Mathematics 921, Springer 1982.
[59] L. M. Morato: On the Dynamics of diffusions and related general electromagnetic potentials. J. Math. Phys. **23** 1020-1924 (1966).
[60] M. Naghdi: Nucleon-Nucleon Interaction: A Typical/Concise Review. arXiv:nucl-th/0702078v2 25 Apr 2013.
[61] E. Nelson: Quantum Fluctuations, Princeton Series in Physics Princeton University Press 1985.
[62] M. M. Nieto: The Titus-Bode Law of Planetary Distances, its History and Theory. Pergamon Press, Oxford 1972.
[63] N. Obata. *White Noise Calculus and Fock Spaces*, volume 1577 of LNM. Springer Verlag, Berlin, Heidelberg, New York, 1994.
[64] G. Parisi and Y.-S. Wu, Sci.Sinica **24**, 483 (1981)
[65] J. Potthoff, L. Streit: Invariant states on random and quantum fields: $\phi-$bounds and white noise analysis. *J. Funct. Anal.* **111**, 295-311 (1993)
[66] L Streit : A generalization of Haag's theorem. Il Nuovo Cimento A · 62, 673-680 (1969)
[67] Streit, L., Energy Forms, Schrödinger Theory, Processes. Phys. Rep. **77** (1981) 363-375.
[68] S. R. S. Varadhan. Appendix to "Euclidean quantum field theory" by K. Symanzik. In R. Jost, editor, *Local Quantum Theory*, New York, 1969. Academic Press.

Chapter 16

On a stability of heat kernel estimates under generalized non-local Feynman-Kac perturbations for stable-like processes

Daehong Kim*† and Kazuhiro Kuwae‡ §

We give a necessary and sufficient condition on the global stability of the fundamental solution for generalized Feynman-Kac semigroup of (relativistic) stable-like processes in terms of the gauge function of Feynman-Kac semigroup under mild conditions for related measures appeared in the perturbations.

Contents

1. Introduction

Let $\mathbf{X} = (X_t, \mathbf{P}_x)$ be a Brownian motion on a complete Riemannian manifold (M, g) with non-negative Ricci curvature. Then, the heat kernel $p_t(x, y)$ associated with $\frac{1}{2}\Delta_M$, the half of Laplace-Beltrami operator on (M, g), satisfies the following global Gaussian lower and upper bound, the so-called *Li-Yau estimate* ([16], see also [10]):

$$\frac{C_1 \exp\left(-c_1 \frac{d^2(x,y)}{t}\right)}{m(B(x, \sqrt{t}))} \leq p_t(x, y) \leq \frac{C_2 \exp\left(-c_2 \frac{d^2(x,y)}{t}\right)}{m(B(x, \sqrt{t}))} \tag{1.1}$$

for all $(t, x, y) \in]0, +\infty[\times M \times M$. Here C_1, c_1, C_2 and c_2 are positive constants and $m(B(x, r))$ denotes the volume of the geodesic ball of radius r centered at $x \in M$. For a signed smooth measure μ satisfying certain Green-tight Kato class condition, let $p_t^\mu(x, y)$ be the heat kernel associated with the Schrödinger operator $\frac{1}{2}\Delta_M + \mu$. In [24], the author gave an analytic condition on the potential μ under which $p_t^\mu(x, y)$ also satisfies the Li-Yau estimate (1.1). To establish the stability for

*partially supported by a Grant-in-Aid for Scientific Research (C) No. 23540147 from Japan Society for the Promotion of Science.
†Department of Mathematics and Engineering, Graduate School of Science and Technology, Kumamoto University, Kumamoto, 860-8555 Japan, daehong@gpo.kumamoto-u.ac.jp
‡partially supported by a Grant-in-Aid for Scientific Research (B) No. 22340036 from Japan Society for the Promotion of Science.
§Department of Mathematics and Engineering, Graduate School of Science and Technology, Kumamoto University, Kumamoto, 860-8555 Japan, kuwae@gpo.kumamoto-u.ac.jp

Li-Yau estimate, he used the uniqueness of Silverstein extensions for local Dirichlet forms and the identification of Girsanov transformed Dirichlet forms ([3]), as well as the analytic characterization for μ being gaugeable, equivalently, $\frac{1}{2}\Delta_M+\mu$ being subcritical ([2, 22]). In [25], the author extended the result for stability of global heat kernel estimates to symmetric jump processes considered in [4] under the additional assumption that μ is a positive smooth measure of finite energy integral.

Let $\mathbf{X}=(X_t,\mathbf{P}_x)$ be a symmetric α-stable-like process on $\mathbb{R}^d$ with $0<\alpha<2$ ([5]). It is well known that $\mathbf{X}$ is conservative ([6, Theorem 1.2]) and admits a locally Hölder continuous transition density function $p_t(x,y)$ on $]0,+\infty[\times\mathbb{R}^d\times\mathbb{R}^d$ ([5, Theorem 1.1], [6, Theorem 1.2]). Moreover, there are constants $C_2>C_1>0$ such that

$$C_1\left(t^{-d/\alpha}\wedge\frac{t}{|x-y|^{d+\alpha}}\right)\le p_t(x,y)\le C_2\left(t^{-d/\alpha}\wedge\frac{t}{|x-y|^{d+\alpha}}\right) \tag{1.2}$$

for all $(t,x,y)\in]0,+\infty[\times\mathbb{R}^d\times\mathbb{R}^d$ (see below for the details). Let $(\mathcal{E},\mathcal{F})$ be the associated Dirichlet form of $\mathbf{X}$. For a bounded function $u\in\dot{\mathcal{F}}_{\rm loc}$ (see §2 for the definition of $\dot{\mathcal{F}}_{\rm loc}$), we consider the transforms by the additive functionals $A_t:=N_t^u+A_t^\mu+A_t^F$ of the form

$$e_A(t):=\exp(A_t),\quad t\ge 0, \tag{1.3}$$

where N_t^u is the continuous additive functional of zero quadratic variation appeared in a Fukushima decomposition of $u(X_t)-u(X_0)$ (see (2.3)), A_t^μ is the continuous additive functional of $\mathbf{X}$ with a signed smooth measure μ in the strict sense as its Revuz measure and $A_t^F:=\sum_{0<s\le t}F(X_{s-},X_s)$ for a bounded symmetric Borel function F on $\mathbb{R}^d\times\mathbb{R}^d$ vanishing on the diagonal. Note that N^u is not necessarily of bounded variation. The transform (1.3) defines a semigroup, namely, the generalized non-local Feynman-Kac semigroup

$$P_t^Af(x):=\mathbf{E}_x[e_A(t)f(X_t)],\quad f\in\mathscr{B}_b(\mathbb{R}^d),\ t\ge 0. \tag{1.4}$$

The purpose of this paper is to give the analytic condition on u, μ and F under which $p_t^A(x,y)$ on $]0,+\infty[\times\mathbb{R}^d\times\mathbb{R}^d$ also satisfies the two-sided estimates (1.2), and the Feynman-Kac semigroup (1.4) is ultracontractive: there exists a constant $C>0$ such that

$$\|P_t^A\|_{1,\infty}\le Ct^{-d/\alpha},\quad t>0. \tag{1.5}$$

Here $\|P_t^A\|_{1,\infty}$ denotes the operator norm of P_t^A from $L^1(\mathbb{R}^d)$ to $L^\infty(\mathbb{R}^d)$. Indeed, we will show that our analytic condition on u, μ and F, the two-sided estimates (1.2) of $p_t^A(x,y)$ for all $(t,x,y)\in]0,\infty[\times\mathbb{R}^d\times\mathbb{R}^d$ and the ultracontractivity (1.5) are equivalent each others (Theorem 4.1). The stability for the Feynman-Kac semigroup $\{P_t^A\}_{t\ge0}$ is proved by applying the analytic characterization of gaugeability for generalized Feynman-Kac transforms studied in [12] to the methods developed by [23–25]. We will furthermore remove the extra assumptions on potentials such that they should be of positive smooth measures of finite energy integrals, by using the generalized Fukushima decomposition studied in [13, Theorem 6.2(2)].

2. Preliminaries

Let $c(x,y)$ be a symmetric measurable function on $\mathbb{R}^d\times\mathbb{R}^d$ which is bounded between two fixed constants $c_2 > c_1 > 0$, that is, $c_1 \le c(x,y) \le c_2$ for $x,y \in \mathbb{R}^d$. For $0<\alpha<2$, define

$$\mathcal{F} = \left\{ f \in L^2(\mathbb{R}^d) : \int_{\mathbb{R}^d}\int_{\mathbb{R}^d} \frac{(f(x)-f(y))^2}{|x-y|^{d+\alpha}} \mathrm{d}x\mathrm{d}y < \infty \right\}$$

$$\mathcal{E}(f,g) = \frac{1}{2}\int_{\mathbb{R}^d}\int_{\mathbb{R}^d} \frac{(f(x)-f(y))(g(x)-g(y))c(x,y)}{|x-y|^{d+\alpha}} \mathrm{d}x\mathrm{d}y, \quad f,g\in\mathcal{F}.$$

It is well known that $(\mathcal{E},\mathcal{F})$ is a regular symmetric Dirichlet form on $L^2(\mathbb{R}^d)$ and hence there is an associated symmetric Hunt process $\mathbf{X}=(X_t,\mathbf{P}_x)$ on $\mathbb{R}^d$ starting from every point in $\mathbb{R}^d$ except for an exceptional set of zero capacity (see, for instance, [5]). The process $\mathbf{X}$ is called an α-stable-like process. Note that $\mathbf{X}$ is nothing but the rotationally symmetric α-stable process on $\mathbb{R}^d$ when

$$c(x,y) = \frac{\alpha\Gamma\left(\frac{d+\alpha}{2}\right)}{2^{1-\alpha}\pi^{d/2}\Gamma\left(1-\frac{\alpha}{2}\right)}.$$

It is shown in [5, 6] that $\mathbf{X}$ is conservative and admits a locally Hölder continuous transition density function $p_t(x,y)$ on $]0,+\infty[\times\mathbb{R}^d\times\mathbb{R}^d$. The latter in particular implies that $\mathbf{X}$ can be modified to start from every point in $\mathbb{R}^d$ as a Feller process. Moreover, there are constants $C_2 > C_1 > 0$ such that (1.2) holds for all $(t,x,y)\in\,]0,1]\times\mathbb{R}^d\times\mathbb{R}^d$. The heat kernel $p_t^{(\delta)}(x,y) := \delta^{-d}p_{\delta^\alpha t}(\delta x,\delta y)$ of the scaled symmetric stable-like process $X_t^\delta := \delta^{-1}X_{\delta^\alpha t}$ also satisfies (1.2) for all $(t,x,y)\in]0,1]\times\mathbb{R}^d\times\mathbb{R}^d$, because the Dirichlet form $(\mathcal{E}^\delta,\mathcal{F}^\delta)$ associated to $\mathbf{X}^\delta=(X_t^\delta,\mathbf{P}_x)$ is given by

$$\mathcal{F}^\delta = \mathcal{F},$$

$$\mathcal{E}^\delta(f,g) = \frac{1}{2}\int_{\mathbb{R}^d}\int_{\mathbb{R}^d} \frac{(f(x)-f(y))(g(x)-g(y))c(\delta x,\delta y)}{|x-y|^{d+\alpha}} \mathrm{d}x\mathrm{d}y, \quad f,g\in\mathcal{F}.$$

Then, one can show that (1.2) is valid for all $(t,x,y)\in]0,+\infty[\times\mathbb{R}^d\times\mathbb{R}^d$ by use of the scaling property of the heat kernel $p_t^\alpha(x,y)$ of symmetric α-stable process: $p_t^\alpha(x,y)=\delta^{-d}p_{\delta^\alpha t}^\alpha(\delta x,\delta y)$, $x,y\in\mathbb{R}^d$.

For $\beta>0$, we define the β-order resolvent kernel

$$R_\beta(x,y) = \int_0^\infty e^{-\beta t}p_t(x,y)\mathrm{d}t, \quad x,y\in\mathbb{R}^d.$$

When $d>\alpha$ the process $\mathbf{X}$ is transient, hence we can define 0-order resolvent kernel $R(x,y) := R_0(x,y) < \infty$ for $x,y\in\mathbb{R}^d$ with $x\ne y$. $R(x,y)$ is called the Green function of $\mathbf{X}$. By virtue of (1.2), we can immediately check that there exist $C_4 > C_3 > 0$ such that

$$\frac{C_3}{|x-y|^{d-\alpha}} \le R(x,y) \le \frac{C_4}{|x-y|^{d-\alpha}}, \quad x,y\in\mathbb{R}^d. \tag{2.1}$$

For a non-negative Borel measure ν, we write $R_\beta\nu(x) := \int_{\mathbb{R}^d} R_\beta(x,y)\nu(dy)$, $R\nu(x) := R_0\nu(x)$ and $R_\beta f(x) = R_\beta\nu(x)$ when $\nu(dx) = f(x)dx$ for any $f \in \mathscr{B}_+(\mathbb{R}^d)$ or $f \in \mathscr{B}_b(\mathbb{R}^d)$. Here $\mathscr{B}_+(\mathbb{R}^d)$ (resp. $\mathscr{B}_b(\mathbb{R}^d)$) denotes the space of non-negative (resp. bounded) Borel functions on $\mathbb{R}^d$. The space of bounded continuous functions on $\mathbb{R}^d$ will be denoted as $C_b(\mathbb{R}^d)$. The process $\mathbf{X}$ has the resolvent strong Feller property (**(RSF)** in abbreviation), that is, $R_\beta(\mathscr{B}_b(\mathbb{R}^d)) \subset C_b(\mathbb{R}^d)$ for any $\beta > 0$. Let $\mathbb{R}^d \cup \{\partial\}$ be the one point compactification of $\mathbb{R}^d$. An increasing sequence $\{F_k\}$ of closed sets is said to be a *strict $\mathcal{E}$-nest* if $\mathbf{P}_x(\lim_{k\to\infty} \sigma_{F_k^c} = \infty) = 1$ a.e. $x \in \mathbb{R}^d$. Here $\sigma_{F_k^c} := \inf\{t > 0 \mid X_t \in \mathbb{R}^d \setminus F_k\}$ is the first hitting time of X_t to $F_k^c := \mathbb{R}^d \setminus F_k$. A function f defined on $\mathbb{R}^d$ (resp. $\mathbb{R}^d \cup \{\partial\}$) is said to be *$\mathcal{E}$-quasi continuous* (resp. *strictly $\mathcal{E}$-quasi continuous*) if there exists an $\mathcal{E}$-nest (resp. a strict $\mathcal{E}$-nest) $\{F_k\}$ of closed sets such that $f|_{F_k}$ (resp. $f|_{F_k\cup\{\partial\}}$) is continuous for each $k \in \mathbb{N}$. Since $\mathbf{X}$ is conservative, any $\mathcal{E}$-nest $\{F_k\}$ of closed sets is automatically a strict $\mathcal{E}$-nest. The regularity of the given Dirichlt form $(\mathcal{E},\mathcal{F})$ for $\mathbf{X}$ tells us that there exists an $\mathcal{E}$-nest of compact sets ([17, Chapter V, Proposition 2.12]), which yields that any $\mathcal{E}$-quasi continuous function on $\mathbb{R}^d$ can be regarded as a strictly $\mathcal{E}$-quasi continuous on $\mathbb{R}^d \cup \{\partial\}$. Denote by $QC(\mathbb{R}^d \cup \{\partial\})$ the family of all strictly $\mathcal{E}$-quasi continuous functions on $\mathbb{R}^d \cup \{\partial\}$. Let $(\mathcal{E},\mathcal{F}_e)$ be the extended Dirichlet space of $(\mathcal{E},\mathcal{F})$ and any element $f \in \mathcal{F}_e$ admits a strictly $\mathcal{E}$-quasi continuous version $\tilde{f}$ with $\tilde{f}(\partial) = 0$ (the proof is similar to the proof of [9, Theorem 2.1.3] if $\mathbf{X}$ is transient, for general case, it can be proved by way of time change method). Throughout this paper, we always take a strictly $\mathcal{E}$-quasi continuous version of the element of $\mathcal{F}_e$, that is, we omit *tilde* from $\tilde{f}$ for $f \in \mathcal{F}_e$.

A positive Radon measure ν on $\mathbb{R}^d$ is said to be *of finite energy integral* if there exists $C > 0$ such that

$$\int_{\mathbb{R}^d} |v(x)|\nu(dx) \le C\sqrt{\mathcal{E}_1(v,v)} \quad \text{for all } v \in \mathcal{F} \cap C_0(\mathbb{R}^d), \tag{2.2}$$

where $\mathcal{E}_1(\cdot,\cdot) := \mathcal{E}(\cdot,\cdot) + (\cdot,\cdot)$. Denote by $S_0(\mathbf{X})$ (resp. $S_{00}(\mathbf{X})$) the family of measures of finite energy integrals (resp. the family of finite measures of finite energy integrals with bounded potentials). A positive Radon measure ν on $\mathbb{R}^d$ is said to be *of 0-order finite energy integral* if there exists $C > 0$ satisfying (2.2) with $\mathcal{E}(\cdot,\cdot)$ in place of $\mathcal{E}_1(\cdot,\cdot)$, and we denote by $S_0^{(0)}(\mathbf{X})$ the family of measures of 0-order finite energy integrals. A positive Borel measure ν on $\mathbb{R}^d$ is said to be *smooth* if there exists an $\mathcal{E}$-nest $\{F_n\}_{n\ge1}$ of compact sets satisfying $\mathbf{1}_{F_n}\nu \in S_{00}(\mathbf{X})$ for each $n \ge 1$. We denote by $S(\mathbf{X})$ the family of all smooth measures. $S(\mathbf{X})$ contains all positive Radon measures on $\mathbb{R}^d$ charging no set of zero capacity. Let $S_1(\mathbf{X})$ be the family of positive smooth measures in the strict sense ([9]).

A measure $\nu \in S_1(\mathbf{X})$ is said to be *of Dynkin class* (resp. *Green-bounded*) *with respect to* $\mathbf{X}$ if $\sup_{x\in\mathbb{R}^d} R_\beta\nu(x) < \infty$ for some $\beta > 0$ (resp. $\sup_{x\in\mathbb{R}^d} R\nu(x) < \infty$). A measure $\nu \in S_1(\mathbf{X})$ is said to be *of Kato class (resp. of extended Kato class) with respect to* $\mathbf{X}$ if $\lim_{\beta\to\infty}\sup_{x\in\mathbb{R}^d} R_\beta\nu(x) = 0$ (resp. $\lim_{\beta\to\infty}\sup_{x\in\mathbb{R}^d} R_\beta\nu(x) < 1$). Denote by $S_D^1(\mathbf{X})$ (resp. $S_{D_0}^1(\mathbf{X})$) the family of measures of Dynkin class (resp. of

Green-bounded measures) and by $S^1_K(\mathbf{X})$ (resp. $S^1_{EK}(\mathbf{X})$) the family of measures of Kato class (resp. of extended Kato class). Clearly, $S^1_K(\mathbf{X}) \subset S^1_{EK}(\mathbf{X}) \subset S^1_D(\mathbf{X})$ and $S^1_{D_0}(\mathbf{X}) \subset S^1_D(\mathbf{X})$. Note that any measure $\nu \in S^1_D(\mathbf{X})$ is a positive Radon measure in view of the Stollmann-Voigt's inequality: $\int_{\mathbb{R}^d} u^2 d\nu \leq \|R_\beta \nu\|_\infty \mathcal{E}_\beta(u,u)$, $u \in \mathcal{F}, \beta \geq 0$ ([20, Theorem 3.1]). Conversely, any positive Radon measure ν satisfying $\sup_{x\in\mathbb{R}^d} R_\beta\nu(x) < \infty$ for some $\beta > 0$ always belongs to $S_1(\mathbf{X})$ in view of [15, Proposition 3.1].

We say that a positive continuous additive functional (PCAF in abbreviation) in the strict sense A^ν of $\mathbf{X}$ and a positive measure $\nu \in S_1(\mathbf{X})$ are in the Revuz correspondence if they satisfy for any bounded $f \in \mathscr{B}_+(\mathbb{R}^d)$,

$$\int_{\mathbb{R}^d} f(x)\nu(\mathrm{d}x) = \uparrow \lim_{t\downarrow 0} \frac{1}{t} \int_{\mathbb{R}^d} \mathbf{E}_x \left[\int_0^t f(X_s)\mathrm{d}A^\nu_s \right] \mathrm{d}x.$$

It is known that the family of equivalence classes of the set of PCAFs in the strict sense and the family of positive measures belonging to $S_1(\mathbf{X})$ are in one to one correspondence under the Revuz correspondence ([9, Theorem 5.1.4]).

It is known that the process $\mathbf{X}$ has a Lévy system (N,H) given by $H_t = t$ and $N(x,\mathrm{d}y) = 2c(x,y)|x-y|^{-(d+\alpha)}\mathrm{d}y$, that is, for any nonnegative Borel function ϕ on $\mathbb{R}^d \times \mathbb{R}^d$ vanishing on the diagonal and any $x \in \mathbb{R}^d$,

$$\mathbf{E}_x \left[\sum_{s\leq t} \phi(X_{s-}, X_s) \right] = \mathbf{E}_x \left[\int_0^t \int_{\mathbb{R}^d} \frac{2c(X_s,y)\phi(X_s,y)}{|X_s - y|^{d+\alpha}} \mathrm{d}y\mathrm{d}s \right].$$

To simplify notation, we will write

$$\mu_\phi(\mathrm{d}x) := \left\{ \int_{\mathbb{R}^d} \frac{2c(x,y)\phi(x,y)}{|x-y|^{d+\alpha}} \mathrm{d}y \right\} \mathrm{d}x$$

and set $\mu_{\langle u \rangle} := \mu_\phi$ for $\phi(x,y) = (u(x)-u(y))^2$.

A function f on $\mathbb{R}^d$ is said to be *locally in $\mathcal{F}$ in the broad sense* (denoted as $f \in \dot{\mathcal{F}}_{\mathrm{loc}}$) if there is an increasing sequence of finely open Borel sets $\{E_n\}$ with $\bigcup_{n=1}^\infty E_n = \mathbb{R}^d$ q.e. and for every $n \geq 1$, there is $f_n \in \mathcal{F}$ such that $f = f_n$ a.e. on E_n. A function f on $\mathbb{R}^d$ is said to be *locally in $\mathcal{F}$ in the ordinary sense* (denoted as $f \in \mathcal{F}_{\mathrm{loc}}$) if for any relatively compact open set G, there exists an element $f_G \in \mathcal{F}$ such that $f = f_G$ a.e. on G. Clearly $\mathcal{F}_{\mathrm{loc}} \subset \dot{\mathcal{F}}_{\mathrm{loc}}$. It is shown in [14, Theorem 4.1], $\mathcal{F}_e \subset \dot{\mathcal{F}}_{\mathrm{loc}}$.

Take a bounded $u \in \dot{\mathcal{F}}_{\mathrm{loc}} \cap QC(\mathbb{R}^d \cup \{\partial\})$. In [13, Theorem 6.2(1)], the authors proved under the condition $\mu_{\langle u \rangle} \in S^1_D(\mathbf{X})$ that the additive functional $u(X_t) - u(X_0)$ admits the following decomposition:

$$u(X_t) - u(X_0) = M^u_t + N^u_t \quad t \in [0, +\infty[\quad \mathbf{P}_x\text{-a.s.} \tag{2.3}$$

for q.e. $x \in \mathbb{R}^d$, where M^u is a square integrable martingale additive functional, $\mu_{\langle u \rangle}$ is the Revuz measure associated with the quadratic variational processe (or the sharp bracket PCAF) $\langle M^u \rangle$ of M^u, and N^u is a continuous additive functional (CAF in abbreviation) locally of zero energy. We moreover note that (2.3) holds for all $x \in \mathbb{R}^d$ as the strict decomposition provided u is (nearly) Borel and finely continuous on $\mathbb{R}^d$ ([13, Theorem 6.2(2)]). Note that N^u is not a process of finite variation in general. The martingale part M^u is characterized by

$$M_t^u = \lim_{n\to\infty} \left\{ \sum_{s\le t} (u(X_s) - u(X_{s-}))\mathbf{1}_{\{|u(X_s)-u(X_{s-})|>1/n\}} \right.$$
$$\left. - \int_0^t \int_{\{y\in\mathbb{R}^d: |u(y)-u(X_s)|>1/n\}} \frac{2c(X_s,y)(u(y)-u(X_s))}{|X_s-y|^{d+\alpha}} dyds \right\}$$

$\mathbf{P}_x$-a.s. for q.e. $x \in \mathbb{R}^d$ (cf. [9, Theorem A.3.9]). Then

$$\mu_{\langle u\rangle}(dx) = \left(\int_{\mathbb{R}^d} \frac{2c(x,y)(u(x)-u(y))^2}{|x-y|^{d+\alpha}} dy \right) dx.$$

Note that $\mathcal{E}(f,f) = \frac{1}{2}\mu_{\langle f\rangle}(\mathbb{R}^d)$ provided $f \in \mathcal{F}_e$.

3. Analytic characterization of gaugeability and subcriticality

Throughout this paper, we treat only transient symmetric α-stable-like process $\mathbf{X}$. In particular, $d > \alpha$ and $\mathbf{X}$ is conservative. Let $\mu := \mu_1 - \mu_2$ be the signed smooth measure in the strict sense for given $\mu_1, \mu_2 \in S_1(\mathbf{X})$ and F a bounded symmetric Borel function on $\mathbb{R}^d \times \mathbb{R}^d$ vanishing on the diagonal. Set $F_+ := F \vee 0$ and $F_- := (-F) \vee 0$. We say that $F := F_+ - F_-$ is in the *class* $J_1(\mathbf{X})$ (resp. $J_D^1(\mathbf{X})$ and $J_{D_0}^1(\mathbf{X})$) if $\mu_{F_+ + F_-}$ belongs to $S_1(\mathbf{X})$ (resp. $S_D^1(\mathbf{X})$ and $S_{D_0}^1(\mathbf{X})$). Any such an F has an expression of the form $F = F_1 - F_2$, where each F_i $(i = 1, 2)$ is a symmetric non-negative bounded Borel function on $\mathbb{R}^d \times \mathbb{R}^d$ vanishing on the diagonal. Indeed, for any non-negative symmetric bounded Borel function ϕ having the same property as for F, $F_1 := F_+ + \phi$ and $F_2 := F_- + \phi$ have those properties. Note that if $F_1 + F_2 \in J_1(\mathbf{X})$ (resp. $J_D^1(\mathbf{X})$ and $J_{D_0}^1(\mathbf{X})$), then $F \in J_1(\mathbf{X})$ (resp. $J_D^1(\mathbf{X})$ and $J_{D_0}^1(\mathbf{X})$). In this case, the following A^F can be defined as an additive functional in the strict sense:

$$A_t^F = A_t^{F_1} - A_t^{F_2}, \quad A_t^{F_i} := \sum_{0<s\le t} F_i(X_{s-}, X_s) \quad (i = 1, 2).$$

For a bounded $u \in \dot{\mathcal{F}}_{\rm loc}$ with $\mu_{\langle u}$ $\in S_1(\mathbf{X})$ and $F_1 + F_2 \in J_1(\mathbf{X})$, we set

$$F^u(x,y) := F(x,y) + \{-u(y) - (-u(x))\} = F(x,y) + u(x) - u(y),$$
$$F_1^u(x,y) := F_1(x,y) + \{-u(y) - (-u(x))\} = F_1(x,y) + u(x) - u(y),$$

and $G^u = e^{F^u} - 1$ (resp. $G_1^u = e^{F_1^u} - 1$) with identifying $F^0 = F$ (resp. $F_1^0 = F_1$) and $G^0 = G := e^F - 1$ (resp. $G_1^0 = G_1 := e^{F_1} - 1$). Since $(F^u)^2 \in J_1(\mathbf{X})$, one can consider a purely discontinuous locally square integrable local martingale additive functional M^{F^u} defined by

$$M_t^{F^u} = M_t^F + M_t^{-u}, \tag{3.1}$$

where $M_t^F = A_t^F - A_t^{\mu_F}$, $t \in [0,+\infty[$. Moreover, since $|G^u(x,y) - F^u(x,y)| \leq \frac{1}{2}e^{\|F^u\|_\infty}|F^u(x,y)|^2$ and

$$|G^u(x,y)|^2 \leq \left(\frac{\|F^u\|_\infty e^{|F^u(x,y)|}}{2}|F^u(x,y)| + |F^u(x,y)|\right)^2$$
$$\leq \left(\frac{\|F^u\|_\infty e^{\|F^u\|_\infty}}{2} + 1\right)^2 |F^u(x,y)|^2,$$

we see that $G^u - F^u \in J_1(\mathbf{X})$ and $(G^u)^2 \in J_1(\mathbf{X})$, respectively. Therefore we can also consider a purely discontinuous locally square integrable local martingale additive functional M^{G^u} defined by

$$M_t^{G^u} = M_t^{F^u} + \sum_{s \leq t}(G^u - F^u)(X_{s-}, X_s) - \int_0^t \int_{\mathbb{R}^d} \frac{2c(X_s,y)(G^u - F^u)(X_s,y)}{|X_s - y|^{d+\alpha}} \mathrm{d}y\mathrm{d}s \tag{3.2}$$

for $t \in [0,+\infty[$. Let $Y_t := \mathrm{Exp}(M^{G^u})_t$ be the Doléans-Dade exponential of $M_t^{G^u}$, that is, Y_t is the unique solution of $Y_t = 1 + \int_0^t Y_{s-}\mathrm{d}M_s^{G^u}$ for $t \in [0,+\infty[$, $\mathbf{P}_x$-a.s. Then Y_t can be represented as

$$Y_t = \exp\left(M_t^{F^u} - \int_0^t \int_{\mathbb{R}^d} \frac{2c(X_s,y)(G^u - F^u)(X_s,y)}{|X_s - y|^{d+\alpha}} \mathrm{d}y\mathrm{d}s\right) \tag{3.3}$$

for $t \in [0,+\infty[$. Note that Y_t is a positive local martingale, therefore it is a supermartingale multiplicative functional for all $t \in [0,+\infty[$.

Let $\mathbf{Y} = (\Omega, \widetilde{\mathscr{F}}_\infty, \widetilde{\mathscr{F}}_t, \widetilde{X}_t, \mathbf{P}_x^Y)$ be the transformed process of $\mathbf{X}$ by Y_t. The transition semigroup $\{P_t^Y\}_{t\geq 0}$ of $\mathbf{Y}$ is defined by $P_t^Y f(x) = \mathbf{E}_x^Y[f(\widetilde{X}_t)] := \mathbf{E}_x[Y_t f(X_t)]$. When $F_2 = 0$, we denote Y_t (resp. $\mathbf{Y}$) by Y_t^1 (resp. $\mathbf{Y}^1$).

The following results are due to [12, Theorem 3.2] (see also [7, Theorem 3.4] for $u \in \mathcal{F}_e$, $\mu \equiv 0$ and $F \equiv 0$ in general settings).

Theorem 3.1 ([12, Theorem 3.2]). *Let $u \in \dot{\mathcal{F}}_{\mathrm{loc}} \cap QC(\mathbb{R}^d \cup \{\partial\})$ be a bounded function satisfying $\mu_{\langle u \rangle} \in S_D^1(\mathbf{X})$ and let $F \in J_1(\mathbf{X})$. Then the transformed process $\mathbf{Y}$ is an $e^{-2u}\mathrm{d}x$-symmetric Hunt process on $\mathbb{R}^d$. Moreover, let $(\mathcal{E}^Y, \mathcal{F}^Y)$ be the Dirichlet form of $\mathbf{Y}$ on $L^2(\mathbb{R}^d; e^{-2u}\mathrm{d}x)$. Then*

$$\mathcal{F} = \mathcal{F}^Y \text{ and } \mathcal{E}^Y(\varphi,\psi) = \frac{1}{2}\int_{\mathbb{R}^d}\int_{\mathbb{R}^d} \frac{(\varphi(x)-\varphi(y))(\psi(x)-\psi(y))e^{F(x,y)-u(x)-u(y)}c(x,y)}{|x-y|^{d+\alpha}}\mathrm{d}x\mathrm{d}y \tag{3.4}$$

for any $\varphi, \psi \in \mathcal{F}$. In particular, $\mathcal{F}_e = \mathcal{F}_e^Y$.

In view of Theorem 3.1 together with the boundedness of u and F, $\mathbf{Y}$ and $\mathbf{Y}^1$ are time changed processes by a bounded positive continuous additive functional of a symmetric α-stable-like process. Thus $\mathbf{Y}$ (resp. $\mathbf{Y}^1$) is also conservative and admits a locally Hölder continuous transition density function $p_t^Y(x,y)$ (resp. $p_t^{Y^1}(x,y)$) on

$]0,+\infty[\times\mathbb{R}^d\times\mathbb{R}^d$ satisfying (1.2) for all $(t,x,y)\in]0,+\infty[\times\mathbb{R}^d\times\mathbb{R}^d$. The Green function $R^Y(x,y)$ of $\mathbf{Y}$ also satisfies the estimate

$$\frac{C_5}{|x-y|^{d-\alpha}}\leq R^Y(x,y)\leq\frac{C_6}{|x-y|^{d-\alpha}},\quad x,y\in\mathbb{R}^d \tag{3.5}$$

for some $C_6>C_5>0$. In particular, $\mathbf{Y}$ (also $\mathbf{Y}^1$) satisfies **(RSF)**.

Now we introduce some notions of Green-tight Kato class measures in the strict sense.

Definition 3.1 (Green-tight Kato class measures). *Let* $\nu\in S_1(\mathbf{X})$.

(1) *ν is said to be* of Green-tight Kato class with respect to $\mathbf{X}$ *if* $\nu\in S^1_K(\mathbf{X})$ *and for any* $\varepsilon>0$ *there exists a compact subset* $K=K(\varepsilon)$ *of* $\mathbb{R}^d$ *such that*

$$\sup_{x\in\mathbb{R}^d}R(\mathbf{1}_{K^c}\nu)(x)<\varepsilon.$$

(2) *ν is said to be* of semi-Green-tight extended Kato class with respect to $\mathbf{X}$ *if* $\nu\in S^1_{EK}(\mathbf{X})$ *and there exists a compact subset* K *of* $\mathbb{R}^d$ *such that*

$$\sup_{x\in\mathbb{R}^d}R(\mathbf{1}_{K^c}\nu)(x)<1.$$

(3) *ν is said to be* of Green-tight Kato class in the sense of Chen with respect to $\mathbf{X}$ *if for any* $\varepsilon>0$ *there exists a Borel subset* $K=K(\varepsilon)$ *of* $\mathbb{R}^d$ *with* $\nu(K)<\infty$ *and a constant* $\delta>0$ *such that for all measurable set* $B\subset K$ *with* $\nu(B)<\delta$,

$$\sup_{x\in\mathbb{R}^d}R(\mathbf{1}_{K^c\cup B}\nu)(x)<\varepsilon.$$

(4) *ν is said to be* of semi-Green-tight extended Kato class in the sense of Chen with respect to $\mathbf{X}$ *if there exists a Borel subset* K *of* $\mathbb{R}^d$ *with* $\nu(K)<\infty$ *and a constant* $\delta>0$ *such that for all measurable set* $B\subset K$ *with* $\nu(B)<\delta$,

$$\sup_{x\in\mathbb{R}^d}R(\mathbf{1}_{K^c\cup B}\nu)(x)<1.$$

We denote by $S^1_{K_\infty}(\mathbf{X})$ (resp. $S^1_{CK_\infty}(\mathbf{X})$, $S^1_{K_1}(\mathbf{X})$, $S^1_{CK_1}(\mathbf{X})$) the family of Green-tight Kato class measures (resp. the family of Green-tight Kato class measures in the sense of Chen, the family of semi-Green-tight extended Kato class measures, the family of semi-Green-tight Kato class measures in the sense of Chen) with respect to $\mathbf{X}$.

Remark 3.1. It is known that $S^1_{CK_\infty}(\mathbf{X})\subset S^1_{CK_1}(\mathbf{X})\subset S^1_{D_0}(\mathbf{X})\cap S^1_{EK}(\mathbf{X})$, $S^1_{CK_\infty}(\mathbf{X})\subset S^1_{K_\infty}(\mathbf{X})\subset S^1_K(\mathbf{X})$ and $S^1_{CK_1}(\mathbf{X})\subset S^1_{K_1}(\mathbf{X})\subset S^1_{EK}(\mathbf{X})$ (see [2], [12]). However, we have $S^1_{K_\infty}(\mathbf{X})=S^1_{CK_\infty}(\mathbf{X})$ and $S^1_{CK_1}(\mathbf{X})\cap S^1_K(\mathbf{X})=S^1_{K_1}(\mathbf{X})\cap S^1_K(\mathbf{X})$ from **(RSF)** for $\mathbf{X}$ ([12, Lemma 4.1]).

Define the measures $\bar{\mu}_1 := \mu_1 + \mu_{G^u - F^u + F_1}$, $\bar{\mu}_2 := \mu_2 + \mu_{F_2}$, $\bar{\mu}_1^* := \mu_1 + \mu_{G_1^u - F_1^u + F_1}$ and $\bar{\mu}_2^* := \mu_2$. Then

$$\begin{aligned}\bar{\mu}(\mathrm{d}x) :=& \ \bar{\mu}_1(\mathrm{d}x) - \bar{\mu}_2(\mathrm{d}x)\\ =& \ \mu(\mathrm{d}x) + \left\{\int_{\mathbb{R}^d} \frac{2c(x,y)(G^u(x,y) - u(x) + u(y))}{|x-y|^{d+\alpha}}\mathrm{d}y\right\}\mathrm{d}x,\\ \bar{\mu}^*(\mathrm{d}x) :=& \ \bar{\mu}_1^*(\mathrm{d}x) - \bar{\mu}_2^*(\mathrm{d}x)\\ =& \ \mu(\mathrm{d}x) + \left\{\int_{\mathbb{R}^d} \frac{2c(x,y)(G_1^u(x,y) - u(x) + u(y))}{|x-y|^{d+\alpha}}\mathrm{d}y\right\}\mathrm{d}x.\end{aligned}$$

Consider the non-local Feynman-Kac transforms by the additive functionals $A_t := N_t^u + A_t^\mu + A_t^F$ of the form (1.3). By (3.3), we see that for all $t \in [0, +\infty[$,

$$e_A(t) = e^{u(X_t) - u(X_0)} Y_t \exp\left(A_t^{\bar{\mu}}\right) = e^{u(X_t) - u(X_0)} Y_t^1 \exp\left(A_t^{\bar{\mu}_1} - A_t^{F_2}\right),$$

which implies that for $x \in \mathbb{R}^d$ and $f \in \mathscr{B}_+(\mathbb{R}^d)$,

$$\begin{aligned}P_t^A f(x) &= e^{-u(x)} \mathbf{E}_x^Y \left[\exp\left(A_t^{\bar{\mu}}\right)(e^u f)(\widetilde{X}_t)\right]\\ &= e^{-u(x)} \mathbf{E}_x^{Y^1} \left[\exp\left(A_t^{\bar{\mu}_1} - A_t^{F_2}\right)(e^u f)(\widetilde{X}_t)\right].\end{aligned}$$

It is easy to see that $\bar{\mu}_1, \bar{\mu}_2 \in S_D^1(\mathbf{X})$ whenever $\mu_1, \mu_2, \mu_{\langle u \rangle} \in S_D^1(\mathbf{X})$ and $F_1 + F_2 \in J_D^1(\mathbf{X})$ hold. Moreover we note that if $\nu \in S_D^1(\mathbf{X})$ (resp. $\nu \in S_{D_0}^1(\mathbf{X})$), then $\int_{\mathbb{R}^d} f^2 d\nu \le \|R_1\nu\|_\infty \mathcal{E}_1(f,f)$ for $f \in \mathcal{F}$ (resp. $\int_{\mathbb{R}^d} f^2 d\nu \le \|R\nu\|_\infty \mathcal{E}(f,f)$ for $f \in \mathcal{F}_e$), hence $\mathcal{F} \subset L^2(\mathbb{R}^d;\nu)$ (resp. $\mathcal{F}_e \subset L^2(\mathbb{R}^d;\nu)$) from [20, Theorem 3.1]. It is shown in [12] that under $\bar{\mu}_1 \in S_{EK}^1(\mathbf{X})$, $\mu_2 + \mu_{F_2} \in S_1(\mathbf{X})$ and $\alpha \ge 0$, the symmetric α-order resolvent kernel $R_\alpha^A(x,y)$ of the Feynman-Kac semigroup P_t^A can be defined for all $x, y \in \mathbb{R}^d$ (possibly infinite), it is α-excessive with respect to $(P_t^A)_{t\ge 0}$ in x (and in y), and it satisfies a resolvent equation as in [9, (4.2.12)]. Moreover, for each $x \in \mathbb{R}^d$, $y \mapsto R_\alpha^A(x,y)$ is finely continuous (see [12, Lemma 6.1]). We write $R^A(x,y) := R_0^A(x,y)$ for $x, y \in \mathbb{R}^d$.

Define the quadratic form $\mathcal{Q}$ by

$$\mathcal{Q}(f,g) := \mathcal{E}(f,g) + \mathcal{E}(u, fg) - \mathcal{H}(f,g), \tag{3.6}$$

where

$$\begin{aligned}\mathcal{H}(f,g) &:= \int_{\mathbb{R}^d} f(x)g(x)\mu(\mathrm{d}x) + \int_{\mathbb{R}^d}\int_{\mathbb{R}^d} \frac{2f(x)g(y)c(x,y)G(x,y)}{|x-y|^{d+\alpha}}\mathrm{d}x\mathrm{d}y,\\ \mathcal{E}(u, fg) &:= \int_{\mathbb{R}^d} f(x)\int_{\mathbb{R}^d} \frac{(g(x)-g(y))(u(x)-u(y))c(x,y)}{|x-y|^{d+\alpha}}\mathrm{d}x\mathrm{d}y\\ &\quad + \int_{\mathbb{R}^d} g(x)\int_{\mathbb{R}^d} \frac{(f(x)-f(y))(u(x)-u(y))c(x,y)}{|x-y|^{d+\alpha}}\mathrm{d}x\mathrm{d}y.\end{aligned}$$

Then it is well-defined for $f, g \in \mathcal{F}$ provided $\mu_{\langle u \rangle} \in S_D^1(\mathbf{X})$, $\mu_1 + \mu_2 \in S_D^1(\mathbf{X})$ and $F_1 + F_2 \in J_D^1(\mathbf{X})$. Moreover, $\mathcal{Q}$ is extended to $\mathcal{F}_e \times \mathcal{F}_e$ with the same expression

(3.6) provided $\mu_{\langle u\rangle} \in S^1_{D_0}(\mathbf{X})$, $\mu_1+\mu_2 \in S^1_{D_0}(\mathbf{X})$ and $F_1+F_2 \in J^1_{D_0}(\mathbf{X})$. Let $\mathbf{F}$ be a non-local linear operator defined by

$$\mathbf{F}f(x) := \int_{\mathbb{R}^d} \frac{2c(x,y)G(x,y)f(y)}{|x-y|^{d+\alpha}}\mathrm{d}y, \quad f \in \mathscr{B}_b(\mathbb{R}^d).$$

The L^2-generator $\mathcal{L}^{\mathcal{Q}}$ associated to $(\mathcal{Q}, D(\mathcal{Q}))$ can be expressed in the formal form $\mathcal{L}^{\mathcal{Q}} := \mathcal{L}^{\mathcal{E}} + \mathcal{L}^{\mathcal{E}}u + \mu + d\mathbf{F}$, where $\mathcal{L}^{\mathcal{E}}$ is the infinitesimal generator for the semigroup of $\mathbf{X}$ and $d\mathbf{F}$ denotes the signed-measure valued operator defined by $d\mathbf{F}f := \mathbf{F}f(x)\mathrm{d}x$. The 0-order resolvent kernel $R^A(x,y)$ associated to $(P_t^A)_{t>0}$ can be regarded as the Green kernel of the L^2-generator $\mathcal{L}^{\mathcal{Q}}$. Under $\mu_{\langle u\rangle} + \mu_1 + \mu_2 + \mu_{F_1+F_2} \in S^1_{D_0}(\mathbf{X})$, we set

$$\lambda^{\mathcal{Q}}(\bar{\mu}_1) := \inf\left\{ \mathcal{Q}(f,f) \;\middle|\; f \in C_0^\infty(\mathbb{R}^d),\ \int_{\mathbb{R}^d} f^2 \mathrm{d}\bar{\mu}_1 = 1 \right\}. \tag{3.7}$$

Let $S^1_{CS_1}(\mathbf{X})$ be the class of semi-conditionally Green-tight measures of extended Kato class (see [12, Definition 6.2] for its definition).

Now we can see from [12, Theorem 2] that the following theorem holds in view of (3.5) and Remark 3.1.

Theorem 3.2. *Let $u \in \mathcal{F}_{\mathrm{loc}} \cap QC(\mathbb{R}^d \cup \{\partial\})$ be a bounded finely continuous (nearly) Borel function on $\mathbb{R}^d$. Assume $\mu_1 \in S^1_{CS_1}(\mathbf{X})$, $\mu_{\langle u\rangle} + \mu_{F_1} \in S^1_{K_\infty}(\mathbf{X})$ and $\mu_2 + \mu_{F_2} \in S^1_{D_0}(\mathbf{X})$. Then the following are equivalent:*

(1) The functional (1.3) *is gaugeable, that is,* $\sup_{x\in\mathbb{R}^d} \mathbf{E}_x[e_A(\zeta)] < \infty$.

(2) The functional (1.3) *is conditionally gaugeable, that is, it holds that* $\sup_{(x,y)\in\mathbb{R}^d\times\mathbb{R}^d, x\neq y} \mathbf{E}_x^y[e_A(\zeta^y)] < \infty$.

(3) For each $x \in \mathbb{R}^d$, $R^A(x,y) < \infty$ for a.e. $y \in \mathbb{R}^d \setminus \{x\}$.

(4) For all $x, y \in \mathbb{R}^d$ with $x \neq y$, $R^A(x,y) < \infty$.

(5) $\lambda^{\mathcal{Q}}(\bar{\mu}_1) > 0$.

4. Stability for generalized Feynman-Kac semigroups

In this section, we state our main results. We prove under some conditions that the analytic characterization $\lambda^{\mathcal{Q}}(\bar{\mu}_1) > 0$ is a necessary and sufficient condition for the heat kernel $p_t^A(x,y)$ (resp. the semigroup $\{P_t^A\}_{t\geq 0}$) also satisfies the two-sided estimates (1.2) for all $(t,x,y) \in]0,+\infty[\times\mathbb{R}^d\times\mathbb{R}^d$ (resp. the ultracontractivity (1.5)), respectively:

Theorem 4.1. *Let $u \in \mathcal{F}_{\mathrm{loc}} \cap QC(\mathbb{R}^d \cup \{\partial\})$ be a bounded finely continuous (nearly) Borel function on $\mathbb{R}^d$. Assume $\mu_1 \in S^1_{CS_1}(\mathbf{X})$, $\mu_{\langle u\rangle} + \mu_{F_1} \in S^1_{K_\infty}(\mathbf{X})$ and $\mu_2 + \mu_{F_2} \in S^1_{D_0}(\mathbf{X})$. Then the following are equivalent each other.*

(1) $\lambda^{\mathcal{Q}}(\bar{\mu}_1) > 0$.

(2) There exists a constant $C = C(\alpha, d, u, \mu, F) > 0$ *such that* $\|P_t^A\|_{1,\infty} \le Ct^{-d/\alpha}$ *for all* $t > 0$.

(3) There exists a constant $C_i = C_i(\alpha, d, u, \mu, F) > 0,\ i = 1, 2$ *such that*

$$C_1\left(t^{-d/\alpha} \wedge \frac{t}{|x-y|^{d+\alpha}}\right) \le p_t^A(x,y) \le C_2\left(t^{-d/\alpha} \wedge \frac{t}{|x-y|^{d+\alpha}}\right)$$

for all $(t, x, y) \in]0, +\infty[\times \mathbb{R}^d \times \mathbb{R}^d$.

Proof. (3) $\Longrightarrow$ (2) : This is clear.

(2) $\Longrightarrow$ (1) : Under the condition, we have $\mu_{\langle u} + \mu_{F_1} \in S^1_{K_\infty}(\mathbf{X}) \subset S^1_K(\mathbf{X})$ and $\bar{\mu}_1^* \in S^1_{CK_1}(\mathbf{X}) \subset S^1_{EK}(\mathbf{X})$. Then one can apply [12, Lemma 4.6(3)] so that $e^{-2u}\bar{\mu}_1^* \in S^1_{EK}(\mathbf{Y}^1)$. For sufficiently small $t_0 > 0$ with $\mathbf{E}_x^{Y^1}[A_{t_0}^{\bar{\mu}_1^*}] < 1$, we have

$$\sup_{x\in\mathbb{R}^d} \mathbf{E}_x^{Y^1}\left[e^{A_{t_0}^{\bar{\mu}_1^*}}\right] \le \frac{1}{1 - \sup_{x\in\mathbb{R}^d}\mathbf{E}_x^{Y^1}[A_{t_0}^{\bar{\mu}_1^*}]} =: C_0 < \infty$$

by Khasminskii's lemma. This yields that there exists $C_1 > 0$ and $\beta > 0$ such that $\sup_{x\in\mathbb{R}^d}\mathbf{E}_x^{Y^1}[e^{A_t^{\bar{\mu}_1^*}}] \le C_1 e^{\beta} t$. Consequently, $\sup_{z\in\mathbb{R}^d}\mathbf{E}_z[e_A(t)] \le C_1 e^{2\|u\|_\infty} e^{\beta t}$, hence $\sup_{z\in\mathbb{R}^d} R_\alpha^A 1(z) \le \frac{C_1 e^{2\|u\|_\infty}}{\alpha-\beta}$ for $\alpha > \beta$. Moreover, Lebesgue's dominated convergence theorem tells us the right continuity of $t \mapsto P_t^A f(x) = e^{-u(x)}\mathbf{E}_x^{Y^1}[e^{u(X_t)}e^{A_t^{\bar{\mu}_1^*} - A_t^{F_2}} f(X_t)]$ for each fixed $x \in \mathbb{R}^d$ and non-negative $f \in C_0(\mathbb{R}^d)$. From this, for each non-negative $f \in C_0(\mathbb{R}^d)$, $P_t^A f(x) \le \|f\|_1 \|P_t^A\|_{1,\infty} \le C\|f\|_1 t^{-d/\alpha}$ a.e. $x \in \mathbb{R}^d$ for each $t > 0$ implies that $P_t^A f(x) \le C\|f\|_1 t^{-d/\alpha}$, $t > 0$ for a.e. $x \in \mathbb{R}^d$. Then, for any non-negative $f \in C_0(\mathbb{R}^d)$, we have

$$\begin{aligned}
R^A f(x) &= \int_0^{t_0} P_t^A f(x)\mathrm{d}t + \int_{t_0}^\infty P_t^A f(x)\mathrm{d}t \\
&\le \|f\|_\infty \int_0^{t_0} \mathbf{E}_x[e_A(t)]\mathrm{d}t + C\|f\|_{L^1}\int_{t_0}^\infty t^{-d/\alpha}\mathrm{d}t \\
&\le \|f\|_\infty e^{2\|u\|_\infty}\int_0^{t_0}\mathbf{E}_x^{Y^1}\left[e^{A_t^{\bar{\mu}^*} - A_t^{F_2}}\right]\mathrm{d}t + C\|f\|_{L^1}\int_{t_0}^\infty t^{-d/\alpha}\mathrm{d}t \\
&\le \|f\|_\infty e^{2\|u\|_\infty}\int_0^{t_0}\mathbf{E}_x^{Y^1}\left[e^{A_t^{\bar{\mu}_1^*}}\right]\mathrm{d}t + C\|f\|_{L^1}\int_{t_0}^\infty t^{-d/\alpha}\mathrm{d}t \\
&\le \|f\|_\infty e^{2\|u\|_\infty} C_0 t_0 + C\|f\|_{L^1}\int_{t_0}^\infty t^{-d/\alpha}\mathrm{d}t =: M < \infty
\end{aligned}$$

for a.e. $x \in \mathbb{R}^d$, that is, $R^A f \le M$ a.e. on $\mathbb{R}^d$. Then $R^A f(x) = R_\alpha^A f(x) + \alpha R_\alpha^A R^A f(x) \le \|f\|_\infty \sup_{z\in\mathbb{R}^d} R_\alpha^A 1(z) + \alpha M \sup_{z\in\mathbb{R}^d} R_\alpha^A 1(z) < \infty$ for each $x \in \mathbb{R}^d$. This implies that for each $x \in \mathbb{R}^d$, $R^A(x, y) < \infty$ for a.e. $y \in \mathbb{R}^d \setminus \{x\}$. Thus we obtain (1) by way of Theorem 3.2 (3)$\Longrightarrow$(1).

(1) $\Longrightarrow$ (3): Suppose $\lambda^{\mathcal{Q}}(\bar{\mu}_1) > 0$, equivalently the gaugeability $\sup_{x\in\mathbb{R}^d}\mathbf{E}_x[e_A(\infty)] < \infty$ holds by Theorem 3.2. This is equivalent to $\sup_{x\in\mathbb{R}^d}\mathbf{E}_x^{Y^1}[e^{A_\infty^{\bar{\mu}^*} - A_\infty^{F_2}}] = \sup_{x\in\mathbb{R}^d} e^{u(x)}\mathbf{E}_x[e^{-u(X_\infty)}e_A(\infty)] < \infty$ in view of [12,

Lemma 4.10]. Note that under the assumptions on measures, we have $\bar{\mu}_1^* \in S^1_{CK_1}(\mathbf{X}) \subset S^1_{D_0}(\mathbf{X})$ and $\bar{\mu}_2^* \in S^1_{D_0}(\mathbf{X})$. So there exists an $\mathcal{E}$-nest $\{F_n\}$ of compact sets such that $\mathbf{1}_{F_n}\bar{\mu}_1^* \in S_0^{(0)}(\mathbf{X})$ and $\bar{\mu}_2^* \in S^1_{D_0}(\mathbf{X})$ for each $n \in \mathbb{N}$, hence $e^{-2u}\mathbf{1}_{F_n}\bar{\mu}_1^* \in S_0^{(0)}(\mathbf{Y}^1)$ and $e^{-2u}\bar{\mu}_2^* \in S^1_{D_0}(\mathbf{Y}^1)$ by [12, Corollary 5.1(1) with Corollary 5.2(2)]. Note that $\lambda^{\mathcal{Q}}(\bar{\mu}_1) > 0$ is equivalent to $\lambda^{\mathcal{Q}}(\bar{\mu}_1^*) > 0$ by [12, Lemma 5.1]. Since $\mathbf{1}_{F_n}\bar{\mu}_1^* \in S^1_{CK_1}(\mathbf{X})$ and $\lambda^{\mathcal{Q}}(\bar{\mu}_1^*) > 0$, we also have $\lambda^{\mathcal{Q}}(\mathbf{1}_{F_n}\bar{\mu}_1^*) > 0$ for all $n \in \mathbb{N}$. Let $\mathbf{U} = (\Omega, \widetilde{\mathscr{F}}_\infty, \widetilde{\mathscr{F}}_t, \widetilde{X}_t, \mathbf{P}_x^U, \zeta^U)$ be the subprocess of $\mathbf{Y}^1$ killed by $\exp(-A_t^{\bar{\mu}_2^*} - A_t^{F_2})$.

Note that the $\mathcal{E}$-quasi notions between Dirichlet forms associated with $\mathbf{X}$ and $\mathbf{U}$, respectively, are equivalent to each other. Indeed, since the expression

$$\mathcal{E}^U(f,f) = \mathcal{E}^Y(f,f) + \int_{\mathbb{R}^d} f^2 e^{-2u} \mathrm{d}\mu_{G_1^u - G^u} + \int_{\mathbb{R}^d} f^2 e^{-2u} \mathrm{d}\bar{\mu}_2^*$$

for $f \in \mathcal{F}$ yields

$$\mathcal{E}^Y(f,f) \le \mathcal{E}^U(f,f) \le \mathcal{E}^Y(f,f) + \|R^{Y^1}(e^{-2u}(\mu_{G_1^u} + \bar{\mu}_2^*))\|_\infty \mathcal{E}^{Y^1}(f,f)$$

for $f \in \mathcal{F}$, and the Dirichlet form $(\mathcal{E}^Y, \mathcal{F}^Y)$ (resp. $(\mathcal{E}^{Y^1}, \mathcal{F}^{Y^1})$) associated to $\mathbf{Y}$ (resp. $\mathbf{Y}^1$) is equivalent to $(\mathcal{E}, \mathcal{F})$ associated to $\mathbf{X}$. The equivalence among both $\mathcal{E}$-quasi notions is now clear because for an increasing sequence $\{F_n\}$ of closed sets, it is an $\mathcal{E}$-nest if and only if $\lim_{n\to\infty} \mathrm{Cap}(K \setminus F_n) = 0$ for each compact subset K of $\mathbb{R}^d$, where Cap denotes the 1-capacity with respect to the Dirichlet form. Consequently, all $\mathcal{E}$-quasi notions among Dirichlet forms associated with $\mathbf{X}$, $\mathbf{Y}^1$ and $\mathbf{U}$ are equivalent to each other. In particular, for an increasing sequence $\{F_n\}$ of closed sets, the following are equivalent:

(a) $\{F_n\}$ is an $\mathcal{E}$-nest. (b) $\{F_n\}$ is a strict $\mathcal{E}$-nest.
(c) $\{F_n\}$ is an $\mathcal{E}^{Y^1}$-nest. (d) $\{F_n\}$ is a strict $\mathcal{E}^{Y^1}$-nest.
(e) $\{F_n\}$ is an $\mathcal{E}^U$-nest. (f) $\{F_n\}$ is a strict $\mathcal{E}^U$-nest.

The equivalence between (a) and (b) (resp. (c) and (d)) follows from the conservativeness of $\mathbf{X}$ (resp. $\mathbf{Y}^1$). The implications (f)$\Longrightarrow$(e)$\Longrightarrow$(a) are clear. Note that $\mathbf{U}$ is not conservative unless $\mu_2 + \mu_{F_2} = 0$. By [19, (62.20)], we see

$$\left.\frac{\mathrm{d}\mathbf{P}_x^U}{\mathrm{d}\mathbf{P}_x^{Y^1}}\right|_{\mathscr{F}_{\sigma_{F_n^c}} \cap \{\sigma_{F_n^c} < \zeta\}} = \exp\left(-A^{\mu_2}_{\sigma_{F_n^c}} - A^{F_2}_{\sigma_{F_n^c}}\right).$$

Then, the implication (d)$\Longrightarrow$(f) is clear from

$$\begin{aligned} \mathbf{E}_x^U[e^{-\sigma_{F_n^c}}] &= \mathbf{E}_x^U[e^{-\sigma_{F_n^c}} : \sigma_{F_n^c} < \zeta] \\ &= \mathbf{E}_x^{Y^1}[e^{-A^{\mu_2}_{\sigma_{F_n^c}} - A^{F_2}_{\sigma_{F_n^c}} - \sigma_{F_n^c}} : \sigma_{F_n^c} < \zeta] \\ &\le \mathbf{E}_x^{Y^1}[e^{-\sigma_{F_n^c}} : \sigma_{F_n^c} < \zeta] = \mathbf{E}_x^{Y^1}[e^{-\sigma_{F_n^c}}], \end{aligned}$$

where we use that $\zeta(\omega) \le \sigma_{F_n^c}(\omega)$ implies $\sigma_{F_n^c}(\omega) = +\infty$.

Set

$$h(x) := \mathbf{E}_x^U\left[e^{A_\infty^{\bar{\mu}_1^*}}\right] \quad \text{and} \quad h_n(x) := \mathbf{E}_x^U\left[e^{\int_0^\infty \mathbf{1}_{F_n}(\widetilde{X}_s)\mathrm{d}A_s^{\bar{\mu}_1^*}}\right]$$

for $x \in \mathbb{R}^d \cup \{\partial\}$. Since any AF C_t of $\mathbf{U}$ satisfies $\mathbf{P}_\partial^U(C_t \equiv 0) = 1$, we see $h(\partial) = h_n(\partial) = 1$. By use of the argument of [19, §62], we have

$$\mathbf{E}_x^U[e^{A_\infty^{\bar{\mu}_1^*}}] = \mathbf{E}_x^{Y^1}\left[\int_0^\infty e^{A_t^{\bar{\mu}_1^*}}\mathrm{d}(e^{-A_t^{\bar{\mu}_2^*}-A_t^{F_2}}) + e^{A_\infty^{\bar{\mu}^*}-A_\infty^{F_2}}\right] \le \mathbf{E}_x^{Y^1}\left[e^{A_\infty^{\bar{\mu}^*}-A_\infty^{F_2}}\right],$$

which shows $1 \le h_n(x) \le h(x) \le \sup_{x\in\mathbb{R}^d\cup\{\partial\}} h(x) < \infty$. For a non-negative Borel measure ν, set $R^U\nu(x) := \int_{\mathbb{R}^d} R^U(x,y)\nu(\mathrm{d}y)$, where $R^U(x,y)$ is the Green kernel of $\mathbf{U}$.

By applying h and h_n to [23, Lemma 3.2], we have

$$h(x) = R^U(he^{-2u}\bar{\mu}_1^*)(x) + 1 \quad \text{and} \quad h_n(x) = R^U(h_n e^{-2u}\mathbf{1}_{F_n}\bar{\mu}_1^*)(x) + 1$$

for all $x \in \mathbb{R}^d$. These expressions yields the fine continuities of h and h_n on $\mathbb{R}^d$ with respect to $\mathbf{U}$. From this observation, both of them are $\mathcal{E}^U$-quasi continuous on $\mathbb{R}^d$, hence $\mathcal{E}$-quasi continuous on $\mathbb{R}^d$. Moreover, we can regard them as strictly $\mathcal{E}$-quasi continuous functions on $\mathbb{R}^d \cup \{\partial\}$, because there exists a common strict $\mathcal{E}$-nest $\{F_l\}$ of compact set such that $\{h, h_n\} \subset C(F_l)$ and ∂ can be thought as an isolated point for each $F_l \cup \{\partial\}$. Thus h and h_n are strictly $\mathcal{E}^U$-quasi continuous on $\mathbb{R}^d \cup \{\partial\}$.

From $e^{-2u}\mathbf{1}_{F_n}\bar{\mu}_1 \in S_0^{(0)}(\mathbf{U})$, [23, Lemma 3.1] tells us

$$h_n - 1 = R^U(h_n e^{-2u}\mathbf{1}_{F_n}\bar{\mu}_1^*) \in \mathcal{F}_e \; \left(= \mathcal{F}_e^{Y^1} = \mathcal{F}_e^U\right)$$

for each $n \in \mathbb{N}$ and each h_n belongs to $QC(\mathbb{R}^d \cup \{\partial\})$, because $h_n - 1$ is strictly $\mathcal{E}$-quasi continuous. Since $R^U(he^{-2u}\bar{\mu}_1^*)$ is a bounded excessive function on $\mathbb{R}^d$ with respect to $\mathbf{U}$, we have

$$h - 1 = R^U(he^{-2u}\bar{\mu}_1^*) \in \dot{\mathcal{F}}_{\mathrm{loc}} \; \left(= \dot{\mathcal{F}}_{\mathrm{loc}}^{Y^1} = \dot{\mathcal{F}}_{\mathrm{loc}}^U\right)$$

by use of [9, Lemma 2.3.2]. Indeed, take any bounded excessive function f and set $f_n := f \wedge \|f\|_\infty(nR_1^U g \wedge 1)$ for a strictly positive bounded function $g \in L^2(\mathbb{R}^d; \mathrm{d}x)$. Then $f_n \le \|f\|_\infty(nR_1^U g \wedge 1)$ and hence, [9, Lemma 2.3.2] yields $f_n \in \mathcal{F}$ and $f = f_n$ on E_n for each $n \in \mathbb{N}$, where $E_n := \{x \in \mathbb{R}^d \mid R_1^U g(x) > 1/n\}$ is a finely open Borel set satisfying $\bigcup_{n=1}^\infty E_n = \mathbb{R}^d$ q.e. Thus we have $h \in \dot{\mathcal{F}}_{\mathrm{loc}}$. In particular, $h \in \mathcal{F}_{\mathrm{loc}}$ if **(RSF)** holds for $\mathbf{U}$.

We easily see that $\{h_n\}$ strictly $\mathcal{E}^U$-quasi uniformly converges to h on $\mathbb{R}^d \cup \{\partial\}$, that is, there exists a common (strict) $\mathcal{E}^U$-nest $\{F_l\}$ of compact sets such that all h_n and h are continuous on each $F_l \cup \{\partial\}$ and $\{h_n\}$ uniformly converges to h on each F_l (hence on each $F_l \cup \{\partial\}$ by $h(\partial) = h_n(\partial) = 1$). In particular $\sup_{0\le t\le T}|h - h_n|(\widetilde{X}_t) \to 0$ as $n \to \infty$ $\mathbf{P}_x^U$-a.s. for q.e. $x \in \mathbb{R}^d$.

We apply Fukushima's decomposition for $h_n - 1 \in \mathcal{F}_e$ (see [8, Theorem 1]) and [9, Lemma 5.4.1] to obtain that

$$\begin{cases} h_n(\widetilde{X}_t) - h_n(\widetilde{X}_0) = M_t^{h_n} + N_t^{h_n}, \\ N_t^{h_n} = -\int_0^t \mathbf{1}_{F_n} h_n(\widetilde{X}_s)\mathrm{d}A_s^{\bar{\mu}_1^*} \end{cases}$$

for all $t \geq 0$ $\mathbf{P}_x^U$-a.s. for q.e. $x \in \mathbb{R}^d$, where $M_t^{h_n}$ (resp. $N_t^{h_n}$) is the martingale additive functional of finite energy (resp. CAF of zero energy) under $\mathbf{U}$. We then see that for $m > n > 0$

$$\begin{aligned}\sup_{0\leq t\leq T}\left|M_t^{h_n} - M_t^{h_m}\right| &\leq 2\sup_{0\leq t\leq T}|h_n - h_m|(\widetilde{X}_t) + \sup_{0\leq t\leq T}\left|N_t^{h_n} - N_t^{h_m}\right| \\ &\leq 2\sup_{0\leq t\leq T}|h - h_m|(\widetilde{X}_t) + \int_0^T |h - \mathbf{1}_{F_n}h_n|(\widetilde{X}_t)\mathrm{d}A_t^{\bar{\mu}_1^*},\end{aligned}$$

which converges to 0 as $m, n \to \infty$, consequently,

$$\mathbf{P}_x^U\left(\sup_{0\leq t\leq T}\left|M_t^{h_n} - M_t^{h_m}\right| \to 0 \ \text{ as } \ m,n\to\infty\right) = 1 \quad \text{for q.e. } x \in \mathbb{R}^d.$$

Moreover, by [11, Lemma 2.2], we have for $p \in \mathbb{N}$

$$\begin{aligned}\mathbf{E}_x^U\left[\left|M_t^{h_n}\right|^p\right] &\leq 2\cdot 4^{p-1}\|h_n\|_\infty^p + 2^{p-1}\mathbf{E}_x^U\left[\left(\int_0^t \mathbf{1}_{F_n}h_n(\widetilde{X}_s)\mathrm{d}A_s^{\bar{\mu}_1^*}\right)^p\right] \\ &\leq 2\cdot 4^{p-1}\|h\|_\infty^p + 2^{p-1}p!\left(\sup_{x\in\mathbb{R}^d}\mathbf{E}_x^U\left[\int_0^t h(\widetilde{X}_s)\mathrm{d}A_s^{\bar{\mu}_1^*}\right]\right)^p \\ &\leq 2\cdot 4^{p-1}\|h\|_\infty^p + 2^{p-1}p!\left\|R^U(he^{-2u}\bar{\mu}_1^*)\right\|_\infty^p < \infty\end{aligned}$$

for q.e. $x \in \mathbb{R}^d$. So $\{M_t^{h_n}\}_{n\in\mathbb{N}}$ is uniformly $\mathbf{P}_x^U$-integrable for q.e. $x \in \mathbb{R}^d$. In particular, we have

$$R^U e^{-2u}\mu_{\langle h_n}(x) = \mathbf{E}_x^U\left[\langle M^{h_n}{}_\infty\right] \leq 8\|h\|_\infty^2 + 4\left\|R^U(he^{-2u}\bar{\mu}_1^*)\right\|_\infty^2 < \infty$$

for q.e. $x \in \mathbb{R}^d$, consequently $\sup_{x\in\mathbb{R}^d} R^U e^{-2u}\mu_{\langle h_n\rangle}(x) < \infty$ because of the fine continuity of $x \mapsto R^U e^{-2u}\mu_{\langle h_n}(x)$. Then we have $e^{-2u}\mu_{\langle h_n} \in S_{D_0}^1(\mathbf{U})$. Therefore, we can redefine M^{h_n} (resp. N^{h_n}) as the martingale additive functional of finite energy (resp. CAF of zero energy) in the strict sense under $\mathbf{U}$, and hence the following Fukushima's decomposition in the strict sense holds: $h_n(X_t) - h_n(X_0) = M_t^{h_n} + N_t^{h_n}$ $t \in [0, +\infty[$ $\mathbf{P}_x^U$-a.s. for all $x \in \mathbb{R}^d$. On the other hand, $M_t^h := \lim_{n\to\infty} M_t^{h_n}$ is a square integrable martingale additive functional under $\mathbf{P}_x^U$-a.s. for q.e. $x \in \mathbb{R}^d$. Combining the same way as above and Fatou's lemma, we can obtain $\sup_{x\in\mathbb{R}^d} R^U e^{-2u}\mu_{\langle h\rangle}(x) \leq 8\|h\|_\infty^2 + 4\|R^U(he^{-2u}\bar{\mu}_1^*)\|_\infty^2 < \infty$. Thus we have $e^{-2u}\mu_{\langle h\rangle} \in S_{D_0}^1(\mathbf{U})$. Set $N_t^h := \lim_{n\to\infty} N_t^{h_n}$. Then we have $N_t^h = -\int_0^t h(\widetilde{X}_s)\mathrm{d}A_s^{\bar{\mu}_1^*}$ under $\mathbf{P}_x^U$-a.s. for q.e. $x \in \mathbb{R}^d$ and

$$h(\widetilde{X}_t) - h(\widetilde{X}_0) = M_t^h + N_t^h \quad \text{for all} \quad t \in [0, +\infty[\quad \mathbf{P}_x^U\text{-a.s.} \tag{4.1}$$

for q.e. $x \in \mathbb{R}^d$. Moreover, by replacing M_t^h (resp. N_t^h) with a martingale additive functional (resp. CAF of zero energy) in the strict sense, (4.1) holds for all $x \in \mathbb{R}^d$ as the generalized Fukushima decomposition for $h \in \dot{\mathcal{F}}_{\mathrm{loc}} \cap QC(\mathbb{R}^d \cup \{\partial\})$ with $e^{-2u}\mu_{\langle h\rangle} \in S_{D_0}^1(\mathbf{U})$ ([13, Theorem 6.2]). Hereafter, we regard (4.1) as a generalized Fukushima decomposition in the strict sense such that M_t^h is a martingale additive functional in the strict sense and $N_t^h = -\int_0^t h(\widetilde{X}_s)\mathrm{d}A_s^{\bar{\mu}_1^*}$ defined under $\mathbf{P}_x^U$.

Define a martingale additive functional of $\mathbf{U}$ by $M_t := \int_0^t h(\widetilde{X}_{s-})^{-1}\mathrm{d}M_s^h$ and denote by $L_t := \mathrm{Exp}(M)_t$ the Doléans-Dade exponential of M_t, that is, L_t is the unique solution of $L_t = 1 + \int_0^t L_{s-}\mathrm{d}M_s$ for $t \in [0, +\infty[$, $\mathbf{P}_x^U$-a.s. for q.e. $x \in \mathbb{R}^d$. Then we see

$$L_t = \exp\left(M_t - \frac{1}{2}\langle M^c\rangle_t\right) \prod_{0<s\le t} \frac{h(\widetilde{X}_s)}{h(\widetilde{X}_{s-})} \exp\left(1 - \frac{h(\widetilde{X}_s)}{h(\widetilde{X}_{s-})}\right),$$

where M_t^c is the continuous part of M_t. By Itô formula applied to the semimartingale $h(X_t)$ with the function $\log x$, we see that L_t has the following expression:

$$L_t = \frac{h(\widetilde{X}_t)}{h(\widetilde{X}_0)} \exp\left(A_t^{\bar{\mu}_1^*}\right).$$

Denote by $\mathbf{U}^h := (\Omega, \widetilde{\mathscr{F}}_\infty, \widetilde{\mathscr{F}}_t, \widetilde{X}_t, \mathbf{P}_x^{U,h}, \zeta^{U,h})$ the transformed process of $\mathbf{U}$ by L_t:

$$\mathbf{E}_x^{U,h}[f(\widetilde{X}_t)] = \mathbf{E}_x^U[L_t f(\widetilde{X}_t)], \quad t > 0, \quad f \in \mathscr{B}_b(\mathbb{R}^d).$$

Set $\ell := -\log h = -\log(R^U(he^{-2u}\bar{\mu}_1^*) + 1)$. Since $R^U(he^{-2u}\bar{\mu}_1^*) \in \dot{\mathcal{F}}_{\mathrm{loc}} \cap QC(\mathbb{R}^d \cup \{\partial\})$, we have $\ell \in \dot{\mathcal{F}}_{\mathrm{loc}} \cap QC(\mathbb{R}^d \cup \{\partial\})$ and $h = e^{-\ell}$. It is easy to see the fine continuity of ℓ on $\mathbb{R}^d$ with respect to $\mathbf{U}$. Moreover since $|\ell(x) - \ell(y)| \le |h(x) - h(y)|$, $x, y \in \mathbb{R}^d$ under $1 \le h \le \|h\|_\infty$ on $\mathbb{R}^d$, we have $e^{-2u}\mu_{\langle \ell} \in S_{D_0}^1(\mathbf{U})$. Then we can see that the transform L_t belongs to the class of Girsanov transforms considered in Section 3 (or in [7]) under $\mathbf{U}$. In particular, $\mathbf{U}^h$ is an $h^2e^{-2u}\mathrm{d}x$-symmetric Hunt process on $\mathbb{R}^d$. From [12, Theorem 3.2] the associated Dirichlet form $(\mathcal{E}^{U,h}, \mathcal{F}^{U,h})$ on $L^2(\mathbb{R}^d; h^2e^{-2u}\mathrm{d}x)$ is identified as follows by:

$$\begin{cases} \mathcal{E}^{U,h}(f,g) \\ \quad = \dfrac{1}{2}\displaystyle\int_{\mathbb{R}^d}\int_{\mathbb{R}^d} \frac{(f(x)-f(y))(g(x)-g(y))h(x)h(y)e^{F(x,y)-u(x)-u(y)}c(x,y)}{|x-y|^{d+\alpha}}\mathrm{d}x\mathrm{d}y \\ \qquad + \displaystyle\int_{\mathbb{R}^d} f(x)g(x)h(x)\int_{\mathbb{R}^d} \frac{(G_1(x,y) - G(x,y))e^{-u(x)-u(y)}c(x,y)}{|x-y|^{d+\alpha}}\mathrm{d}x\mathrm{d}y \\ \qquad\quad + \displaystyle\int_{\mathbb{R}^d} f(x)g(x)h(x)e^{-2u(x)}\bar{\mu}_2^*(\mathrm{d}x), \\ \mathcal{F}^{U,h} = \mathcal{F}. \end{cases}$$

By way of Stollmann-Voigt's inequality, we can see that $(\mathcal{E}^{U,h}, \mathcal{F}^{U,h})$ is equivalent to $(\mathcal{E}, \mathcal{F})$ under $\mu_2 + \mu_{F_2} \in S_{D_0}^1(\mathbf{X})$, that is, $\mathcal{F}^{U,h} = \mathcal{F}$ and

$$\begin{aligned} e^{-\|F_2\|_\infty - 2\|u\|_\infty}\mathcal{E}(f,f) &\le \mathcal{E}^{U,h}(f,f) \\ &\le \|h\|_\infty^2 e^{\|F_1\|_\infty + 2\|u\|_\infty}(\|R\mu_2\|_\infty + \|R\mu_{F_2}\|_\infty)\mathcal{E}(f,f) \end{aligned}$$

for $f \in \mathcal{F} = \mathcal{F}^{U,h}$. Here we use the estimate $G_1 - G = e^{F_1}(1 - e^{-F_2}) \le e^{\|F_1\|_\infty}F_2$. Note that $(\mathcal{E}^{U,h}, \mathcal{F}^{U,h})$ is not necessarily the Dirichlet form associated to a symmetric stable like process except the case $\mu_2 + \mu_{F_2} = 0$. From this equivalence of the Dirichlet forms, we see that there exists an integral kernel $p_t^{U,h}(x,y)$ of $(\mathcal{E}^{U,h}, \mathcal{F}^{U,h})$ associated to the Feynman-Kac semigroup $\{P_t^{U,h}\}_{t>0}$ of $\mathbf{U}^h$ defined by

$$P_t^{U,h}f(x) := \mathbf{E}_x^U\left[L_t f(\widetilde{X}_t)\right]$$

$$= h(x)^{-1}\mathbf{E}_x^U\left[h(\widetilde{X}_t)e^{A_t^{\bar{\mu}_1^*}}f(\widetilde{X}_t)\right] =: h(x)^{-1}P_t^{U,\bar{\mu}_1^*}(hf)(x)$$

and it also satisfies the two-sided estimates (1.2) for all $(t,x,y)\in]0,+\infty[\times\mathbb{R}^d\times\mathbb{R}^d$ under $\mu_2+\mu_{F_2}=0$. Moreover,

$$p_t^{U,\bar{\mu}_1^*}(x,y) = p_t^{U,h}(x,y)h(x)h(y), \tag{4.2}$$

where $p_t^{U,\bar{\mu}_1^*}(x,y)$ is the integral kernel of the semigroup $P_t^{U,\bar{\mu}_1^*}g(x) := \mathbf{E}_x^U[e^{A_t^{\bar{\mu}_1^*}}g(\widetilde{X}_t)]$. On the other hand, since

$$\begin{aligned}P_t^A f(x) &= \mathbf{E}_x\left[e^{N_t^u+A_t^\mu+A_t^F}\right] = \mathbf{E}_x\left[e^{u(X_t)-u(x)-M_t^u+A_t^\mu+A_t^F}f(X_t)\right]\\ &= e^{-u(x)}\mathbf{E}_x^U\left[e^{A_t^{\bar{\mu}_1^*}}e^{u(\widetilde{X}_t)}f(\widetilde{X}_t)\right],\end{aligned}$$

the integral kernel $p_t^A(x,y)$ of the Feynman-Kac semigroup $\{P_t^A\}_{t>0}$ can be represented as

$$h(x)h(y)p_t^{U,h}(x,y)e^{-u(x)-u(y)}$$

in view of (4.2). Now we can conclude that if $\lambda^{\mathcal{Q}}(\bar{\mu}_1^*)>0$ and $\mu_2+\mu_{F_2}=0$, then $p_t^A(x,y)$ satisfies the two-sided estimates (1.2) for all $(t,x,y)\in]0,+\infty[\times\mathbb{R}^d\times\mathbb{R}^d$ by the boundedness of u and h. Thus Theorem 4.1 is proved for the case $\mu_2+\mu_{F_2}=0$. Finally we prove the implication (1) $\Longrightarrow$ (3) for general case. Let $\nu\in S_{D_0}^1(\mathbf{X})$ be a non-trivial smooth measure in the strict sense. For sufficiently large $p>1+\sup_{x\in\mathbb{R}^d}\mathbf{E}_x[A_\infty^\nu]$, we see $\frac{\nu}{p-1}\in S_{CS_1}^1(\mathbf{X})$ and

$$\begin{aligned}&\lambda\left(\frac{\nu}{p-1}\right)\\ &:= \inf\left\{\mathcal{E}(f,f)-\frac{1}{p-1}\int_{\mathbb{R}^d}f^2\mathrm{d}\nu \;\middle|\; f\in C_0^\infty(\mathbb{R}^d),\ \frac{1}{p-1}\int_{\mathbb{R}^d}f^2\mathrm{d}\nu=1\right\}\\ &\geq \inf\left\{\left(1-\frac{1}{p-1}\sup_{x\in\mathbb{R}^d}\mathbf{E}_x[A_\infty^\nu]\right)\mathcal{E}(f,f)\;\middle|\; f\in C_0^\infty(\mathbb{R}^d),\right.\\ &\qquad\qquad\left.\frac{1}{p-1}\int_{\mathbb{R}^d}f^2\mathrm{d}\nu=1\right\}\\ &\geq \left(1-\frac{1}{p-1}\sup_{x\in\mathbb{R}^d}\mathbf{E}_x[A_\infty^\nu]\right)\frac{p-1}{\sup_{x\in\mathbb{R}^d}\mathbf{E}_x[A_\infty^\nu]}>0.\end{aligned}$$

Therefore, one can apply Theorem 4.1 to $\mu_1:=\frac{\nu}{p-1}$ and $\mu_{\langle u\rangle}+\mu_2+\mu_{F_1+F_2}=0$. In this case, the integral kernel $p_t^B(x,y)$ for $B_t:=\frac{1}{p-1}A_t^\nu$ enjoys the two-sided estimates (1.2) for all $(t,x,y)\in]0,+\infty[\times\mathbb{R}^d\times\mathbb{R}^d$ by applying the assertion of our Theorem 4.1 under $\mu_2+\mu_{F_2}=0$. The heat kernel $p_t^\nu(x,y)$ of the process killed by $e^{-A_t^\nu}$ can be constructed as an upper semi continuous function on $]0,+\infty[\times\mathbb{R}^d\times\mathbb{R}^d$ by way of the method of Blanchard-Ma [1]. Though the underlying Markov process of [1] is assumed to be the d-dimensional transient Brownian motion and the measures appeared in the perturbation are of Kato classes, the method of the proof of [1, Theorem 3.1] remains valid for Markov processes admitting heat kernel having an on-diagonal upper estimate and for measures of extended Kato class. Indeed, let $\{F_k\}$ be an $\mathcal{E}$-nest of compact sets such that $\mathbf{1}_{F_k}\nu \in S^1_K(\mathbf{X})$ for each $k \in \mathbb{N}$. As noted above, there exists a heat kernel $p_t^{\mathbf{1}_{F_k}\nu}(x,y)$ which is jointly continuous on $]0,+\infty[\times\mathbb{R}^d \times \mathbb{R}^d$ satisfying

$$P_t^{\mathbf{1}_{F_k}\nu} f(x) = \int_{\mathbb{R}^d} p_t^{\mathbf{1}_{F_k}\nu}(x,y) f(y)\mathrm{d}y$$

for all $(t,x) \in]0,+\infty[\times\mathbb{R}^d$ by [1, Theorem 3.1]. It is easy to see that

$$p_t^{\mathbf{1}_{F_{k+1}}\nu}(x,y) \le p_t^{\mathbf{1}_{F_k}\nu}(x,y) \le p_t(x,y)$$

for all $(t,x,y) \in]0,+\infty[\times\mathbb{R}^d \times \mathbb{R}^d$ and $k \in \mathbb{N}$ because of the continuity of the heat kernels. We set $p_t^{\nu}(x,y) :=\downarrow \lim_{k\to\infty} p_t^{\mathbf{1}_{F_k}\nu}(x,y)$. Then $p_t^{\nu}(x,y)$ is the upper semi continuous heat kernel associated to the Feynman-Kac semigroup $P_t^{\nu} f(x) := \mathbf{E}_x[e^{-A_t^{\nu}} f(X_t)]$.

Using the method described in [1, pp. 180], we have the following estimates:

$$\frac{p_t(x,y)^p}{p_t^B(x,y)^{p-1}} \le p_t^{\nu}(x,y) \le p_t(x,y) \quad \text{for all } t>0, \quad x,y \in \mathbb{R}^d, \tag{4.3}$$

where we use the continuities of $p_t(x,y)$ and the integral kernel $p_t^B(x,y)$ for $B := \frac{1}{p-1}A^{\nu}$. Then, (4.3) shows that $p_t^{\nu}(x,y)$ satisfies the two-sided estimates (1.2) for all $(t,x,y) \in]0,+\infty[\times\mathbb{R}^d \times \mathbb{R}^d$. Applying this fact to the symmetric stable like process $\mathbf{U}^{h,j}$ associated to the the Dirichlet form $(\mathcal{E}^{U,h,j},\mathcal{F})$, which is the pure jump part of $(\mathcal{E}^{U,h},\mathcal{F}^{U,h})$ on $L^2(\mathbb{R}^d; h^2e^{-2u}\mathrm{d}x)$, one can conclude that the integral kernel $p_t^{U,h}(x,y)$ defined above, regarded as the heat kernel of the killed process by $\exp(-A_t^{\mu_2} - A_t^{\mu_{G_1}-G})$, enjoys the two-sided estimates (1.2) for all $(t,x,y) \in]0,+\infty[\times\mathbb{R}^d \times \mathbb{R}^d$ without assuming $\mu_2 + \mu_{F_2} = 0$. Here we use that $\mu_2 + \mu_{F_2} \in S^1_{D_0}(\mathbf{X})$ implies $he^{-2u}\mu_2 + he^{-u}\mu_{F_2e^{-u}} \in S^1_{D_0}(\mathbf{U}^{h,j})$ by applying [12, Corollary 5.1(1) with Corollary 5.2(2)] twice, because $\mathbf{U}^{h,j}$ is the Girsanov transformed process obtained from $\mathbf{Y}^1$ by L_t. Therefore, in the same way as described above, we can obtain the two-sided estimates (1.2) of the integral kernel $p_t^A(x,y)$ for all $(t,x,y) \in]0,+\infty[\times\mathbb{R}^d \times \mathbb{R}^d$ without assuming $\mu_2 + \mu_{F_2} = 0$. □

Remark 4.1.

(1) The stability of the heat kernel estimates for Feynman-Kac semigroup $\{P_t^A\}_{t\ge 0}$ given by $A_t := N_t^u + A_t^{\mu}$ also can be proved for a diffusion process on $\mathbb{R}^d$ satisfying a uniform elliptic condition by applying [12, Theorem 1.2] to the equivalence among weak Poincaré inequality with uniform volume doubling condition, parabolic Harnack inequality and heat kernel estimates proved by [18, 21] (cf. [24]).

(2) In [25], the author proved the stability of the fundamental solution for local Feynman-Kac semigroup in the framework of symmetric stable-like processes or symmetric relativistic stable-like processes considered in [4], and discovered good Green kernel estimates yielding the $3G$-inequality in these underlying processes (see also [12, Example 7.2]). One can easily check that a result similar to Theorem 4.1 is also true in the framework of symmetric relativistic stable-like processes.

References

[1] P. Blanchard and Z.-M. Ma, *Semigroup of Schrödinger operators with potentials given by Radon measures*, Stochastic processes, physics and geometry, 160–195, World Sci. Publishing, Teaneck, NJ, 1990.

[2] Z.-Q. Chen, *Gaugeability and conditional gaugeability*, Trans. Amer. Math. Soc. **354** (2002), no. 11, 4639–4679.

[3] Z.-Q. Chen, P. J. Fitzsimmons, M. Takeda, J. Ying and T.-S. Zhang, *Absolute continuity of symmetric Markov processes*, Ann. Probab. **32** (2004), no. 3A, 2067–2098.

[4] Z.-Q. Chen, P. Kim and T. Kumagai, *Global heat kernel estimates for symmetric jump processes*, Trans. Amer. Math. Soc. **363** (2011), no. 9, 5021–5055.

[5] Z.-Q. Chen and T. Kumagai, *Heat kernel estimates for stable-like processes on d-sets*, Stochastic processes and their Applications **108** (2003), no. 1, 23–62.

[6] Z.-Q. Chen and T. Kumagai, *Heat kernel estimates for jump processes of mixed types on metric measure spaces*, Probab. Theory Related Fields **140** (2008), no. 1-2, 277–317.

[7] Z.-Q. Chen and T.-S. Zhang, *Girsanov and Feynman-Kac type transformations for symmetric Markov processes*, Ann. Inst. H. Poincaré Probab. Statist. **38** (2002), no. 4, 475–505.

[8] M. Fukushima, *On a decomposition of additive functionals in the strict sense for a symmetric Markov process*, Dirichlet forms and stochastic processes (Beijing, 1993), 155–169, de Gruyter, Berlin, 1995.

[9] M. Fukushima, Y. Oshima and M. Takeda, *Dirichlet forms and symmetric Markov processes*, Second revised and extended edition. de Gruyter Studies in Mathematics, **19**. Walter de Gruyter & Co., Berlin, 2011.

[10] A. Grigoryan, *Heat kernels on weighted manifolds and applications*, Cont. Math. **398** (2006), 93–191.

[11] G. De Leva, D. Kim and K. Kuwae, *L^p-independence of spectral bounds of Feynman-Kac semigroups by continuous additive functionals*, J. Funct. Anal. **259** (2010), no. 3, 690–730.

[12] D. Kim and K. Kuwae, *Analytic characterizations of gaugeability for generalized Feynman-Kac functionals*, preprint (2012).

[13] D. Kim, K. Kuwae and Y. Tawara, *Large deviation principle for generalized Feynman-Kac functionals and its applications*, preprint (2012).

[14] K. Kuwae, *Functional calculus for Dirichlet forms*, Osaka J. Math. **35** (1998), no. 3, 683–715.

[15] K. Kuwae and M. Takahashi, *Kato class functions of Markov processes under ultracontractivity*, Potential theory in Matsue, 193–202, Adv. Stud. Pure Math. **44**, Math. Soc. Japan, Tokyo, 2006.

[16] P. Li and S.-T. Yau, *On the parabolic kernel of the Schrödinger operator*, Acta Math. **156** (1986), no. 3-4, 153–201.
[17] Z.-M. Ma and M. Röckner, *Introduction to the Theory of (Non-Symmetric) Dirichlet Forms*, Springer Universitext, 1992.
[18] L. Saloff-Coste, *Uniformly elliptic operators on Riemannian manifolds*, (English summary) J. Differential Geom. **36** (1992), no. 2, 417–450.
[19] M. Sharpe, *General Theory of Markov Processes*, Academic Press, Boston, 1988.
[20] P. Stollmann and J. Voigt, *Perturbation of Dirichlet forms by measures*, Potential Anal. **5** (1996), no. 2, 109–138.
[21] K.-Th. Sturm, *Analysis on local Dirichlet spaces. III. The parabolic Harnack inequality*, J. Math. Pures Appl. (9) **75** (1996), no. 3, 273–297.
[22] M. Takeda, *Conditional gaugeability and subcriticality of generalized Schrödinger operators*, J. Funct. Anal. **191** (2002), no. 2, 343–376.
[23] ———, *Gaugeability for Feynman-Kac functionals with applications to symmetric α-stable processes*, Proc. Amer. Math. Soc. **134** (2006), no. 9, 2729–2738.
[24] ———, *Gaussian bounds of heat kernels for Schrödinger operators on Riemannian manifolds*, Bull. Lond. Math. Soc. **39** (2007), no. 1, 85–94.
[25] M. Wada, *Perturbation of Dirichlet forms and the stablility of fundamental solutions*, to appear in Tohoku Math. J. (2013).

Chapter 17

Martin boundary for some symmetric Lévy processes

Panki Kim*, Renming Song† and Zoran Vondraček‡

In this paper we study the Martin boundary of open sets with respect to a large class of purely discontinuous symmetric Lévy processes in $\mathbb{R}^d$. We show that, if $D \subset \mathbb{R}^d$ is an open set which is κ-fat at a boundary point $Q \in \partial D$, then there is exactly one Martin boundary point associated with Q and this Martin boundary point is minimal.

Contents

1. Introduction

The Martin boundary of an open set D is an abstract boundary introduced in 1941 by Martin [26] so that every nonnegative classical harmonic function in D can be written as an integral of the Martin kernel with respect to a finite measure on the Martin boundary. This integral representation is called a Martin representation. The concepts of Martin boundary and Martin kernel were extended to general Markov processes by Kunita and Watanabe [25] in 1965. In order for the Martin representation to be useful, one needs to have a better understanding of the Martin boundary, for instance, its relation with the Euclidean boundary. In 1970, Hunt and Wheeden [14] proved that, in the classical case, the Martin boundary of a bounded Lipschitz domain coincides with its Euclidean boundary. Subsequently, a lot of progress has been made in studying the Martin boundary in the classical case.

With the help of the boundary Harnack principle for rotationally invariant α-stable ($\alpha \in (0, 2)$) processes established in [2], it was proved in [3, 11, 27] that the Martin boundary, with respect to the rotationally invariant α-stable process,

*Department of Mathematical Sciences and Research Institute of Mathematics, Seoul National University, Building 27, 1 Gwanak-ro, Gwanak-gu Seoul 151-747, Republic of Korea. The research of this author is suppoerted by the National Research Foundation of Korea(NRF) grant funded by the Korea government(MEST) (NRF-2013R1A2A2A01004822).
†Department of Mathematics, University of Illinois, Urbana, IL 61801, USA. The research of this author is supported in part by a grant from the Simons Foundation (208236).
‡Department of Mathematics, University of Zagreb, Zagreb, Croatia

of a bounded Lipschitz domain D coincides with its Euclidean boundary and that any nonnegative harmonic function with respect to the killed rotationally invariant α-stable process in D can be written uniquely as an integral of the Martin kernel with respect to a finite measure on ∂D. In [29] this result was extended to bounded κ-fat open sets. The Martin boundary, with respect to truncated stable processes, of any roughly connected κ-fat open set was shown in [16] to coincide with its Euclidean boundary. In [17], the results of [29] were extended to a large class of purely discontinuous subordinate Brownian motions. For Martin boundary at infinity with respect to subordinate Brownian motions, see [23].

In this paper, we study the Martin boundary of open set $D \subset \mathbb{R}^d$ with respect to a large class of symmetric, not necessarily rotationally invariant, (transient) Lévy processes killed upon exiting D. We show that if D is an open set and D is κ-fat at a single point $Q \in \partial D$, then the Martin boundary associated with Q consists of exactly one point and the corresponding Martin kernel is a minimal harmonic function. Another point is that, unlike [3, 11, 16, 17, 27, 29], the set D is not necessarily bounded. In the case when D is unbounded, we do not study the Martin boundary associated with infinite boundary points.

Now we describe the class of processes we are going to work with.

Throughout this paper, $r \mapsto j(r)$ is a strictly positive and non-increasing function on $(0, \infty)$ satisfying

$$j(r) \le c j(r+1) \qquad \text{for } r \ge 1, \tag{1.1}$$

and $X = (X_t, \mathbb{P}_x)$ is a purely discontinuous symmetric Lévy process with Lévy exponent $\Psi_X(\xi)$ so that

$$\mathbb{E}_x\left[e^{i\xi\cdot(X_t - X_0)}\right] = e^{-t\Psi_X(\xi)}, \qquad t > 0, x \in \mathbb{R}^d, \xi \in \mathbb{R}^d.$$

We assume that the Lévy measure of X has a density J_X such that

$$\gamma_1^{-1} j(|y|) \le J_X(y) \le \gamma_1 j(|y|), \quad \text{for all } y \in \mathbb{R}^d, \tag{1.2}$$

for some $\gamma_1 > 1$. Since $\int_0^\infty j(r)(1 \wedge r^2) r^{d-1} dr < \infty$ by (1.2), the function $x \to j(|x|)$ is the Lévy density of an isotropic unimodal Lévy process whose characteristic exponent is

$$\Psi(|\xi|) = \int_{\mathbb{R}^d} (1 - \cos(\xi \cdot y)) j(|y|) dy. \tag{1.3}$$

The Lévy exponent Ψ_X can be written as

$$\Psi_X(\xi) = \int_{\mathbb{R}^d} (1 - \cos(\xi \cdot y)) J_X(y) dy$$

and, clearly by (1.2), it satisfies

$$\gamma_1^{-1} \Psi(|\xi|) \le \Psi_X(\xi) \le \gamma_1 \Psi(|\xi|), \quad \text{for all } \xi \in \mathbb{R}^d. \tag{1.4}$$

The function Ψ may be not increasing. However, if we put $\Psi^*(r) := \sup_{s\le r}\Psi(s)$, then, by [4, Proposition 2] (cf. also [13, Proposition 1]), we have

$$\Psi(r) \le \Psi^*(r) \le \pi^2\Psi(r).$$

Thus by (1.4),

$$(\pi^2\gamma_1)^{-1}\Psi^*(|\xi|) \le \Psi_X(\xi) \le \gamma_1\Psi^*(|\xi|), \quad \text{for all } \xi \in \mathbb{R}^d\,. \tag{1.5}$$

We will always assume that Ψ satisfies the following scaling condition at infinity:

(**H**): There exist constants $0<\delta_1\le\delta_2<1$ and $a_1, a_2>0$ such that

$$a_1\lambda^{2\delta_1}\Psi(t) \le \Psi(\lambda t) \le a_2\lambda^{2\delta_2}\Psi(t), \quad \lambda\ge 1, t\ge 1\,. \tag{1.6}$$

Then by [4, (15) and Corollary 22], for every $R>0$, there exists $c=c(R)>1$ such that

$$c^{-1}\frac{\Psi(r^{-1})}{r^d} \le j(r) \le c\frac{\Psi(r^{-1})}{r^d} \quad \text{for } r\in(0,R]. \tag{1.7}$$

Note that the class of purely discontinuous symmetric Lévy processes considered in this paper contains some of the purely discontinuous isotropic unimodal Lévy processes dealt with in [4].

Let us now formulate precisely the main result of this paper.

Definition 1.1. Let $D\subset\mathbb{R}^d$ be an open set and $Q\in\mathbb{R}^d$. We say that D is κ-fat at Q for some $\kappa\in(0,\frac{1}{2})$, if there exists $R>0$ such that for all $r\in(0,R]$, there is a ball $B(A_r(Q),\kappa r)\subset D\cap B(Q,r)$. The pair (R,κ) is called the characteristics of the κ-fat open set D at Q. We say that an open set D is κ-fat with characteristics (R,κ) if D is κ-fat at $Q\in\partial D$ with characteristics (R,κ) for all $Q\in\partial D$.

For $D\subset\mathbb{R}^d$ we denote by $\partial_M D$ the Martin boundary of D. A point $w\in\partial_M D$ is said to be associated with Q if there is a sequence $(y_n)_{n\ge1}\subset D$ converging to w in the Martin topology and to Q in the Euclidean topology. The set of Martin boundary points associated with Q is denoted by $\partial_M^Q D$.

Theorem 1.1. *Suppose that the assumption* (**H**) *is satisfied. Let X be a symmetric Lévy process with a Lévy density satisfying* (1.2) *and let D be an open subset of $\mathbb{R}^d$ which is κ-fat at $Q\in\partial D$. If D is bounded, then $\partial_M^Q D$ consists of exactly one point and this point is a minimal Martin boundary point. If D is unbounded and the process X is transient, the same conclusion is true.*

In the case when D is unbounded, a natural assumption would be that D is Greenian, that is, the killed process X^D is transient. Unfortunately, under the assumption (**H**), which governs the behavior of the process in small space, it seems difficult to obtain estimates of the Green function $G_D(x,y)$ when either x or y becomes large. This is why in the case of unbounded D we impose the transience assumption on X which gives the asymptotic behavior of the free Green function $G(x,y)$, cf. Lemma 2.10.

We begin the paper by showing that only small modifications are needed to extend some results from the isotropic case studied in [20] to the symmetric Lévy processes X considered in this paper. These results include exit time estimates, Poisson kernel estimates and Harnack inequality. A little more work is needed to establish the upper and the lower bound on the Green function G_D. Those are used to obtain sharp bounds on the Poisson kernel and the boundary Harnack principle in the same way as in [20]. In Section 3 we follow the well-established route, see [2, 17, 19, 23], to identify the Martin boundary point associated to Q. After preliminary estimates about harmonic functions, we first show that the oscillation reduction lemma, see [2, Lemma 16], is valid in our setting (with essentially the same proof). The lemma almost immediately implies that $\partial_M^Q D$ consists of exactly one point. We then show that this point is a minimal Martin boundary point. We end the paper by giving the Martin representation for bounded κ-fat open sets.

We finish this introduction by setting up some notation and conventions. We use ":=" to denote a definition, which is read as "is defined to be"; we denote $a \wedge b := \min\{a, b\}$, $a \vee b := \max\{a, b\}$; we denote by $B(x, r)$ the open ball centered at $x \in \mathbb{R}^d$ with radius $r > 0$; for any two positive functions f and g, $f \asymp g$ means that there is a positive constant $c \geq 1$ so that $c^{-1} g \leq f \leq c g$ on their common domain of definition; for any Borel subset $E \subset \mathbb{R}^d$ and $x \in E$, $\text{diam}(E)$ stands for the diameter of E and $\delta_E(x)$ stands for the Euclidean distance between x and E^c; $\mathbb{N}$ is the set of natural numbers.

In this paper, we use the following convention: The values of the constants $R, \delta_1, \delta_2, C_1, C_2, C_3, C_4$ remain the same throughout this paper, while $c, c_0, c_1, c_2, \ldots$ represent constants whose values are unimportant and may change. All constants are positive finite numbers. The labeling of the constants $c_0, c_1, c_2, \ldots$ starts anew in the statement and proof of each result. The dependence of the constants on dimension d is not mentioned explicitly.

2. Green function estimates

Let $S = (S_t : t \geq 0)$ be a subordinator with no drift. The Laplace exponent ϕ of S is a Bernstein function and admits the following representation

$$\phi(\lambda) = \int_0^\infty (1 - e^{-\lambda t})\mu(dt), \qquad \lambda > 0,$$

where μ satisfies $\int_0^\infty (1 \wedge t)\mu(dt) < \infty$. μ is called the Lévy measure of S or ϕ. The function ϕ is called a complete Bernstein function if μ has a completely monotone density.

The following elementary result observed in [19, 22] will be used several times later in this paper.

Lemma 2.1. *If ϕ is a Bernstein function, then for all $\lambda, t > 0$, $1 \wedge \lambda \leq \phi(\lambda t)/\phi(t) \leq 1 \vee \lambda$.*

Recall that, throughout this paper, we assume that $X = (X_t, \mathbb{P}_x)$ is a purely discontinuous symmetric Lévy process in $\mathbb{R}^d$ with Lévy exponent $\Psi_X(\xi)$ and a Lévy density J_X satisfying (1.2).

It follows from [4, (28)], (1.4), (1.6) and (1.7) that there exist a constant $\gamma_2 > 1$ and a complete Bernstein function ϕ such that

$$\gamma_2^{-1}\phi(|\xi|^2) \le \Psi(|\xi|) \le \gamma_2\phi(|\xi|^2), \quad \text{for all } \xi \in \mathbb{R}^d, \tag{2.1}$$

and j enjoys the following property: for every $R > 0$,

$$j(r) \asymp \frac{\phi(r^{-2})}{r^d} \quad \text{for } r \in (0, R]. \tag{2.2}$$

Furthermore, there exist $b_1, b_2 > 0$ such that

$$b_1\lambda^{\delta_1}\phi(t) \le \phi(\lambda t) \le b_2\lambda^{\delta_2}\phi(t), \quad \lambda \ge 1, t \ge 1. \tag{2.3}$$

Throughout this paper, we assume that ϕ is the above complete Bernstein function.

From Lemma 2.1 and (2.2), we also get that for every $R > 0$,

$$j(r) \le cj(2r), \quad r \in (0, R]. \tag{2.4}$$

The infinitesimal generator $\mathbf{L}$ of X is given by

$$\mathbf{L}f(x) = \int_{\mathbb{R}^d} \left(f(x+y) - f(x) - y \cdot \nabla f(x)\mathbf{1}_{\{|y|\le 1\}}\right) J_X(y)dy \tag{2.5}$$

for $f \in C_b^2(\mathbb{R}^d)$. Furthermore, for every $f \in C_b^2(\mathbb{R}^d)$, $f(X_t) - f(X_0) - \int_0^t \mathbf{L}f(X_s)\, ds$ is a $\mathbb{P}_x$-martingale for every $x \in \mathbb{R}^d$.

The following two results are valid without assuming (**H**). The next lemma is a special case of [13, Corollary 1].

Lemma 2.2. *There exists a constant* $c > 0$ *depending only on* d *such that*

$$c^{-1}\Psi(r^{-1}) \le \frac{1}{r^2}\int_0^r s^{d+1}j(s)ds + \int_r^\infty s^{d-1}j(s)ds \le c\Psi(r^{-1}), \quad \forall r > 0. \tag{2.6}$$

Lemma 2.3. *There exists a constant* $c = c(\Psi, \gamma_1, \gamma_2) > 0$ *such that for every* $f \in C_b^2(\mathbb{R}^d)$ *with* $0 \le f \le 1$,

$$\mathbf{L}f_r(x) \le c\,\phi(r^{-2})\left(2 + \sup_y \sum_{j,k} |(\partial^2/\partial y_j \partial y_k) f(y)|\right), \quad \forall x \in \mathbb{R}^d,\ r > 0,$$

where $f_r(y) := f(y/r)$.

Proof. Using (1.2) and Lemma 2.2 or [4, Corollary 3], this result can be obtained by following the proof of [21, Lemma 4.2]. We omit the details. □

For any open set D, we use τ_D to denote the first exit time of D, i.e., $\tau_D = \inf\{t > 0 : X_t \notin D\}$.

Using Lemmas 2.1 and 2.3 and (2.1), the proof of the next result is the same as those of [18, Lemmas 13.4.1 and 13.4.2]. Thus we omit the proof.

Lemma 2.4. *There exists a constant $c = c(\Psi, \gamma_1, \gamma_2) > 0$ such that for every $r > 0$ and every $x \in \mathbb{R}^d$,*

$$\inf_{z \in B(x,r/2)} \mathbb{E}_z \left[\tau_{B(x,r)}\right] \geq \frac{c}{\phi(r^{-2})}.$$

The idea of the following key result comes from [31].

Lemma 2.5. *There exists a constant $c = c(\Psi, \gamma_1, \gamma_2) > 0$ such that for any $r > 0$ and $x_0 \in \mathbb{R}^d$,*

$$\mathbb{E}_x[\tau_{B(x_0,r)}] \leq c\,(\phi(r^{-2})\phi((r - |x - x_0|)^{-2}))^{-1/2}, \qquad x \in B(x_0, r).$$

Proof. Without loss of generality, we may assume that $x_0 = 0$. We fix $x \neq 0$ and put $Z_t = \frac{X_t \cdot x}{|x|}$. Then, Z_t is a one dimensional symmetric Lévy process in $\mathbb{R}$ with Lévy exponent

$$\Psi_Z(\theta) = \int_{\mathbb{R}^d} \left(1 - \cos(\frac{\theta x}{|x|} \cdot y)\right) J_X(y)dy, \qquad \theta \in \mathbb{R}.$$

By (1.2),

$$\Psi_Z(\theta) \asymp \int_{\mathbb{R}^d} \left(1 - \cos(\frac{\theta x}{|x|} \cdot y)\right) j(|y|)dy = \Psi(\theta).$$

It is easy to see that, if $X_t \in B(0, r)$, then $|Z_t| < r$, hence $\mathbb{E}_x[\tau_{B(0,r)}] \leq \mathbb{E}_{|x|}[\tilde{\tau}]$, where $\tilde{\tau} = \inf\{t > 0 : |Z_t| \geq r\}$. By [5, (2.17)], the proof of [5, Proposition 2.4] and Lemma 2.2.

$$\mathbb{E}_x[\tau_{B(0,r)}] \leq \mathbb{E}_{|x|}[\tilde{\tau}] \leq c\,(\Psi(r^{-1})\Psi((r - |x|)^{-1}))^{-1/2}.$$

Now the assertion of the lemma follows immediately by (2.1). □

Given an open set $D \subset \mathbb{R}^d$, we define $X_t^D(\omega) = X_t(\omega)$ if $t < \tau_D(\omega)$ and $X_t^D(\omega) = \partial$ if $t \geq \tau_D(\omega)$, where ∂ is a cemetery state. We now recall the definitions of harmonic functions with respect to X and with respect to X^D.

Definition 2.1. Let D be an open subset of $\mathbb{R}^d$. A nonnegative function u on $\mathbb{R}^d$ is said to be

(1) harmonic in D with respect to X if

$$u(x) = \mathbb{E}_x \left[u(X_{\tau_U})\right], \qquad x \in U,$$

for every open set U whose closure is a compact subset of D;

(2) regular harmonic in D with respect to X if for each $x \in D$,

$$u(x) = \mathbb{E}_x \left[u(X_{\tau_D}); \tau_D < \infty\right].$$

Definition 2.2. Let D be an open subset of $\mathbb{R}^d$. A nonnegative function u on D is said to be harmonic with respect to X^D if

$$u(x) = \mathbb{E}_x \left[u(X_{\tau_U}^D)\right], \qquad x \in U,$$

for every open set U whose closure is a compact subset of D.

Obviously, if u is harmonic with respect to X^D, then the function which is equal to u in D and zero outside D is harmonic with respect to X in D.

Since our X satisfies [7, (1.6), **(UJS)**], by [7, Theorem 1.4] and using the standard chain argument one has the following form of Harnack inequality.

Theorem 2.1. *For every $a \in (0,1)$, there exists $c = c(a, \Psi, \gamma_1, \gamma_2) > 0$ such that for every $r \in (0,1)$, $x_0 \in \mathbb{R}^d$, and every function u which is nonnegative on $\mathbb{R}^d$ and harmonic with respect to X in $B(x_0, r)$, we have*

$$u(x) \le c\, u(y), \quad \text{for all } x, y \in B(x_0, ar)\,.$$

Let $D \subset \mathbb{R}^d$ be an open set. Since J_X satisfies the assumption [7, (1.6)], by [7, Theorem 3.1], bounded functions that are harmonic in D with respect to X are Hölder continuous. Suppose that u is a nonnegative function which is harmonic with respect to X in D. For any ball $B := B(x_0, r)$ with $B \subset \overline{B} \subset D$, the functions u_n, $n \ge 1$, defined by

$$u_n(x) := \mathbb{E}_x\left[(u \wedge n)(X_{\tau_B})\right], \qquad x \in \mathbb{R}^d,$$

are bounded functions which are harmonic with respect to X in B. Applying Theorem 2.1 to $v_n(x) := u(x) - u_n(x) = \mathbb{E}_x\left[(u - (u \wedge n))(X_{\tau_B})\right]$, it is easy to see that u_n converges to u uniformly in $B(x_0, r/2)$. Thus u is continuous in D. This implies that all nonnegative functions that are harmonic in D with respect to X are continuous.

A subset D of $\mathbb{R}^d$ is said to be Greenian (for X) if X^D is transient. By [7, Theorem 3.1] X^D has Hölder continuous transition densities $p_D(t, x, y)$. For any Greenian open set D in $\mathbb{R}^d$, let $G_D(x, y) = \int_0^\infty p_D(t, x, y)dt$ be the Green function of X^D. Then $G_D(x, y)$ is finite off the diagonal of $D \times D$. Furthermore, $x \mapsto G_D(x, y)$ is harmonic in $D \setminus \{y\}$ with respect to X and therefore continuous. Using the Lévy system for X, we know that for every Greenian open subset D and every $f \ge 0$ and $x \in D$,

$$\mathbb{E}_x\left[f(X_{\tau_D});\, X_{\tau_D -} \ne X_{\tau_D}\right] = \int_{\overline{D}^c} \int_D G_D(x, z) J_X(z - y) dz f(y) dy. \tag{2.7}$$

We define the Poisson kernel

$$K_D(x, y) := \int_D G_D(x, z) J_X(z - y) dz, \qquad (x, y) \in D \times \overline{D}^c. \tag{2.8}$$

Thus (2.7) can be simply written as

$$\mathbb{E}_x\left[f(X_{\tau_D});\, X_{\tau_D -} \ne X_{\tau_D}\right] = \int_{\overline{D}^c} K_D(x, y) f(y) dy.$$

The following result will be used later in this paper.

Lemma 2.6. *There exist $c_1 = c_1(\Psi, \gamma_1, \gamma_2) > 0$ and $c_2 = c_2(\Psi, \gamma_1, \gamma_2) > 0$ such that for every $r \in (0,1]$ and $x_0 \in \mathbb{R}^d$,*

$$K_{B(x_0, r)}(x, y) \le c_1 j(|y - x_0| - r)(\phi(r^{-2})\phi((r - |x - x_0|)^{-2}))^{-1/2}$$

$$\leq c_1 j(|y-x_0|-r)\phi(r^{-2})^{-1} \tag{2.9}$$

for all $(x,y) \in B(x_0,r) \times \overline{B(x_0,r)}^c$ *and*

$$K_{B(x_0,r)}(x_0,y) \geq c_2 j(|y-x_0|)\phi(r^{-2})^{-1}, \qquad \text{for all } y \in \overline{B(x_0,r)}^c. \tag{2.10}$$

Proof. This proof is exactly the same as that of [18, Proposition 13.4.10]. We provide the proof to show that only the monotonicity of j, (1.1), (2.4) and Lemmas 2.4 and 2.5 are used.

Without loss of generality, we assume $x_0 = 0$. For $z \in B(0,r)$ and $r < |y| < 2$,

$$|y| - r \leq |y| - |z| \leq |y-z| \leq |z| + |y| + r + |y| \leq 2|y|,$$

and for $z \in B(0,r)$ and $y \in B(0,2)^c$,

$$|y| - r \leq |y| - |z| \leq |y-z| \leq |z| + |y| + r + |y| \leq |y| + 1.$$

Thus by the monotonicity of j, (1.1) and (2.4), there exists a constant $c > 0$ such that

$$cj(|y|) \leq j(|z-y|) \leq j(|y|-r), \qquad (z,y) \in B(0,r) \times \overline{B(0,r)}^c.$$

Applying the above inequalities, Lemmas 2.4 and 2.5 to (2.8), we immediately get the assertion of the lemma. □

As in [20], to deal with κ-fat open set, we need the following form of Harnack inequality.

Theorem 2.2. *Let $L > 0$. There exists a positive constant $c = c(L, \Psi, \gamma_1, \gamma_2) > 1$ such that the following is true: If $x_1, x_2 \in \mathbb{R}^d$ and $r \in (0,1)$ are such that $|x_1 - x_2| < Lr$, then for every nonnegative function u which is harmonic with respect to X in $B(x_1,r) \cup B(x_2,r)$, we have*

$$c^{-1}u(x_2) \leq u(x_1) \leq cu(x_2).$$

Proof. Let $r \in (0,1]$, $x_1, x_2 \in \mathbb{R}^d$ be such that $|x_1 - x_2| < Lr$ and let u be a nonnegative function which is harmonic in $B(x_1,r) \cup B(x_2,r)$ with respect to X. If $|x_1 - x_2| < \frac{1}{4}r$, then since $r < 1$, the theorem is true by Theorem 2.1. Thus we only need to consider the case when $\frac{1}{4}r \leq |x_1 - x_2| \leq Lr$ with $L > \frac{1}{4}$.

Let $w \in B(x_1, \frac{r}{8})$. Because $|x_2 - w| \leq |x_1 - x_2| + |w - x_1| < (L + \frac{1}{8})r \leq 2Lr$, first using the monotonicity of j and (2.10), then using (2.2) and Lemma 2.1, we get

$$K_{B(x_2,\frac{r}{8})}(x_2,w) \geq c_1\, j(2Lr)\phi(r^{-2})^{-1} \geq c_2\, r^{-d}\frac{\phi((2Lr)^{-2})}{\phi(r^{-2})} \geq c_3 r^{-d}. \tag{2.11}$$

For any $y \in B(x_1, \frac{r}{8})$, u is regular harmonic in $B(y, \frac{7r}{8}) \cup B(x_1, \frac{7r}{8})$. Since $|y - x_1| < \frac{r}{8}$, by the already proven part of this theorem,

$$u(y) \geq c_2 u(x_1), \quad y \in B(x_1, \frac{r}{8}), \tag{2.12}$$

for some constant $c_2 > 0$. Therefore, by (2.7) and (2.11)–(2.12),

$$\begin{aligned} u(x_2) &= \mathbb{E}_{x_2}\left[u(X_{\tau_{B(x_2,\frac{r}{8})}})\right] \geq \mathbb{E}_{x_2}\left[u(X_{\tau_{B(x_2,\frac{r}{8})}}); X_{\tau_{B(x_2,\frac{r}{8})}} \in B(x_1,\frac{r}{8})\right] \\ &\geq c_2\, u(x_1)\, \mathbb{P}_{x_2}\left(X_{\tau_{B(x_2,\frac{r}{8})}} \in B(x_1,\frac{r}{8})\right) \\ &= c_2\, u(x_1) \int_{B(x_1,\frac{r}{8})} K_{B(x_2,\frac{r}{8})}(x_2,w)\, dw \\ &\geq c_3\, u(x_1) \left|B(x_1,\frac{r}{8})\right| r^{-d} = c_4\, u(x_1). \end{aligned}$$

Thus we have proved the right-hand side inequality in the conclusion of the theorem. The inequality on the left-hand side follows by symmetry. □

For notational convenience, we define

$$\Phi(r) = \frac{1}{\phi(r^{-2})}, \qquad r > 0. \tag{2.13}$$

The inverse function of Φ will be denoted by the usual notation $\Phi^{-1}(r)$.

Our process X belongs to the class of Markov processes considered in [7]. Thus we have the following two-sided estimates for $p(t,x,y)$ from [7]. The proof is the same as that of [9, Proposition 2.2].

Proposition 2.1. *For any $T > 0$, there exists $c_1 = c_1(T,\Psi,\gamma_1,\gamma_2) > 0$ such that*

$$p(t,x,y) \leq c_1\, (\Phi^{-1}(t))^{-d} \qquad \text{for all } (t,x,y) \in (0,T] \times \mathbb{R}^d \times \mathbb{R}^d. \tag{2.14}$$

For any $T, R > 0$, there exists $c_2 = c_2(T,R,\Psi,\gamma_1,\gamma_2) > 1$ such that for all $(t,x,y) \in [0,T] \times \mathbb{R}^d \times \mathbb{R}^d$ with $|x-y| < R$,

$$\begin{aligned} c_2^{-1}\left((\Phi^{-1}(t))^{-d} \wedge \frac{t}{|x-y|^d\,\Phi(|x-y|)}\right) &\leq p(t,x,y) \\ \leq c_2\left((\Phi^{-1}(t))^{-d} \wedge \frac{t}{|x-y|^d\,\Phi(|x-y|)}\right). \end{aligned} \tag{2.15}$$

Our argument to obtain upper bound on the Green functions of bounded open sets is similar to those in [8, 15]. We give the details here for the completeness.

Lemma 2.7. *For every bounded open set D, the Green function $G_D(x,y)$ is finite and continuous off the diagonal of $D \times D$ and there exists $c = c(\mathrm{diam}(D),\Psi,\gamma_1,\gamma_2) \geq 1$ such that for all $x,y \in D$,*

$$G_D(x,y) \leq c\frac{\Phi(|x-y|)}{|x-y|^d} = \frac{c}{|x-y|^d\phi(|x-y|^{-2})}. \tag{2.16}$$

Proof. Put $L := \mathrm{diam}(D)$. By (2.15), for every $x \in D$ we have

$$\mathbb{P}_x(\tau_D \leq 1) \geq \mathbb{P}_x(X_1 \in \mathbb{R}^d \setminus D) = \int_{\mathbb{R}^d\setminus D} p(1,x,y)dy$$

$$\geq c_1 \int_{\mathbb{R}^d \setminus D} \left(1 \wedge \frac{1}{|x-y|^d\, \Phi(|x-y|)}\right) dy$$
$$\geq c_1 \int_{\{\,|z| \geq L\,\}} \left(1 \wedge \frac{1}{|z|^d\, \Phi(|z|)}\right) dz = c_2 > 0.$$

Thus

$$\sup_{x \in D} \int_D p_D(1,x,y)dy \;=\; \sup_{x \in D} \mathbb{P}_x(\tau_D > 1) \;<\; 1.$$

Now the Markov property of X implies that there exist positive constants c_3 and c_4 such that

$$\int_D p_D(t,x,y)dy \;\leq\; c_3 e^{-c_4 t} \quad \text{for all } (t,x) \in (0,\infty) \times D.$$

Thus combining this, (2.15) and the semigroup property, we have that for any $(t,x,y) \in (1,\infty) \times D \times D$,

$$p_D(t,x,y) = \int_D p_D(t-1,x,z) p_D(1,z,y) dz$$
$$\leq c_5 \int_D p_D(t-1,x,z)dz \leq c_6 e^{-c_4 t}.$$

This with (2.15) implies that such that for any $(x,y) \in D \times D$,

$$G_D(x,y) = \int_0^\infty p_D(t,x,y)dt$$
$$\leq c_7 \int_0^1 \left((\Phi^{-1}(t))^{-d} \wedge \frac{t}{|x-y|^d\, \Phi(|x-y|)} \right) dt + c_7. \tag{2.17}$$

By the proof of [7, Theorem 6.1],

$$\int_0^1 \left((\Phi^{-1}(t))^{-d} \wedge \frac{t}{|x-y|^d\, \Phi(|x-y|)} \right) dt \asymp \frac{\Phi(|x-y|)}{|x-y|^d}.$$

Therefore the Green function $G_D(x,y) = \int_0^\infty p_D(t,x,y)dt$ is finite and continuous off the diagonal of $D \times D$. Furthermore, by (2.3), we have $\inf_{a \leq \text{diam}(D)} \Phi(a) a^{-d} > 0$. Consequently, (2.16) holds. □

For interior lower bound on the Green function, we use some recent results from [9]. The next result is an analog of [9, Proposition 3.6], which is the main result of [9, Section 3]. Even though it is assumed in [9] that X is rotationally symmetric and its Lévy density satisfies a little stronger assumption than in this paper, all the arguments of [9, Section 3] only use the results in [7], (1.1), (1.2), (2.2), (2.3), and the semigroup property. Thus by following the same arguments line by line, one can prove the next proposition. We omit the details. We note in passing that D is not necessarily bounded in the next proposition.

Proposition 2.2. *Let T and a be positive constants. There exists $c = c(T, a, \Psi, \gamma_1, \gamma_2) > 0$ such that for any open set D,*

$$p_D(t, x, y) \ge c\left((\Phi^{-1}(t))^{-d} \wedge tj(|x - y|)\right)$$

for every $(t, x, y) \in (0, T] \times D \times D$ with $\delta_D(x) \wedge \delta_D(y) \ge a\Phi^{-1}(t)$.

Lemma 2.8. *For every $L, T > 0$, there exists $c = c(T, L, \Psi, \gamma_1, \gamma_2) > 0$ such that for any bounded open set D with* $\mathrm{diam}(D) \le T$, *$x, y \in D$ with $|x-y| \le L(\delta_D(x) \wedge \delta_D(y))$,*

$$G_D(x, y) \ge c\frac{\Phi(|x - y|)}{|x - y|^d} = \frac{c}{|x - y|^d \phi(|x - y|^{-2})}. \tag{2.18}$$

Proof. Without loss of generality, we assume $L \ge 1$, $\delta_D(y) \le \delta_D(x)$ and $\mathrm{diam}(D) \le T$. By Proposition 2.2 and (2.2), there exists $c_1 = c_1(T, \Psi, \gamma_1, \gamma_2) > 0$ such that for all $(t, z, w) \in (0, T] \times D \times D$ with $\delta_D(z) \wedge \delta_D(w) \ge \Phi^{-1}(t)$,

$$p_D(t, z, w) \ge c_1 \left((\Phi^{-1}(t))^{-d} \wedge \frac{t}{|z - w|^d \Phi(|z - w|)}\right). \tag{2.19}$$

Using this we have

$$\begin{aligned} G_D(x, y) &\ge \int_0^T p_D(t, x, y)dt \ge \int_0^{\Phi(\delta_D(y))} p_D(t, x, y)dt \\ &\ge c_1 \int_0^{\Phi(\delta_D(y))} (\Phi^{-1}(t))^{-d} \wedge \frac{t}{|x - y|^d \Phi(|x - y|)}dt \\ &\ge c_1 \int_0^{\Phi(L^{-1}|x-y|)} (\Phi^{-1}(t))^{-d} \wedge \frac{t}{|x - y|^d \Phi(|x - y|)}dt. \end{aligned} \tag{2.20}$$

Let $r = |x - y|$. By the change of variable $u = \frac{\Phi(r)}{t}$ and the fact that $t \to \Phi(t)$ is increasing,

$$\begin{aligned} &\int_0^{\Phi(L^{-1}r)} (\Phi^{-1}(t))^{-d} \wedge \frac{t}{r^d \Phi(r)}dt \\ &= \frac{\Phi(r)}{r^d} \int_{\Phi(r)/\Phi(L^{-1}r)}^{\infty} u^{-2} \left(\left(\frac{r}{\Phi^{-1}(u^{-1}\Phi(r))}\right)^d \wedge u^{-1}\right) du \\ &= \frac{\Phi(r)}{r^d} \int_{\Phi(r)/\Phi(L^{-1}r)}^{\infty} u^{-3}du. \end{aligned} \tag{2.21}$$

Since by (2.3) and (2.13)

$$\int_{\Phi(r)/\Phi(L^{-1}r)}^{\infty} u^{-3}du \ge \int_{c_2 L^{2\delta_1}}^{\infty} u^{-3}du > 0, \tag{2.22}$$

we conclude from (2.20) and (2.21) that (2.18) holds. □

It follows from (2.3), Lemmas 2.8 and 2.5 that, for all $r \in (0,1)$ and all $\alpha \in (0,\pi)$, there exists $c = c(\alpha) > 0$ such that for all cones V of angle α with vertex at the origin,

$$\begin{aligned} &\mathbb{E}_0 \int_0^{\tau_{B(0,r)}} 1_V(X_s)ds \geq \int_{V\cap B(0,r/2)} G_{B(0,r)}(0,y)dy \\ &\geq c_1 \int_{V\cap B(0,r/2)} \frac{1}{|y|^d \phi(|y|^{-2})} dy \\ &\geq c_2 \int_0^{r/2} \frac{1}{r\phi(r^{-2})} dr \geq c_3 \phi(r^{-2}) \geq c_4 \mathbb{E}_0 \tau_{B(0,r)}. \end{aligned}$$

Thus X satisfies hypothesis **H** in [30]. It follows from (1.2), (2.2) and Lemma 2.4 that for every cone V with vertex at the origin,

$$\int_{V\cap B(0,1)} \mathbb{E}_0 \tau_{B(0,|y|)} J_X(y)dy \geq c \int_{V\cap B(0,1)} |y|^{-d} dy = \infty.$$

Therefore it follows from [30, Theorem 1] that if V is a Lipschitz open set in $\mathbb{R}^d$ and D is an open subset of V,

$$\mathbb{P}_x(X_{\tau_D} \in \partial V) = 0, \qquad x \in D.$$

Using Theorem 2.1, the proof of the next result is the same as that of [18, Proposition 13.4.11]. So we omit the proof.

Proposition 2.3. *For every $a \in (0,1)$, there exists $c = c(\Psi, \gamma_1, \gamma_2, a) > 0$ such that for every $r \in (0,1]$, $x_0 \in \mathbb{R}^d$ and $x_1, x_2 \in B(x_0, ar)$,*

$$K_{B(x_0,r)}(x_1,y) \leq cK_{B(x_0,r)}(x_2,y), \qquad a.e.\ y \in \overline{B(x_0,r)}^c.$$

Proposition 2.4. *For every $a \in (0,1)$, there exists $c = c(\Psi, \gamma_1, \gamma_2, a) > 0$ such that for every $r \in (0,1]$ and $x_0 \in \mathbb{R}^d$,*

$$K_{B(x_0,r)}(x,y) \leq c\, r^{-d} \left(\frac{\phi((|y-x_0|-r)^{-2})}{\phi(r^{-2})} \right)^{1/2}$$

for all $x \in B(x_0, ar)$ and a.e. y such that $r < |x_0 - y| < 2r$.

Proof. Using (2.2), Lemmas 2.2, 2.5 and Proposition 2.3, the proof is exactly the same as that of [21, Proposition 4.9]. We omit it. □

Lemma 2.9. *For every $a \in (0,1)$, there exists a constant $c = c(\Psi, \gamma_1, \gamma_2, a) > 0$ such that for any $r \in (0,1)$ and any open set D with $D \subset B(0,r)$ we have*

$$\mathbb{P}_x\left(X_{\tau_D} \in B(0,r)^c\right) \leq c\,\phi(r^{-2}) \int_D G_D(x,y)dy, \qquad x \in D \cap B(0,ar).$$

Proof. Using Lemma 2.3, the proof is exactly the same as that of [21, Lemma 4.10]. We omit it. □

With these preparations in hand, we can repeat the argument of [21, Section 5] to get the following form of the boundary Harnack principle established in [21]. We omit the details. Note that the open set D in the next result is not necessarily bounded.

Theorem 2.3. *There exists $C_1 = C_1(\Psi, \gamma_1, \gamma_2) \geq 1$ such that the following hold for all $r \in (0,1)$.*

(i) For every $z_0 \in \mathbb{R}^d$, every open set $U \subset B(z_0, r)$ and for any nonnegative function u in $\mathbb{R}^d$ which is regular harmonic in U with respect to X and vanishes a.e. in $U^c \cap B(z_0, r)$ it holds that

$$C_1^{-1}\mathbb{E}_x[\tau_U] \int_{B(z_0,r/2)^c} j(|y-z_0|)u(y)dy \leq u(x)$$
$$\leq C_1\, \mathbb{E}_x[\tau_U] \int_{B(z_0,r/2)^c} j(|y-z_0|)u(y)dy$$

for every $x \in U \cap B(z_0, r/2)$.

(ii) For every $z_0 \in \mathbb{R}^d$, every open set $D \subset \mathbb{R}^d$, every $r \in (0,1)$ and for any nonnegative functions u, v in $\mathbb{R}^d$ which are regular harmonic in $D \cap B(z_0, r)$ with respect to X and vanish a.e. in $D^c \cap B(z_0, r)$, we have

$$\frac{u(x)}{v(x)} \leq C_1^4 \frac{u(y)}{v(y)}, \qquad x, y \in D \cap B(z_0, r/2).$$

(iii) For every $z_0 \in \mathbb{R}^d$, every Greenian open set $D \subset \mathbb{R}^d$, every $r \in (0,1)$, we have

$$K_D(x_1, y_1)K_D(x_2, y_2) \leq C_1^4 K_D(x_1, y_2)K_D(x_2, y_1)$$

for all $x_1, x_2 \in D \cap B(z_0, r/2)$ and a.e. $y_1, y_2 \in \overline{D}^c \cap B(z_0, r)^c$.

In the next two results, we will assume that X is transient and will use $G(x, y)$ to denote the Green function of X. Note that $G(x, y) = G(y, x)$ by symmetry and $G(x, y) = G(0, y - x)$ by translation invariance. Since we only assume that X is symmetric, rather than unimodal, the next result does not follow from [13].

Theorem 2.4. *For every $M \geq 1$ there exists a constant $C_2(M) = C_2(M, \Psi, \gamma_1, \gamma_2) > 0$ such that for all $x \in B(0, M)$,*

$$C_2(M)^{-1} \frac{\Phi(|x|)}{|x|^d} \leq G(x, 0) \leq C_2(M) \frac{\Phi(|x|)}{|x|^d}.$$

Proof. In this proof, we always assume that $x \in B(0, M)$. It follows from Lemma 2.8 that

$$G(x, 0) \geq G_{B(0,2M)}(x, 0) \geq c_1 \frac{\Phi(|x|)}{|x|^d}.$$

On the other hand, by the strong Markov property and Lemma 2.7,

$$G(x, 0) = G_{B(0,2M)}(x, 0) + \mathbb{E}_x[G(X_{\tau_{B(0,2M)}}, 0)] \leq c_2 \frac{\Phi(|x|)}{|x|^d} + \mathbb{E}_x[G(X_{\tau_{B(0,2M)}}, 0)].$$

Choose $x_0 = (0, \dots, 0, M/2)$. By Theorem 2.2 and the strong Markov property, we have

$$\mathbb{E}_x[G(X_{\tau_{B(0,2M)}}, 0)] \le c_3 \mathbb{E}_{x_0}[G(X_{\tau_{B(0,2M)}}, 0)] \le c_3 G(x_0, 0) \le c_4 < \infty.$$

Furthermore, by (2.3), we have $\inf_{a \le 2M} \Phi(a) a^{-d} > 0$. Thus we conclude that

$$G(x,0) = G_{B(0,2M)}(x,0) + \mathbb{E}_x[G(X_{\tau_{B(0,2M)}}, 0)] \le c_2 \frac{\Phi(|x|)}{|x|^d} + c_4 \le c_5 \frac{\Phi(|x|)}{|x|^d}.$$

□

The following result will be needed in the proof of Theorem 3.2.

Lemma 2.10. *If X is transient, then*

$$\lim_{x \to \infty} G(x,0) = 0.$$

Proof. It follows from the transience assumption and (1.5) that $1/\Psi_X$ is locally integrable in $\mathbb{R}^d$, thus by [1, Proposition 13.23] that the semigroup of X is integrable (in the sense of [1]). Therefore it follows from [1, Proposition 13.21] that, for any nonnegative continuous function f on $\mathbb{R}^d$ with compact support, $\lim_{x \to \infty} Gf(x) = 0$.

By Theorem 2.1, for any $x \in B(0,4)^c$,

$$G(x,0) \le c\, G(x,y), \qquad y \in B(0,2).$$

Take a nonnegative function f with support in $\overline{B(0,2)}$ such that f is identically 1 on $B(0,1)$. Then

$$\begin{aligned} G(x,0) \int_{B(0,2)} f(y)dy &= \int_{B(0,2)} G(x,0) f(y) dy \\ &\le c \int_{B(0,2)} G(x,y) f(y) dy = Gf(x), \end{aligned}$$

which yields

$$G(x,0) \le c \frac{Gf(x)}{\int_{B(0,2)} f(y)dy}.$$

Therefore $\lim_{x \to \infty} G(x,0) = 0$. □

Remark 2.1. Note that several results of this section are stated only for small radii r, namely $r \in (0,1]$. This is, of course, a consequence of the scaling condition (**H**) at infinity which governs the behavior of the process for small time and small space. If we want to study the large time and large space behavior of X, we would need to add the following scaling condition on Ψ near the origin too:

(**H2**): There exist constants $0 < \delta_3 \le \delta_4 < 1$ and $a_3, a_4 > 0$ such that

$$a_3 \lambda^{2\delta_4} \Psi(t) \le \Psi(\lambda t) \le a_4 \lambda^{2\delta_3} \phi(t), \quad \lambda \le 1, t \le 1\,.$$

One consequence of adding condition **(H2)** to condition (**H**) is that many results that were valid for small r only will hold true for all $r > 0$. For future reference, we list below precisely which statements are true.

First note that if both (**H**) and (**H2**) hold, there exist $a_5, a_6 > 0$ such that

$$a_5 \left(\frac{R}{r}\right)^{2(\delta_1 \wedge \delta_3)} \leq \frac{\Psi(R)}{\Psi(r)} \leq a_6 \left(\frac{R}{r}\right)^{2(\delta_2 \vee \delta_4)}, \quad a > 0,\ 0 < r < R < \infty, \tag{2.23}$$

cf. [22, (2.6)].

In the remainder of this remark, in addition to all the conditions from Section 1, namely (1.1), (1.2) and **(H)**, we also assume (**H2**). Then X satisfies the assumptions in [10]. Furthermore, it follows from [4, (15), Corollary 23, Proposition 28] and (2.23) that (2.2) and (2.4) hold for all $r > 0$. That is, under both (**H**) and (**H2**)

$$j(r) \asymp \frac{\phi(r^{-2})}{r^d} \quad \text{for } r > 0, \tag{2.24}$$

and

$$j(r) \leq cj(2r)\,, \quad r > 0\,. \tag{2.25}$$

The following results are now true for all $r > 0$: Theorem 2.1, Lemma 2.6, Theorem 2.2, Proposition 2.3, Proposition 2.4, Lemma 2.9 and Theorem 2.3. In order to prove Theorem 2.1, we also have to use [10, Theorem 4.12], while in the proof of Theorem 2.2 we use (2.24) instead of (2.2). Furthermore, using results in [10], Proposition 2.2 and Lemma 2.8 hold with $T = \infty$. Proofs of all other listed results stay the same.

3. Martin Boundary

In this section we will use the boundary Harnack principle (Theorem 2.3) and follow the well-established route, see [2, 17, 19, 23], to study the Martin boundary of an open set D with respect to X. The key ingredient is the oscillation reduction technique used in the proof of Lemma 3.4.

Throughout this section we assume that $D \subset \mathbb{R}^d$ is a Greenian open set, $Q \in \overline{D}$ and D is κ-fat at Q for some $\kappa \in (0, \frac{1}{2})$, that is, there is $R > 0$ such that for all $r \in (0, R]$, there is a ball $B(A_r(Q), \kappa r) \subset D \cap B(Q, r)$. Without loss of generality, we assume that $R \leq 1/2$. After Lemma 3.4, we will further assume that $Q \in \partial D$.

Lemma 3.1. *There exist $C_3 = C_3(\Psi, \kappa, \gamma_1, \gamma_2) > 0$ and $\xi = \xi(\Psi, \kappa, \gamma_1, \gamma_2) \in (0, 1)$ such that for every $r \in (0, R)$ and any non-negative function h in $\mathbb{R}^d$ which is harmonic in $D \cap B(Q, r)$ it holds that*

$$\begin{aligned} &h(A_r(Q)) \\ &\leq C_3(\phi(r^{-2}))^{-1}\xi^k \phi((\kappa/2)^{-2k} r^{-2}) h(A_{(\kappa/2)^k r}(Q))\,, \quad k = 0, 1, 2, \dots\,. \end{aligned} \tag{3.1}$$

Proof. Without loss of generality, we may assume that $Q = 0$. Fix $r \in (0, R)$. For $k = 0, 1, 2, \dots$, let $\eta_k = (\kappa/2)^k r$, $A_k = A_{\eta_k}(0)$ and $B_k = B(A_k, \eta_{k+1})$. Note that the balls B_k are pairwise disjoint. By harmonicity of h, for every $k = 0, 1, 2 \dots$,

$$h(A_k) = \mathbb{E}_{A_k}\left[h(X_{\tau_{B_k}})\right] \geq \sum_{l=0}^{k-1} \mathbb{E}_{A_k}\left[h(X_{\tau_{B_k}}) : X_{\tau_{B_k}} \in B_l\right]$$

$$= \sum_{l=0}^{k-1} \int_{B_l} K_{B_k}(A_k, z) h(z)\, dz\,.$$

By Theorem 2.2, there exists $c_1 = c_1(\Psi, \gamma_1, \gamma_2) > 0$ such that for every $l = 0, 1, 2, \dots$, $h(z) \geq c_1 h(A_l)$ for all $z \in B_l$. Hence

$$\int_{B_l} K_{B_k}(A_k, z) h(z)\, dz \geq c_1 h(A_l) \int_{B_l} K_{B_k}(A_k, z)\, dz\,, \quad 0 \leq l \leq k-1\,.$$

By (2.10), we have

$$\int_{B_l} K_{B_k}(A_k, z) dz \geq c_2 \phi(\eta_k^{-2})^{-1} \int_{B_l} j(|2(A_k - z)|) dz\,, \quad 0 \leq l \leq k-1\,.$$

For $l = 0, 1, \cdots, k-1$ and $z \in B_l$, it holds that $|z| \leq \kappa(\kappa/2)^l r + (\kappa/2)^{(l+1)} r = (3\kappa/2)(\kappa/2)^l r$. Since $|A_k| \leq \kappa\eta_k$, we have that $|A_k - z| \leq |A_k| + |z| < 3\eta_l$. Together with (2.2) and Lemma 2.1, this implies that $j(|2(A_k - z)|) \geq c_3 |\eta_l|^{-d} \phi(\eta_k^{-2})$ for every $z \in B_l$ and $0 \leq l \leq k-1$. Therefore,

$$\int_{B_l} K_{B_k}(A_k, z)\, dz$$
$$\geq c_4 |\eta_l|^{-d} \phi(\eta_k^{-2}) \phi(\eta_k^{-2})^{-1} |B_l| \geq c_5 \phi(\eta_l^{-2}) / \phi(\eta_k^{-2})\,, \quad 0 \leq l \leq k-1\,.$$

Hence,

$$\phi(\eta_k^{-2}) h(A_k) \geq c_5 \sum_{l=0}^{k-1} \phi(\eta_l^{-2}) h(A_l)\,, \quad \text{for all } k = 1, 2, \dots,$$

where $c_5 = c_5(\Psi, \gamma_1, \gamma_2, \kappa)$. Let $a_k := \phi(\eta_k^{-2}) h(A_k)$ so that $a_k \geq c_5 \sum_{l=0}^{k-1} a_l$. Using the identity $1 + c_5 \sum_{l=0}^{k-2} (1 + c_5)^l = (1 + c_5)^{k-1}$ for $k \geq 3$, by induction it follows that $a_k \geq c_5 (1 + c_5)^{k-1} a_0$. Let $\xi := (1 + c_5)^{-1} \in (0, 1)$ so that

$$\phi(r^{-2}) h(A_0) = a_0 \leq (1 + c_5) c_5^{-1} \xi^k a_k = (1 + c_5) c_5^{-1} \xi^k \phi(\eta_k^{-2}) h(A_k).$$

Then,

$$h(A_r(0)) \leq (1 + c_5) c_5^{-1} (\phi(r^{-2}))^{-1} \xi^k \phi(\eta_k^{-2}) h(A_{(\kappa/2)^k r}(0))\,.$$

□

Lemma 3.2. *There exists $c = c(\Psi, \kappa, \gamma_1, \gamma_2) > 0$ such that for every $r \in (0, R)$ and every non-negative function u in $\mathbb{R}^d$ which is regular harmonic in $D \cap B(Q, r)$ with respect to X,*

$$u(A_r(Q)) \geq c \phi(r^{-2})^{-1} \int_{B(Q,r)^c} j(|z - Q|) u(z) dz.$$

Proof. Without loss of generality, we may assume that $Q = 0$. Fix $r \in (0, R)$ and let $A := A_r(0)$. Since u is regular harmonic in $D \cap B(0, (1-\kappa/2)r)$ with respect to X, we have

$$\begin{aligned} u(A) &\geq \mathbb{E}_A\left[u(X_{\tau_{D\cap B(0,(1-\kappa/2)r)}}) : X_{\tau_{D\cap B(0,(1-\kappa/2)r)}} \in B(0,r)^c\right] \\ &= \int_{B(0,r)^c} K_{D\cap B(0,(1-\kappa/2)r)}(A,z)u(z)dz \\ &= \int_{B(0,r)^c}\int_{D\cap B(0,(1-\kappa/2)r)} G_{D\cap B(0,(1-\kappa/2)r)}(A,y)J_X(y-z)dyu(z)dz. \end{aligned}$$

Since $B(A, \kappa r/2) \subset D \cap B(0, (1-\kappa/2)r)$, by the domain monotonicity of Green functions,

$$G_{D\cap B(0,(1-\kappa/2)r)}(A,y) \geq G_{B(A,\kappa r/2)}(A,y), \qquad y \in B(A,\kappa r/2).$$

Thus

$$\begin{aligned} u(A) &\geq \int_{B(0,r)^c}\int_{B(A,\kappa r/2)} G_{B(A,\kappa r/2)}(A,y)J_X(y-z)dyu(z)dz \\ &= \int_{B(0,r)^c} K_{B(A,\kappa r/2)}(A,z)u(z)dz \\ &\geq c_1\phi((\kappa r/2)^{-2})^{-1}\int_{B(0,r)^c} j(|A-z|)u(z)dz, \end{aligned}$$

where in the last line we used (2.10). Note that $|A - z| \leq 2|z|$ for $z \in A(0, r, 1)$ and $|A - z| \leq |z| + 1$ for $z \in B(0, 1)^c$. Hence by (1.1), (2.4) and Lemma 2.1,

$$u(A) \geq c_2\phi(r^{-2})^{-1}\int_{B(0,r)^c} j(|z|)u(z)dz.$$

□

Lemma 3.3. *There exist $C_4 = C_4(\Psi, \kappa, \gamma_1, \gamma_2) \geq 1$ and $\xi = \xi(\Psi, \kappa, \gamma_1, \gamma_2) \in (0, 1)$ such that for any $r \in (0, 1)$, and any non-negative function u on $\mathbb{R}^d$ which is regular harmonic in $D \cap B(Q, r)$ with respect to X and vanishes in $D^c \cap B(Q, r)$ we have*

$$\mathbb{E}_x\left[u(X_{\tau_{D\cap B_k}}) : X_{\tau_{D\cap B_k}} \in B(Q,r)^c\right] \leq C_4\xi^k u(x), \qquad x \in D \cap B_k,$$

where $B_k := B(Q, (\kappa/2)^k r)$, $k = 0, 1, 2, \cdots$.

Proof. Without loss of generality, we may assume that $Q = 0$. Fix $r \in (0, R)$. Let $\eta_k := (\kappa/2)^k r$, $B_k := B(0, \eta_k)$ and

$$u_k(x) := \mathbb{E}_x\left[u(X_{\tau_{D\cap B_k}}) : X_{\tau_{D\cap B_k}} \in B(0,r)^c\right], \qquad x \in D \cap B_k.$$

Note that for $x \in D \cap B_{k+1}$,

$$\begin{aligned} u_{k+1}(x) &= \mathbb{E}_x\left[u(X_{\tau_{D\cap B_{k+1}}}) : X_{\tau_{D\cap B_{k+1}}} \in B(0,r)^c\right] \\ &= \mathbb{E}_x\left[u(X_{\tau_{D\cap B_{k+1}}}) : \tau_{D\cap B_{k+1}} = \tau_{D\cap B_k}, X_{\tau_{D\cap B_{k+1}}} \in B(0,r)^c\right] \end{aligned}$$

$$= \mathbb{E}_x \left[u(X_{\tau_{D\cap B_k}}) : \tau_{D\cap B_{k+1}} = \tau_{D\cap B_k}, X_{\tau_{D\cap B_k}} \in B(0,r)^c \right]$$
$$\leq \mathbb{E}_x \left[u(X_{\tau_{D\cap B_k}}) : X_{\tau_{D\cap B_k}} \in B(0,r)^c \right].$$

Thus

$$u_{k+1}(x) \leq u_k(x), \qquad x \in D \cap B_{k+1}. \tag{3.2}$$

Let $A_k := A_{\eta_k}(0)$. Similarly we have

$$\begin{aligned} u_k(A_k) &= \mathbb{E}_{A_k} \left[u(X_{\tau_{D\cap B_k}}) : X_{\tau_{D\cap B_k}} \in B(0,r)^c \right] \\ &\leq \mathbb{E}_{A_k} \left[u(X_{\tau_{B_k}}) : X_{\tau_{D\cap B_k}} \in B(0,r)^c \right] \\ &\leq \int_{B(0,r)^c} K_{B_k}(A_k, z) u(z) dz. \end{aligned}$$

By (2.9), we have

$$K_{B_k}(A_k, z) \leq c_1 j(|z| - \eta_k) \phi(\eta_k^{-2})^{-1}, \qquad z \in B(0,r)^c.$$

Note that $|z| - \eta_k \geq |z|/2$ for $z \in A(0,r,2)$ and $|z| - \eta_k \geq |z| - 1$ for $z \geq 2$. Thus by (1.1), (2.4) and the monotonicity of j,

$$u_k(A_k) \leq c_2 \phi(\eta_k^{-2})^{-1} \int_{B(0,r)^c} j(|z|) u(z) dz, \qquad k = 1, 2, \cdots. \tag{3.3}$$

By Lemma 3.2, we have

$$u(A_0) \geq c_3 \phi(\eta_1^{-2})^{-1} \int_{B(0,r)^c} j(|z|) u(z) dz. \tag{3.4}$$

Thus (3.3)–(3.4) imply that

$$u_k(A_k) \leq c_4 \phi(\eta_1^{-2}) / \phi(\eta_k^{-2}) u(A_0).$$

On the other hand, using Lemma 3.1, we get

$$u(A_0) \leq c_5 (\phi(r^{-2}))^{-1} \xi^k \phi(\eta_k^{-2}) u(A_k).$$

Combining the last two displays and using Lemma 2.1, we get

$$u_k(A_k) \leq c_6 \xi^k \frac{\phi(\eta_1^{-2})}{\phi(r^{-2})} u(A_k) \leq c_7 \xi^k u(A_k).$$

By the boundary Harnack principle, Theorem 2.3 (i), we have

$$\frac{u_k(x)}{u(x)} \leq \frac{u_{k-1}(x)}{u(x)} \leq c_8 \frac{u_{k-1}(A_{k-1})}{u(A_{k-1})} \leq c_9 \xi^k$$

for $k = 1, 2, \cdots$. The proof is now complete. □

Using the boundary Harnack principle and Lemma 3.3 (instead of Lemmas 13 and 14 in [2]), we can repeat the argument of [2, Lemma 16] (which dealt with isotropic stable process) to get the following result. We include the proof from [2] to show that it does not depend on scaling property of stable processes, and on the way make some constants explicit and provide the detailed computation in the end of the induction argument.

Lemma 3.4. *There exist $c = c(\Psi, \gamma_1, \gamma_2, \kappa) > 0$ and $\beta = \beta(\Psi, \gamma_1, \gamma_2, \kappa) > 0$ such that for all $r \in (0, R/2)$ and non-negative functions u and v on $\mathbb{R}^d$ which are regular harmonic in $D \cap B(Q, 2r)$ with respect to X, vanish on $D^c \cap B(Q, 2r)$ and satisfy $u(A_r(Q)) = v(A_r(Q))$, the limit $g = \lim_{D\ni x\to Q} \frac{u(x)}{v(x)}$ exists, and we have*

$$\left|\frac{u(x)}{v(x)} - g\right| \le c\left(\frac{|x-Q|}{r}\right)^{\beta}, \qquad x \in D \cap B(Q, r). \tag{3.5}$$

Proof. Fix $r \in (0, R/2)$. Without loss of generality, we assume that $u(A_r(Q)) = v(A_r(Q)) = 1$. Let $D_0 = D \cap B(Q, r)$. We start the proof by fixing several constants. We first choose $c_1 = c_1(\Psi, \gamma_1, \gamma_2) \ge 10$ such that

$$\sup_{x\in D_0} \frac{u(x)}{v(x)} \le \left(1 + \frac{c_1}{2}\right) \inf_{x\in D_0} \frac{u(x)}{v(x)}. \tag{3.6}$$

This is possible because of Theorem 2.3 (it suffices to choose $1 + c_1/2 \ge C_1^2$). Let $\delta = \delta(\Psi, \gamma_1, \gamma_2) := 1 - \frac{1}{2}C_1^{-1} \in (1/2, 1)$, where C_1 is the constant from Theorem 2.3. We further define $\epsilon = \epsilon(\Psi, \gamma_1, \gamma_2) \in (0, 1/4)$ by

$$\epsilon := \frac{1-\delta}{20c_1},$$

and choose $k_0 = k_0(\Psi, \gamma_1, \gamma_2, \kappa) \in \mathbb{N}$ large enough so that $(1 - \xi^{k_0})^{-1} \le 2$ and $C_4\xi^{k_0} \le \varepsilon(4C_1)^{-1}$ where $C_4 \ge 1$ and $\xi \in (0, 1)$ are the constants from Lemma 3.3.

For $k = 0, 1, \ldots,$ define

$$r_k = (\kappa/2)^{k_0 k} r, \quad B_k = B(Q, r_k), \quad D_k = D \cap B_k, \quad \Pi_k = D_k \setminus D_{k+1}, \quad \Pi_{-1} = B_0^c.$$

For $l = -1, 0, 1, \ldots, k-1$, let

$$u_k^l(x) := \mathbb{E}_x\left[u(X_{\tau_{D_k}}); X_{\tau_{D_k}} \in \Pi_l\right], \qquad x \in \mathbb{R}^d, \tag{3.7}$$

$$v_k^l(x) := \mathbb{E}_x\left[v(X_{\tau_{D_k}}); X_{\tau_{D_k}} \in \Pi_l\right], \qquad x \in \mathbb{R}^d. \tag{3.8}$$

Apply Lemma 3.3 with $\tilde{r} = r_{l+1}$ instead of r. Then $r_k = (\kappa/2)^{k_0 k} r = (\kappa/2)^{k_0(k-l-1)}\tilde{r}$, hence for $k = 0, 1, 2, \ldots$ and $x \in D_k$

$$u_k^l(x) \le C_4(\xi^{k_0})^{k-l-1} u(x), \qquad l = -1, 0, 1, \ldots, k-2. \tag{3.9}$$

Then since $\sum_{l=-1}^{k-2} (\xi^{k_0})^{k-1-l} = \xi^{k_0} \sum_{n=0}^{k-1} (\xi^{k_0})^n \le \xi^{k_0}(1 - \xi^{k_0})^{-1} \le 2\xi^{k_0}$, we have that for $k = 1, 2, \ldots$ and $x \in D_k$,

$$\sum_{l=-1}^{k-2} u_k^l(x) \le 2C_4\xi^{k_0} u(x). \tag{3.10}$$

Since $2C_4\xi^{k_0} \le \varepsilon/2 < 1/2$, for $k = 1, 2, \ldots, l = -1, 0, 1, \ldots, k-2$, we have $\sum_{l=-1}^{k-2} u_k^l(x) \le u_k^{k-1}(x)$ for all $x \in D_k$ and, by (3.10)

$$u_k^l(x) \le \epsilon^{k-1-l} u_k^{k-1}(x), \qquad x \in D_k. \tag{3.11}$$

By symmetry we also have that for $k = 1, 2, \ldots, l = -1, 0, 1, \ldots, k-2$,

$$v_k^l(x) \le \epsilon^{k-1-l} v_k^{k-1}(x), \qquad x \in D_k. \tag{3.12}$$

Define $\zeta = \zeta(\Psi, \gamma_1, \gamma_2) \in (1/2, 1)$ by

$$\zeta = \sqrt{\frac{1+\delta}{2}}.$$

We claim that for all $l = 0, 1, \ldots$

$$\sup_{x \in D_l} \frac{u(x)}{v(x)} \le (1 + c_1\zeta^l) \inf_{x \in D_l} \frac{u(x)}{v(x)}. \tag{3.13}$$

By Theorem 2.3 with $y = A_r(Q)$ we have that

$$C_1^{-1} \le \frac{u(x)}{v(x)} \le C_1, \qquad x \in D_0 = D \cap B(Q, r). \tag{3.14}$$

We show that (3.13) and (3.14) imply the statement of the lemma. Indeed,

$$\sup_{x \in D_l} \frac{u(x)}{v(x)} - \inf_{x \in D_l} \frac{u(x)}{v(x)} \le c_1\zeta^l \inf_{x \in D_l} \frac{u(x)}{v(x)} \le C_1 c_1 \zeta^l, \qquad l = 0, 1, \ldots.$$

Since the right-hand side goes to zero as $l \to \infty$, the same is valid for the left-hand side proving that the limit $g = \lim_{D \ni x \to Q} \frac{u(x)}{v(x)}$ exists. Further, for $x \in D \cap B(Q, r)$ there is a unique $l \ge 1$ such that $x \in \Pi_{l-1} \subset D_{l-1}$. Thus $x \notin D_l$ implying $|x - Q| \ge r_l = (\kappa/2)^{k_0 l} r$, that is

$$l \ge \frac{\log \frac{r}{|x-Q|}}{k_0 \log \frac{2}{\kappa}}.$$

Let $\beta := -\frac{\log \zeta}{k_0 \log \frac{2}{\kappa}}$. Then $\beta = \beta(\Psi, \gamma_1, \gamma_2, \kappa)$ and

$$\begin{aligned}\left|\frac{u(x)}{v(x)} - g\right| &\le C_1 c_1 \zeta^l \le C_1 c_1 \zeta^{\frac{\log \frac{r}{|x-Q|}}{k_0 \log \frac{2}{\kappa}}} \\ &= C_1 c_1 \left(\frac{r}{|x-Q|}\right)^{\frac{\log \zeta}{k_0 \log \frac{2}{\kappa}}} = C_1 c_1 \left(\frac{|x-Q|}{r}\right)^{\beta}.\end{aligned}$$

We now prove (3.13) by induction. By (3.6) we wee that (3.13) holds for $l = 0$. Again by (3.6) and the fact that $\zeta > 1/2$ we have

$$\begin{aligned}\sup_{x \in D_1} \frac{u(x)}{v(x)} &\le \sup_{x \in D_0} \frac{u(x)}{v(x)} \le \left(1 + \frac{c_1}{2}\right) \inf_{x \in D_0} \frac{u(x)}{v(x)} \\ &\le (1 + c_1\zeta) \inf_{x \in D_0} \frac{u(x)}{v(x)} \le (1 + c_1\zeta) \inf_{x \in D_1} \frac{u(x)}{v(x)},\end{aligned}$$

hence (3.13) holds also for $l = 1$. Let $k = 0, 1, \ldots$ and assume that (3.13) holds for $l = 0, 1, 2, \ldots, k$. By the definitions (3.7)–(3.8), and the regular harmonicity of u and v, we have

$$u(x) = \sum_{l=-1}^{k} u_{k+1}^{l}(x), \qquad x \in D_{k+1}, \tag{3.15}$$

$$v(x) = \sum_{l=-1}^{k} v_{k+1}^{l}(x), \qquad x \in D_{k+1}. \tag{3.16}$$

For any function f on a set A we define

$$\mathrm{Osc}_A f = \sup_{x \in A} f(x) - \inf_{x \in A} f(x).$$

Let $g(x) := u_{k+1}^{k}(x)/v_{k+1}^{k}(x)$, $x \in D_k$. We claim that

$$\mathrm{Osc}_{D_{k+2}} g \le \delta\, \mathrm{Osc}_{D_k} g. \tag{3.17}$$

Recall that $\delta = 1 - \frac{1}{2}C_1^{-1} \in (1/2, 1)$. Let $m_1 := \inf_{x \in D_k} g(x)$ and $m_2 := \sup_{x \in D_k} g(x)$. By (3.7), (3.8) and Theorem 2.3, it holds that $0 < m_1 \le m_2 < \infty$. If $m_1 = m_2$, then both $\mathrm{Osc}_{D_{k+2}} g$ and $\mathrm{Osc}_{D_k} g$ are zero so (3.17) holds trivially. Otherwise, let

$$\widetilde{g}(x) := \frac{g(x) - m_1}{m_2 - m_1} = \frac{u_{k+1}^{k}(x) - m_1 v_{k+1}^{k}(x)}{(m_2 - m_1) v_{k+1}^{k}(x)}, \qquad x \in D_k.$$

Note that $\widetilde{g}$ is the quotient of two non-negative functions regular harmonic in D_{k+1} with respect to X. Clearly $\mathrm{Osc}_{D_k} \widetilde{g} = 1$. Furthermore,

$$\mathrm{Osc}_{D_{k+2}} g = \mathrm{Osc}_{D_{k+2}} \widetilde{g} \cdot \mathrm{Osc}_{D_k} g. \tag{3.18}$$

This is clear from $g(x) = (m_2 - m_1)\widetilde{g}(x) + m_1 = (\mathrm{Osc}_{D_k} g)\widetilde{g}(x) + m_1$. If $\sup_{D_{k+2}} \widetilde{g}(x) \le 1/2$, then $\mathrm{Osc}_{D_{k+2}} \widetilde{g} \le 1/2$, and it follows from (3.18) that

$$\mathrm{Osc}_{D_{k+2}} g \le \frac{1}{2} \mathrm{Osc}_{D_k} g. \tag{3.19}$$

If, on the other hand, $\sup_{D_{k+2}} \widetilde{g}(x) > 1/2$, we apply Theorem 2.3 to the functions $\widetilde{u}(x) = u_{k+1}^{k}(x) - m_1 v_{k+1}^{k}(x)$ and $\widetilde{v}(x) = v_{k+1}^{k}(x)$ to conclude that

$$C_1^{-1} \frac{\widetilde{u}(y)}{\widetilde{v}(y)} \le \frac{\widetilde{u}(x)}{\widetilde{v}(x)} \le C_1 \frac{\widetilde{u}(y)}{\widetilde{v}(y)}, \qquad x, y \in D_{k+1}.$$

This can be written as

$$C_1^{-1} \widetilde{g}(y) \le \widetilde{g}(x) \le C_1 \widetilde{g}(y), \qquad x, y \in D_{k+1}.$$

Hence, for all $x \in D_{k+2}$, we have

$$\widetilde{g}(x) \ge C_1^{-1} \sup_{y \in D_{k+2}} \widetilde{g}(y) \ge \frac{1}{2} C_1^{-1}.$$

Therefore, $\inf_{y\in D_{k+2}}\widetilde{g}(y)\ge \frac{1}{2}C_1^{-1}$, and since $\widetilde{g}\le 1$, we get that

$$\mathrm{Osc}_{D_{k+2}}\widetilde{g}\le 1-\frac{1}{2}C_1^{-1}=\delta. \tag{3.20}$$

By (3.18)–(3.20) we get (3.17).

We claim that

$$\inf_{x\in D_{k+2}}\frac{u_{k+1}^k(x)}{v_{k+1}^k(x)}\ge \inf_{x\in D_{k+1}}\frac{u_{k+1}^k(x)}{v_{k+1}^k(x)}\ge \inf_{x\in D_k}\frac{u(x)}{v(x)}\ge \inf_{x\in D_i}\frac{u(x)}{v(x)},\qquad i=0,\dots k. \tag{3.21}$$

Indeed, let $\eta:=\inf_{x\in D_k}\frac{u(x)}{v(x)}$ so that $u(x)\ge \eta v(x)$ on D_k. Then for $x\in D_{k+1}$,

$$\begin{aligned} u_{k+1}^k(x) &= \mathbb{E}_x[u(X_{\tau_{D_{k+1}}});X_{\tau_{D_{k+1}}}\in \Pi_k] \\ &\ge \eta\mathbb{E}_x[v(X_{\tau_{D_{k+1}}});X_{\tau_{D_{k+1}}}\in \Pi_k]=\eta v_{k+1}^k(x), \end{aligned}$$

so the second inequality of (3.21) holds. The first and third equalities of (3.21) are trivial. Similarly we have

$$\sup_{x\in D_{k+2}}\frac{u_{k+1}^k(x)}{v_{k+1}^k(x)}\le \sup_{x\in D_{k+1}}\frac{u_{k+1}^k(x)}{v_{k+1}^k(x)}\le \sup_{x\in D_k}\frac{u(x)}{v(x)}. \tag{3.22}$$

Combining (3.21), (3.22) and (3.17) we get

$$\frac{\sup_{x\in D_{k+2}}\frac{u_{k+1}^k(x)}{v_{k+1}^k(x)}}{\inf_{x\in D_{k+2}}\frac{u_{k+1}^k(x)}{v_{k+1}^k(x)}}-1\le \delta\left(\frac{\sup_{x\in D_k}\frac{u(x)}{v(x)}}{\inf_{x\in D_k}\frac{u(x)}{v(x)}}-1\right). \tag{3.23}$$

Using (3.23) and (3.13) with $l=k$ (the induction hypothesis) we obtain

$$\sup_{x\in D_{k+2}}\frac{u_{k+1}^k(x)}{v_{k+1}^k(x)}=(1+c_1\delta\rho\zeta^k)\inf_{x\in D_{k+2}}\frac{u_{k+1}^k(x)}{v_{k+1}^k(x)} \tag{3.24}$$

with a suitably chosen $\rho\in[0,1]$ (independent of $x\in D_{k+2}$). Indeed,

$$M:=\frac{\sup_{x\in D_{k+2}}\frac{u_{k+1}^k(x)}{v_{k+1}^k(x)}}{\inf_{x\in D_{k+2}}\frac{u_{k+1}^k(x)}{v_{k+1}^k(x)}}\le 1+\delta\left(\frac{\sup_{x\in D_k}\frac{u(x)}{v(x)}}{\inf_{x\in D_k}\frac{u(x)}{v(x)}}-1\right)\le 1+c_1\delta\zeta^k.$$

Thus, $1\le M\le 1+c_1\delta\zeta^k$ which implies that there exists $\rho=\rho(c_1,\zeta,k)\in[0,1]$ such that $M=1+c_1\delta\rho\zeta^k$. Next, by symmetry and (3.24) we have

$$\sup_{x\in D_{k+2}}\frac{v_{k+1}^k(x)}{u_{k+1}^k(x)}=(1+c_1\delta\rho\zeta^k)\inf_{x\in D_{k+2}}\frac{v_{k+1}^k(x)}{u_{k+1}^k(x)}. \tag{3.25}$$

Combining (3.21) and (3.13) we get

$$\sup_{x\in D_l}\frac{u(x)}{v(x)}\le (1+c_1\zeta^l)\inf_{x\in D_{k+2}}\frac{u_{k+1}^k(x)}{v_{k+1}^k(x)},\qquad l=0,1,\dots,k. \tag{3.26}$$

Now we fix $x \in D_{k+2}$. Then by (3.15) and (3.16),

$$\begin{aligned} \frac{u(x)}{v(x)} &= \frac{\sum_{l=0}^k u_{k+1}^l(x) + u_{k+1}^{-1}(x)}{\sum_{l=0}^k v_{k+1}^l(x) + v_{k+1}^{-1}(x)} \\ &\le \frac{\sum_{l=0}^k u_{k+1}^l(x) + \epsilon^{k+1} u_{k+1}^k(x)}{\sum_{l=0}^k v_{k+1}^l(x)} \\ &\le (1+\epsilon^{k+1}) \frac{\sum_{l=0}^k u_{k+1}^l(x)}{\sum_{l=0}^k v_{k+1}^l(x)}, \end{aligned} \tag{3.27}$$

where in the first inequality we used (3.11) with $l=-1$. Now we apply [2, Lemma 15] with

$$\begin{aligned} &u_0 = u_{k+1}^k(x), \quad v_0 = v_{k+1}^k(x), \quad a = (1+c_1\delta\rho\zeta^k) \inf_{y \in D_{k+2}} \frac{u_{k+1}^k(y)}{v_{k+1}^k(y)}, \\ &u_i = u_{k+1}^{k-i}(x), \quad v_i = v_{k+1}^{k-i}(x), \quad b_i = (1+c_1\zeta^{k-i}) \inf_{y \in D_{k+2}} \frac{u_{k+1}^k(y)}{v_{k+1}^k(y)}, \\ &\epsilon_i = \epsilon^i, \quad i = 1, \dots, k. \end{aligned}$$

We need to check the conditions of [2, Lemma 15]. [2, (5.16)], $a \le b_i$, $i = 1, \dots, k$, is immediate. [2, (5.18)], $u_0 \le a v_0$, is true by (3.24). [2, (5.17)] consists of two parts: $u_i \le b_i v_i$ and $v_i \le \epsilon_i v_0$ for $i = 1, \dots, k$. The first part follows from

$$\sup_{y \in D_{k+1}} \frac{u_{k+1}^{k-i}(y)}{v_{k+1}^{k-i}(y)} \le \sup_{y \in D_{k-i}} \frac{u(y)}{v(y)} \le (1+c_1\zeta^{k-i}) \inf_{y \in D_{k+2}} \frac{u_{k+1}^{k-i}(y)}{v_{k+1}^{k-i}(y)},$$

where the first inequality is (3.22) and the second inequality is (3.26). The second part of [2, (5.17)] is precisely (3.12). Now we apply [2, Lemma 15] to conclude that

$$\begin{aligned} &\sum_{i=0}^k u_{k+1}^{k-i}(x) \\ &\le \Bigg[(1+c_1\delta\rho\zeta^k) \inf_{y \in D_{k+2}} \frac{u_{k+1}^k(y)}{v_{k+1}^k(y)} \\ &\qquad + \sum_{i=1}^k c_1(\zeta^{k-i} - \delta\rho\zeta^k) \inf_{y \in D_{k+2}} \frac{u_{k+1}^k(y)}{v_{k+1}^k(y)} \cdot \epsilon^i \Bigg] \cdot \sum_{i=0}^k v_{k+1}^{k-i}(x) \\ &\le \left[1 + c_1\delta\rho\zeta^k + c_1 \sum_{i=1}^k \zeta^{k-i}\epsilon^i \right] \inf_{y \in D_{k+2}} \frac{u_{k+1}^k(y)}{v_{k+1}^k(y)} \cdot \sum_{i=0}^k v_{k+1}^{k-i}(x). \end{aligned}$$

Hence, letting $\tau := (1+\epsilon^{k+1})(1 + c_1\delta\rho\zeta^k + c_1 \sum_{i=1}^k \zeta^{k-i}\epsilon^i)$ and applying (3.27) we get

$$\frac{u(x)}{v(x)} \le (1+\epsilon^{k+1}) \frac{\sum_{i=0}^k u_{k+1}^{k-i}(x)}{\sum_{i=0}^k v_{k+1}^{k-i}(x)} \le \tau \cdot \inf_{y \in D_{k+2}} \frac{u_{k+1}^k(y)}{v_{k+1}^k(y)}. \tag{3.28}$$

By symmetry,

$$\frac{v(x)}{u(x)} \le \tau \cdot \inf_{y\in D_{k+2}} \frac{v^k_{k+1}(y)}{u^k_{k+1}(y)}. \tag{3.29}$$

Now, (3.28), (3.24) and (3.29) imply that

$$\begin{aligned}\sup_{y\in D_{k+2}} \frac{u(y)}{v(y)} &\le \tau \cdot \inf_{y\in D_{k+2}} \frac{u^k_{k+1}(y)}{v^k_{k+1}(y)} \\ &= \frac{\tau}{1+c_1\delta\rho\zeta^k} \sup_{y\in D_{k+2}} \frac{u^k_{k+1}(y)}{v^k_{k+1}(y)} \\ &\le \frac{\tau^2}{1+c_1\delta\rho\zeta^k} \inf_{y\in D_{k+2}} \frac{u(y)}{v(y)}.\end{aligned}$$

The proof of the induction step will be finished by showing that

$$\frac{\tau^2}{1+c_1\delta\rho\zeta^k} \le 1 + c_1\zeta^{k+2}. \tag{3.30}$$

First note that

$$\tau = (1+\epsilon^{k+1})\left(1 + c_1\left(\delta\rho + \frac{\epsilon}{\zeta-\epsilon}\left(1-\left(\frac{\epsilon}{\zeta}\right)^k\right)\right)\zeta^k\right).$$

Let

$$\widetilde{\tau} := (1+\epsilon^{k+1})\left(1 + c_1\left(\delta\rho + \frac{\epsilon}{\zeta-\epsilon}\right)\zeta^k\right),$$

so that $\tau \le \widetilde{\tau}$. Then

$$\begin{aligned}\tau^2 \le \widetilde{\tau}^2 &= (1+\epsilon^{k+1})^2\left(1 + 2c_1\left(\delta\rho + \frac{\epsilon}{\zeta-\epsilon}\right)\zeta^k + c_1^2\left(\delta\rho + \frac{\epsilon}{\zeta-\epsilon}\right)^2\zeta^{2k}\right) \\ &\le (1+3\epsilon^{k+1})A,\end{aligned}$$

where

$$A := 1 + c_1\left(2\delta\rho + \frac{2\epsilon}{\zeta-\epsilon}\right)\zeta^k + c_1^2\left(\delta\rho + \frac{\epsilon}{\zeta-\epsilon}\right)^2\zeta^{2k}.$$

Further, let

$$B := (1+c_1\delta\rho\zeta^k)(1+c_1\zeta^{k+2}) = 1 + c_1(\delta\rho+\zeta^2)\zeta^k + c_1^2(\delta\rho\zeta^2)\zeta^{2k}.$$

We will show that $B-(1+3\epsilon^{k+1})A = B-A-3\epsilon^{k+1}A \ge 0$. Since $B-(1+3\epsilon^{k+1})A \le B-\tau^2$, this will prove what want. We now prove $B-(1+3\epsilon^{k+1})A = B-A-3\epsilon^{k+1}A \ge 0$. Note that $\epsilon \in (0, 1/200) \subset (0, 1/4)$ and thus $\zeta - \epsilon > 1/2 - 1/4 = 1/4$. We will need the following estimate:

$$\frac{\epsilon}{\zeta-\epsilon} \le 4\epsilon < 1. \tag{3.31}$$

Since $\delta \in (1/2, 1)$, $\rho \in [0,1]$ and $\zeta < 1$ we have

$$A \le 1 + c_1(2+2) + c_1^2(1+1)^2 = 1 + 4c_1 + 4c_1^2 = (1+2c_1)^2 \le (3c_1)^2 = 9c_1^2\,. \quad (3.32)$$

Note that it follows from (3.31) that

$$\zeta^2 - \delta - 8\epsilon = \frac{1+\delta}{2} - \delta - \frac{2(1-\delta)}{5c_1} = (1-\delta)\left(\frac{1}{2} - \frac{2}{5c_1}\right) \ge \frac{2}{5}(1-\delta)\,,$$

where we used that $1/2 - 2/(5c_1) \ge 1/2 - 2/50 = 23/50 \ge 2/5$ which follows immediately from $c_1 \ge 10$. Thus,

$$\begin{aligned}
B - A &= c_1\left(\delta\rho + \zeta^2 - 2\delta\rho - \frac{2\epsilon}{\zeta-\epsilon}\right)\zeta^k \\
&\quad + c_1^2\left(\delta\rho\zeta^2 - \delta^2\rho^2 - 2\delta\rho\frac{\epsilon}{\zeta-\epsilon} - \frac{\epsilon^2}{(\zeta-\epsilon)^2}\right)\zeta^{2k} \\
&= c_1\left(\zeta^2 - \delta\rho - \frac{2\epsilon}{\zeta-\epsilon}\right)\zeta^k + c_1^2\left[\delta\rho\left(\zeta^2 - \delta\rho - \frac{2\epsilon}{\zeta-\epsilon}\right) - \frac{\epsilon^2}{(\zeta-\epsilon)^2}\right]\zeta^{2k} \\
&> c_1(\zeta^2 - \delta\rho - 8\epsilon)\zeta^k - c_1^2\frac{\epsilon^2}{(\zeta-\epsilon)^2}\zeta^{2k} \\
&\ge c_1(\zeta^2 - \delta - 8\epsilon)\zeta^k - c_1^2(4\epsilon)^2\zeta^{2k} \\
&\ge \frac{2}{5}c_1(1-\delta)\left(\frac{1+\delta}{2}\right)^{k/2} - c_1^2\left(\frac{1-\delta}{5c_1}\right)^2\left(\frac{1+\delta}{2}\right)^k \\
&= \frac{2}{5}c_1(1-\delta)\left(\frac{1+\delta}{2}\right)^{k/2} - \frac{(1-\delta)^2}{25}\left(\frac{1+\delta}{2}\right)^k \\
&= (1-\delta)\left(\frac{1+\delta}{2}\right)^{k/2}\left(\frac{2}{5}c_1 - \frac{1-\delta}{25}\left(\frac{1+\delta}{2}\right)^{k/2}\right) \\
&\ge (1-\delta)\left(\frac{1+\delta}{2}\right)^{k/2}\left(4 - \frac{1}{25}\left(\frac{1+\delta}{2}\right)^{k/2}\right), \qquad (3.33)
\end{aligned}$$

where in the third line above we used (3.31) and neglected the first term in the square brackets in the line above, in the fourth line we used that $\rho \le 1$ and again (3.31), and in the last line we used that $c_1 \ge 10$.

By combining (3.32) and (3.33) we get

$$\begin{aligned}
B - A - 3\epsilon^{k+1}A &\ge B - A - 3\epsilon^{k+1}\cdot 9c_1^2 \\
&\ge (1-\delta)\left(\frac{1+\delta}{2}\right)^{k/2}\left(4 - \frac{1}{25}\left(\frac{1+\delta}{2}\right)^{k/2}\right) \\
&\quad -27\left(\frac{1-\delta}{20c_1}\right)^{k-1}\left(\frac{1-\delta}{20c_1}\right)^2 c_1^2 \\
&= (1-\delta)\left(\frac{1+\delta}{2}\right)^{k/2}\left(4 - \frac{1}{25}\left(\frac{1+\delta}{2}\right)^{k/2}\right)
\end{aligned}$$

$$
\begin{aligned}
&-\frac{27}{20^2}\left(\frac{1-\delta}{20c_1}\right)^{k-1}(1-\delta)^2\\
&\geq (1-\delta)\left(\frac{1+\delta}{2}\right)^{k/2}\left(4-\frac{1}{25}\left(\frac{1+\delta}{2}\right)^{k/2}\right)\\
&\quad -27\left(\frac{1-\delta}{20}\right)^{k+1}\\
&\geq 3(1-\delta)\left(\frac{1+\delta}{2}\right)^{k/2}-27\left(\frac{1-\delta}{20}\right)^{k+1}\\
&\geq 3\frac{1}{2}\left(\frac{1}{2}\right)^{k/2}-27\left(\frac{1}{20}\right)^{k+1}\\
&= 3\left(\frac{1}{2}\right)^{k/2+1}-27\left(\frac{1}{20}\right)^{k+1}>0
\end{aligned}
$$

for all $k \geq 1$. In the penultimate line we used that $1-\delta \geq 1/2$, $(1+\delta)/2 \geq 1/2$ and $1-\delta < 1$.

The proof is now complete. □

An alternative approach to the oscillation reduction in case of rotationally stable processes is given in [6, Lemma 8]. We note that for non-local operators the oscillation reduction technique seems much harder than for the Laplacian, because the subtraction used in this process may destroy global nonnegativity. In this regard, we note that there is a gap of this nature in the proof of [11, Lemma 3.3].

In the remainder of this section, we assume that $Q \in \partial D$. Since D is Greenian, the Green function $G_D(x,y)$, $x,y \in D$, is well defined. Fix $x_0 \in D$ and set

$$
M_D(x,y) = \frac{G_D(x,y)}{G_D(x_0,y)}, \qquad x,y \in D\,, y \neq x_0.
$$

Let $r < \frac{1}{2}\min\{\mathrm{dist}(x,Q), \mathrm{dist}(x_0,Q)\}$. Since $G_D(x,\cdot)$ and $G_D(x_0,\cdot)$ are regular harmonic in $D \cap B(Q,2r)$ and vanish in $D^c \cap B(Q,2r)$, using Lemma 3.4, one immediately gets the following.

Corollary 3.1. *The limit $M_D(x,Q) := \lim_{D\ni y\to Q} M_D(x,y)$ exists. Furthermore, there exist positive constants c and β depending on $(\Psi,\gamma_1,\gamma_2,\kappa)$ such that for any $r \in (0, \frac{1}{2}(R \wedge \mathrm{dist}(x_0,Q))$, any $y \in D \cap B(Q,r)$ and any $x \in D \setminus B(Q,2r)$,*

$$
|M_D(x,y) - M_D(x,Q)| \leq cM_D(x, A_r(Q))\left(\frac{|y-Q|}{r}\right)^{\beta}. \tag{3.34}
$$

Proof. Put

$$
u(y) := \frac{G_D(x,y)}{G_D(x,A_r(Q))}, \qquad v(y) := \frac{G_D(x_0,y)}{G_D(x_0,A_r(Q))}.
$$

Since u and v satisfy the assumptions of Lemma 3.4, there exists

$$g := \lim_{D\ni y\to Q} \frac{u(y)}{v(y)} = \lim_{D\ni y\to Q} \frac{M_D(x,y)}{M_D(x,A_r(Q))}\,.$$

This implies the existence of $M_D(x,Q) := \lim_{D\ni y\to Q} M_D(x,y)$. Furthermore, by (3.5) we have that

$$\left|\frac{u(y)}{v(y)} - g\right| \le c\left(\frac{|y-Q|}{r}\right)^{\beta}, \quad \text{for all } y\in D\cap B(Q,r)\,.$$

Equivalently,

$$\left|\frac{M_D(x,y)}{M_D(x,A_r(Q))} - \frac{M_D(x,Q)}{M_D(x,A_r(Q))}\right| \le c\left(\frac{|y-Q|}{r}\right)^{\beta}, \quad \text{for all } y\in D\cap B(Q,r)\,,$$

which is (3.34). □

Recall that X^D is the process X killed upon exiting D. As the process X^D satisfies Hypothesis (B) in [25], D has a Martin boundary $\partial_M D$ with respect to X^D satisfying the following properties:

(M1) $D\cup\partial_M D$ is a compact metric space (with the metric denoted by d);
(M2) D is open and dense in $D\cup\partial_M D$, and its relative topology coincides with its original topology;
(M3) $M_D(x,\,\cdot\,)$ can be uniquely extended to $\partial_M D$ in such a way that
  (a) $M_D(x,y)$ converges to $M_D(x,w)$ as $y\to w\in\partial_M D$ in the Martin topology;
  (b) for each $w\in D\cup\partial_M D$ the function $x\to M_D(x,w)$ is excessive with respect to X^D;
  (c) the function $(x,w)\to M_D(x,w)$ is jointly continuous on $D\times(D\cup\partial_M D)$ in the Martin topology and
  (d) $M_D(\cdot,w_1)\neq M_D(\cdot,w_2)$ if $w_1\neq w_2$ and $w_1,w_2\in\partial_M D$.

We will say that a point $w\in\partial_M D$ is a finite Martin boundary point if there exists a bounded sequence $(y_n)_{n\ge1}\subset D$ converging to w in the Martin topology. The finite part of the Martin boundary will be denoted by $\partial_M^f D$. Recall that a point w on the Martin boundary $\partial_M D$ of D is said to be associated with $Q\in\partial D$ if there is a sequence $(y_n)_{n\ge1}\subset D$ converging to w in the Martin topology and to Q in the Euclidean topology. The set of Martin boundary points associated with Q is denoted by $\partial_M^Q D$.

Proposition 3.1. *$\partial_M^Q D$ consists of exactly one point.*

Proof. We first note that $\partial_M^Q D$ is not empty. Indeed, let $(y_n)_{n\ge1}\subset D$ converge to Q in the Euclidean topology. Since $D\cup\partial_M D$ is a compact metric space with the Martin metric, there exist a subsequence $(y_{n_k})_{k\ge1}$ and $w\in D\cup\partial_M D$ such that $\lim_{k\to\infty} d(y_{n_k},w)=0$. Clearly, $w\notin D$ (since relative topologies on D are equivalent). Thus we have found a sequence $(y_{n_k})_{k\ge1}\subset D$ which converges to $w\in\partial_M D$ in the Martin topology and to Q in the Euclidean topology.

Let $w \in \partial_M^Q D$ and let $M_D(\cdot, w)$ be the corresponding Martin kernel. If $(y_n)_{n\geq 1} \subset D$ is a sequence converging to w in the Martin topology and to Q in the Euclidean topology, then, by (M3)(a), $M_D(x, y_n)$ converge to $M_D(x, w)$. On the other hand, $|y_n - Q| \to 0$, thus by Corollary 3.1,

$$\lim_{n\to\infty} M_D(x, y_n) = M_D(x, Q).$$

Hence, for each $w \in \partial_M^Q D$ it holds that $M_D(\cdot, w) = M_D(\cdot, Q)$. Since, by (M3)(d), for two different Martin boundary points $w^{(1)}$ and $w^{(2)}$ it always holds that $M_D(\cdot, w^{(1)}) \neq M_D(\cdot, w^{(2)})$, we conclude that $\partial_M^Q D$ consists of exactly one point. □

Because of the proposition above, we will also use Q to denote the point on the Martin boundary $\partial_M^Q D$ associated with Q. Note that it follows from the proof of Proposition 3.1 that if $(y_n)_{n\geq 1}$ converges to Q in the Euclidean topology, then it also converges to Q in the Martin topology.

For $\epsilon > 0$ let

$$K_\epsilon := \{w \in \partial_M D : d(w, Q) \geq \epsilon\} \tag{3.35}$$

be a closed subset of $\partial_M D$. By the definition of Martin boundary, for each $w \in K_\epsilon$ there exists a sequence $(y_n^w)_{n\geq 1} \subset D$ such that $\lim_{n\to\infty} d(y_n^w, w) = 0$. Without loss of generality we may assume that $d(y_n^w, w) < \frac{\epsilon}{2}$ for all $n \geq 1$.

Lemma 3.5. *There exists $c = c(\epsilon) > 0$ such that $|y_n^w - Q| \geq c$ for all $w \in K_\epsilon$ and all $n \geq 1$.*

Proof. Suppose the lemma is not true. Then $\{y_n^w : w \in K_\epsilon, n \in \mathbb{N}\}$ contains a sequence $(y_{n_k}^{w_k})_{k\geq 1}$ such that $\lim_{k\to\infty} |y_{n_k}^{w_k} - Q| = 0$. Then also $\lim_{k\to\infty} d(y_{n_k}^{w_k}, Q) = 0$. On the other hand,

$$d(y_{n_k}^{w_k}, Q) \geq d(w_k, Q) - d(y_{n_k}^{w_k}, w_k) \geq \epsilon - \frac{\epsilon}{2} = \frac{\epsilon}{2}.$$

This contradiction proves the claim. □

We continue by showing that $M_D(\cdot, Q)$ is harmonic in D with respect to X.

Lemma 3.6. *For every bounded open $U \subset \overline{U} \subset D$ and every $x \in D$, $M_D(X_{\tau_U}, Q)$ is $\mathbb{P}_x$-integrable.*

Proof. Let $(y_m)_{m\geq 1}$ be a sequence in $D\setminus\overline{U}$ such that $|y_m - Q| \to 0$. Then $M_D(\cdot, y_m)$ is regular harmonic in U. Hence, by Fatou's lemma and Corollary 3.1,

$$\begin{aligned}\mathbb{E}_x[M_D(X_{\tau_U}, Q)] &= \mathbb{E}_x[\lim_{m\to\infty} M_D(X_{\tau_U}, y_m)] \leq \liminf_{m\to\infty} \mathbb{E}_x[M_D(X_{\tau_U}, y_m)] \\ &= \liminf_{m\to\infty} M_D(x, y_m) = M_D(x, Q) < \infty.\end{aligned}$$

□

Using the results above, we can get the following result.

Lemma 3.7. *Suppose that D is either (i) a bounded open set or (ii) an unbounded open set and X is transient. For any $x \in D$ and $\rho \in (0, R \wedge (\delta_D(x)/2)]$,*

$$M_D(x,Q) = \mathbb{E}_x[M_D(X_{\tau_{B(x,\rho)}}, Q)]\,.$$

Proof. Fix $x \in D$ and a positive $r < R \wedge \frac{\delta_D(x)}{2}$. Let

$$\eta_m := \left(\frac{\kappa}{2}\right)^m r \quad \text{and} \quad z_m = A_{\eta_m}(Q), \quad m = 0, 1, \ldots.$$

Note that

$$B(z_m, \eta_{m+1}) \subset D \cap B(Q, \frac{\eta_m}{2}) \subset D \cap B(Q, \eta_m) \subset D \cap B(Q, r) \subset D \setminus B(x, r)$$

for all $m \geq 0$. Thus by the harmonicity of $M_D(\cdot, z_m)$, we have

$$M_D(x, z_m) = \mathbb{E}_x\left[M_D(X_{\tau_{B(x,r)}}, z_m)\right].$$

On the other hand, by Theorem 2.3, there exist $m_0 = m_0(\kappa) \geq 2$ and $c_1 = c_1(\Psi, \gamma_1, \gamma_2, \kappa) > 0$ such that for every $w \in D \setminus B(Q, \eta_m)$ and $y \in D \cap B(Q, \eta_{m+1})$,

$$M_D(w, z_m) = \frac{G_D(w, z_m)}{G_D(x_0, z_m)} \leq c_1 \frac{G_D(w, y)}{G_D(x_0, y)} = c_1 M_D(w, y), \qquad m \geq m_0.$$

Letting $y \to Q$ we get

$$M_D(w, z_m) \leq c_1 M_D(w, Q), \qquad m \geq m_0, w \in D \setminus B(Q, \eta_m). \tag{3.36}$$

To prove this lemma, it suffices to show that $\{M_D(X_{\tau_{B(x,r)}}, z_m) : m \geq m_0\}$ is $\mathbb{P}_x$-uniformly integrable. Since $M_D(X_{\tau_{B(x,r)}}, Q)$ is integrable by Lemma 3.6, for any $\epsilon > 0$, there is an $N_0 > 1$ such that

$$\mathbb{E}_x\left[M_D(X_{\tau_{B(x,r)}}, Q); M_D(X_{\tau_{B(x,r)}}, Q) > N_0/c_1\right] < \frac{\epsilon}{2c_1}. \tag{3.37}$$

Note that by (3.36) and (3.37),

$$\begin{aligned}
&\mathbb{E}_x\left[M_D(X_{\tau_{B(x,r)}}, z_m); M_D(X_{\tau_{B(x,r)}}, z_m) > N_0 \text{ and } X_{\tau_{B(x,r)}} \in D \setminus B(Q, \eta_m)\right] \\
&\leq c_1 \mathbb{E}_x\left[M_D(X_{\tau_{B(x,r)}}, Q); c_1 M_D(X_{\tau_{B(x,r)}}, Q) > N_0\right] < c_1 \frac{\epsilon}{2c_1} = \frac{\epsilon}{2}.
\end{aligned}$$

By (2.9), we have for $m \geq m_0$,

$$\begin{aligned}
&\mathbb{E}_x\left[M_D(X_{\tau_{B(x,r)}}, z_m); X_{\tau_{B(x,r)}} \in D \cap B(Q, \eta_m)\right] \\
&= \int_{D \cap B(Q,\eta_m)} M_D(w, z_m) K_{B(x,r)}(x, w) dw \\
&\leq c_2 \phi(r^{-2})^{-1} \int_{D \cap B(Q,\eta_m)} M_D(w, z_m) j(|w - x| - r) dw
\end{aligned}$$

for some $c_2 = c_2(\Psi, \gamma_1, \gamma_2) > 0$. Since $|w - x| \geq |x - Q| - |Q - w| \geq \delta_D(x) - \eta_m \geq \frac{7}{4}r$, using (2.4) and (2.2), we get that

$$\mathbb{E}_x\left[M_D(X_{\tau_{B(x,r)}}, z_m); X_{\tau_{B(x,r)}} \in D \cap B(Q, \eta_m)\right]$$

$$\begin{aligned}
&\le c_3 j(r)\phi(r^{-2})^{-1}\int_{D\cap B(Q,\eta_m)} M_D(w,z_m)dw \\
&\le c_4 r^{-d}\int_{D\cap B(Q,\eta_m)} M_D(w,z_m)dw \\
&= c_4 r^{-d} G_D(x_0,z_m)^{-1}\int_{D\cap B(Q,\eta_m)} G_D(w,z_m)dw \qquad (3.38)
\end{aligned}$$

for some $c_3 = c_3(\Psi, \gamma_1, \gamma_2) > 0$ and $c_4 = c_4(\Psi, \gamma_1, \gamma_2) > 0$. Note that, by Lemma 3.1,

$$G_D(x_0, z_m)^{-1} \le C_3(\phi(\eta_{m_0}^{-2}))^{-1}\xi^{m-m_0}\phi(\eta_m^{-2})G_D(x_0, z_{m_0})^{-1}. \qquad (3.39)$$

By (2.3), there exists $c_5 = c_5(\Psi, \gamma_1, \gamma_2) > 0$ such that for any $\eta < 1$,

$$\int_0^\eta \frac{1}{s\phi(s^{-2})}ds \le c_5\phi(\eta^{-2})^{-1}.$$

Thus by Lemma 2.7 in case D is bounded and by Theorem 2.4 in case of unbounded D,

$$\begin{aligned}
\int_{B(Q,\eta_m)} G_D(w,z_m)dw &\le c_6\int_{B(z_m,2\eta_m)} \frac{1}{|w-z_m|^d\phi(|w-z_m|^{-2})}dw \\
&\le c_7\phi((2\eta_m)^{-2})^{-1} \qquad (3.40)
\end{aligned}$$

for some constants $c_6 = c_6(\Psi, \gamma_1, \gamma_2) > 0$ and $c_7 = c_7(\Psi, \gamma_1, \gamma_2) > 0$. It follows from (3.38)–(3.40) that

$$\begin{aligned}
&\mathbb{E}_x\left[M_D(X_{\tau_{B(x,r)}}, z_m); X_{\tau_{B(x,r)}} \in D\cap B(Q,\eta_m)\right] \\
&\le c_8 r^{-d}(\phi(\eta_{m_0}^{-2}))^{-1}G_D(x_0,z_{m_0})^{-1}\frac{\phi(\eta_m^{-2})}{\phi((2\eta_m)^{-2})}\xi^{m-m_0}.
\end{aligned}$$

Applying Lemma 2.1, we get

$$\begin{aligned}
&\mathbb{E}_x\left[M_D(X_{\tau_{B(x,r)}}, z_m); X_{\tau_{B(x,r)}} \in D\cap B(Q,\eta_m)\right] \\
&\quad \le c_9 r^{-d}(\phi(\eta_{m_0}^{-2}))^{-1}G_D(x_0,z_{m_0})^{-1}\xi^{m-m_0}.
\end{aligned}$$

Thus there exists $N > 0$ such that for all $m \ge N$,

$$\mathbb{E}_x\left[M_D(X_{\tau_{B(x,r)}}, z_m); X_{\tau_{B(x,r)}} \in D\cap B(Q,\eta_m)\right] \le \frac{\epsilon}{2}.$$

Consequently, for all $m \ge N$,

$$\mathbb{E}_x\left[M_D(X_{\tau_{B(x,r)}}, z_m); M_D(X_{\tau_{B(x,r)}}, z_m) > N\right] \le \epsilon,$$

which implies that $\{M_D(X_{\tau_{B(x,r)}}, z_m) : m \ge m_0\}$ is $\mathbb{P}_x$-uniformly integrable. □

Using this, we can easily get the following

Theorem 3.1. *The function $M_D(\cdot, Q)$ is harmonic in D with respect to X.*

Proof. The proof is basically the same as that of [19, Theorem 3.9]. We write the details here for completeness. Let $h(x) := M_D(x, Q)$. Consider a relatively compact open set $D_1 \subset \overline{D_1} \subset D$, and put $r(x) = R \wedge (\frac{1}{3}\delta_D(x))$ and $B(x) = B(x, r(x))$. Define a sequence $\{T_m : m \geq 1\}$ of stopping times as follows: $T_1 := \inf\{t > 0 : X_t \notin B(X_0)\}$, and for $m \geq 2$,

$$T_m := \begin{cases} T_{m-1} + \tau_{B(X_{T_{m-1}})} \cdot \theta_{T_{m-1}} & \text{if } X_{T_{m-1}} \in D_1 \\ \tau_{D_1} & \text{otherwise.} \end{cases}$$

Note that $X_{\tau_{D_1}} \in \partial D_1$ on $\cap_{n=1}^{\infty}\{T_n < \tau_{D_1}\}$. Thus, since $\lim_{m\to\infty} T_m = \tau_{D_1}$ $\mathbb{P}_x$-a.s. and h is continuous in D, using the quasi-left continuity of X^D, we have $\lim_{m\to\infty} h(X^D_{T_m}) = h(X^D_{\tau_{D_1}})$ on $\cap_{n=1}^{\infty}\{T_n < \tau_{D_1}\}$. Now by the dominated convergence theorem and Lemma 3.7,

$$\begin{aligned} h(x) &= \lim_{m\to\infty} \mathbb{E}_x[h(X^D_{T_m}); \cup_{n=1}^{\infty}\{T_n = \tau_{D_1}\}] \\ &\quad + \lim_{m\to\infty} \mathbb{E}_x[h(X^D_{T_m}); \cap_{n=1}^{\infty}\{T_n < \tau_{D_1}\}] \\ &= \mathbb{E}_x[h(X^D_{\tau_{D_1}}); \cup_{n=1}^{\infty}\{T_n = \tau_{D_1}\}] + \mathbb{E}_x[h(X^D_{\tau_{D_1}}); \cap_{n=1}^{\infty}\{T_n < \tau_{D_1}\}] \\ &= \mathbb{E}_x[h(X^D_{\tau_{D_1}})]. \end{aligned}$$

□

Part (b) of the following result is proved in [23, Lemma 4.18]. Part (a) is even simpler.

Lemma 3.8. *(a) Let D be a bounded open set and suppose that u is a bounded nonnegative harmonic function for X^D. If there exists a polar set $N \subset \partial D$ such that for any $z \in \partial D \setminus N$*

$$\lim_{D\ni x\to z} u(x) = 0\,, \tag{3.41}$$

then u is identically equal to zero.
(b) Let D be an unbounded open set and suppose that u is a bounded nonnegative harmonic function for X^D. If there exists a polar set $N \subset \partial D$ such that for any $z \in \partial D \setminus N$ (3.41) holds true and additionally

$$\lim_{D\ni x\to \infty} u(x) = 0\,,$$

then u is identically equal to zero.

The next result completes the proof of Theorem 1.1. Recall that a point $z \in \partial D$ is said to regular boundary point of D if $\mathbb{P}_z(\tau_D = 0) = 1$band an irregular boundary point if $\mathbb{P}_z(\tau_D = 0) = 0$. The set of irregular boundary points is polar.

Theorem 3.2. *Assume that either D is bounded, or D is unbounded and X is transient. Then Q is a minimal Martin boundary point, that is, $M_D(\cdot, Q)$ is a minimal harmonic function.*

Proof. Let h be a positive harmonic function for X^D such that $h \le M_D(\cdot, Q)$. By the Martin representation in [25], there is a finite measure on $\partial_M D$ such that

$$h(x) = \int_{\partial_M D} M_D(x,w)\,\mu(dw) = \int_{\partial_M D\setminus\{Q\}} M_D(x,w)\,\mu(dw) + M_D(x,Q)\mu(\{Q\}).$$

In particular, $\mu(\partial_M D) = h(x_0) \le M_D(x_0, Q) = 1$ (because of the normalization at x_0). Hence, μ is a sub-probability measure.

For $\epsilon > 0$, K_ϵ is the compact subset of $\partial_M D$ defined in (3.35). Define

$$u(x) := \int_{K_\epsilon} M_D(x,w)\,\mu(dw). \tag{3.42}$$

Then u is a positive harmonic function with respect to X^D satisfying

$$u(x) \le h(x) - \mu(\{Q\})M_D(x,Q) \le \big(1-\mu(\{Q\})\big)M_D(x,Q). \tag{3.43}$$

Let $c = c(\epsilon) > 0$ be the constant from Lemma 3.5. Hence, for $w \in K_\epsilon$ and $(y_n^w)_{n\ge 1}$ a sequence such that $\lim_{n\to\infty} d(y_n^w, w) = 0$, it holds that $|y_n^w - Q| \ge c$. Fix $x_1 \in D\cap B(Q, c/2)$ and choose arbitrary $y_0 \in D\setminus B(Q,c)$. For any $x \in D\cap B(Q,c/2)$ and any $y \in D\setminus B(Q,c)$ we have that

$$\frac{G_D(x,y)}{G_D(x_0,y)} = \frac{G_D(x,y)}{G_D(x_1,y)}\,\frac{G_D(x_1,y)}{G_D(x_0,y)} \le c_1 \frac{G_D(x,y_0)}{G_D(x_1,y_0)}\,\frac{G_D(x_1,y)}{G_D(x_0,y)}.$$

Here the inequality follows from Theorem 2.3 applied to functions $G_D(\cdot, y)$ and $G_D(\cdot, y_0)$ which are regular harmonic in $D\cap B(Q,c)$ and vanish in $D\setminus B(Q,c)$. Now fix $w \in K_\epsilon$ and apply the above inequality to y_n^w to get

$$\begin{aligned} M_D(x,w) &= \lim_{n\to\infty} \frac{G_D(x,y_n^w)}{G_D(x_0,y_n^w)} \le c_1 \frac{G_D(x,y_0)}{G_D(x_1,y_0)} \lim_{n\to\infty} \frac{G_D(x_1,y_n^w)}{G_D(x_0,y_n^w)} \\ &= c_1 \frac{G_D(x,y_0)}{G_D(x_1,y_0)} M_D(x_1,w) \le c_1 \frac{G_D(x,y_0)}{G_D(x_1,y_0)} \sup_{w\in K_\epsilon} M_D(x_1,w) \\ &\le c_2 \frac{G_D(x,y_0)}{G_D(x_1,y_0)} = c_3 G_D(x,y_0). \end{aligned}$$

In the last line we used property (M3) (c) of the Martin kernel. Thus,

$$M_D(x,w) \le c_3 G_D(x,y_0), \qquad x \in D\cap B(Q,c/2), w \in K_\epsilon. \tag{3.44}$$

Choose $r < c/4$. For any $x \in D\setminus B(Q,2r)$ and $y \in D\cap B(Q,r)$ with r small enough, by Theorem 2.3 applied to $G_D(x,\cdot)$ and $G_D(x_0,\cdot)$, we have

$$\frac{G_D(x,y)}{G_D(x_0,y)} \le c_4 \frac{G_D(x, A_r(Q))}{G_D(x_0, A_r(Q))}.$$

Letting $D \ni y \to Q$, we get

$$M_D(x,Q) \le c_4 \frac{G_D(x,A_r(Q))}{G_D(x_0,A_r(Q))} = c_5 G_D(x, A_r(Q)), \quad x \in D\setminus B(Q,2r). \tag{3.45}$$

Recall that $\lim_{D\ni x\to z} G_D(x,y)=0$ for every regular $z\in\partial D$. Since $r<c/4$ can be arbitrarily small, we see from (3.45) and (3.43) that $\lim_{D\ni x, x\to z} u(x)=0$ for every regular $z\in\partial D$, $z\neq Q$.

Assume D is bounded. Fix $r<c/4$. It follows from Lemma 2.7 that for all $x\in D\setminus B(Q,2r)$,

$$G_D(x,A_r(Q))\le c_6\frac{\Phi(|x-A_r(Q)|)}{|x-A_r(Q)|^d}\le c_6\sup_{a\ge r}\frac{\Phi(a)}{a^d}\le c_7\,. \tag{3.46}$$

From (3.45) and (3.43) we conclude that u is bounded in $x\in D\setminus B(Q,2r)$. Similarly, for every $x\in D\cap B(Q,c/2)$ we have that $G_D(x,y_0)\le c_8\sup_{a\ge c/2}\Phi(c)c^{-d}=: c_9$ (recall $y_0\in D\setminus B(Q,c)$). Hence by (3.44) and (3.43) we see that u is bounded on $D\cap B(Q,c/2)$. Thus u is bounded on D. Now it follows from Lemma 3.8 (a) that $u\equiv 0$ in D.

If D is unbounded, we argue as follows. Since $G_D(x,A_r(Q))\le G(x,A_r(Q))$, it follows from (3.44) and Lemma 2.10 that $\lim_{D\ni x\to\infty} M_D(x,Q)=0$. Hence by (3.43) $\lim_{D\ni x\to\infty} u(x)=0$. Thus, there exists $R\ge 2$ such that $u(x)\le 1$ for all $x\in D\setminus B(Q,R)$. Fix $r<c/4\wedge 1$ and let $x\in D\cap(B(Q,R)\setminus B(Q,2r))$. By (3.45) and Theorem 2.4,

$$M_D(x,Q)\le c_5G(x,A_r(Q))\le c_5C_2(R)\frac{\Phi(|x-A_r(Q)|)}{|x-A_r(Q)|^d}\le c_{10}\sup_{a\ge r}\frac{\Phi(a)}{a^d}\le c_{11}\,.$$

It follows that u is bounded in $D\cap(B(Q,R)\setminus B(Q,2r))$. The proof that u is bounded on $D\cap B(Q,c/2)$ is the same as in the case of a bounded D. Hence, u is bounded, and again we conclude from Lemma 3.8 (b) that $u\equiv 0$ in D.

We see from (3.42) that $\nu=\mu_{|K_\epsilon}=0$. Since $\epsilon>0$ was arbitrary and $\partial_M D\setminus\{Q\}=\cup_{\epsilon>0}K_\epsilon$, we see that $\mu_{|\partial_M D\setminus\{Q\}}=0$. Hence $h=\mu(\{Q\})M_D(\cdot,Q)$ showing that $M_D(\cdot,Q)$ is minimal. □

In the next two results we assume that D is a κ-fat set. Then one can define $\Xi:\partial D\to\partial_M^f D$ so that $\Xi(Q)$ is the unique element of $\partial_M^Q D$, cf. Proposition 3.1.

Theorem 3.3. *Suppose that either D is bounded, or D is unbounded and X in transient. If D is a κ-fat set, then the finite part of the minimal Martin boundary of D and the finite part of the Martin boundary of D both coincide with the Euclidean boundary ∂D of D. More precisely, Ξ is 1-1- and onto.*

Proof. Since every finite Martin boundary point is associated with some $Q\in\partial D$, we see that Ξ is onto. We show now that Ξ is 1-1. If not, there are $Q,Q'\in\partial D$, $Q\neq Q'$, such that $\Xi(Q)=\Xi(Q')=w$. Then $M_D(\cdot,Q)=M_D(\cdot,w)=M_D(\cdot,Q')$. Choose $r>0$ small enough and satisfying $r<|Q-Q'|/4$. By (3.45) and (3.46) we see that there exists a constant $c_1=c_1(Q)$ such that $M_D(x,Q)\le c_1$ for all $x\in D\setminus B(Q,2r)$. Similarly, there exists $c_2=c_2(Q')$ such that $M_D(x,Q')\le c_2$ for all $x\in D\setminus B(Q',2r)$. Since $B(Q,2r)$ and $B(Q',2r)$ are disjoint, we conclude that $M_D(\cdot,Q)=M_D(\cdot,Q')$ is bounded on D by $c_1\vee c_2$. Again by (3.45), $\lim_{D\ni x\to z} M_D(x,Q)=0$ for all regular

$z \in \partial D$. In case of unbounded D, we showed in the proof of Theorem 3.2 that $\lim_{x\to\infty} M_D(x, Q) = 0$. Hence by Lemma 3.8 we conclude that $M_D(\cdot, Q) \equiv 0$. This is a contradiction with $M_D(x_0, Q) = 1$.

The statement about the minimal Martin boundary follows from Theorem 3.2. □

As a consequence of the result above and the general result of [25], we have the following Martin representation for nonnegative harmonic functions with respect to the killed process X^D.

Theorem 3.4. *Suppose that D is a bounded κ-fat set. Then $\Xi : \partial D \to \partial_M D$ is a homeomorphism. Furthermore, for any nonnegative function u which is harmonic with respect to X^D, there exists a unique finite measure μ on ∂D such that*

$$u(x) = \int_{\partial D} M_D(x, z)\mu(dz), \qquad x \in D.$$

Proof. Let $Q \in \partial D$ and $x \in D$. Choose $r < \frac{1}{2}\min\{R, \mathrm{dist}(x, Q), \mathrm{dist}(x_0, Q)\}$ so that $x \in D\backslash B(Q, 2r)$. Let $Q' \in \partial D \cap B(Q, r/2)$. Since D is κ-fat at Q', by Corollary 3.1 there exists $M_D(x, Q') = \lim_{D\ni y\to Q'} M_D(x, y)$. Further, by letting $y \to Q'$ in (3.34) we get that

$$|M_D(x, Q') - M_D(x, Q)| \le cM_D(x, A_r(Q)) \left(\frac{|Q' - Q|}{r}\right)^{\beta}.$$

This shows that if $(Q_n)_{n\ge1}$ is a sequence of points in ∂D converging to $Q \in \partial D$, then $M_D(\cdot, Q) = \lim_{n\to\infty} M_D(\cdot, Q_n)$.

In order to show that Ξ is continuous we proceed as follows. Let $Q_n \to Q$ in ∂D. Since $\partial_M D$ is compact, $(\Xi(Q_n))_{n\ge1}$ has a subsequence $(\Xi(Q_{n_k}))_{k\ge1}$ converging in the Martin topology to some $w \in \partial_M D$. By property (M3), $M_D(\cdot, \Xi(Q_{n_k})) \to M_D(\cdot, w)$. On the other hand, by the first part of the proof, $M_D(\cdot, \Xi(Q_{n_k})) = M_D(\cdot, Q_{n_k}) \to M_D(\cdot, Q)$, implying that $w = \Xi(Q)$. This shows in fact that $(\Xi(Q_n))_{n\ge1}$ is convergent with the limit $\Xi(Q)$. Using the fact that ∂D is compact, the proof of the continuity of the inverse is similar.

The Martin representation for nonnegative harmonic functions is now a consequence of the general result form [25]. □

Acknowledgements. We thank the referee for helpful comments on the first version of this paper.

References

[1] C. Berg and G. Forst: *Potential Theory on Locally Compact Abelian Groups.* Springer, 1975.

[2] K. Bogdan: The boundary Harnack principle for the fractional Laplacian. *Studia Math.* **123(1)**(1997), 43–80.

[3] K. Bogdan: Representation of α-harmonic functions in Lipschitz domains. *Hiroshima Math. J.* **29** (1999), 227–243.
[4] K. Bogdan, T. Grzywny and M. Ryznar: Density and tails of unimodal convolution semigroups. *J. Funct. Anal.* **266** (2014), 3543–3571.
[5] K. Bogdan, T. Grzywny and M. Ryznar: Barriers, exit time and survival probability for unimodal Lévy processes. Preprint. arXiv:1307.0270 [math.PR]
[6] K. Bogdan, T. Kulczycki, M. Kwaśnicki: Estimates and structure of α-harmonic functions. *Probab. Theory Rel. Fields.* **140** (2008), 345–381.
[7] Z.-Q. Chen, P. Kim and T. Kumagai: On heat kernel estimates and parabolic Harnack inequality for jump processes on metric measure spaces. *Acta Mathematica Sinica, English Series* **25** (2009), 1067–1086.
[8] Z.-Q. Chen, P. Kim and R. Song: Dirichlet heat kernel estimates for fractional Laplacian with gradient perturbation. *Ann. Probab.* **40** (2012), 2483–2538.
[9] Z.-Q. Chen, P. Kim and R. Song: Dirichlet heat kernel estimates for rotationally symmetric Lévy processes. To appear in *Proc. London Math. Soc.*
[10] Z.-Q. Chen and T. Kumagai: Heat kernel estimates for jump processes of mixed types on metric measure spaces. *Probab. Theory Relat. Fields*, **140** (2008), 277–317.
[11] Z.-Q. Chen and R. Song: Martin boundary and integral representation for harmonic functions of symmetric stable processes. *J. Funct. Anal.* **159** (1998), 267–294.
[12] K.-L. Chung: *Lectures from Markov Processes to Brownian Motion.* Springer, New York (1982)
[13] T. Grzywny: On Harnack inequality and Hölder regularity for isotropic unimodal Lévy processes. To appear in *Potential Anal.*
[14] R. A. Hunt and R. L. Wheeden: Positive harmonic functions on Lipschitz domains. *Trans. Amer. Math. Soc.* **147** (1970), 507–527.
[15] P. Kim and R. Song: Two-sided estimates on the density of Brownian motion with singular drift. *Illinois J. Math.*, **50 (3)**, (2006), 635–688.
[16] P. Kim and R. Song: Boundary behavior of harmonic functions for truncated stable processes. *J. Theoret. Probab.* **21** (2008), 287–321.
[17] P. Kim, R. Song and Z. Vondraček: Boundary Harnack principle for subordinate Brownian motions. *Stoch. Processee Appl.* **119** (2009), 1601–1631.
[18] P. Kim, R. Song and Z. Vondraček: Potential theory of subordinate Brownian motions revisted. *Analysis applications to finance, essays in honour of Jia-an Yan.* Interdisciplinary Mathematical Sciences - Vol. 13, World Scientific, 2012, 277–317.
[19] P. Kim, R. Song and Z. Vondraček: Minimal thinness for subordinate Brownian motion in half space. *Ann. Inst. Fourier* **62 (3)** (2012), 1045–1080.
[20] P. Kim, R. Song and Z. Vondraček: Two-sided Green function estimates for killed subordinate Brownian motions. *Proc. London Math. Soc.* **104** (2012), 927–958.
[21] P. Kim, R. Song and Z. Vondraček: Uniform boundary Harnack principle for rotationally symmetric Lévy processes in general open sets. *Science China Math.* **55** (2012), 2317–2333.
[22] P. Kim, R. Song and Z. Vondraček: Global uniform boundary Harnack principle with explicit decay rate and its application. *Stoch. Proc. Appl.* **124(1)** (2014), 235–267.
[23] P. Kim, R. Song and Z. Vondraček: Boundary Harnack principle and Martin boundary at infinity for subordinate Brownian motions. To appear in *Potential Analysis*
[24] M. Kwaśnicki, J. Małecki, M. Ryznar: Suprema of Lévy processes. *Ann. Probab.* **41** (2013) 2047–2065.
[25] H. Kunita and T. Watanabe: Markov processes and Martin boundaries I. *Illinois J. Math.* **9(3)** (1965) 485–526.
[26] R.S. Martin: Minimal harmonic functions. *Trans. Amer. Math. Soc.* **49** (1941), 137–

172.

[27] K. Michalik and K. Samotij: Martin representation for α-harmonic functions. *Probab. Math. Statist.* **20** (2000), 75–91.

[28] R. Song, Z. Vondraček: *Potential theory of subordinate Brownian motion.* In: Potential Analysis of Stable Processes and its Extensions, P. Graczyk, A. Stos, editors, Lecture Notes in Mathematics 1980, (2009) 87–176.

[29] R. Song and J. Wu: Boundary Harnack principle for symmetric stable processes. *J. Funct. Anal.* **168** (1999), 403–427.

[30] P. Sztonyk: On harmonic measure for Lévy processes. *Probab. Math. Statist.* **20** (2000), 383–390.

[31] P. Sztonyk: Boundary potential theory for stable Lévy processes. *Colloq. Math.* **95** (2003), 191–206.

Chapter 18

Level statistics of one-dimensional Schrödinger operators with random decaying potential

Shinichi Kotani* and Fumihiko Nakano†

We study the level statistics of one-dimensional Schrödinger operator with random potential decaying like $x^{-\alpha}$ at infinity. We consider the point process ξ_L consisting of the rescaled eigenvalues and show that: (i) (ac spectrum case) for $\alpha > \frac{1}{2}$, ξ_L converges to a clock process, and the fluctuation of the eigenvalue spacing converges to Gaussian. (ii) (critical case) for $\alpha = \frac{1}{2}$, ξ_L converges to the limit of the circular β-ensemble.

Contents

*Kwanseigakuin University, Sanda, 669-1337, Japan, kotani@kwansei.ac.jp
†Gakushuin University, Tokyo, 171-8588, Japan, fumihiko@math.gakushuin.ac.jp

1. Introduction

1.1. *Background*

In this paper, we study the following Schrödinger operator

$$H := -\frac{d^2}{dt^2} + a(t)F(X_t) \quad \text{on } L^2(\mathbf{R})$$

where $a \in C^\infty$ is real valued, $a(-t) = a(t)$, non-increasing for $t \geq 0$, and satisfies

$$C_1 t^{-\alpha} \leq a(t) \leq C_2 t^{-\alpha}, \quad t \geq 1$$

for some positive constants C_1, C_2 and $\alpha > 0$. F is a real-valued, smooth, and non-constant function on a compact Riemannian manifold M such that

$$\langle F \rangle := \int_M F(x)dx = 0.$$

$\{X_t\}$ is a Brownian motion on M. Since the potential $a(t)F(X_t)$ is $-\frac{d^2}{dt^2}$-compact, we have $\sigma_{ess}(H) = [0, \infty)$. Kotani-Ushiroya [4] proved that the spectrum of H in $[0, \infty)$ is :

(i) for $\alpha < \frac{1}{2}$: pure point with exponentially decaying eigenfunctions,

(ii) for $\alpha = \frac{1}{2}$: pure point on $[0, E_c]$ and purely singular continuous on $[E_c, \infty)$ with some explicitly computable E_c, and

(iii) for $\alpha > \frac{1}{2}$: purely absolutely continuous.

In this paper we study the level statistics of this operator. For that purpose, let $H_L := H|_{[0,L]}$ be the local Hamiltonian with Dirichlet boundary condition and let $\{E_n(L)\}_{n=1}^\infty$ be its eigenvalues in increasing order. Let $n(L) \in \mathbf{N}$ be s.t. $\{E_n(L)\}_{n \geq n(L)}$ coincides with the set of positive eigenvalues of H_L. We take the reference energy $E_0 > 0$ arbitrarily and consider the point process

$$\xi_L := \sum_{n \geq n(L)} \delta_{L(\sqrt{E_n(L)} - \sqrt{E_0})}$$

in order to study the local fluctuation of eigenvalues near E_0. Our aim is to identify the limit of ξ_L as $L \to \infty$. Here we consider the scaling of $\sqrt{E_n(L)}$'s instead of $E_n(L)$'s, which corresponds to the unfolding with respect to the density of states. This problem was first studied by Molchanov [7]. He proved that, when $a(t)$ is constant, ξ_L converges to the Poisson process. It was extended to the multidimensional Anderson model by Minami [8]. Killip-Stoiciu [3] studied the CMV matrices whose matrix elements decay like $n^{-\alpha}$. They showed that ξ_L converges to

(i) $\alpha > \frac{1}{2}$: the clock process,

(ii) $\alpha = \frac{1}{2}$: the limit of the circular β-ensemble,

(iii) $0 < \alpha < \frac{1}{2}$: the Poisson process.

Krichevski-Valkó-Virág [6] studied the one-dimensional discrete Schrödinger operator with the random potential decaying like $n^{-1/2}$, and proved that ξ_L converges to the Sine$_\beta$-process, which is the limit of the Gaussian β-ensemble found by Valkó-Virág [10].

The aim of our work is to do the analogue of that by Killip-Stoiciu [3] for the one-dimensional Schrödinger operator in the continuum. In subsection 1.2 (resp. subsection

1.3), we state our results for ac-case : $\alpha > \frac{1}{2}$ (resp. critical-case : $\alpha = \frac{1}{2}$). We have not obtained results for pp-case : $\alpha < \frac{1}{2}$.

1.2. *AC-case*

Definition 1.1. Let μ be a probability measure on $[0, \pi)$. We say that ξ is the clock process with spacing π with respect to μ if and only if

$$\mathbf{E}[e^{-\xi(f)}] = \int_0^\pi d\mu(\phi) \exp\left(- \sum_{n \in \mathbf{Z}} f(n\pi - \phi) \right)$$

where $f \in C_c(\mathbf{R})$ and $\xi(f) := \int_{\mathbf{R}} f d\xi$.

We set

$$[x]_{\pi \mathbf{Z}} := \max\{y \in \pi \mathbf{Z} \,|\, y \leq x\}, \quad (x)_{\pi \mathbf{Z}} := x - [x]_{\pi \mathbf{Z}}.$$

We study the limit of ξ_L under the following assumption

(A)

(1) $\alpha > \frac{1}{2}$,

(2) A sequence $\{L_j\}_{j=1}^\infty$ *satisfies* $\lim_{j \to \infty} L_j = \infty$, *and*

$$(\sqrt{E_0} L_j)_{\pi \mathbf{Z}} = \beta + o(1), \quad j \to \infty.$$

for some $\beta \in [0, \pi)$.

Condition A(2) is set to guarantee the convergence of ξ_L to a point process. If $a \equiv 0$ for instance, A(2) is indeed necessary.

Theorem 1.1. *Assume (A). Then* ξ_{L_j} *converges in distribution to the clock process with spacing* π *with respect to a probability measure* μ_β *on* $[0, \pi)$.

Remark 1.1. Let x_t be the solution to the eigenvalue equation: $H_L x_t = \kappa^2 x_t$ ($\kappa > 0$). Let $\tilde{\theta}(\kappa)$ be the one defined in (2.2). Then $\tilde{\theta}_t(\kappa)$ has a limit as t goes to infinity [4]: $\lim_{t \to \infty} \tilde{\theta}_t(\kappa) = \tilde{\theta}_\infty(\kappa)$, a.s.; μ_β is the distribution of the random variable $(\beta + \tilde{\theta}_\infty(\sqrt{E_0}))_{\pi \mathbf{Z}}$. In some special cases, we can show that $(\tilde{\theta}_\infty(\sqrt{E_0}))_{\pi \mathbf{Z}}$ is not uniformly distributed on $[0, \pi)$ for large E_0, implying that μ_β really depends on β.

Remark 1.2. We can consider point processes with respect to two reference energies $E_0, E_0' (E_0 \neq E_0')$ simultaneously : suppose a sequence $\{L_j\}_{j=1}^\infty$ satisfies $(\sqrt{E_0} L_j)_{\pi \mathbf{Z}} = \beta + o(1)$, $(\sqrt{E_0'} L_j)_{\pi \mathbf{Z}} = \beta' + o(1)$, $j \to \infty$ for some $\beta, \beta' \in [0, \pi)$. We set $\xi_L := \sum_{n \geq n(L)} \delta_{L(\sqrt{E_n(L)} - \sqrt{E_0})}$, $\xi_L' := \sum_{n \geq n(L)} \delta_{L(\sqrt{E_n(L)} - \sqrt{E_0'})}$. Then the joint distribution of ξ_{L_j}, ξ_{L_j}' converges, for $f, g \in C_c(\mathbf{R})$,

$$\lim_{j \to \infty} \mathbf{E}\left[\exp\left(-\xi_{L_j}(f) - \xi_{L_j}(g) \right) \right]$$
$$= \int_0^\pi d\mu(\phi, \phi') \exp\left(- \sum_{n \in \mathbf{Z}} (f(n\pi - \phi) + g(n\pi - \phi')) \right)$$

where $\mu(\phi, \phi')$ is the joint distribution of $(\beta + \tilde{\theta}_\infty(\sqrt{E_0}))_{\pi \mathbf{Z}}$ and $(\beta' + \tilde{\theta}_\infty(\sqrt{E_0'}))_{\pi \mathbf{Z}}$. We are unable to identify $\mu(\phi, \phi')$ but it may be possible that ϕ and ϕ' are correlated.

Remark 1.3. Suppose we renumber the eigenvalues near the reference energy E_0 so that

$$\cdots < E'_{-2}(L) < E'_{-1}(L) < E_0 \le E'_0(L) < E'_1(L) < E'_2(L) < \cdots.$$

Then an argument similar to the proof of Theorem 2.4 in [5] proves the following fact: for any $n \in \mathbf{Z}$ we have

$$\lim_{L\to\infty} L(\sqrt{E'_{n+1}(L)} - \sqrt{E'_n(L)}) = \pi, \quad a.s. \tag{1.1}$$

which is called the strong clock behavior [1]. We note that the integrated density of states is equal to $\sqrt{E}/\pi$.

We next study the finer structure of the eigenvalue spacing, under the following assumption.

(B)
(1) $a(t) = t^{-\alpha}(1+o(1))$, $t \to \infty$, $\frac{1}{2} < \alpha < 1$,
(2) A sequence $\{L_j\}_{j=1}^{\infty}$ *satisfies* $\lim_{j\to\infty} L_j = \infty$ *and*

$$\sqrt{E_0}L_j = m_j\pi + \beta + \epsilon_j, \quad j \to \infty$$

for some $\{m_j\}_{j=1}^{\infty}(\subset \mathbf{N})$, $\beta \in [0,\pi)$ *and* $\{\epsilon_j\}_{j=1}^{\infty}$ *with* $\lim_{j\to\infty}\epsilon_j = 0$.
Roughly speaking, $E_{m_j}(L_j)$ is the eigenvalue closest to E_0. In view of (1.1), we set

$$X_j(n) := \left\{\left(\sqrt{E_{m_j+n+1}(L_j)} - \sqrt{E_{m_j+n}(L_j)}\right)L_j - \pi\right\} L_j^{\alpha-\frac{1}{2}}, \quad n \in \mathbf{Z}.$$

Theorem 1.2. *Assume (B). Then* $\{X_j(n)\}_{n\in\mathbf{Z}}$ *converges in distribution to the Gaussian system with covariance*

$$C(n,n') = \frac{C(E_0)}{8E_0} Re \int_0^1 s^{-2\alpha} e^{2i(n-n')\pi s} 2(1-\cos 2\pi s)ds, \quad n,\ n' \in \mathbf{Z},$$

where $C(E) := \int_M \left|\nabla(L + 2i\sqrt{E})^{-1}F\right|^2 dx$ *and* L *is the generator of* (X_t).

Remark 1.4. Lemma 2.1 in [4] and Lemma 4.1 imply that

$$\sqrt{E_{m_j}(L_j)} = \sqrt{E_0} - \frac{\beta + \tilde{\theta}_\infty(\sqrt{E_0})}{L_j} + Y_j$$

where $Y_j = O(L_j^{-\alpha-\frac{1}{2}+\epsilon}) + O(\epsilon_j L_j^{-1})$, a.s. for any $\epsilon > 0$. Furthermore by definition of $\{X_j(n)\}$ we have

$$\sqrt{E_{m_j+n}(L_j)} = \begin{cases} \sqrt{E_{m_j}(L_j)} + \frac{n\pi}{L_j} + \frac{1}{L_j^{\alpha+\frac{1}{2}}}\sum_{l=0}^{n-1} X_j(l) \ (n \ge 1) \\ \sqrt{E_{m_j}(L_j)} + \frac{n\pi}{L_j} - \frac{1}{L_j^{\alpha+\frac{1}{2}}}\sum_{l=n}^{-1} X_j(l) \ (n \le -1). \end{cases}$$

Theorem 1.2 thus describes the behavior of eigenvalues near $E_{m_j}(L_j)$ in the second order.

Remark 1.5. Suppose we consider two reference energies $E_0, E_0'(E_0 \neq E_0')$ simultaneously and suppose a sequence $\{L_j\}_{j=1}^{\infty}$ satisfies $\lim_{j\to\infty} L_j = \infty$ and $\sqrt{E_0}L_j = m_j\pi + \beta + o(1)$, $\sqrt{E_0'}L_j = m_j'\pi + \beta' + o(1)$, $j \to \infty$ for some $m_j, m_j' \in \mathbf{N}$, and $\beta, \beta' \in [0, \pi)$. Then $\{X_j(n)\}_n$ and $\{X_j'(n)\}_n$ converge jointly to the mutually independent Gaussian systems.

1.3. *Critical Case*

We set the following assumption.

(C) $\quad a(t) = t^{-\frac{1}{2}}(1 + o(1)), \quad t \to \infty.$

Theorem 1.3. *Assume (C). Then*

$$\lim_{L\to\infty} \mathbf{E}[e^{-\xi_L(f)}] = \mathbf{E}\left[\int_0^{2\pi} \frac{d\theta}{2\pi} \exp\left(-\sum_{n\in\mathbf{Z}} f(\Psi_1^{-1}(2n\pi + \theta))\right)\right] \tag{1.2}$$

where $\{\Psi_t(\cdot)\}_{t\geq 0}$ is the strictly-increasing function valued process such that for any $c_1, \cdots, c_m \in \mathbf{R}$, $\{\Psi_t(c_j)\}_{j=1}^m$ is the unique solution to the following SDE:

$$d\Psi_t(c_j) = 2c_j dt + D(E_0) Re\left\{(e^{i\Psi_t(c_j)} - 1)\frac{dZ_t}{\sqrt{t}}\right\} \tag{1.3}$$

$$\Psi_0(c_j) = 0, \quad j = 1, 2, \cdots, m$$

where $C(E_0) := \int_M \left|\nabla(L + 2i\sqrt{E_0})^{-1}F\right|^2 dx$, $D(E_0) := \sqrt{\frac{C(E_0)}{2E_0}}$ and Z_t is a complex Browninan motion.

Definition 1.2. For $\beta > 0$, the circular β-ensemble with n-points is given by

$$\mathbf{E}_n^{\beta}[G] := \frac{1}{Z_{n,\beta}} \int_{-\pi}^{\pi} \frac{d\theta_1}{2\pi} \cdots \int_{-\pi}^{\pi} \frac{d\theta_n}{2\pi} G(\theta_1, \cdots, \theta_n) |\triangle(e^{i\theta_1}, \cdots, e^{i\theta_n})|^{\beta}$$

where $Z_{n,\beta}$ is the normalization constant, $G \in C(\mathbf{T}^n)$ is bounded and $\triangle$ is the Vandermonde determinant. The limit ξ_β of the circular β-ensemble is defined by

$$\mathbf{E}[e^{-\xi_\beta(f)}] = \lim_{n\to\infty} \mathbf{E}_n^{\beta}\left[\exp\left(-\sum_{j=1}^{n} f(n\theta_j)\right)\right], \quad f \in C_c^+(\mathbf{R})$$

Killip-Stoiciu [3] proved that the limit ξ_β exists and satisfies (1.2), (1.3) where $D(E_0)$ is replaced by $\frac{2}{\sqrt{\beta}}$ and $2c_j$ is replaced by c_j. Therefore the limit of ξ_L coincides with that of the circular β-emsemble modulo a scaling.

Corollary 1.1. *Assume (C). Writing $\xi_\beta = \sum_n \delta_{\lambda_n}$, let $\xi_\beta' := \sum_n \delta_{\lambda_n/2}$. Then $\xi_L \xrightarrow{d} \xi_\beta'$ with $\beta = \beta(E_0) := \frac{8E_0}{C(E_0)}$.*

Remark 1.6. The corresponding $\beta = \beta(E_0) = \frac{8E_0}{C(E_0)}$ depends on the reference energy E_0, so that the spacing distribution may change if we look at the different point in the spectrum. In fact we have $\beta(E) = \gamma(E)^{-1}$ where $\gamma(E)$ is the Lyapunov exponent defined in [4] such that the generalized eigenfunction ψ_E of H satisfies $\psi_E \simeq |x|^{-\gamma(E)}$, $|x| \to \infty$. It then follows that $E < E_c$ (resp. $E > E_c$) if and only if $\beta(E) < 2$ (resp. $\beta(E) > 2$) and $\beta(E_c) = 2$ (Figure 1.). Similar statement also holds for discrete Hamiltonian, the Jacobi matrix arising from the β-ensemble, and CMV matrices studied respectively by [2, 3, 6].

This is consistent with our general belief that in the point spectrum (resp. in the continuous spectrum) the level repulsion is weak (resp. strong).

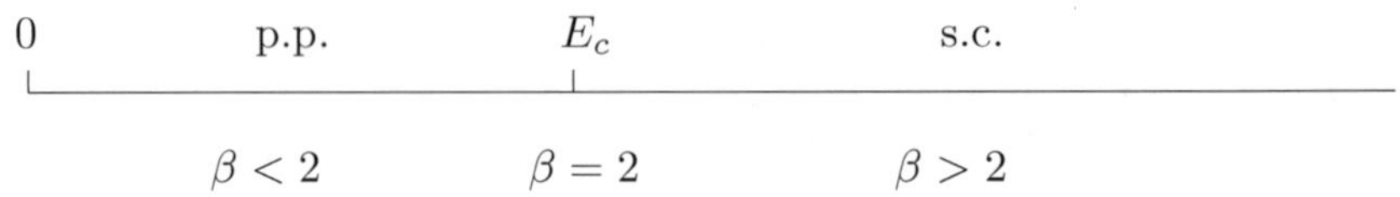

Fig. 1. Spectrum and corresponding β.

We note that, for $\beta = 2$, the circular β-ensemble with n-points coincides with the eigenvalue distribution of the unitary ensemble with the Haar measure on $U(n)$. In [10], Valkó-Virág showed that Sine_β process has a phase transition at $\beta = 2$.

Remark 1.7. If we consider two reference energies $E_0, E_0'(E_0 \neq E_0')$, then the corresponding point process ξ_L, ξ_L' converges jointly to the independent $\xi_\beta, \xi_{\beta'}'$.

In later sections, we prove theorems mentioned above based on the argument in [3–5]: The main ingredient of the proof is to study the limiting behavior of the relative Prüfer phase Ψ_L, by which the Laplace transform of ξ_L is represented(Lemma 2.1). The major difference from the argument in [3] is that $\phi(E_0, L)$, which is defined in Section 2 to be the projection to the torus of the Prüfer phase associated to E_0, is not uniformly distributed and is not independent of Ψ_L. Hence our additional task is to show that, the joint limit of $(\Psi_L, \phi(E_0, L))$ is independent each other, the convergence of Ψ_L is stronger, and that the limit of which is strictly monotone and continuous. In Section 2 we prepare some notations and basic facts. In Sections 3, 4, we consider the ac-case and prove Theorems 1.1, 1.2. In Sections 6-9, we consider the critical case and prove Theorem 1.3 which is outlined in Section 5. In what follows, C denotes general positive constant which is subject to change from line to line in each argument.

2. Preliminaries

Let x_t be the solution to the equation $H_L x_t = \kappa^2 x_t$ $(\kappa > 0)$ which we set in the following form

$$\begin{pmatrix} x_t \\ x_t'/\kappa \end{pmatrix} = r_t \begin{pmatrix} \sin\theta_t \\ \cos\theta_t \end{pmatrix}, \quad \theta_0 = 0. \tag{2.1}$$

We define $\tilde{\theta}_t(\kappa)$ by

$$\theta_t(\kappa) = \kappa t + \tilde{\theta}_t(\kappa). \tag{2.2}$$

Then it follows that

$$r_t(\kappa) = \exp\left(\frac{1}{2\kappa} Im \int_0^t a(s)F(X_s)e^{2i\theta_s(\kappa)}ds\right) \tag{2.3}$$

$$\tilde{\theta}_t(\kappa) = \frac{1}{2\kappa}\int_0^t Re(e^{2i\theta_s(\kappa)} - 1)a(s)F(X_s) \tag{2.4}$$

$$\frac{\partial\theta_t(\kappa)}{\partial\kappa} = \int_0^t \frac{r_s^2}{r_t^2}ds + \frac{1}{2\kappa^2}\int_0^t \frac{r_s^2}{r_t^2}a(s)F(X_s)(1 - Re\, e^{2i\theta_s(\kappa)})ds. \tag{2.5}$$

By using the behavior of solutions x_t [4], we can show the following fact: let $I \subset (0,\infty)$ be an interval. Then for sufficiently large $t > 0$ we have $\inf_{\kappa\in I} \frac{\partial\theta_t(\kappa)}{\partial\kappa} > 0$, so that $\theta_t(\kappa)$ is increasing as a function of κ on I. Here and henceforth, for simplicity, we say f is increasing if and only if $x < y$ implies $f(x) < f(y)$. Set

$$m(E_0, L)\pi := [\theta_L(\sqrt{E_0})]_{\pi\mathbf{Z}}, \quad \phi(E_0, L) := (\theta_L(\sqrt{E_0}))_{\pi\mathbf{Z}} \in [0, \pi). \tag{2.6}$$

Moreover we define the relative Prüfer phase

$$\Phi_L(x) = \theta_L(\sqrt{E_0} + \frac{x}{L}) - \theta_L(\sqrt{E_0})$$

which is continuous and increasing. As in [3] we use the following representation of Laplace transform of ξ_L in terms of Φ_L.

Lemma 2.1. *For $f \in C_c^+(\mathbf{R})$ we have*

$$\mathbf{E}[e^{-\xi_L(f)}] = \mathbf{E}\left[\exp\left(-\sum_{n=n(L)-m(E_0,L)}^{\infty} f\left(\Phi_L^{-1}(n\pi - \phi(E_0, L)\right)\right)\right].$$

3. Convergence to a clock process

In what follows, for simplicity, we set

$$\kappa := \sqrt{E_0}$$

3.1. *The behavior of Ψ_L*

Proposition 3.1. *If $\alpha > \frac{1}{2}$, following fact holds for a.s.:*

$$\lim_{L\to\infty} \Phi_L(x) = x$$

pointwise and this holds compact uniformly with respect to κ.

Proof. By (2.4) we have

$$\Phi_L(x) = x + \frac{1}{2\kappa} Re \int_0^L a(s)F(X_s)\left(e^{2i\theta_s(\kappa+\frac{x}{L})} - e^{2i\theta_s(\kappa)}\right) ds + O(L^{-\alpha}).$$

We set

$$A_t(\kappa, \beta) := \int_0^t a(s)F(X_s)e^{i\beta\theta_s(\kappa)} ds.$$

Take $\delta > 0$ such that $\int_0^\infty a(s)^2 s^\delta ds < \infty$. Then by [4] Lemma 2.2, for any compact set $K \subset (0, \infty)$ and for any $\epsilon < \frac{\delta}{2}$, $\beta \in \mathbf{R}$ we have

$$\sup_{t\geq 0,\ \kappa,\ \kappa_1\in K} \frac{|A_t(\kappa,\beta) - A_t(\kappa_1,\beta)|}{|\kappa - \kappa_1|^\epsilon} < \infty, \quad a.s..$$

Hence for fixed x, we have $\Phi_L(x) = x + O(L^{-\epsilon})$, a.s.. Since the function $f(x) = x$ is continuous, the proof is complete. □

3.2. *Proof of Theorem 1.1*

We sometimes use the following elementary lemma.

Lemma 3.1. *Let $\Psi_n, n = 1, 2, \cdots$, and Ψ are continuous and increasing functions on an open interval I such that $\lim_{n\to\infty} \Psi_n(x) = \Psi(x)$ pointwise. If $y_n \in Ran\,\Psi_n$, $y \in Ran\,\Psi$ and $y_n \to y$, then it holds that $\Psi_n^{-1}(y_n) \stackrel{n\to\infty}{\to} \Psi^{-1}(y)$.*

Proof of Theorem 1.1
By the fact that $\tilde{\theta}_t(\kappa) \stackrel{t\to\infty}{=} \tilde{\theta}_\infty(\kappa) + o(1)$ ([4] Proposition 2.1) and by (A)(2), $\lim_{j\to\infty} \phi(\kappa^2, L_j) = \left(\tilde{\theta}_\infty(\kappa) + \beta\right)_{\pi\mathbf{Z}}$, a.s.. Together with Proposition 3.1, the assumption for Lemma 3.1 is satisfied. □

4. Second Limit Theorem

4.1. *Behavior of eigenvalues near E_0*

Lemma 4.1. *Assume (B) and let $n \in \mathbf{Z}$. Then for $j \to \infty$ we have*

$$(1) \qquad \sqrt{E_{m_j+n}(L_j)} = \kappa + o(1)$$

$$(2) \qquad \sqrt{E_{m_j+n}(L_j)} = \kappa + \frac{n\pi - \beta - \tilde{\theta}_\infty(\kappa)}{L_j} + o(L_j^{-1}).$$

Lemma 4.1 follows from the fact that $\tilde{\theta}_L(\kappa) \stackrel{L\to\infty}{\to} \tilde{\theta}_\infty(\kappa)$ holds compact uniformly w.r.t. κ. By definition we see that

$$X_j(n) = -L_j^{\alpha-\frac{1}{2}} \left(\tilde{\theta}_{L_j}(\sqrt{E_{m_j+n+1}(L_j)}) - \tilde{\theta}_{L_j}(\sqrt{E_{m_j+n}(L_j)})\right).$$

By Lemma 4.1(2)

$$\sqrt{E_{m_j+n+1}(L_j)} = \kappa + \frac{c_1}{L_j}, \quad \sqrt{E_{m_j+n}(L_j)} = \kappa + \frac{c_2}{L_j}$$
$$c_1 = (n+1)\pi - \beta - \tilde{\theta}_\infty(\kappa) + o(1), \quad c_2 = n\pi - \beta - \tilde{\theta}_\infty(\kappa) + o(1), \quad j \to \infty.$$

We set

$$\Theta_t^{(n)}(c_1, c_2) := \left(\tilde{\theta}_{nt}(\kappa + \frac{c_1}{n}) - \tilde{\theta}_{nt}(\kappa + \frac{c_2}{n})\right) n^{\alpha-\frac{1}{2}}$$
$$l_t((c_1, c_2), (c_1', c_2')) := \frac{C(\kappa^2)}{8\kappa^2} \int_0^t s^{-2\alpha} Re\left(e^{2ic_1 s} - e^{2ic_2 s}\right) \overline{\left(e^{2ic_1' s} - e^{2ic_2' s}\right)} ds.$$

When c_1, c_2 are constant, the following fact is proved in [5] Lemma 3.1.

Proposition 4.1. $\{\Theta_t^{(n)}(c_1, c_2)\}_{t\geq 0,\, c_1, c_2 \in \mathbf{R}} \stackrel{d}{\to} \{Z(t, c_1, c_2)\}_{t\geq 0,\, c_1, c_2 \in \mathbf{R}}$ *as $n \to \infty$ where $\{Z(t, c_1, c_2)\}_{t\geq 0,\, c_1, c_2 \in \mathbf{R}}$ is the Gaussian system with covariance $l_{t\wedge t'}((c_1, c_2), (c_1', c_2'))$.*

4.2. *Independence of the limits*

To finish the proof of Theorem 1.2, it is then sufficient to prove that $(\tilde{\theta}_{nt}(\kappa), \{\Theta_t^{(n)}(c_1, c_2)\}_{c_1,c_2})$ converges jointly to the independent ones. Let $0 < \kappa_1 < \kappa_2$ and $I := [\kappa_1, \kappa_2]$. In the lemma below, we regard $\tilde{\theta}_t, \tilde{\theta}_\infty$ are $C(I)$-valued random elements.

Lemma 4.2. *For $t > 0$ fixed, we have*

$$(\tilde{\theta}_{nt}, \{\Theta_t^{(n)}(c_1, c_2)\}_{c_1,c_2}) \overset{d}{\to} (\tilde{\theta}_\infty, \{Z(t, c_1, c_2)\}_{c_1,c_2})$$

as $n \to \infty$ where $\tilde{\theta}_\infty$ and $\{Z(t, c_1, c_2)\}_{c_1,c_2}$ are independent.

Proof. Let $A(\subset C(I))$ be a $\tilde{\theta}_\infty$-continuity set (i.e., $\mathbf{P}(\tilde{\theta}_\infty \in \partial A) = 0$) and set $A_\epsilon := \{f \in C(I) \,|\, d(f, A) < \epsilon\}$. Since $\tilde{\theta}_t(\kappa) \overset{a.s.}{\to} \tilde{\theta}_\infty(\kappa)$ compact uniformly in κ, for any $\epsilon > 0$ $\mathbf{P}\left(\tilde{\theta}_{nt} \in A,\ \tilde{\theta}_T \notin A_\epsilon\right) = o(1)$ for sufficiently large T, n. Here we recall eq. (3.3) in [5].

$$\begin{aligned} \Theta_t^{(n)}(c_1, c_2) &= T_t^{(n)}(c_1, c_2) + O(n^{\frac{1}{2}-\alpha}) \\ \text{where} \quad T_t^{(n)}(c_1, c_2) &:= n^{\alpha - \frac{1}{2}} Re \left(S_t^{(n)}\left(\kappa + \frac{c_1}{n}\right) - S_t^{(n)}\left(\kappa + \frac{c_2}{n}\right)\right) \\ S_t^{(n)}(\kappa) &:= \frac{1}{2\kappa} \int_0^{nt} a(s) e^{2i\tilde{\theta}_s(\kappa)} dM_s(\kappa) \end{aligned}$$

$M_s(\kappa)$ is the complex martingale defined in subsection 6.2. Let $m \in \mathbf{N}$. For $\mathbf{c}_1 = (c_1^{(1)}, \cdots, c_1^{(m)})$, $\mathbf{c}_2 = (c_2^{(1)}, \cdots, c_2^{(m)})$, we use the following convention : $\Theta_t^{(n)}(\mathbf{c}_1, \mathbf{c}_2) = \left(\Theta_t^{(n)}(c_1^{(1)}, c_2^{(1)}), \cdots, \Theta_t^{(n)}(c_1^{(m)}, c_2^{(m)})\right)$ and similarly for $T_t^{(n)}(\mathbf{c}_1, \mathbf{c}_2)$ and $Z(t, \mathbf{c}_1, \mathbf{c}_2)$. Let $B \in \mathcal{B}(\mathbf{R}^m)$ be a $Z(t, \mathbf{c}_1, \mathbf{c}_2)$-continuity set and let $B_\epsilon := \{x \in \mathbf{R}^m \,|\, d(x, B) < \epsilon\}$. Writing $\Theta_t^{(n)} = \Theta_t^{(n)}(\mathbf{c}_1, \mathbf{c}_2)$, $T_t^{(n)} = T_t^{(n)}(\mathbf{c}_1, \mathbf{c}_2)$ we have, for sufficiently large n,

$$\begin{aligned} \mathbf{P}\left(\tilde{\theta}_{nt} \in A, \Theta_t^{(n)} \in B\right) &\le \mathbf{P}\left(\tilde{\theta}_T \in A_\epsilon, T_t^{(n)} \in B_\epsilon\right) + o(1) \\ &= \mathbf{P}\left(\tilde{\theta}_T \in A_\epsilon, T_t^{(n)} - T_{T/n}^{(n)} + T_{T/n}^{(n)} \in B_\epsilon\right) + o(1) \\ &= \mathbf{P}\left(\tilde{\theta}_T \in A_\epsilon, T_t^{(n)} - T_{T/n}^{(n)} \in B_{2\epsilon}\right) + o(1). \end{aligned}$$

Here we used $T_{T/n}^{(n)} \overset{P}{\to} 0$. By the Markov property

$$= \mathbf{E}\left[1_{\{\tilde{\theta}_T \in A_\epsilon\}} \mathbf{E}_{X_T}\left[1_{\{\widetilde{T}_{t-T/n}^{(n)} \in B_{2\epsilon}\}}\right]\right] + o(1)$$

where $\widetilde{T}_t^{(n)}$ is the suitable "time-shift" of $T_t^{(n)}$. Because $\widetilde{T}_t^{(n)}$ converges in distribution to $Z(t, \mathbf{c}_1, \mathbf{c}_2)$ as $n \to \infty$ being irrespective of X_T,

$$\begin{aligned} &= \mathbf{P}\left(\tilde{\theta}_T \in A_\epsilon\right) \mathbf{P}\left(Z(t, \mathbf{c}_1, \mathbf{c}_2) \in B_{2\epsilon}\right) + o(1) \\ &\le \mathbf{P}\left(\tilde{\theta}_\infty \in A_{2\epsilon}\right) \mathbf{P}\left(Z(t, \mathbf{c}_1, \mathbf{c}_2) \in B_{2\epsilon}\right) + o(1). \end{aligned}$$

Since A is a $\tilde{\theta}_\infty$-continuity set and $B \in \mathcal{B}(\mathbf{R}^m)$ is a $Z(t, \mathbf{c}_1, \mathbf{c}_2)$-continuity set,

$$\limsup_{n\to\infty} \mathbf{P}\left(\tilde{\theta}_{nt} \in A, \Theta_t^{(n)} \in B\right) \le \mathbf{P}(\tilde{\theta}_\infty \in A)\mathbf{P}(Z(t, \mathbf{c}_1, \mathbf{c}_2) \in B).$$

The opposite inequality can be proved similarly. □

5. SC-case: outline of proof of Theorem 1.3

In this section we overview the proof of Theorem 1.3. First of all, set

$$[x]_{2\pi\mathbf{Z}} := \max\{y \in 2\pi\mathbf{Z} \mid y \le x\}, \quad (x)_{2\pi\mathbf{Z}} := x - [x]_{2\pi\mathbf{Z}},$$
$$2m(\kappa^2, L)\pi := [2\theta_L(\kappa)]_{2\pi\mathbf{Z}}, \quad \phi(\kappa^2, L) := (2\theta_L(\kappa))_{2\pi\mathbf{Z}} \in [0, 2\pi).$$

We also set the relative Prüfer phase by $\Psi_L(x) := 2\theta_L(\kappa + \frac{x}{L}) - 2\theta_L(\kappa)$. Then we have a variant of Lemma 2.1.

Lemma 5.1. *For* $f \in C_c^+(\mathbf{R})$

$$\mathbf{E}[e^{-\xi_L(f)}] = \mathbf{E}\left[\exp\left(-\sum_{n=n(L)-m(\kappa^2,L)}^{\infty} f\left(\Psi_L^{-1}(2n\pi - \phi(\kappa^2, L))\right)\right)\right].$$

So our task is to study the limit of the joint distribution of $(\Psi_L, \phi(\kappa^2, L))$ as $L \to \infty$. Following [3] we consider

$$\Psi_t^{(n)}(x) := 2\theta_{nt}(\kappa + \frac{x}{n}) - 2\theta_{nt}(\kappa), \tag{5.1}$$

regard it as an increasing function-valued process, and find a process $\Psi_t(x)$ such that for any fixed $c_1, \cdots, c_m \in \mathbf{R}$ $\{\Psi_t^{(n)}(c_j)\}_{j=1}^m \stackrel{d}{\to} \{\Psi_t(c_j)\}_{j=1}^m$ (Theorem 6.1). Ψ_t is characterized as the unique solution to the SDE (1.3). Moreover, $\Psi_t(c)$ is continuous and increasing with respect to c (Lemma 6.10). On the other hand we have $(\{\Psi_1^{(n)}(c_j)\}_{j=1}^m, \phi(\kappa^2, n)) \stackrel{d}{\to} (\{\Psi_1(c_j)\}_{j=1}^m, \phi_1)$ jointly, where ϕ_1 is uniformly distributed on $[0, 2\pi)$ and independent of Ψ_1 (Proposition 9.1). Moreover $\Psi^{(n)}$ converges to Ψ also as a sequence of increasing function-valued process (Lemma 9.1), so that we can find a coupling such that for a.s. $((\Psi_1^{(n)})^{-1}(x), \phi(\kappa^2, n)) \to (\Psi_1^{-1}(x), \phi_1)$ for any $x \in \mathbf{R}$ (Proposition 9.2). Therefore we obtain (1.2).

6. Convergence of Ψ

6.1. *Preliminaries*

We recall the basic tool used in [4, 5]. For $f \in C^\infty(M)$ let $R_\beta f := (L + i\beta)^{-1} f$ $(\beta > 0)$, $R_0 f := L^{-1}(f - \langle f \rangle)$. Then by Ito's formula,

$$\int_0^t e^{i\beta s} f(X_s) ds = \left[e^{i\beta s}(R_\beta f)(X_s)\right]_0^t + \int_0^t e^{i\beta s} dM_s(f, \beta)$$
$$\int_0^t f(X_s) ds = \langle f \rangle t + [(R_0 f)(X_s)]_0^t + M_t(f, 0)$$

where $M_s(f, \beta), M_s(f, 0)$ are the complex martingales whose variational process satisfy

$$\langle M(f, \beta), M(f, \beta) \rangle_t = \int_0^t [R_\beta f, R_\beta f](X_s) ds,$$

$$\langle M(f,\beta), \overline{M(f,\beta)}\rangle_t = \int_0^t [R_\beta f, \overline{R_\beta f}](X_s)ds$$

where $[f_1, f_2](x) := L(f_1 f_2)(x) - (Lf_1)(x)f_2(x) - f_1(x)(Lf_2)(x) = (\nabla f_1, \nabla f_2)(x)$. Then the integration by parts gives us the following formulas to be used frequently.

Lemma 6.1.

$$\begin{aligned}
(1) \int_0^t & b(s)e^{i\beta s}e^{i\gamma\tilde{\theta}_s} f(X_s)ds \\
&= \left[b(s)e^{i\gamma\tilde{\theta}_s}e^{i\beta s}(R_\beta f)(X_s)\right]_0^t - \int_0^t b'(s)e^{i\gamma\tilde{\theta}_s}e^{i\beta s}(R_\beta f)(X_s)ds \\
&- \frac{i\gamma}{2\kappa}\int_0^t b(s)a(s)Re(e^{2i\theta_s}-1)e^{i\gamma\tilde{\theta}_s}e^{i\beta s}F(X_s)(R_\beta f)(X_s)ds \\
&+ \int_0^t b(s)e^{i\beta s}e^{i\gamma\tilde{\theta}_s}dM_s(f,\beta).
\end{aligned}$$

$$\begin{aligned}
(2) \int_0^t & b(s)e^{i\gamma\tilde{\theta}_s} f(X_s)ds \\
&= \langle f\rangle \int_0^t b(s)e^{i\gamma\tilde{\theta}_s}ds \\
&+ \left[b(s)e^{i\gamma\tilde{\theta}_s}(R_0 f)(X_s)\right]_0^t - \int_0^t b'(s)e^{i\gamma\tilde{\theta}_s}(R_0 f)(X_s)ds \\
&- \frac{i\gamma}{2\kappa}\int_0^t a(s)b(s)Re(e^{2i\theta_s}-1)e^{i\gamma\tilde{\theta}_s}F(X_s)(R_0 f)(X_s)ds \\
&+ \int_0^t b(s)e^{i\gamma\tilde{\theta}_s}dM_s(f,0).
\end{aligned}$$

We will also use following notation for simplicity.

$$g_\kappa := (L+2i\kappa)^{-1}F, \quad g := L^{-1}(F - \langle F\rangle),$$
$$M_s(\kappa) := M_s(F, 2\kappa), \quad M_s := M_s(F,0).$$

6.2. *A priori estimates*

In this section we derive a priori estimate for (5.1). We set

$$\begin{aligned}
Y_t(\kappa) &:= \int_0^t a(s)e^{2i\theta_s(\kappa)}dM_s(\kappa), \\
\delta_t(\kappa) &:= \left[a(s)e^{2i\theta_s(\kappa)}g_\kappa(X_s)\right]_0^t - \int_0^t a'(s)e^{2i\theta_s(\kappa)}g_\kappa(X_s)ds \\
&\quad - \frac{i}{\kappa}\int_0^t a(s)^2 e^{2i\theta_s(\kappa)}\left(\frac{e^{2i\theta_s(\kappa)}}{2} - 1\right)g_\kappa(X_s)F(X_s)ds, \\
V_t^{(n)}(c) &:= Y_{nt}\left(\kappa + \frac{c}{n}\right) - Y_{nt}(\kappa).
\end{aligned}$$

Lemma 6.2. *Suppose* $\int_0^\infty a(s)^3 ds < \infty$. *We then have*

(1)

$$\int_0^t a(s)e^{2i\theta_s(\kappa)}F(X_s)ds = -\frac{i}{2\kappa}\int_0^t a(s)^2 g_\kappa(X_s)F(X_s)ds + Y_t(\kappa) + \delta_t(\kappa)$$

(2) For a.s., $\delta_t(\kappa)$ *has the limit as* $t \to \infty$ F $\lim_{t\to\infty}\delta_t(\kappa) = \delta_\infty(\kappa)$, *a.s.*
(3) For any $0 < T < \infty$, *we have*

$$\mathbf{E}\left[\max_{0\le t\le T}\left|\delta_{nt}(\kappa + \frac{c}{n}) - \delta_{nt}(\kappa)\right|^2\right] \overset{n\to\infty}{\to} 0.$$

Proof. (1) It follows directly from Lemma 6.1(1).
(2) We further decompose the remainder term $\delta_t(\kappa)$:

$$\begin{aligned}
\delta_t(\kappa) &= \delta_t^{(1)}(\kappa) + \delta_t^{(2)}(\kappa) &(6.1)\\
\delta_t^{(1)}(\kappa) &:= \Big[a(s)e^{2i\theta_s(\kappa)}g_\kappa(X_s)\Big]_0^t - \int_0^t a'(s)e^{2i\theta_s(\kappa)}g_\kappa(X_s)ds &(6.2)\\
\delta_t^{(2)}(\kappa) &:= -\frac{i}{\kappa}\int_0^t a(s)^2\left(\frac{e^{2i\theta_s(\kappa)}}{2} - 1\right)e^{2i\theta_s(\kappa)}g_\kappa(X_s)F(X_s)ds.
\end{aligned}$$

It is easy to see $\lim_{t\to\infty}\delta_t^{(1)}(\kappa) = \delta_\infty^{(1)}(\kappa)$, a.s.. To see the convergence of $\delta_t^{(2)}(\kappa)$ we write

$$\begin{aligned}
\delta_t^{(2)}(\kappa) &= -\frac{i}{2\kappa}D_t^{(4)}(\kappa) + \frac{i}{\kappa}D_t^{(2)}(\kappa) &(6.3)\\
D_t^{(\beta)}(\kappa) &:= \int_0^t a(s)^2 e^{i\beta\theta_s(\kappa)}F(X_s)g_\kappa(X_s)ds, \quad \beta = 2, 4.
\end{aligned}$$

We use Lemma 6.1(1) to decompose $D_t^{(\beta)}(\kappa)$ into martingale part and the remainder: Setting $h_{\kappa,\beta} = R_{\beta\kappa}(Fg_\kappa)$ and $\widetilde{M_s}^{(\beta)}(\kappa) = M_s(Fg_\kappa, \beta\kappa)$, we have

$$\begin{aligned}
D_t^{(\beta)}(\kappa) &= I_t^{(\beta)}(\kappa) + N_t^{(\beta)}(\kappa) &(6.4)\\
I_t^{(\beta)}(\kappa) &:= \Big[a(s)^2 e^{i\beta\theta_s(\kappa)}h_{\kappa,\beta}(X_s)\Big]_0^t - \int_0^t (a(s)^2)' e^{i\beta\theta_s(\kappa)}h_{\kappa,\beta}(X_s)ds\\
&\quad -\frac{i\beta}{2\kappa}\int_0^t a(s)^3 Re(e^{2i\theta_s(\kappa)} - 1)e^{i\beta\theta_s(\kappa)}F(X_s)h_{\kappa,\beta}(X_s)ds\\
N_t^{(\beta)}(\kappa) &:= \int_0^t a(s)^2 e^{i\beta\theta_s(\kappa)}d\widetilde{M_s}^{(\beta)}(\kappa).
\end{aligned}$$

$I_t^{(\beta)}(\kappa)$ is easily seen to be convergent : $\lim_{t\to\infty} I_t^{(\beta)}(\kappa) = I_\infty^{(\beta)}(\kappa)$, a.s.. Since

$$|\langle N^{(\beta)}, N^{(\beta)}\rangle_t|, \quad |\langle N^{(\beta)}, \overline{N^{(\beta)}}\rangle_t| \le (const.)\int_0^t a^4(s)ds < \infty.$$

Re N, Im N can be represented by the time-change of a Brownian motion and thus have limit a.s..
(3) We consider $\delta_t^{(1)}(\kappa)$, $\delta_t^{(2)}(\kappa)$ separately. For $\delta_t^{(1)}(\kappa)$, we have

$$\begin{aligned}
\delta_{nt}^{(1)}(\kappa + \frac{c}{n}) - \delta_{nt}^{(1)}(\kappa) &= a(nt)\left(e^{2i\theta_{nt}(\kappa+\frac{c}{n})} - e^{2i\theta_{nt}(\kappa)}\right)g_{\kappa+\frac{c}{n}}(X_{nt})\\
&\quad - \int_0^{nt} a'(s)\left(e^{2i\theta_s(\kappa+\frac{c}{n})} - e^{2i\theta_s(\kappa)}\right)g_{\kappa+\frac{c}{n}}(X_s)ds + O(n^{-1}) &(6.5)
\end{aligned}$$

by (6.2). The second term of (6.5) is $o(1)$ as $n \to \infty$ due to Lebesgue's dominated convergence theorem. For the first term, we note

$$\max_{0 \le t \le M} |e^{2i\theta_t(\kappa + \frac{c}{n})} - e^{2i\theta_t(\kappa)}| \le \frac{C_M}{n} \tag{6.6}$$

for some positive constant C_M depending on M, which follows from (2.3)-(2.5). We can then show that the first term of (6.5) vanishes uniformly w.r.t. $t \in [0, T]$ so that $\max_{0 \le t \le T} \mathbf{E}[|\delta_{nt}^{(1)}(\kappa + \frac{c}{n}) - \delta_{nt}^{(1)}(\kappa)|^2] \overset{n\to\infty}{\to} 0$. Similar argument shows $\max_{0 \le t \le T} \left| I_{nt}^{(\beta)}(\kappa + \frac{c}{n}) - I_{nt}^{(\beta)}(\kappa) \right| \to 0$ so that we have only to show

$$\mathbf{E}\left[\max_{0 \le t \le T} \left| N_{nt}^{(\beta)}(\kappa + \frac{c}{n}) - N_{nt}^{(\beta)}(\kappa) \right|^2 \right] \overset{n\to\infty}{\to} 0, \quad \beta = 2, 4$$

to finish the proof of Lemma 6.2(3). By the martingale inequality,

$$\mathbf{E}\left[\max_{0 \le t \le T} \left| N_{nt}^{(\beta)}(\kappa + \frac{c}{n}) - N_{nt}^{(\beta)}(\kappa) \right|^2 \right] \le C\mathbf{E}\left[\int_0^{nt} a(s)^4 \left[H_{\beta,\kappa}, \overline{H_{\beta,\kappa}} \right] ds \right]$$
$$\text{where} \quad H_{\beta,\kappa}(s) := e^{i\beta\theta_s(\kappa + \frac{c}{n})} h_{\beta, \kappa + \frac{c}{n}} - e^{i\beta\theta_s(\kappa)} h_{\beta,\kappa}$$

which converges to 0 due to the fact that $\int_0^\infty a(s)^4 ds < \infty$ and Lebesgue's theorem. □

We assume in what follows $a(t) = t^{-1/2}(1 + o(1))$.

Lemma 6.3.

$$\Psi_t^{(n)}(c) = 2ct + Re\ \epsilon_t^{(n)} + \frac{1}{\kappa} Re\ V_t^{(n)}(c) + \frac{1}{\kappa} Re \left(\delta_{nt}(\kappa + \frac{c}{n}) - \delta_{nt}(\kappa) \right) \tag{6.7}$$

for some $\epsilon_t^{(n)}$ *satisfying*

$$|\epsilon_t^{(n)}| \le Ct + C\sqrt{\frac{t}{n}}.$$

Proof. By Lemma 6.2(1) we have (6.7) with

$$\begin{aligned} \epsilon_t^{(n)} &:= -\frac{\frac{c}{n}}{\kappa(\kappa + \frac{c}{n})} \int_0^{nt} (e^{2i\theta_s(\kappa)} - 1) a(s) F(X_s) ds \\ &+ \frac{1}{\kappa} \left\{ \frac{i}{2} \cdot \frac{\frac{c}{n}}{\kappa(\kappa + \frac{c}{n})} \int_0^{nt} a(s)^2 g_{\kappa + \frac{c}{n}}(X_s) F(X_s) ds \right. \\ &\left. + \frac{i}{2\kappa} \int_0^{nt} a(s)^2 \left(g_\kappa(X_s) - g_{\kappa + \frac{c}{n}}(X_s) \right) F(X_s) ds \right\}. \end{aligned}$$

It then suffices to see $|\epsilon_t^{(n)}| \le \frac{C}{n} \int_0^{nt} a(s) ds \le Ct + C\sqrt{\frac{t}{n}}$. □

Lemma 6.4.

$$\mathbf{E}[|\Psi_t^{(n)}(c)|] \le C \left(t + \sqrt{\frac{t}{n}} + \frac{1}{\sqrt{n}} \right), \quad t \ge 0,\ n > 0.$$

Proof. We decompose $\delta_t(\kappa)$ as is done in (6.1) to estimate $\delta_t(\kappa)$ further. Let

$$\Lambda_t^{(n)}(c) := e^{2i\theta_{nt}(\kappa+\frac{c}{n})} - e^{2i\theta_{nt}(\kappa)}$$

then

$$\begin{aligned}\delta_{nt}^{(1)}(\kappa+\frac{c}{n}) - \delta_{nt}^{(1)}(\kappa) &= \Lambda_t^{(n)}(c)a(nt)g_{\kappa+\frac{c}{n}}(X_{nt}) \\ &- \int_0^{nt} a'(s)g_{\kappa+\frac{c}{n}}(X_s)\Lambda_{s/n}^{(n)}(c)ds + O(n^{-1}).\end{aligned}$$

$\delta_t^{(2)}$ is also decomposed, as in (6.3), (6.4). The $I_{nt}^{(\beta)}$-term can be written as

$$\begin{aligned}&I_{nt}^{(\beta)}(\kappa+\frac{c}{n}) - I_{nt}^{(\beta)}(\kappa) = a(nt)^2 h_{\kappa,n}^{(\beta)}(nt)\Lambda_t^{(n)}(c) \\ &- \int_0^{nt}(a(s)^2)' f_{\kappa,n}^{(\beta)}(s)\Lambda_{s/n}^{(n)}(c)ds - \int_0^{nt} a(s)^3 g_{\kappa,n}^{(\beta)}(s)\Lambda_{s/n}^{(n)}(c)ds\end{aligned}$$

for some bounded functions $f_{\kappa,n}^{(\beta)}, g_{\kappa,n}^{(\beta)}, h_{\kappa,n}^{(\beta)}$. Putting together we have

$$\begin{aligned}\delta_{nt}(\kappa+\frac{c}{n}) - \delta_{nt}(\kappa) &= \Lambda_t^{(n)}(c)\Big(a(nt)g_{\kappa+\frac{c}{n}}(X_{nt}) + a(nt)^2 h_{\kappa,n}(nt)\Big) \\ &+ \int_0^{nt}\Lambda_{s/n}^{(n)}(c)b_{\kappa,n}(s)ds + N_{nt}(\kappa+\frac{c}{n}) - N_{nt}(\kappa) + O(n^{-1})\end{aligned}$$

for some bounded functions $h_{\kappa,n}, b_{\kappa,n}$ and a martingale N_t. $b_{\kappa,n}(s)$ is a linear combination of $a'(s)g_{\kappa+\frac{c}{n}}$, $(a(s)^2)'f_{\kappa,n}^{(\beta)}$, and $a(s)^3 g_{\kappa,n}^{(\beta)}$, so that it is integrable : $\int_0^\infty b_{\kappa,n}(s)ds < \infty$. Taking expectations, the martingale terms vanish and it follows that

$$\begin{aligned}\mathbf{E}\left[\delta_{nt}(\kappa+\frac{c}{n}) - \delta_{nt}(\kappa)\right] &= \mathbf{E}\left[\Lambda_t^{(n)}(c)\Big(a(nt)g_{\kappa+\frac{c}{n}}(X_{nt}) + a(nt)^2 h_{\kappa,n}(nt)\Big)\right] \\ &+ \int_0^{nt}\mathbf{E}\left[\Lambda_{s/n}^{(n)}(c)b_{\kappa,n}(s)\right]ds + O(n^{-1}).\end{aligned}$$

Therefore we can find a non-random function

$$b(s) = C(a'(s) + (a(s)^2)' + a(s)^3)$$

for some $C > 0$ such that $\int_0^\infty b(s)ds < \infty$ and

$$\left|\mathbf{E}\left[\delta_{nt}(\kappa+\frac{c}{n}) - \delta_{nt}(\kappa)\right]\right| \le Ca(nt)\mathbf{E}[|\Lambda_t^{(n)}(c)|] + \int_0^{nt}\mathbf{E}[|\Lambda_{s/n}^{(n)}(c)|]b(s)ds + \frac{C}{n}.$$

Here without loss of generality, we may suppose $c \ge 0$. We use $\Psi_t^{(n)}(c) \ge 0$ for $c \ge 0$ and take expectation in (6.7).

$$\begin{aligned}&\mathbf{E}[|\Psi_t^{(n)}(c)|] = \mathbf{E}[\Psi_t^{(n)}(c)] \\ &= 2ct + \mathbf{E}[Re\,\epsilon_t^{(n)}] + \frac{1}{\kappa}\mathbf{E}\left[Re\left(\delta_{nt}(\kappa+\frac{c}{n}) - \delta_{nt}(\kappa)\right)\right] \\ &\le Ct + C\sqrt{\frac{t}{n}} + Ca(nt)\mathbf{E}[|\Lambda_t^{(n)}(c)|] + C\int_0^{nt}\mathbf{E}\left[\left|\Lambda_{s/n}^{(n)}(c)\right|\right]b(s)ds + \frac{C}{n}.\end{aligned}$$

Let

$$\rho_n(t) := C\left(t + \sqrt{\frac{t}{n}} + \frac{1}{n}\right).$$

Since $|\Lambda_t^{(n)}(c)| \le |\Psi_t^{(n)}(c)|$ we have

$$\mathbf{E}\left[\left|\Psi_t^{(n)}(c)\right|\right] \le \rho_n(t) + Ca(nt)\mathbf{E}\left[\left|\Psi_t^{(n)}(c)\right|\right] + C\int_0^{nt} \mathbf{E}\left[\left|\Psi_{s/n}^{(n)}(c)\right|\right] b(s)ds.$$

Fix $M > 0$ arbitrary. We may suppose $nt > M$ since otherwise Lemma 6.4 holds true by (6.6). (6.6) also implies $\int_0^M \mathbf{E}\left[\left|\Psi_{s/n}^{(n)}\right|\right] b(s)ds \le \frac{C}{n}$ which gives us

$$\mathbf{E}\left[\left|\Psi_t^{(n)}(c)\right|\right] \le \rho_n(t) + Ca(M)\mathbf{E}\left[\left|\Psi_t^{(n)}(c)\right|\right]$$
$$+C\int_M^{nt} \mathbf{E}\left[\left|\Psi_{s/n}^{(n)}(c)\right|\right] b(s)ds + \frac{C}{n}.$$

Take M large enough such that $Ca(M) < 1$ and renew the positive constant C in the definition of $\rho_n(t)$. Then we have

$$\mathbf{E}\left[\left|\Psi_t^{(n)}(c)\right|\right] \le \rho_n(t) + C\int_{M/n}^{t} \mathbf{E}\left[\left|\Psi_s^{(n)}(c)\right|\right] nb(ns)ds.$$

By Grownwall's inequality,

$$\mathbf{E}\left[\left|\Psi_t^{(n)}(c)\right|\right] \le \rho_n(t) + C\int_{M/n}^{t} \rho_n(s)nb(ns)\exp\left(C\int_s^t nb(nu)du\right)ds.$$

Since b is integrable, $\exp\left(C\int_s^t nb(nu)du\right)$ is bounded so that

$$\mathbf{E}\left[\left|\Psi_t^{(n)}(c)\right|\right] \le \rho_n(t) + C\int_{M/n}^{t} \rho_n(s)nb(ns)ds. \tag{6.8}$$

Substituting

$$\int_{M/n}^{t} \rho_n(s)nb(ns)ds = C\int_M^{nt}\left(\frac{s}{n} + \sqrt{\frac{s}{n^2}} + \frac{1}{n}\right)b(s)ds \le \frac{C}{\sqrt{n}}$$

into (6.8) yields the conclusion. □

Lemma 6.5. *For $t > 0$, we have*

$$\mathbf{E}[\langle V^{(n)}(c), \overline{V^{(n)}(c)}\rangle_t] \le Ct + o(1)$$

as $n \to \infty$. In particular, $\sup_n \mathbf{E}[\langle V^{(n)}(c), \overline{V^{(n)}(c)}\rangle_t] < \infty$.

Proof. A straightforward computation using Lemma 6.1(2) yields

$$\langle V^{(n)}(c), \overline{V^{(n)}(c)}\rangle_t = \int_0^{nt} a(s)^2\left|e^{2i(\theta_s(\kappa+\frac{c}{n})-\theta_s(\kappa))} - 1\right|^2 [g_\kappa, \overline{g_\kappa}](X_s)ds + o(1)$$

$$= \langle [g_\kappa, \overline{g_\kappa}] \rangle \int_0^{nt} a(s)^2 \left| e^{2i(\theta_s(\kappa + \frac{c}{n}) - \theta_s(\kappa))} - 1 \right|^2 ds + o(1)$$

as $n \to \infty$. We take expectations and use Lemma 6.4.

$$\begin{aligned}
\mathbf{E}[\langle V^{(n)}(c), \overline{V^{(n)}(c)} \rangle_t] &= Cn \int_0^t a(ns)^2 \mathbf{E}\left[|e^{i\Psi_s^{(n)}(c)} - 1|^2 \right] ds + o(1) \\
&\leq Cn \int_0^t a(ns)^2 \mathbf{E}\left[\left| \Psi_s^{(n)}(c) \right| \right] ds + o(1) \\
&\leq C \left(t + \sqrt{\frac{t}{n}} + \frac{\log(nt)}{\sqrt{n}} \right) + o(1).
\end{aligned}$$

□

Lemma 6.6. *For each $c > 0$, $T > 0$ fixed we have*

$$\begin{aligned}
&\mathbf{E}\left[\sup_{0 \leq t \leq T} \Psi_t^{(n)}(c) \right] \\
&\leq C \left(T + \sqrt{\frac{T}{n}} \right) + CT^{\frac{1}{2}} + o(1) + C\mathbf{E}\left[\max_{0 \leq t \leq T} |\delta_{nt}(\kappa + \frac{c}{n}) - \delta_{nt}(\kappa)| \right].
\end{aligned}$$

as $n \to \infty$.

Proof. We estimate the third term of (6.7) by the martingale inequality and use Lemma 6.5: $\mathbf{E}\left[\sup_{0 \leq t \leq T} |V_t^{(n)}(\kappa)| \right] \leq C\mathbf{E}\left[|V_T^{(n)}(\kappa)|^2 \right]^{1/2} \leq C(T + o(1))^{\frac{1}{2}}$. □

Lemma 6.7. *For each $0 < t_0 < t_1 < \infty$, we can find $C = C(t_0, t_1)$ such that for large n, we have*

$$\mathbf{E}\left[\left| V_t^{(n)}(c) - V_s^{(n)}(c) \right|^4 \right] \leq C(t-s)^2$$

for any $s, t \in [t_0, t_1]$.

Proof. By martingale inequality,

$$\begin{aligned}
&\mathbf{E}\left[\left| V_t^{(n)}(c) - V_s^{(n)}(c) \right|^4 \right] \leq C\mathbf{E}\left[\left| V_t^{(n)}(c) - V_s^{(n)}(c) \right|^2 \right]^2 \\
&\leq C\mathbf{E}\left[\int_{ns}^{nt} a(u)^2 \left[G_\kappa(u), \overline{G_\kappa(u)} \right] (X_u) du \right]^2 \\
&\leq C \left(\int_{ns}^{nt} a(u)^2 du \right)^2.
\end{aligned}$$

where $G_\kappa(u) := e^{2i\theta_u(\kappa + \frac{c}{n})} g_{\kappa + \frac{c}{n}} - e^{2i\theta_u(\kappa)} g_\kappa$.

We can find $N = N(t_0)$ such that for $n \geq N$

$$C \left(\int_{ns}^{nt} a(u)^2 du \right)^2 \leq C \log \left(1 + \frac{t-s}{t_0} \right)^2 \leq C(t-s)^2.$$

□

6.3. *Tightness of* Ψ

Lemma 6.8. *For any* $c = (c_1, c_2, \cdots, c_m) \in \mathbf{R}^m$, *the sequence of* $\mathbf{R}^m$*-valued process* $\{\Psi_t^{(n)}(c)\}_{n\ge1} = \{(\Psi_t^{(n)}(c_1), \cdots, \Psi_t^{(n)}(c_m))\}_{n\ge1}$ *is tight as a family in* $C([0,T] \to \mathbf{R}^m)$.

Proof. It is sufficient to show

$$(1)\ \lim_{A\to\infty} \sup_n \mathbf{P}(|\Psi_t^{(n)}(c)| \ge A) = 0$$

$$(2)\ \lim_{\delta\downarrow 0} \limsup_{n\to\infty} \mathbf{P}\left(\sup_{0\le s,t\le T,\ |t-s|<\delta} |\Psi_t^{(n)}(c) - \Psi_s^{(n)}(c)| > \rho \right) = 0, \quad T, \rho > 0.$$

(1) follows from Lemma 6.4. To prove (2), we fix $M > 0$ arbitrary and decompose

$$\begin{aligned}
&\mathbf{P}\left(\sup_{0\le s,t\le T,\ |t-s|<\delta} |\Psi_t^{(n)}(c) - \Psi_s^{(n)}(c)| > \rho \right) \\
\le\ &\mathbf{P}\left(\sup_{0\le s,t\le M,\ |t-s|<\delta} |\Psi_t^{(n)}(c) - \Psi_s^{(n)}(c)| > \rho \right) \\
&\quad + \mathbf{P}\left(\sup_{M\le s,t\le T,\ |t-s|<\delta} |\Psi_t^{(n)}(c) - \Psi_s^{(n)}(c)| > \rho \right) =: I + II.
\end{aligned}$$

Since $\Psi_0^{(n)}(c) = 0$ we have

$$I \le \mathbf{P}\left(\sup_{t\le M} |\Psi_t^{(n)}(c)| > \frac{\rho}{2} \right) + \mathbf{P}\left(\sup_{s\le M} |\Psi_s^{(n)}(c)| > \frac{\rho}{2} \right)$$

and we use Lemma 6.6

$$\begin{aligned}
&\mathbf{P}\left(\sup_{t\le M} |\Psi_t^{(n)}(c)| > \frac{\rho}{2} \right) \le \frac{2}{\rho} \mathbf{E}\left[\sup_{0\le t\le M} |\Psi_t^{(n)}(c)| \right] \\
\le\ &C\left(M + \sqrt{\frac{M}{n}} \right) + CM^{\frac{1}{2}} + o(1) + C\mathbf{E}\left[\max_{0\le t\le M} \left| \delta_{nt}(\kappa + \frac{c}{n}) - \delta_{nt}(\kappa) \right| \right]
\end{aligned}$$

as $n \to \infty$. By Lemma 6.2(3) the third term vanishes as $n \to \infty$ and it holds that $\limsup_{n\to\infty} I \le CM^{1/2}$. Thus following estimate will be sufficient

$$\lim_{\delta\downarrow 0} \limsup_{n\to\infty} II = 0. \tag{6.9}$$

By Lemmas 6.2, 6.3, eq.(6.9) will follow from the following equation

$$\lim_{\delta\downarrow 0} \limsup_{n\to\infty} \mathbf{P}\left(\sup_{M\le t,s\le T,\ |t-s|<\delta} \left| V_t^{(n)}(c) - V_s^{(n)}(c) \right| > \rho \right) = 0$$

which, in turn, follows from Lemma 6.7 and Kolmogorov's theorem. □

6.4. *SDE satisfied by* Ψ

In this subsection we show that $\Psi^{(n)}$ has a limit Ψ which satisfies (1.3).

Lemma 6.9. *For any* $c_1, \cdots, c_m \in \mathbf{R}$, *the solution of the following martingale problem is unique:*

$$W_t(c_j) = \Psi_t(c_j) - 2c_j t, \quad j = 1, 2, \cdots, m$$

are martingales whose variational process satisfy

$$\langle W(c_i), W(c_j)\rangle_t = D^2 \int_0^t s^{-1} Re\left\{\left(e^{i\Psi_s(c_i)} - 1\right)\left(e^{-i\Psi_s(c_j)} - 1\right)\right\} ds.$$

Moreover $\Psi_t(c_j)$ *can be characterized by the unique solution to the following SDE.*

$$d\Psi_t(c_j) = 2c_j dt + Dt^{-1/2} Re[(e^{i\Psi_t(c_j)} - 1)dZ_t], \quad \Psi_0(c_j) = 0.$$

The proof of Lemma 6.9 is similar to that of Lemma 6.10 except that we consider the regularized p-th power and use Lemma 6.4.

Theorem 6.1. *For any* $c_1, \cdots, c_m \in \mathbf{R}$, $(\Psi_t^{(n)}(c_1), \cdots, \Psi_t^{(n)}(c_m)) \stackrel{d}{\to} (\Psi_t(c_1), \cdots, \Psi_t(c_m))$, *where* $\{\Psi_t(c_j)\}$ *satisfies (1.3).*

Proof. By Lemma 6.8, the sequence $\{(\Psi_t^{(n)}(c_1), \cdots, \Psi_t^{(n)}(c_m))\}_{n\geq 1}$ has a limit point $(\Psi_t(c_1), \cdots, \Psi_t(c_m))$. Since Lemmas 6.2, 6.3 imply

$$\Psi_t^{(n)}(c) = 2ct + \frac{1}{\kappa} Re\, V_t^{(n)}(c) + o(1)$$

in probability, we study $V_t^{(n)}(c)$. By a computation using Lemma 6.1, $\langle V^{(n)}(c), V^{(n)}(c')\rangle_t \stackrel{n\to\infty}{\to} 0$ in mean square. Similarly,

$$\begin{aligned}&\langle V^{(n)}(c), \overline{V^{(n)}(c')}\rangle_t \\ &= \langle [g_\kappa, \overline{g_\kappa}]\rangle \int_0^t na(nu)^2 \left(e^{i\Psi_u^{(n)}(c)} - 1\right)\overline{\left(e^{i\Psi_u^{(n)}(c')} - 1\right)} du + o(1).\end{aligned}$$

By Skorohod's theorem, we can suppose $\Psi_t^{(n)}(c) \to \Psi_t(c)$ compact uniformly with respect to t. Hence for $0 < s < t$,

$$\begin{aligned}&\langle V^{(n)}(c), \overline{V^{(n)}(c')}\rangle_t - \langle V^{(n)}(c), \overline{V^{(n)}(c')}\rangle_s \\ &\stackrel{n\to\infty}{\to} \langle [g_\kappa, \overline{g_\kappa}]\rangle \int_s^t u^{-1}\left(e^{i\Psi_u(c)} - 1\right)\overline{(e^{i\Psi_u(c')} - 1)} du.\end{aligned}$$

On the other hand by Lemma 6.4 we have

$$\int_0^t \mathbf{E}\left[\left|e^{i\Psi_s(c)} - 1\right|^2\right] \frac{ds}{s} \leq C \int_0^t \mathbf{E}\left[|\Psi_s(c)|\right] \frac{ds}{s} < \infty$$

so that $V_t(c) = \lim_{n\to\infty} V_t^{(n)}(c)$ is a square integrable continuous martingale whose variational process satisfy

$$\langle V(c), V(c')\rangle_t = 0$$

$$\langle V(c), \overline{V(c')} \rangle_t = \langle [g_\kappa, \overline{g_\kappa}] \rangle \int_0^t \left(e^{i\Psi_s(c)} - 1 \right) \overline{(e^{i\Psi_s(c')} - 1)} \frac{ds}{s}.$$

Therefore

$$W_t(c) = \Psi_t(c) - 2ct = \frac{1}{\kappa} Re\, V_t(c)$$

is a square integrable continuous martingale whose variational process is equal to

$$\langle W(c), W(c') \rangle_t = \frac{\langle [g_\kappa, \overline{g_\kappa}] \rangle}{2\kappa^2} \int_0^t Re \left[(e^{i\Psi_s(c)} - 1)(e^{-i\Psi_s(c')} - 1) \frac{ds}{s} \right].$$

Lemma 6.9 yields the conclusion. □

Lemma 6.10. *For a.s., $\Psi_t(c)$ is continuous on $[0, \infty) \times \mathbf{R}$ and is increasing with respect to c.*

Proof. We shall show the following inequality: for $p > 1$ sufficiently close to 1,

$$\mathbf{E}[|\Psi_t(c_1) - \Psi_t(c_2)|^p] \leq \frac{2^p (c_1 - c_2)^p}{1 - \frac{1}{2}(p-1) D^2} t^p. \tag{6.10}$$

Hence by Kolmogorov's theorem, for any fixed $t > 0$, $\Psi_t(c)$ has a continuous version with respect to $c \in \mathbf{R}$ a.s.. We first note that $\Psi_t(c)$ satisfies

$$d\Psi_t(c) = 2c dt + \frac{D}{2\sqrt{t}} \left\{ (e^{i\Psi_t} + e^{-i\Psi_t} - 2) dB_t^1 + i(e^{i\Psi_t} - e^{-i\Psi_t}) dB_t^2 \right\}.$$

Here we note that if $c_1 > c_2$ then $\Psi_t(c_1) > \Psi_t(c_2)$ by the comparison theorem of SDE which proves the desired monotonicity of $\Psi_t(c)$. We set

$$\Gamma_t := \Psi_t(c_1) - \Psi_t(c_2), \quad \Xi_t := e^{i\Psi_t(c_1)} - e^{i\Psi_t(c_2)}.$$

For $c_1 > c_2$, we see

$$d\Gamma_t = 2(c_1 - c_2) dt + \frac{D}{2\sqrt{t}} \left\{ (\Xi_t + \overline{\Xi}_t) dB_t^1 + i(\Xi_t - \overline{\Xi}_t) dB_t^2 \right\}.$$

Hence

$$(d\Gamma_t)^2 = \frac{D^2}{4t} \left\{ (\Xi_t + \overline{\Xi}_t)^2 - (\Xi_t - \overline{\Xi}_t)^2 \right\} dt = \frac{D^2}{t} |\Xi_t|^2 dt.$$

Then for $p > 1$

$$\begin{aligned} d\Gamma_t^p &= p\Gamma_t^{p-1} d\Gamma_t + \frac{p(p-1)}{2} \Gamma_t^{p-2} (d\Gamma_t)^2 \\ &= 2(c_1 - c_2) p \Gamma_t^{p-1} dt + \frac{p(p-1)}{2} \Gamma_t^{p-2} \frac{D^2}{t} |\Xi_t|^2 dt \\ &\quad + p\Gamma_t^{p-1} \frac{D}{2\sqrt{t}} \left\{ (\Xi_t + \overline{\Xi_t}) dB_t^1 + i(\Xi_t - \overline{\Xi_t}) dB_t^2 \right\}. \end{aligned}$$

Taking expectation yields

$$\mathbf{E}[\Gamma_t^p] = 2(c_1 - c_2)p \int_0^t \mathbf{E}[\Gamma_s^{p-1}]ds + \frac{p(p-1)}{2} D^2 \int_0^t \mathbf{E}[\Gamma_s^{p-2}|\Xi_s|^2]\frac{ds}{s}. \tag{6.11}$$

We have $|\Xi_t|^2 \le C\Gamma_t^\gamma$, $0 < \gamma < 2$ for some positive constant C and some $0 < \gamma < 2$. Hence

$$\int_0^t \mathbf{E}[\Gamma_s^{p-2}|\Xi_s|^2]\frac{ds}{s} \le C \int_0^t \mathbf{E}[\Gamma_s^{p-2+\gamma}]\frac{ds}{s}.$$

We use $\mathbf{E}[|X|^r] \le \mathbf{E}[|X|]^r$ for $r \le 1$ and the fact that $\mathbf{E}[\Psi_t(c)] = 2ct$. Assuming $p - 1 \le 1$ and $0 < p - 2 + \gamma \le 1$ yields

$$\begin{aligned}\mathbf{E}[\Gamma_t^p] &\le 2(c_1 - c_2)p \int_0^t \mathbf{E}[\Gamma_s]^{p-1}ds + C \int_0^t \mathbf{E}[\Gamma_s]^{p-2+\gamma}\frac{ds}{s} \\ &= 2^p(c_1 - c_2)^p t^p + C(c_1 - c_2)^{p-2+\gamma} t^{p-2+\gamma}\end{aligned}$$

so that for $0 \le t \le T$ we have

$$f(t) := \mathbf{E}[\Gamma_t^p] \le C_T t^{p-2+\gamma}$$

and hence

$$h(t) := \int_0^t \frac{f(s)}{s} ds \le C_T t^{p-2+\gamma}.$$

Thus for any $p > 1$ sufficiently close to 1, we take γ satisfying $1+(p-1)(\frac{p}{2}D^2-1) < \gamma \le 3-p$ so that

$$h(t) \le C t^{\frac{p}{2}(p-1)D^2+\delta} \tag{6.12}$$

for some $\delta > 0$. On the other hand by using $|\Xi_s|^2 \le \Gamma_s^2$ in (6.11) we have

$$\begin{aligned}\mathbf{E}[\Gamma_t^p] &\le 2(c_1 - c_2)p \int_0^t 2^{p-1}(c_1 - c_2)^{p-1} s^{p-1} ds + \frac{p}{2}(p-1)D^2 \int_0^t \mathbf{E}[\Gamma_s^p]\frac{ds}{s} \\ &= 2^p(c_1 - c_2)^p t^p + \frac{p}{2}(p-1)D^2 \int_0^t \mathbf{E}[\Gamma_s^p]\frac{ds}{s}.\end{aligned}$$

Hence if $\frac{1}{2}(p-1)D^2 < 1$, (6.12) and a Grownwall type argument give the desired inequality (6.10).

Having established the continuity of $\Psi_{t_0}(c)$ with respect to c, the joint continuity of $\Psi_t(c)$ on $[t_0, \infty) \times \mathbf{R}$ is valid due to the absence of singularity in this time domain. The continuity of $\Psi_t(c)$ at $t = 0$ follows from the monotonicity of $\Psi_t(c)$ with respect to c. □

Remark 6.1. $\{\Psi_t(c)\}_{t\ge 0, c\in\mathbf{R}}$ satisfies the following properties:
(1) The process has invariance

$$\{\Psi_t(c)\}_{t\ge 0, c\in\mathbf{R}} \stackrel{law}{=} \{\Psi_t(c + c_0) - \Psi_t(c_0)\}_{t\ge 0, c\in\mathbf{R}}$$

for any $c_0 \in \mathbf{R}$.
(2) For each fixed c there exists a 1-D Brownian motion $\{B_t(c)\}$ such that

$$\frac{\partial \Psi_t}{\partial c} = 2\int_0^t \exp\left(\int_s^t \frac{D}{\sqrt{u}} dB_u - \int_s^t \frac{D^2}{2u} du\right) ds$$

where $\{B_t(c)\}$ is a family of martingales satisfying

$$\langle B_\cdot(c), B_\cdot(c')\rangle_t = \int_0^t \cos\left(\Psi_s(c) - \Psi_s(c')\right) ds.$$

7. Convergence of $\theta_t(\kappa)$ mod π

Proposition 7.1. *As $t \to \infty$ $(2\theta_t(\kappa))_{2\pi\mathbf{Z}}$ converges to the uniform distribution on $[0, 2\pi)$.*

Proof. Letting $\xi_t(\kappa) := e^{2mi\tilde{\theta}_t(\kappa)}, m \in \mathbf{Z}$, it suffices to show $\mathbf{E}[\xi_t(\kappa)] \overset{t\to\infty}{\to} 0, m \neq 0$. We omit the κ-dependence of θ_t. By (2.4) we decompose

$$\begin{aligned}
\xi_t &= 1 + \frac{mi}{2\kappa}\int_0^t e^{2i\kappa s + 2(m+1)i\tilde{\theta}_s} a(s) F(X_s) ds \\
&\quad + \frac{mi}{2\kappa}\int_0^t e^{-2i\kappa s + 2(m-1)i\tilde{\theta}_s} a(s) F(X_s) ds - \frac{mi}{\kappa}\int_0^t e^{2mi\tilde{\theta}_s} a(s) F(X_s) ds \\
&=: 1 + I + II + III.
\end{aligned}$$

We use Lemma 6.1(1) and decompose I further:

$$\begin{aligned}
I = \frac{mi}{2\kappa}\Bigg(& -\frac{2i(m+1)}{4\kappa}\int_0^t a(s)^2 e^{2mi\tilde{\theta}_s} F(X_s) g_\kappa(X_s) ds \\
& + \int_0^t a(s) e^{2i\kappa s} e^{2i(m+1)\tilde{\theta}_s} dM_s(\kappa) + \delta_{1,1}(t)\Bigg)
\end{aligned} \tag{7.1}$$

where

$$\begin{aligned}
&\delta_{1,1}(t) \\
&:= \left[a(s) e^{2i(m+1)\tilde{\theta}_s} e^{2i\kappa s} g_\kappa(X_s)\right]_0^t - \int_0^t a'(s) e^{2i(m+1)\tilde{\theta}_s} e^{2i\kappa s} g_\kappa(X_s) ds \\
&\quad -\frac{2i(m+1)}{2\kappa}\int_0^t a(s)^2 \left(\frac{e^{2(m+2)i\tilde{\theta}_s} e^{4i\kappa s}}{2} - e^{2(m+1)i\tilde{\theta}_s} e^{2i\kappa s}\right) F(X_s) g_\kappa(X_s) ds.
\end{aligned}$$

We further compute the third term of $\delta_{1,1}$ by Lemma 6.1(1) and see that $\delta_{1,1}(t)$ has a limit as $t \to \infty$. Taking expectation, martingale term vanishes and we have

$$\mathbf{E}[\delta_{1,1}(t)] - \mathbf{E}[\delta_{1,1}(\infty)] = O(a(t)), \quad t \to \infty. \tag{7.2}$$

By Lemma 6.1(2), the first term of (7.1) satisfies

$$\int_0^t a(s)^2 e^{2mi\tilde{\theta}_s} F(X_s) g_\kappa(X_s) ds = \langle F g_\kappa \rangle \int_0^t a(s)^2 e^{2mi\tilde{\theta}_s} ds + \delta_{1,2}(t)$$

where $\delta_{1,2}(t)$ has a limit as $t \to \infty$ and satisfies the same estimate as (7.2). We substitute it into (7.1) and let $\delta_1 = \delta_{1,1} + \delta_{1,2}$. Then

$$I = \frac{mi}{2\kappa}\left(-\frac{2i(m+1)}{4\kappa}\langle Fg_\kappa \rangle \int_0^t a(s)^2 e^{2mi\tilde{\theta}_s} ds \right.$$
$$\left. + \int_0^t a(s) e^{2i\kappa s} e^{2i(m+1)\tilde{\theta}_s} dM_s(\kappa) + \delta_1(t) \right). \tag{7.3}$$

We compute II, III in a similar manner and consequently

$$\xi_t = 1 - \frac{mi}{\kappa}\langle F \rangle \int_0^t a(s) e^{2mi\tilde{\theta}_s} ds + \langle G_m \rangle \int_0^t a(s)^2 e^{2mi\tilde{\theta}_s} ds + N_t + \delta(t) \tag{7.4}$$

where

$$\begin{aligned} G_m &= \left(\frac{m(m+1)}{4\kappa^2} g_\kappa + \frac{m(m-1)}{4\kappa^2} g_{-\kappa} + \frac{m^2}{\kappa^2} g \right) F \\ N_t &= \frac{mi}{2\kappa} \int_0^t a(s) e^{2i\kappa s} e^{2i(m+1)\tilde{\theta}_s} dM_s(\kappa) + \frac{mi}{2\kappa} \int_0^t a(s) e^{-2i\kappa s} e^{2i(m-1)\tilde{\theta}_s} dM_s(-\kappa) \\ &\quad - \frac{mi}{\kappa} \int_0^t a(s) e^{2mi\tilde{\theta}_s} dM_s \end{aligned}$$

where $\delta(\infty) = \lim_{t\to\infty} \delta(t)$ exists a.s. and

$$\mathbf{E}[\delta(t)] - \mathbf{E}[\delta(\infty)] = O(a(t)), \quad t \to \infty.$$

Let $\sigma_F(d\lambda)$ be the spectral measure of L with respect to F. Then by noting

$$Re\langle Fg_\kappa \rangle = Re\langle Fg_{-\kappa} \rangle = \int_{-\infty}^0 \frac{\lambda \sigma_F(d\lambda)}{\lambda^2 + 4\kappa^2} < 0, \quad Re\langle Fg \rangle = \int_{-\infty}^0 \frac{\sigma_F(d\lambda)}{\lambda} < 0$$

we have $-\gamma := Re\langle G_m \rangle < 0$. Set

$$\rho(t) := \mathbf{E}[\xi_t], \quad b(t) := -\frac{mi}{\kappa}\langle F \rangle a(t) + \langle G_m \rangle a(t)^2.$$

Then (7.4) turns to

$$\rho(t) = 1 + \int_0^t b(s)\rho(s) ds + \mathbf{E}[\delta(t)]$$

and hence

$$\begin{aligned} \rho(t) &= \exp\left(\int_0^t b(u) du \right) + \mathbf{E}[\delta(t)] + \int_0^t \mathbf{E}[\delta(s)] b(s) \exp\left(\int_s^t b(u) du \right) ds \\ &= \exp\left(\int_0^t b(u) du \right) + \mathbf{E}[\delta(t)] + \mathbf{E}[\delta(\infty)] \int_0^t b(s) \exp\left(\int_s^t b(u) du \right) ds \\ &\quad + \int_0^t \left(\mathbf{E}[\delta(s)] - \mathbf{E}[\delta(\infty)] \right) b(s) \exp\left(\int_s^t b(u) du \right) ds \\ &=: I + II + III + IV. \end{aligned}$$

Noting $Re\, b(t) = Re\, \langle G_m \rangle a(t)^2 = -\gamma a(t)^2$, we compute I, III

$$|I| \le \exp\left(\int_0^t Re\, b(s)ds\right) \le C \exp\left(-\gamma \int_1^t \frac{1}{s} ds\right) \overset{t\to\infty}{\to} 0$$

$$III = \mathbf{E}[\delta(\infty)] \left(-1 + \exp\left(\int_0^t b(u)du\right)\right) \overset{t\to\infty}{\to} -\mathbf{E}[\delta(\infty)].$$

We further decompose IV:

$$\begin{aligned}|IV| &= \left|\int_0^t (\mathbf{E}[\delta(s)] - \mathbf{E}[\delta(\infty)])\, b(s) \exp\left(\int_s^t b(u)du\right) ds\right| \\ &\le C\left(\int_0^M + \int_M^t\right) a(s)|b(s)| \exp\left(Re \int_s^t b(u)du\right) ds \\ &=: IV_1 + IV_2.\end{aligned}$$

It is easy to see that $IV_1 \overset{t\to\infty}{\to} 0$. For IV_2 we use $\langle F \rangle = 0$ and compute, for large M,

$$\begin{aligned}|IV_2| &\le C \int_M^t a(s)^3 \exp\left(\int_s^t Re\, b(u)du\right) ds \\ &\le C \int_M^t s^{-3/2} \left(\frac{t}{s}\right)^{-\gamma} ds = \begin{cases} Ct^{-\gamma} \log \frac{t}{M} & (\gamma = \frac{1}{2}) \\ Ct^{-\gamma} \frac{t^{\gamma-\frac{1}{2}} - M^{\gamma-\frac{1}{2}}}{\gamma - \frac{1}{2}} & (\gamma \ne \frac{1}{2}) \end{cases} \quad \overset{t\to\infty}{\to} 0.\end{aligned}$$

□

8. Limiting behavior of $\tilde{\theta}_t$

To study the limiting behavior of $(2\tilde{\theta}_t)_{2\pi\mathbf{Z}}$ we set

$$\tilde{\xi}_t := e^{2i\tilde{\theta}_t(\kappa)}.$$

8.1. *Estimate of integral equation*

As in the proof of Proposition 7.1, we can show the following lemma.

Lemma 8.1. *Let $0 < t_0 < t$. Then we have*

$$\begin{aligned}\tilde{\xi}_t &= \tilde{\xi}_{t_0} + \frac{1}{2\kappa^2} \langle F \cdot (g_\kappa + 2g) \rangle \int_{t_0}^t a(s)^2 e^{2i\tilde{\theta}_s} ds - \frac{i}{\kappa} \langle F \rangle \int_{t_0}^t a(s) e^{2i\tilde{\theta}_s} ds \\ &\quad + \frac{i}{2\kappa}\left(Y_t + \widetilde{Y}_t - 2\widehat{Y}_t\right) + O(a(t_0)), \quad t_0 \to \infty.\end{aligned}$$

where $Y_t := \int_{t_0}^t a(s) e^{2i\kappa s + 4i\tilde{\theta}_s} dM_s(\kappa), \quad \widetilde{Y}_t := \int_{t_0}^t a(s) e^{-2i\kappa s} dM_s(-\kappa)$

$$\widehat{Y}_t := \int_{t_0}^t a(s) e^{2i\tilde{\theta}_s} dM_s.$$

The variational process of $Y, \tilde{Y}$, and $\hat{Y}$ satisfy, as $t_0 \to \infty$,

$$\langle Y, Y \rangle_t = O(a(t_0)),\ \langle Y, \overline{Y} \rangle_t = \langle [g_\kappa, \overline{g}_\kappa] \rangle \int_{t_0}^t a(s)^2 ds + O(a(t_0))$$

$$\langle \widetilde{Y}, \widetilde{Y} \rangle_t = O(a(t_0)),\ \langle \widetilde{Y}, \overline{\widetilde{Y}} \rangle_t = \langle [g_\kappa, \overline{g}_\kappa] \rangle \int_{t_0}^t a(s)^2 ds + O(a(t_0))$$

$$\langle \widehat{Y}, \widehat{Y} \rangle_t = \langle [g, g] \rangle \int_{t_0}^t a(s)^2 e^{4i\tilde{\theta}_s} ds + O(a(t_0)),$$

$$\langle \widehat{Y}, \overline{\widehat{Y}} \rangle_t = \langle [g, g] \rangle \int_{t_0}^t a(s)^2 ds + O(a(t_0)).$$

8.2. *Tightness of* η

Let

$$\eta_t^{(n)} := \tilde{\xi}_{nt} = e^{2i\tilde{\theta}_{nt}(\kappa)}, \quad \mathbf{U} := \{ z \in \mathbf{C} \mid |z| = 1 \}.$$

Lemma 8.2. *$\{\eta_t^{(n)}\}_{n \ge 1}$ is tight as a family in $C((0,\infty) \to \mathbf{U})$.*

Proof. It suffices to show, for any $t_0 > 0, \rho > 0$,

$$\lim_{\delta \downarrow 0} \limsup_{n \to \infty} \mathbf{P} \left(\sup_{t_0 < s < t, t-s < \delta} |\eta_t^{(n)} - \eta_s^{(n)}| > \rho \right) = 0.$$

Noting $\langle F \rangle = 0$, Lemma 8.1 implies

$$\tilde{\xi}_{nt} - \tilde{\xi}_{ns} = \frac{1}{2\kappa^2} \langle F \cdot (g_\kappa + 2g) \rangle \int_{ns}^{nt} a(u)^2 e^{2i\tilde{\theta}_u} du + \frac{i}{2\kappa} W_{t,s}^{(n)} + o(1), \tag{8.1}$$

$$\text{where} \quad W_{t,s}^{(n)} := \left(Y_{nt} + \tilde{Y}_{nt} - 2\hat{Y}_{nt} \right) - \left(Y_{ns} + \tilde{Y}_{ns} - 2\hat{Y}_{ns} \right),$$

as $n \to \infty$. We note that $W_{t,s}^{(n)}$ satisfies the estimate in Lemma 6.7 and the rest of the argument is the same as that in Lemma 6.8. □

8.3. *Identification of* η_t

Let η_t be a limit point of $\eta_t^{(n)}$ which is uniformly distributed on $\mathbf{U}$ for each fixed $t > 0$ by Lemma 7.1. In this subsection we show that the distribution of the process η_t is uniquely determined.

Lemma 8.3. *(1) For any $0 < t_0 < t$,*

$$\lim_{n \to \infty} \mathbf{E}[e^{2mi(\tilde{\theta}_{nt} - \tilde{\theta}_{nt_0})} | \mathcal{F}_{nt_0}] = \left(\frac{t}{t_0} \right)^{\langle G_m \rangle}$$

where $\mathcal{F}_t$ is the σ-algebra generated by $\{X_s\}_{0 \le s \le t}$.
(2) For any $0 < t_0 < t_1 < \cdots < t_k$, the family of random variables $\{\eta_{t_0}, \eta_{t_1}/\eta_{t_0}, \cdots, \eta_{t_k}/\eta_{t_{k-1}}\}$ are independent.

Proof. (1) Let $m, m' \in \mathbf{Z}$. By a argument similar to deduce (7.4), we have

$$e^{2mi(\tilde{\theta}_{nt} - \tilde{\theta}_{nt_0})} = 1 + \langle G_m \rangle \int_{t_0}^t na(nu)^2 e^{2mi(\tilde{\theta}_{nu} - \tilde{\theta}_{nt_0})} du$$
$$+ N_{nt,nt_0} e^{-2mi\tilde{\theta}_{nt_0}} + \delta_n(t) e^{-2mi\tilde{\theta}_{nt_0}}$$

where

$$\begin{aligned} G_m &= \left(\frac{m(m+1)}{4\kappa^2} g_\kappa + \frac{m(m-1)}{4\kappa^2} g_{-\kappa} + \frac{m^2}{\kappa^2} g \right) F \\ N_{nt,nt_0} &= \frac{mi}{2\kappa} \int_{nt_0}^{nt} a(s) e^{2i\kappa s} e^{2i(m+1)\tilde{\theta}_s} dM_s(\kappa) \\ &\quad + \frac{mi}{2\kappa} \int_{nt_0}^{nt} a(s) e^{-2i\kappa s} e^{2i(m-1)\tilde{\theta}_s} dM_s(-\kappa) - \frac{mi}{\kappa} \int_{nt_0}^{nt} a(s) e^{2mi\tilde{\theta}_s} dM_s \\ \mathbf{E}[\delta_n(t)|\ \mathcal{F}_{nt_0}\ ] &\overset{n\to\infty}{\to} 0, \quad a.s.. \end{aligned}$$

Taking a conditional expectation and letting

$$\rho_n(t) := \mathbf{E}[e^{2mi(\tilde{\theta}_{nt} - \tilde{\theta}_{nt_0})} | \mathcal{F}_{nt_0}]$$

we have

$$\rho_n(t) = 1 + \langle G_m \rangle \int_{t_0}^t na(nu)^2 \rho_n(u) du + \mathbf{E}[\delta_n(t) | \mathcal{F}_{nt_0}] e^{-2mi\tilde{\theta}_{nt_0}}.$$

Therefore

$$\begin{aligned} \rho_n(t) &= \exp\left(\langle G_m \rangle \int_{t_0}^t na(nu)^2 du \right) + \mathbf{E}[\delta_n(t)|\mathcal{F}_{nt_0}] e^{-2mi\tilde{\theta}_{nt_0}} \\ &+ \textstyle\int_{t_0}^t \mathbf{E}[\delta_n(s)|\mathcal{F}_{nt_0}] e^{-2mi\tilde{\theta}_{nt_0}} \langle G_m \rangle na(nu)^2 \exp\left(\langle G_m \rangle \displaystyle\int_s^t na(nu)^2 du \right) ds \\ &\overset{n\to\infty}{\to} \exp\left(\langle G_m \rangle \int_{t_0}^t \frac{du}{u} \right) = \left(\frac{t}{t_0} \right)^{\langle G_m \rangle}. \end{aligned}$$

(2) The required independence easily follows from (1) and the fact that $e^{2i\tilde{\theta}_{nt}}$ converges to the uniform distribution on $\mathbf{U}$ as $n \to \infty$. In fact, for $k = 1$,

$$\begin{aligned} &\mathbf{E}\left[\mathbf{E}[e^{2mi(\tilde{\theta}_{nt} - \tilde{\theta}_{nt_0})} | \mathcal{F}_{nt_0}] e^{2m' i\tilde{\theta}_{nt_0}} \right] \\ &= \mathbf{E}\left[\left(\mathbf{E}[e^{2mi(\tilde{\theta}_{nt} - \tilde{\theta}_{nt_0})} | \mathcal{F}_{nt_0}] - \left(\frac{t}{t_0} \right)^{\langle G_m \rangle} \right) e^{2m' i\tilde{\theta}_{nt_0}} \right] \\ &\quad + \mathbf{E}\left[\left(\frac{t}{t_0} \right)^{\langle G_m \rangle} e^{2m' i\tilde{\theta}_{nt_0}} \right] \\ &\to \begin{cases} 0 & (m' \neq 0) \\ \left(\frac{t}{t_0} \right)^{\langle G_m \rangle} & (m' = 0) \end{cases} \end{aligned}$$

For $k \geq 2$, the proof is similar. □

Lemma 8.4. *For each fixed $t_0 > 0$, η_t satisfies the following SDE on $t \geq t_0$:*

$$d\eta_t = C_1 \frac{\eta_t}{t} dt + C_2 \frac{\eta_t}{\sqrt{t}} dB_t, \tag{8.2}$$

where $C_1 := \dfrac{\langle (g_\kappa + 2g) F \rangle}{2\kappa^2}, \quad C_2 := \dfrac{i}{2\kappa} \sqrt{\langle 2[g_\kappa, \overline{g_\kappa}] + 4[g, g] \rangle}.$

Proof. Letting $s = t_0 > 0$ in (8.1) yields, as $n \to \infty$,

$$\int_{nt_0}^{nt} a(u)^2 e^{2i\tilde{\theta}_u} du \to \int_{t_0}^{t} \frac{\eta_u}{u} du$$

$$\langle W^{(n)}_{\cdot,t_0}, W^{(n)}_{\cdot,t_0} \rangle_t \to \langle 2[g_\kappa, \overline{g_\kappa}] + 4[g,g] \rangle \int_{t_0}^{t} \frac{\eta_u^2}{u} du$$

$$\langle W^{(n)}_{\cdot,t_0}, \overline{W^{(n)}_{\cdot,t_0}} \rangle_t \to \langle 2[g_\kappa, \overline{g_\kappa}] + 4[g,g] \rangle \int_{t_0}^{t} \frac{du}{u}.$$

We then proceed as in the proof of Theorem 6.1. □

Remark 8.1. Z_t, B_t which appear in SDE's (1.3), (8.2) of Ψ, η are not independent. In fact, $W_t = \lim_{n\to\infty} W_t^{(n)}$, $V_t = \lim_{n\to\infty} V_t^{(n)}$ satisfy

$$\begin{aligned} dW_t &= \sqrt{2\langle [g_\kappa, \overline{g_\kappa}] \rangle + 4[g,g]} \frac{\eta_t}{\sqrt{t}} dB_t \\ dV_t &= \sqrt{\langle [g_\kappa, \overline{g_\kappa}] \rangle} \left(e^{i\Psi_t(c)} - 1 \right) \frac{dZ_t}{\sqrt{t}} \\ d\langle W, V \rangle &= \langle [g_\kappa, \overline{g_\kappa}] \rangle \left(e^{i\Psi_t(c)} - 1 \right) \eta_t \frac{dt}{t} \\ d\langle W, \overline{V} \rangle &= \langle [g_\kappa, \overline{g_\kappa}] \rangle \left(e^{-i\Psi_t(c)} - 1 \right) \eta_t \frac{dt}{t} \end{aligned}$$

which imply

$$dZdB = \sqrt{\frac{\langle [g_\kappa, \overline{g_\kappa}] \rangle}{2\langle [g_\kappa, \overline{g_\kappa}] + 4[g,g] \rangle}} dt.$$

Here we note the following fact. By the time change $u = \log t$, $\zeta_u := \log \eta_{e^u}$ satisfies the following SDE which is stationary in time.

$$\begin{aligned} d\zeta_u &= iC_3 du + iC_4 d\tilde{B}_u \qquad (8.3) \\ \text{where} \quad C_3 &:= -\frac{1}{\kappa} \langle |g_\kappa|^2 \rangle \in \mathbf{R}, \quad C_4 := \frac{1}{2\kappa} \sqrt{2[g_\kappa, \overline{g}_\kappa] + 4[g,g]} \in \mathbf{R} \end{aligned}$$

To summarize, the following facts have been proved.
(i) For any $t > 0$, η_t is uniformly distributed(Lemma 7.1).
(ii) For any $0 < t_0 < t_1 < t_2 < \cdots < t_n$, random variables $\{\eta_{t_0}, \eta_{t_1}/\eta_{t_0}, \cdots \eta_{t_n}/\eta_{t_{n-1}}\}$ are independent(Lemma 8.3).
(iii) For any $t_0 > 0$, $x_t = \eta_t/\eta_{t_0}$ satisfies an SDE on $t \geq t_0$ (Lemma 8.4):

$$dx_t = C_1 \frac{x_t}{t} dt + C_2 \frac{x_t}{\sqrt{t}} dB_t, \quad x_{t_0} = 1.$$

These facts determines (in distribution) the process η_t uniquely. In fact, for any $0 < t_0 < t_1 < \cdots < t_n$, the distribution of $\{\eta_{t_0}, \eta_{t_1}, \cdots, \eta_{t_n}\}$ can be computed from that of $\{\eta_{t_0}, \eta_{t_1}/\eta_{t_0}, \cdots, \eta_{t_n}/\eta_{t_{n-1}}\}$ and the latter distribution can be determined uniquely from (ii) and (iii). Therefore the distribution of $\{\eta_t\}$ is characterized by the constants C_1, C_2. More concretely, if we prepare 1D Brownian motion $\{B_t\}_{t\in\mathbf{R}}$ with $B_0 = 0$ and independent random variable $X \in \mathbf{C}$ with uniform distribution on $\mathbf{U}$, a process $X \exp[i(C_3 u + C_4 B_u)]$ has the same distribution as $\{\eta_{e^u}\}$ by (8.2), (8.3).

9. Convergence of the joint distribution

We finish the proof of Theorem 1.3.

9.1. *Behavior of the joint distribution*

Proposition 9.1. *For any* $c_1, \cdots, c_m \in \mathbf{R}$, $t > 0$,

$$(\Psi_t^{(n)}(c_1), \cdots, \Psi_t^{(n)}(c_m), (\theta_{nt}(\kappa))_{2\pi\mathbf{Z}}) \xrightarrow{d} (\Psi_t(c_1), \cdots, \Psi_t(c_m), \phi_t), \tag{9.1}$$

as $n \to \infty$, *where* $(\Psi_t(c_1), \cdots, \Psi_t(c_m))$ *and* ϕ_t *are independent and* ϕ_t *is uniformly distributed on* $[0, 2\pi)$.

Proof. For simplicity, we use the following notation. $\mathbf{c} := (c_1, \cdots, c_m)$, $\Psi_t^{(n)}(\mathbf{c}) := (\Psi_t^{(n)}(c_1), \cdots, \Psi_t^{(n)}(c_m))$, and $\Psi_t(\mathbf{c}) := (\Psi_t(c_1), \cdots, \Psi_t(c_m))$. It suffices to show (9.1) with $(\theta_{nt}(\kappa))_{2\pi\mathbf{Z}}$ being replaced by $(\tilde{\theta}_{nt}(\kappa))_{2\pi\mathbf{Z}}$, since $(\tilde{\theta}_{nt}(\kappa))_{2\pi\mathbf{Z}}$ converges to the uniform distribution by Lemma 7.1. By Lemmas 6.8, 8.2, for any fixed $t_0 > 0$, the process $\{(\Psi_t^{(n)}(\mathbf{c}), \eta_t^{(n)})\}_{n\geq 1}$ on $[t_0, \infty)$ is a tight family. Hence we can assume $(\Psi_t^{(n)}(\mathbf{c}), \eta_t^{(n)})_{t>0} \xrightarrow{d} (\Psi_t(\mathbf{c}), \eta_t)_{t>0}$. By Lemma 8.3 $\eta_{1/n}$ and $\eta_t/\eta_{1/n}$ are independent.

We next consider a process $\tilde{\Psi}_t^{(n)}(\mathbf{c})$ which is defined on $[\frac{1}{n}, \infty)$ and is the solution to (1.3) with initial value $\tilde{\Psi}_{\frac{1}{n}}^{(n)}(\mathbf{c}) = \frac{\mathbf{c}}{n}$. [3] Proposition 4.5 proves the following fact

$$\sup_{n^{-1}<t<n} |\tilde{\Psi}_t^{(n)}(\mathbf{c}) - \Psi_t(\mathbf{c})| \xrightarrow{P} 0, \quad n \to \infty.$$

Since $\eta_{\frac{1}{n}}$ and $(\eta_t/\eta_{1/n}, \tilde{\Psi}_t^{(n)}(\mathbf{c}))$ are independent, by letting $n \to \infty$, it follows that $\eta_0 := \lim_{t\downarrow 0} \eta_t$ and $(\eta_t/\eta_0, \Psi_t(\mathbf{c}))$ are independent. Since η_0 is uniformly distributed on $\mathbf{U}$, $\tilde{\phi}_t := \arg \eta_t = \arg\left(\eta_0 \cdot \frac{\eta_t}{\eta_0}\right)$ and Ψ_t are independent. □

9.2. *Convergence of* $\Psi_t^{(n)}$ *as increasing functions*

Proposition 9.2. *Fix any* $t > 0$. *Then we can find a coupling such that the following statement is valid for a.s.*

$$\lim_{n\to\infty} (\Psi_t^{(n)})^{-1}(x) = \Psi_t^{-1}(x), \quad \lim_{n\to\infty} (2\theta_{nt}(\kappa))_{2\pi\mathbf{Z}} = \phi_t$$

for any $x \in \mathbf{R}$ *where* ϕ_t *is uniformly distributed and independent of* Ψ_t.

As is explained in section 5, Proposition 9.2 completes the proof of Theorem 1.3. To prove Proposition 9.2 we shall show below that the convergence $\Psi_t^{(n)} \to \Psi_t$ holds in the sense of increasing function-valued process. Let $\mathcal{M}$ be the set of non-negative measures on $[a, b]$. Fix $\{f_j\}_{j\geq 1}$ a family of smooth functions on $[a, b]$ satisfying the property

$$\text{for } \omega \in \mathcal{M} \text{ if } \int_a^b f_j(x) d\omega(x) = 0 \text{ for any } j \geq 1 \Rightarrow \omega = 0.$$

We define a metric ρ on $\mathcal{M}$ by

$$\rho(\omega_1, \omega_2) := \sum_{j=1}^{\infty} \frac{1}{2^j} \left(\left| \int_a^b f_j(x) d(\omega_1(x) - \omega_2(x)) \right| \wedge 1 \right).$$

Let

$$\Omega := C([0,T] \to \mathcal{M})$$

for $T < \infty$. We further define for a smooth function f on $[a,b]$ a map $\Phi_f : \Omega \to C([0,T] \to \mathbf{R})$ by

$$\Phi_f(\omega)(t) := \int_a^b f(x) d\omega_t(x) = [f(x)\omega_t(x)]_a^b - \int_a^b f'(x)\omega_t(x)dx. \tag{9.2}$$

Lemma 9.1. *Let $\{\mu_n\}_{n\geq 1}$ be a family of probability measures on Ω. Suppose for each smooth function f on $[a,b]$ a family of probability measures $\{\Phi_f^{-1}\mu_n\}_{n\geq 1}$ on $C([0,T] \to \mathbf{R})$ is tight. Assume further there exists a constant C such that*

$$\mathbf{E}_{\mu_n}\left[\sup_{0\leq t\leq T}\int_a^b d\omega_t(x)\right] \leq C \tag{9.3}$$

holds for any $n \geq 1$. Then $\{\mu_n\}_{n\geq 1}$ is tight.

Proof. From (9.3) we see that for any $\epsilon > 0$, there exists a $M > 0$ such that

$$\mu_n\left(\sup_{0\leq t\leq T}\int_a^b d\omega_t(x) \leq M\right) \geq 1 - \epsilon.$$

Set

$$\Omega_0 := \left\{\omega \in \Omega \,\middle|\, \sup_{0\leq t\leq T}\int_a^b d\omega_t(x) \leq M\right\}.$$

Since $\{\Phi_{f_j}^{-1}\mu_n\}_{n\geq 1}$ is tight for each $j \geq 1$, there exists a compact set $\mathcal{K}_j$ in $C([0,T] \to \mathbf{R})$ such that $\mu_n\left(\Phi_{f_j}^{-1}(\mathcal{K}_j)\right) > 1 - \frac{\epsilon}{2^j}$. Set

$$\mathcal{K} := \bigcap_{j=1}^{\infty} \Phi_{f_j}^{-1}(\mathcal{K}_j) \cap \Omega_0 \subset \Omega.$$

Then

$$\mu_n(\mathcal{K}^c) \leq \sum_{j=1}^{\infty}\mu_n\left(\Phi_{f_j}^{-1}(\mathcal{K}_j^c)\right) + \mu_n(\Omega_0^c) \leq \epsilon + \sum_{j=1}^{\infty}\frac{\epsilon}{2^j} = 2\epsilon. \tag{9.4}$$

We show $\mathcal{K}$ is compact in Ω. Let $\{\omega_n\}_{n\geq 1}$ be a sequence in $\mathcal{K}$. Since $\mathcal{K}_1$ is compact, there exists a subsequence $\{n_i^1\}$ along which $\Phi_{f_1}\left(\omega_{n_i^1}\right)$ is uniformly convergent in $C([0,T] \to \mathbf{R})$. Then, using the compactness of $\mathcal{K}_2$ we can find a subsequence $\{n_i^2\}$ of $\{n_i^1\}$ along which $\Phi_{f_2}\left(\omega_{n_i^2}\right)$ is uniformly convergent in $C([0,T] \to \mathbf{R})$. Continuing this procedure for each j we find a subsequence $\{n_i^j\}$ of $\{n_i^{j-1}\}$ along which $\Phi_{f_j}\left(n_i^j\right)$ is uniformly convergent in $C([0,T] \to \mathbf{R})$. Let $m_i = n_i^i$. Then for each $j \geq 1$, $\Phi_{f_j}(\omega_{m_i})$ converges uniformly in $C([0,T] \to \mathbf{R})$. Since, for any $f \in C[a,b]$ and $\epsilon' > 0$, there exists a finite linear combination g of $\{f_j\}$ such that $\sup_{x\in[a,b]}|f(x) - g(x)| < \epsilon'$. We easily have

$$\sup_{t\in[0,T]}|\Phi_f(\omega_{m_i})(t) - \Phi_g(\omega_{m_i})(t)| \leq \epsilon' M$$

where we have used $\int_a^b d\omega_t(x) \leq M$ for any $\omega \in \mathcal{K}$. Therefore we see that the limit $\lim_{i\to\infty} \Phi_f(\omega_{m_i})(t)$ exists uniformly with respect to $t \in [0,T]$, which implies that there exists a $\omega \in \Omega$ satisfying

$$\int_a^b d\omega_t(x) \leq M \text{ and } \lim_{i\to\infty} \Phi_f(\omega_{m_i}) = \Phi_f(\omega)(t)$$

for any $t \in [0,T]$ and $f \in C([a,b])$. Consequently we have the compactness of $\mathcal{K}$ which together with (9.4) shows the tightness of $\{\mu_n\}_{n\geq 1}$. □

We shall check that the conditions for Lemma 9.1 are satisfied for $\Psi_t^{(n)}(\cdot)$. The inequality (9.3) follows from Lemma 6.6. In view of (9.2), the required tightness is implied by the following lemma.

Lemma 9.2. *For $f \in C^\infty(a,b)$ let*

$$g_n(t) := \int_a^b f(x)\Psi_t^{(n)}(x)dx.$$

Then, as a family of probability measures on $C([0,T] \to \mathbf{R})$, $\{g_n\}_{n\geq 1}$ is tight.

Proof. It is sufficient to show that following two equations.
(1) $\lim_{A\to\infty} \sup_n \mathbf{P}\left(|g_n(0)| \geq A\right) = 0$,
(2) For any $\rho > 0$,

$$\lim_{\delta\downarrow 0} \limsup_{n\to\infty} \mathbf{P}\left(\sup_{|t-s|<\delta} |g_n(t) - g_n(s)| > \rho \right) = 0.$$

(1) follows from Lemma 6.6. By bounding f, the following equation implies (2).

$$\lim_{\delta\downarrow 0} \limsup_{n\to\infty} \mathbf{P}\left[\sup_{|t-s|<\delta} \int_a^b \left|\Psi_t^{(n)}(x) - \Psi_s^{(n)}(x)\right| dx > \rho \right] = 0.$$

Here we borrow an argument in [3] Proposition 2.5: We divide $[a,b]$ into N-intervals $x_j = a + \frac{b-a}{N}x_j$, $j = 0, 1, \cdots, N-1$, and have

$$\int_a^b |\Psi_t^{(n)}(x) - \Psi_s^{(n)}(x)|dx = \sum_{j=0}^{N-1} \int_{x_j}^{x_{j+1}} |\Psi_t^{(n)}(x) - \Psi_s^{(n)}(x)|dx. \tag{9.5}$$

Since $\Psi_t^{(n)}(x)$ is increasing with respect to x, for $x \in [x_j, x_{j+1}]$ the integrand is bounded from above by

$$\begin{aligned} &|\Psi_t^{(n)}(x) - \Psi_s^{(n)}(x)| \\ &\leq \Psi_t^{(n)}(x_{j+1}) - \Psi_t^{(n)}(x_j) + |\Psi_t^{(n)}(x_j) - \Psi_s^{(n)}(x_j)| + \Psi_s^{(n)}(x_{j+1}) - \Psi_s^{(n)}(x_j). \end{aligned}$$

Substituting it into (9.5) yields

$$J := \int_a^b |\Psi_t^{(n)}(x) - \Psi_s^{(n)}(x)|dx$$

$$\leq \frac{1}{N}(b-a)\left(\Psi_t^{(n)}(b)-\Psi_t^{(n)}(a)\right)+(t\leftrightarrow s)$$
$$+\sum_{j=0}^{N}\frac{1}{N}(b-a)|\Psi_t^{(n)}(x_j)-\Psi_s^{(n)}(x_j)| =: I+II.$$

Thus we decompose the probability in question into two terms.

$$\mathbf{P}\left(\sup_{|t-s|<\delta} J>\rho\right) \leq \mathbf{P}\left(\sup_{|t-s|<\delta} I>\rho/2\right)+\mathbf{P}\left(\sup_{|t-s|<\delta} II>\rho/2\right)$$
$$=: III+IV.$$

The III-term can be estimated by Lemma 6.6.

$$III \leq \mathbf{P}\left(\frac{b-a}{N}\left(\Psi_t^{(n)}(b)-\Psi_t^{(n)}(a)\right)>\rho/4\right)+(t\leftrightarrow s)$$
$$\leq \frac{4}{\rho}\cdot\frac{b-a}{N}\mathbf{E}\left[\left(\Psi_t^{(n)}(b)-\Psi_t^{(n)}(a)\right)\right]+(t\leftrightarrow s)$$
$$\leq 2\cdot\frac{4}{\rho}\cdot\frac{b-a}{N}\mathbf{E}\left[\sup_{0\leq t\leq T}\left(\Psi_t^{(n)}(b)\right)\right]\leq\frac{C}{N}.$$

Thus for any $\epsilon>0$ we take N large enough independently of δ to have $III<\frac{\epsilon}{2}$. For such fixed N, we have

$$IV \leq \sum_{j=0}^{N}\mathbf{P}\left(\frac{b-a}{N}\sup_{|t-s|<\delta}|\Psi_t^{(n)}(x_j)-\Psi_s^{(n)}(x_j)|>\frac{\rho}{2N}\right)$$
$$=\sum_{j=0}^{N}\mathbf{P}\left(\sup_{|t-s|<\delta}|\Psi_t^{(n)}(x_j)-\Psi_s^{(n)}(x_j)|>\frac{\rho}{2(b-a)}\right).$$

Since $\{\Psi_t^{(n)}(x_j)\}_{j=0}^{N}$ is tight by Lemma 6.8, we can let $IV<\epsilon/2$ by taking n large and then taking $\delta>0$ small. □

We identify an element of $\mathcal{M}$ with a non-decreasing and right continuous function ω on $[a,b]$ satisfying $\omega(a)=0$. Then ω_n converges to $\omega\in\Omega$ if and only if $\omega_n(x)\to\omega(x)$ at any point of continuity of ω.

Lemma 9.3. *Suppose $\{\omega_n\}_{n\geq1}\subset\mathcal{M}$ converges to ω of $\mathcal{M}$. Assume ω is continuous. Then the convergence is uniform.*

Proof. Assume $\{\omega_n\}_{n\geq1}$ does not converge to ω uniformly. Then there exists a sequence $n_1<n_2<\cdots$, $\{t_k\}_{k\geq1}$ and a positive number ϵ_0 such that

$$|\omega_{n_k}(t_k)-\omega(t_k)|\geq\epsilon_0 \tag{9.6}$$

is valid for any $k=1,2,\cdots$. We can assume $t_k\to t_0\in[a,b]$ keeping $t_1<t_2<\cdots<t_0$. Then

$$\omega_{n_k}(t_l)-\omega(t_k)\leq\omega_{n_k}(t_k)-\omega(t_k)\leq\omega_{n_k}(t_0)-\omega(t_k)$$

for any $l < k$, hence letting $k \to \infty$, we have

$$\begin{aligned}\omega(t_l) - \omega(t_0) &\leq \liminf_{k\to\infty} (\omega_{n_k}(t_k) - \omega(t_k)) \\ &\leq \limsup_{k\to\infty} (\omega_{n_k}(t_k) - \omega(t_k)) \leq \omega(t_0) - \omega(t_0) = 0.\end{aligned}$$

Consequently, letting $l \to \infty$, we see $\lim_{k\to\infty} (\omega_{n_k}(t_k) - \omega(t_k)) = 0$, which contradicts (9.6). □

Proof of Proposition 9.2

By Lemma 9.1, the sequence of increasing function-valued process $\{\Psi_t^{(n)}(\cdot)\}_n$ is tight. Hence $(\Psi_t^{(n_k)}, (2\theta_{n_k t})_{2\pi\mathbf{Z}}) \stackrel{d}{\to} (\Psi_t, \phi_t)$ for some subsequence $\{n_k\}$. By Skorohod's theorem, we can suppose $(\Psi_t^{(n_k)}, (2\theta_{n_k t})_{2\pi\mathbf{Z}}) \stackrel{a.s.}{\to} (\Psi_t, \phi_t)$. Hence in particular we fix any $t > 0$ and obtain

$$\rho(\Psi_t^{(n_k)}, \Psi_t) = \sum_{j\geq 1} \frac{1}{2^j} \left(\left| \int_a^b f_j(x) d(\Psi_t^{(n_k)}(x) - \Psi_t(x)) \right| \wedge 1 \right) \stackrel{n\to\infty}{\to} 0, \quad a.s.$$

By Lemma 6.10 Ψ_t is continuous and increasing. Hence for a.s., $\Psi_t^{(n_k)}(x) \to \Psi_t(x)$ holds for any x. Moreover by Lemma 3.1 $(\Psi_t^{(n_k)})^{-1}(x) \stackrel{a.s.}{\to} \Psi_t^{-1}(x)$. Therefore Proposition 9.2 is proved.

Acknowledgement This work is partially supported by JSPS grants Kiban-C no.22540163(S.K.) and no.22540140(F.N.).

References

[1] Avila, A., Last, Y., and Simon, B., Bulk Universality and Clock Spacing of zeros for Ergodic Jacobi Matrices with A.C. spectrum, Anal. PDE **3**(2010), 81-108.

[2] Breuer, J., Forrester, P., and Smilansky, U., Random discrete Schrödinger operators from random matrix theory, J. Phys. **A40**(2007), no.5, F161-168.

[3] Killip, R., Stoiciu, M., Eigenvalue statistics for CMV matrices: from Poisson to clock via random matrix ensembles, Duke Math. **146**(2009), 361-399.

[4] Kotani, S. Ushiroya, N., One-dimensional Schrödinger operators with random decaying potentials, Comm. Math. Phys. **115**(1988), 247-266.

[5] Kotani, S., On limit behavior of eigenvalues spacing for 1-D random Schrödinger operators, Kokyuroku Bessatsu **B27**(2011), 67-78.

[6] Kritchevski, E., Valkó, B. and Virág, B., The scaling limit of the critical one-dimensional random Schrödinger operators, Comm. Math. Phys. **314**(2012), 775-806.

[7] Molchanov, S. A., The local structure of the spectrum of the one-dimensional Schrödinger operator, Comm. Math. Phys. **78**(1981), 429-446.

[8] Minami, N., Local fluctuation of the spectrum of a multidimensional Anderson tight-binding model, Comm. Math. Phys. **177**(1996), 709-725.

[9] Nakano, F., Level statistics for one-dimensional Schrödinger operators and Gaussian beta ensemble, to appear in J. Stat. Phys.

[10] Valkó, B. and Virág, V., Continuum limits of random matrices and the Brownian carousel, Invent. Math. **177**(2009), 463-508.

Chapter 19

Perturbation of the loop measure

Yves Le Jan* and Jay Rosen†

The loop measure is associated with a Markov generator. We compute the variation of the loop measure induced by an infinitesimal variation of the generator affecting the killing rates or the jumping rates.

Contents

1. Introduction

Professor Fukushima's contribution to probabilistic potential theory was seminal. His book on Dirichlet spaces [7] was the first complete exposition in which the functional analytic, potential theoretic, and probabilistic aspects of the theory were considered jointly, as different aspects of the same mathematical object. In particular, transformations of the energy form, such as restriction to an open set, trace on a closed set, change of the equilibrium (or killing) measure, change of reference measure, etc... have probabilistic counterparts. In the second chapter, the possibility of superposing different Dirichlet forms was recognized. The present paper follows this line of research. But we will focus on Markov loops and bridges rather than Markov paths. We will present them in the next section. Let us stress however that the existence of loop or bridge measures seems to require more than the assumption of a Markov process defined up to polar sets, which is the basic assumption for a Markov process associated with a Dirichlet form. The existence of a Green function seems to be required. Our purpose is to compute the variation of the loop measure induced by an infinitesimal variation of the generator. This variation may in particular affect the killing rates or the jumping rates. In the case of symmetric continuous time Markov chains, the question has been addressed in chapter 6 of [9]. The results are formally close to formulas used in conformal field theory for operator insertions

*Equipe Probabilités et Statistiques, Université Paris-Sud, Bâtiment 425, 91405 Orsay Cedex, France, yves.lejan@math.u-psud.fr
†Department of Mathematics, College of Staten Island, CUNY, Staten Island, NY 10314, U.S.A, jrosen30@optimum.net

(see for example [8]). We try here to extend them to a more general situation, to show in particular that the bridge measures can be derived from the loop measure.

2. Background on loop measures

We begin by introducing loop measures for Borel right processes (such as Feller processes) on a rather general state space S, which we assume to be locally compact with a countable base. Let $X=(\Omega, \mathcal{F}_t, X_t, \theta_t, P^x)$ be a transient Borel right process [12] with cadlag paths (such as a standard Markov process [1]) with state space S and jointly measurable transition densities $p_t(x,y)$ with respect to some σ-finite measure m on S. We assume that the potential densities

$$u(x,y) = \int_0^\infty p_t(x,y)\,dt$$

are finite off the diagonal, but allow them to be infinite on the diagonal. We do not require that $p_t(x,y)$ is symmetric.

We assume furthermore that $0 < p_t(x,y) < \infty$ for all $0 < t < \infty$ and $x, y \in S$, and that there exists another Borel right process $\widehat{X}$ in duality with X (Cf [1]), relative to the measure m, so that its transition probabilities $\widehat{P}_t(x,dy) = p_t(y,x)\,m(dy)$. These conditions allow us to use material on bridge measures in [6]. In particular, for all $0 < t < \infty$ and $x, y \in S$, there exists a finite measure $Q_t^{x,y}$ on $\mathcal{F}_{t-}$, of total mass $p_t(x,y)$, such that

$$Q_t^{x,y}\left(1_{\{\zeta > s\}}\, F_s\right) = P^x\left(F_s\, p_{t-s}(X_s, y)\right), \tag{2.1}$$

for all $F_s \in \mathcal{F}_s$ with $s < t$. (We use the letter Q for measures which are not necessarily of mass 1, and reserve the letter P for probability measures.) $Q_t^{x,y}$ should be thought of as a measure for paths which begin at x and end at y at time t. When normalized, this gives the bridge measure $P_t^{x,y}$ of [6].

We use the canonical representation of X in which Ω is the set of cadlag paths ω in $S_\Delta = S \cup \Delta$ with $\Delta \notin S$, and is such that $\omega(t) = \Delta$ for all $t \geq \zeta = \inf\{t > 0\,|\,\omega(t) = \Delta\}$. Set $X_t(\omega) = \omega(t)$. We define a σ-finite measure μ on $(\Omega, \mathcal{F})$ by

$$\int F\,d\mu = \int_0^\infty \frac{1}{t} \int Q_t^{x,x}\left(F \circ k_t\right)\,dm(x)\,dt \tag{2.2}$$

for all $\mathcal{F}$ measurable functions F on Ω. Here k_t is the killing operator defined by $k_t\omega(s) = \omega(s)$ if $s < t$ and $k_t\omega(s) = \Delta$ if $s \geq t$, so that $k_t^{-1}\mathcal{F} \subset \mathcal{F}_{t-}$. As usual, if F is a function, we often write $\mu(F)$ for $\int F\,d\mu$. μ is σ-finite as any set of loops in a compact set with lifetime bounded away from zero and infinity has finite measure.

We call μ the loop measure of X because, when X has continuous paths, μ is concentrated on the set of continuous loops with a distinguished starting point (since $Q_t^{x,x}$ is carried by loops starting at x). Moreover, it is shift invariant. More precisely let ρ_u denote the loop rotation defined by

$$\rho_u\omega(s) = \begin{cases} \omega(s + u \bmod \zeta(\omega)), & \text{if } 0 \leq s < \zeta(\omega) \\ \Delta, & \text{otherwise.} \end{cases}$$

Here, for two positive numbers a, b we define $a \bmod b = a - mb$ as the unique positive integer m such that $0 \leq a - mb < b$. μ is invariant under ρ_u, for any u. We let $\mathcal{F}_\rho$ denote the σ-algebra of $\mathcal{F}$ measurable functions F on Ω which are invariant under ρ, that is $F \circ \rho_u = F$ for all $u \geq 0$. Loop functionals of interest are mostly $\mathcal{F}_\rho$ - measurable.

Recall that Poisson processes of intensity μ appear naturally as produced by loop erasure in the construction of random spanning trees through the Wilson algorithm (see chapter 8 in [9]). Although the definition of μ in (2.2), especially the $\frac{1}{t}$, may look forbidding, μ often has a nice form when applied to specific functions in $\mathcal{F}_\rho$. A particular function in $\mathcal{F}_\rho$ is given by

$$\phi(f) = \int_0^\infty f(X_t)\,dt, \tag{2.3}$$

where f is any measurable function on S. If f_j, $j = 1, \ldots, k \geq 2$ are non-negative functions on S, then by [11, Lemma 2.1]

$$\mu\left(\prod_{j=1}^k \phi(f_j)\right) \tag{2.4}$$

$$= \sum_{\pi \in \mathcal{P}_k^{\odot}} \int u(y_{\pi(1)}, y_{\pi(2)}) \cdots u(y_{\pi(k-1)}, y_{\pi(k)}) u(y_{\pi(k)}, y_{\pi(1)}) \prod_{j=1}^k f_j(y_j)\,dy_j$$

where $\mathcal{P}_k^{\odot}$ denotes the set of permutations of $[1, k]$ on the circle. (For example, $(1, 3, 2)$, $(2, 1, 3)$ and $(3, 2, 1)$ are considered to be one permutation $\pi \in \mathcal{P}_3^{\odot}$.) We note however that in general when u is infinite on the diagonal

$$\mu\left(\phi(f_j)\right) = \infty.$$

For $k \geq 2$, the integral (2.4) can be finite if the f_i satisfy certain integrability conditions: see the beginning of section 3. Consider more generally the multiple integral

$$\sum_{\pi \in \mathcal{T}_k^{\odot}} \int_{0 \leq r_1 \leq \cdots \leq r_k \leq t} f_{\pi(1)}(X_{r_1}) \cdots f_{\pi(k)}(X_{r_k}) \tag{2.5}$$

where $\mathcal{T}_k^{\odot}$ denotes the set of translations π of $[1, k]$ which are cyclic mod k, that is, for some i, $\pi(j) = j + i, \mod k$, for all $j = 1, \ldots, k$.

Finite sums of multiple integrals such as these form an algebra (see exercise 11, p. 21 in [9]) which generates $\mathcal{F}_\rho$, [3].

Finally, let $a(x)$ be a bounded, strictly positive function on S. Define the time changed process $Y_t = X_{\tau(t)}$ where $\tau(t) = \int_0^t a(Y_s)\,ds$ is the inverse of the CAF $A_t = \int_0^t 1/a(X_s)\,ds$. It satisfies the duality assumption relative to the measure $a^{-1} \cdot m$. It then follows as in [5, section 7.3] that if $u_X(x, y), u_Y(x, y)$ denote the potential densities of X, Y respectively with respect to $m, a^{-1} \cdot m$ respectively, then

$$u_Y(x, y) = u_X(x, y). \tag{2.6}$$

It follows that μ_Y is the image of μ_X by the time change.

3. Multiple CAF's and perturbation of loop measures

Let $\mathcal{M}(S)$ be the set of finite signed Radon measures on $\mathcal{B}(S)$. We say that a norm $\|\cdot\|$ on $\mathcal{M}(S)$ is a **proper norm** with respect to a kernel u if for all $n \geq 2$ and $\nu_1, \ldots, \nu_n$ in $\mathcal{M}(S)$

$$\left| \int \prod_{j=1}^n u(x_j, x_{j+1}) \prod_{i=1}^n d\nu_j(x_i) \right| \leq C^n \prod_{j=1}^n \|\nu_j\|, \tag{3.1}$$

(with $x_{n+1} = x_1$) for some universal constant $C < \infty$. In Section 6 of [10] we present several examples of proper norms which depend upon various hypotheses about the kernel u.

In particular, the following norm is related to the square root of the generator of X, which defines the Dirichlet space in the m-symmetric case:

$$\|\nu\|_w := \left(\iint \left(\int w(x,y)w(y,z)\,d\nu(y)\right)^2 dm(x)\,dm(z)\right)^{1/2}, \tag{3.2}$$

where

$$w(x,y) = \int_0^\infty \frac{p_s(x,y)}{\sqrt{\pi s}}\,ds. \tag{3.3}$$

To see that $\|\nu\|_w$ is a proper norm we first note that

$$u(x,z) = \int w(x,y)w(y,z)\,dm(y). \tag{3.4}$$

(It is interesting to note that w is the potential density of the process X_{T_t} where T_t is the stable subordinator of index 1/2. In operator notation (3.4) says that $W^2 = U$ where W and U are operators with kernels w and u respectively.)

Using (3.4)

$$\begin{aligned}\prod_{j=1}^n u(z_j, z_{j+1}) &= \prod_{j=1}^n \int w(z_j,\lambda_j)w(\lambda_j, z_{j+1})\,dm(\lambda_j) \\ &= \prod_{j=1}^n \int w(z_j,\lambda_j)w(\lambda_{j-1}, z_j)\,dm(\lambda_j)\end{aligned} \tag{3.5}$$

in which $z_{n+1} = z_1$ and $\lambda_0 = \lambda_n$. It follows from this that

$$\begin{aligned}&\Big|\int \prod_{j=1}^n u(z_j, z_{j+1}) \prod_{j=1}^n d\nu_j(z_j)\Big| \\ &= \left|\int \prod_{j=1}^n \left(\int w(z_j,\lambda_j)w(\lambda_{j-1}, z_j)\,d\nu_j(z_j)\right) \prod_{j=1}^n dm(\lambda_j)\right| \\ &\le \prod_{j=1}^n \left(\iint \left(\int w(z_j,s)w(t,z_j)\,d\nu_j(z_j)\right)^2 dm(s)\,dm(t)\right)^{1/2} \le \prod_{j=1}^n \|\nu_j\|_w,\end{aligned} \tag{3.6}$$

where, for the first inequality, we use repeatedly the Cauchy-Schwarz inequality.

Lastly, set

$$M_\nu(x,z) = \int w(x,y)w(y,z)\,d\nu(y). \tag{3.7}$$

Since $\|\nu\|_w$ is the L^2 norm of M_ν, and $M_{\nu+\nu'} = M_\nu + M_{\nu'}$, we see that $\|\nu\|_w$ is a norm. (This can also be viewed as the Hilbert-Schmidt norm of the operator defined by the kernel M_ν). □

We denote by $\mathcal{R}^+$ the set of positive bounded Revuz measures ν on S that are associated with X. This is explained in detail in Section 2.1 of [10]. We use L_t^ν to denote the CAF with Revuz measure ν.

Let $\|\cdot\|$ be a proper norm on $\mathcal{M}(S)$ with respect to the kernel u. Set

$$\mathcal{M}^+_{\|\cdot\|} = \{\text{positive } \nu \in \mathcal{M}(S) \,|\, \|\nu\| < \infty\}, \tag{3.8}$$

and

$$\mathcal{R}^+_{\|\cdot\|} = \mathcal{R}^+ \cap \mathcal{M}^+_{\|\cdot\|}. \tag{3.9}$$

Let $\mathcal{M}_{\|\cdot\|}$ and $\mathcal{R}_{\|\cdot\|}$ denote the set of measures of the form $\nu = \nu_1 - \nu_2$ with $\nu_1, \nu_2 \in \mathcal{M}^+_{\|\cdot\|}$ or $\mathcal{R}^+_{\|\cdot\|}$ respectively. We often omit saying that both $\mathcal{R}_{\|\cdot\|}$ and $\|\cdot\|$ depend on the kernel u.

Let $\|\cdot\|$ be a proper norm for u. For $\nu_j \in \mathcal{R}_{\|\cdot\|}$, $j = 1, \ldots, k$, let

$$M_t^{\nu_1,\ldots,\nu_k} = \sum_{\pi \in \mathcal{T}_k^{\odot}} \int_{0 \le r_1 \le \cdots \le r_k \le t} dL_{r_1}^{\nu_{\pi(1)}} \cdots dL_{r_k}^{\nu_{\pi(k)}}. \tag{3.10}$$

We refer to $M_t^{\nu_1,\ldots,\nu_k}$ as a multiple CAF. Let

$$Q^{x,y}(F) = \int_0^\infty Q_t^{x,y}(F \circ k_t)\, dt. \tag{3.11}$$

We have the following analogue of [9, Proposition 5] and [10, Lemma 2.1].

Lemma 3.1. *For any measures $\nu_j \in \mathcal{R}^+_{\|\cdot\|}$, $j = 1, \ldots, k \ge 2$,*

$$\mu(M_\infty^{\nu_1,\ldots,\nu_k}) = \int u(y_1, y_2) \cdots u(y_{k-1}, y_k) u(y_k, y_1) \prod_{j=1}^k d\nu_j(y_j), \tag{3.12}$$

$$\begin{aligned} &Q^{x,y}(M_\infty^{\nu_1,\ldots,\nu_k}) \\ &= \sum_{\pi \in \mathcal{T}_k^{\odot}} \int u(x, y_1) u(y_1, y_2) \cdots u(y_{k-1}, y_k) u(y_k, y) \prod_{j=1}^k d\nu_{\pi(j)}(y_j), \end{aligned} \tag{3.13}$$

and if $\nu \in \mathcal{R}^+_{\|\cdot\|}$

$$\begin{aligned} &\mu(M_\infty^{\nu_1,\ldots,\nu_k} L_\infty^\nu) \\ &= \sum_{i=1}^k \int \left(\prod_{j=1}^{i-1} u(y_j, y_{j+1}) \right) \\ &\qquad u(y_i, x) u(x, y_{i+1}) \left(\prod_{j=i+1}^k u(y_j, y_{j+1}) \right) \prod_{j=1}^k d\nu_j(y_j)\, d\nu(x) \\ &= \int Q^{x,x}(M_\infty^{\nu_1,\ldots,\nu_k})\, d\nu(x), \end{aligned} \tag{3.14}$$

with $y_{k+1} = y_1$.

The proof of (3.12) follows that of [10, Lemma 2.1], noticing that the crucial step [10, (2.23)-(2.28)] used the fact that the set of permutations of $[1, k]$ is invariant under translation mod k. Since $\mathcal{T}_k^{\odot}$ is invariant under translation mod k, the same proof will work here. The proof of (3.13), which is much easier, follows that of [10, Lemma 4.2]. The first equality in (3.14) follows from (3.12) and the fact that

$$M_\infty^{\nu_1,\ldots,\nu_k}\, L_\infty^\nu = \sum_{i=1}^k M_\infty^{\nu_1,\ldots,\nu_{i-1},\nu,\nu_i,\ldots,\nu_k}, \tag{3.15}$$

which is easily verified using $L_\infty^\nu = \int_0^\infty dL_t^\nu$. The second equality in (3.14) then follows by comparing with (3.13). □

Let now $X_{(\epsilon)}$, $\epsilon \geq 0$, $X_{(0)} = X$, be a family of Markov processes with potential densities $u_{(\epsilon)}(x,y)$, and let $\mu_{(\epsilon)}$ denote the loop measure for $X_{(\epsilon)}$. Assume that we can use the same proper norm $\|\cdot\|$ for all $u_{(\epsilon)}$. Let

$$u'_{(0)}(x,y) = \frac{du_{(\epsilon)}(x,y)}{d\epsilon}\,|_{\epsilon=0}, \tag{3.16}$$

and assume that $\|\cdot\|$ is also a proper norm for $u'_{(0)}$. Then using the last Lemma we have formally, that is, assuming we can justify interchanging derivative and integral in the second equality,

$$\begin{aligned}
&\frac{d}{d\epsilon}\,\mu_{(\epsilon)}(M_\infty^{\nu_1,\ldots,\nu_k})\Big|_{\epsilon=0} \\
&= \frac{d}{d\epsilon}\int u_{(\epsilon)}(y_1,y_2)\cdots u_{(\epsilon)}(y_{k-1},y_k)u_{(\epsilon)}(y_k,y_1)\prod_{j=1}^k d\nu_j(y_j)\Big|_{\epsilon=0} \\
&= \sum_{i=1}^k \int u(y_1,y_2)\cdots u'_{(0)}(y_i,y_{i+1})\cdots u(y_k,y_1)\prod_{j=1}^k d\nu_j(y_j) \\
&= \sum_{\pi\in\mathcal{T}_k^{\odot}} \int u(y_1,y_2)\cdots u(y_{k-1},y_k)u'_{(0)}(y_k,y_1)\prod_{j=1}^k d\nu_{\pi(j)}(y_j)
\end{aligned} \tag{3.17}$$

where we have set $y_{k+1} = y_1$.

Assume now that for some distribution F on $S\times S$ we have

$$u'_{(0)}(y_k,y_1) = \int_{S\times S} u(y_k,x)F(x,y)\,u(y,y_1)\,dm(x)\,dm(y). \tag{3.18}$$

Let $\mathcal{A}_{\|\cdot\|}$ denote the space spanned by the multiple CAF's with $\nu_j \in \mathcal{R}_{\|\cdot\|}$. Note it is an algebra. Then comparing (3.17) and (3.13) we would obtain

$$\frac{d\mu_{(\epsilon)}(A)}{d\epsilon}\,|_{\epsilon=0} = \int_{S\times S} F(x,y)Q^{y,x}(A)\,dm(x)\,dm(y), \tag{3.19}$$

for all $A \in calA_{\|\cdot\|}$.

In the following sections we present specific examples where this heuristic approach is made rigorous.

4. Perturbation of Lévy processes

Let X be a transient Lévy process in R^d with characteristic exponent ψ so that, as distributions

$$u(x,y)=\int \frac{e^{i\lambda(y-x)}}{\psi(\lambda)}\,d\lambda. \tag{4.1}$$

In [10] we showed that $\|\cdot\|_{\psi,2}$ is a proper norm for u where

$$\|\nu\|_{\psi,2}^2=\int \left(\frac{1}{|\psi|}*\frac{1}{|\psi|}(\lambda)\right)\,|\widehat{\nu}(\lambda)|^2\,d\lambda. \tag{4.2}$$

Let κ be a Lévy characteristic exponent, so that the same is true for $\psi+\epsilon\kappa$, and let $X_{(\epsilon)}$ be the Lévy process with characteristic exponent $\psi+\epsilon\kappa$. We let $u_{(\epsilon)}(x,y)$ denote the potential of $X_{(\epsilon)}$ so that, as distributions

$$u_{(\epsilon)}(x,y)=\int \frac{e^{i\lambda(y-x)}}{\psi(\lambda)+\epsilon\kappa(\lambda)}\,d\lambda. \tag{4.3}$$

If we assume that

$$|\kappa(\lambda)|\le C|\psi(\lambda)| \tag{4.4}$$

for some $C<\infty$, then for $\epsilon>0$ sufficiently small

$$\|\nu\|_{\psi+\epsilon\kappa,2}\le C'\|\nu\|_{\psi,2}, \tag{4.5}$$

for some $C'<\infty$. Thus $\|\cdot\|_{\psi,2}$ is a proper norm for $u_{(\epsilon)}$.

Let F be the distribution given by

$$F(x,y)=\int e^{i\lambda(y-x)}\kappa(\lambda)\,d\lambda, \tag{4.6}$$

and let $\widehat{Q}^{\lambda_1,\lambda_2}(A)$ denote the Fourier transform of $Q^{x,y}(A)$ in x,y.

Theorem 4.1. *If (4.4) holds, then*

$$\begin{aligned}\frac{d\mu_{(\epsilon)}(A)}{d\epsilon}\,\Big|_{\epsilon=0}&=-\int_{R^d\times R^d} Q^{y,x}(A)F(x,y)\,dm(x)\,dm(y)\\ &=-\int \widehat{Q}^{\lambda,-\lambda}(A)\;\kappa(\lambda)\,d\lambda\end{aligned} \tag{4.7}$$

for all $A\in\mathcal{A}_{\|\cdot\|_{\psi,2}}$.

Proof of Theorem 4.1: It suffices to show that for any $\nu_1,\ldots,\nu_k\in\mathcal{R}^+_{\|\cdot\|_{\psi,2}}$

$$\begin{aligned}&\frac{d}{d\epsilon}\int\prod_{j=1}^k u_{(\epsilon)}(y_j,y_{j+1})\prod_{j=1}^k d\nu_j(y_j)\;\Big|_{\epsilon=0}\\ &=-\sum_{i=1}^k\int\left(\prod_{j=1}^{i-1}u(y_j,y_{j+1})\right)\\ &\qquad\left(\int_{R^d\times R^d}u(y_i,x)\,F(x,y)\,u(y,y_{i+1})\,dm(x)\,dm(y)\right)\end{aligned} \tag{4.8}$$

$$\left(\prod_{j=i+1}^{k} u(y_j, y_{j+1})\right) \prod_{j=1}^{k} d\nu_j(y_j),$$

with $y_{k+1} = y_1$.

Using (4.3) we see that

$$\begin{aligned} I(\epsilon) =: & \int \prod_{j=1}^{k} u_{(\epsilon)}(y_j, y_{j+1}) \prod_{j=1}^{k} d\nu_j(y_j) \qquad (4.9) \\ = & \iint \prod_{j=1}^{k} e^{i(y_{j+1}-y_j)\cdot\lambda_j} \frac{1}{\psi(\lambda_j) + \epsilon\kappa(\lambda_j)} d\lambda_j d\nu_j(y_j) \\ = & \int \left(\prod_{j=1}^{k} \int e^{-i(\lambda_j-\lambda_{j-1})\cdot y_j} d\nu_j(y_j)\right) \prod_{j=1}^{k} \frac{1}{\psi(\lambda_j) + \epsilon\kappa(\lambda_j)} d\lambda_j \end{aligned}$$

where $\lambda_0 = \lambda_n$. We take the Fourier transforms of the $\{\nu_j\}$ to see that

$$\begin{aligned} & \int \prod_{j=1}^{k} u_{(\epsilon)}(y_j, y_{j+1}) \prod_{j=1}^{k} d\nu_j(y_j) \qquad (4.10) \\ & = \int \prod_{j=1}^{k} \hat{\nu}_j(\lambda_j - \lambda_{j-1}) \frac{1}{\psi(\lambda_j) + \epsilon\kappa(\lambda_j)} d\lambda_j. \end{aligned}$$

We have

$$\frac{1}{\psi(\lambda_j) + \epsilon\kappa(\lambda_j)} = \frac{1}{\psi(\lambda_j)} - \epsilon\frac{\kappa(\lambda_j)}{\psi^2(\lambda_j)} + \epsilon^2 \frac{\kappa^2(\lambda_j)}{\psi^2(\lambda_j)(\psi(\lambda_j) + \epsilon\kappa(\lambda_j))}. \qquad (4.11)$$

Substituting this into (4.11) we can write the result as

$$I(\epsilon) = I(0) - \epsilon J + K(\epsilon), \qquad (4.12)$$

where

$$J = \sum_{i=1}^{k} \int \left(\prod_{\substack{j=1 \\ j\neq i}}^{k} \frac{1}{\psi(\lambda_j)}\right) \frac{\kappa(\lambda_i)}{\psi^2(\lambda_i)} \prod_{j=1}^{k} \hat{\nu}_j(\lambda_j - \lambda_{j-1})\, d\lambda_j, \qquad (4.13)$$

and $K(\epsilon)$ is the sum of all remaining terms.

We now show that J is precisely the right hand side of (4.8), and that

$$|K(\epsilon)| = O(\epsilon^2), \qquad (4.14)$$

which will complete the proof of our Proposition.

For the first point, using the relation between Fourier transforms and convolutions we have

$$\begin{aligned} & \int_{R^d \times R^d} u(y_i, x)\, F(x, y)\, u(y, y_{i+1})\, dm(x)\, dm(y) \qquad (4.15) \\ & = \int e^{i\lambda_i(y_{i+1}-y_i)} \frac{\kappa(\lambda_i)}{\psi^2(\lambda_i)} d\lambda_i. \end{aligned}$$

Using this in the right hand side of (4.8) and proceeding as in (4.9)-(4.10) we indeed obtain J. As for (4.14), $K(\epsilon)$ is the sum many terms each of which has a factor of ϵ^m for some $m \geq 2$. We need only show that the corresponding integrals are bounded uniformly in ϵ. For example, consider the term which arises when using the last term in (4.11) for all j. This term is ϵ^{2k} times

$$\widetilde{K}(\epsilon) =: \int \prod_{j=1}^{k} \hat{\nu}_j(\lambda_j - \lambda_{j-1}) \frac{\kappa^2(\lambda_j)}{\psi^2(\lambda_j)(\psi(\lambda_j) + \epsilon\kappa(\lambda_j))} \, d\lambda_j. \tag{4.16}$$

By our assumption (4.4), for sufficiently small ϵ

$$\begin{aligned} |\widetilde{K}(\epsilon)| &\leq C' \int \prod_{j=1}^{k} |\hat{\nu}_j(\lambda_j - \lambda_{j-1})| \frac{1}{|\psi(\lambda_j)|} \, d\lambda_j \\ &\leq C'' \prod_{j=1}^{k} \|\nu_j\|_{\psi,2}, \end{aligned} \tag{4.17}$$

by repeated use of the Cauchy-Schwarz inequality, as in our proof in [10] that $\|\cdot\|_{\psi,2}$ is a proper norm for u. □

5. Perturbation by multiplicative functionals

Let m_t be a continuous decreasing multiplicative functional of X, with $m_t \leq 1$ for all t and $m_\zeta = 0$. By [12, Theorem 61.5], there is a right process $\widetilde{X}_t$ with transition semigroup

$$\widetilde{P}_t f(x) = P^x(f(X_t)m_t) = \int Q_t^{x,y}(m_t) f(y) \, dm(y), \tag{5.1}$$

where $Q_t^{x,y}$ are the bridge measures (2.1) for our original process X and the second equality is [6, (2.8)]. Thus $\widetilde{X}$ has transition densities

$$\widetilde{p}_t(x,y) =: Q_t^{x,y}(m_t), \tag{5.2}$$

and using [6, (2.8)] once more we can verify that these satisfy the Chapman-Kolmogorov equations. It follows from the construction in [12, Theorem 61.5] that if X has cadlag paths so will $\widetilde{X}$. Let $r_t(\omega)(s) = \omega(t-s)^-$, $0 \leq s \leq t$ be the time reversal mapping. If we set $\widehat{m}_t = m_t \circ r_t$, then it is easy to check that $\widehat{m}_t$ is a multiplicative functional as above, see [2, p. 359]. If $\widehat{X}$ is the dual process for X as described in Section 2, with bridge measures $\widehat{Q}_t^{x,y}$ then as above there exists a process Y with transition densities $\widehat{Q}_t^{x,y}(\widehat{m}_t)$. By [6, Corollary 1]

$$\widehat{Q}_t^{x,y}(\widehat{m}_t) = \widehat{Q}_t^{x,y}(m_t \circ r_t) = Q_t^{y,x}(m_t), \tag{5.3}$$

which shows that Y is dual to $\widetilde{X}$.

We now show that if $\widetilde{Q}_t^{x,y}$ are the bridge measures for $\widetilde{X}$, then

$$\widetilde{Q}_t^{x,y}(F) = Q_t^{x,y}(F\,m_t), \text{ for } F \in \mathcal{F}_s,\ s < t. \tag{5.4}$$

To see this, using the fact that m_t is continuous and decreasing, we have for $F \in \mathcal{F}_s$, $s < t$

$$\begin{aligned}
\widetilde{Q}_t^{x,y}(F) &= \widetilde{P}^x \left(F Q_{t-s}^{X_s,y}(m_{t-s}) \right) \\
&= \lim_{t^* \uparrow t} \widetilde{P}^x \left(F Q_{t-s}^{X_s,y}(m_{t^*-s}) \right) \\
&= \lim_{t^* \uparrow t} \widetilde{P}^x \left(F P^{X_s} \left(m_{t^*-s}\, p_{t-t^*}(X_{t^*-s}, y) \right) \right) \\
&= \lim_{t^* \uparrow t} P^x \left(F m_s P^{X_s} \left(m_{t^*-s}\, p_{t-t^*}(X_{t^*-s}, y) \right) \right) \\
&= \lim_{t^* \uparrow t} P^x \left(F m_s\, m_{t^*-s} \circ \theta_s\, p_{t-t^*}(X_{t^*}, y) \right) \\
&= \lim_{t^* \uparrow t} P^x \left(F m_{t^*}\, p_{t-t^*}(X_{t^*}, y) \right) \\
&= \lim_{t^* \uparrow t} Q_t^{x,y}(F\, m_{t^*}) = Q_t^{x,y}(F\, m_t),
\end{aligned} \tag{5.5}$$

which proves (5.4).

If A_t is a CAF, then $m_t = e^{-A_t}$ is a continuous decreasing multiplicative functional of X. Let $X_{(\epsilon)}$ denote the Markov process $\widetilde{X}$ with $m_t = e^{-\epsilon L_t^\nu}$. If $\mu_{(\epsilon)}$ denotes the loop measure for $X_{(\epsilon)}$, and μ the loop measure for X, it now follows from (2.2) and (5.4) that

$$\mu_{(\epsilon)}(A) = \mu \left(A e^{-\epsilon L_\infty^\nu} \right). \tag{5.6}$$

Theorem 5.1. *If $\nu \in \mathcal{R}^+_{\|\cdot\|}$, for some proper norm $\|\cdot\|$, then*

$$\frac{d\mu_{(\epsilon)}(A)}{d\epsilon} \Big|_{\epsilon=0} = -\mu(L_\infty^\nu A) = -\int_S Q^{x,x}(A)\, d\nu(x) \tag{5.7}$$

for all $A \in \mathcal{A}_{\|\cdot\|}$.

Proof: It suffices to prove this for A of the form $M_\infty^{\nu_1,\ldots,\nu_k}$ with $\nu_j \in \mathcal{R}^+_{\|\cdot\|}$, $1 \le j \le k$. Since $0 \le e^{-x} - 1 + x \le x^2/2$ for $x \ge 0$, it follows from (5.6) that

$$|\mu_{(\epsilon)}(M_\infty^{\nu_1,\ldots,\nu_k}) - \mu(M_\infty^{\nu_1,\ldots,\nu_k}) - \epsilon\mu(L_\infty^\nu M_\infty^{\nu_1,\ldots,\nu_k})| \le \epsilon^2 \mu \left((L_\infty^\nu)^2 M_\infty^{\nu_1,\ldots,\nu_k} \right). \tag{5.8}$$

$\mu\left((L_\infty^\nu)^2 A\right)$ is bounded by our assumption about the proper norm $\|\cdot\|$, so the first equality in (5.7) follows. The second equality is (3.14). □

It follows from (5.2) that $X_{(\epsilon)}$ has potential densities

$$u_{(\epsilon)}(x,y) = \int_0^\infty Q_t^{x,y}(e^{-\epsilon L_t^\nu})\, dt \le u(x,y). \tag{5.9}$$

Then, using [6, Lemma 1]

$$\begin{aligned}
u'_{(0)}(x,y) &= -\int_0^\infty Q_t^{x,y}(L_t^\nu)\, dt \\
&= -\int_0^\infty \int_0^t \int p_s(x,z) p_{t-s}(z,y)\, d\nu(z)\, ds\, dt \\
&= -\int u(x,z) u(z,y)\, d\nu(z).
\end{aligned} \tag{5.10}$$

In view of this, Theorem 5.1 is another example of our heuristic formula (3.19), where the distribution F on $S \times S$ of (3.18) is $\delta(x-y)\, d\nu(x)$.

For use in the next section we now recast Theorem 5.1. For $\nu_j \in \mathcal{R}^+_{\|\cdot\|}$, $1 \le j \le k$ let

$$I(\epsilon) = \int \prod_{j=1}^{k} u_{(\epsilon)}(y_j, y_{j+1}) \prod_{j=1}^{k} d\nu_j(y_j). \tag{5.11}$$

Since $\int \prod_{j=1}^k u_{(\epsilon)}(y_j, y_{j+1}) \prod_{j=1}^k d\nu_j(y_j) = \mu_{(\epsilon)}(M_\infty^{\nu_1,\ldots,\nu_k})$ by (3.12), it follows from Theorem 5.1 and (3.14) that

$$\begin{aligned} &\lim_{\epsilon\to 0} \frac{I(\epsilon) - I(0)}{\epsilon} \qquad (5.12)\\ &= -\sum_{i=1}^{k} \int \left(\prod_{j=1}^{i-1} u(y_j, y_{j+1}) \right) u(y_i, x) u(x, y_{i+1}) \\ &\qquad \left(\prod_{j=i+1}^{k} u(y_j, y_{j+1}) \right) \prod_{j=1}^{k} d\nu_j(y_j)\, d\nu(x). \end{aligned}$$

In the next section we will also use the following observation. If $m_t = e^{-\epsilon \int_0^t c(s)\, ds}$ then $\widehat{m}_t = m_t \circ r_t = m_t$. Hence if $\widehat{X}$ is the dual process for X as described in Section 2, it follows from (5.3) that Y, the dual process to $X_{(\epsilon)}$, is the process $(\widehat{X})_{(\epsilon)}$ obtained by perturbing $\widehat{X}$ by the same multiplicative functional m_t. In particular

$$\widehat{u}_{(\epsilon)}(x, y) = u_{(\epsilon)}(y, x). \tag{5.13}$$

6. Perturbation by addition of jumps

Let $j(x, y)$ be a nonnegative $m \otimes m$-integrable function on $S \times S$. We will assume that

$$c(x) =: \int j(x, y) m(dy) = \int j(y, x) m(dy). \tag{6.1}$$

$c(x)$ is integrable and we will assume moreover that it is bounded and strictly positive. Then

$$G(x, dy) =: \frac{1}{c(x)} j(x, y) m(dy) \tag{6.2}$$

is a probability kernel on $S \times \mathcal{B}(S)$. $c(x)$ will govern the rate at which we will add jumps to the process, which may depend on the position x of the process, and $G(x, dy)$ will describe the distribution of the jumps from position x.

In more detail, define the CAF

$$A_t = \int_0^t c(X_s)\, ds, \tag{6.3}$$

and let τ_t be the right continuous inverse of A_t. Let λ be an independent mean 1 exponential. We define a new process Y_t to be equal to X_t for $t < \tau_\lambda$, and then re-birthed at a random point independent of λ, distributed according to $G(X_{\tau_\lambda}, dy)$, with this process being iterated. We use $U_{c,G}$ to denote the potential operator of Y.

Let

$$\|\nu\|_{u^2,\infty} := |\nu|(S) \vee \sup_x \int \left(u^2(x, y) + u^2(y, x)\right) d|\nu|(y), \tag{6.4}$$

where $|\nu|$ is the total variation of the measure ν. This is a proper norm for u, see [10, (3.25)].

We use $\mu_{(\epsilon)}$ to denote the loop measure associated to Y, where we have replaced c by ϵc.

Theorem 6.1. *Assume that*

$$\sup_z \int u(z,y)dm(y) < \infty, \qquad \sup_z \int u(y,z)dm(y) < \infty. \tag{6.5}$$

Then $\mu_{(\epsilon)}$ is well defined for ϵ small enough and

$$\frac{d\mu_{(\epsilon)}(A)}{d\epsilon}\,|_{\epsilon=0} = \int_{S\times S} \left(Q^{y,x}(A) - Q^{x,x}(A)\right) c(x)G(x,\,dy)\,dm(x), \tag{6.6}$$

for all $A \in \mathcal{A}_{\|\cdot\|_{u^2,\infty}}$.

Before proving this theorem, we first show that $U_{\epsilon c,G}$ has densities $u_{\epsilon c,G}(x,y)$ for ϵ sufficiently small. Note first that by (6.1)-(6.2)

$$\int c(z)G(z,dy)\,dm(z) = \int j(z,y)\,dm(y)\,dm(z) = c(y)\,dm(y), \tag{6.7}$$

and since we assumed that c is bounded it follows from (6.5) that

$$\sup_x \int c(z)G(z,dy)u(y,x)\,dm(z) = \sup_x \int c(y)u(y,x)\,dm(y) < \infty. \tag{6.8}$$

Let $\lambda_1, \lambda_2, \ldots$ be a sequence of independent mean 1 exponentials, and set $T_j = \sum_{i=1}^{j} \lambda_j$. Using the fact that $\tau_{t+u} = \tau_t + \tau_u \circ \theta_{\tau_t}$ we see that

$$\begin{aligned} W_{c,G,n}f(x) &=: P^x\left(\int_{\tau(T_n)}^{\tau(T_{n+1})} f(Y_t)\,dt\right) \\ &= P^x\left(\left(\int_0^{\tau(\lambda_{n+1})} f(Y_t)\,dt\right)\circ\theta_{\tau(T_n)}\right) \\ &= P^x\left(P^{Y_{\tau(T_n)}}\left(\int_0^{\tau(\lambda_{n+1})} f(X_t)\,dt\right)\right) \\ &= P^x\left(\int G(Y^-_{\tau(T_n)},dz)P^z\left(\int_0^{\tau(\lambda_{n+1})} f(X_t)\,dt\right)\right). \end{aligned} \tag{6.9}$$

We have

$$\begin{aligned} &P^x_\lambda\left(\int_0^{\tau_\lambda} f(X_t)\,dt\right) \\ &= P^x_\lambda\left(\int_0^\infty 1_{\{t<\tau_\lambda\}} f(X_t)\,dt\right) \\ &= P^x_\lambda\left(\int_0^\infty 1_{\{A_t<\lambda\}} f(X_t)\,dt\right) \\ &= P^x\left(\int_0^\infty e^{-A_t} f(X_t)\,dt\right). \end{aligned} \tag{6.10}$$

Hence setting

$$V_c f(x) = P^x \left(\int_0^\infty e^{-A_t} f(X_t)\, dt \right), \tag{6.11}$$

and writing $Gh(x) = \int_S G(z, dy)h(y)$ for any nonnegative or bounded function h, we have shown that

$$\begin{aligned} W_{c,G,n} f(x) &= P^x \left(\int G(Y^-_{\tau(T_n)}, dz) V_c f(z) \right) \\ &= P^x \left(G V_c f(Y^-_{\tau(T_n)}) \right). \end{aligned} \tag{6.12}$$

Using once again the fact that $\tau_{t+u} = \tau_t + \tau_u \circ \theta_{\tau_t}$ and the Markov property, we see that for any h

$$\begin{aligned} P^x \left(h(Y^-_{\tau(T_n)}) \right) &= P^x \left(\left(h(Y^-_{\tau(\lambda_n)}) \right) \circ \theta_{\tau(T_{n-1})} \right) \\ &= P^x \left(P^{Y_{\tau(T_{n-1})}} \left(h(Y^-_{\tau(\lambda_n)}) \right) \right) \\ &= P^x \left(\int G(Y^-_{\tau(T_{n-1})}, dz) P^z \left(h(X^-_{\tau(\lambda_n)}) \right) \right). \end{aligned} \tag{6.13}$$

Using the change of variables formula, [4, Chapter 6, (55.1)], and the fact that $X^-_t = X_t$ for a.e. t, we see that

$$\begin{aligned} P^z_\lambda \left(h(X^-_{\tau_\lambda}) \right) = P^z_\lambda \left(h(X_{\tau_\lambda}) \right) &= E^x \left(\int_0^\infty e^{-t} h(X_{\tau_t})\, dt \right) \\ &= P^z \left(\int_0^{A_\zeta} e^{-t} h(X_{\tau_t})\, dt \right) \\ &= P^z \left(\int_0^\infty e^{-A_s} h(X_s)\, dA_s \right) \\ &= P^z \left(\int_0^\infty e^{-A_s} h(X_s)\, c(X_s)\, ds \right) \\ &= V_c(ch)(z). \end{aligned} \tag{6.14}$$

Thus we can write (6.13) as

$$P^x \left(h(Y^-_{\tau(T_n)}) \right) = P^x \left(G V_c\, c\, h(Y^-_{\tau(T_{n-1})}) \right). \tag{6.15}$$

Iterating this we obtain

$$\begin{aligned} P^x \left(h(Y^-_{\tau(T_n)}) \right) &= P^x \left(G V_c\, c\, h(Y^-_{\tau(T_{n-1})}) \right) \\ &= P^x \left((G V_c\, c)^2\, h(Y^-_{\tau(T_{n-2})}) \right) \\ &= \cdots \\ &= P^x \left((G V_c\, c)^{n-1}\, h(Y^-_{\tau(\lambda)}) \right) \\ &= V_c\, c (G V_c\, c)^{n-1}\, h(x), \end{aligned} \tag{6.16}$$

where the last step used (6.14). Applying this to (6.12) we have that

$$W_{c,G,n}f(x) = V_c\, c(GV_c\, c)^{n-1}\, GV_c f(x) = V_c\,(cGV_c)^n\, f(x). \tag{6.17}$$

It follows from (5.9) that V_c has a density which we write as $v_c(x,y)$, and therefore $W_{c,G,n}$ has the density

$$\begin{aligned} w_{c,G,n}(x,y) = \int v_c(x,z_1)G(z_1,\, dz_2)v_c(z_2,z_3)\cdots \\ \cdots G(z_{2n-3},\, dz_{2n-2})v_c(z_{2n-2},z_{2n-1})G(z_{2n-1},\, dz_{2n})v_c(z_{2n},y) \\ \prod_{j=1}^n c(z_{2j-1})\, dm(z_{2j-1}). \end{aligned} \tag{6.18}$$

By (6.5) it follows from (5.9) that

$$\sup_z \int v_c(z,y)dm(y) \leq M, \tag{6.19}$$

and thus by (6.18) that

$$\sup_z \int w_{c,G,n}(z,y)dm(y) \leq C^n M^{n+1}. \tag{6.20}$$

Replacing c by ϵc we have shown that for ϵ sufficiently small

$$U_{\epsilon c,G}f(x) = \sum_{n=0}^\infty \epsilon^n\; V_{\epsilon c}\,(cGV_{\epsilon c})^n\, f(x) \tag{6.21}$$

for all bounded measurable f. Hence $U_{\epsilon c,G}$ has a density

$$\begin{aligned} u_{\epsilon c,G}(x,y) = \sum_{n=0}^\infty \epsilon^n \int v_{\epsilon c}(x,z_1)G(z_1,\, dz_2)v_{\epsilon c}(z_2,z_3)\cdots \\ \cdots G(z_{2n-3},\, dz_{2n-2})v_{\epsilon c}(z_{2n-2},z_{2n-1})G(z_{2n-1},\, dz_{2n})v_{\epsilon c}(z_{2n},y) \\ \prod_{j=1}^n c(z_{2j-1})\, dm(z_{2j-1}), \end{aligned} \tag{6.22}$$

with

$$\sup_z \int u_{\epsilon c,G}(z,y)dm(y) < \infty. \tag{6.23}$$

We can also write this as

$$\begin{aligned} u_{\epsilon c,G}(x,y) = \sum_{n=0}^\infty \epsilon^n \int v_{\epsilon c}(x,z_1)j(z_1,z_2)v_{\epsilon c}(z_2,z_3)\cdots \\ \cdots j(z_{2n-3},z_{2n-2})v_{\epsilon c}(z_{2n-2},z_{2n-1})j(z_{2n-1},z_{2n})v_{\epsilon c}(z_{2n},y)\prod_{j=1}^{2n} dm(z_j). \end{aligned} \tag{6.24}$$

Similar expansions can be given for the semigroup which has therefore a density:

$$p_{\epsilon c,G,t}(x,y)=\sum_{n=0}^{\infty}\epsilon^n\int_{0\le t_1\le\cdots\le t_n\le t}q_{t_1}(x,z_1)c(z_1)G(z_1,\,dz_2)q_{t_2-t_1}(z_2,z_3)\cdots$$

$$\cdots c(z_{2n-1})G(z_{2n-1},\,dz_{2n})q_{t-t_n}(z_{2n},y)\prod_{j=1}^{n}dm(z_{2j-1})\,dt_j, \tag{6.25}$$

where q_t denotes the kernel of the semigroup associated with the process killed at rate ϵc.

To see this let

$$W_{c,G,n,t}f(x)=:P^x\left(f\left(Y_t\right);\,\tau(T_n)<t<\tau(T_{n+1})\right). \tag{6.26}$$

Using the fact that $\tau_{t+u}=\tau_t+\tau_u\circ\theta_{\tau_t}$ we have

$$\begin{aligned}
&=P^x\left(f\left(Y_t\right);\,\tau(T_n)<t<\tau(T_{n+1})\Big|\tau(T_n)\right) \qquad (6.27)\\
&=P^x\left(\left(f\left(Y_{t-\tau(T_n)}\right);\,t-\tau(T_n)<\tau(\lambda_{n+1})\right)\circ\theta_{\tau(T_n)}\Big|\tau(T_n)\right)\\
&=P^x\left(P^{Y_{\tau(T_n)}}\left(f\left(X_{t-\tau(T_n)}\right);\,t-\tau(T_n)<\tau(\lambda_{n+1})\right)\Big|\tau(T_n)\right)\\
&=P^x\Bigg(\int G(Y^-_{\tau(T_n)},\,dz)\\
&\qquad P^z_{\lambda_{n+1}}\left(f\left(X_{t-\tau(T_n)}\right);\,t-\tau(T_n)<\tau(\lambda_{n+1})\right)\Big|\tau(T_n)\Bigg).
\end{aligned}$$

Conditionally on $\tau(T_n)$ we have

$$\begin{aligned}
&P^x_\lambda\left(f\left(X_{t-\tau(T_n)}\right);\,t-\tau(T_n)<\tau(\lambda)\right)\\
&=P^x_\lambda\left(f\left(X_{t-\tau(T_n)}\right);\,A_{t-\tau(T_n)}<\lambda\right)\\
&=P^x\left(e^{-A_{t-\tau(T_n)}}f\left(X_{t-\tau(T_n)}\right)\right).
\end{aligned} \tag{6.28}$$

Hence setting

$$P_{c,t}f(x)=P^x\left(e^{-A_t}f\left(X_t\right)\right), \tag{6.29}$$

we have shown that

$$\begin{aligned}
W_{c,G,n,t}f(x)&=P^x\left(\int G(Y^-_{\tau(T_n)},\,dz)P_{c,t-\tau(T_n)}f(z)\right)\\
&=P^x\left(GP_{c,t-\tau(T_n)}f(Y^-_{\tau(T_n)})\right).
\end{aligned} \tag{6.30}$$

Using once again the fact that $\tau_{t+u}=\tau_t+\tau_u\circ\theta_{\tau_t}$ and the strong Markov property, we see that for any h

$$\begin{aligned}
&P^x\left(h(Y^-_{\tau(T_n)},t-\tau(T_n))\Big|\tau(T_{n-1})\right) \qquad (6.31)\\
&=P^x\left(\left(h(Y^-_{\tau(\lambda_n)},t-\tau(T_{n-1})-\tau(\lambda_n))\right)\circ\theta_{\tau(T_{n-1})}\Big|\tau(T_{n-1})\right)\\
&=P^x\left(P^{Y_{\tau(T_{n-1})}}\left(h(Y^-_{\tau(\lambda_n)},t-\tau(T_{n-1})-\tau(\lambda_n))\right)\Big|\tau(T_{n-1})\right)
\end{aligned}$$

$$= P^x \left(\int G(Y^-_{\tau(T_{n-1})}, dz) P^z \left(h(X^-_{\tau(\lambda_n)}, t - \tau(T_{n-1}) - \tau(\lambda_n)) \right) \Big| \tau(T_{n-1}) \right).$$

Using the change of variables formula, [4, Chapter 6, (55.1)], and the fact that $X^-_t = X_t$ for a.e. t, we see that conditionally on $\tau(T_{n-1})$

$$\begin{aligned}
& P^z_{\lambda_n} \left(h(X^-_{\tau_{\lambda_n}}, t - \tau(T_{n-1}) - \tau(\lambda_n)) \right) \\
&= P^z_{\lambda_n} \left(h(X_{\tau_{\lambda_n}}, t - \tau(T_{n-1}) - \tau(\lambda_n)) \right) \\
&= E^x \left(\int_0^\infty e^{-r} h\left(X_{\tau_r}, t - \tau(T_{n-1}) - \tau_r\right) dr \right) \\
&= P^z \left(\int_0^{A_\zeta} e^{-r} h\left(X_{\tau_r}, t - \tau(T_{n-1}) - \tau_r\right) dr \right) \\
&= P^z \left(\int_0^\infty e^{-A_s} h\left(X_s, t - \tau(T_{n-1}) - s\right) dA_s \right) \\
&= \int_0^\infty P^z \left(e^{-A_s} h\left(X_s, t - \tau(T_{n-1}) - s\right) c(X_s) \right) ds \\
&= \int_0^\infty P_{c,s}(ch(\cdot, t - s - \tau(T_{n-1})))(z)\, ds.
\end{aligned} \tag{6.32}$$

Combining this with (6.31) we have

$$\begin{aligned}
& P^x \left(h(Y^-_{\tau(T_n)}, t - \tau(T_n)) \right) \\
&= \int P^x \left(GP_{c,s_n} ch(Y^-_{\tau(T_{n-1})}, t - s_n - \tau(T_{n-1})) \right) ds_n.
\end{aligned} \tag{6.33}$$

Iterating this we obtain, conditionally on $\tau(T_{n-1})$, then $\tau(T_{n-2}), \cdots$

$$\begin{aligned}
& P^x \left(h(Y^-_{\tau(T_n)}), t - \tau(T_n)) \right) \\
&= \int P^x \left(GP_{c,s_n} ch(Y^-_{\tau(T_{n-1})}, t - s_n - \tau(T_{n-1})) \right) ds_n \\
&= \int P^x \left(GP_{c,s_{n-1}} cGP_{c,s_n} ch(Y^-_{\tau(T_{n-2})}, t - s_n - s_{n-1} - \tau(T_{n-2})) \right) ds_{n-1}\, ds_n \\
&= \cdots \\
&= \int P^x \left(GP_{c,s_2} c \cdots GP_{c,s_n} ch(Y^-_{\tau(\lambda)}, t - \sum_{j=2}^n s_j - \tau(\lambda)) \right) \prod_{j=2}^n ds_j \\
&= \int P_{c,s_1} cGP_{c,s_2} c \cdots GP_{c,s_n} ch(x, t - \sum_{j=1}^n s_j) \prod_{j=1}^n ds_j
\end{aligned} \tag{6.34}$$

where the last step used (6.32). Applying this to (6.30) we have that

$$W_{c,G,n,t} f(x) = \int P_{c,s_1} cGP_{c,s_2} c \cdots GP_{c,s_n} cGP_{c,t-\sum_{j=1}^n s_j} f(x) \prod_{j=1}^n ds_j. \tag{6.35}$$

(6.25) then follows as in the proof of (6.22).

Let

$$\hat{G}(x,dy)=\frac{1}{c(x)}j(y,x)m(dy). \tag{6.36}$$

By (6.1), $\hat{G}(x,dy)$ is a probability kernel on $S\times\mathcal{B}(S)$. We now add jumps to the dual process $\widehat{X}$, where $c(x)$ will again govern the rate of jumps but we use $\hat{G}(x,dy)$ to describe the distribution of the jumps from position x. Performing the same calculation as before, but with the dual process and using (6.24) and (5.13), we see that the two processes obtained by adding jumps have dual potential kernels:

$$\hat{u}_{\epsilon c,\hat{G}}(x,y)=u_{\epsilon c,G}(y,x). \tag{6.37}$$

The same will be true for the associated resolvents and semigroups. The duality assumptions are verified and we can therefore define a loop measure associated with these processes.

We use $\mu_{(\epsilon)}$ to denote the loop measure associated to Y, where we have replaced c by ϵc.

The next Lemma is needed for the proof of Theorem 6.1.

Lemma 6.1. *Assume (6.5) (which implies (6.8)). Then for any positive measure ν and ϵ sufficiently small.*

$$\sup_z\int u_{\epsilon c,G}(z,y)d\nu(y)\le C\sup_z\int u(z,y)d\nu(y), \tag{6.38}$$

and

$$\|\nu\|_{u^2_{\epsilon c,G},\infty}\le C\|\nu\|_{u^2,\infty}, \tag{6.39}$$

Proof of Lemma 6.1: (6.38) follows immediately from (6.22) and (5.9). It also follows from (6.22) that for a positive measure ν

$$\begin{aligned}&\int u^2_{\epsilon c,G}(x,y)\,d\nu(y) \\ &=\sum_{m,n=0}^{\infty}\epsilon^{m+n}\int V_{\epsilon c}\left(cGV_{\epsilon c}\right)^m(x,y)V_{\epsilon c}\left(cGV_{\epsilon c}\right)^n(x,y)\,d\nu(y) \\ &=\sum_{m,n=0}^{\infty}\epsilon^{m+n}\int V_{\epsilon c}\left(cGV_{\epsilon c}\right)^{m-1}cG(x,\,dz_1)V_{\epsilon c}\left(cGV_{\epsilon c}\right)^{n-1}cG(x,\,dz_2) \\ &\qquad\qquad\left(\int v_{\epsilon c}(z_1,y)v_{\epsilon c}(z_2,y)\,d\nu(y)\right).\end{aligned} \tag{6.40}$$

Hence for ϵ small enough

$$\begin{aligned}&\sup_x\int u^2_{\epsilon c,G}(x,y)\,d\nu(y) \\ &\le\sum_{m,n=0}^{\infty}\epsilon^{m+n}\sup_x\int V_{\epsilon c}\left(cGV_{\epsilon c}\right)^{m-1}cG(x,\,dz_1)V_{\epsilon c}\left(cGV_{\epsilon c}\right)^{n-1}cG(x,\,dz_2) \\ &\qquad\qquad\sup_{z_1,z_2}\left(\int v_{\epsilon c}(z_1,y)v_{\epsilon c}(z_2,y)\,d\nu(y)\right) \\ &\le C\|\nu\|_{u^2,\infty}.\end{aligned} \tag{6.41}$$

Similarly

$$\begin{aligned}&\int u^2_{\epsilon c,G}(y,x)\,d\nu(y) \qquad (6.42)\\&=\sum_{m,n=0}^{\infty}\epsilon^{m+n}\int V_{\epsilon c}\left(cGV_{\epsilon c}\right)^m(y,x)V_{\epsilon c}\left(cGV_{\epsilon c}\right)^n(y,x)\,d\nu(y)\\&=\sum_{m,n=0}^{\infty}\epsilon^{m+n}\int\left(cGV_{\epsilon c}\right)^m(z_1,x)\left(cGV_{\epsilon c}\right)^n(z_2,x)\,dm(z_1)\,dm(z_2)\\&\qquad\qquad\left(\int v_{\epsilon c}(y,z_1)v_{\epsilon c}(y,z_2)\,d\nu(y)\right).\end{aligned}$$

Using (6.8) it follows that for ϵ small enough

$$\begin{aligned}&\sup_x\int u^2_{\epsilon c,G}(x,y)\,d\nu(y) \qquad (6.43)\\&\le\sum_{m,n=0}^{\infty}\epsilon^{m+n}\sup_x\int\left(cGV_{\epsilon c}\right)^m(z_1,x)\left(cGV_{\epsilon c}\right)^n(z_2,x)\,dm(z_1)\,dm(z_2)\\&\qquad\qquad\sup_{z_1,z_2}\left(\int v_{\epsilon c}(y,z_1)v_{\epsilon c}(y,z_2)\,d\nu(y)\right)\\&\le C\|\nu\|_{u^2,\infty}.\end{aligned}$$

□

It follows from [10, Lemma 3.3] that $\|\nu\|_{u^2,\infty}$ is a proper norm for $u_{\epsilon c,G,n}$. Theorem 6.1 follows from the next Lemma.

Lemma 6.2. *Under the assumptions of Lemma 6.1, for any* $\nu_1,\ldots,\nu_k\in\mathcal{R}^+_{\|\cdot\|_{u^2,\infty}}$

$$\begin{aligned}&\frac{d}{d\epsilon}\int\prod_{j=1}^k u_{\epsilon c,G}(y_j,y_{j+1})\prod_{j=1}^k d\nu_j(y_j)\,\Big|_{\epsilon=0} \qquad (6.44)\\&=\sum_{i=1}^k\int\left(\prod_{j=1}^{i-1}u(y_j,y_{j+1})\right)\left(-\int u(y_i,x)u(x,y_{i+1})c(x)\,dm(x)\right.\\&\qquad\left.+\int_{S\times S}u(y_i,z_1)c(z_1)G(z_1,\,dz_2)u(z_2,y_{i+1})\,dm(z_1)\right)\\&\qquad\qquad\left(\prod_{j=i+1}^k u(y_j,y_{j+1})\right)\prod_{j=1}^k d\nu_j(y_j),\end{aligned}$$

with $y_{k+1}=y_1$.

Proof of Lemma 6.2: Set

$$\mathcal{I}(\epsilon)=\int\prod_{j=1}^k u_{\epsilon c,G}(y_j,y_{j+1})\prod_{j=1}^k d\nu_j(y_j) \qquad (6.45)$$

with $y_{k+1}=y_1$.

We can write (6.22) as

$$
\begin{aligned}
&u_{\epsilon c,G}(x,y) \qquad (6.46)\\
&= v_{\epsilon c}(x,y) + \epsilon \int v_{\epsilon c}(x,z_1)c(z_1)G(z_1,\,dz_2)v_{\epsilon c}(z_2,y)\,dm(y_1)\\
&+\epsilon^2 \int v_{\epsilon c}(x,z_1)c(z_1)G(z_1,\,dz_2)v_{\epsilon c}(z_2,y)c(z_2)G(z_2,\,dz_3)u_{\epsilon c,G}(z_3,y)\,dm(y_1)\,dm(y_3)\\
&= v_{\epsilon c}(x,y) + \epsilon V_{\epsilon c}cGV_{\epsilon c}(x,y) + \epsilon^2 V_{\epsilon c}cGV_{\epsilon c}cGU_{\epsilon c,G}(x,y),
\end{aligned}
$$

with operator notation.

We substitute this in (6.45) and collect terms to obtain

$$
\mathcal{I}(\epsilon) = I(\epsilon) + \epsilon \sum_{i=1}^{k} \mathcal{J}_i(\epsilon) + \mathcal{K}(\epsilon), \qquad (6.47)
$$

where

$$
I(\epsilon) = \int \prod_{j=1}^{k} v_{\epsilon c}(y_j, y_{j+1}) \prod_{j=1}^{k} d\nu_j(y_j) \qquad (6.48)
$$

$$
\begin{aligned}
&\mathcal{J}_i(\epsilon) \qquad (6.49)\\
&= \int \left(\prod_{j=1}^{i-1} v_{\epsilon c}(y_j, y_{j+1})\right) V_{\epsilon c}cGV_{\epsilon c}(y_i, y_{i+1}) \left(\prod_{j=i+1}^{k} v_{\epsilon c}(y_j, y_{j+1})\right) \prod_{j=1}^{k} d\nu_j(y_j),
\end{aligned}
$$

and $\mathcal{K}(\epsilon)$ represents all the remaining terms. Noting that $\mathcal{I}(0) = I(0)$, we can write (6.47) as

$$
\mathcal{I}(\epsilon) - \mathcal{I}(0) = I(\epsilon) - I(0) + \epsilon \sum_{i=1}^{k} \mathcal{J}_i(\epsilon) + \mathcal{K}(\epsilon). \qquad (6.50)
$$

Note also that $I(\epsilon)$ of (6.48) is a special case of the $I(\epsilon)$ of (5.11) with $\nu(dx) = c(x)\,dm(x)$. Hence by (5.12)

$$
\begin{aligned}
&\lim_{\epsilon \to 0} \frac{I(\epsilon) - I(0)}{\epsilon} \qquad (6.51)\\
&= -\sum_{i=1}^{k} \int \left(\prod_{j=1}^{i-1} u(y_j, y_{j+1})\right) \left(\int u(y_i, x)u(x, y_{i+1})c(x)\,dm(x)\right)\\
&\qquad\qquad \left(\prod_{j=i+1}^{k} u(y_j, y_{j+1})\right) \prod_{j=1}^{k} d\nu_j(y_j).
\end{aligned}
$$

Since $v_{\epsilon c}(x,y) \uparrow u(x,y)$ as $\epsilon \downarrow 0$, it follows by the Monotone Convergence Theorem that

$$
\begin{aligned}
&\lim_{\epsilon \to 0} \mathcal{J}_i(\epsilon) \qquad (6.52)\\
&= \int \left(\prod_{j=1}^{i-1} u(y_j, y_{j+1})\right)
\end{aligned}
$$

$$\left(\int_{S\times S} u(y_i, z_1)c(z_1)G(z_1,\, dz_2)u(z_2, y_{i+1})\, dm(z_1)\right)$$
$$\left(\prod_{j=i+1}^{k} u(y_j, y_{j+1})\right)\prod_{j=1}^{k} d\nu_j(y_j).$$

To complete the proof of our Lemma it remains to show that

$$\lim_{\epsilon\to 0}\frac{\mathcal{K}(\epsilon)}{\epsilon} = 0. \tag{6.53}$$

However every term in $\mathcal{K}(\epsilon)$ comes with a pre-factor of ϵ^m for some $m \geq 2$, so we need only bound the integrals uniformly in ϵ. For this we will use Lemma 6.1. We illustrate this with the most complicated term, which has the pre-factor ϵ^{2k}:

$$\begin{aligned}
&\int \prod_{j=1}^{k} V_{\epsilon c} cGV_{\epsilon c} cGU_{\epsilon c,G}(y_j, y_{j+1}) \prod_{j=1}^{k} d\nu_j(y_j) \qquad (6.54)\\
&= \int v_{\epsilon c}(y_1, x_1) cGV_{\epsilon c} cGU_{\epsilon c,G}(x_1, y_2)\cdots\\
&\quad v_{\epsilon c}(y_{k-1}, x_{k-1}) cGV_{\epsilon c} cGU_{\epsilon c,G}(x_{k-1}, y_k)\\
&\qquad v_{\epsilon c}(y_k, x_k) cGV_{\epsilon c} cGU_{\epsilon c,G}(x_k, y_1)\prod_{j=1}^{k} d\nu_j(y_j)\, dm(x_j)\\
&= \int cGV_{\epsilon c} cGU_{\epsilon c,G}(x_1, y_2) v_{\epsilon c}(y_2, x_2)\cdots\\
&\quad cGV_{\epsilon c} cGU_{\epsilon c,G}(x_{k-1}, y_k)\, v_{\epsilon c}(y_k, x_k)\\
&\qquad cGV_{\epsilon c} cGU_{\epsilon c,G}(x_k, y_1) v_{\epsilon c}(y_1, x_1)\prod_{j=1}^{k} d\nu_j(y_j)\, dm(x_j),
\end{aligned}$$

where the last step is just a rearrangement. We can rewrite this as

$$\begin{aligned}
&\int \left(\int cGV_{\epsilon c} cGU_{\epsilon c,G}(x_1, y_2) v_{\epsilon c}(y_2, x_2)\, d\nu_2(y_2)\right)\cdots \qquad (6.55)\\
&\quad \left(\int cGV_{\epsilon c} cGU_{\epsilon c,G}(x_{k-1}, y_k)\, v_{\epsilon c}(y_k, x_k)\, d\nu_k(y_k)\right)\\
&\qquad \left(\int cGV_{\epsilon c} cGU_{\epsilon c,G}(x_k, y_1) v_{\epsilon c}(y_1, x_1)\, d\nu_1(y_1)\right)\prod_{j=1}^{k} dm(x_j).
\end{aligned}$$

Then by Lemma 6.1, and the fact that $v_{\epsilon c} \leq u$

$$\begin{aligned}
&\int cGV_{\epsilon c} cGU_{\epsilon c,G}(x_1, y_2) v_{\epsilon c}(y_2, x_2)\, d\nu_2(y_2) \qquad (6.56)\\
&= \int c(x_1)G(x_1, dz_1) v_{\epsilon c}(z_1, z_2) c(z_2) G(z_2, dz_3) u_{\epsilon c,G}(z_3, y_2) v_{\epsilon c}(y_2, x_2)\, d\nu_2(y_2)\, dm(z_2)\\
&\leq \int c(x_1)G(x_1, dz_1) v_{\epsilon c}(z_1, z_2) c(z_2) G(z_2, dz_3)\, dm(z_2)
\end{aligned}$$

$$\sup_{z_3,x_2} \int u_{\epsilon c,G}(z_3,y_2)v_{\epsilon c}(y_2,x_2)\,d\nu_2(y_2)$$

$$\leq C\|\nu_2\|_{u^2,\infty} \int c(x_1)G(x_1,dz_1)v_{\epsilon c}(z_1,z_2)c(z_2)G(z_2,dz_3)\,dm(z_2). \tag{6.57}$$

Using (6.19), the fact that c is bounded and the fact that $G(\cdot,dz)$ is a probability density

$$\int c(x_1)G(x_1,dz_1)v_{\epsilon c}(z_1,z_2)c(z_2)G(z_2,dz_3)\,dm(z_2) \leq Cc(x_1). \tag{6.58}$$

Thus (6.55) is bounded independently of ϵ by

$$C \int \prod_{j=1}^{k} c(x_j)\; dm(x_j) < \infty \tag{6.59}$$

since $c(x)$ is integrable.

The other terms of $\mathcal{K}(\epsilon)$ can be bounded similarly. □

References

1. R. Blumenthal and R. Getoor, *Markov Processes and Potential Theory*, Academic Press, New York, 1968.
2. K. L. Chung and J. Walsh, *Markov Processes, Brownian Motion and Time Symmetry*, second edition, Springer, New York, 2005.
3. Y. Chang, Multi-occupation field generates the Borel sigma-field of loops. arXiv:1309.1558 [math.PR]
4. C. Dellacherie, and P.-A. Meyer, (1982). *Probabilities and Potential B.* North Holland Publishing Company, Amsterdam.
5. P. J. Fitzsimmons and J, Rosen, Markovian loop soups: permanental processes and isomorphism theorems, arxiv.org/pdf/1211.5163.pdf
6. P. Fitzsimmons, J. Pitman, and M. Yor, Markovian bridges: construction, Palm interpretation, and splicing. *Seminar on Stochastic Processes*, 1992, E. Cinlar and K.L. Chung and M.J. Sharpe editors, 101-134, Birkhauser, Boston (1993).
7. M. Fukushima, Y. Oshima and M. Takeda, *Dirichlet Forms and Symmetric Markov Processes.* Walter de Gruyter, New York, 1994.
8. K. Gawedski, *Lectures on conformal field theory*, IAS, Princeton, (2009). www.math.ias.edu/QFT/fall/
9. Y. Le Jan, *Markov paths, loops and fields.* École d'Été de Probabilités de Saint-Flour XXXVIII - 2008. Lecture Notes in Mathematics 2026. (2011) Springer-Verlag, Berlin-Heidelberg.
10. Y. Le Jan, M. B. Marcus and J. Rosen, Permanental fields, loop soups and continuous additive functionals, arxiv.org/pdf/1209.1804.pdf, Version 1.
11. Y. Le Jan, M. B. Marcus and J. Rosen, Intersection local times, loop soups and permanental Wick powers, arxiv.org/pdf/1308.2701.pdf
12. M. Sharpe, *General theory of Markov processes*, Acad. Press, New York, 1988.

Chapter 20

Regular subspaces of Dirichlet forms

Liping Li* and Jiangang Ying†

In honor of Professor Masatoshi Fukushima on occasion of his 80th birthday

The regular subspaces of a Dirichlet form are the regular Dirichlet forms that inherit the original form but possess smaller domains. The two problems we are concerned are: (1) the existence of regular subspaces of a fixed Dirichlet form, (2) the characterization of the regular subspaces if exists. In this paper, we will first research the structure of regular subspaces for a fixed Dirichlet form. The main results indicate that the jumping and killing measures of each regular subspace are just equal to that of the original Dirichlet form. By using the independent coupling of Dirichlet forms and some celebrated probabilistic transformations, we will study the existence and characterization of the regular subspaces of local Dirichlet forms.

Contents

1. Introduction

Let E be a locally compact separable metric space, m a Radon measure fully supported on E, then $L^2(E,m)$ is a Hilbert space. A non-negative definite symmetric bilinear form $\mathcal{E}$ with domain $\mathcal{F}$ densely defining on $L^2(E,m)$ is called a Dirichlet form and denoted by

*School of Mathematical Sciences, Fudan University, Shanghai, China 200433, lipingli10@fudan.edu.cn
†School of Mathematical Sciences, Fudan University, Shanghai, China 200433, jgying@fudan.edu.cn

$(\mathcal{E},\mathcal{F})$ if it is closed and Markovian. $(T_t)_{t>0}, (G_\alpha)_{\alpha>0}$ are its semigroup and resolvent. Define:

$$\mathcal{E}_\alpha(u,v) := \mathcal{E}(u,v) + \alpha\cdot(u,v)_m, \quad u,v\in\mathcal{F}, \alpha\geq 0.$$

We then denote the space of all real continuous functions on E by $C(E)$ and its subspace of continuous functions with compact support (resp. bounded continuous functions, one order continuous differentiable functions with compact supports, the continuous functions which converge to zero at infinite) by $C_c(E)$ (resp. $C_b(E), C_c^1(E), C_0(E)$). A Dirichlet form $(\mathcal{E},\mathcal{F})$ is called regular if $\mathcal{F}\cap C_c(E)$ is dense in $\mathcal{F}$ with $\mathcal{E}_1^{\frac{1}{2}}$-norm and dense in $C_c(E)$ with uniform norm. A core of $\mathcal{E}$ is by definition a subset $\mathcal{C}$ of $\mathcal{F}\cap C_c(E)$ such that $\mathcal{C}$ is dense in $\mathcal{F}$ with $\mathcal{E}_1^{\frac{1}{2}}$-norm and dense in $C_c(E)$ with uniform norm. We refer the definitions of standard core and special standard core to Sec. 1.1 of Ref. 1. The 1-capacity of regular Dirichlet form $(\mathcal{E},\mathcal{F})$ is always denoted by **Cap**, see Eq. (2.1.2) and Eq. (2.1.3) of Ref. 1. An increasing sequence $\{F_k, k\geq 1\}$ of closed sets of E is an $\mathcal{E}$-*nest* if $\cup_{k\geq 1}\mathcal{F}_{F_k}$ is $\mathcal{E}_1^{\frac{1}{2}}$-dense in $\mathcal{F}$ where $\mathcal{F}_{F_k} := \{f\in\mathcal{F} : f = 0\ m\text{-a.e. on } F_k\}$. A subset N of E is $\mathcal{E}$-*polar* if there is an $\mathcal{E}$-nest $\{F_k, k\geq 1\}$ such that $N\subset\cap_{k\geq 1}(E\setminus F_k)$. A statement depending on $x\in A$ is said to hold $\mathcal{E}$-quasi-everywhere ($\mathcal{E}$-q.e. in abbreviation) on A if there is an $\mathcal{E}$-polar set $N\subset A$ such that the statement is true for every $x\in A\setminus N$. A function f on E is said to be $\mathcal{E}$-quasi-continuous if there is an $\mathcal{E}$-nest $\{F_k, k\geq 1\}$ such that $f|_{F_k}$ is finite and continuous on F_k for each $k\geq 1$, which will be denoted in abbreviation as $f\in C(\{F_k\})$. In fact, a set N is $\mathcal{E}$-polar if and only if N is of zero 1-capacity, i.e. $\text{Cap}(N) = 0$. An increasing sequence of closed sets $\{F_k\}$ is an $\mathcal{E}$-nest if and only if $\lim_{k\to\infty}\text{Cap}(K\setminus F_k) = 0$ for any compact set $K\subset E$. If there is no confusion with the fixed regular Dirichlet form $(\mathcal{E},\mathcal{F})$, for convenience we drop "$\mathcal{E}$-" from the terminology, then "$\mathcal{E}$-polar", an "$\mathcal{E}$-nest" and "$\mathcal{E}$-quasi-continuous" will simply be called *polar*, a *nest* and *quasi continuous* respectively. We call an increasing sequence $\{F_k\}$ of closed sets a *Cap-nest* if $\lim_{k\to\infty}\text{Cap}(E\setminus F_k) = 0$. Any Cap-nest is a ($\mathcal{E}$-)nest but not vice versa. However, a function f is quasi continuous if and only if there is a Cap-nest $\{F_k\}$ such that $f\in C(\{F_k\})$, i.e. $f|_{F_k}$ is finite and continuous on F_k for each $k\geq 1$. By Theorem 2.1.7 of Ref. 1, any function $u\in\mathcal{F}$ admits a quasi continuous version $\tilde{u}$, i.e. $\tilde{u}$ is quasi continuous and $\tilde{u} = u$ m-a.e.

For an m-measurable function u, the support $\text{supp}[|u|\cdot m]$ of the measure $|u|\cdot m$ is simply denoted by $\text{supp}[u]$. We say that $(\mathcal{E},\mathcal{F})$ possesses the local property if $\mathcal{E}(u,v) = 0$ for any $u,v\in\mathcal{F}$ with disjoint compact supports. Moreover $(\mathcal{E},\mathcal{F})$ is called strongly local if $\mathcal{E}(u,v) = 0$ for any $u,v\in\mathcal{F}$ with compact supports and v is constant on a neighbourhood of $\text{supp}[u]$. For more details, see Ref. 1.

The basic concept of this paper is the regular subspace of a Dirichlet form.

Definition 1.1. Let $(\mathcal{E},\mathcal{F}), (\mathcal{E}',\mathcal{F}')$ be two regular Dirichlet forms on $L^2(E,m)$, we say $(\mathcal{E}',\mathcal{F}')$ is a *regular subspace* of $(\mathcal{E},\mathcal{F})$ if

$$\mathcal{F}'\subset\mathcal{F}, \qquad \mathcal{E}(u,v) = \mathcal{E}'(u,v) \quad u,v\in\mathcal{F}'. \tag{1.1}$$

We denote it by $(\mathcal{E}',\mathcal{F}')\prec(\mathcal{E},\mathcal{F})$ or just $\mathcal{E}'\prec\mathcal{E}$ for simple. If in addition $\mathcal{F}'\neq\mathcal{F}$, $(\mathcal{E}',\mathcal{F}')$ is said to be a proper subspace of $(\mathcal{E},\mathcal{F})$.

If Y and Z are the corresponding Hunt processes of $(\mathcal{E},\mathcal{F})$ and $(\mathcal{E}',\mathcal{F}')$, we say Z is an *R-subprocess* of Y and denote it by $Z\prec Y$.

Remark 1.1. Let $(\mathcal{E}',\mathcal{F}')\prec(\mathcal{E},\mathcal{F})$ and Cap, Cap$'$ the 1-capacity of $(\mathcal{E},\mathcal{F})$, $(\mathcal{E}',\mathcal{F}')$ respectively. Clearly

$$\text{Cap}'(A)\geq\text{Cap}(A),$$

for any $A \subset E$ by the definition of 1-capacity. Hence any $\mathcal{E}'$-polar set is also $\mathcal{E}$-polar, any $\mathcal{E}'$-(resp. Cap$'$-)nest is also $\mathcal{E}$-(resp. Cap-)nest and any $\mathcal{E}'$-quasi-continuous function is also $\mathcal{E}$-quasi-continuous.

Two problems that we are concerned in this paper are the existence of regular subspaces and the characterization of regular subspaces if they exist for a fixed regular Dirichlet form. These problems were raised firstly in Ref. 3 which showed that all regular subspaces of one-dimensional Brownian motion can be achieved by a time change of additive functional firstly and a spatial transform secondly. It is a pity that they hold only for one-dimensional local cases, in other words, the diffusions on $\mathbf{R}$, due to the method used in that paper, see Ref. 4.

The structure of this paper is as follows. We will first research the structure of regular subspaces for a fixed Dirichlet form. In Sec. 2.1, we will prove that the jumping and killing measures of every regular subspace are just equal to that of the original Dirichlet form, see Theorem 2.1. In Sec. 2.2, we will discuss several probabilistic transformations such as time changes, killing and resurrected transformations. It is shown that if $(\tilde{\mathcal{E}}, \tilde{\mathcal{F}})$ is a regular Dirichlet form which is transformed from another one, say $(\mathcal{E}, \mathcal{F})$, then every regular subspace of $(\tilde{\mathcal{E}}, \tilde{\mathcal{F}})$ can be produced by the same transformation on some regular subspace of $(\mathcal{E}, \mathcal{F})$. If $(\mathcal{E}, \mathcal{F})$ and $(\tilde{\mathcal{E}}, \tilde{\mathcal{F}})$ have the above relationship, we say that **they are two Dirichlet forms with the same structure of regular subspaces**. These transformations are powerful tools to deal with what we are concerned. For example, the killing part of a Dirichlet form doesn't play a role in generating a regular subspace, we could always assume that it has no killing inside (by a resurrection if necessary), see Sec. 2.2.4. On the other hand, by spatial homeomorphic transformation and time changes, we can easily extend the results of one-dimensional Brownian motion to one-dimensional diffusions. This is the main topic of Ref. 4, see Example 2.1.

In Sec. 3, we will study the regular subspaces of a local Dirichlet form. By introducing the independent coupling of Dirichlet forms, we will prove that the associated Dirichlet forms of multidimensional Brownian motions always have proper regular subspaces. On the other hand, these corresponding R-subprocesses of multidimensional Brownian motions could be acted as the independent products of R-subprocess of one-dimensional Brownian motion, see Proposition 3.1 and Theorem 3.2. In Sec. 3.3, we will extend the similar results to the planar reflecting and absorbing Brownian motion on any domain of $\mathbf{R}^2$ by using the transformation method introduced in Sec. 2.2. It indicates that the associated Dirichlet forms of absorbing or reflecting Brownian motions on any planar domains always possess the same structure of regular subspaces. In addition, the associated Dirichlet forms of the reflecting and absorbing Brownian motions on arbitrary domains of $\mathbf{R}^d$ always possess proper regular subspaces not only for the two dimensional cases, see Sec. 3.4. At last, we will also prove that the uniform-ellipticity-type strongly local Dirichlet forms on arbitrary domains always possess proper regular subspaces, see Proposition 3.6.

2. Structure of regular subspaces

In this section, we will research the structure of regular subspaces. We first conclude that the jumping and killing measures of every regular subspace are equal to that of the original Dirichlet form. Then we will discuss several probabilistic transformations which could become powerful tools to deal with what we concerned about some specific Dirichlet forms.

2.1. *Jumping and killing measures of regular subspaces*

The following theorem gives a basic property about regular subspaces that the jumping and killing measures of regular subspace must be the same as that of the original Dirichlet form. This is the essential reason for Theorem 4.1(1) of Ref. 4 which is just the diffusion cases.

Let us first review the Beurling-Deny formulae[a] which state that any regular Dirichlet form $(\mathcal{E},\mathcal{F})$ on $L^2(E,m)$ can be expressed for any $u,v\in\mathcal{F}$ as follows:

$$\begin{aligned}\mathcal{E}(u,v) =&\mathcal{E}^{(c)}(u,v) + \int_{E\times E\setminus d}(\tilde{u}(x)-\tilde{u}(y))(\tilde{v}(x)-\tilde{v}(y))J(dx,dy)\\ &+\int_E \tilde{u}(x)\tilde{v}(x)k(dx)\end{aligned} \tag{2.1}$$

where $\mathcal{E}^{(c)}$ is a symmetric form, satisfying the strongly local property and Markovian property[b], J is a symmetric positive Radon measure on the product space $E\times E$ off the diagonal d called the *jumping measure* and k is a positive Radon measure on E called the *killing measure*. They both charge no set of zero capacity, i.e. if A is measurable and $\mathrm{Cap}(A)=0$, then $k(A)=0$, $J(A\times E)=0$. Note that the following equation is satisfied:

$$\mathcal{E}(u,v) = -2\int_{E\times E} u(x)v(y)J(dx,dy) \tag{2.2}$$

for any $u,v\in\mathcal{F}\cap C_c(E)$ with disjoint supports. The triple $(\mathcal{E}^{(c)},J,k)$ is uniquely determined by $\mathcal{E}$ and we call it the Beurling-Deny triple.

We refer the definitions of the energy measure $\mu_{<u>}$ as well as its strongly local part measure $\mu^c_{<u>}$ of $u\in\mathcal{F}_b$ to §3.2 of Ref. 1. Moreover, for $u,v\in\mathcal{F}_b$ put

$$\mu^c_{<u,v>} := \frac{1}{2}(\mu^c_{<u+v>} - \mu^c_{<u>} - \mu^c_{<v>}) \tag{2.3}$$

which is a bounded signed measure. We then have

$$\int_E f d\mu^c_{<u,v>} = \mathcal{E}^{(c)}(uf,v) + \mathcal{E}^{(c)}(vf,u) - \mathcal{E}^{(c)}(uv,f) \tag{2.4}$$

for any $f\in\mathcal{F}\cap C_c(E)$.

Theorem 2.1. *Let $(\mathcal{E},\mathcal{F}),(\mathcal{E}',\mathcal{F}')$ be two regular Dirichlet forms on $L^2(E,m)$ with Beurling-Deny triples $(\mathcal{E}^{(c)},J,k)$ and $(\mathcal{E}^{(c)'},J',k')$ respectively. Then $(\mathcal{E}',\mathcal{F}')\prec(\mathcal{E},\mathcal{F})$ if and only if $\mathcal{F}'\cap C_c\subset\mathcal{F}\cap C_c$, $J=J'$, $k=k'$ and $\mu^c_{<u,v>}={\mu'}^c_{<u,v>}$ for any $u,v\in\mathcal{F}'\cap C_c$ where ${\mu'}^c_{<u,v>}$ is defined similarly as Eq. (2.3) with respect to $(\mathcal{E}',\mathcal{F}')$.*

Proof. We first prove the "if" part. Note that the quasi-continuous versions of an appropriate function for $\mathcal{E}$ and $\mathcal{E}'$ are different, but for any $u,v\in\mathcal{F}'\cap C_c$, it is easy to verify that $\mathcal{E}(u,v)=\mathcal{E}'(u,v)$ under these assumptions. We only need to prove $\mathcal{F}'\subset\mathcal{F}$ and $\mathcal{E}(u,u)=\mathcal{E}'(u,u)$ for any $u\in\mathcal{F}'$ (then we have $\mathcal{E}(u,v)=\mathcal{E}'(u,v)$ for any $u,v\in\mathcal{F}'$ by using polarization formula).

For any $u\in\mathcal{F}'$, as $\mathcal{E}'$ is regular there exists a sequence $\{u_n\}\subset\mathcal{F}'\cap C_c$ such that $u_n\to u$ in ${\mathcal{E}'}_1^{\frac{1}{2}}$-norm. Because $\mathcal{E}_1(u_n,u_n)=\mathcal{E}'_1(u_n,u_n)$ is convergent to $\mathcal{E}'_1(u,u)$, u_n is

[a]See Theorem 3.2.1 of Ref. 1.
[b]See Sec. 1.4 of Ref. 1.

$\mathcal{E}_1$-Cauchy. Since $\mathcal{E}$ is closed, there exists $v \in \mathcal{F}$ such that $u_n \to v$ in $\mathcal{E}_1^{\frac{1}{2}}$-norm, especially $u_n \to v$ in $L^2(E,m)$. But $u_n \to u$ in $L^2(E,m)$ and hence $u = v \in \mathcal{F}$ and $\mathcal{E}(u,u) = \mathcal{E}'(u,u)$.

For the other direction, clearly $\mathcal{F}' \cap C_c(E) \subset \mathcal{F} \cap C_c(E)$. We only need to prove $J = J', k = k'$. In fact, if we have $J = J', k = k'$, then for any $u, v \in \mathcal{F}' \cap C_c(E)$, it holds that

$$\mathcal{E}^{(c)}(u,v) = \mathcal{E}^{(c)'}(u,v)$$

because of Eq. (2.1). Then we can deduces from Eq. (2.4) that

$$\int_E f d\mu^c_{<u,v>} = \int_E f d\mu'^c_{<u,v>}$$

for any $f \in \mathcal{F}' \cap C_c(E)$. Hence $\mu^c_{<u,v>} = \mu'^c_{<u,v>}$ by using the regularity of $(\mathcal{E}', \mathcal{F}')$.

To prove $J = J'$, let G_1, G_2 be two disjoint relatively compact open sets. For any $u, v \in \mathcal{F}' \cap C_c$ with $\text{supp}[u] \subset G_1, \text{supp}[v] \subset G_2$, we know from Eq. (1.1) and Eq. (2.2) that

$$\begin{aligned} -2\int u(x)v(y)J(dx,dy) &= \mathcal{E}(u,v) \\ &= \mathcal{E}'(u,v) = -2\int u(x)v(y)J'(dx,dy). \end{aligned}$$

As $\mathcal{E}'$ is regular, $\mathcal{F}' \cap C_c$ is dense in C_c. Then one can easily check that

$$J(G_1 \times G_2) = J'(G_1 \times G_2).$$

On the other hand, for any two disjoint compact sets K_1, K_2, one can find two sequences of relatively compact open sets $\{G_n^1\}, \{G_n^2\}$ such that $G_n^1 \cap G_n^2 = \varnothing$ for any n and $G_n^1 \downarrow K_1$, $G_n^2 \downarrow K_2$. It follows from $J(G_n^1 \times G_n^2) = J'(G_n^1 \times G_n^2)$ that $J(K_1 \times K_2) = J(K_1 \times K_2)$. We conclude that, for any two disjoint Borel sets B_1, B_2 we have $J(B_1 \times B_2) = J'(B_1 \times B_2)$. Then a simple argument gains directly that $J = J'$.

We now prove $k = k'$.

For any relatively compact open set G, there exists $u \in \mathcal{F}' \cap C_c$ such that $u|_G \equiv 1$. In fact, we can choose a function $u_0 \in C_c(E)$, such that $u_0|_{\bar{G}} \geqslant 2$. Since $\mathcal{F} \cap C_c(E)$ is dense in $C_c(E)$ with uniform norm, for ε small enough there exists $u_{0,\varepsilon} \in \mathcal{F} \cap C_c(E)$ such that $\|u_0 - u_{0,\varepsilon}\| < \varepsilon$. Let $\varphi_\varepsilon(t) := t - (\varepsilon \wedge t) \vee (-\varepsilon)$, and $\psi(t) := (0 \vee t) \wedge 1$, which are normal contractions. It is easy to check that $u = \psi(\varphi_\varepsilon(u_{0,\varepsilon})) \in \mathcal{F} \cap C_c(E)$ and equals to 1 on $\bar{G}$. Let v be any function in $\mathcal{F}' \cap C_c$ with $\text{supp}[v] \subset G$. Because $\mathcal{E}^{(c)}$ and $\mathcal{E}^{(c)'}$ are strongly local, $\mathcal{E}^{(c)}(u,v) = \mathcal{E}^{(c)'}(u,v) = 0$. As we have proved $J = J'$, it follows from Eq. (1.1) and Eq. (2.1) that

$$\int u(x)v(x)k(dx) = \int u(x)v(x)k'(dx),$$

for any $v \in \mathcal{F}' \cap C_c$ with $\text{supp}[v] \subset G$, $u|_G \equiv 1$. Then we have $\int v(x)k(dx) = \int v(x)k'(dx)$. By using the regularity of $\mathcal{E}'$, it is easy to get $k = k'$ which completes the proof. □

The following corollaries are interesting and not difficult to verify.

Corollary 2.1. *Let $(\mathcal{E}, \mathcal{F})$, $(\mathcal{E}', \mathcal{F}')$ be two regular Dirichlet forms on $L^2(E,m)$ and $(\mathcal{E}', \mathcal{F}') \prec (\mathcal{E}, \mathcal{F})$.*

(1) If $(\mathcal{E}, \mathcal{F})$ is local, so is $(\mathcal{E}', \mathcal{F}')$. Moreover if $(\mathcal{E}, \mathcal{F})$ is strongly local, so is $(\mathcal{E}', \mathcal{F}')$ too.

(2) *If* $(\mathcal{E},\mathcal{F})$ *is a pure-jumping type Dirichlet form which means* $\mathcal{E}^{(c)}=0$, $k=0$, *then* $(\mathcal{E}',\mathcal{F}')$ *is pure-jumping type too.*

(3) *If* $(\mathcal{E},\mathcal{F})$ *is a pure-killing Dirichlet form which means* $\mathcal{E}^{(c)}=0$, $J=0$, *then it must be* $\mathcal{F}=L^2(E,m+k)$ *and* $(\mathcal{E}',\mathcal{F}')=(\mathcal{E},\mathcal{F})$.

(4) *If the jumping part or killing part of* $(\mathcal{E},\mathcal{F})$ *does not disappear, it also does not disappear for that of* $(\mathcal{E}',\mathcal{F}')$.

Proof. Clearly (1)(2) and (4) hold by Theorem 2.1.

(3) When $\mathcal{E}^{(c)}=0$ and $J=0$ it is obvious that $\mathcal{E}(u,u)=\int u^2dk$ and $\|u\|_{\mathcal{E}_1}:=\mathcal{E}(u,u)^{\frac{1}{2}}=\|u\|_{L^2(m+k)}$ for any $u\in\mathcal{F}$. By the regularity of $\mathcal{E}$, $\mathcal{F}\cap C_c$ is dense in C_c with uniform norm, then it is also dense in $L^2(E,m+k)$ with norm $\|\cdot\|_{L^2(m+k)}=\|\cdot\|_{\mathcal{E}_1}$. For any $u\in L^2(E,m+k)$, there exists a sequence u_n in $\mathcal{F}\cap C_c$ such that $\|u_n-u\|_{\mathcal{E}_1}\to 0$. Hence $u\in\mathcal{F}$ and $\mathcal{F}=L^2(E,m+k)$. It is also easy to check that $\mathcal{F}'=L^2(E,m+k)$ which means $(\mathcal{E}',\mathcal{F}')=(\mathcal{E},\mathcal{F})$. That completes the proof. □

Before showing another corollary, we should give some necessary statements about the Lévy-type Dirichlet forms. A Lévy-type Dirichlet form on $L^2(\mathbf{R}^d,dx)$ is generated by a symmetric convolution semigroup $\{\nu_t,t>0\}$ on $\mathbf{R}^d$. Its corresponding Markov process is just the so-called symmetric Lévy process. The celebrated Lévy-Khinchin formula under the symmetric assumption reads as follows:

$$\hat{\nu}_t(x)=\exp\{-t\psi(x)\} \tag{2.5}$$

where

$$\psi(x)=\frac{1}{2}(Sx,x)+\int_{\mathbf{R}^d}(1-\cos(x,y))j(dy), \tag{2.6}$$

S is a non-negative definite $d\times d$ symmetric matrix and j is a symmetric Radon measure on $\mathbf{R}^d\setminus\{0\}$ such that $\int_{\mathbf{R}^d\setminus\{0\}}1\wedge|x|^2j(dx)<\infty$. The corresponding regular Dirichlet form is

$$\begin{aligned}\mathcal{D}[\mathcal{E}]&=\{u\in L^2(\mathbf{R}^d):\int_{\mathbf{R}^d}|\hat{u}|(x)^2\psi(x)dx<\infty\},\\ \mathcal{E}(u,v)&=\int_{\mathbf{R}^d}\hat{u}(x)\bar{\hat{v}}(x)\psi(x)dx\quad\forall u,v\in\mathcal{D}[\mathcal{E}],\end{aligned} \tag{2.7}$$

where $\hat{u}$ means the Fourier transform of u. Note that the symmetric convolution semigroup, as well as the regular Dirichlet form, are characterized by the pair (S,j).

The strongly local part of $(\mathcal{E},\mathcal{D}[\mathcal{E}])$ which is determinate by S is:

$$\mathcal{E}^{\mathrm{S}}(u,u)=\frac{1}{2}\int(S\nabla u,\nabla u)dx \tag{2.8}$$

and the jumping part can be rewritten as

$$\mathcal{E}^j(u,u)=\frac{1}{2}\int_{\mathbf{R}^d\times(\mathbf{R}^d\setminus\{0\})}(u(x+y)-u(x))^2j(dy)dx. \tag{2.9}$$

Of course $\mathcal{E}=\mathcal{E}^{\mathrm{S}}+\mathcal{E}^j$. On the other hand, as S is a non-negative definite symmetric matrix there exists an orthogonal matrix P, i.e. $P' = P^{-1}$, such that

$$P'SP = \begin{pmatrix} \lambda_1 & 0 & \cdots & \cdots & \cdots & 0 \\ 0 & \cdots & 0 & \cdots & \cdots & 0 \\ 0 & 0 & \lambda_r & 0 & \cdots & 0 \\ 0 & \cdots & \cdots & 0 & \cdots & 0 \\ 0 & \cdots & \cdots & \cdots & \cdots & 0 \\ 0 & \cdots & \cdots & \cdots & \cdots & 0 \end{pmatrix}_{d\times d} = \mathrm{diag}\{\lambda_1, \cdots, \lambda_r, 0, \cdots, 0\}. \tag{2.10}$$

It is just the diagonalization of S with the rank r and $\lambda_1, \cdots, \lambda_r$ are its positive eigenvalues, i.e. $\lambda_i > 0$ for any $1 \leqslant i \leqslant r$. Note that $S \neq 0$ means $r \geq 1$. For any $u \in \mathcal{D}[\mathcal{E}]$, suppose $u(x) = \varphi(Px)$ for a unique function φ. Then

$$\begin{aligned}\mathcal{E}^{\mathrm{S}}(u,u) =& \frac{1}{2}\int (S\nabla\varphi(Px), \nabla\varphi(Px))dx \\ =& \frac{1}{2}\int (P'SP\nabla_y\varphi(y), \nabla_y\varphi(y))dP'y \\ =& \frac{1}{2}\int (P'SP\nabla_y\varphi(y), \nabla_y\varphi(y))dy,\end{aligned}$$

hence,

$$\mathcal{E}^{\mathrm{S}}(u,u) = \frac{1}{2}\sum_{i=1}^{r}\lambda_i \int (\frac{\partial\varphi}{\partial y_i})^2 dy. \tag{2.11}$$

Corollary 2.2. *If $(\mathcal{E}, \mathcal{F})$ is a non-pure-jumping Lévy-type Dirichlet form, i.e. the scatter coefficient $S \neq 0$ and $(\mathcal{E}', \mathcal{F}') \prec (\mathcal{E}, \mathcal{F})$. Then $(\mathcal{E}', \mathcal{F}')$ mustn't be pure-jumping type, in other words, the strongly local part of $(\mathcal{E}', \mathcal{F}')$ never disappears.*

Proof. Suppose that $(\mathcal{E}, \mathcal{F})$ is just the Lévy Dirichlet form Eq. (2.7) such that $S \neq 0$, i.e. $r \geq 1$ and $(\mathcal{E}', \mathcal{F}') \prec (\mathcal{E}, \mathcal{F})$. If $(\mathcal{E}', \mathcal{F}')$ is a pure-jumping Dirichlet form, then $\mathcal{E}^{\mathrm{S}}(u,u) = 0$ for any $u \in \mathcal{F}'$. For any $u(x) = \varphi(Px) \in \mathcal{F}' \cap C_{\mathrm{c}}$, it follows from Eq. (2.11) that

$$\int (\frac{\partial\varphi}{\partial y_1})^2 dy = 0$$

and hence $\frac{\partial\varphi}{\partial y_1} = 0$, a.e. Note that $\frac{\partial\varphi}{\partial y_1}$ is the weak derivative of φ.

Set $y = (y_1, \bar{y}) \in \mathbf{R}^d$, i.e. $\bar{y} \in \mathbf{R}^{d-1}$ and $\varphi_{\bar{y}}(y_1) := \varphi(y_1, \bar{y})$. Take any $\psi_1(y_1) \in C_{\mathrm{c}}^\infty(\mathbf{R}), \psi_2(\bar{y}) \in C_{\mathrm{c}}^\infty(\mathbf{R}^{d-1})$, and write $\psi(y) := \psi_1(y_1)\psi_2(\bar{y}) \in C_{\mathrm{c}}^\infty(\mathbf{R}^d)$. By the definition of weak derivative,

$$0 = (\frac{\partial\varphi}{\partial y_1}, \psi) = -(\varphi, \psi_1'\psi_2) = -\int \psi_2(\bar{y})d\bar{y}\int \varphi_{\bar{y}}(y_1)\psi_1'(y_1)dy_1.$$

Because $\int \varphi_{\bar{y}}(y_1)\psi_1'(y_1)dy_1$ is continuous with respect to $\bar{y}$ and $\psi_2(\bar{y})$ is arbitrary in $C_{\mathrm{c}}^\infty(\mathbf{R}^{d-1})$, we have

$$\int \varphi_{\bar{y}}(y_1)\psi_1'(y_1)dy_1 = 0, \text{ for any fix } \bar{y} \in \mathbf{R}^{d-1}.$$

Then $0 = \int \varphi_{\bar{y}}(y_1)\psi_1'(y_1)dy_1 = -(\varphi_{\bar{y}}'(y_1), \psi_1(y_1))$. It follows from Lemma 3.31 and Corollary 3.32 of Ref. 8 that $\varphi_{\bar{y}}(y_1)$ is independent of y_1. But $\varphi_{\bar{y}}$ is continuous, hence

$\varphi_{\bar{y}}(y_1) \equiv$ const. for each $\bar{y}$ and we have

$$\varphi(y_1, \bar{y}) = \bar{\varphi}(\bar{y}),$$

for some $\bar{\varphi} \in C_c(\mathbf{R}^{d-1})$, which implies $\varphi \equiv 0$ because of $\varphi \in C_c(\mathbf{R}^d)$. This contradicts that $\mathcal{F}' \cap C_c(\mathbf{R}^d)$ is dense in $C_c(\mathbf{R}^d)$ with uniform norm. We then conclude that $(\mathcal{E}', \mathcal{F}')$ mustn't be pure-jumping type. □

2.2. *Probabilistic transformations*

An m-measurable set $A \subset E$ is said to be (T_t)-invariant if $T_t(1_A f) = 1_A \cdot T_t f$, m-a.e for any $f \in L^2(E)$ and $t > 0$. A Dirichlet form with semigroup $(T_t)_{t>0}$ is called *m-irreducible* if every (T_t)-invariant set A is m-trivial, i.e. $m(A) = 0$, or $m(E \backslash A) = 0$. In this section we will discuss several celebrated probabilistic transformations.

2.2.1. *Spatial homeomorphous transformation*

Let $(\mathcal{E}, \mathcal{F})$ be a regular Dirichlet form on $L^2(E, m)$. $\hat{E}$ is another topological space and there exists a homeomorphism mapping $j : E \to \hat{E}$. Then $\hat{E}$ is also a locally compact separable metric space. Set $\hat{m} := m \circ j^{-1}$ which is a Radon measure fully supported on $\hat{E}$. Define

$$\begin{aligned} \hat{\mathcal{F}} &:= \{\hat{u} \in L^2(\hat{E}, \hat{m}) : \hat{u} \circ j \in \mathcal{F}\}, \\ \hat{\mathcal{E}}(\hat{u}, \hat{v}) &:= \mathcal{E}(\hat{u} \circ j, \hat{v} \circ j), \quad \forall u, v \in \hat{\mathcal{F}}. \end{aligned} \tag{2.12}$$

We denote $(\hat{\mathcal{E}}, \hat{\mathcal{F}})$ by $j(\mathcal{E}, \mathcal{F})$. Note that $(\mathcal{E}, \mathcal{F}) = j^{-1}(\hat{\mathcal{E}}, \hat{\mathcal{F}})$.

Lemma 2.1. *Let $L^2(E, m)$, $L^2(\hat{E}, \hat{m})$, $(\mathcal{E}, \mathcal{F})$ and $(\hat{\mathcal{E}}, \hat{\mathcal{F}})$ be above.*

(1) $(\hat{\mathcal{E}}, \hat{\mathcal{F}})$ is a regular Dirichlet form on $L^2(\hat{E}, \hat{m})$.

(2) $(\hat{\mathcal{E}}, \hat{\mathcal{F}})$ is irreducible if and only if $(\mathcal{E}, \mathcal{F})$ is irreducible.

(3) B is $\mathcal{E}$-polar if and only if $\hat{B} = j(B)$ is $\hat{\mathcal{E}}$-polar, u is $\mathcal{E}$-quasi-continuous if and only if $\hat{u} = u \circ j^{-1}$ is $\hat{\mathcal{E}}$-quasi-continuous.

Proof. The proofs of (1)(3) are trivial and we omit them.

For (2) we need only to prove that if A is $(\mathcal{E}, \mathcal{F})$-invariant, $\hat{A} = j(A)$ is $(\hat{\mathcal{E}}, \hat{\mathcal{F}})$-invariant. This can be checked by Propsiotion 2.1.6 of Ref. 2 and for any $\hat{u} \in \hat{\mathcal{F}}$,

$$1_{\hat{A}} \cdot \hat{u} = 1_A \circ j^{-1} \cdot u \circ j^{-1} = (1_A \cdot u) \circ j^{-1},$$

where $u = \hat{u} \circ j \in \mathcal{F}$. □

Proposition 2.1. *$(\mathcal{E}, \mathcal{F})$ and $(\hat{\mathcal{E}}, \hat{\mathcal{F}})$ are two Dirichlet forms with the same structure of regular subspaces with respect to the spatial homeomorphic transformation j. Then for any $(\hat{\mathcal{E}}', \hat{\mathcal{F}}') \prec (\hat{\mathcal{E}}, \hat{\mathcal{F}})$, there exists a unique $(\mathcal{E}', \mathcal{F}') \prec (\mathcal{E}, \mathcal{F})$ such that $(\hat{\mathcal{E}}', \hat{\mathcal{F}}') = j(\mathcal{E}', \mathcal{F}')$.*

Proof. Note that j is reversible. So the conclusions are trivial because $(\mathcal{E}', \mathcal{F}') \prec (\mathcal{E}, \mathcal{F})$ if and only if $(\hat{\mathcal{E}}', \hat{\mathcal{F}}') := j(\mathcal{E}', \mathcal{F}') \prec (\hat{\mathcal{E}}, \hat{\mathcal{F}})$ □

Combining with Lemma 2.1(2), we have

Corollary 2.3. *Let $(\mathcal{E}, \mathcal{F})$ and $(\hat{\mathcal{E}}, \hat{\mathcal{F}})$ be above. For any irreducible $(\hat{\mathcal{E}}', \hat{\mathcal{F}}') \prec (\hat{\mathcal{E}}, \hat{\mathcal{F}})$, there exists a unique irreducible $(\mathcal{E}', \mathcal{F}') \prec (\mathcal{E}, \mathcal{F})$ such that $(\hat{\mathcal{E}}', \hat{\mathcal{F}}') = j(\mathcal{E}', \mathcal{F}')$.*

2.2.2. *Time changes with full quasi support*

Assume that $(\mathcal{E},\mathcal{F})$ is a regular Dirichlet form on $L^2(E,m)$ and X is its corresponding Hunt process. Denote the set of Radon smooth measures on E with full quasi support by S_E. Let $\mu \in S_E$, $(A_t)_{t\geq 0}$ its corresponding positive continuous additive functional and $(\tau_t)_{t\geq 0}$ the inverse of $(A_t)_{t\geq 0}$. Then the time-changed process $(X_{\tau_t})_{t\geq 0}$ is μ-symmetric and its Dirichlet form $(\check{\mathcal{E}},\check{\mathcal{F}})$ on $L^2(E,\mu)$ can be presented as

$$\begin{aligned}\check{\mathcal{F}} &= \mathcal{F}_{\mathrm{e}} \cap L^2(E,\mu),\\ \check{\mathcal{E}}(u,u) &= \mathcal{E}(u,u), \quad u \in \check{\mathcal{F}},\end{aligned} \tag{2.13}$$

where $\mathcal{F}_{\mathrm{e}}$ is the extended space of $(\mathcal{E},\mathcal{F})$. Moreover $(\check{\mathcal{E}},\check{\mathcal{F}})$ is regular as μ is Radon and its extended space $\check{\mathcal{F}}_{\mathrm{e}} = \mathcal{F}_{\mathrm{e}}$. Note that the quasi-notions (nests, polar sets and quasi-continuous functions) of $(\mathcal{E},\mathcal{F})$ and $(\check{\mathcal{E}},\check{\mathcal{F}})$ are equivalent. We denote $(\check{\mathcal{E}},\check{\mathcal{F}})$ by $\mu(\mathcal{E},\mathcal{F})$ if $\mu \in S_E$. The above procedures are reversible because m is a Radon smooth measure with quasi support E with respect to $(\check{\mathcal{E}},\check{\mathcal{F}})$. In other words, $(\mathcal{E},\mathcal{F}) = m(\check{\mathcal{E}},\check{\mathcal{F}})$.

Similarly $\check{S}_E$ (resp. S'_E) represents the set of all Radon smooth measures on E with full quasi support with respect to $(\check{\mathcal{E}},\check{\mathcal{F}})$ (resp. $(\mathcal{E}',\mathcal{F}')$ which is another Dirichlet form).

Lemma 2.2. *Let $(\mathcal{E},\mathcal{F})$ and $(\mathcal{E}',\mathcal{F}')$ be two Dirichlet forms on $L^2(E,m)$.*

(1) It holds that $\mathcal{F}' \subset \mathcal{F}$ if and only if $\mathcal{F}'_{\mathrm{e}} \subset \mathcal{F}_{\mathrm{e}}$. In particular, $\mathcal{F}' \subsetneqq \mathcal{F}$ if and only if $\mathcal{F}'_{\mathrm{e}} \subsetneqq \mathcal{F}_{\mathrm{e}}$.

(2) If $(\mathcal{E}',\mathcal{F}') \prec (\mathcal{E},\mathcal{F})$ and $\mu \in S_E$, then $\mu \in S'_E$.

(3) Assume $(\mathcal{E},\mathcal{F})$ is regular on $L^2(E,m)$ and $\mu \in S_E$. Then $(\mathcal{E},\mathcal{F})$ is irreducible if and only if $\mu(\mathcal{E},\mathcal{F})$ is irreducible.

Proof. (1) The second assertion follows directly from the first one. For the first one, the if part is because $\mathcal{F} = \mathcal{F}_{\mathrm{e}} \cap L^2(E,m)$ (resp. $\mathcal{F}' = \mathcal{F}'_{\mathrm{e}} \cap L^2(E,m)$). The only if part is obvious by the definition of extended space.

(2) By Remark 1.1, any Cap$'$-zero-capacity set is also of Cap-zero-capacity. As μ is Radon and smooth with respect to $(\mathcal{E},\mathcal{F})$, μ is also smooth with respect to $(\mathcal{E}',\mathcal{F}')$. It suffices to prove that the quasi support of μ with respect to $(\mathcal{E}',\mathcal{F}')$ is E. As $\mu \in S_E$, it holds that $u = 0$ μ-a.e. if and only if $u = 0$ $\mathcal{E}$-q.e. for any $\mathcal{E}$-quasi continuous function $u \in \mathcal{F}$ by Theorem 3.3.5 of Ref. 2. For any $\mathcal{E}'$-quasi continuous $v \in \mathcal{F}'$, it also holds that $v \in \mathcal{F}$ and v is $\mathcal{E}$-quasi continuous. Therefore if $v = 0$ μ-a.e., then $u = 0$ $\mathcal{E}$-q.e, and hence also m-a.e. But v is $\mathcal{E}'$-quasi continuous, we have $v = 0$ $\mathcal{E}'$-q.e. On the other hand, it follows from $v = 0$ $\mathcal{E}'$-q.e. and $\mu \in S_E$ that $v = 0$ μ-a.e. In other words, $v = 0$ μ-a.e. if and only if $v = 0$ $\mathcal{E}'$-q.e for any $\mathcal{E}'$-quasi continuous function $v \in \mathcal{F}'$. Hence $\mu \in S'_E$ by using Theorem 3.3.5 of Ref. 2 again.

(3) It is clear by using Eq. (2.13), $\check{\mathcal{F}}_{\mathrm{e}} = \mathcal{F}_{\mathrm{e}}$ and Proposition 2.1.6 of Ref. 2. □

Remark 2.1. By Lemma 2.2(1), the Dirichlet spaces $\mathcal{F}$ and $\mathcal{F}'$ appeared in Definition 1.1 can be replaced by their extended spaces $\mathcal{F}_{\mathrm{e}}$ and $\mathcal{F}'_{\mathrm{e}}$ respectively.

Proposition 2.2. *If $(\mathcal{E},\mathcal{F})$ is a regular Dirichlet form on $L^2(E,m)$ and $\mu \in S_E$, then $(\check{\mathcal{E}},\check{\mathcal{F}}) = \mu(\mathcal{E},\mathcal{F})$ has the same structure of regular subspaces as $(\mathcal{E},\mathcal{F})$ with respect to μ in the meaning that for any $(\check{\mathcal{E}}',\check{\mathcal{F}}') \prec (\check{\mathcal{E}},\check{\mathcal{F}})$, there exists a unique $(\mathcal{E}',\mathcal{F}') \prec (\mathcal{E},\mathcal{F})$ such that $(\check{\mathcal{E}}',\check{\mathcal{F}}') = \mu(\mathcal{E}',\mathcal{F}')$.*

Proof. Clearly $m \in \check{S}_E$, hence it follows from Lemma 2.2(2) that $m \in \check{S}'_E$. Set $(\mathcal{E}', \mathcal{F}') := m(\check{\mathcal{E}}', \check{\mathcal{F}}')$. As

$$\mathcal{F}_e = \check{\mathcal{F}}_e, \quad \mathcal{F}'_e = \check{\mathcal{F}}'_e,$$

it follows from Lemma 2.2(1) that $(\mathcal{E}', \mathcal{F}') \prec (\mathcal{E}, \mathcal{F})$. By the similar method, it is easy to check that if another $(\mathcal{E}'', \mathcal{F}'') \prec (\mathcal{E}, \mathcal{F})$ such that $(\check{\mathcal{E}}', \check{\mathcal{F}}') = \mu(\mathcal{E}'', \mathcal{F}'')$, it must hold that $(\mathcal{E}'', \mathcal{F}'') = m(\check{\mathcal{E}}', \check{\mathcal{F}}') = (\mathcal{E}', \mathcal{F}')$. The uniqueness is proved. □

Combining with Lemma 2.2(3), we have

Corollary 2.4. *Under the conditions of Proposition 2.2, if $(\check{\mathcal{E}}', \check{\mathcal{F}}')$ is irreducible, then the corresponding $(\mathcal{E}', \mathcal{F}')$ is also irreducible.*

2.2.3. *Killing and resurrected transformations*

Let $(\mathcal{E}, \mathcal{F})$ be a regular Dirichlet form on $L^2(E, m)$ with no killing inside (or zero killing measure) and X its corresponding Hunt process. Assume that the functions in $\mathcal{F}$ always take their quasi-continuous versions. Take a positive Radon measure k on E charging no $\mathcal{E}$-zero-capacity sets and the perturbed (killing) Dirichlet form $(\mathcal{E}^k, \mathcal{F}^k)$ of $(\mathcal{E}, \mathcal{F})$ with respect to k on $L^2(E, m)$ is defined by

$$\begin{aligned} \mathcal{F}^k &:= \mathcal{F} \cap L^2(E, k), \\ \mathcal{E}^k(u, v) &:= \mathcal{E}(u, v) + (u, v)_k, \quad u, v \in \mathcal{F}^k. \end{aligned} \tag{2.14}$$

By Theorem 6.1.2 of Ref. 1, $(\mathcal{E}^k, \mathcal{F}^k)$ is a regular Dirichlet form whose corresponding symmetric Markov process is the canonical subprocess X^k of X relative to the multiplicative functional $(e^{-A_t})_{t\geq 0}$ where A_t is the corresponding PCAF of smooth measure k. In particular, the killing measure of $(\mathcal{E}^k, \mathcal{F}^k)$ is just the Radon measure k. Note that the perturbed Dirichlet form has the equivalent $\mathcal{E}$-quasi notions as the original one by Theorem 5.1.4 of Ref. 2, i.e. $\mathcal{E}$-polar (resp. $\mathcal{E}$-nest, $\mathcal{E}$-quasi-continuous) and $\mathcal{E}^k$-polar (resp. $\mathcal{E}^k$-nest, $\mathcal{E}^k$-quasi-continuous) are equivalent. Moreover, any special standard core of $(\mathcal{E}, \mathcal{F})$ remains to be a special standard core of $(\mathcal{E}^k, \mathcal{F}^k)$.

The opposite procedure is called "resurrection". Let $(\mathcal{E}, \mathcal{F})$ be a regular Dirichlet form on $L^2(E, m)$ with killing measure $k \neq 0$. For any $u, v \in \mathcal{F}_e$, define

$$\mathcal{E}^{\mathrm{res}}(u, v) := \mathcal{E}^{(c)}(u, v) + \frac{1}{2}\int_{E\times E\setminus d} (u(x) - u(y))(v(x) - v(y))J(dxdy) \tag{2.15}$$

and $m_k := m + k$ where $(\mathcal{E}^{(c)}, J, k)$ is the Beurling-Deny triple of $(\mathcal{E}, \mathcal{F})$. Then it follows from

$$\begin{aligned} \mathcal{F} &= \mathcal{F}_e \cap L^2(E, m_k), \\ \mathcal{E}_1(u, v) &= \mathcal{E}^{\mathrm{res}}(u, v) + (u, v)_{m_k}, \quad u, v \in \mathcal{F}, \end{aligned}$$

that $(\mathcal{E}^{\mathrm{res}}, \mathcal{F})$ is a regular Dirichlet form on $L^2(E, m_k)$. Let $(\mathcal{E}^{\mathrm{res}}, \mathcal{F}^{\mathrm{res}}_e)$ be the extended Dirichlet space of $(\mathcal{E}^{\mathrm{res}}, \mathcal{F})$. Define

$$\mathcal{F}^{\mathrm{res}} = \mathcal{F}^{\mathrm{res}}_e \cap L^2(E, m), \tag{2.16}$$

then $(\mathcal{E}^{\mathrm{res}}, \mathcal{F}^{\mathrm{res}})$ is the so-called resurrected Dirichlet form of $(\mathcal{E}, \mathcal{F})$ which has no killing inside. It is regular on $L^2(E, m)$ and shares the same $\mathcal{E}$-quasi notions with $(\mathcal{E}, \mathcal{F})$. Moreover any special standard core of $(\mathcal{E}, \mathcal{F})$ is a core of $(\mathcal{E}^{\rm res}, \mathcal{F}^{\rm res})$. Note that the extended Dirichlet space of $(\mathcal{E}^{\rm res}, \mathcal{F}^{\rm res})$ on $L^2(E,m)$ is just $(\mathcal{E}^{\rm res}, \mathcal{F}^{\rm res}_{\rm e})$. For more details, see Ref. 1, 2.

Killing and resurrected transformations are reciprocal in the following sense: for a regular Dirichlet form with no killing inside, kill it firstly by a Radon smooth measure and then resurrect. The final Dirichlet form is the same as the original one. On the contrary, for a regular Dirichlet form with killing measure $k \neq 0$, resurrect it firstly and then kill it by k. The final Dirichlet form is also the same as the original one.

We are now ready to prove that the killing and resurrected transformations are both equivalent relationships between Dirichlet forms, in the meaning that they have the same structures holding regular subspaces between the transformed Dirichlet form by killing or resurrection and the original Dirichlet form. The following lemma is trivial and we omit its proof.

Lemma 2.3. *If $(\mathcal{E}', \mathcal{F}') \prec (\mathcal{E}, \mathcal{F})$ and k is a Radon smooth measure with respect to $(\mathcal{E}, \mathcal{F})$, then k is also a Radon smooth measure with respect to $(\mathcal{E}', \mathcal{F}')$.*

Lemma 2.4. *The following assertions about the irreducibility hold.*

(1) Let $(\mathcal{E}, \mathcal{F})$ be a regular Dirichlet form on $L^2(E,m)$ with no killing inside and k a Radon smooth measure on E. Then the perturbed Dirichlet form $(\mathcal{E}^k, \mathcal{F}^k)$ is irreducible if and only if $(\mathcal{E}, \mathcal{F})$ is irreducible.

(2) Let $(\mathcal{E}, \mathcal{F})$ be a regular Dirichlet form on $L^2(E,m)$ with killing measure $k \neq 0$. Then $(\mathcal{E}, \mathcal{F})$ is irreducible if and only if $(\mathcal{E}^{\rm res}, \mathcal{F}^{\rm res})$ is irreducible.

Proof. (1) We need only to prove $\mathcal{E}$-invariant sets are equivalent to $\mathcal{E}^k$-invariant sets. In fact, let A be an $\mathcal{E}$-invariant set, then $1_A \cdot u \in \mathcal{F}$ for any $u \in \mathcal{F}^k \subset \mathcal{F}$. Hence $1_A \cdot u \in \mathcal{F}^k$ because $1_A \cdot u \in L^2(E,k)$. By Proposition 2.1.6 of Ref. 2 and

$$(u,u)_k = (1_A \cdot u, 1_A \cdot u)_k + (1_{A^c} \cdot u, 1_{A^c} \cdot u)_k,$$

we have A is an $\mathcal{E}^k$-invariant set. On the other hand, assume B is an $\mathcal{E}^k$-invariant set, then $u \in \mathcal{F} \cap C_c(E)$ implies $u \in \mathcal{F}^k$, hence $1_B \cdot u \in \mathcal{F}^k \subset \mathcal{F}$ and

$$\mathcal{E}^k(u,u) = \mathcal{E}^k(1_B \cdot u, 1_B \cdot u) + \mathcal{E}^k(1_{B^c} \cdot u, 1_{B^c} \cdot u).$$

This shows that $u \in \mathcal{F} \cap C_c(E)$ implies $1_B \cdot u \in \mathcal{F}$ and

$$\mathcal{E}(u,u) = \mathcal{E}(1_B \cdot u, 1_B \cdot u) + \mathcal{E}(1_{B^c} \cdot u, 1_{B^c} \cdot u)$$

because of $(u,u)_k = (1_B \cdot u, 1_B \cdot u)_k + (1_{B^c} \cdot u, 1_{B^c} \cdot u)_k$. From this equation and the regularity of $(\mathcal{E}, \mathcal{F})$, we can deduce that $u \in \mathcal{F}$ implies $1_B \cdot u \in \mathcal{F}$ and

$$\mathcal{E}(u,u) = \mathcal{E}(1_B \cdot u, 1_B \cdot u) + \mathcal{E}(1_{B^c} \cdot u, 1_{B^c} \cdot u).$$

(2) They are just the former conclusions in an opposite direction. □

Proposition 2.3. *The following conclusions about the structure of the regular subspaces hold.*

(1) Let $(\mathcal{E}, \mathcal{F})$ be a regular Dirichlet form on $L^2(E,m)$ with no killing inside and k a Radon smooth measure on E, then the perturbed Dirichlet form $(\mathcal{E}^k, \mathcal{F}^k)$ and $(\mathcal{E}, \mathcal{F})$ are two Dirichlet forms with the same structure of regular subspaces with respect to k in the

meaning that, for any $(\mathcal{E}'^k, \mathcal{F}'^k) \prec (\mathcal{E}^k, \mathcal{F}^k)$, there exists a unique $(\mathcal{E}', \mathcal{F}') \prec (\mathcal{E}, \mathcal{F})$ such that $(\mathcal{E}'^k, \mathcal{F}'^k)$ is the perturbed Dirichlet form of $(\mathcal{E}, \mathcal{F})$ with respect to k.

(2) If $(\mathcal{E}, \mathcal{F})$ is a regular Dirichlet form on $L^2(E, m)$ with killing measure $k \neq 0$, then $(\mathcal{E}^{\text{res}}, \mathcal{F}^{\text{res}})$ and $(\mathcal{E}, \mathcal{F})$ are two Dirichlet forms with the same structure of regular subspaces with respect to the resurrection in the meaning that, for any $(\mathcal{E}^{\text{res}'}, \mathcal{F}^{\text{res}'}) \prec (\mathcal{E}^{\text{res}}, \mathcal{F}^{\text{res}})$, there exists a unique $(\mathcal{E}', \mathcal{F}') \prec (\mathcal{E}, \mathcal{F})$ such that $(\mathcal{E}^{\text{res}'}, \mathcal{F}^{\text{res}'})$ is the resurrected Dirichlet form of $(\mathcal{E}', \mathcal{F}')$.

Proof. (1) Take any $(\mathcal{E}', \mathcal{F}') \prec (\mathcal{E}, \mathcal{F})$, then k is also a Radon smooth measure with respect to $(\mathcal{E}', \mathcal{F}')$ by Lemma 2.3. It follows from Eq. (2.14) that

$$(\mathcal{E}'^k, \mathcal{F}'^k) \prec (\mathcal{E}^k, \mathcal{F}^k).$$

We have known that $(\mathcal{E}^k, \mathcal{F}^k)$ is regular with killing measure k and its resurrected Dirichlet form is just $(\mathcal{E}, \mathcal{F})$. It suffices to prove that for any $(\mathcal{E}^{k'}, \mathcal{F}^{k'}) \prec (\mathcal{E}^k, \mathcal{F}^k)$, the resurrected Dirichlet form of $(\mathcal{E}^{k'}, \mathcal{F}^{k'})$ is a regular subspace of $(\mathcal{E}, \mathcal{F})$. In fact, $(\mathcal{E}^{k',\text{res}}, \mathcal{F}^{k'}) \prec (\mathcal{E}^{k,\text{res}}, \mathcal{F}^k)$ on $L^2(E, m_k)$, where $\mathcal{E}^{k,\text{res}}$ and $\mathcal{E}^{k',\text{res}}$ are the resurrected forms of $\mathcal{E}^k$ and $\mathcal{E}^{k'}$ defined by Eq. (2.15) respectively. By Lemma 2.2(1) and Eq. (2.16)

$$\mathcal{F}_{\text{e}}^{k',\text{res}} \subset \mathcal{F}_{\text{e}}^{k,\text{res}}, \quad \mathcal{F}^{k',\text{res}} \subset \mathcal{F}^{k,\text{res}} = \mathcal{F}.$$

Hence

$$(\mathcal{E}^{k',\text{res}}, \mathcal{F}^{k',\text{res}}) \prec (\mathcal{E}, \mathcal{F}).$$

(2) We need only to change the order of the two steps in the first assertion. □

2.2.4. *Conclusions and examples*

From the above discussions, we could summarize that:

Theorem 2.2. *Let $(\mathcal{E}, \mathcal{F})$ be a regular Dirichlet form on $L^2(E, m)$ and T a transformation described above, i.e. T is one of the transformations including spatial homeomorphic transformation, time change with full quasi support, killing and resurrection. Define $(\tilde{\mathcal{E}}, \tilde{\mathcal{F}}) := T(\mathcal{E}, \mathcal{F})$, which is the transformed Dirichlet form of $(\mathcal{E}, \mathcal{F})$ by T. Then $(\tilde{\mathcal{E}}, \tilde{\mathcal{F}})$ and $(\mathcal{E}, \mathcal{F})$ have the same structure of regular subspaces with respect to the transformation T in the meaning that for any $(\tilde{\mathcal{E}}', \tilde{\mathcal{F}}') \prec (\tilde{\mathcal{E}}, \tilde{\mathcal{F}})$, there exists a unique $(\mathcal{E}', \mathcal{F}') \prec (\mathcal{E}, \mathcal{F})$ such that $(\tilde{\mathcal{E}}', \tilde{\mathcal{F}}') = T(\mathcal{E}', \mathcal{F}')$. Conversely, it holds that $(\mathcal{E}, \mathcal{F}) = T^{-1}(\tilde{\mathcal{E}}, \tilde{\mathcal{F}})$ and they also have the same structure of regular subspaces with respect to the inverse transformation T^{-1} in the similar sense.*

In Corollary 2.1(3), we have proved that if the Dirichlet form $(\mathcal{E}, \mathcal{F})$ is pure-killing type, i.e. $\mathcal{E}(u, u) = (u, u)_k$ for some Radon measure k, then it has no proper regular subspaces. In addition, for general Dirichlet forms, the killing part does not play a role in generating a proper regular subspaces because it follows from Sec. 2.2.3 that the resurrected Dirichlet form is equivalent to the original one up to the resurrection. In other words, adding or erasing the killing part of a Dirichlet form does not really change the structure of its regular subspaces. Hence it is not necessary to consider the killing part for the discussions about the regular subspaces of a Dirichlet form. We assume all the Dirichlet forms appeared in the sequel have no killing inside unless otherwise stated.

It is discussed in Ref. 3 that how to characterize the regular subspaces of one-dimensional (reflecting) Brownian motion. By Theorem 2.1 every corresponding Markov process of its regular subspace must be a diffusion without killing inside. If the diffusion is irreducible, it can be characterized by a scale function s which is uniquely determined up to a linear transformation. The additional condition on s is that $s' = 0$ or 1 a.e. Using the above theorem we can give a simple characterization to the regular subspaces of one-dimensional diffusions.

Example 2.1. Let $I \subset \mathbf{R}$ be an interval and $(\mathcal{E}, \mathcal{F})$ an irreducible strongly local Dirichlet form on $L^2(I, m)$ whose corresponding Markov process is an irreducible diffusion $(X_t)_{t\geq 0}$. Write $I = \langle a, b\rangle$ with two endpoints, $-\infty \leq a < b \leq \infty$, which may or may not be in I (X is reflecting or absorbing at the endpoints respectively). Clearly m is the speed measure of X (unique up to a constant) and let s be its scale function which is a strictly increasing and continuous function on I. Hence

$$\mathcal{E}(u, u) = \frac{1}{2}\int_I (\frac{du}{ds})^2 ds, \quad u \in \mathcal{F}.$$

Set $J := s(I) \subset \mathbf{R}$, then $s^{-1} : J \to I$ is a homeomorphism mapping. Let $(B_t)_{t\geq 0}$ be the Brownian motion on J which is reflecting or absorbing at the finite endpoint of J if X is absorbing or reflecting at the corresponding endpoint of I. Then $(X_t)_{t\geq 0}$ is the time-changed Markov process of $(s^{-1}(B_t))_{t\geq 0}$ by m, i.e. a homeomorphism transformation s^{-1} firstly and a time change by m next on $(B_t)_{t\geq 0}$. Applying Theorem 2.2, we know that the associated Dirichlet forms of the one-dimensional irreducible diffusion X on I and the corresponding Brownian motion B on J have the same structure of regular subspaces with respect to the transformation s^{-1} and the time change by Revuz measure m. In particular, by the characterization of the regular subspaces of one-dimensional Brownian motion[c], any irreducible regular subspace $(\mathcal{E}', \mathcal{F}')$ of $(\mathcal{E}, \mathcal{F})$ can be characterized by the speed measure m and another scale function s' on I, i.e. a strictly increasing and continuous function on I, such that

$$ds' \ll ds \text{ and } \frac{ds'}{ds} = 0 \text{ or } 1, \quad ds\text{-a.e.}$$

Note that this is the main results of Ref. 4.

More applications of Theorem 2.2 will be found in the following text, especially in dealing with the planar reflecting or absorbing Brownian motions in Sec. 3.3.

3. Regular subspaces of local Dirichlet forms

The authors proved in Ref. 3 that one-dimensional Brownian motion always possesses proper regular subspaces which are characterized by a special scale function. We shall study the similar problems for the symmetric diffusions on multidimensional space. Before that let's review the results of one-dimensional Brownian motion.

3.1. *One-dimensional Brownian motion*

It is known that $(\frac{1}{2}\mathbf{D}, H^1(\mathbf{R}))$ is the corresponding Dirichlet form of one-dimensional Brownian motion $(B_t)_{t\geq 0}$ where

$$\begin{aligned} H^1(\mathbf{R}) &= \{u \in L^2(\mathbf{R}) : u \text{ is absolutely continuous, } u' \in L^2(\mathbf{R})\}, \\ \mathbf{D}(u, v) &= \int u'(x)v'(x)dx. \end{aligned} \tag{3.1}$$

[c]See Theorem 3.1.

An irreducible diffusion on $\mathbf{R}$ associated with a local Dirichlet form can be characterized by a scale function s, a speed measure m and a killing measure k where s is a strictly increasing and continuous function on $\mathbf{R}$ and m, k are Radon measures on $\mathbf{R}$. We may and will assume $k = 0$ because of the notes after Theorem 2.2. Define

$$\begin{aligned}\mathbf{S}(\mathbf{R}) := \{s : \mathbf{R} \to \mathbf{R} \,|\, s \text{ is strictly increasing and continuous, } s(0) = 0 \\ \text{and } s'(x) = 0 \text{ or } 1 \text{ a.e. }\},\end{aligned} \tag{3.2}$$

$$\hat{\mathbf{S}}(\mathbf{R}) := \{s \in \mathbf{S}(\mathbf{R}) : \ s(\pm\infty) = \pm\infty\}. \tag{3.3}$$

For any $s \in \mathbf{S}(\mathbf{R})$, define

$$\begin{aligned}\mathcal{F}^{(s)} &:= \{u \in L^2(\mathbf{R}) : \ u \ll s, \ \int_{\mathbf{R}} (\frac{du}{ds})^2 ds < \infty\} \\ \mathcal{E}^{(s)}(u, v) &:= \frac{1}{2} \int_{\mathbf{R}} \frac{du}{ds} \frac{dv}{ds} ds, \quad u, v \in \mathcal{F}^{(s)},\end{aligned} \tag{3.4}$$

where $u \ll s$ means u is absolutely continuous with respect to s, in other words, there exists an absolutely continuous function ϕ such that $u = \phi \circ s$. Although s^{-1} may not be absolutely continuous, ds^{-1} is a smooth measure of (B_t, P^x) because the only polar set of (B_t) is null set. Let (A_t) be the PCAF associated with ds^{-1} and (τ_t) the inverse of (A_t). Then we have:

Theorem 3.1 (Ref. 3, 2005). *The following assertions about the regular subspaces of one-dimensional Brownian motion hold.*

(1) The Dirichlet form $(\mathcal{F}^{(s)}, \mathcal{E}^{(s)})$ is strongly local and regular on $L^2(\mathbf{R})$ whose associated diffusion has the scale function s up to a linear transform, speed measure dx. Hence the generator of $(\mathcal{F}^{(s)}, \mathcal{E}^{(s)})$ is

$$A^{(s)} = \frac{1}{2} \frac{d}{dx} \frac{d}{ds}.$$

(2) The class $C_c^1(s) := \{\varphi \circ s : \varphi \in C_c^1(J) \text{ where } J = s(\mathbf{R})\}$ is a special standard core of $(\mathcal{F}^{(s)}, \mathcal{E}^{(s)})$.

(3) For any $s \in \mathbf{S}(\mathbf{R})$, $(\mathcal{F}^{(s)}, \mathcal{E}^{(s)}) \prec (\frac{1}{2}\mathbf{D}, H^1(\mathbf{R}))$. Moreover $\mathcal{F}^{(s)} \subsetneqq H^1(\mathbf{R})$ if and only if $|E_s| > 0$ where $E_s = \{x \in \mathbf{R} : s'(x) = 0\}$ and $|E_s|$ means the Lebesgue measure of E_s.

(4) The regular subspace $(\mathcal{E}', \mathcal{F}')$ of $(\frac{1}{2}\mathbf{D}, H^1(\mathbf{R}))$ is recurrent if and only if $(\mathcal{E}', \mathcal{F}') = (\mathcal{F}^{(s)}, \mathcal{E}^{(s)})$ for some $s \in \hat{\mathbf{S}}(\mathbf{R})$. Moreover its associated diffusion is $(X_t^s, P^{s(x)})$ where $X_t^s = s^{-1}(B_{\tau_t})$.

Note that when $s^{-1}(x) = c(x) + x$, where $c(x)$ is the standard Cantor function, it holds that $|E_s| > 0$. Moreover, if the Brownian motion lies only on an proper interval I of $\mathbf{R}$, set $I = \langle a, b\rangle$ where two endpoints a, b may or may not be in I, accordingly the Brownian motion is reflecting or absorbing at the endpoints. Let $\mathbf{S}(I)$ be defined as the similar way as Eq. (3.2), replacing $\mathbf{R}$ by I and $s(0) = 0$ by $s(a) = 0$ if $a > -\infty$. And the domain of $\mathcal{E}^{(s)}$ should be restricted to a subspace of $\mathcal{F}^{(s)}$ satisfying that the value at the absorbing endpoint of every function in this subspace is zero. The results about the regular subspaces of the reflecting or absorbing Brownian motion on I are similar as Theorem 3.1, see Ref. 3.

We are going to give a brief explanation of $\mathbf{S}(\mathbf{R})$ (or $\mathbf{S}(I)$). Let $(\mathcal{E}', \mathcal{F}')$ be a regular subspace of $(\frac{1}{2}\mathbf{D}, H^1(\mathbf{R}))$ whose associated Markov process is denoted by X. By

Corollary 2.1, X is a diffusion on $\mathbf{R}$ and hence it can be characterized by a strictly increasing and continuous scale function s on $\mathbf{R}$ and a speed measure which is just the Lebesgue measure. Moreover it holds that

$$\mathcal{E}'(u,u) = \frac{1}{2}\int (\frac{du}{ds})^2 ds, \quad u \in \mathcal{F}'.$$

However $\mathcal{E}'$ is equal to $\frac{1}{2}\mathbf{D}$ on $\mathcal{F}'$ and

$$\frac{1}{2}\mathbf{D}(u,u) = \frac{1}{2}\int (u'(x))^2 dx = \frac{1}{2}\int (\frac{du}{ds}\cdot\frac{ds}{dx})^2 dx = \frac{1}{2}\int (\frac{du}{ds})^2 (s'(x))^2 dx,$$

therefore we must assume that $s'(x)^2 = s'(x)$ a.e., in other words, $s'(x) = 0$ or 1 a.e, to ensure that

$$\mathcal{E}'(u,u) = \frac{1}{2}\mathbf{D}(u,u), \quad \forall u \in \mathcal{F}'.$$

This is the essential reason for the definition of $\mathbf{S}(\mathbf{R})$.

3.2. *Multidimensional Brownian motions*

In this section, we will study the regular subspaces of multidimensional Brownian motions. We first introduce the independent coupling of Dirichlet forms.

Let X be an m^X-symmetric Markov process on a locally compact separable metric space E^X whose corresponding Dirichlet form on $L^2(E^X, m^X)$ is $(\mathcal{E}^X, \mathcal{F}^X)$. Similarly, the Markov process Y is m^Y-symmetric on another space E^Y whose Dirichlet form is $(\mathcal{E}^Y, \mathcal{F}^Y)$ on $L^2(E^Y, m^Y)$. Note that m^X and m^Y are Radon measures on E^X and E^Y respectively. Our target is the Dirichlet form of the independent coupling process $Z = (X, Y)$ on $E := E^X \times E^Y$ with $m := m^X \times m^Y$. Before that we need to give some notes. Let $(\Omega, \mathcal{M}, \mu)$ be a measurable space and H a real Hilbert space with the inner product $(\cdot,\cdot)_H$ and the norm $\|\cdot\|_H$. For any $p \geq 1$,

$$L^p(\Omega, H) := \{u : \Omega \to H \mid \|u(\cdot)\|_H \in L^p(\Omega, \mu)\}$$

is a Banach space with the norm $(\int_\Omega \|u(\omega)\|_H^p \mu(d\omega))^{\frac{1}{p}}$. Write:

$$\begin{aligned} L^2(E^Y, \mathcal{F}^X) = \{u \in L^2(E,m) : u(\cdot, y) \in \mathcal{F}^X \; m^Y\text{-a.e. } y, \\ \|u(\cdot,y)\|_{\mathcal{E}_1^X} \in L^2(E^Y, m^Y)\}, \end{aligned} \tag{3.5}$$

$$\begin{aligned} L^2(E^X, \mathcal{F}^Y) = \{u \in L^2(E,m) : u(x, \cdot) \in \mathcal{F}^Y \; m^X\text{-a.e. } x, \\ \|u(x,\cdot)\|_{\mathcal{E}_1^Y} \in L^2(E^X, m^X)\}. \end{aligned} \tag{3.6}$$

We have:

Proposition 3.1. *Let* X, Y, $L^2(E^X, m^X)$, $L^2(E^Y, m^Y)$ *and* $(\mathcal{E}^X, \mathcal{F}^X)$, $(\mathcal{E}^Y, \mathcal{F}^Y)$ *be above and* $Z = (X, Y)$ *the independent coupling process of* X *and* Y. *Set* $E = E^X \times E^Y$ *and*

$m = m^X \times m^Y$. Then Z is m-symmetric on E and

$$\begin{cases} \mathcal{F} = L^2(E^Y, \mathcal{F}^X) \cap L^2(E^X, \mathcal{F}^Y), \\ \mathcal{E}(u,v) = \int_Y \mathcal{E}^X(u(\cdot,y), v(\cdot,y)) m^Y(dy) \\ \qquad\qquad + \int_X \mathcal{E}^Y(u(x,\cdot), v(x,\cdot)) m^X(dx), \quad u,v \in \mathcal{F}, \end{cases} \tag{3.7}$$

is a regular Dirichlet form on $L^2(E,m)$ whose corresponding Markov process is Z. Moreover if $\mathcal{C}^X$ (resp. $\mathcal{C}^Y$) is a core of $(\mathcal{E}^X, \mathcal{F}^X)$ (resp. $(\mathcal{E}^Y, \mathcal{F}^Y)$), then the tensor product

$$\mathcal{C} := \mathcal{C}^X \otimes \mathcal{C}^Y$$

is a core of $(\mathcal{E}, \mathcal{F})$.

Proof. Let $(P_t^X), (P_t^Y), (P_t^Z)$ be the transition functions of X, Y and Z respectively. For any $f_1, f_2 \in bL^2(E^X, m^X)$ and $g_1, g_2 \in bL^2(E^Y, m^Y)$,

$$\begin{aligned} (P_t^Z(f_1 \otimes g_1), f_2 \otimes g_2)_m &= (P_t^X f_1 \otimes P_t^Y g_1, f_2 \otimes g_2) m \\ &= (P_t^X f_1, f_2)_{m^X} \cdot (P_t^Y g_1, g_2)_{m^Y} \\ &= (f_1, P_t^X f_2)_{m^X} \cdot (g_1, P_t^Y g_2)_{m^Y} \\ &= (f_1 \otimes g_1, P_t^Z(f_2 \otimes g_2))_m. \end{aligned}$$

Therefore $(P_t^Z)_{t \geq 0}$ (i.e. Z) is m-symmetric.

We are going to prove that $(\mathcal{E}, \mathcal{F})$ is a Dirichlet form on $L^2(E,m)$. Clearly $\mathcal{F} \subset L^2(E,m)$ and $\mathcal{C} \subset \mathcal{F}$, hence $\mathcal{F}$ is dense in $L^2(E,m)$. The bilinearity and Markov property are trivial. We need only to prove that $(\mathcal{E}, \mathcal{F})$ is closed. For any $\mathcal{E}_2$-Cauchy sequence $\{u_n\} \subset \mathcal{F}$ and any n, m,

$$\|u_n - u_m\|_{\mathcal{E}_2}^2 = \int \|(u_n - u_m)(\cdot, y)\|_{\mathcal{E}_1^X}^2 m^Y(dy) + \int \|(u_n - u_m)(x, \cdot)\|_{\mathcal{E}_1^Y}^2 m^X(dx),$$

hence u_n is also $L^2(E^Y, \mathcal{F}^X)$-Cauchy and $L^2(E^X, \mathcal{F}^Y)$-Cauchy. Then there exist $u \in L^2(E^Y, \mathcal{F}^X)$ and $v \in L^2(E^X, \mathcal{F}^Y)$ such that u_n is convergent to u (resp. v) in $L^2(E^Y, \mathcal{F}^X)$ (resp. $L^2(E^X, \mathcal{F}^Y)$). But u_n is $L^2(E,m)$-convergent to u (resp. v), hence $u = v \in \mathcal{F}$ and u_n is convergent to u in $\mathcal{E}_2$.

We shall now give some notations for convenience. The semigroups, resolvents and generators of X, Y, Z are denoted by (T_t^X, G_α^X, A^X), (T_t^Y, G_α^Y, A^Y) and (T_t^Z, G_α^Z, A^Z) respectively. We have proved that $(\mathcal{E}, \mathcal{F})$ is a Dirichlet form, whose semigroup, resolvent and generator are denoted by T_t, G_α and A. As $\mathcal{C} \subset \mathcal{F}$ is dense in $C_c(E)$ by Stone-Weierstrass theorem, its closure $\overline{\mathcal{C}}^{\mathcal{E}_1} \subset \mathcal{F}$ is a regular Dirichlet space. The semigroup, resolvent and generator of $(\overline{\mathcal{C}}^{\mathcal{E}_1}, \mathcal{E})$ are denoted by $T_t^{\mathcal{C}}, G_\alpha^{\mathcal{C}}$ and $A^{\mathcal{C}}$. Note that for any $f \in L^2(E^X, m^X), g \in L^2(E^Y, m^Y), t > 0$,

$$T_t^Z(f \otimes g) = T_t^X f \otimes T_t^Y g. \tag{3.8}$$

In addition, one can easily check that $\mathcal{F}^X \otimes \mathcal{F}^Y \subset \mathcal{F}^{\mathcal{C}}$.

We are now going to prove that for any $\alpha > 0$, $G_\alpha = G_\alpha^Z = G_\alpha^{\mathcal{C}}$ which implies that $(\mathcal{E}, \mathcal{F})$ is the associated Dirichlet form of Z with the core $\mathcal{C}$. In particular, $(\mathcal{E}, \mathcal{F})$ is regular.

Firstly we claim that $\mathcal{D}(A^X)\otimes\mathcal{D}(A^Y)\subset\mathcal{D}(A^Z)\cap\mathcal{D}(A^C)$ and for any $f\in\mathcal{D}(A^X)$, $g\in\mathcal{D}(A^Y)$, it holds that

$$A^Z(f\otimes g)=A^C(f\otimes g)=A^Xf\otimes g+f\otimes A^Yg. \tag{3.9}$$

In fact, $f\otimes g\in\mathcal{F}^X\otimes\mathcal{F}^Y\subset\mathcal{F}^C$ and for any $h\in\mathcal{F}^C$

$$\begin{aligned}\mathcal{E}(f\otimes g,h)&=\int\mathcal{E}^X(f,h(\cdot,y))g(y)m^Y(dy)+\int\mathcal{E}^Y(g,h(x,\cdot))f(x)m^X(dx)\\&=\int(-A^Xf,h(\cdot,y))_{m^X}g(y)m^Y(dy)\\&\qquad+\int(-A^Yg,h(x,\cdot))_{m^Y}f(x)m^X(dx)\\&=(-A^Xf\cdot g-f\cdot A^Yg,h)_m.\end{aligned}$$

It is similar to see that $f\otimes g\in\mathcal{F}$ and

$$\mathcal{E}(f\otimes g,\tilde{h})=(-A^Xf\cdot g-f\cdot A^Yg,\tilde{h})_m,\quad \tilde{h}\in\mathcal{F}.$$

Hence Eq. (3.9) is proved.

On the other hand, $f\in\mathcal{D}(A^X),g\in\mathcal{D}(A^Y)$ implies that $T_t^Xf\in\mathcal{D}(A^X),T_t^Yg\in\mathcal{D}(A^Y)$. By Eq. (3.8) and Eq. (3.9), $T_t^Z(f\otimes g)\in\mathcal{D}(A^X)\otimes\mathcal{D}(A^Y)$ and

$$\begin{aligned}A^CT_t^Z(f\otimes g)&=A^C(T_t^Xf\otimes T_t^Yg)\\&=A^XT_t^Xf\cdot T_t^Yg+T_t^Xf\cdot A^YT_t^Yg\\&=\frac{d}{dt}T_t^Xf\cdot T_t^Yg+T_t^Xf\cdot\frac{d}{dt}T_t^Yg\\&=\frac{d}{dt}T_t^Z(f\otimes g)\\&=A^ZT_t^Z(f\otimes g).\end{aligned}$$

As A^C is a closed linear operator and $\int_0^\infty A^ZT_t^Z(f\otimes g)e^{-\alpha t}dt$ is Bochner integrable,

$$\begin{aligned}A^CG_\alpha^Z(f\otimes g)&=A^C\int_0^\infty e^{-\alpha t}T_t^Z(f\otimes g)dt\\&=\int_0^\infty e^{-\alpha t}A^CT_t^Z(f\otimes g)dt\\&=\int_0^\infty e^{-\alpha t}A^ZT_t^Z(f\otimes g)dt\\&=A^Z\int_0^\infty e^{-\alpha t}T_t^Z(f\otimes g)dt\\&=A^ZG_\alpha^Z(f\otimes g).\end{aligned}$$

Then it follows that

$$(\alpha-A^C)G_\alpha^Z(f\otimes g)=(\alpha-A^Z)G_\alpha^Z(f\otimes g)=f\otimes g.$$

Making use of $G_\alpha^{\mathcal{C}} = (\alpha - A^{\mathcal{C}})^{-1}$, we have

$$G_\alpha^Z(f \otimes g) = G_\alpha^{\mathcal{C}}(f \otimes g).$$

Similarly we can prove $G_\alpha^Z(f \otimes g) = G_\alpha(f \otimes g)$. Therefore for any $\alpha > 0$,

$$G_\alpha = G_\alpha^{\mathcal{C}} = G_\alpha^Z$$

on $\mathcal{D}(A^X) \otimes \mathcal{D}(A^Y)$ which is dense in $L^2(E, m)$. As they are bounded operators on $L^2(E, m)$, it also holds

$$G_\alpha h = G_\alpha^{\mathcal{C}} h = G_\alpha^Z h, \quad h \in L^2(E, m).$$

That completes the proof. □

This proposition can be easily extended to the higher dimensional independent coupling of Markov processes. Let X^i be an m_i-symmetric Markov process on E^i whose associated Dirichlet form is $(\mathcal{E}^i, \mathcal{F}^i)$ where $1 \leq i \leq d$ and d is an arbitrary natural number. For any $1 \leq i \leq d$, set

$$\begin{aligned} E :&= E^1 \times \cdots \times E^d, \\ \hat{E}^i :&= E^1 \times \cdots \times E^{i-1} \times E^{i+1} \cdots \times E^d, \end{aligned}$$

and for any $x = (x_1, \cdots, x_d) \in E$, $\hat{x}_i \in \hat{E}^i$ is the projection of x on $\hat{E}^i$. For any function f on E, set $f_{\hat{x}_i}(x_i) := f(x)$ for any $x \in E$. Similarly $m := m_1 \times \cdots \times m_d$ and $\hat{m}_i$ is its projection on $\hat{E}^i$. Note that if $X^1, \cdots, X^d$ are independent, X is symmetric if and only if X^i is m_i-symmetric for any $1 \leq i \leq d$ and X is Markov if and only if X^i is Markov for any $1 \leq i \leq d$. The following corollary is direct by Proposition 3.1 and we omit its proof.

Corollary 3.1. *Let $X = (X^1, \cdots, X^d)$ be the independent coupling process of $X^1, \cdots, X^d$. Then X is m-symmetric on E and its associated Dirichlet form is*

$$\begin{aligned} \mathcal{F} &= \cap_{i=1}^d L^2(\hat{E}^i, \mathcal{F}^i), \\ \mathcal{E}(u, v) &= \sum_{i=1}^d \int \mathcal{E}^i(f_{\hat{x}_i}(x_i), g_{\hat{x}_i}(x_i))\hat{m}^i(d\hat{x}_i) \quad f, g \in \mathcal{F}. \end{aligned} \tag{3.10}$$

Note that $L^2(\hat{E}^i, \mathcal{F}^i)$ is defined similar as Eq. (3.5). Moreover, $(\mathcal{E}, \mathcal{F})$ is regular and if $\mathcal{C}^i$ is a core of $(\mathcal{E}^i, \mathcal{F}^i)$, $1 \leq i \leq d$, then the tensor product

$$\mathcal{C} := \mathcal{C}^1 \otimes \cdots \otimes \mathcal{C}^d$$

is a core of $(\mathcal{E}, \mathcal{F})$.

Since the d-dimensional Brownian motion is an independent coupling process of one-dimensional Brownian motions, we could conclude some assertions about the regular subspaces of d-mensional Brownian motions. It is well known that $(\frac{1}{2}\mathbf{D}, H^1(\mathbf{R}^d))$ (or equivalently the Lévy-type Dirichlet form with $S = I_d, j = 0$) is the associated Dirichlet form of d-dimensional Brownian motion. Choose $s_i \in \mathbf{S}(\mathbf{R})$ for any $1 \leq i \leq d$ and let X^{s_i} be the R-subprocess of one-dimensional Brownian motion with respect to s_i appeared in Theorem 3.1. Denote the independent coupling process of $X^{s_1}, \cdots, X^{s_d}$ by $(X^{s_1}, \cdots, X^{s_d})$ and its corresponding Dirichlet form by $(\mathcal{E}^{s_1,\cdots,s_d}, \mathcal{F}^{s_1,\cdots,s_d})$. Define

$$\mathcal{C}^{s_1,\cdots,s_d} := C_c^1(s_1) \otimes \cdots \otimes C_c^1(s_d),$$

where $C_c^1(s_i) = \{\phi \circ s_i : \phi \in C_c^1(J_i) \text{ where } J_i = s_i(\mathbf{R})\}$ for any $1 \leq i \leq d$.

Theorem 3.2. *For any $s_1, \cdots, s_d \in \mathbf{S}(\mathbf{R})$, $X := (X^{s_1}, \cdots, X^{s_d})$ is an R-subprocess of d-dimensional Brownian motion, in other words, the associated Dirichlet form $(\mathcal{E}^{s_1,\cdots,s_d}, \mathcal{F}^{s_1,\cdots,s_d})$ of X is a regular subspace of $(\frac{1}{2}\mathbf{D}, H^1(\mathbf{R}^d))$ and $\mathcal{C}^{s_1,\cdots,s_d}$ is a core of $(\mathcal{E}^{s_1,\cdots,s_d}, \mathcal{F}^{s_1,\cdots,s_d})$. In particular, if $|E_{s_i}| > 0$ for some $1 \leq i \leq d$, then the regular subspace $(\mathcal{E}^{s_1,\cdots,s_d}, \mathcal{F}^{s_1,\cdots,s_d})$ is a proper one.*

The proof is obvious by Theorem 3.1 and Corollary 3.1. Note that $|E_{s_i}| > 0$ means s_i is not the scale function of one-dimensional Brownian motion. Usually the R-subprocess of d-dimensional Brownian motion is not an independent coupling, but the following corollary is not difficult to prove. So we omit its proof.

Corollary 3.2. *If $X = (X^1, \cdots, X^d)$ is an R-subprocess of d-dimensional Brownian motion such that $X^1, \cdots, X^d$ are independent and X^i is irreducible for $1 \leq i \leq d$, then there exist $s_1, \cdots, s_d \in \mathbf{S}(\mathbf{R})$ such that $X^i = X^{s_i}$, for any $1 \leq i \leq d$.*

3.3. *Planar absorbing and reflecting Brownian motions*

The results in Theorem 3.2 can be extended to the absorbing or reflecting Brownian motion on a special domain of $\mathbf{R}^d$, such as the rectangle $G = I_1 \times \cdots \times I_d \subset \mathbf{R}^d$, where I_i is an interval of $\mathbf{R}$. It is because that on these rectangles the Brownian motions are independent couplings of one dimensional absorbing or reflecting Browinan motions. However on other usual domains the independent coupling method is invalid.

When $d = 2$, it is known that planar Brownian motion is conformal invariant. In fact the absorbing (or reflecting) Brownian motions on some domains are also conformal invariant. Let φ be a conformal mapping from domain $U \subset \mathbf{R}^2$ to domain $V \subset \mathbf{R}^2$ and B^U the absorbing Brownian motion on U. Then up to a time change, $\varphi(B^U)$ is an absorbing Brownian motion on V[d]. In addition, by the Riemann mapping theorem, for any non-trivial simply connected open subsets U, V of $\mathbf{R}^2$, there exists a biholomorphic (bijective and holomorphic) mapping f from U to V. Obviously f is homeomorphic and conformal mapping. Thus by Proposition 2.1 and 2.2 we have

Proposition 3.2. *Let U, V be two arbitrary non-trivial (non-empty and not $\mathbf{R}^2$) simply connected domains of $\mathbf{R}^2$, then the associated Dirichlet forms of absorbing Brownian motions on U and V are with the same structure of regular subspaces with respect to the corresponding conformal transformation f and time change. In other words, their regular subspaces also have a similar correspondence with respect to these transformations.*

In particular, it is known that the absorbing Brownian motion on the upper half plane is the independent coupling process of a one-dimensional Brownian motion on $\mathbf{R}$ and a Brownian motion on $(0, \infty)$ which is absorbing at 0. By Theorem 3.1, Proposition 3.1 and 3.2, we have

[d]See Theorem 5.3.1 of Ref. 2.

Corollary 3.3. *For any non-trivial simply connected domain U of $\mathbf{R}^2$, the associated Dirichlet form of absorbing Brownian motion on U always has proper regular subspaces.*

The results about reflecting Brownian motion are similar, but the domains should be the Jordan domains in the meaning that they are simply connected and their boundaries are non-self-intersecting continuous loops in the plane. When U, V are Jordan domains the biholomorphic mapping $f : U \to V$ can be extended to a topological homeomorphism from $\bar{U}$ to $\bar{V}$. If $B^{\bar{U}}$ is the reflecting Brownian motion on $\bar{U}$, then up to a time change, $f(B^{\bar{U}})$ is a reflecting Brownian motion on $\bar{V}$[e]. Hence we have

Proposition 3.3. *If U, V are two Jordan domains of $\mathbf{R}^2$, then the associated Dirichlet forms of reflecting Brownian motions on $\bar{U}$ and $\bar{V}$ are with the same structure of regular subspaces with respect to the corresponding biholomorphic mapping and time change. In particular, the associated Dirichlet form of any reflecting Brownian motion on a Jordan domain always has proper regular subspaces.*

In the next section, we will prove that the existence of the proper regular subspaces of absorbing or reflecting Brownian motions will be held not only for these two dimensional cases but for that on arbitrary domains of $\mathbf{R}^d$.

3.4. *Absorbing and reflecting Brownian motions on arbitrary domains*

Let Γ be an arbitrary domain of $\mathbf{R}^d$. Set

$$\begin{aligned} H^1(\Gamma) &:= \{u \in L^2(\Gamma) : \int_\Gamma |\nabla u(x)|^2 dx < \infty\}, \\ \mathbf{D}_\Gamma(u,v) &:= \int_\Gamma \nabla u(x) \cdot \nabla v(x) dx, \quad \forall u, v \in H^1(\Gamma), \end{aligned} \tag{3.11}$$

and denote the closure of $C_c^\infty(\Gamma)$ in $H^1(\Gamma)$ by $H_0^1(\Gamma)$. By regarding $H_0^1(\Gamma) \subset H_0^1(\mathbf{R}^d) = H^1(\mathbf{R}^d)$, we can write $\mathbf{D}_\Gamma(u,v) = \mathbf{D}(u,v)$ for $u, v \in H_0^1(\Gamma)$. It is well known that $(\frac{1}{2}\mathbf{D}, H_0^1(\Gamma))$ is a regular Dirichlet form on $L^2(\Gamma)$ with the core $C_c^\infty(\Gamma)$ and its associated Markov process is just the absorbing Brownian motion B^Γ on Γ.

Proposition 3.4. *The associated Dirichlet form $(\frac{1}{2}\mathbf{D}, H_0^1(\Gamma))$ on $L^2(\Gamma)$ of the absorbing Brownian motion B^Γ on Γ always possesses proper regular subspaces.*

Proof. Clearly, there exists a rectangle $G = I_1 \times \cdots I_d \subset \Gamma$ where I_i is an open interval of $\mathbf{R}$ for any $1 \le i \le d$. Choose $s_1, \cdots s_d \in \mathbf{S}(\mathbf{R})$ such that for some $1 \le i \le d$, the Lebesgue measure of the set

$$E_{s_i}^{I_i} := \{x \in I_i : s_i'(x) = 0\}$$

is positive. Define

$$\mathcal{C} := C_c^\infty(s_1) \otimes \cdots \otimes C_c^\infty(s_d) \tag{3.12}$$

where $C_c^\infty(s_i) = \{\phi \circ s_i : \phi \in C_c^\infty(J_i) \text{ where } J_i = s_i(\mathbf{R})\}$. Clearly, $\mathcal{C}$ is closable in $(\frac{1}{2}\mathbf{D}, H^1(\mathbf{R}^d))$[f] and its closure $(\mathcal{E}, \mathcal{F}) := (\mathcal{E}^{s_1,\cdots,s_d}, \mathcal{F}^{s_1,\cdots,s_d})$ is the corresponding regular Dirichlet form of $X = (X^{s_1}, \cdots, X^{s_d})$ which is defined in the notes before Theorem 3.2. Since $|E_{s_i}^{I_i}| > 0$ for some i, X is different to B and $(\mathcal{E}, \mathcal{F})$ is a proper regular subspace of $(\frac{1}{2}\mathbf{D}, H^1(\mathbf{R}^d))$.

[e]See Sec. 5.3(2°) of Ref. 2.
[f]Note that $\mathbf{D}$ is the corresponding integral on $\mathbf{R}^d$.

As usual, the part Dirichlet form $(\mathcal{E}_\Gamma, \mathcal{F}_\Gamma)$ of $(\mathcal{E}, \mathcal{F})$ on open set Γ can be written as:

$$\begin{aligned} \mathcal{F}_\Gamma &= \{u \in \mathcal{F} : \tilde{u} = 0 \text{ q.e. on } \Gamma^c\} \\ \mathcal{E}_\Gamma(u,v) &= \mathcal{E}(u,v), \quad u, v \in \mathcal{F}_\Gamma. \end{aligned} \tag{3.13}$$

It is well known that $(\mathcal{E}_\Gamma, \mathcal{F}_\Gamma)$ is a regular Dirichlet form on $L^2(\Gamma)$ with the core

$$\mathcal{C}_\Gamma := \{u \in \mathcal{C} : \text{supp}[u] \subset \Gamma\}.$$

On the other hand, $(\frac{1}{2}\mathbf{D}, H_0^1(\Gamma))$ is just the part Dirichlet form of $(\frac{1}{2}\mathbf{D}, H^1(\mathbf{R}^d))$ on Γ. Hence it is easy to check that

$$(\mathcal{E}_\Gamma, \mathcal{F}_\Gamma) \prec (\frac{1}{2}\mathbf{D}, H_0^1(\Gamma)).$$

Note that the part Dirichlet form $(\mathcal{E}_G, \mathcal{F}_G)$ of $(\mathcal{E}, \mathcal{F})$ on G is a regular Dirichlet form with the core

$$\mathcal{C}_G := \{u \in \mathcal{C} : \text{supp}[u] \subset G\}.$$

On the other hand,

$$\mathcal{C}_G = C_c^\infty(s_i|_{I_i}) \otimes \cdots \otimes C_c^\infty(s_d|_{I_d})$$

where $C_c^\infty(s_i|_{I_i}) = \{\phi(s_i|_{I_i}) : \phi \in C_c^\infty(J_i') \text{ where } J_i' = s_i(I_i)\}$. As $|E_{s_i}^{I_i}| > 0$ for some i, $(\mathcal{E}_G, \mathcal{F}_G)$ is a proper regular subspace of $(\frac{1}{2}\mathbf{D}, H_0^1(G))$.

Now we are going to prove that $\mathcal{F}_\Gamma \subsetneqq H_0^1(\Gamma)$. If not, for any $f \in C_c^\infty(G) \subset H_0^1(\Gamma) = \mathcal{F}_\Gamma$, there exists $\{u_n\} \subset \mathcal{C}_\Gamma$ such that $u_n \to f$ in the norm $||\cdot||_{H^1(\Gamma)}$. Let $K = \text{supp}[f] \subset G$. Clearly K is compact. Then there exists a $\varphi \in \mathcal{F}_G \cap C_c(G)$ such that $\varphi|_K \equiv 1$. Let $v_n := u_n \cdot \varphi \in \mathcal{F}_G$, it is easy to prove that $||v_n - f||_{H^1(G)} \to 0$ as $n \to \infty$. Hence $H_0^1(G) \subset \mathcal{F}_G$ which contradicts that $\mathcal{F}_G \subsetneqq H_0^1(G)$. □

Usually $(\frac{1}{2}\mathbf{D}_\Gamma, H^1(\Gamma))$ is a Dirichlet form on $L^2(\Gamma)$ but not regular. However under some conditions, for example, when the domain Γ has **continuous boundary** in the following sense: any $x \in \partial D$ has a neighborhood U such that

$$\Gamma \cap U = \{(x_1, \cdots, x_d) : x_d > F(x_1, \cdots, x_{d-1})\} \cap U$$

in some coordinate $(x_1, \cdots, x_d)$ and with a continuous function F, then

$$C_c^\infty(\bar{\Gamma}) := \{u|_{\bar{\Gamma}} : u \in C_c^\infty(\mathbf{R}^d)\}$$

is dense in $H^1(\Gamma)$ where $u|_{\bar{\Gamma}}$ is the restriction of u on $\bar{\Gamma}$. Hence $(\frac{1}{2}\mathbf{D}_\Gamma, H^1(\Gamma))$ is a regular Dirichlet form on $L^2(\bar{\Gamma})$, where $L^2(\bar{\Gamma}) := L^2(\bar{\Gamma}, 1_{\bar{\Gamma}}(x) \cdot dx)$. Its corresponding Markov process is just the reflecting Browninan motion $B^{\bar{\Gamma}}$ on $\bar{\Gamma}$.

Proposition 3.5. *Let Γ be a domain of $\mathbf{R}^d$ with continuous boundary. Then the associated Dirichlet form $(\frac{1}{2}\mathbf{D}_\Gamma, H^1(\Gamma))$ on $L^2(\bar{\Gamma})$ of the reflecting Brownian motion $B^{\bar{\Gamma}}$ on $\bar{\Gamma}$ always possesses proper regular subspaces.*

Proof. Similarly as in the proof of Proposition 3.4, choose an open rectangle $G = I_1 \times \cdots I_d \subset \Gamma$ and $s_1, \cdots s_d \in \mathbf{S}(\mathbf{R})$ such that for some $1 \leq i \leq d$, the Lebesgue measure of the set $E_{s_i}^{I_i}$ is positive. We also define $\mathcal{C}$ as Eq. (3.12). Define:

$$\mathcal{C}(\bar{\Gamma}) := \{u|_{\bar{\Gamma}} : u \in \mathcal{C}\}.$$

Since $\mathcal{C}$ is dense in $C_c(\mathbf{R}^d)$, it also follows that $\mathcal{C}(\bar{\Gamma})$ is dense in $C_c(\bar{\Gamma}) := \{u|_{\bar{\Gamma}} : u \in C_c(\mathbf{R}^d)\}$. It is easy to check that $(\frac{1}{2}\mathbf{D}_\Gamma, \mathcal{C}(\bar{\Gamma}))$ is closable on $L^2(\bar{\Gamma})$. Its closure denoted by $(\mathcal{E}_{\bar{\Gamma}}, \mathcal{F}_{\bar{\Gamma}})$, i.e. $\mathcal{F}_{\bar{\Gamma}} = \overline{\mathcal{C}(\bar{\Gamma})}^{H^1(\Gamma)}, \mathcal{E}_{\bar{\Gamma}}(u,v) = \frac{1}{2}\mathbf{D}_\Gamma(u,v)$ for any $u, v \in \mathcal{F}_{\bar{\Gamma}}$, is a regular Dirichlet form on $L^2(\bar{\Gamma})$. Clearly,

$$(\mathcal{E}_{\bar{\Gamma}}, \mathcal{F}_{\bar{\Gamma}}) \prec (\frac{1}{2}\mathbf{D}_\Gamma, H^1(\Gamma)). \tag{3.14}$$

Note that $\mathcal{C}(\bar{G}) := \{u|_{\bar{G}} : u \in \mathcal{C}\} = C_c^\infty(s_i|_{\bar{I}_i}) \otimes \cdots \otimes C_c^\infty(s_d|_{\bar{I}_d})$ where $C_c^\infty(s_i|_{\bar{I}_i}) = \{\phi(s_i|_{\bar{I}_i}) : \phi \in C_c^\infty(\bar{J}_i')$ where $\bar{J}_i' = s_i(\bar{I}_i)\}$. Clearly G has the continuous boundary. It follows from $|E_{s_i}^{I_i}| > 0$ for some i and Corollary 3.1 that the closure $(\mathcal{E}_{\bar{G}}, \mathcal{F}_{\bar{G}})$ of $\mathcal{C}(\bar{G})$ is a proper regular subspace of $(\frac{1}{2}\mathbf{D}_G, H^1(G))$ on $L^2(\bar{G})$.

Now we are going to prove $\mathcal{F}_{\bar{\Gamma}} \subsetneqq H^1(\Gamma)$. If $\mathcal{F}_{\bar{\Gamma}} = H^1(\Gamma)$, for any $f \in C_c^\infty(\mathbf{R}^d)$, $f|_{\bar{\Gamma}} \in H^1(\Gamma)$, there exists $\{u_n\} \subset \mathcal{C}(\bar{\Gamma})$ such that $u_n \to f|_{\bar{\Gamma}}$ in $H^1(\Gamma)$. Hence it also holds that $u_n|_{\bar{G}} \to f|_{\bar{G}}$ in $H^1(G)$. It follows from $u_n|_{\bar{G}} \in \mathcal{C}(\bar{G}) \subset \mathcal{F}_{\bar{G}}$ that $C_c^\infty(\bar{G}) \subset \mathcal{F}_{\bar{G}}$ and hence $\mathcal{F}_{\bar{G}} = H^1(G)$ which contradicts that $(\mathcal{E}_{\bar{G}}, \mathcal{F}_{\bar{G}})$ is a proper regular subspace of $(\frac{1}{2}\mathbf{D}_G, H^1(G))$ on $L^2(\bar{G})$. □

Remark 3.1. It is well known that the active reflected Dirichlet form of $(\frac{1}{2}\mathbf{D}, H_0^1(\Gamma))$ is just $(\frac{1}{2}\mathbf{D}_\Gamma, H^1(\Gamma))$. The proper regular subspace $(\mathcal{E}_\Gamma, \mathcal{F}_\Gamma)$ of $(\frac{1}{2}\mathbf{D}, H_0^1(\Gamma))$ and the proper regular subspace $(\mathcal{E}_{\bar{\Gamma}}, \mathcal{F}_{\bar{\Gamma}})$ of $(\frac{1}{2}\mathbf{D}_\Gamma, H^1(\Gamma))$ appeared in Eq. (3.13) and Eq. (3.14) were introduced by the same class $\mathcal{C}$ defined by Eq. (3.12). Let $(\mathcal{E}_\Gamma^{\text{ref}}, (\mathcal{F}_\Gamma^{\text{ref}})_\text{a})$ be the active reflected Dirichlet form of $(\mathcal{E}_\Gamma, \mathcal{F}_\Gamma)$. It holds that

$$\begin{aligned}
&\mathcal{F}_{\bar{\Gamma}} \subset (\mathcal{F}_\Gamma^{\text{ref}})_\text{a} \subsetneqq H^1(\Gamma), \\
&\mathcal{E}_\Gamma^{\text{ref}}(u,v) = \frac{1}{2}\mathbf{D}_\Gamma(u,v), \quad u, v \in (\mathcal{F}_\Gamma^{\text{ref}})_\text{a},
\end{aligned}$$

which is followed by Eq. (4.3.13), Theorem 6.2.13 and Theorem 6.4.2 of Ref. 2.

3.5. *Symmetric uniformly elliptic diffusions*

Let Γ be an arbitrary domain of $\mathbf{R}^d$. In this section we pay special attention to the Markov symmetric form on $L^2(\Gamma)$:

$$\begin{aligned}
&\mathcal{D}[\mathcal{E}] := C_c^\infty(\Gamma), \\
&\mathcal{E}(u,v) := \sum_{i,j=1}^d \int_\Gamma a_{ij}(x) \frac{\partial u(x)}{\partial x_i} \frac{\partial v(x)}{\partial x_j} dx,
\end{aligned} \tag{3.15}$$

where a_{ij} are locally bounded Borel functions on Γ for any $1 \le i, j \le d$ such that for any $\xi \in \mathbf{R}^d, x \in \Gamma$

$$\sum_{i,j=1}^d a_{ij}(x)\xi_i\xi_j \ge 0, \quad a_{ij}(x) = a_{ji}(x), \ 1 \le i, j \le d. \tag{3.16}$$

Moreover, $(a_{ij})_{d\times d}$ is said to satisfy the *uniform ellipticity* if there exists $\delta > 0$ such that

$$\sum_{i,j=1}^d a_{ij}(x)\xi_i\xi_j \ge \delta|\xi|^2, \quad \xi \in \mathbf{R}^d, x \in \Gamma. \tag{3.17}$$

If (a_{ij}) satisfies the uniform ellipticity, then $(\mathcal{E}, \mathcal{D}[\mathcal{E}])$ is closable. The closure of $\mathcal{D}[\mathcal{E}]$ is denoted by $\mathcal{F}$. Clearly $(\mathcal{E}, \mathcal{F})$ is a regular Dirichlet form on $L^2(\Gamma)$. For more details, see Sec. 3.1 of Ref. 1.

Proposition 3.6. *Let $(\mathcal{E}, \mathcal{F})$ be above where (a_{ij}) are locally bounded Borel functions on Γ and satisfy Eq. (3.16) and Eq. (3.17). For any proper $(\frac{1}{2}\mathbf{D}, \mathcal{G}) \prec (\frac{1}{2}\mathbf{D}, H_0^1(\Gamma))$ with a core $\mathcal{C}$, it holds that $(\mathcal{E}, \mathcal{C})$ is closable on $L^2(\Gamma)$ and its closure $(\mathcal{E}, \mathcal{F}^{\mathcal{C}})$ is a proper regular subspace of $(\mathcal{E}, \mathcal{F})$. In particular, $(\mathcal{E}, \mathcal{F})$ always possesses proper regular subspaces.*

Proof. We need only to prove $\mathcal{C} \subset \mathcal{F}$. In fact, if we have proved $\mathcal{C} \subset \mathcal{F}$, then clearly $(\mathcal{E}, \mathcal{C})$ is closable and its closure $(\mathcal{E}, \mathcal{F}^{\mathcal{C}})$ is a regular Dirichlet form with $\mathcal{F}^{\mathcal{C}} \subset \mathcal{F}$. If $(\mathcal{E}, \mathcal{F}^{\mathcal{C}})$ is not a proper subspace, in other words, $\mathcal{F}^{\mathcal{C}} = \mathcal{F}$, then for any $u \in C_c^\infty(\Gamma)$ there exists $\{u_n\} \subset \mathcal{C}$ such that $u_n \to u$ in the norm $||\cdot||_{\mathcal{E}_1}$. It follows Eq. (3.17) that $u_n \to u$ also in $H^1(\Gamma)$, hence $C_c^\infty(\Gamma) \subset \mathcal{G}$ which contradits that $(\frac{1}{2}\mathbf{D}, \mathcal{G})$ is a proper regular subspace of $(\frac{1}{2}\mathbf{D}, H_0^1(\Gamma))$.

We are now ready to prove $\mathcal{C} \subset \mathcal{F}$. For any $u \in \mathcal{C}$, let $K := \text{supp}[u]$. Clearly $K \subset \Gamma$ is compact. Choose a $\phi \in C_c^\infty(\mathbf{R}^d)$ such that ϕ supports on $\{x : |x| \leq 1\}$ and let $\phi_n(x) := n^d\phi(nx)$. Define $u_n := \phi_n * u \in C_c^\infty(\mathbf{R}^d)$. It is easy to check that there exists another compact set $\tilde{K} \supset K$ and $\tilde{K} \subset \Gamma$ such that $\text{supp}[u_n] \subset \tilde{K}$, and hence $u_n \in C_c^\infty(\Gamma)$, for enough large n. As (a_{ij}) is locally bounded, there exists a constant $M > 0$ such that $|a_{ij}(x)| < M$ for any $1 \leq i, j \leq d$ and $x \in \tilde{K}$. On the other hand, $u_n \to u$ in $H^1(\Gamma)$. Hence we have:

$$\begin{aligned}
\mathcal{E}(u_n - u, u_n - u) &= \int_\Gamma \sum_{i,j=1}^d a_{ij}(x) \frac{\partial(u_n - u)}{\partial x_i} \frac{\partial(u_n - u)}{\partial x_j} dx \\
&= \int_{\tilde{K}} \sum_{i,j=1}^d a_{ij}(x) \frac{\partial(u_n - u)}{\partial x_i} \frac{\partial(u_n - u)}{\partial x_j} dx \\
&\leq M \cdot \int_{\tilde{K}} \sum_{i,j=1}^d \left| \frac{\partial(u_n - u)}{\partial x_i} \frac{\partial(u_n - u)}{\partial x_j} \right| dx \\
&\leq d \cdot M \cdot \sum_{i=1}^d \int_{\tilde{K}} \left| \frac{\partial(u_n - u)}{\partial x_i} \right|^2 dx \\
&\to 0
\end{aligned}$$

as $n \to \infty$. It follows that $u \in \mathcal{F}$ which complete the proof. □

Acknowledgements

The authors would like to thank Prof. Zhen-Qing Chen for stimulating discussions and many helpful suggestions.

References

1. M. Fukushima, Y. Oshima and M. Takeda, Dirichlet forms and symmetric Markov processes. Second revised and extended edition ed. *Walter de Gruyter* Vol. 19. Berlin (2011).
2. M. Fukushima, and Z.-Q. Chen, Symmetric Markov processes, Time change and Boundary Theory. *Princeton University Press*, (2011).

3. X. Fang, M. Fukushima and J. Ying, On regular Dirichlet subspaces of $H^1(I)$ and associated linear diffusions. *Osaka Journal of Mathematics.* 42(1): p. 27–41, (2005).
4. X. Fang, P. He and J. Ying, Dirichlet Forms Associated with Linear Diffusions. *Chi. Ann. Math. Ser. B*, 31(4): p. 507-518, (2010).
5. U. Mosco, Composite media and asymptotic Dirichlet forms. *Journal of Functional Analysis*, 123(2): p. 368–421, (1994).
6. P.A. Meyer, Intégrales stochastiques. I, II, III, IV, Séminaire de Probabilitiés I, *Springer-Verlag*, New York, (1967).
7. M.T. Barlow, Non-Local Dirichlet Forms and Symmetric Jump Processes. *Trans. Amer. Math. Soc.*, 361(4): p. 1963–1999, (2009).
8. R.A. Adams and J.J.F. Fournier, Sobolev spaces. 2ed. Vol. 140. *Elsevier/Academic Press*, Amsterdam. xiv+305, (2003).
9. K. Yosida, Functional analysis. *Springer*, (1978).
10. R.M.C. Blumenthal and R.K. Getoor, Markov processes and potential theory. *Dover Publications*, (1968).
11. J. Bertoin, Lévy Processes: XD-US. *Cambridge University Press*, (1998).
12. L. C. G. Rogers and D. Williams, Diffusions, Markov processes and martingales: Volume 2, Itô calculus. *Cambridge University Press*, (2000).
13. K. Itô and H.P. McKean, Difussion Processes and Their Sample Paths. *Springer*, (1996).
14. Y. Le Jan, Mesures associées à une forme de Dirichlet, Applications (French) *Bull. Soc. Math. France* 106 (1978), no. 1, 61–112.

Chapter 21

Quasi-regular semi-Dirichlet forms and beyond

Zhi-Ming Ma*, Wei Sun† and Li-Fei Wang‡

In this survey paper we review the concept of quasi-regularity and report some of our recent progress on quasi-regular semi-Dirichlet forms. We start with an introduction to the concept of quasi-regularity and then review three results closely related to this concept. Afterwards we report our recent results on representations of semi-Dirichlet forms and the Fukushima type decomposition for Markov processes associated with quasi-regular semi-Dirichlet forms. We report also our results on semi-h-transform of semi-Dirichlet forms, some applications of semi-Dirichlet forms in the study of nonlinear filtering, and our recent investigation concerning Hunt's hypothesis (H).

Contents

*Institute of Applied Mathematics, Academy of Mathematics and Systems Science, Chinese Academy of Sciences, Beijing 100190, China, mazm@amt.ac.cn
†Department of Mathematics and Statistics, Concordia University, Montreal H3G 1M8, Canada, wei.sun@concordia.ca
‡College of Mathematics and Information Science, Hebei Normal University, Shijiazhuang 050024, China, flywit1986@163.com

1. Introduction

The fundamental work of Professor Masatoshi Fukushima [36, 37] in 1971 on the construction of Hunt processes associated with regular Dirichlet forms brought a rapid development of the modern theory of Dirichlet forms. The analytic part of the theory goes back to the pioneering papers of A. Beurling and J. Deny [16, 17]. Because of the connection between the analytic part and the probabilistic part, since Refs. 36 and 37, the theory of Dirichlet forms has been very active and has applied to a wide range of related areas of mathematics and physics.

Historically, the connection between Dirichlet forms and Markov processes was initially established between regular Dirichlet forms and symmetric Markov processes living on locally compact separable metric spaces [36, 37, 39, 80], and then extended to include non-symmetric Markov processes [21, 61–63, 73]. Meanwhile there were various investigations on Dirichlet forms and Markov processes in infinite dimensional spaces and/or in singular cases, see e.g. Refs. 3, 58, 40, 13, 1, 2, 14, 15, 20, 79. We refer the readers to Ref. 12 in this volume for more references in this connection.

The concept of quasi-regular Dirichlet forms was developed in the 90s of the last century for the study of Dirichlet forms in connection with Markov processes on more general state spaces, including infinite dimensional state spaces [4–9, 70] (see Ref. 70 for a systematic study of quasi-regular Dirichlet forms). This concept together with the quasi-homeomorphism method [27] enable us to study general Markov processes via regular Dirichlet forms. We refer the readers to the recent book Ref. 25 for very nice developments in this aspect.

In Refs. 69, 11 and 32, Markov processes associated with quasi-regular semi-Dirichlet forms are studied. In the recent book Ref. 74, various topics on the connection between regular semi-Dirichlet forms and Markov processes are discussed.

In this survey paper, we shall briefly review the concept of quasi-regularity and report some of our recent progress on quasi-regular semi-Dirichlet forms. The paper is organized as follows.

In Section 2, we first recall the concept of quasi-regularity for semi-Dirichlet forms in Subsection 2.1. In the subsequent three subsections we review some results closely related to the concept of quasi-regularity. These include the nice correspondence between Markov processes and semi-Dirichlet forms, local compactification and quasi-homeomorphism, and quasi-regular topology of semi-Dirichlet forms. In the latter we review some results in Ref. 29 by Z. Dong and in Ref. 18 by L. Beznea et al., which enable us to study semi-Dirichlet forms on arbitrary measurable Lusin spaces by means of quasi-regular semi-Dirichlet forms. In Section 3, we present our results in recent years on representations of semi-Dirichlet forms. We start with recalling the notion of SPV integrability. Then we present respectively the Beurling-Deny type decomposition, LeJan type transformation rule and Lévy-Khintchine type representation of semi-Dirichlet forms. In Section 4, we briefly report our recent result on the Fukushima type decomposition for semi-Dirichlet forms. Apart from the main result, in Subsection 4.2 we provide some remarks on the uniqueness of the Fukushima type decomposition, which expose some oversights in the literature. The oversights may be traced back even to Theorem 8.26 of the book Ref. 47, concerning a question about the uniqueness of the Doob-Meyer decomposition on optional sets of interval type. With the title "Semi-h-transform of semi-Dirichlet forms", in Section 5 we briefly report some of our results in Ref. 46 (published in 2011) on $h\hat{h}$-transforms of positivity preserving semigroups. In Section 6, we report some applications of semi-Dirichlet forms in the study of nonlinear filtering. In Section 7, we report some of our recent investigation concerning Hunt's hypothesis (H).

We refer the readers to Refs. 41, 70 and 69 for notations and terminologies of this paper. We also refer the readers to the recently published book Ref. 74 for a self-contained introduction to regular semi-Dirichlet forms.

We would like to take this opportunity to thank all our colleagues and friends with whom we have shared pleasant discussions and collaborations in various stages. Especially, we thank X.F. Han, Z.C. Hu, L. Ma and J. Zhang for their permission to present our recent joint work here. Z.M. Ma is grateful to S. Albeverio and M. Röckner for the long standing pleasant collaboration. We all thank Z.Q. Chen and an anonymous referee for helpful comments that improved the presentation of this paper. We are grateful to the support of 973 project (2011CB808000), NCMIS, NSFC (11021161) and NSERC (Grant No. 311945-2013).

2. Quasi-regularity of semi-Dirichlet forms

2.1. *Introduction to the concept of quasi-regularity*

Let $(E;\mathcal{B};m)$ be a measure space. We shall work on the (real) Hilbert space $L^2(E;m) := L^2(E;\mathcal{B};m)$ with the usual inner product $(\ ,\ )$. As usual, we shall not distinguish a function on E from its m-equivalence class unless otherwise stated. For $u, v : E \to \mathbb{R}$, we set $u \vee v := \sup(u,v)$, $u \wedge v := \inf(u,v)$, $u^+ := u \vee 0$. For a bilinear form $\mathcal{E}$ with domain $D(\mathcal{E}) \subset L^2(E;m)$, we define for $u, v \in D(\mathcal{E})$,

$$\tilde{\mathcal{E}}(u,v) := \frac{1}{2}(\mathcal{E}(u,v) + \mathcal{E}(v,u)),$$
$$\check{\mathcal{E}}(u,v) := \frac{1}{2}(\mathcal{E}(u,v) - \mathcal{E}(v,u)),$$
$$\hat{\mathcal{E}}(u,v) := \mathcal{E}(v,u).$$

$\tilde{\mathcal{E}}$ is called the *symmetric part* of $\mathcal{E}$, $\check{\mathcal{E}}$ the *antisymmetric part* of $\mathcal{E}$ and $\hat{\mathcal{E}}$ the *dual form* of $\mathcal{E}$. For $\alpha > 0$ we write

$$\mathcal{E}_\alpha(u,v) = \mathcal{E}(u,v) + \alpha(u,v), \quad \forall u, v \in D(\mathcal{E}).$$

Definition 2.1. Let $(\mathcal{E}, D(\mathcal{E}))$ be a bilinear form on $L^2(E;m)$. Then $(\mathcal{E}, D(\mathcal{E}))$ is called a *semi-Dirichlet form* if the following three conditions are satisfied.

(a) $D(\mathcal{E})$ is dense in $L^2(E;m)$, and the symmetric part $(\tilde{\mathcal{E}}, D(\mathcal{E}))$ is positive definite and closed on $L^2(E;m)$.

(b) *(weak sector condition)* There exists $K > 0$ (called *continuity constant*) such that

$$|\mathcal{E}_1(u,v)| \le K\mathcal{E}_1(u,u)^{1/2}\mathcal{E}_1(v,v)^{1/2}, \quad \forall u, v \in D(\mathcal{E}).$$

(c) *(semi-Dirichlet property)* If $u \in D(\mathcal{E})$, then

$$u^+ \wedge 1 \in D(\mathcal{E}) \text{ and } \mathcal{E}(u + u^+ \wedge 1, u - u^+ \wedge 1) \ge 0.$$

Remark 2.1. (i) In the literature $(\mathcal{E}, D(\mathcal{E}))$ is called a *Dirichlet form* if both $\mathcal{E}$ and its dual form $\hat{\mathcal{E}}$ are semi-Dirichlet forms, $(\mathcal{E}, D(\mathcal{E}))$ is called a *coercive closed form* if it satisfies only the above conditions (a) and (b).

(ii) If $(\mathcal{E}, D(\mathcal{E}))$ is a coercive closed form, then $D(\mathcal{E})$ equipped with the inner product $\tilde{\mathcal{E}}_1$ is a Hilbert space. Below we shall always implicitly assume that the topology on $D(\mathcal{E})$ is induced by this Hilbert structure.

(iii) Given a coercive closed form $(\mathcal{E}, D(\mathcal{E}))$. Let $(T_t)_{t\geq 0}$ and $(G_\alpha)_{\alpha\geq 0}$ be the semigroup and the resolvent associated with $(\mathcal{E}, D(\mathcal{E}))$, respectively. Then the semi-Dirichlet property (c) is equivalent to the sub-Markov property for $(T_t)_{t\geq 0}$ or for $(G_\alpha)_{\alpha\geq 0}$, respectively. The sub-Markov property of $(T_t)_{t\geq 0}$ and $(G_\alpha)_{\alpha\geq 0}$ is defined as:

$$f \in L^2(E;m), 0 \leq f \leq 1 \ m\text{-}a.e. \Rightarrow 0 \leq T_t f \leq 1 \ m\text{-}a.e., \ \forall t > 0$$

and

$$f \in L^2(E;m), 0 \leq f \leq 1 \ m\text{-}a.e. \Rightarrow 0 \leq \alpha G_\alpha f \leq 1 \ m\text{-}a.e., \ \forall \alpha > 0.$$

To introduce the concept of quasi-regularity, from now on we assume that E is a Hausdorff topological space such that $\mathcal{B} := \mathcal{B}(E)$, the Borel-$\sigma$-algebra of E, is generated by continuous functions on E, and m is a σ-finite positive Borel measure with full support on E. Let $(\mathcal{E}, D(\mathcal{E}))$ be a semi-Dirichlet form on $L^2(E;m)$. For a closed set $F \subset E$, we set

$$D(\mathcal{E})_F := \{u \in D(\mathcal{E}) \,|\, u = 0 \ m\text{-}a.e. \ on \ F^c\},$$

here and henceforth $F^c := E \backslash F$.

Definition 2.2. (i) An increasing sequence $(F_k)_{k\in\mathbb{N}}$ of closed subsets of E is called an *$\mathcal{E}$-nest* if $\bigcup_{k\geq 1} D(\mathcal{E})_{F_k}$ is dense in $D(\mathcal{E})$.

(ii) A subset $N \subset E$ is called *$\mathcal{E}$-exceptional* if $N \subset \bigcap_{k\geq 1} F_k^c$ for some $\mathcal{E}$-nest $(F_k)_{k\in\mathbb{N}}$. We say that a property of points in E holds *$\mathcal{E}$-quasi-everywhere* (abbreviated *$\mathcal{E}$-q.e.*), if the property holds outside some $\mathcal{E}$-exceptional set.

(iii) An $\mathcal{E}$-q.e. defined function f on E is called *$\mathcal{E}$-quasi-continuous* if there exists an $\mathcal{E}$-nest $(F_k)_{k\in\mathbb{N}}$ such that $f \in C(\{F_k\})$, where

$$C(\{F_k\}) := \{f : A \to \mathbb{R} \,|\, \bigcup_{k\geq 1} F_k \subset A \subset E, f|_{F_k} \text{ is continuous for every } k \in \mathbb{N}\}.$$

We now state the concept of quasi-regularity in the definition below.

Definition 2.3. A semi-Dirichlet form $(\mathcal{E}, D(\mathcal{E}))$ on $L^2(E;m)$ is called *quasi-regular* if

(Q.1) There exists an $\mathcal{E}$-nest $(E_k)_{k\in\mathbb{N}}$ consisting of metrizable compact sets.

(Q.2) There exists an $\tilde{\mathcal{E}}_1^{1/2}$-dense subset of $D(\mathcal{E})$ whose elements have $\mathcal{E}$-quasi-continuous m-versions.

(Q.3) There exist $u_n \in D(\mathcal{E})$, $n \in \mathbb{N}$, having $\mathcal{E}$-quasi-continuous m-versions $\tilde{u}_n$, $n \in \mathbb{N}$, and an $\mathcal{E}$-exceptional set $N \subset E$ such that $\{\tilde{u}_n \,|\, n \in \mathbb{N}\}$ separates the points of $E\backslash N$.

The concept of quasi-regular Dirichlet forms was initially introduced and developed in the 90s of the last century in its various stages in a series of publications, see Refs. 4, 5, 6, 7, 70, 8, 9, 27, 76. In particular, the terminology of quasi-regularity appeared for the first time in Refs. 6 and 70, in the latter a systematic study of quasi-regular Dirichlet forms is given. Later, the concept of quasi-regularity was adopted for the study of semi-Dirichlet forms and positivity forms in Refs. 69, 11, 71, 32, respectively. It was also adopted in the study of generalized Dirichlet forms in Ref. 82.

In the following three subsections we review some results closely related to the concept of quasi-regularity.

2.2. *Correspondence between Markov processes and semi-Dirichlet forms*

The correspondence between nice Markov processes and Dirichlet forms was first established by M. Fukushima in 1971. In the celebrated article [37], M. Fukushima constructed for the first time Hunt processes associated with regular symmetric Dirichlet forms on locally compact separable metric spaces. Since then the theory of Dirichlet forms has been rapidly developed as an important branch of stochastic analysis. An original motivation for formulating the quasi-regularity condition was attempting to extend the correspondence between Markov processes and Dirichlet forms to cover also the infinite dimensional case and/or the non-regular case.

We now assume that E is a metrizable Lusin space (i.e., E is topologically isomorphic to a Borel subset of a complete separable metric space) and m is a σ-finite positive measure on its Borel σ-algebra $\mathcal{B}(E)$. We adjoin an extra point Δ (the cemetery) to E and write $E_\Delta := E \cup \{\Delta\}$. When E is locally compact, we regard E_Δ as a one point compactification of E. Otherwise we regard Δ as an isolated point of E_Δ. Denote by $\mathcal{B}(E_\Delta)$ the Borel sets of E_Δ, and denote by $\mathcal{B}^*(E)$ (resp. $\mathcal{B}^*(E_\Delta)$) the σ-algebra consisting of all universally measurable subsets of E (resp. E_Δ). Any function defined on E is extended to E_Δ by declaring its value at Δ to be 0. Any measure defined on $\mathcal{B}(E)$ is extended to $\mathcal{B}^*(E_\Delta)$ by declaring its value on $\{\Delta\}$ to be 0 and by a standard method of completing. Below for the definition of right processes, we refer to e.g. Chapter IV of Ref. 70 or Appendix A of Ref. 25. We note that the notion of right processes was originally introduced in Ref. 43 and intensively studied in the book Ref. 78.

For a right process $\mathbf{M} = (\Omega, \mathcal{F}, (\mathcal{F}_t)_{t\geq 0}, (X_t)_{t\geq 0}, (P_x)_{x\in E_\Delta})$ with state space E and cemetery Δ, we use $(P_t)_{t\geq 0}$ to denote its transition semigroup. A right process $\mathbf{M} = (\Omega, \mathcal{F}, (\mathcal{F}_t)_{t\geq 0}, (X_t)_{t\geq 0}, (P_x)_{x\in E_\Delta})$ is said to be *(properly) associated* with a semi-Dirichlet form $(\mathcal{E}, D(\mathcal{E}))$ on $L^2(E;m)$ if and only if $P_t f$ is an ($\mathcal{E}$-quasi-continuous) m-version of $T_t f$ for all $f \in \mathcal{B}_b(E) \cap L^2(E;m)$ and all $t > 0$. Hereafter $\mathcal{B}_b(E)$ denotes the set of all bounded Borel measurable functions on E.

Theorem 2.1. *Let $(\mathcal{E}, D(\mathcal{E}))$ be a semi-Dirichlet form on $L^2(E;m)$. Then there exists a right process $\mathbf{M} = (\Omega, \mathcal{F}, (\mathcal{F}_t)_{t\geq 0}, (X_t)_{t\geq 0}, (P_x)_{x\in E_\Delta})$ associated with $(\mathcal{E}, D(\mathcal{E}))$ if and only if $(\mathcal{E}, D(\mathcal{E}))$ is quasi-regular. In this case $\mathbf{M}$ is always properly associated with $(\mathcal{E}, D(\mathcal{E}))$. Moreover, in this case $\mathbf{M}$ is always m-special standard and m-tight in the sense of the following definition.*

Definition 2.4. Let $\mathbf{M} = (\Omega, \mathcal{F}, (\mathcal{F}_t)_{t\geq 0}, (X_t)_{t\geq 0}, (P_x)_{x\in E_\Delta})$ be a right process with state space E, life time ζ, and cemetery Δ. We take an arbitrary probability measure μ on $(E_\Delta, \mathcal{B}(E_\Delta))$ such that μ is equivalent to m.

(i) $\mathbf{M}$ is said to be *m-special standard* if it enjoys the following additional properties:

(a) *(left limits up to ζ)* $X_{t-} := \lim_{s\uparrow t, s<t} X_s$ exists in E for all $t \in (0,\zeta)$ P_μ-a.s.

(b) *(quasi-left continuity up to ζ and special)* If τ, τ_n, $n \in \mathbb{N}$, are $(\mathcal{F}_t^{P_\mu})$-stopping times such that $\tau_n \uparrow \tau$, then $X_{\tau_n} \to X_\tau$ as $n \to \infty$ P_μ-a.s. on $\{\tau < \zeta\}$ and X_τ is $\bigvee_{n\in\mathbb{N}} \mathcal{F}_{\tau_n}^{P_\mu}$-measurable. Hereafter $\mathcal{F}_t^{P_\mu}$ denotes the completion of $\mathcal{F}_t$ w.r.t. $P_\mu(\cdot) := \int_{E_\Delta} P_x(\cdot)\mu(dx)$.

(ii) $\mathbf{M}$ is said to be *m-tight* if there exists an increasing sequence $(K_n)_{n\in\mathbb{N}}$ of compact metrizable sets in E such that

$$P_\mu[\lim_{n\to\infty} \sigma_{E\setminus K_n} < \zeta] = 0,$$

where $\sigma_A := \inf\{t > 0 \,|\, X_t \in A\}$ is the *first hitting time* of a subset A of E_Δ.

Remark 2.2. The result as stated in the current version of Theorem 2.1 was developed in the 90s of the last century along with the concept of quasi-regularity. The general correspondence between quasi-regular (semi-)Dirichlet forms and m-special Borel standard processes was developed in various stages in Refs. 4, 5, 7, 70, 8, 9, 69, 11. P. J. Fitzsimmons proved that any right process associated with a semi-Dirichlet form is indistinguishable from an m-special standard process and the Borel measurability can be weakened to the universal measurability (Refs. 31 and 32), which enables us to state the result in the current version.

Remark 2.3. The notion of m-special standard processes was introduced in Ref. 44. The notion of m-tightness was introduced in Chapter IV of Ref. 70. In fact it was proved in Ref. 10 (see also Theorem IV. 1.15 in Ref. 70) that any process with values in a metrizable Lusin (or co-Souslin) space and with trajectories almost surely càdlàg up to its life time is always m-tight. The main idea of Ref. 10 is taken from Ref. 66, but generalizes the corresponding result of Ref. 66 considerably. Theorem 2.1 is valid for more general state space E provided we impose a priori the m-tightness condition on the underlying process (cf. Chapter IV of Ref. 70).

Remark 2.4. For a detailed proof of the correspondence between quasi-regular Dirichlet forms and m-tight m-special standard processes we refer to Chapter IV of Ref. 70, of which the last section contains detailed notes in connection with references in the literature. Among other things, we mention that a modification of an idea of Y. LeJan [64] plays an important role in checking the necessity of (Q.2).

2.3. *Local compactification and quasi-homeomorphism*

An important feature of the quasi-regularity concept is that it is not only sufficient but also necessary for the existence of a right process associated with a semi-Dirichlet form. As asserted in Theorem 2.1, if a right process is associated with a semi-Dirichlet form $(\mathcal{E}, D(\mathcal{E}))$, then $(\mathcal{E}, D(\mathcal{E}))$ is always quasi-regular. This feature turns out to be very useful due to the result of quasi-homeomorphism, which enables us to transfer the fruitful results known for the regular case to the quasi-regular case. The quasi-homeomorphism between quasi-regular Dirichlet forms and regular Dirichlet forms was established in Ref. 27 (see also Chapter VI of Ref. 70). We refer the readers to the recent book Ref. 25 in which fruitful results for general symmetric Markov processes were obtained based on the framework of regular Dirichlet forms by virtue of the quasi-homeomorphism method. In this subsection we present the result of quasi-homeomorphism for semi-Dirichlet forms, of which the proof can be found in Ref. 49.

Following the definition of regular Dirichlet forms [37, 39], we stated in Section 2 of Ref. 49 the definition of regular semi-Dirichlet forms as below. The readers can find fruitful results about regular semi-Dirichlet forms in the recently published book Ref. 74.

Definition 2.5. A semi-Dirichlet form $(\mathcal{E}, D(\mathcal{E}))$ on $L^2(E;m)$ is called *regular* if the following conditions hold:

(R.1) E is a locally compact separable metric space and m is a positive Radon measure on E with $\mathrm{supp}[m] = E$.

(R.2) $C_0(E) \cap D(\mathcal{E})$ is dense in $D(\mathcal{E})$ w.r.t. the $\tilde{\mathcal{E}}_1^{1/2}$-norm.

(R.3) $C_0(E) \cap D(\mathcal{E})$ is dense in $C_0(E)$ w.r.t. the uniform norm $\| \ \|_\infty$.

Hereafter $C_0(E)$ denotes the set of all continuous functions on E with compact supports.

In Section 2 of Ref. 49, we introduced also the notion of core and special core for a regular semi-Dirichlet form. For our further discussion we quote it as the definition below.

Definition 2.6. A subset $D \subset C_0(E) \cap D(\mathcal{E})$ is called a *core* if the following conditions hold:

(C.1) D is dense in $D(\mathcal{E})$ w.r.t. the $\tilde{\mathcal{E}}_1^{1/2}$-norm.

(C.2) D is dense in $C_0(E)$ w.r.t. the uniform norm $\| \|_\infty$.

(C.3) D is a linear lattice.

D is called a *special core* if in addition to (C.1)-(C.3), it holds that

(C.4) For any compact set K and relatively compact open set G with $K \subset G$, there exists a $u \in D$ such that $0 \le u \le 1, u|_K = 1$ and $u|_{E\backslash G} = 0$.

Let $(\mathcal{E}, D(\mathcal{E}))$ be a semi-Dirichlet form on $L^2(E;m)$ and $E^\sharp$ be another Hausdorff topological space with Borel σ-algebra $\mathcal{B}(E^\sharp)$. Suppose that N is an $\mathcal{E}$-exceptional set. Set $Y = E\backslash N$. Suppose that j is a $\mathcal{B}(Y)/\mathcal{B}(E^\sharp)$-measurable map from Y into $E^\sharp$. Let $m \circ j^{-1}$ be the image measure of m on $(E^\sharp, \mathcal{B}(E^\sharp))$. If $u^\sharp$ is $m \circ j^{-1}$-a.e. defined on $E^\sharp$, then $u^\sharp \circ j$ is m-a.e. defined on E since $m(N) = 0$. Define $j^* u^\sharp = u^\sharp \circ j$ m-a.e. for $u^\sharp \in L^2(E^\sharp; m \circ j^{-1})$. Then, j^* is an isometric map from $L^2(E^\sharp, m \circ j^{-1})$ into $L^2(E;m)$.

We define

$$\begin{cases} D(\mathcal{E}^j) = \{u^\sharp \in L^2(E^\sharp; m \circ j^{-1}) \,|\, j^* u^\sharp \in D(\mathcal{E})\}, \\ \mathcal{E}^j(u^\sharp, v^\sharp) = \mathcal{E}(j^* u^\sharp, j^* v^\sharp), \ \forall u^\sharp, v^\sharp \in D(\mathcal{E}^j). \end{cases}$$

Then $(\mathcal{E}^j, D(\mathcal{E}^j))$ is called the image of $(\mathcal{E}, D(\mathcal{E}))$ under j. If j^* is onto then one can check that $(\mathcal{E}^j, D(\mathcal{E}^j))$ is a semi-Dirichlet form.

Now let $m^\sharp$ be a σ-finite Borel measure on $E^\sharp$, and $(\mathcal{E}^\sharp, D(\mathcal{E}^\sharp))$ be another semi-Dirichlet form on $L^2(E^\sharp; m^\sharp)$. All the notations w.r.t. $(\mathcal{E}^\sharp, D(\mathcal{E}^\sharp))$ will be marked by "$\sharp$".

Definition 2.7. $(\mathcal{E}, D(\mathcal{E}))$ is said to be *quasi-homeomorphic* to $(\mathcal{E}^\sharp, D(\mathcal{E}^\sharp))$, if there exists a map $j : \bigcup_{k\ge1} F_k \to \bigcup_{k\ge1} F_k^\sharp$, where $\{F_k\}_{k\in\mathbb{N}}$ is an $\mathcal{E}$-nest in E and $\{F_k^\sharp\}_{k\in\mathbb{N}}$ an $\mathcal{E}^\sharp$-nest in $E^\sharp$, such that:

(i) j is a topological homeomorphism from F_k onto $F_k^\sharp$ for each $k \in \mathbb{N}$.

(ii) $m^\sharp = m \circ j^{-1}$.

(iii) $(\mathcal{E}^\sharp, D(\mathcal{E}^\sharp)) = (\mathcal{E}^j, D(\mathcal{E}^j))$, where $(\mathcal{E}^j, D(\mathcal{E}^j))$ is the image of $(\mathcal{E}, D(\mathcal{E}))$ under j.

The map j is called a *quasi-homeomorphism* from $(\mathcal{E}, D(\mathcal{E}))$ to $(\mathcal{E}^\sharp, D(\mathcal{E}^\sharp))$.

The theorem below extends the corresponding result of [27, Theorem 3.7] or [70, VI.1] to the setting of semi-Dirichlet forms.

Theorem 2.2. (local compactification, Hu et al. [49, Theorem 3.5]) *Let $(\mathcal{E}, D(\mathcal{E}))$ be a quasi-regular semi-Dirichlet form on $L^2(E;m)$. Then there exist an $\mathcal{E}$-nest $\{F_k\}_{k\in\mathbb{N}}$ consisting of compact metrizable subsets of E and a locally compact separable metric space $Y^\sharp$ such that:*

(i) $Y^\sharp$ is a local compactification of $Y := \bigcup_{k\ge1} F_k$ in the sense that $Y^\sharp$ is a locally compact space containing Y as a dense subset and $\mathcal{B}(Y) = \{A \in \mathcal{B}(Y^\sharp) \,|\, A \subset Y\}$.

(ii) The trace topologies on F_k induced by E and $Y^\sharp$ coincide for each $k \in \mathbb{N}$.

(iii) The image $(\mathcal{E}^\sharp, D(\mathcal{E}^\sharp))$ of $(\mathcal{E}, D(\mathcal{E}))$ under the inclusion map $i : Y \subset Y^\sharp$ is a regular semi-Dirichlet form on $L^2(Y^\sharp; m^\sharp)$, where $m^\sharp := m \circ i^{-1}$ is the image measure of m on $(Y^\sharp, \mathcal{B}(Y^\sharp))$.

With the above result, we can obtain the following theorem establishing quasi-homeomorphism for semi-Dirichlet forms, which extends the corresponding result of [27, Theorem 3.7].

Theorem 2.3. (Hu et al. [49, Theorem 3.8]) *A semi-Dirichlet form $(\mathcal{E}, D(\mathcal{E}))$ on $L^2(E;m)$ is quasi-regular if and only if it is quasi-homeomorphic to a regular semi-Dirichlet form $(\mathcal{E}^\sharp, D(\mathcal{E}^\sharp))$ on $L^2(E^\sharp;m^\sharp)$.*

2.4. *Quasi-regular topology of semi-Dirichlet forms*

As we have seen from the above two subsections that the quasi-regularity condition plays the role of a bridge connecting on the one side right processes and on the other side regular semi-Dirichlet forms, which enables us to study quasi-regular semi-Dirichlet forms employing both fruitful results of Markov processes (e.g. Ref. 78 and references therein) and fruitful results of regular semi-Dirichlet forms (e.g. Ref. 74 and references therein). In order to be placed in this pleasant situation, an interesting question arises naturally. Suppose that we are working on a semi-Dirichlet form $(\mathcal{E}, D(\mathcal{E}))$ on $L^2(E;m)$ where $(E;\mathcal{B};m)$ is only a measure space with no topological structure, or the original topology on E does not ensure $(\mathcal{E}, D(\mathcal{E}))$ to be quasi-regular, can we impose a suitable topology τ compatible with the measurable structure, such that $(\mathcal{E}, D(\mathcal{E}))$ becomes quasi-regular with this topology? We call such a topology τ an *$\mathcal{E}$-quasi-regular topology.*

This question was first studied by Z. Dong. In Ref. 29 he obtained the following result.

Theorem 2.4. (Dong, [29, Theorem 1.2]) *Let $(E;\mathcal{B};m)$ be a Hausdorff measure space and $(\mathcal{E}, D(\mathcal{E}))$ be a Dirichlet form on $L^2(E;m)$. Then a necessary and sufficient condition for the existence of an $\mathcal{E}$-quasi-topology τ is that there exists a countable dense Q-linear algebra $D \subset D(\mathcal{E}) \cap \mathcal{B}(E)$ (here Q denotes all the rational numbers) and $E_0 \in \mathcal{B}$ with $m(E \setminus E_0) = 0$, satisfying the following conditions:*

(a) There exists $u \in D$ such that $u > 0$ on E_0;

(b) $(E_0;\mathcal{B} \cap E_0)$ is a Lusin measurable space and D separates points of E_0;

(c) For any bounded α-excessive function $g \in D(\mathcal{E})$, there exists a σ-finite measure μ on $(E_0;\mathcal{B} \cap E_0)$ such that $\mathcal{E}_\alpha(g,u) = \int_{E_0} u d\mu$ for all $u \in D$.

Z. Dong extended also the above results to positivity preserving forms (hence including semi-Dirichlet forms) with slight modifications.

More interestingly, L. Beznea, N. Boboc and M. Röckner obtained the following theorem in Ref. 18, which enables us to study semi-Dirichlet forms on arbitrary measurable Lusin spaces by means of quasi-regular semi-Dirichlet forms.

Theorem 2.5. (Beznea et al. [18, Corollary 3.4]) *Let $(\mathcal{E}, D(\mathcal{E}))$ be a semi-Dirichlet form on $L^2(E;m)$, where m is a σ-finite measure on the measurable Lusin space $(E;\mathcal{B})$. Then there exists a (larger) topological Lusin space E_1 such that $E \subset E_1$, E belongs to $\mathcal{B}(E_1)$, $\mathcal{B} = \mathcal{B}(E_1)|_E$, and $(\mathcal{E}, D(\mathcal{E}))$ regarded as a semi-Dirichlet form on $L^2(E_1;\bar{m})$ is quasi-regular, where $\bar{m}$ is the measure on $(E_1;\mathcal{B}(E_1))$ extending m by setting $\bar{m}(E_1 \backslash E) = 0$.*

3. Representations of semi-Dirichlet forms

3.1. *Beurling-Deny type decomposition for semi-Dirichlet forms*

The Beurling-Deny formula plays a fundamental role in studying symmetric Dirichlet forms and their associated Markov processes. To extend the classical Beurling-Deny formula (cf. Ref. 17 and [41, Theorem 3.2.1]) to the settings of non-symmetric Dirichlet forms and semi-Dirichlet forms, we introduced in Refs. 48 and 49 the notion of *SPV integrability* as stated below.

Definition 3.1. Let E be a metrizable Lusin space and Q be a σ-finite positive Borel measure on $E \times E \backslash d$, where d denotes the diagonal of the product space $E \times E$. A measurable function f on $E \times E \backslash d$ is said to be integrable w.r.t. Q in the sense of *symmetric principle value* (abbreviated by *SPV integrable*) if there exists an increasing sequence $\{A_n\}_{n \in \mathbb{N}}$ of subsets of $E \times E \backslash d$ satisfying $Q((E \times E \backslash d) \setminus (\bigcup_{n \geq 1} A_n)) = 0$, $I_{A_n}(x,y) = I_{A_n}(y,x)$ for all $x, y \in E$, f is integrable on each A_n, $n \geq 1$, and for any sequence $\{A_n\}_{n \in \mathbb{N}}$ with these properties, the limit

$$SPV \int_{E \times E \backslash d} f(x,y) Q(dxdy) := \lim_{n \to \infty} \int_{A_n} f(x,y) Q(dxdy)$$

exists and is independent of the specific choice of the sequence $\{A_n\}_{n \in \mathbb{N}}$.

Below we use $\tilde{D}(\mathcal{E})$ to denote the set of all $\mathcal{E}$-quasi-continuous versions of elements in $D(\mathcal{E})$. The following version of Beurling-Deny type decomposition was obtained in Refs. 48 and 49. Note that the SPV integrable condition cannot be dropped in the decomposition (3.1) below. See Ref. 52 for a counter example.

Theorem 3.1. (Hu et al. [49, Theorem 4.1]) *Let $(\mathcal{E}, D(\mathcal{E}))$ be a quasi-regular semi-Dirichlet form on $L^2(E;m)$. Then the following assertions hold.*

(i) There exist a unique σ-finite positive Borel measure J (called jumping measure) on $E \times E \backslash d$ and a unique σ-finite positive Borel measure K (called killing measure) on E satisfying the following properties:

(a) $J(N \times E \backslash d) = J(E \times N \backslash d) = 0$ and $K(N) = 0$ for any $\mathcal{E}$-exceptional set N.

(b) For $v \in \tilde{D}(\mathcal{E})$ and $u \in I_q(v)$,

$$\mathcal{E}(u,v) = \int_{E \times E \backslash d} 2(u(y) - u(x))v(y) J(dxdy) + \int_E u(x)v(x)K(dx),$$

where

$$I_q(v) := \{u \in \tilde{D}(\mathcal{E}) \mid u \text{ is constant } \mathcal{E}\text{-q.e. on a quasi-open set containing } \operatorname{supp}_q[v]\}.$$

(ii) Define

$$\tilde{\mathcal{A}}(v) := \{u \in \tilde{D}(\mathcal{E}) \mid (u(y) - u(x))v(y) \text{ is SPV integrable w.r.t. } J(dxdy) \text{ and } uv \text{ is integrable w.r.t. } K\}.$$

Then we have the following unique decomposition:

$$\mathcal{E}(u,v) = \mathcal{E}^c(u,v) + SPV \int_{E \times E \backslash d} 2(u(y) - u(x))v(y) J(dxdy) + \int_E u(x)v(x)K(dx) \tag{3.1}$$

for $v \in \tilde{D}(\mathcal{E})$ and $u \in \tilde{\mathcal{A}}(v)$, where $\mathcal{E}^c$ satisfies the left strong local property in the sense that $I_q[v] \subset \tilde{\mathcal{A}}(v)$ and $\mathcal{E}^c(u,v) = 0$ whenever $v \in \tilde{D}(\mathcal{E})$ and $u \in I_q(v)$.

Remark 3.1. In Ref. 74, Y. Oshima derived a version of Beurling-Deny type decomposition for regular semi-Dirichlet forms under his condition $(\mathcal{E}.5)$ and (J). See Theorem 5.2.1

of Ref. 74 for his result. See also Remark 3 on page 177 of Ref. 74 for a connection of his result with ours.

In Ref. 49 we derived also a version of Beurling-Deny type decomposition which is available for some functions getting rid of the SPV integrable condition. Theorem 3.2 below is implied by Proposition 4.4 and Theorem 4.5 (and their proofs) of Ref. 49. We now state it in terms of quasi-homeomorphism to avoid involving more notations.

Theorem 3.2. *Let $(\mathcal{E}, D(\mathcal{E}))$ be a quasi-regular semi-Dirichlet form on $L^2(E;m)$. Then there exist a locally compact separable metric space E^* quipped with a metric ρ and a Radon measure m^*, and a regular semi-Dirichlet form $(\mathcal{E}^*, D(\mathcal{E}^*))$ on $L^2(E^*;m^*)$ having the following properties.*

(i) $(\mathcal{E}^*, D(\mathcal{E}^*))$ is quasi-homeomorphic to $(\mathcal{E}, D(\mathcal{E}))$.

(ii) $(\mathcal{E}^*, D(\mathcal{E}^*))$ possesses a special core (cf. Definition 2.6) D^* satisfying the following conditions.

(a) For $u \in D^*$ we have

$$\int_{E^*\times\{u\neq 0\}\setminus d}(u(y)-u(x))^2 J^*(dxdy) + \int_E u^2(x)K^*(dx) < \infty,$$

where J^* and K^* are the jumping and killing measures of $(\mathcal{E}^*, D(\mathcal{E}^*))$ specified by Theorem 3.1, respectively.

(b) For any $\varepsilon > 0$ and any $u, v \in D^*$, we have the following decomposition:

$$\mathcal{E}^*(u,v) = \mathcal{E}^{\rho,\varepsilon}(u,v) + \int_{\rho(x,y)>\varepsilon} 2(u(y)-u(x))v(y)J^*(dxdy) + \int_E u(x)v(x)K^*(dx), \tag{3.2}$$

where $\mathcal{E}^{\rho,\varepsilon}$ is a bilinear form with domain D^* and satisfies

$$\mathcal{E}^{\rho,\varepsilon}(u,v) = \int_{\rho(x,y)<\varepsilon} 2(u(y)-u(x))v(y)J^*(dxdy) \quad \text{if} \quad u \in D^*\cap I_q(v).$$

Moreover, if $(u(y)-u(x))v(y)$ is SPV integrable w.r.t. $J^*(dxdy)$ then $\lim_{\varepsilon\downarrow 0}\mathcal{E}^{\rho,\varepsilon}(u,v) = (\mathcal{E}^*)^c(u,v)$, where $(\mathcal{E}^*)^c(u,v)$ is specified by (3.1).

Remark 3.2. The regular semi-Dirichlet form $(\mathcal{E}^*, D(\mathcal{E}^*))$ in the above theorem is not unique, but the decomposition (3.2) is quasi-homeomorphism invariant in the sense that if we denote by j the quasi-homeomorphism map from $(\mathcal{E}, D(\mathcal{E}))$ to $(\mathcal{E}^*, D(\mathcal{E}^*))$, then we have always $J^* = J\circ j^{-1}$ and $K^* = K\circ j^{-1}$.

3.2. *LeJan type transformation rule and Lévy-Khintchine type representation of semi-Dirichlet forms*

In this subsection, we investigate further the structure of semi-Dirichlet forms. By the quasi-homeomorphism method, without loss of generality, below we shall concentrate on the representation of regular semi-Dirichlet forms.

In Refs. 51 and 52, the structure of non-symmetric Dirichlet forms has been investigated and the diffusion part of non-symmetric Dirichlet forms has been characterized.

Note that if $(\mathcal{E}, D(\mathcal{E}))$ is only a semi-Dirichlet form, its dual form $(\hat{\mathcal{E}}, D(\mathcal{E}))$ generally does not satisfy the semi-Dirichlet property. So we do not have the decomposition obtained in Refs. 51 and 52. In particular, the existence of the dual killing measure $\hat{K}$ is not ensured. Also, the symmetric part $\tilde{\mathcal{E}}$ of $\mathcal{E}$ is only a symmetric positivity preserving form but not a symmetric Dirichlet form, which causes extra difficulty in characterizing the structure of $\mathcal{E}$. In the recent paper Ref. 84, the LeJan type transformation rule of Ref. 52 has been generalized to the setting of semi-Dirichlet forms.

Let $(\mathcal{E}, D(\mathcal{E}))$ be a regular semi-Dirichlet form on $L^2(E;m)$. We make the following assumption.

Assumption 1. For $f, g \in C_0(E) \cap D(\mathcal{E})$, we have $fg \in C_0(E) \cap D(\mathcal{E})$ and $(f(y)-f(x))g(y)$ is integrable w.r.t. $J(dxdy)$.

It is shown in [84, §4] that under Assumption 1, for $u, v \in C_0(E) \cap D(\mathcal{E})$, there exists a unique signed Radon measure $\mu^c_{<u,v>}$ on E such that

$$\int_E f d\mu^c_{<u,v>} = \mathcal{E}^c(uf, v) + \mathcal{E}^c(vf, u) - \mathcal{E}^c(uv, f), \quad f \in C_0(E) \cap D(\mathcal{E}).$$

Further, define $\hat{J}(dxdy) := J(dydx)$ and let $\chi \in C_0(E) \cap D(\mathcal{E})$ satisfying $\chi = 1$ on a neighbourhood of $\mathrm{supp}[u] \cup \mathrm{supp}[v]$. We can well-define

$$\hat{\mathcal{E}}^c(u, v) := \hat{\mathcal{E}}(u, v) - \int_{E\times E\setminus d} 2(u(y) - u(x))v(y)\chi(x)\hat{J}(dxdy) - \hat{\mathcal{E}}(\chi, uv).$$

We introduce a linear functional $L(u, v)$ on $C_0(E) \cap D(\mathcal{E})$ by

$$< L(u, v), f >:= \frac{1}{2}(\mathcal{E}^c(u, vf) - \hat{\mathcal{E}}^c(u, vf)), \quad f \in C_0(E) \cap D(\mathcal{E}).$$

Theorem 3.3. (Sun et al. [84, Theorems 4.2 and 4.3]) *Suppose Assumption 1 holds. Let $u, v \in C_0(E) \cap D(\mathcal{E})$ and $\chi \in C_0(E) \cap D(\mathcal{E})$ satisfying $\chi = 1$ on a neighbourhood of* $\mathrm{supp}[u] \cup \mathrm{supp}[v]$. *Then*

$$\mathcal{E}^c(u, v) = \frac{1}{2}\int_E \chi d\mu^c_{<u,v>} + < L(u, v), \chi > .$$

Moreover, for $u, v, w, f \in C_0(E) \cap D(\mathcal{E})$, we have

(i) $d\mu^c_{<uv,w>} = ud\mu^c_{<v,w>} + vd\mu^c_{<u,w>}$.

(ii) $< L(u, vw), f >=< L(u, v), wf >$.

(iii) $< L(uv, w), f >=< L(u, w), vf > + < L(v, w), uf >$.

If the state space E is equipped with a differential structure then the semi-Dirichlet form admits more explicit expression. Let U be an open set of $\mathbb{R}^n$ and m be a positive Radon measure on U such that $\mathrm{supp}[m] = U$. Suppose that $(\mathcal{E}, D(\mathcal{E}))$ is a semi-Dirichlet form on $L^2(U;m)$ such that $C_0^\infty(U) \subset D(\mathcal{E})$. For $\delta > 0$, we define

$$U^\delta := \{x \in U \mid \inf_{y\in\partial U} |x - y| > \delta\}.$$

Hereafter, for $B \subset \mathbb{R}^n$, we denote by ∂B its boundary in $\mathbb{R}^n$.

Assumption 2. Let O be a relatively compact open set of U. Suppose that $\{f_n\}_{n=1}^{\infty} \subset C_0^{\infty}(O)$ and $f \in C_0^{\infty}(O)$ satisfying f_n and all of its partial derivatives converge uniformly to f and its corresponding partial derivatives as $n \to \infty$. Then, $\mathcal{E}(f,g) = \lim_{n\to\infty} \mathcal{E}(f_n, g)$ and $\mathcal{E}(g,f) = \lim_{n\to\infty} \mathcal{E}(g, f_n)$ for any $g \in C_0^{\infty}(U)$.

Theorem 3.4. (Sun et al. [84, Theorem 1.4]) *Under Assumption 2, let $\delta > 0$ be a constant such that $U^\delta \neq \emptyset$. Then, we have the decomposition:*

$$\begin{aligned}\mathcal{E}(u,v) &= \frac{1}{2}\sum_{i,j=1}^{n}\int_U \frac{\partial u}{\partial x_i}\frac{\partial v}{\partial x_j}\nu_{ij}(dx) + \sum_{i=1}^{n}\left\langle \Psi_i^\delta, \frac{\partial u}{\partial x_i}v\right\rangle \\ &+ \int_{U\times U\backslash d}\sum_{i=1}^{n}(y_i - x_i)\left(\frac{\partial u}{\partial y_i}(y)v(y) - \frac{\partial u}{\partial x_i}(x)v(x)\right) I_{\{|x-y|\leq\delta\}}(x,y)J(dxdy) \\ &+ \int_{U\times U\backslash d} 2\left(u(y) - u(x) - \sum_{i=1}^{n}(y_i - x_i)\frac{\partial u}{\partial y_i}(y)I_{\{|x-y|\leq\delta\}}(x,y)\right)v(y)J(dxdy) \\ &+ \int_U u(x)v(x)K(dx), \qquad \forall u,v\in C_0^\infty(U^\delta), \end{aligned} \tag{3.3}$$

where J and K are the jumping and killing measures, respectively, $\{\nu_{ij}\}_{i,j=1}^n$ are signed Radon measures on U such that for any compact set $G \subset U$, $\nu_{ij}(G) = \nu_{ji}(G)$ and $\sum_{i,j=1}^n \xi_i\xi_j\nu_{ij}(G) \geq 0$ for all $(\xi_1,\ldots,\xi_n) \in \mathbb{R}^n$, and $\{\Psi_i^\delta\}_{i=1}^n$ are generalized functions on U^δ.

If Assumption 2 is replaced by the assumption that $(\mathcal{E}, D(\mathcal{E}))$ is locally controlled by Dirichlet forms, then we can obtain a clearer characterization of the generalized functions $\{\Psi_i^\delta\}_{i=1}^n$ given in Theorem 3.4.

Assumption 3. Let $\Omega_l \uparrow U$ be a sequence of relatively compact open sets. There exist a sequence of Dirichlet forms $(\mathcal{Q}^l, D(\mathcal{Q}^l))$ on $L^2(\Omega_l; m)$ and a sequence of positive constants C_l such that

$$C_0^\infty(\Omega_l) \subset D(\mathcal{Q}_l),$$

and

$$\mathcal{E}_1(g,g) \leq C_l\mathcal{Q}_1^l(g,g), \quad \forall g \in C_0^\infty(\Omega_l).$$

Corollary 3.1. *Assume Assumption 3 holds. Then, we have the decomposition given in Theorem 3.4. Moreover, for any relatively compact open set O of U^δ, there exist signed Radon measures $\{\mu_i^O\}_{i=1}^n$ and $\{\mu_{ij}^O\}_{i,j=1}^n$ on O such that for $1 \leq i \leq n$,*

$$< \Psi_i^\delta, g >= \int_O g(x)\mu_i^O(dx) + \sum_{j=1}^n \int_O \frac{\partial g}{\partial x_j}(x)\mu_{ij}^O(dx), \quad \forall g \in C_0^\infty(O).$$

Our Theorem 3.4 show that, under suitable conditions, any semi-Dirichlet form on $\mathbb{R}^n$ has the representation (3.3). On the other hand, many authors have studied Markov processes on $\mathbb{R}^n$ by using the Lévy-Khintchine type representation. Their results provide non-trivial examples for Theorem 3.4. Before ending this subsection, we give such an example.

Example 3.1. (see Ref. Uemura, [85]) Let U be an open set of $\mathbb{R}^n$. Suppose that the following conditions hold:

(A.I) There exist $0 < \lambda \leq \Lambda$ such that

$$\lambda|\xi|^2 \leq \sum_{i,j=1}^{n} a_{ij}(x)\xi_i\xi_j \leq \Lambda|\xi|^2 \quad \text{for } x \in U,\ \xi \in \mathbb{R}^n.$$

(A.II) $b_i \in L^n(U;dx)$, $i = 1,\dots,n$.
(A.III) $c \in L_+^{n/2}(U;dx)$.
(A.IV) $k_s(x,y)$ is a nonnegative measurable function defined on $U \times U\backslash d$ satisfying $x \to \int_{y\neq x}(1 \wedge |x-y|^2)k_s(x,y)dy \in L^1_{loc}(U;dx)$.
(A.V) $k_a(x,y)$ is a measurable function defined on $U \times U\backslash d$ satisfying $\sup_{x\in U}\int_{\{|x-y|\geq 1, y\in U\}}|k_a(x,y)|dy < \infty$, $\sup_{x\in U}\int_{\{|x-y|<1,y\in U\}}|k_a(x,y)|^\gamma dy < \infty$ for some $0 < \gamma \leq 1$, and $|k_a(x,y)|^{2-\gamma} \leq Ck_s(x,y)$, $x,y \in U$ with $0 < |x-y| < 1$ for some constant $C > 0$.

Define for $u,v \in C_0^1(U)$,

$$\begin{aligned}\mathcal{Q}(u,v) = &\frac{1}{2}\sum_{i=1}^{n}\int_U \frac{\partial u}{\partial x_i}(x)\frac{\partial v}{\partial x_i}(x)dx \\ &+\frac{1}{2}\iint_{x\neq y}(u(x)-u(y))(v(x)-v(y))k_s(x,y)dxdy\end{aligned}$$

and

$$\begin{aligned}\mathcal{E}(u,v) = &\frac{1}{2}\sum_{i,j=1}^{n}\int_U a_{ij}(x)\frac{\partial u}{\partial x_i}(x)\frac{\partial v}{\partial x_j}(x)dx + \sum_{i=1}^{n}\int_U b_i(x)u(x)\frac{\partial v}{\partial x_i}(x)dx \\ &+\int_U u(x)v(x)c(x)dx \\ &+\frac{1}{2}\iint_{x\neq y}(u(x)-u(y))(v(x)-v(y))k_s(x,y)dxdy \\ &+\iint_{x\neq y}(u(x)-u(y))v(x)k_a(x,y)dxdy.\end{aligned}$$

Then, when λ is sufficiently large, there exists a positive constant β such that $(\mathcal{E}_\beta, C_0^1(U))$ is closable on $L^2(U;dx)$ and its closure $(\mathcal{E}_\beta, D(\mathcal{E}_\beta))$ is a regular semi-Dirichlet form on $L^2(U;dx)$ (see [85, Theorem 3.1]). Note that Assumption 3 (hence Assumption 2) is satisfied for $(\mathcal{E}_\beta, D(\mathcal{E}_\beta))$ since there exists a constant $C' > 1$ such that for any $u \in C_0^1(U)$,

$$\frac{1}{C'}\mathcal{Q}_1(u,u) \leq \mathcal{E}_\beta(u,u) \leq C'\mathcal{Q}_1(u,u).$$

4. Fukushima type decomposition for semi-Dirichlet forms

In this section, we briefly report our recent result on the Fukushima type decomposition for $u \in D(\mathcal{E})_{loc}$ in the general quasi-regular semi-Dirichlet forms setting. The material of this section is mainly taken from Ref. 72. We refer the readers to Ref. 72 for the proofs and more details.

4.1. *Main result on Fukushima type decomposition for semi-Dirichlet forms*

Let E be a metrizable Lusin space and m be a σ-finite positive measure on its Borel σ-algebra $\mathcal{B}(E)$. We consider a quasi-regular semi-Dirichlet form $(\mathcal{E}, D(\mathcal{E}))$ on $L^2(E;m)$. Denote by $(T_t)_{t\geq 0}$ and $(G_\alpha)_{\alpha\geq 0}$ (resp. $(\hat{T}_t)_{t\geq 0}$ and $(\hat{G}_\alpha)_{\alpha\geq 0}$) the semigroup and resolvent (resp. co-semigroup and co-resolvent) associated with $(\mathcal{E}, D(\mathcal{E}))$. Let $\mathbf{M} = (\Omega, \mathcal{F}, (\mathcal{F}_t)_{t\geq 0}, (X_t)_{t\geq 0}, (P_x)_{x\in E_\Delta})$ be an m-tight special standard process which is properly associated with $(\mathcal{E}, D(\mathcal{E}))$.

Similar to the symmetric case, in the semi-Dirichlet forms setting we can introduce the notion of additive functional (abbreviated by AF), continuous additive functional (abbreviated by CAF), and positive continuous additive functional (abbreviated by PCAF). We can define also the family of smooth measures (denoted by S), and establish a one-to-one correspondence between all equivalent classes of PCAFs and S. The lemma below is taken from Ref. 67. We remark that the readers can now find more detailed descriptions and discussions in Ref. 74 on the potential theory of semi-Dirichlet forms including the correspondence between PCAFs and smooth measures.

Lemma 4.1. (cf. Ma et al. [67, Theorem A.8], see also Oshima, [74, Section 4.1]) *Let A be a PCAF. Then there exists a unique $\mu \in S$, which is referred to as the Revuz measure of A and is denoted by μ_A, such that:*

For any γ-co-excessive function g ($\gamma \geq 0$) in $D(\mathcal{E})$ and $f \in \mathcal{B}^+(E)$,

$$\lim_{t\downarrow 0}\frac{1}{t}E_{g\cdot m}((fA)_t) = < f\cdot\mu, \tilde{g} > .$$

Conversely, let $\mu \in S$, then there exists a unique (up to the equivalence) PCAF A such that $\mu = \mu_A$.

Below we fix a function $\phi \in L^2(E;m)$ with $0 < \phi \leq 1$ m-a.e. and set $h = G_1\phi$, $\hat{h} = \hat{G}_1\phi$. Denote $\tau_B := \inf\{t > 0 \mid X_t \notin B\}$ for $B \subset E$. Let V be a quasi-open subset of E. We denote by $X^V = (X_t^V)_{t\geq 0}$ the part process of X on V and denote by $(\mathcal{E}^V, D(\mathcal{E})_V)$ the part form of $(\mathcal{E}, D(\mathcal{E}))$ on $L^2(V;m)$. It is known that X^V is a standard process and $(\mathcal{E}^V, D(\mathcal{E})_V)$ is a quasi-regular semi-Dirichlet form (cf. Ref. 59). Denote by $(T_t^V)_{t\geq 0}$, $(\hat{T}_t^V)_{t\geq 0}$, $(G_\alpha^V)_{\alpha\geq 0}$ and $(\hat{G}_\alpha^V)_{\alpha\geq 0}$ the semigroup, co-semigroup, resolvent and co-resolvent associated with $(\mathcal{E}^V, D(\mathcal{E})_V)$, respectively. One can check that $\hat{h}|_V$ is 1-co-excessive w.r.t. $(\mathcal{E}^V, D(\mathcal{E})_V)$. Define $\bar{h}^V := \hat{h}|_V \wedge \hat{G}_1^V\phi$. Then $\bar{h}^V \in D(\mathcal{E})_V$ and $\bar{h}^V$ is 1-co-excessive. Denote $D(\mathcal{E})_{V,b} := \mathcal{B}_b(E) \cap D(\mathcal{E})_V$.

For an AF $A = (A_t)_{t\geq 0}$ of X^V, we define

$$e^V(A) := \lim_{t\downarrow 0}\frac{1}{2t}E_{\bar{h}^V\cdot m}(A_t^2)$$

whenever the limit exists in $[0,\infty]$. Define

$$\dot{\mathcal{M}}^V := \{M \mid M \text{ is an AF of } X^V,\ E_x(M_t^2) < \infty, E_x(M_t) = 0 \\ \text{for all } t \geq 0 \text{ and } \mathcal{E}\text{-}q.e.\ x \in V, e^V(M) < \infty\},$$

$$\mathcal{N}_c^V := \{N \mid N \text{ is a CAF of } X^V, E_x(|N_t|) < \infty \text{ for all } t \geq 0 \\ \text{and } \mathcal{E}\text{-}q.e.\ x \in V, e^V(N) = 0\},$$

$$\Theta := \{\{V_n\} \mid V_n \text{ is } \mathcal{E}\text{-quasi-open}, \ V_n \subset V_{n+1} \ \mathcal{E}\text{-}q.e. \ \forall\, n \in \mathbb{N}, \text{ and } E = \cup_{n=1}^{\infty} V_n \ \mathcal{E}\text{-}q.e.\},$$

and

$$D(\mathcal{E})_{loc} := \{u \mid \exists\, \{V_n\} \in \Theta \text{ and } \{u_n\} \subset D(\mathcal{E}) \text{ such that } \ u = u_n \ m\text{-}a.e. \text{ on } V_n, \ \forall\, n \in \mathbb{N}\}.$$

In what follows we shall employ the notion of local AFs introduced in Ref. 41 as follows.

Definition 4.1. (cf. Fukushima et al. [41, page 271]) A family $A = (A_t)_{t\geq 0}$ of functions on Ω is called a *local AF* of $\mathbf{M}$, if A satisfies all the requirements for an AF, except that the additivity property is required only for $s, t \geq 0$ with $t + s < \zeta(\omega)$, hereafter ζ denotes the life time of $(X_t)_{t\geq 0}$.

Two local AFs $A^{(1)}$, $A^{(2)}$ are said to be equivalent if for $\mathcal{E}$-$q.e.$ $x \in E$, it holds that

$$P_x(A_t^{(1)} = A_t^{(2)}; t < \zeta) = P_x(t < \zeta), \quad \forall\, t \geq 0.$$

We now define

$$\begin{aligned} \dot{\mathcal{M}}_{loc} :=\ & \{M \mid M \text{ is a local AF of } \mathbf{M}, \ \exists\, \{V_n\}, \{E_n\} \in \Theta \text{ and} \\ & \{M^n \mid M^n \in \dot{\mathcal{M}}^{V_n}\} \ \text{ such that } E_n \subset V_n, \\ & M_{t\wedge\tau_{E_n}} = M^n_{t\wedge\tau_{E_n}}, \ t \geq 0, \ n \in \mathbb{N}\} \end{aligned}$$

and

$$\begin{aligned} \mathcal{L}_c := \{N \mid\ & N \text{ is a local AF of } \mathbf{M}\ , \ \exists\, \{E_n\} \in \Theta \text{ such that } t \to N_{t\wedge\tau_{E_n}} \\ & \text{is continuous and of zero quadratic variation}, \ n \in \mathbb{N}\}. \end{aligned}$$

Proposition 4.1. (Ma et al. [72, Proposition 2.4]) *(i) There exists an $\{\mathcal{F}_t\}$-stopping time ζ_i (may be identically equal to ∞) which is the totally inaccessible part of ζ w.r.t. P_x for $\mathcal{E}$-q.e. $x \in E$. Such a ζ_i is unique in the sense of P_x-a.s. for $\mathcal{E}$-q.e. $x \in E$.*

(ii) Denote by $I(\zeta) := [\![0, \zeta[\![\cup [\![\zeta_i]\!]$. Then $I(\zeta)$ is a predictable set of interval type, and there exists a sequence $\{V_n\} \in \Theta$ such that for any $\{U_n\} \in \Theta$, $I(\zeta) = \cup_n [\![0, \tau_{V_n \cap U_n}]\!]$ P_x-a.s. for *$\mathcal{E}$-q.e. $x \in E$.*

We write $I(\zeta) := [\![0, \zeta[\![\cup [\![\zeta_i]\!]$. By Proposition 4.1, $I(\zeta)$ is a predictable set of interval type (cf. [47, Theorem 8.18]). In this section a local AF M is called a locally square integrable martingale additive functional (abbreviated by MAF) on $I(\zeta)$, denoted by $M \in \mathcal{M}_{loc}^{I(\zeta)}$, if $M \in (\mathcal{M}^2_{loc})^{I(\zeta)}$ in the sense of [47, Definition 8.19].

We put the following assumption:

Assumption 4. (Ma et al. [72, Assumption 1.3]) There exist $\{V_n\} \in \Theta$ and locally bounded function $\{C_n\}$ on $\mathbb{R}$ such that for each $n \in \mathbb{N}$, if $u, v \in D(\mathcal{E})_{V_n,b}$ then $uv \in D(\mathcal{E})$ and

$$\mathcal{E}(uv, uv) \leq C_n(\|u\|_\infty + \|v\|_\infty)(\mathcal{E}_1(u, u) + \mathcal{E}_1(v, v)).$$

Now we can state our main result on the Fukushima type decomposition. In condition (S) below, J denotes the jumping measure of $(\mathcal{E}, D(\mathcal{E}))$ (cf. Theorem 3.1 in the above Subsection 3.1).

Theorem 4.1. (Ma et al. [72, Theorem 1.4]) *Suppose that $(\mathcal{E}, D(\mathcal{E}))$ is a quasi-regular semi-Dirichlet form on $L^2(E;m)$ satisfying Assumption 4. Then for $u \in D(\mathcal{E})_{loc}$ the following two assertions are equivalent to each other.*

(i) u admits a Fukushima type decomposition. That is, there exist $M^{[u]} \in \mathcal{M}_{loc}^{I(\zeta)}$ and $N^{[u]} \in \mathcal{L}_c$ such that

$$\tilde{u}(X_t) - \tilde{u}(X_0) = M_t^{[u]} + N_t^{[u]}, \quad t \geq 0, \quad P_x\text{-a.s. for } \mathcal{E}\text{-q.e. } x \in E. \tag{4.1}$$

(ii) u satisfies condition (S) specified below.

$$(S): \quad \mu_u(dx) := \int_E (\tilde{u}(x) - \tilde{u}(y))^2 J(dydx) \text{ is a smooth measure.}$$

Moreover, if u satisfies condition (S), then the decomposition (4.1) is unique up to the equivalence of local AFs, and the continuous part of $M^{[u]}$ belongs to $\dot{\mathcal{M}}_{loc}$.

The above theorem extends our previous result in Ref. 67 from the setting of local semi-Dirichlet forms to that of general semi-Dirichlet forms. It is also an extension of [74, Theorem 5.1.5]. We take this opportunity to thank Y. Oshima for sending us the manuscript of his new book (Ref. 74). Among other things, in Ref. 74 Y. Oshima proved Fukushima's decomposition for $u \in D(\mathcal{E})_b$ in the setting of regular semi-Dirichlet forms satisfying his condition ($\mathcal{E}$.5). The Assumption 4 stated above was partly stimulated by the condition ($\mathcal{E}$.5) in Ref. 74. We would like to point out that the main techniques employed in [74, Theorem 5.1.5] and in [72, Theorem 1.4] are different. In proving [74, Theorem 5.1.5] Y. Oshima employed his ingenious auxiliary bilinear forms and introduced a weak sense energy, while in proving [72, Theorem 1.4] we employed mainly the localization and pasting technique.

Theorem 4.1 extends also the corresponding results of [41, Theorem 5.5.1] and [60, Theorem 4.2] from the symmetric case to the semi-Dirichlet form case. Note that for a symmetric Dirichlet form $(\mathcal{E}, D(\mathcal{E}))$, Assumption 4 is satisfied automatically. Also, $u \in D(\mathcal{E})_{loc}$ satisfies condition (S) trivially if $(\mathcal{E}, D(\mathcal{E}))$ is local. When $(\mathcal{E}, D(\mathcal{E}))$ is non-local, condition (S) is necessary even in the symmetric case. In developing stochastic calculus with Nakao's integral, K. Kuwae obtained in Ref. 60 a generalized Fukushima decomposition in the symmetric case for a subclass of $D(\mathcal{E})_{loc}$, which is equivalent to impose condition (S) for $u \in D(\mathcal{E})_{loc}$. In Ref. 72 when dealing with the purely discontinuous part of $M^{[u]}$, we adopted some idea from Ref. 60 without making use of Nakao's integral.

4.2. *Remark on the uniqueness of Fukushima type decomposition*

The reader might notice that in the above Theorem 4.1 we used $I(\zeta)$ instead of $[\![0, \zeta[\![$, the latter is customarily used in the literature. The reason of this variation is that we discovered that the decomposition on $I(\zeta)$ is unique, but it may fail to be unique on $[\![0, \zeta[\![$. This difference is essentially due to the fact that $I(\zeta)$ is a predictable set of interval type while $[\![0, \zeta[\![$ is not necessarily predictable. This discovery exposes not only an oversight in the previous paper Ref. 67, but also similar oversights in the literature e.g. Refs. 22 and 60. The oversight may be traced back even to Theorem 8.26 of the book Ref. 47, which exposes a question about the uniqueness of the Doob-Meyer decomposition on optional sets of interval type. We would like to thank Z.Q. Chen for bringing the work in Ref.

24 to our attention. They were aware of the situation $I(\zeta)$ vs $[\![0,\zeta[\![$ and made such a point in [24, Section 3] (see in particular the first paragraph of Section 3, Theorem 3.1 and Lemma 3.2 in Ref. 24).

Let B be an optional set of interval type. We denote respectively by $(\mathcal{S}_p)^B$, $(\mathcal{M}_{loc})^B$ and $(\mathcal{A}_{loc})^B$ the sets of special semi-martingales, local martingales and adapted processes with locally integrable variation on B. The assertion below, which is referred as the Doob-Meyer decomposition on sets of interval type, was stated in [47, Theorem 8.26].

Assertion. Let B be an optional set of interval type and $Y \in (\mathcal{S}_p)^B$. Then Y can be uniquely decomposed as: $Y = M + A$, where $M \in (\mathcal{M}_{loc})^B$ and $A \in (\mathcal{A}_{loc,0})^B$ is a predictable process (i.e., A is the restriction of a predictable process on B).

Although the above assertion has been employed by several papers (including our previous paper Ref. 67), during the course of our research we observed the following remark.

Remark 4.1. In the above assertion if B is not a predictable set of interval type, then the uniqueness of the decomposition $Y = M + A$ may fail to be true.

The above remark reveals that the Doob-Meyer decomposition may fail to be unique on an optional set of interval type. In the same manner, we observe that the Fukushima type decomposition may fail to be unique on an optional set of interval type. Note that with the notation of Theorem 4.1, $[\![0,\zeta[\![$ is an optional set of interval type but is not necessarily a predictable set.

Remark 4.2. In Theorem 4.1 if we use $\mathcal{M}_{loc}^{[\![0,\zeta[\![}$ instead of $\mathcal{M}_{loc}^{I(\zeta)}$, then the uniqueness of the decomposition (4.1) may fail to be true.

We refer the readers to Ref. 72 for the proofs of the above two remarks.

4.3. *Transformation formula for MAFs*

As an application of Theorem 4.1, we give a transformation formula for MAFs.

Theorem 4.2. (Ma et al. [72, Theorem 3.1]) *Suppose that $(\mathcal{E}, D(\mathcal{E}))$ is a quasi-regular semi-Dirichlet form on $L^2(E;m)$ satisfying Assumption 4. Let $m \in \mathbb{N}$, $\Phi \in C^1(\mathbb{R}^m)$, and $u = (u_1, u_2, \ldots, u_m)$ with $u_i \in D(\mathcal{E})_{loc}$, $1 \le i \le m$. Then $\Phi(u) \in D(\mathcal{E})_{loc}$ and*

$$M^{[\Phi(u)],c} = \sum_{i=1}^{m} \Phi_{x_i}(u) \cdot M^{[u_i],c} \text{ on } I(\zeta), \quad P_x\text{-a.s. for } \mathcal{E}\text{-q.e. } x \in E.$$

Below we consider a concrete example.

Example 4.1. (see Ref. Fukushima et al. [42] and cf. also Ref. Schilling et al. [77]) Let (E,d) be a locally compact separable metric space, m be a positive Radon Measure on E with full support, and $k(x,y)$ be a nonnegative Borel measurable function on $\{(x,y) \in E \times E \,|\, x \neq y\}$. Set $k_s(x,y) = \frac{1}{2}(k(x,y) + k(y,x))$ and $k_a(x,y) = \frac{1}{2}(k(x,y) - k(y,x))$. Denote by $C_0^{lip}(E)$ the family of all uniformly Lipschitz continuous functions on E with compact support. Suppose that the following conditions hold:

(B.I) $x \to \int_{y \neq x} (1 \wedge d(x,y)^2) k_s(x,y) m(dy) \in L^1_{loc}(E;m)$.

(B.II) $\sup_{x \in E} \int_{\{y:\, k_s(x,y) \neq 0\}} \frac{k_a^2(x,y)}{k_s(x,y)} m(dy) < \infty$.

Define for $u, v \in C_0^{lip}(E)$,

$$\eta(u,v) = \iint_{x \neq y} (u(x) - u(y))(v(x) - v(y))k_s(x,y)m(dx)m(dy)$$

and

$$\mathcal{E}(u,v) = \frac{1}{2}\eta(u,v) + \iint_{x \neq y} (u(x) - u(y))v(y)k_a(x,y)m(dx)m(dy).$$

Then, there exists $\alpha > 0$ such that $(\mathcal{E}_\alpha, C_0^{lip}(E))$ is closable on $L^2(E;m)$ and its closure $(\mathcal{E}_\alpha, D(\mathcal{E}_\alpha))$ is a regular semi-Dirichlet form on $L^2(E;m)$. Moreover, there exists $C > 1$ such that for any $u \in D(\mathcal{E}_\alpha)$,

$$\frac{1}{C}\eta_\alpha(u,u) \leq \mathcal{E}_\alpha(u,u) \leq C\eta_\alpha(u,u).$$

Therefore, our Theorems 4.1 and 4.2 hold for any $u \in D(\mathcal{E})_{loc}$ which satisfies condition (S), in particular, for any $u \in D(\mathcal{E})$ by noting that $|k_a(x,y)| \leq k_s(x,y)$.

5. Semi-h-transform of semi-Dirichlet forms

In Ref. 46 (published in 2011) we discussed $h\hat{h}$-transforms of positivity preserving semigroups. We call it semi-h-transform when the semigroup is associated with a semi-Dirichlet form, because in this case we may take $h = 1$. Among other things, in Ref. 46 we found that by changing the reference measure using a coexcessive function, we can obtain a dual sub-Markov strongly continuous semigroup (cf. Theorem 5.2 below). Moreover, if $(\mathcal{E}, D(\mathcal{E}))$ is a quasi-regular semi-Dirichlet form, then by changing the reference measure using a coexcessive function, we can obtain a pair $(\mathbf{M}, \hat{\mathbf{M}})$ of m-tight special standard processes which are in duality with respect to the changed measure (cf. Theorem 5.4 below).

We are very glad that in the recent book Ref. 74, Y. Oshima obtained a dual Markov resolvent (using an argument different from ours), and introduced a very useful auxiliary bilinear form by changing the reference measure with a coexcessive function. Moreover, he obtained a dual Hunt process relative to the changed measure, and introduced a dual pseudo Hunt process by interpolating the dual process. Employing his ingenious auxiliary bilinear form and pseudo Hunt process, Y. Oshima obtained fruitful results for semi-Dirichlet forms.

From the Markov processes point of view, there are several authors having studied the issue when a dual process can be constructed, including M. Liao (dual Hunt process) and Fitzsimmons-Getoor (general Borel right process), see Refs. 65 and 33. We thank Z.Q. Chen for bringing these works to our attention.

In this section, we shall report briefly some of the results on $h\hat{h}$-transforms of positivity preserving semigroups. We refer the readers to Ref. 46 for the proofs and more details.

Let $(E;\mathcal{B})$ be a measurable space and m be a σ-finite measure on $(E;\mathcal{B})$. We consider a strongly continuous contraction semigroup $(T_t)_{t\geq 0}$ on $L^2(E;m)$. Recall that $(T_t)_{t\geq 0}$ is called positivity preserving if for any $f \in L^2(E;m)$ with $f \geq 0$, we have $T_t f \geq 0$ for all $t > 0$; $(T_t)_{t\geq 0}$ is called sub-Markovian if for any $f \in L^2(E;m)$ with $0 \leq f \leq 1$, we have $0 \leq T_t f \leq 1$ for all $t > 0$. In what follows, we assume that $(T_t)_{t\geq 0}$ is a positivity preserving semigroup on $L^2(E;m)$. Denote by $(\hat{T}_t)_{t\geq 0}$ the dual semigroup of $(T_t)_{t\geq 0}$ on $L^2(E;m)$. One finds that $(\hat{T}_t)_{t\geq 0}$ is also a positivity preserving semigroup on $L^2(E;m)$.

For given $\alpha > 0$, let $h > 0$ (m-a.e.) be an α-excessive function in $L^2(E;m)$ and $\hat{h} > 0$ (m-a.e.) be an α-coexcessive function in $L^2(E;m)$. We define for $f \in L^\infty(E;m)$ and $t \geq 0$,

$$T_t^h f := h^{-1}e^{-\alpha t}T_t(hf), \quad \hat{T}_t^{\hat{h}} f := \hat{h}^{-1}e^{-\alpha t}\hat{T}_t(\hat{h}f).$$

One finds that T_t^h and $\hat{T}_t^{\hat{h}}$ are contraction operators on $L^\infty(E;h\hat{h}\cdot m)$, and both of them can be (uniquely) continuously extended to be contraction operators on $L^1(E;h\hat{h}\cdot m)$. Therefore, by virtue of the Riesz-Thorin interpolation theorem or the representation of operators by kernels, one can show that T_t^h and $\hat{T}_t^{\hat{h}}$ can be (uniquely) continuously extended to be contraction operators on $L^2(E;h\hat{h}\cdot m)$ (cf. Refs. 46 and 86). We use still T_t^h and $\hat{T}_t^{\hat{h}}$ to denote their continuous extensions on $L^2(E;h\hat{h}\cdot m)$.

Theorem 5.1. (Hu et al. [46, Theorem 2.2]) *$(T_t^h)_{t\geq 0}$ and $(\hat{T}_t^{\hat{h}})_{t\geq 0}$ are a pair of dual sub-Markovian strongly continuous contraction semigroups on $L^2(E;h\hat{h}\cdot m)$.*

Remark 5.1. Theorem 5.1 can be applied to both symmetric and non-symmetric positivity preserving semigroups. An interesting non-symmetric positivity preserving semigroup that Theorem 5.1 can be applied to is the one induced by the generalized Feynman-Kac transformation for nearly symmetric Markov processes, see [68, Theorems 1.1 and 1.2].

Suppose that $(T_t)_{t\geq 0}$ is sub-Markovian but $(\hat{T}_t)_{t\geq 0}$ is not sub-Markovian, e.g. $(T_t)_{t\geq 0}$ is associated with a semi-Dirichlet form but not a Dirichlet form. One may perform an h-transform to $(\hat{T}_t)_{t\geq 0}$ only while keeping $(T_t)_{t\geq 0}$ unchanged. Theorem 5.2 below will be useful in this situation.

Theorem 5.2. (Hu et al. [46, Theorem 2.5]) *Suppose that $(T_t)_{t\geq 0}$ is a sub-Markovian strongly continuous contraction semigroup on $L^2(E;m)$. Then $(e^{-\alpha t}T_t)_{t\geq 0}$ and $(\hat{T}_t^{\hat{h}})_{t\geq 0}$ are a pair of dual sub-Markovian strongly continuous contraction semigroups on $L^2(E;\hat{h}\cdot m)$.*

We now assume that E is a Hausdorff topological space, $\mathcal{B} := \mathcal{B}(E)$ is the Borel σ-algebra of E, and m is a σ-finite measure on $(E;\mathcal{B})$. Let $(T_t)_{t\geq 0}$ be an analytic positivity preserving semigroup on $L^2(E;m)$. Then, by [70, I.2] and [71, Theorem 1.5], $(T_t)_{t\geq 0}$ is associated with a positivity preserving coercive (closed) form $(\mathcal{E}, D(\mathcal{E}))$ on $L^2(E;m)$. Let $\alpha > 0$ and $h > 0$ (m-a.e.) be an α-excessive function in $D(\mathcal{E})$. We define

$$\begin{cases} D(\mathcal{E}^h) := \{u \in L^2(E;h^2\cdot m) \,|\, hu \in D(\mathcal{E})\}, \\ \mathcal{E}^h(u,v) := \mathcal{E}(hu,hv) \quad \text{for } u,v \in D(\mathcal{E}^h). \end{cases}$$

$(\mathcal{E}^h, D(\mathcal{E}^h))$ is called the h-transform of $(\mathcal{E}, D(\mathcal{E}))$.

We say that $(T_t)_{t\geq 0}$, or equivalently $(\mathcal{E}, D(\mathcal{E}))$, is h-associated with a special standard process $\mathbf{M} = ((X_t)_{t\geq 0}, (P_x)_{x\in E_\Delta})$ if for all $t > 0$, $he^{\alpha t}E_\cdot[(\tilde{f}h^{-1})(X_t)]$ is an m-version of $T_t f$ for any m-version $\tilde{f}$ of $f \in L^2(E;m)$. We say that $(T_t)_{t\geq 0}$, or equivalently $(\mathcal{E}, D(\mathcal{E}))$, is properly h-associated with $\mathbf{M}$ if, in addition, for all $t > 0$, $he^{\alpha t}E_\cdot[(\tilde{f}h^{-1})(X_t)]$ is $\mathcal{E}$-quasi-continuous for any m-version $\tilde{f}$ of $f \in L^2(E;m)$. Here E_x denotes the expectation w.r.t. P_x for $x \in E$.

Let $\tilde{h}$ be a strictly positive m-version of some m-a.e. strictly positive and finite valued α-excessive function $h \in D(\mathcal{E})$. By [71, Theorem 5.2], a necessary and sufficient condition for $(\mathcal{E}, D(\mathcal{E}))$ to be properly $\tilde{h}$-associated with an m-tight special standard process $\mathbf{M}$ is that $(\mathcal{E}, D(\mathcal{E}))$ is quasi-regular. In this case $(\mathcal{E}_\alpha^h, D(\mathcal{E}^h))$ is the quasi-regular semi-Dirichlet form associated with $\mathbf{M}$ on $L^2(E;h^2\cdot m)$. However, in general, $(\mathcal{E}_\alpha^h, D(\mathcal{E}^h))$ is not necessarily a Dirichlet form. We are interested in giving a pair $(\mathbf{M}, \hat{\mathbf{M}})$ of standard processes which are associated with $(\mathcal{E}, D(\mathcal{E}))$ and are in duality in some sense. We have the following result.

Theorem 5.3. (Hu et al. [46, Theorem 3.1]) *Let $(\mathcal{E}, D(\mathcal{E}))$ be a quasi-regular positivity preserving coercive form on $L^2(E;m)$ and $\alpha > 0$. Then, there exist m-a.e. strictly positive α-excessive function $h \in D(\mathcal{E})$, m-a.e. strictly positive α-coexcessive function $\hat{h} \in D(\mathcal{E})$, and a pair $(\mathbf{M}, \hat{\mathbf{M}})$ of m-tight special standard processes satisfying the following properties:*

(i) $\mathbf{M}$ is properly h-associated with $(\mathcal{E}, D(\mathcal{E}))$ and $\hat{\mathbf{M}}$ is properly $\hat{h}$-associated with $(\hat{\mathcal{E}}, D(\mathcal{E}))$.

(ii) $\mathbf{M}$ and $\hat{\mathbf{M}}$ are in weak duality w.r.t. $h\hat{h} \cdot m$.

(iii) All semipolar sets of $\mathbf{M}$ and $\hat{\mathbf{M}}$ are $\mathcal{E}$-exceptional.

Remark 5.2. By Theorem 5.3, any h-associated process of a positivity preserving coercive form can be studied in the framework of Markov processes in weak duality. Hence many existing literatures such as [19, Chapter VI], Ref. 38, [28, Chapter 13] and Ref. 26 can be applied. In Ref. 23 the general perturbation of symmetric Markov processes is studied, which is a combination of Girsanov transforms, continuous and discontinuous Feynman-Kac transforms, and generalized Feynman-Kac transforms. We may use Theorem 5.3 to construct a pair of dual standard processes $(\mathbf{M}, \hat{\mathbf{M}})$ which are $h\hat{h}$-associated with the positivity preserving semigroup $(T_t)_{t\geq 0}$ given in [23, Theorem 3.1]. The dual processes $(\mathbf{M}, \hat{\mathbf{M}})$ can help us analyze further the perturbed semigroup $(T_t)_{t\geq 0}$.

Theorem 5.4. (Hu et al. [46, Theorem 3.3]) *Let $(\mathcal{E}, D(\mathcal{E}))$ be a quasi-regular semi-Dirichlet form on $L^2(E;m)$ and $\alpha > 0$. Then, there exist m-a.e. strictly positive α-coexcessive function $\hat{h} \in D(\mathcal{E})$, and a pair $(\mathbf{M}, \hat{\mathbf{M}})$ of m-tight special standard processes satisfying the following properties:*

(i) $\mathbf{M}$ is properly associated with $(\mathcal{E}_\alpha, D(\mathcal{E}))$ and $\hat{\mathbf{M}}$ is properly $\hat{h}$-associated with $(\hat{\mathcal{E}}, D(\mathcal{E}))$.

(ii) $\mathbf{M}$ and $\hat{\mathbf{M}}$ are in weak duality w.r.t. $\hat{h} \cdot m$.

(iii) All semipolar sets of $\mathbf{M}$ and $\hat{\mathbf{M}}$ are $\mathcal{E}$-exceptional.

6. Nonlinear filtering of semi-Dirichlet processes

The optimal filtering problem consists of computing the conditional distributions of the *signal* process $(X_t)_{t\geq 0}$ based on the collection $(Y_t)_{t\geq 0}$ of partial and noisy *observations*

$$Y_t = \int_0^t h(X_s)ds + W_t, \tag{6.1}$$

where $h : E \to \mathbb{R}^d$ is measurable and W is the standard d-dimensional Brownian motion independent of X. Under the assumption that X is the solution of a classical martingale problem or satisfies an stochastic differential equation (SDE), the well-known FKK (Fujisaki-Kallianpur-Kunita) (also known as Kushner-Stratonovich) and DMZ (Duncan-Mortensen-Zakai) equations have been established respectively for the normalized and unnormalized filtering processes. However, these fundamental results cannot be applied directly to the Markov processes that are not characterized by solutions of classical martingale problems or SDEs, which arise naturally in many applications.

The framework of semi-Dirichlet forms provides a useful tool to consider filtering of Markov processes associated with singular generators living possibly in infinite-dimensional state spaces. In this section, we shall report our results on nonlinear filtering of semi-Dirichlet processes. The material is taken from Ref. 50 to which the readers are referred for more details.

Let $(X_t)_{t\geq 0}$ be a right process taking values in a metrizable Lusin space E, $\mathcal{B}(E)$ be the Borel σ-algebra of E, and m be a σ-finite measure on $(E;\mathcal{B}(E))$. Suppose that $(X_t)_{t\geq 0}$ is associated with a semi-Dirichlet form $(\mathcal{E}, D(\mathcal{E}))$ on $L^2(E;m)$. By Ref. 32, $(\mathcal{E}, D(\mathcal{E}))$ must be quasi-regular. Then, every element $u \in D(\mathcal{E})$ admits an $\mathcal{E}$-quasi-continuous m-version, which we denote by $\tilde{u}$. Note that $\tilde{u}$ is uniquely determined except for an $\mathcal{E}$-exceptional set. We use $\tilde{D}(\mathcal{E})$ to denote the family of $\mathcal{E}$-quasi-continuous versions of elements in $D(\mathcal{E})$. Without loss of generality, we assume that every element $u \in \tilde{D}(\mathcal{E})$ is Borel measurable. We use the same notation for a function f (m-a.e. defined) on E and for the m-equivalence class of functions represented by f.

Let $(L, D(L))$ be the L^2-generator of $(\mathcal{E}, D(\mathcal{E}))$. Define

$$\tilde{D}(L) := \{G_1 f \mid f \in \mathcal{Q}_b(E) \cap L^2(E;m)\},$$

where $(G_\alpha)_{\alpha\geq 0}$ denotes the resolvent corresponding to $(\mathcal{E}, D(\mathcal{E}))$ and $\mathcal{Q}_b(E)$ denotes the family of bounded measurable functions on E that have $\mathcal{E}$-quasi-continuous m-versions. Denote $D_b(\mathcal{E}) := \mathcal{B}_b(E) \cap D(\mathcal{E})$. Note that $LG_\alpha f = \alpha G_\alpha f - f$ for $f \in L^2(E;m)$ and $\alpha > 0$. Then $G_\alpha(D_b(\mathcal{E})) \subset \tilde{D}(L)$ and thus $\tilde{D}(L)$ is dense in $D(L)$ w.r.t. the norm on $D(L)$. By [69, Proposition 2.19], any $\mathcal{E}$-quasi-continuous function in $\mathcal{B}_b(E) \cap L^2(E;m)$ is bounded $\mathcal{E}$-q.e. Hence any function $f \in \tilde{D}(L)$ is bounded $\mathcal{E}$-q.e. Moreover, since X is a right process, $(s,\omega) \to \widetilde{Lf}(X_s)(\omega)$ is jointly measurable.

Assume that the *initial distribution* ν of X charges no $\mathcal{E}$-exceptional sets. We denote

$$P(\cdot) = \int_E P_x(\cdot)\nu(dx)$$

and $\mathcal{Y}_t := \sigma(\{Y_s, 0 \leq s \leq t\}, \mathcal{N})$, where Y is defined in (6.1) and $\mathcal{N}$ is the collection of P-null sets. Our goal is to estimate the conditional distributions

$$\pi_t(\cdot) := P(X_t \in \cdot \mid \mathcal{Y}_t), \quad t \geq 0.$$

We assume that the observation function h satisfies

$$P\left(\int_0^T |h(X_t)|^2 dt < \infty\right) = 1 \tag{6.2}$$

and

$$\int_0^T E[|h(X_t)|]dt < \infty. \tag{6.3}$$

Hereafter $|\cdot|$ denotes the Euclidean distance. Note that conditions (6.2) and (6.3) assume weaker moment conditions on X than the following standard finite energy condition

$$\int_0^T E[|h(X_t)|^2]dt < \infty.$$

Denote by $\mathcal{P}(E)$ the family of probability measures on $(E, \mathcal{B}(E))$. We fix a $T > 0$. Since X is a right process, there exists a version of $(\pi_t)_{0\leq t\leq T}$ which belongs to $D([0,T], \mathcal{P}(E))$ (see Ref. 87). In the sequel, we will use this version without further comment.

Set $\mathcal{X}_t := \sigma\{\{X_s, 0 \le s \le t\}, \mathcal{N}\}$. For $f \in \tilde{D}(L)$ and $t \ge 0$, define

$$M_t(f) := \tilde{f}(X_t) - \tilde{f}(X_0) - \int_0^t \widetilde{Lf}(X_s)ds.$$

Note that different choices of $\mathcal{E}$-quasi-continuous m-versions of f and Lf give indistinguishable versions of the process $(M_t)_{t\ge 0}$. Denote by $(P_t)_{t\ge 0}$ and $(T_t)_{t\ge 0}$ the transition semigroup and L^2-semigroup of $(X_t)_{t\ge 0}$, respectively. Note that $P_t\tilde{f} - \int_0^t P_s\widetilde{Lf}ds = \tilde{f}$ $\mathcal{E}$-q.e. for $f \in \tilde{D}(L)$ and $t \ge 0$. By [30, Page 162], $(M_t(f), \mathcal{X}_t, P)$ is a martingale.

We define the innovation process $I_t = Y_t - \int_0^t \pi_s(h)ds$. Then, by [35, Lemma 2.2 and Remark 2.1], we know that $(I_t, \mathcal{Y}_t, P)$ is a d–dimensional continuous martingale under condition (6.3). Hereafter, we use $\langle , \rangle$ to denote the standard inner product of $\mathbb{R}^d$. Now we define

$$A_t = \exp\left\{\int_0^t \langle h(X_s), dY_s\rangle - \frac{1}{2}\int_0^t |h(X_s)|^2 ds\right\} \tag{6.4}$$

and

$$\mathcal{XY}_t := \sigma\{\{X_t, 0 \le t < \infty\}, \mathcal{Y}_t, \mathcal{N}\}.$$

Note that by (6.1) and (6.4), we get

$$A_t^{-1} = \exp\left\{\int_0^t \langle -h(X_s), dW_s\rangle - \frac{1}{2}\int_0^t |h(X_s)|^2 ds\right\}.$$

By the independence of X and W, one finds that $(A_t^{-1}, \mathcal{XY}_t, P)$ is a martingale and $(W_t, \mathcal{XY}_t, P)$ is a d-dimensional Brownian motion (cf. [55, Lemma 11.3.1]). Define a new probability measure $\overline{P}$ via the Girsanov transformation

$$\left.\frac{d\overline{P}}{dP}\right|_{\mathcal{XY}_t} = A_t^{-1} = E[A_T^{-1}|\mathcal{XY}_t], \quad 0 \le t \le T.$$

Then, $\overline{P} \circ (X, Y)^{-1} = P \circ (X, W)^{-1}$, $(A_t, \mathcal{XY}_t)_{0\le t\le T}$ is a martingale under $\overline{P}$, and one has the Kallianpur-Striebel formula

$$\pi_t(f) = E[\tilde{f}(X_t)|\mathcal{Y}_t] = \frac{\overline{E}[\tilde{f}(X_t)A_t|\mathcal{Y}_t]}{\overline{E}[A_t|\mathcal{Y}_t]} := \frac{\sigma_t(\tilde{f})}{\sigma_t(1)}, \quad f \in \tilde{D}(L).$$

Now we can state the filtering equations in the setting of quasi-regular semi-Dirichlet forms.

Theorem 6.1. (Hu et al. [50, Theorem 2.2]) *Suppose that h satisfies (6.2) and (6.3). Then, the DMZ and FKK equations*

$$\sigma_t(\tilde{f}) = \sigma_0(\tilde{f}) + \int_0^t \sigma_s(\widetilde{Lf})ds + \int_0^t \langle \sigma_s(h\tilde{f}), dY_s\rangle, \quad \overline{P}\text{-a.s.}, \tag{6.5}$$

$$\pi_t(\tilde{f}) = \pi_0(\tilde{f}) + \int_0^t \pi_s(\widetilde{Lf})ds + \int_0^t \langle \pi_s(h\tilde{f}) - \pi_s(h)\pi_s(\tilde{f}), dI_s\rangle, \quad P\text{-a.s.}$$

hold for all $f \in \tilde{D}(L)$ and $t \in [0, T]$.

We next consider the uniqueness of solutions of the filtering equations. To simplify notation, from now on we consider only the one dimensional case, i.e., $d = 1$. But the similar results hold for general d. We denote by $\mathcal{M}_b(E)$ the family of finite signed measures on $(E; \mathcal{B}(E))$. In what follows, we set $\widetilde{T_t f}(x) = P_t f(x) = E_x[f(X_t)]$ whenever the latter is well-defined. We have the following equivalence of weak solution and mild solution.

Lemma 6.1. (Hu et al. [50, Lemma 3.1]) *Let $(v_t)_{0\le t\le T}$ be an $\mathcal{M}_b(E)$-valued càdlàg process. Suppose that*

$$\overline{P}\left\{\int_0^T |v_s|^2(|h|)ds < \infty\right\} = 1.$$

Then the following two assertions are equivalent.
(i) For any $f \in \tilde{D}(L)$,

$$v_t(\tilde{f}) = v_0(\tilde{f}) + \int_0^t v_s(\widetilde{Lf})ds + \int_0^t v_s(h\tilde{f})dY_s, \quad \overline{P}\text{-a.s.}$$

(ii) For any $f \in \mathcal{B}_b(E)$,

$$v_t(f) = v_0(\widetilde{T_t f}) + \int_0^t v_s(h\widetilde{T_{t-s}f})dY_s, \quad \overline{P}\text{-a.s.}$$

By virtue of Lemma 6.1, we can obtain the following result.

Theorem 6.2. (Hu et al. [50, Corollary 3.7]) *Suppose that h satisfies (6.2), (6.3) and*

$$\sum_{n=1}^{\infty}\int_0^T\int_0^{t_1}\cdots\int_0^{t_n}\nu^2(T_{t_n}(|h|\widetilde{T_{t_{n-1}-t_n}}(|h|\cdots T_{t_1}|h|)))dt_{n-1}\cdots dt_2dt_1 < \infty.$$

Then $\{\sigma_t\}_{0\le t\le T}$, the unnormalized filtering process, is the unique $\mathcal{M}_b(E)$-valued solution of the DMZ equation (6.5). Moreover, we have the unique Wiener chaos expansion

$$\begin{aligned}\sigma_t(f) &= \nu(\widetilde{T_t f}) + \int_0^t \nu(T_{t_1}(\widetilde{hT_{t-t_1}}f))dY_{t_1}\\ &+\int_0^t\int_0^{t_1}\nu(T_{t_2}(hT_{t_1}\widetilde{-t_2(hT_{t-t_1}}f)))dY_{t_2}dY_{t_1}+\cdots, \quad \overline{P}\text{-a.s.}, \quad \forall f\in\mathcal{B}_b(E).\end{aligned}$$

Finally, we consider the density equations of the filtering processes. Suppose that ν is absolutely continuous w.r.t. m and denote $u_0 := d\nu/dm$. Note that

$$\sigma_t(f) = \overline{E}[f(X_t)A_t|\mathcal{Y}_t] = \overline{E}\{\overline{E}[A_t|(\sigma(X_t)\vee\mathcal{Y}_t)]f(X_t)|\mathcal{Y}_t\}.$$

Since $\overline{E}[A_t|(\sigma(X_t)\vee\mathcal{Y}_t)] = g(X_t, Y_{[0,t]})$ for some positive measurable function g on $E\times(\mathbb{R}^d)^{[0,t]}$,

$$\sigma_t(f) = \int_E P_t\{g(\cdot, Y_{[0,t]})f\}d\sigma_0$$

by the independence of X and Y under $\overline{P}$. Since X is associated with a semi-Dirichlet form, $f = 0$ m-a.e. implies that $P_t\{g(\cdot, Y_{[0,t]})f\} = 0$ m-a.e., hence $\{\sigma_t\}_{0\le t\le T}$ is absolutely continuous w.r.t. m.

Theorem 6.3. (Hu et al. [50, Theorem 4.1]) *Suppose that h satisfies (6.2), (6.3) and*

$$\sum_{n=1}^{\infty}\int_0^T\int_0^{t_1}\cdots\int_0^{t_n}\nu^2(T_{t_n}(|h|\widetilde{T_{t_{n-1}-t_n}}(|h|\cdots T_{t_1}|h|)))dt_{n-1}\cdots dt_2dt_1<\infty.$$

Let $(v_t)_{0\le t\le T}$ be an $L^1(E;m)$-valued càdlàg process satisfying the following conditions.
(i) $v_0=u_0$.
(ii) $\int_0^T\overline{E}[(\int_E|v_th|dm)^2]dt<\infty$.
(iii) For any $f\in\tilde{D}(L)$, $\{(v_t,f)_m\}_{0\le t\le T}$ is a $\{\mathcal{Y}_t\}_{0\le t\le T}$ semi-martingale with

$$(v_t,f)_m=(v_0,f)_m+\int_0^t(v_s,Lf)_mds+\int_0^t(v_s,hf)_mdY_s,\quad \overline{P}\text{-a.s.},$$

where $(v_t,f)_m:=\int_E v_tfdm$.
Denote $u_t=d\sigma_t/dm$ for $t\in[0,T]$. Then $v_t=u_t$ for all $t\in[0,T]$.

Remark 6.1. By virtue of Poisson chaos expansion, we may also consider nonlinear filtering of semi-Dirichlet processes with marked point process observations. We refer the interested reader to the recent paper Ref. 83 for this connection.

7. Hunt's hypothesis (H) and Getoor's conjecture

In this section, we shall discuss Hunt's hypothesis (H), which is an important and powerful principle in the potential theory of Markov processes.

Let X be a time-homogeneous Markov process. Hunt's hypothesis (H) says that "every semipolar set of X is polar". To illustrate its importance, let us recall some potential-theoretic principles. Assume that E is a locally compact space with a countable base. Let (X,P^x) and $(\hat{X},\hat{P}^x)$ be a pair of dual standard Markov processes on E as described in [19, VI]. Denote by $\mathcal{B}^n$ the family of all nearly Borel measurable subsets of E. For $D\subset E$, we define the first hitting time of D by

$$\sigma_D:=\inf\{t>0\,|\,X_t\in D\}.$$

A set $D\subset E$ is called polar if there exists a set $C\in\mathcal{B}^n$ such that $D\subset C$ and $P^x(\sigma_C<\infty)=0$ for every $x\in E$. D is called a thin set if there exists a set $C\in\mathcal{B}^n$ such that $D\subset C$ and $P^x(\sigma_C=0)=0$ for every $x\in E$. D is called semipolar if $D\subset\bigcup_{n=1}^{\infty}D_n$ for some thin sets $\{D_n\}_{n=1}^{\infty}$.

Denote by E^x the expectation w.r.t. P^x. Let $\alpha>0$. Denote by $(U^\alpha)_{\alpha>0}$ the resolvent operators for X.

- **Bounded positivity principle**: If ν is a finite signed measure such that $U^\alpha\nu$ is bounded, then $\nu U^\alpha\nu\ge 0$, where $\nu U^\alpha\nu:=\int_E U^\alpha\nu(x)\nu(dx)$.
- **Bounded energy principle**: If ν is a finite measure with compact support such that $U^\alpha\nu$ is bounded, then ν does not charge semipolar sets.
- **Bounded maximum principle**: If ν is a finite measure with compact support K such that $U^\alpha\nu$ is bounded, then $\sup\{U^\alpha\nu(x)\,|\,x\in E\}=\sup\{U^\alpha\nu(x)\,|\,x\in K\}$.
- **Bounded regularity principle**: If ν is a finite measure with compact support such that $U^\alpha\nu$ is bounded, then $U^\alpha\nu$ is regular.

It has been proved in literatures that (see Ref. 54 for a summary of the proofs), if all 1-excessive (equivalently, all α-excessive, $\alpha > 0$) functions are lower semicontinuous, then all the above principles are equivalent to Hunt's hypothesis (H). In addition, (H) holds if and only if the fine and cofine topologies differ by polar sets.

It is well-known that (H) holds for Markov processes associated with Dirichlet forms, or more generally, semi-Dirichlet forms (see Refs. 81 and 32). (H) has many important consequences in the theory of Dirichlet forms. For example, it implies that a function in $D(\mathcal{E})$ is quasi-continuous if and only if it is finely continuous q.e.

In spite of its importance, Hunt's hypothesis (H) is still not fully understood. It remains an open problem to characterize when a Markov process satisfies (H). Below we introduce two recent results on this topic.

Theorem 7.1. (see Ref. Wang, [86]) *Let $(G_n)_{n\geq 1}$ be an increasing sequence of nearly Borel measurable finely open sets satisfying $E = \bigcup_{n\geq 1} G_n$. Then, X satisfies (H) if and only if for each $n \geq 1$, the part process X^{G_n} on G_n satisfies (H).*

Theorem 7.2. (Sun et al. [84, Theorem 1.2]) *Let U be an open set of $\mathbb{R}^n$ and m be a positive Radon measure on U such that* $\mathrm{supp}[m] = U$. *Suppose that $(A, D(A))$ is a generator on $L^2(U; m)$ such that A is a Dirichlet operator and $C_0^\infty(U) \subset D(A) \cap D(\hat{A})$. Let $\delta > 0$ be a constant such that $U^\delta := \{x \in U \mid \inf_{y\in\partial U} |x - y| > \delta\} \neq \emptyset$. Then, we have the decomposition:*

$$
\begin{aligned}
(-Au, v) &= \frac{1}{2}\sum_{i,j=1}^{n} \int_U \frac{\partial u}{\partial x_i}\frac{\partial v}{\partial x_j}\nu_{ij}(dx) + \sum_{i=1}^{n}\int_{U^\delta} \frac{\partial u}{\partial x_i}(x)v(x)\nu_i^\delta(dx) \\
&+ \int_{U\times U\backslash d} \sum_{i=1}^{n}(y_i - x_i)\left(\frac{\partial u}{\partial y_i}(y)v(y) - \frac{\partial u}{\partial x_i}(x)v(x)\right) I_{\{|x-y|\leq\delta\}}(x, y)J(dxdy) \\
&+ \int_{U\times U\backslash d} 2\left(u(y) - u(x) - \sum_{i=1}^{n}(y_i - x_i)\frac{\partial u}{\partial y_i}(y)I_{\{|x-y|\leq\delta\}}(x, y)\right) v(y)J(dxdy) \\
&+ \int_U u(x)v(x)K(dx), \qquad \forall u, v \in C_0^\infty(U^\delta),
\end{aligned}
$$

where J and K are the jumping and killing measures, respectively, $\{\nu_{ij}\}_{i,j=1}^n$ are signed Radon measures on U such that for any compact set $G \subset U$, $\nu_{ij}(G) = \nu_{ji}(G)$ and $\sum_{i,j=1}^n \xi_i\xi_j\nu_{ij}(G) \geq 0$ for all $(\xi_1, \ldots, \xi_n) \in \mathbb{R}^n$, and $\{\nu_i^\delta\}_{i=1}^n$ are signed Radon measures on U^δ.

Recall that an L^2-generator $(A, D(A))$ on $L^2(U; m)$ is a Dirichlet operator if $(Au, (u - 1)\vee 0) \leq 0$ for all $u \in D(A)$, where $(\, , \,)$ denotes the inner product of $L^2(U; m)$. Note that in Theorem 7.2, the weak sector condition is not assumed for $(A, D(A))$ and the co-generator $(\hat{A}, D(\hat{A}))$ is not assumed to be a Dirichlet operator. Let X be a Markov process living on an open set of $\mathbb{R}^n$ without jump (i.e. $J = 0$). Theorem 7.2 implies that X consists of three parts: a symmetric diffusion, a deterministic drift and a symmetric killing part. If the diffusion part is non-degenerate or, more generally, the drift part is locally well-controlled by the diffusion part and the killing part, then X is locally associated with a semi-Dirichlet form. In this case, Theorem 7.1 implies that X satisfies (H). If the drift part is not well-controlled by the other two symmetric parts, we cannot expect that (H) always holds.

If a Markov process has jumps, things become much more complicated. It is not hard to construct a pure jump process that does not satisfy (H). For example, we consider a

Markov process X living on the real line. For any $x < 0$, we let the process stay in x for an exponential distributed time with mean 1 and then jump to 0. For any $x \geq 0$, we let the process move to the right according to a one-sided α-stable process for some $0 < \alpha < 1$. It is easy to see that $\{0\}$ is a thin set but not a polar set for X. Hence (H) is not satisfied for X. Note that in this example, the Markov process X is not spatially homogeneous. Motivated by this example, when we consider (H) for Markov processes with jumps, we may first concentrate on Lévy processes.

Some forty years ago, R. K. Getoor conjectured that essentially all Lévy processes satisfy (H). Let $(\Omega, \mathcal{F}, P)$ be a probability space and $X = (X_t)_{t \geq 0}$ be an $\mathbb{R}^n$-valued Lévy process on $(\Omega, \mathcal{F}, P)$ with Lévy-Khintchine exponent ψ, i.e.,

$$E[\exp\{i\langle z, X_t\rangle\}] = \exp\{-t\psi(z)\},\ \ z \in \mathbb{R}^n, t \geq 0,$$

where E denotes the expectation with respect to P and $\langle \cdot, \cdot \rangle$ denotes the Euclidean inner product of $\mathbb{R}^n$. The classical Lévy-Khintchine formula tells us that

$$\psi(z) = i\langle a, z\rangle + \frac{1}{2}\langle z, Qz\rangle + \int_{\mathbb{R}^n} \left(1 - e^{i\langle z,x\rangle} + i\langle z, x\rangle 1_{\{|x|<1\}}\right) \mu(dx),$$

where $a \in \mathbb{R}^n, Q$ is a symmetric nonnegative definite $n \times n$ matrix, and μ is a measure (called the Lévy measure) on $\mathbb{R}^n \backslash \{0\}$ satisfying $\int_{\mathbb{R}^n \backslash \{0\}} (1 \wedge |x|^2)\mu(dx) < \infty$. Hereafter, we use $\mathrm{Re}(\psi)$ and $\mathrm{Im}(\psi)$ to denote the real and imaginary parts of ψ, respectively, and use (a, Q, μ) to denote ψ.

We now recall some known results obtained so far for Getoor's conjecture. When $n = 1$, H. Kesten [57] showed that if X is not a compound Poisson process, then every $\{x\}$ is non-polar if and only if

$$\int_0^\infty \mathrm{Re}([1 + \psi(z)]^{-1})dz < \infty.$$

Blumenthal and Getoor [19] showed that all stable processes with index $\alpha \in (0, 2)$ on the line satisfy (H). M. Kanda [56] and G. Forst [34] proved that (H) holds if X has bounded continuous transition densities and the Lévy-Khintchine exponent ψ satisfies

$$|\mathrm{Im}(\psi)| \leq M(1 + \mathrm{Re}(\psi)) \tag{7.1}$$

for some positive constant M. Note that a Lévy process satisfies the weak sector condition (w.r.t. the Lebesgue measure) if and only if (7.1) holds. In Ref. 81, M. L. Silverstein extended the result of Kanda and Forst and showed that the sector condition implies that semipolar sets are quasi-polar.

Let X be a Lévy process on $\mathbb{R}^n$ with Lévy-Khintchine exponent (a, Q, μ). Hu and Sun [53] showed that if Q is non-degenerate then X satisfies (H); if Q is degenerate then, under the assumption that $\mu(\mathbb{R}^n \backslash \sqrt{Q}\mathbb{R}^n) < \infty$, X satisfies (H) if and only if the equation

$$\sqrt{Q}y = -a - \int_{\{x \in \mathbb{R}^n \backslash \sqrt{Q}\mathbb{R}^n :\, |x|<1\}} x\mu(dx)$$

has at least one solution $y \in \mathbb{R}^n$.

In Ref. 75, M. Rao relaxed the Kanda-Forst condition. He proved that if all 1-excessive functions of a Lévy process X are lower semicontinuous and $|\mathrm{Im}(\psi)| \leq (1 + \mathrm{Re}(\psi))f(1 + \mathrm{Re}(\psi))$, where f is an increasing function on $[1, \infty)$ such that $\int_N^\infty (\lambda f(\lambda))^{-1} d\lambda = \infty$ for every $N \geq 1$, then X satisfies (H). Rao's condition is one of the weakest sufficient conditions obtained so far for the validity of (H). However, a type of truncated α-stable processes has been constructed recently in Ref. 54 for which Rao's condition is not satisfied. It is unknown whether (H) holds for this type of subordinators.

It is shown in Ref. 53 that if a subordinator satisfies (H) then its drift coefficient must be 0. To the best of our knowledge, which subordinators satisfy (H) is unknown in general. We refer the readers to Ref. 54 for some examples of subordinators that satisfy Rao's condition and hence (H). Finally, we use the following interesting result obtained by Glover and Rao to end this section.

Theorem 7.3. (see Ref. Glover et al. [45]) *Let* $(X_t)_{t\geq 0}$ *be a standard process on a locally compact space with a countable base and* $(T_t)_{t\geq 0}$ *be an independent subordinator satisfying Hunt's hypothesis (H). Then* $(X_{T_t})_{t\geq 0}$ *satisfies (H).*

References

1. S. Albeverio, T. Hida, J. Potthoff, M. Röckner, and L. Streit, Dirichlet forms in terms of white noise analysis I-construction and QFT examples, *Rev. Math. Phys.* **1**, 291-312, (1990).
2. S. Albeverio, T. Hida, J. Potthoff, M. Röckner, and L. Streit, Dirichlet forms in terms of white noise analysis II-closability and diffusion processes, *Rev. Math. Phys.* **1**, 313-323, (1990).
3. S. Albeverio and R. Høegh-Krohn, Dirichlet forms and diffusion processes on rigged Hilbert spaces, *Z. Wahrsch. verw. Geb.* **40**, 1-57, (1977).
4. S. Albeverio and Z. M. Ma, Necessary and sufficient conditions for the existence of m-perfect processes associated with Dirichlet forms, *Séminaire de Probabilités, XXV*, Lect. Notes in Math. **1485**, 374–406, (1991). Springer, Berlin.
5. S. Albeverio and Z. M. Ma, A general correspondence between Dirichlet forms and right processes, *Bull. Amer. Math. Soc. (N.S.)* **26**, 245-252, (1991).
6. S. Albeverio, Z. M. Ma, and M. Röckner, A Beurling-Deny type structure theorem for Dirichlet forms on general state spaces. In *Memorial Volume for R. Hϕegh-Krohn, vol. 1. Ideals and Methods in Math. Anal. Stochastics and Appl.*, (Ed. S. Albeverio et al.), 115-123, (1992). Cambridge Univ. Press, Cambridge.
7. S. Albeverio, Z. M. Ma, and M. Röckner, Non-symmetric Dirichlet forms and Markov processes on general state space, *C. R. Acad. Sci. Paris Sér. I Math.* **314**, 77-82, (1992).
8. S. Albeverio, Z. M. Ma, and M. Röckner, Quasi-regular Dirichlet forms and Markov processes, *J. Funct. Anal.* **111**, 118–154, (1993).
9. S. Albeverio, Z. M. Ma, and M. Röckner, Local property of Dirichlet forms and diffusions on general state spaces, *Math. Ann.* **296**, 677-686, (1993).
10. S. Albeverio, Z. M. Ma, and M. Röckner, A remark on the support of cadlag processes, Stochastic Processes. A festschrift in honour of G. Kallianpur (S. Cambanis et al., eds.), 1-5, (1993). Springer, Berlin.
11. S. Albeverio, Z. M. Ma, and M. Röckner, Characterization of (non-symmetric) semi-Dirichlet forms associated with Hunt processes, *Random Oper. and Stoch. Equ.* **3**, 161-179, (1995).
12. S. Albeverio, Z. M. Ma, and M. Röckner, Quasi regular Dirichlet forms and the stochastic quantization problem, in this volume, (2014).
13. S. Albeverio and M. Röckner, Classical Dirichlet forms on topological vector spaces-the construction of the associated diffusion process, *Probab. Theory Relat. Fields* **83**, 405-434, (1989).

14. S. Albeverio and M. Röckner, Classical Dirichlet forms on topological vector spaces-closability and Cameron-Martin formula, *J. Funct. Anal.* **88**, 395-436, (1990).
15. S. Albeverio and M. Röckner, Stochastic differential equations in infinite dimensions: solution via Dirichlet forms, *Probab. Theory Relat. Fields* **89**, 347-386, (1991).
16. A. Beurling and J. Deny, Espaces de Dirichlet, *Acta Math.* **99**, 203-224, (1958).
17. A. Beurling and J. Deny, Dirichlet spaces, *Proc. Nat. Acad. Sci. U.S.A.* **45**, 208-215, (1959).
18. L. Beznea, N. Boboc, and M. Röckner, Quasi-regular topologies for L^p-resolvents and semi-Dirichlet forms, *Potential Anal.* **25**, 269-282, (2006).
19. R. M. Blumenthal and R. K. Getoor, *Markov Processes and Potential Theory*, (1968). Academic Press, New York.
20. N. Bouleau and F. Hirsch, *Dirichlet Forms and Analysis on Wiener Space*, de Gruyter Studies in Math. **14** (1991). Walter de Gruyter, Berlin.
21. S. Carillo-Menendez, Processus de Markov associé à une forme de Dirichlet non symétrique, *Z. Wahrsch. verw. Geb.* **33**, 139-154, (1975).
22. Z. Q. Chen, P. J. Fitzsimmons, K. Kuwae, and T. S. Zhang, Stochastic calculus for symmetric Markov processes, *Ann. Probab.* **36**, 931-970, (2008).
23. Z. Q. Chen, P. J. Fitzsimmons, K. Kuwae, and T. S. Zhang, Perturbation of symmetric Markov Processes, *Probab. Theory Relat. Fields.* **140**, 239-275, (2008).
24. Z. Q. Chen, P. J. Fitzsimmons, M. Takeda, J. Ying, and T. S. Zhang, Absolute continuity of symmetric Markov processes. *Ann. Probab.* **32**, 2067-2098, (2004).
25. Z. Q. Chen and M. Fukushima, *Symmetric Markov Processes, Time Change, and Boundary Theory*, (2012). Princeton University Press.
26. Z. Q. Chen, M. Fukushima, and J. G. Ying, Entrance law, exit system and Levy system of time-changed processes, *Illinois J. Math.* **50**, 269-312, (2006).
27. Z. Q. Chen, Z. M. Ma, and M. Röckner, Quasi-homeomorphisms of Dirichlet forms, *Nagoya Math. J.* **136**, 1-15, (1994).
28. K. L. Chung and J. B. Walsh, *Markov Processes, Brownian Motion, and Time Symmetry*, (2005). Springer, Berlin.
29. Z. Dong, Quasi-regular topologies for Dirichlet forms on arbitrary Hausdorff spaces, *Acta Math. Sinica, New Series Supplement.* **14**, 683-690, (1998).
30. S. N. Ethier and T. G. Kurtz, *Markov Processes, Characterization and Convergence*, (1986). Wiley.
31. P. J. Fitzsimmons, Markov processes and nonsymmetric Dirichlet forms without regularity, *J. Funct. Anal.* **85**, 287-306, (1989).
32. P. J. Fitzsimmons, On the quasi-regularity of semi-Dirichlet forms, *Potential Anal.* **15**, 151-185, (2001).
33. P. J. Fitzsimmons and R. K. Getoor, Potential theory of moderate Markov dual processes, *Potential Anal.* **31**, 275-310, (2009).
34. G. Forst, The definition of energy in non-symmetric translation invariant Dirichlet spaces, *Math. Ann.* **216**, 165-172, (1975).
35. M. Fujisaki, G. Kallianpur, and H. Kunita, Stochastic differential equations for the non linear filtering problem, *Osaka J. Math.* **9**, 19-40, (1972).
36. M. Fukushima, Regular representations of Dirichlet spaces, *Trans. Amer. Math. Soc.* **155**, 455-473, (1971).
37. M. Fukushima, Dirichlet spaces and strong Markov processes, *Trans. Amer. Math. Soc.* **162**, 185-224, (1971).
38. M. Fukushima, Almost polar sets and an ergodic theorem, *J. Math. Soc. Japan*, **26**, 17-32, (1974).
39. M. Fukushima, *Dirichlet Forms and Markov Processes*, (1980). North Holland,

Amsterdam-Oxford-New York.
40. M. Fukushima, A Dirichlet form on the Wiener space and properties of Brownian motion, Lect. Notes in Math. **1096**, 290-300, (1984). Springer, Berlin.
41. M. Fukushima, Y. Oshima, and M. Takeda, *Dirichlet Forms and Symmetric Markov Processes*, second revised and extended edition, (2011). Walter de Gruyter, Berlin.
42. M. Fukushima and T. Uemura, Jump-type Hunt processes generated by lower bounded semi-Dirichlet forms, *Ann.Probab.* **40**, 858-889, (2012).
43. R. K. Getoor, *Markov Processes: Ray Processes and Right Processes*, Lect. Notes in Math. **440**, (1975). Springer, Berlin.
44. R. K. Getoor and M. T. Sharpe, Naturality, standardness and weak duality for Markov processes, *Z. Wahrsch. verw. Geb.* **64**, 1-62, (1984).
45. J. Glover and M. Rao, Hunt's hypothesis (H) and Getoor's conjecture, *Ann. Probab.* **14**, 1085-1087, (1986).
46. X. F. Han, Z. M. Ma, and W. Sun, $h\hat{h}$-transforms of positivity preserving semigroups and associated Markov processes, *Acta Math. Sin. (Engl. Ser.)* **27**, 369-376, (2011).
47. S. W. He, J. G. Wang, and J. A. Yan, *Semimartingale Theory and Stochastic Calculus*, (1992). Science Press, Beijing.
48. Z. C. Hu and Z. M. Ma, Beurling-Deny formula of semi-Dirichlet forms, *C. R. Acad. Sci. Paris Sér. I Math.* **338**, 521-526, (2004).
49. Z. C. Hu, Z. M. Ma, and W. Sun, Extensions of Lévy-Khintchine formula and Beurling-Deny formula in semi-Dirichlet forms setting, *J. Funct. Anal.* **239**, 179-213, (2006).
50. Z. C. Hu, Z. M. Ma, and W. Sun, Nonlinear filtering of semi-Dirichlet processes, *Stoch. Proc. Appl.* **119**, 3890-3913, (2009).
51. Z. C. Hu, Z. M. Ma, and W. Sun, Some remarks on representations of non-symmetric local Dirichlet forms, *Potential Theory and Stochastics in Albac, Aurel Cornea Memorial Volume.* 145-156, (2009).
52. Z. C. Hu, Z. M. Ma, and W. Sun, On representations of non-symmetric Dirichlet forms, *Potential Anal.* **32**, 101-131, (2010).
53. Z. C. Hu and W. Sun, Hunt's hypothesis (H) and Getoor's conjecture for Lévy processes, *Stoch. Proc. Appl.*, **122**, 2319-2328, (2012).
54. Z. C. Hu, W. Sun, and J. Zhang, New results on Hunt's hypothesis (H) for Lévy processes, (2013). Preprint.
55. G. Kallianpur, *Stochastic Filtering Theory*, (1980). Springer-Verlag, Berlin-Heidelberg-New York.
56. M. Kanda, Two theorems on capacity for Markov processes with stationary independent increments, *Z. Wahrsch. verw. Gebiete* **35**, 159-165, (1976).
57. H. Kesten, Hitting probabilities of single points for processes with stationary independent increments, *Memoirs of AMS*, No. 93, (1969). American Mathematical Society, Providence, R.I.
58. S. Kusuoka, Dirichlet forms and diffusion processes on Banach spaces, *J. Fac. Sc. Univ. Tokyo* **29**, 79-95, (1982).
59. K. Kuwae, Maximum principles for subharmonic functions via local semi-Dirichlet forms, *Canad. J. Math.* **60**, 822-874, (2008).
60. K. Kuwae, Stochastic calculus over symmetric Markov processes without time reversal, *Ann. Probab.* **38**, 1532-1569, (2010).
61. Y. LeJan, Balayage et formes de Dirichlet, *Z. Wahrsch. verw. Geb.* **37**, 297-319, (1977).
62. Y. LeJan, Mesures associées à une forme de Dirichlet. Applications, *Bull. Soc. Math. France* **106**, 61-112, (1978).
63. Y. LeJan, Dual Markovian semigroups and processes, In *Functional Analysis in*

Markov Processes (Ed. M. Fukushima), Lect. Notes in Math. **923**, 47-75, (1982). Springer-Verlag, Berlin-Heidelberg-New York.

64. Y. LeJan, Quasi-continuous functions and Hunt processes, *J. Math. Soc. Japan* **35**, 37-42, (1983).
65. M. Liao, Riesz representation and duality of Markov processes, Lect. Notes in Math. **1123**, 366-396, (1985). Springer, Berlin.
66. T. Lyons and M. Röckner, A note on tightness of capacities associated with Dirichlet forms, *Bull. London Math. Soc.* **24**, 181-184, (1992).
67. L. Ma, Z. M. Ma, and W. Sun, Fukushima's decomposition for diffusions associated with semi-Dirichlet forms, *Stoch. Dyn.* **12**, 1250003, (2012).
68. L. Ma and W. Sun, On the generalized Feynman-Kac transformation for nearly symmetric Markov processes, *J. Theoret. Probab.* **25**, 733-755, (2012).
69. Z. M. Ma, L. Overbeck, and M. Röckner, Markov processes associated with semi-Dirichlet forms, *Osaka J. Math.* **32**, 97-119, (1995).
70. Z. M. Ma and M. Röckner, *Introduction to the Theory of (Non-Symmetric) Dirichlet Forms*, (1992). Springer-Verlag, Berlin-Heidelberg-New York.
71. Z. M. Ma and M. Röckner, Markov processes associated with positivity preserving coercive forms, *Canad. J. Math.* **47**, 817-840, (1995).
72. Z. M. Ma, W. Sun, and L. F. Wang, Fukushima type decomposition for semi-Dirichlet forms, (2014). Preprint.
73. Y. Oshima, *Lecture on Dirichlet Spaces*, (1988). Univ. Erlangen-Nürnberg.
74. Y. Oshima, *Semi-Dirichlet Forms and Markov Processes*, (2013). Walter de Gruyter, Berlin.
75. M. Rao, Hunt's hypothesis for Lévy processes, *Proc. Amer. Math. Soc.* **104**, 621-624, (1988).
76. M. Röckner and B. Schmuland, Quasi-regular Dirichlet forms: examples and counterexamples, *Canad. J. Math.* **47**, 165-200 (1995).
77. R. L. Schilling and J. Wang, Lower bounded semi-Dirichlet forms associated with Lévy type operators, *arXiv:1108.3499*, (2012).
78. M. T. Sharpe, *General Theory of Markov Processes*, (1988). Academic Press, New York.
79. I. Shigekawa and S. Taniguchi, Dirichlet forms on separable metric spaces, In *Probability Theory and Mathematical Statistics* (Ed. A.N. Shiryaev et al.), 324-353, (1992). World Sci., Singapore.
80. M. L. Silverstein, *Symmetric Markov Processes*, Lect. Notes in Math. **516**, (1974). Springer-Verlag, Berlin-Heidelberg-New York.
81. M. L. Silverstein, The sector condition implies that semipolar sets are quasi-polar, *Z. Wahrsch. verw. Gebiete* **41**, 13-33, (1977).
82. W. Stannat, The theory of generalized Dirichlet forms and its applications in analysis and stochastics, *Memoirs of AMS*, No. 678, (1999). American Mathematical Society, Providence, R.I.
83. W. Sun, Y. Zeng, and S. Zhang, Filtering with marked point process observations via Poisson chaos expansion, *Appl. Math. Opt.*, **67**, 323-351, (2013).
84. W. Sun and J. Zhang, Lévy-Khintchine type representation of Dirichlet generators and semi-Dirichlet forms, *Forum Math.*, to appear.
additive functionals of Brownian motion, *Ann. Probab.* **25**, 940-952, (1997).
85. T. Uemura, On multidimensional diffusion processes with jumps, *Osaka J. Math.*, to appear.
86. L. F. Wang, Fukushima's decomposition of semi-Dirichlet forms and some related research, (2013). Ph.D. thesis, Chinese Academy of Sciences.

87. M. Yor, Sur les théories du filtrage et de la prédiction, *Séminaire de Probabilités XI*, Lect. Notes in Math. **581**, 257-297, (1977). Springer, Berlin.

Chapter 22

Large deviation estimates for controlled semi-martingales

Hideo Nagai*

Motivated by downside risk minimization on the wealth process in an incomplete market model, we consider minimizing the probability that the empirical mean of a controlled semi-martingale falls below a certain level on a time horizon T, and then study the asymptotic behavior of the minimizing probability as $T \to \infty$. This asymptotic behavior relates to an risk-sensitive stochastic control problem in the risk-averse case. Indeed, we can obtain an expression of the decay rate of the probability by the Legendre transform of the limit value of the value function of the stochastic control problem, which is characterized as the solution to an H-J-B equation of ergodic type.

Contents

1. Introduction

Let X_t be a solution to the following stochastic differential equation:

$$dX_t = \lambda(X_t)dW_t + \beta(X_t)dt, \quad X_0 = x \in R^N, \tag{1.1}$$

with $\lambda(x) \in C_{lip}(R^N; R^N \otimes R^M)$, $\beta(x) \in C_{lip}(R^N; R^N)$ and consider the following asymptotic behavior:

$$J(\kappa) := \varliminf_{T\to\infty} \frac{1}{T} \inf_{h.} \log P\left(\frac{1}{T} F_T(X_., h_.) \le \kappa\right), \tag{1.2}$$

where

$$F_T(X_., h_.) = F_0 + \int_0^T f(X_s, h_s)ds + \int_0^T \varphi(X_s, h_s)^* dW_s,$$

*Department of Mathematics, Faculty of Engineering Science, Kansai University, Suita, 564-8680 and Center for the Study of Finance and Insurance, Osaka University, Toyonaka, 560-8531, Japan, nagaih@kansai-u.ac.jp, nagai@sigmath.es.osaka-u.ac.jp

W_t is an M- dimensional Brownian motion process defined on a probability space $(\Omega, \mathcal{F}, P)$ with filtration $\mathcal{F}_t$, F_0 an $\mathcal{F}_0$ - measurable random variable, and $h_t = h(t, X_t)$ an $\mathcal{F}_t$ - progressively measurable R^m -valued process with $m, N \leq M$. In particular, we consider the case where

$$f(x,h) := -\frac{1}{2}h^*S(x)h + h^*g(x) + U(x), \quad \varphi(x,h) = \delta(x)h,$$

with

$$S(x) \in C(R^N; R^m \otimes R^m), \; g(x) \in C(R^N; R^m), \; \delta(x) \in C(R^N; R^M \otimes R^m).$$

To study (1.2), we relate it to the following limit value:

$$\chi(\theta) := \varliminf_{T\to\infty} \frac{1}{T} \inf_{h.} \log E[e^{\theta F_T(X_\cdot, h_\cdot)}]$$

and are going to show the duality relationship:

$$J(\kappa) = -\sup_{\theta<0}\{\theta\kappa - \chi(\theta)\}$$

for $\chi'(-\infty) < \kappa < \chi'(-0)$ and construct an asymptotically optimal strategy. Note that extensive studies of large deviation principle for Markov process expectations for large time have been done in the case where $F_T(X_\cdot, h_\cdot)$ does not include control parameter h and the stochastic integral term $\int_0^T \varphi(X_s, h_s)^* dW_s$ since the works by Donsker and Varadhan [11], [12], or Gärtner [13]. See the volumes, for example, [10] or [9] on large deviation principle and applications. The current work is considered an extension of the large deviation estimates for additive functionals of bounded variation of Markov processes to semi-martingales including control parameters. The motivated examples coming from mathematical finance are seen in the following.

Example 1.1. Consider the following market model with the price process S_t^0 of the riskless asset whose dynamics is defined by:

$$dS_t^0 = r(X_t)S_t^0 dt, \quad S_0^0 = s^0$$

and the price processes $S_t^1, S_t^2, \ldots, S_t^m$ of risky assets governed by the stochastic differential equations:

$$dS_t^i = S_t^i\{\alpha^i(X_t)dt + \sum_{k=1}^{n+m} \sigma_k^i(X_t)dW_t^k\}, \quad S_0^i = s^i, \; i = 1, \ldots, m, \tag{1.3}$$

where X_t is a factor process defined by the stochastic differential equation:

$$dX_t = \tilde{\beta}(X_t)dt + \tilde{\lambda}(X_t)dW_t, \quad X_0 = x.$$

Here, $W_t = (W_t^1, \ldots, W_t^{n+m})$ is an $n+m$ dimensional standard Brownian motion process. Note that this market model is an incomplete one called a "factor model". Then, we set the wealth process

$$V_t := \sum_{i=0}^m N_t^i S_t^i,$$

where N_t^i is the number of shares invested to i-th security, $i = 0, 1, \dots, m$. When defining the portfolio proportion $h_t^i = \frac{N_t^i S_t^i}{V_t}$ invested to i -th security, under the condition of self-financing the wealth process V_t has the stochastic differential:

$$\frac{dV_t}{V_t} = \{r(X_t) + h_t^*(\alpha(X_t) - r(X_t)\mathbf{1}\}dt + h_t^*\sigma(X_t)dW_t,$$

where $h_t = (h^1, \dots, h_t^m)$. The investor may be naturally concerned with asymptotic behavior of the probability of minimizing downside risk with the wealth process V_t:

$$J_0(\kappa) := \varliminf_{T\to\infty} \frac{1}{T} \inf_{h.} \log P(\frac{1}{T}\log V_T(h) \le \kappa)$$

for a given target growth rate κ. In this example, when setting $n + m = M$, $n = N$, we have a controlled semi-martingale:

$$\log V_T(h) = \log v_0 + \int_0^T f(X_s, h_s)ds + \int_0^T \varphi(X_s, h_s)^* dW_s,$$

where

$$f(x,h) = -\frac{1}{2}h^*\sigma\sigma^*(x)h + h^*\hat{\alpha}(x) + r(x), \quad \varphi(x,h) = \sigma^*(x)h,$$

$\hat{\alpha}(x) = \alpha(x) - r(x)\mathbf{1}$, and arrive at the above problem (1.2) with $S(x) = \sigma\sigma^*(x)$ and $\delta(x) = \sigma^*(x)$. Concerning this problem, we have actually obtained duality relationship between $J_0(\kappa)$ and the corresponding risk-sensitive portfolio optimization on infinite time horizon in [16], [28].

We further note that Markovian complete market cases are also discussed in this framework. Indeed, in those cases, we set $n = 0$, $m = N = M$, namely there are no factor processes. Instead, processes X_t^i, $i = 1, \dots, m$, governed by the stochastic differential equation:

$$dX_t^i = \{\alpha^i(X_t) - \frac{1}{2}(\sigma\sigma^*(X_t))^{ii}\}dt + \sum_{j=1}^m \sigma_j^i(X_t)dW_t^j, \quad X_0^i = x^i$$

are regarded as "factor processes". Then, one can see that $S_t^i = e^{X_t^i}$ satisfies (1.3) with $n = 0$ and $s^i = e^{x^i}$. So, setting $S_t^0 = s^0 e^{\int_0^t r(X_s)ds}$, we have a complete market model with the security prices $(S^0, \dots, S^m)$ if $\{\sigma_k^i(x)\}$ is non-degenerate. For such a model, we have discussed the above mentioned duality relationship in [27], and reformulating the problem in the current situation, we have an expository work [29]. As for related works, see [15], [17], [18], [31], [32], etc.. Concerning further financial motivation, see [6], [7], [34], etc..

In this paper we shall give the proofs of the results announced in [29]. Let us come back to (1.2) above mentioned. In what follows, we always assume that $F_0 \equiv 0$ for simplicity. Then, for each control h such that

$$E[e^{\theta\int_0^T \varphi(X_s,h_s)^* dW_s - \frac{\theta^2}{2}\int_0^T |\varphi(X_s,h_s)|^2 ds}] = 1, \quad \forall T > 0, \tag{1.4}$$

introduce a probability measure P^h satisfying

$$\left.\frac{dP^h}{dP}\right|_{\mathcal{F}_T} := e^{\theta\int_0^T \varphi(X_s,h_s)^* dW_s - \frac{\theta^2}{2}\int_0^T |\varphi(X_s,h_s)|^2 ds}.$$

Then, $W_t^h := W_t - \theta\int_0^t \varphi(X_s,h_s)ds$ is a Brownian motion process under P^h and X_t satisfies

$$dX_t = \lambda(X_t)dW_t^h + \{\beta(X_t) + \theta\lambda(X_t)\delta(X_t)h_t\}dt, \quad X_0 = x \in R^N. \tag{1.5}$$

Thus, we have the risk-sensitive stochastic control problem on a finite time horizon with the value

$$\begin{aligned}\hat{v}(x,0;T) &:= \inf_{h_\cdot\in\mathcal{A}_T} \log E[e^{\theta F_T(X_\cdot,h_\cdot)}] \\ &= \inf_{h_\cdot\in\mathcal{A}_T} \log E^h[e^{\int_0^T\{\theta f(X_s,h_s)+\frac{\theta^2}{2}\varphi(X_s,h_s)\}ds}]\end{aligned} \tag{1.6}$$

and the controlled process X_t governed by stochastic differential equation (1.5). Here, $\mathcal{A}_T$ is the totality of progressively measurable processes $h_t = h(t,X_t)$ with a Borel function $h(t,x)$ such that $|h(t,x)| \le C(1+|x|)$ satisfying (1.4). The H-J-B equation of this stochastic control problem is seen to be

$$\begin{cases} \frac{\partial v}{\partial t} + \frac{1}{2}\mathrm{tr}[\lambda\lambda^* D^2 v] + \frac{1}{2}(Dv)^*\lambda\lambda^* Dv \\ \qquad + \inf_h\{[\beta(x) + \theta\lambda(x)\delta(x)h]^* Dv + \theta f(x,h) + \frac{\theta^2}{2}|\varphi(x,h)|^2\} = 0, \\ v(x,T) = 0, \end{cases} \tag{1.7}$$

which is written as

$$\begin{cases} \frac{\partial v}{\partial t} + \frac{1}{2}\mathrm{tr}[\lambda\lambda^* D^2 v] + \beta_\theta^* Dv + \frac{1}{2}(Dv)^*\lambda(I + \theta\delta Q_\theta^{-1}\delta^*)\lambda^* Dv \\ \qquad\qquad + \frac{\theta}{2}g^* Q_\theta^{-1} g + \theta U = 0, \\ v(T,x) = 0. \end{cases} \tag{1.8}$$

Here,

$$\beta_\theta = \beta + \theta\lambda\delta Q_\theta^{-1} g, \quad Q_\theta = S - \theta\delta\delta^*. \tag{1.9}$$

Under the suitable conditions (cf. (2.1) - (2.4)) such that

$$N_\theta := I + \theta\delta Q_\theta^{-1}\delta^* \tag{1.10}$$

is uniformly positive definite, we shall see that H-J-B equation (1.7) has a structure similar to the one studied in [28], and so does the H-J-B equation of ergodic type corresponding to the risk-sensitive stochastic control problem on infinite time horizon. Thus, an analysis similar to [28] works well from the analytical view points. Although the difference between the current H-J-B equations and the ones discussed in [28] is not serious, we need further analysis to obtain the duality theorem for the controlled semi-martingales treated here. In the current paper, in proving the duality theorem, we newly introduce an "equivalent" stochastic differential game in section 5 having the same H-J-B equation (1.8) as the above risk-sensitive stochastic control problem. The relevant H-J-B equation of ergodic type to its infinite time horizon counter part naturally leads the proof of the duality theorem as is presented in section 6. Further, in proving the convexity of $\chi(\theta)$, which is the solution to the H-J-B equation of ergodic type, we regard it as the averaging limit of stochastic control problem (2.15) of classical type as was done in [28]. Thus, we give the proof that the solution $\chi(\theta)$ of the equation can be obtained as the averaging limit of the solution of H-J-B equation (2.14) of parabolic type which characterizes (2.15), by showing refinement of the results on asymptotic behavior of the solution in section 3. We here slightly generalize the results recently obtained by Ichihara and Sheu [21] by adding diffusion coefficients and the proofs become simpler because of the specified nonlinear term. Convexity of $\chi(\theta)$ plays a crucial role in proving the duality theorem and thus we arrive at the natural proofs under the current situation (cf. Corollary 3.1).

We further mention that the robust estimates of the large deviation estimates for controlled semi-martingales is also given under admitting model uncertainty (cf. [30]).

2. Studies of the problems on a finite time horizon

In the following we always assume that $F_0 = 0$ for simplicity and so

$$F_T(X_\cdot, h_\cdot) = \int_0^T f(X_s, h_s)ds + \int_0^T \varphi(X_s, h_s)^* dW_s$$

is considered. Let us assume the following assumptions.

$$\begin{cases} \lambda,\ \beta,\ S,\ g,\ \delta \text{ are smooth and globally Lipschitz,} \\ U \text{ is a smooth function bounded below such that} \\ |U(x)|,\ |DU| \leq M_1|x|^2 + M_2, \quad M_1,\ M_2 > 0, \end{cases} \tag{2.1}$$

$$c_0\delta^*\delta(x) \leq S(x) \leq c_1\delta^*\delta(x), \quad x \in R^N, \ c_0,\ c_1 > 0, \tag{2.2}$$

$$c_\delta I \leq \delta^*\delta(x) \leq c_\delta' I, \quad c_\delta,\ c_\delta' > 0, \tag{2.3}$$

$$c_2|\xi|^2 \leq \xi^*\lambda\lambda^*(x)\xi \leq c_3|\xi|^2, \ c_2,\ c_3 > 0, \ \xi \in R^n. \tag{2.4}$$

Note that, under these assumptions, Q_θ defined by (1.9) satisfies

$$(c_0 - \theta)\delta^*\delta(x) \leq Q_\theta(x) \leq (c_1 - \theta)\delta^*\delta(x) \tag{2.5}$$

and

$$\frac{\theta}{c_0 - \theta}(\delta^*\delta(x))^{-1} \leq \theta Q_\theta^{-1}(x) \leq \frac{\theta}{c_1 - \theta}(\delta^*\delta(x))^{-1}. \tag{2.6}$$

Moreover,

$$\frac{c_0}{c_0 - \theta} I \leq I + \theta\delta Q_\theta^{-1}\delta^* \leq I \tag{2.7}$$

holds. Indeed, (2.5) follows directly from (2.4) and thus (2.6) is obtained from (2.5). The left-hand side of (2.7) is seen since

$$\frac{\theta}{c_0 - \theta} I \leq \frac{\theta}{c_0 - \theta}\delta(\delta^*\delta)^{-1}\delta^* \leq \theta\delta Q_\theta^{-1}\delta^*,$$

which follows from (2.6). The right hand side of (2.7) is obvious.
Introduce the value function

$$\hat{v}(t,x) = \inf_{h_\cdot \in \mathcal{A}(T-t)} \log E[e^{\theta\{\int_0^{T-t} f(X_s,h_s)ds + \int_0^{T-t}\varphi(X_s,h_s)^* dW_s\}}], \tag{2.8}$$

which is written by P^h as

$$\hat{v}(t,x) = \inf_{h_\cdot \in \mathcal{A}(T-t)} \log E^h[e^{\theta\int_0^{T-t}\{f(X_s,h_s)+\frac{\theta}{2}h_s^*\delta^*\delta(X_s)h_s\}ds}]. \tag{2.9}$$

The corresponding H-J-B equation is seen to be (1.7). We see from (2.5) we see that N_θ defined by (1.10) satisfies

$$\frac{c_0}{c_0-\theta} I \le N_\theta \le I. \tag{2.10}$$

Thus, we can see that H-J-B equation (1.7) has a sufficiently smooth solution satisfying nice gradient estimates, similarly to Theorem 2.1 in [27]. Indeed, we have the the following proposition (cf. [4], [25]).

Proposition 2.1. *Under assumptions (2.1) - (2.4), H-J-B equation (1.7) has a solution such that*

$$v(t,x) \le K_0,$$

$$v,\ \tfrac{\partial v}{\partial t},\ \tfrac{\partial v}{\partial x_i},\ \tfrac{\partial^2 v}{\partial x_i \partial x_j}, \in L^p(0,T;L^p_{loc}(R^n)), \quad \forall p>1,$$

$$\tfrac{\partial v}{\partial t} \ge -C,$$

$$\tfrac{\partial^2 v}{\partial^2 t},\ \tfrac{\partial^2 v}{\partial x_i \partial t},\ \tfrac{\partial^3 v}{\partial x_i \partial x_j \partial x_k},\ \tfrac{\partial^3 v}{\partial x_i \partial x_j \partial t} \in L^p(0,T;L^p_{loc}(R^n)), \quad \forall p>1,$$

$$\begin{aligned}|Dv|^2 + \tfrac{(c_0-\theta)(1+c)}{c_0c_2}(\tfrac{\partial v}{\partial t}+C) &\le c'(|DN_\theta|^2_{2r} + |N_\theta|^2_{2r} + |D(\lambda\lambda^*)|^2_{2r} + |D\beta_\theta|_{2r} \\ &+|\beta_\theta|^2_{2r} + |\theta U|_{2r} + |\theta DU|_{2r} + |\theta g^* Q_\theta^{-1} g|_{2r} + |\theta D g^* Q_\theta^{-1} g|_{2r} + 1), \\ &x \in B_r, \quad t\in[0,T).\end{aligned}$$

Here, we note that $\frac{\partial v}{\partial t} \ge -C$ because of assumption (2.1) and the gradient estimates have a minor difference from those in [[28]: Theorem 2.1].

To have (1.8) from (1.7) the following is to be noted. It follows that

$$\begin{aligned}&\inf_{h\in R^m}\{\theta[\lambda\delta h]^* Dv + \theta f(x,h) + \frac{\theta^2}{2}|\varphi(x,h)|^2\} \\ &= \inf_{h\in R^m}\{-\frac{\theta}{2}h^*(S(x)-\theta\delta\delta^*(x))h + \theta h^*(\delta^*\lambda^* Dv + g) + \theta U\} \\ &= \inf_{h\in R^m}[-\frac{\theta}{2}\{h - Q_\theta^{-1}(\delta^*\lambda^* Dv+g)\}Q_\theta\{h - Q_\theta^{-1}(\delta^*\lambda^* Dv+g)\} \\ &\quad + \frac{\theta}{2}(\delta^*\lambda^* Dv + g)^* Q_\theta^{-1}(\delta^*\lambda^* Dv+g) + \theta U].\end{aligned}$$

Then, we have the following verification theorem.

Proposition 2.2. *Assume assumptions (2.1)-(2.4) and set*

$$\hat{h}(t,x) := Q_\theta^{-1}(\delta^*\lambda^* Dv(t,x) + g(x)),$$

where $v(t,x)$ is the solution to (1.7). Then, $\hat{h}_t^{(T)} \equiv \hat{h}_t^{(\theta,T)} := \hat{h}(t,X_t)$ is the optimal strategy:

$$\begin{aligned} v(0,x;T) &= \log E[e^{\theta\{\int_0^T f(X_s,\hat{h}_s^{(T)})ds + \int_0^T \varphi(X_s,\hat{h}_s^{(T)})^* dW_s\}}] \\ &= \inf_{h_\cdot \in \mathcal{A}(T)} \log E[e^{\theta\{\int_0^T f(X_s,h_s)ds + \int_0^T \varphi(X_s,h_s)^* dW_s}]. \end{aligned}$$

To prove this proposition, the following lemma is useful.

Lemma 2.1. *Let Y_t be a solution to the stochastic differential equation*

$$dY_t = \sigma(t,Y_t)dW_t + \mu(t,Y_t)dt, \quad Y_0 = x,$$

where σ, μ are locally Lipschitz continuous functions such that

$$|r(t,x)| \leq K(1+|x|) \tag{2.11}$$

for $r = \sigma$, μ and $H(t,x)$ be a continuous function satisfying (2.11). We further assume that σ is bounded. Then, setting

$$\rho_t := e^{\int_0^t H(s,Y_s)^* dW_s - \frac{1}{2}\int_0^t |H(s,X_s)|^2 ds},$$

we have

$$E[\rho_t] = 1, \quad \forall t.$$

The proof of this lemma is similar to that of Lemma 4.1.1 in [2]. Further, we see that the proof of Proposition 2.2 is given by using this lemma when noting that the solution v of the H-J-B equation (1.7) satisfies the estimate $|Dv(t,x)| \leq K(1+|x|)$ for some positive constant K owing to Proposition 2.1 (cf. [28]).

Now, let us rewrite H-J-B equation (1.8). Since $N_\theta - I = \theta\delta Q_\theta^{-1}\delta^*$ we have

$$\begin{aligned} \theta Q_\theta^{-1} &= (\delta^*\delta)^{-1}\delta^*(N_\theta - I)\delta(\delta^*\delta)^{-1} \\ &= (\delta^*\delta)^{-1}\delta^* N_\theta \delta(\delta^*\delta)^{-1} - (\delta^*\delta)^{-1} \end{aligned} \tag{2.12}$$

and thus (1.8) is changed to

$$\begin{cases} \frac{\partial v}{\partial t} + \frac{1}{2}\mathrm{tr}[\lambda\lambda^* D^2 v] + [\beta - \lambda\delta(\delta^*\delta)^{-1}g]^* Dv + \frac{1}{2}(Dv)^*\lambda N_\theta \lambda^* Dv \\ \qquad + \{\lambda\delta(\delta^*\delta)^{-1}\delta^* N_\theta \delta(\delta^*\delta)^{-1}g\}^* Dv - \frac{1}{2}g^*(\delta^*\delta)^{-1}g \\ \qquad\qquad + \frac{1}{2}g^*(\delta^*\delta)^{-1}\delta^* N_\theta\delta(\delta^*\delta)^{-1}g + \theta U = 0, \\ v(T,x) = 0. \end{cases} \tag{2.13}$$

Further, from $N_\theta = I + \theta\delta Q_\theta^{-1}\delta^*$ it follows that $I = N_\theta^{-1} + \theta N_\theta^{-1}\delta Q_\theta^{-1}\delta^*$ and that

$$\delta(\delta^*\delta)^{-1} = N_\theta^{-1}\delta(\delta^*\delta)^{-1} + \theta N_\theta^{-1}\delta Q_\theta^{-1}.$$

Therefore

$$\begin{aligned}N_\theta^{-1}\delta(\delta^*\delta)^{-1}\delta^* N_\theta\delta(\delta^*\delta)^{-1} &= N_\theta^{-1}\delta(\delta^*\delta)^{-1}\delta^*(I+\theta\delta Q_\theta^{-1}\delta^*)\delta(\delta^*\delta)^{-1}\\ &= N_\theta^{-1}\delta(\delta^*\delta)^{-1}+\theta N_\theta^{-1}\delta Q_\theta^{-1}\\ &= \delta(\delta^*\delta)^{-1}.\end{aligned}$$

Thus, (2.13) can be rewritten as

$$\begin{cases}\frac{\partial v}{\partial t}+\frac{1}{2}\mathrm{tr}[\lambda\lambda^* D^2 v]+\{\beta-\lambda\delta(\delta^*\delta)^{-1}g\}^* Dv\\ \qquad +\frac{1}{2}[\lambda^* Dv+\delta(\delta^*\delta)^{-1}g]^* N_\theta[\lambda^* Dv+\delta(\delta^*\delta)^{-1}g]\\ \qquad\qquad -\frac{1}{2}g^*(\delta^*\delta)^{-1}g+\theta U=0,\\ v(T,x)=0.\end{cases} \tag{2.14}$$

Then, by seeing that

$$\begin{aligned}&\tfrac{1}{2}[\lambda^* Dv+\delta(\delta^*\delta)^{-1}g]^* N_\theta[\lambda^* Dv+\delta(\delta^*\delta)^{-1}g]\\ &\quad=\sup_{z\in R^M}\{-\tfrac{1}{2}z^* N_\theta^{-1}z+z^*(\delta(\delta^*\delta)^{-1}g+\lambda^* Dv)\},\end{aligned}$$

equation (2.14) is regarded as the H-J-B equation of the stochastic control problem

$$v_*(0,x;T;\theta)=\sup_{z_\cdot} E[\int_0^T \Phi(Y_s,z_s)ds] \tag{2.15}$$

subject to

$$dY_t=\lambda(Y_t)dB_t+\{G(Y_t)+\lambda(Y_t)z_t\}dt,\quad Y_0=x, \tag{2.16}$$

$$G(y)=\beta(y)-\lambda\delta(\delta^*\delta)^{-1}g,$$

$$\Phi(y,z;\theta)=-\frac{1}{2}z^* N_\theta^{-1}z+g^*(\delta^*\delta)^{-1}\delta^*(y)z-\frac{1}{2}g^*(\delta^*\delta)^{-1}g(y)+\theta U(y).$$

Here control process $z_\cdot$ ranges over the set of all progressively measurable process such that

$$E[\int_0^T |z_s|^2 ds]<\infty.$$

Indeed, we have the following verification theorem by using Proposition 2.1.

Proposition 2.3. *Assume assumptions (2.1)-(2.4) and set*

$$\hat{z}(t,y):=N_\theta\{\lambda^* Dv(t,x)+\delta(\delta^*\delta)^{-1}g(x)\}$$

for the solution $v(t,x)$ to (1.8). Then, the stochastic differential equation

$$dY_t=\lambda(Y_t)dB_t+\{G(Y_t)+\lambda(Y_t)\hat{z}(t,Y_t)\}dt,\quad Y_0=x$$

has a unique solution and the verification

$$v(0,x;T)=E[\int_0^T \Phi(Y_s,\hat{z}_s)ds]=\sup_{z_\cdot} E[\int_0^T \Phi(Y_s,z_s)ds]$$

holds, where $\hat{z}_t = \hat{z}(t, Y_t)$.

The following lemma is useful in the later arguments.

Lemma 2.2. *Assume (2.1)-(2.4). Then, function* $\Phi(y, z; \theta)$ *is linear with respect to* θ.

Proof. Since $\frac{\partial Q_\theta}{\partial \theta} = -\delta^* \delta$ we have

$$\frac{\partial Q_\theta^{-1}}{\partial \theta} = -Q_\theta^{-1} \frac{\partial Q_\theta}{\partial \theta} Q_\theta^{-1} = Q_\theta^{-1} \delta^* \delta Q_\theta^{-1}.$$

Therefore,

$$\frac{\partial N_\theta}{\partial \theta} = \delta Q_\theta^{-1} \delta^* + \theta \delta Q_\theta^{-1} \delta^* \delta Q_\theta^{-1} \delta^* = \delta Q_\theta^{-1} \delta^* N_\theta = N_\theta \delta Q_\theta^{-1} \delta^*$$

holds and thus we have

$$\frac{\partial^2 N_\theta}{\partial \theta^2} = \frac{\partial N_\theta}{\partial \theta} \delta Q_\theta^{-1} \delta^* + N_\theta \delta \frac{\partial Q_\theta^{-1}}{\partial \theta} \delta^* = 2 N_\theta (\delta Q_\theta^{-1} \delta^*)^2.$$

Therefore,

$$\frac{\partial N_\theta^{-1}}{\partial \theta} = -N_\theta^{-1} \frac{\partial N_\theta}{\partial \theta} N_\theta^{-1} = -\delta Q_\theta^{-1} \delta^* N_\theta^{-1} = -N_\theta^{-1} \delta Q_\theta^{-1} \delta^*$$

holds and hence we obtain

$$\begin{aligned}\frac{\partial^2 N_\theta^{-1}}{\partial \theta^2} =& - \frac{\partial N_\theta^{-1}}{\partial \theta} \frac{\partial N_\theta}{\partial \theta} N_\theta^{-1} - N_\theta^{-1} \frac{\partial^2 N_\theta}{\partial \theta^2} N_\theta^{-1} - N_\theta^{-1} \frac{\partial N_\theta}{\partial \theta} \frac{\partial N_\theta^{-1}}{\partial \theta} \\ =& N_\theta^{-1} (\delta Q_\theta^{-1} \delta^*)^2 - 2 N_\theta^{-1} (\delta Q_\theta^{-1} \delta^*)^2 + N_\theta^{-1} (\delta Q_\theta^{-1} \delta^*)^2 \\ =& 0,\end{aligned}$$

which implies that Φ is linear with respect to θ. □

Corollary 2.1. *The value function* $v_*(0, x; T)$ *(and hence* $v(0, x; T)$*) is a convex function of* θ.

3. H-J-B equations of ergodic type

Let us set $\bar{v} = -v$. Then,

$$\begin{cases} \frac{\partial \bar{v}}{\partial t} + \frac{1}{2} \mathrm{tr}[\lambda \lambda^* D^2 \bar{v}] + \beta_\theta^* D \bar{v} - \frac{1}{2} (D \bar{v})^* \lambda N_\theta \lambda^* D \bar{v} \\ \qquad\qquad\qquad\qquad - \frac{\theta}{2} g^* Q_\theta^{-1} g - \theta U = 0, \\ \bar{v}(T, x) = 0. \end{cases} \tag{3.1}$$

Now let us consider the infinite horizon counterpart of H-J-B equation (1.8) (or (2.14)), which is called H-J-B equation of ergodic type. We have the following proposition according to [3] and Proposition 3.2 in [28].

Proposition 3.1.

i) Assume that

$$\lim_{r\to\infty}\inf_{|x|\geq r}\{g^*(\delta^*\delta)^{-1}g(x)+U(x)\}=\infty \tag{3.2}$$

besides assumptions (2.1) - (2.4). Then, we have a solution $(\bar{\chi}(\theta),\bar{w})$ *of :*

$$\begin{aligned}\bar{\chi}(\theta)=\tfrac{1}{2}tr[\lambda\lambda^*D^2\bar{w}]+\beta_\theta^*D\bar{w}-\tfrac{1}{2}(D\bar{w})^*\lambda N_\theta\lambda^*D\bar{w}\\ -\tfrac{\theta}{2}g^*Q_\theta^{-1}g-\theta U\end{aligned} \tag{3.3}$$

such that $\bar{w}(x)$ *is bounded below. Moreover, such a solution* $(\bar{\chi},\bar{w})$ *is unique up to additive constants with respect to* $\bar{w}$ *and satisfies the following estimate*

$$|\nabla\bar{w}(x)|^2\leq C_w|x|^2+C_w'. \tag{3.4}$$

Furthermore, if we assume stronger assumption

$$c_4|x|^2-c_5\leq\frac{1}{c_1-\theta}g^*(\delta^*\delta)^{-1}g(x)+U(x) \tag{3.5}$$

than (3.2), then we have

$$c_w|x|^2-c_w'\leq\bar{w}(x),\quad c_w,\ c_w'>0. \tag{3.6}$$

ii) Assume that

$$\beta(x)^*x\leq-c_\beta|x|^2+c_\beta',\quad c_\beta>0\,,c_\beta'>0 \tag{3.7}$$

besides assumptions (2.1) - (2.4). Then, there exists a positive constant $b_*>0$ *such that* $\psi_{b^*}(x):=-b_*|x|^2$ *satisfies*

$$F(\psi_{b^*})(x)\to\infty,\ as\ |x|\to\infty,$$

where

$$F(\psi)=\frac{1}{2}tr[\lambda\lambda^*D^2\psi]+\beta_\theta^*D\psi-\frac{1}{2}(D\psi)^*\lambda N_\theta\lambda^*D\psi-\frac{\theta}{2}g^*Q_\theta^{-1}g-\theta U.$$

Moreover, we have a solution $(\bar{\chi}(\theta),\bar{w})$ *to (3.3) such that* $\bar{w}-\psi_b(x)$ *with* $0<b\leq b_*$ *is bounded below.*

Here we note that c_w in (3.6) is a positive constant such that

$$2c_3c_w^2+C_\beta c_w<\frac{-c_4\theta}{4}$$

and C_β is the one such that $\beta_\theta(x)\leq C_\beta|x|+C$ (cf. Proposition 3.2 in [28]).

We further note that when considering $\bar{w}-\psi_b$ in place of $\bar{w}$ we have

$$\begin{aligned}\bar{\chi}(\theta)=\tfrac{1}{2}\mathrm{tr}[\lambda\lambda^*D^2(\bar{w}-\psi_b)]+(\beta_\theta-\lambda N_\theta\lambda^*D\psi_b)^*D(\bar{w}-\psi_b)\\ -\tfrac{1}{2}(D\bar{w}-\psi_b)^*\lambda N_\theta\lambda^*D(\bar{w}-\psi_b)+F(\psi_b),\end{aligned}$$

and thus, the case of ii) is reduced to i) (cf. also Ichihara [20]). In the following we shall proceed assuming the assumptions of Proposition 3 i) with (3.4). We can develop parallel arguments as well in the case of ii) of the proposition.

Let us write H-J-B equation (3.1) as

$$\begin{cases} \frac{\partial \bar{v}}{\partial t} + \frac{1}{2}\mathrm{tr}[\lambda\lambda^* D^2\bar{v}] + H(x, D\bar{v}) = 0, \\ \bar{v}(T, x) = 0, \end{cases} \tag{3.8}$$

by introducing

$$H(x, p) = \beta_\theta^* p - \frac{1}{2}p^*\lambda N_\theta \lambda^* p - \frac{\theta}{2}g^* Q_\theta^{-1} g - \theta U,$$

which is concave with respect to p. Here we note that the optimal diffusion process for stochastic control problem (2.15) turns out to be governed by stochastic differential equation

$$d\bar{Y}_t = \lambda(\bar{Y}_t)dW_t + D_p H(\bar{Y}_t, \bar{v}(t, \bar{Y}_t))dt, \quad \bar{Y}_0 = x,$$

where

$$D_p H(x, p) = \frac{\partial H(x, p)}{\partial p} = \beta_\theta + \lambda N_\theta \lambda^* p.$$

Now let us extend the solution to (3.1)' to $(-\infty, 0]$. It is indeed possible because the coefficients of the H-J-B equation do not depend on t and for each $-s < 0$, the solution $\bar{v}(-s, x; T)$ is identical to $\bar{v}(0, x; T + s)$, which is the solution of (3.1)' with the terminal condition $\bar{v}(T + s, x; T + s) = 0$. Thus, we have $\bar{v}(-s, x; T)$ defined on $-T \leq s < \infty$. In what follows we specify the solution $\bar{w}(x)$ of H-J-B equation (3.3) such that $\bar{w}(x) \geq 0$ by suitably taking an additive constant. Then, from the maximum principle it follows that

$$\bar{v}(-s, x; T) \leq \bar{w}(x) + \bar{\chi}(T + s), \quad -T \leq s < \infty \tag{3.9}$$

since $\bar{w}(x) + \bar{\chi}(T + s)$ is a super-solution to (3.1). Now we have the following proposition.

Proposition 3.2. *Under the assumptions of Proposition 3.1 i) with (3.5), we have*

$$\bar{v}(-s, x; T) - (\bar{w}(x) + \bar{\chi}(T + s)) \geq -c_E|x|^2 - c_E', \tag{3.10}$$

where c_E and c_E' are positive constants independent of s.

Proof. For given $-s \leq 0$, let $\bar{Y}_t$ be a solution to

$$d\bar{Y}_t = \lambda(\bar{Y}_t)dW_t + D_p H(\bar{Y}_t, \bar{v}(-s + t, \bar{Y}_t))dt, \quad \bar{Y}_0 = x. \tag{3.11}$$

Then, we see that

$$\bar{v}(-s, x; T) = \inf_{z.} E_x[\int_0^{T+s} -\Phi(Y_s, z_s)ds] = E_x[\int_0^{T+s} -\Phi(\bar{Y}_t, \hat{z}(-s + t, \bar{Y}_t)dt]$$

and also

$$\bar{w}(x) + \bar{\chi}(T + s) \leq E_x[\int_0^{T+s} -\Phi(Y_t, z_t)dt + \bar{w}(Y_{T+s})]$$

for each control z_t. Therefore,

$$\begin{aligned}\bar{v}(-s,x;T)-(\bar{w}(x)+\bar{\chi}(T+s)) &\geq -E_x[\bar{w}(\bar{Y}_{T+s})]\\ &\geq -c_w E[|\bar{Y}_{T+s}|^2]+c'_w.\end{aligned}$$

We conclude the present proposition since we see that $E[|\bar{Y}_{T+s}|^2] \leq C|x|^2 + C'$ for some positive constants C and C' independent of s from the following lemma and the estimate $|\bar{w}(x)| \leq C|x|^2 + C'$ obtained from (3.4) . □

Lemma 3.1. *Assume the assumptions of Proposition 3.2 and set $\psi(x) = -c_*|x|^2$ with a positive constant c_* such that*

$$c_\beta c_* + \frac{c_3}{2}c_*^2 \leq -c_4\theta. \tag{3.12}$$

Then,

$$E[c_*|\bar{Y}_{T+s}|^2] \leq e^{-\alpha(T+s)}(\bar{w}(x)-\psi(x)) + \frac{C}{\alpha}(1-e^{-\alpha(T+s)}), \quad -T \leq s < \infty \tag{3.13}$$

holds for sufficiently small $\alpha > 0$ and a positive constant C.

Proof. First note that there exist positive constants c_ψ and c'_ψ such that

$$\frac{1}{2}\mathrm{tr}[\lambda\lambda^* D^2\psi] + H(x, D\psi) \geq c_\psi|x|^2 - c'_\psi \tag{3.14}$$

because of the estimates $|\beta_\theta^* D\psi| \leq c_\beta c_*|x|^2 + C$, $\frac{1}{2}(D\psi)^*\lambda N_\theta\lambda^* D\psi \leq \frac{c_3}{2}c_*^2$, and assumption (3.5). Further, since $H(x,p)$ is concave in p we have

$$H(x,D\psi) - H(x,D\bar{v}) \leq D_pH(x,D\bar{v})\cdot(D\psi - D\bar{v}).$$

Therefore, by setting

$$\hat{v}(-s+t,x;T) = \bar{v}(-s+t,x;T) - \{\psi(x) + \bar{\chi}(T-t+s)\}, \qquad 0 \leq t \leq T+s,$$

and applying Itô's formula to $e^{\alpha t}\hat{v}(-s+t,x;T)$, we see that

$$\begin{aligned}&e^{\alpha t}\hat{v}(-s+t,\bar{Y}_t,;T) - \hat{v}(-s,x;T)\\ &\quad= \int_0^t e^{\alpha r}\{L\hat{v}(-s+r,\bar{Y}_r) + D_pH(\bar{Y}_r, D\bar{v}(-s+r,\bar{Y}_r))^* D\hat{v}(-s+r,\bar{Y}_r)\}dr\\ &\qquad\qquad + \alpha\int_0^t e^{\alpha r}\hat{v}(-s+r,\bar{Y}_r)dr + \text{local mart.},\end{aligned}$$

where $Lv := \frac{\partial v}{\partial t} + \frac{1}{2}\mathrm{tr}[\lambda\lambda^* D^2 v]$. Thus, by using the stopping time arguments, we obtain

$$\begin{aligned}&e^{\alpha(T+s)}E[\hat{v}(T,\bar{Y}_{T+s};T)] - \hat{v}(-s,x;T)\\ &\quad\leq \int_0^{T+s} e^{\alpha r}E[L\bar{v}(-s+r,\bar{Y}_r) + H(\bar{Y}_r, D\bar{v}(-s+r,\bar{Y}_r))]dr\\ &\qquad - \int_0^{T+s} e^{\alpha r}E[\tfrac{1}{2}\mathrm{tr}[\lambda\lambda^* D^2\psi(\bar{Y}_r) + H(\bar{Y}_r, D\psi(\bar{Y}_r))]dr\\ &\qquad\qquad + \alpha\int_0^{T+s} e^{\alpha r}E[\hat{v}(-s+r,\bar{Y}_r)]dr\\ &\quad\leq \int_0^{T+s} e^{\alpha r}E[\alpha(\bar{w}-\psi)(\bar{Y}_r) - c_\psi|\bar{Y}_r|^2 + c'_\psi]dr,\end{aligned}$$

from (3.14) and (3.9). Since $\alpha(\bar{w}-\psi)(x) \le c_\psi|x|^2 + C'$ for sufficiently small $\alpha > 0$, we have (3.13) from (3.9). □

Remark 3.1. Let $\bar{X}_t$ be the solution to

$$d\bar{X}_t = \lambda(\bar{X}_t)dW_t + D_pH(\bar{X}_t, D\bar{w}(\bar{X}_t)dt, \quad \bar{X}_0 = x$$

and set

$$L^{\bar{w}}u := \frac{1}{2}\mathrm{tr}[\lambda\lambda^* D^2u] + D_pH(x, D\bar{w})^* Du.$$

Then, the diffusion prices $\bar{X}_t$ is ergodic under the assumptions of Proposition 3.1. Indeed, we see that $\bar{w}(x) \to \infty$, and also

$$L^{\bar{w}}\bar{w}(x) = -\frac{1}{2}(D\bar{w})^*\lambda N_\theta\lambda^* D\bar{w} + \frac{\theta}{2}g^* Q_\theta^{-1} g + \theta U \equiv -K(x;\bar{w}) \to -\infty$$

as $|x| \to \infty$.

Now set

$$\bar{u}(0,x;T) := \bar{v}(0,x;T) - \{\bar{w}(x) + \chi T\}$$

for a specified solution $\bar{w}(x)$ to H-J-B equation (3.3) such that $\bar{w}(x) \ge 0$. Then, we have the following theorem.

Theorem 3.1. *Under the assumptions of Proposition 3.2, as $T \to \infty$, $\bar{u}(0,x;T)$ converges to a constant $c_\infty \in R$ uniformly on each compact set.*

Proof. Let us first note that $\bar{u}(0,x;T)$ is nonpositive and bounded below by $-c_E|x|^2 - c'_E$ owing to (3.9) and (3.10). Further, its bounds do not depend on T. Moreover, $|\nabla\bar{u}(0,x;T)|^2 \le C_R, \quad x \in B_R$, where C_R is a positive constant depending only on R. Thus, we see that $\{\bar{u}(0,x;T)\}_{T\ge 1}$ is relatively compact in $C(R^N)$ and so, there exists a sequence T_j such that $\bar{u}(0,x;T_j)$ converges to a function $u_\infty \in C(R^N)$ as $T_j \to \infty$. Set

$$\mathcal{U} := \{u_\infty \in C(R^N); \lim_{T_j\to\infty} u(0,x;T_j) = u_\infty \text{ in } C(R^N), \text{ for some } \{T_j\}\}.$$

Then, proving that $\mathcal{U} = \{c_\infty\}$ for some constant $c_\infty \in R$ makes the proof of the theorem complete. Apply Itô's formula to $\bar{u}(-s+t,x;T)$, $0 \le t \le T+s$ and we have

$$\begin{aligned}
&\bar{u}(-s+t,\bar{X}_t;T) - \bar{u}(-s,X_0;T) \\
&= \textstyle\int_0^t \{L\bar{u} + D_pH(\cdot, D\bar{w})^* D\bar{u}\}(-s+r,\bar{X}_r)dr \\
&\qquad + \textstyle\int_0^t D\bar{u}(-s+r,\bar{X}_r)^*\lambda(\bar{X}_r)dW_r \\
&= \textstyle\int_0^t \frac{1}{2}(D\bar{u})^*\lambda N_\theta\lambda^* D\bar{u}(-s+r,\bar{X}_r)dr + \int_0^t D\bar{u}(-s+r,\bar{X}_r)^*\lambda(\bar{X}_r)dW_r \\
&\equiv \textstyle\int_0^t Q(\bar{X}_r, D\bar{u}(-s+r,\bar{X}_r))dr + \int_0^t D\bar{u}(-s+r,\bar{X}_r)^*\lambda(\bar{X}_r)dW_r
\end{aligned}$$

when noting that

$$H(x, D\bar{w}+p) - H(x, D\bar{w}) = D_pH(x, D\bar{w})^* p - \frac{1}{2}p^*\lambda N_\theta\lambda^* p. \tag{3.15}$$

Thus, setting $\kappa := \frac{c_0 c_2}{c_0 - \theta}$,

$$e^{-\kappa\bar{u}(-s,x;T)-\kappa\int_0^t (D\bar{u}(-s+r,\bar{X}_r)^*\lambda(\bar{X}_r)dW_r - \frac{\kappa^2}{2}\int_0^t (D\bar{u})^*\lambda\lambda^* D\bar{u}(-s+r,\bar{X}_r)dr}$$

$$= e^{-\kappa\bar{u}(-s+t,X_t;T)+\kappa\int_0^t Q(\bar{X}_r,D\bar{u}(-s+r,\bar{X}_r)dr - \frac{\kappa^2}{2}\int_0^t (D\bar{u})^*\lambda\lambda^* D\bar{u}(-s+r,\bar{X}_r)dr}$$

$$\geq e^{-\kappa\bar{u}(-s+t,X_t;T)}$$

since $Q(x,p) \geq \frac{\kappa}{2}|p|^2$. Therefore we have

$$E[e^{-\kappa\bar{u}(-s+t,X_t;T)}] \leq e^{-\kappa\bar{u}(-s,x;T)}, \quad 0 \leq t \leq T+s.$$

Choose a sequence T_j such that $\{\bar{u}(T-T_j,x;T)\} = \{\bar{u}(0,x;T_j)\}$ converges to some $u_\infty \in C(R^N)$ as $T_j \to \infty$ and set $s = t = T_j - T$. Then, we obtain

$$E[e^{-\kappa\bar{u}(0,X_{T_j-T};T)}] \leq e^{-\kappa\bar{u}(T-T_j,x;T)},$$

which implies that

$$\int e^{-\kappa\bar{u}(0,x;T)} m_\theta(dx) \leq e^{-\kappa u_\infty(x)},$$

where $m_\theta(dx)$ is the invariant measure of the diffusion process with the generator $L^{\bar{w}}$. Taking $T = T_j$ sending it to ∞ again, we have

$$\int e^{-\kappa u_\infty(y)} m_\theta(dy) \leq \varliminf_{T_j\to\infty} \int e^{-\kappa\bar{u}(0,y;T_j)} m_\theta(dy) \leq e^{-\kappa u_\infty(x)}$$

for each $x \in R^N$. By taking the infimum of both sides with respect to x we obtain

$$\int (e^{-\kappa u_\infty(y)} - \inf_{x\in R^n} e^{-\kappa u_\infty(x)}) m_\theta(dy) \leq 0.$$

Since supp $m_\theta = R^N$ we see that u_∞ is a constant function.

Suppose that $\bar{u}(0,x;T_j)$ (respectively $\bar{u}(0,x;T_j')$) converges to a constant c_∞ (resp. c'_∞), as T_j and $T_j\ ' \to \infty$. Then, we see that

$$\int e^{-\kappa\bar{u}(0;y;T)} m_\theta(dy) \leq e^{-\kappa c_\infty}.$$

By setting $T = T_j'$ and sending T_j' to ∞ we have

$$e^{-\kappa c'_\infty} \leq e^{-\kappa c_\infty}$$

and thus we obtain $c'_\infty \geq c_\infty$. By replacing the roles of the both we also see that $c_\infty \geq c'_\infty$ and hence $c_\infty = c'_\infty$ □

Now the following is the direct consequence of Theorem 3.1 and Corollary 2.1.

Corollary 3.1. *Under the assumptions of Theorem 3.1, for some constant c_∞,*

$$\lim_{T\to\infty} \{v(0,x;T) - \chi(\theta)T - c_\infty - w(x)\} = 0,$$

uniformly on each compact set, which implies that

$$\lim_{T\to\infty} \frac{v(0,x;T)}{T} = \chi(\theta),$$

where $(\chi(\theta), w(x))$ *is the solution to H-J-B equation of ergodic type:*

$$\chi(\theta) = \tfrac{1}{2}tr[\lambda\lambda^* D^2 w] + \beta_\theta^* Dw + \tfrac{1}{2}(Dw)^*\lambda N_\theta \lambda^* Dw \\ +\tfrac{\theta}{2} g^* Q_\theta^{-1} g + \theta U. \tag{3.16}$$

Further, $\chi(\theta)$ *is convex.*

Moreover, similarly to Lemma 4.2 in [28], we have the following useful lemma.

Lemma 3.2. *Under the assumptions of Theorem 3.1, for each* $\theta_1 \le \theta \le \theta_0$ *there exist positive constants* $\delta > 0$ *and* $C > 0$ *independent of* T *and* $\theta \in [\theta_1, \theta_0]$ *such that*

$$E[e^{\delta |X_T|^2)}] \le C. \tag{3.17}$$

4. Differentiability of H-J-B equation

Let us take a solution $\bar{w}$ to (3.3) such that $\bar{w}(x) > 0$. Noting that $\frac{c_0}{c_0-\theta} I \le N_\theta$, we see that for sufficiently large $R_0 > 0$

$$K(x;\bar{w}) \equiv -L^{\bar{w}}\bar{w}(x) > \frac{c_0}{(c_0-\theta)\bar{w}}(D\bar{w})^*\lambda\lambda^* D\bar{w}(x), \quad x \in B_{R_0}^c.$$

and condition (A.3) of Appendix in [28] is satisfied for $L^{\bar{w}}$ and $\bar{w}$. The other conditions are also satisfied. Thus, setting

$$F_{\bar{w}} = \{u \in W_{loc}^{2,p};\ \text{esssup}_{x\in B_{R_0}^c} \frac{|u(x)|}{\bar{w}(x)} < \infty\}$$

and

$$F_K = \{f \in L_{loc}^\infty;\ \text{esssup}_{x\in B_{R_0}^c} \frac{|f(x)|}{K(x;\bar{w})} < \infty\},$$

we see that

$$-\zeta(\theta) = L^{\bar{w}}u - U - \frac{1}{2}(\lambda^* D\bar{w} - \delta(\delta^*\delta)^{-1}g)^* \frac{\partial N_\theta}{\partial\theta}(\lambda^* D\bar{w} - \delta(\delta^*\delta)^{-1}g) \tag{4.1}$$

has a solution $u \in F_{\bar{w}}$ such that

$$\zeta(\theta) = \int \{U + \frac{1}{2}(\lambda^* D\bar{w} - \delta(\delta^*\delta)^{-1}g)^* \frac{\partial N_\theta}{\partial\theta}(\lambda^* D\bar{w} - \delta(\delta^*\delta)^{-1}g)\} m_\theta(dx)$$

(cf. Corollary 5.1 in [28]). Here we note that

$$0 \le \frac{\partial N_\theta}{\partial\theta} = \delta Q_\theta^{-1}\delta^* + \theta(\delta Q_\theta^{-1}\delta^*)^2 \le \frac{1}{c_0-\theta} I$$

since $Q_\theta > -\theta\delta^*\delta$ and so $-\frac{1}{\theta}Q_\theta > \delta^*\delta$. Equation (4.1) is the one the derivative of the solution $(\bar{\chi}, \bar{w})$ of H-J-B equation of ergodic type should satisfy and we can see that $\bar{\chi}(\theta)$ is differentiable and the solution satisfies

$$\bar{\chi}'(\theta) = L^{\bar{w}}\bar{w}' - U - \frac{1}{2}(\lambda^* D\bar{w} - \delta(\delta^*\delta)^{-1}g)^* \frac{\partial N_\theta}{\partial \theta}(\lambda^* D\bar{w} - \delta(\delta^*\delta)^{-1}g), \tag{4.2}$$

where $\bar{w}' = \frac{\partial \bar{w}}{\partial \theta} \equiv -\frac{\partial w}{\partial \theta}$ (cf. Lemma 6.4 in [28]).

5. Interpretation by a stochastic differential game

Let us introduce a stochastic differential game with the value w

$$\begin{aligned}\tilde{J}(0,x;T) = \inf_{h.} \sup_{\nu.} E^{h,\nu}[\theta\{\textstyle\int_0^T f(X_s,h_s)ds + \int_0^T \varphi(X_s,h_s)^* dW_s\} \\ -\tfrac{1}{2}\textstyle\int_0^T |\nu_s + \theta\delta(\tilde{X}_s)h_s|^2 ds],\end{aligned} \tag{5.1}$$

where X_t is a solution to the stochastic differential equation

$$dX_t = \{\beta(X_t) + \lambda(X_t)(\nu_t + \theta\delta(X_t)h_t)\}dt + \lambda(X_t)d\tilde{W}_t, \quad X_0 = x, \tag{5.2}$$

$P^{h,\nu}$ is a probability measure defined by

$$P^{h,\nu}(A) = E[e^{\int_0^T (\nu_s+\theta\delta(X_s)h_s)^* dW_s - \frac{1}{2}\int_0^T |\nu_s+\theta\delta(X_s)h_s|^2 ds}; A]$$

and

$$\tilde{W}_t = W_t - \int_0^t (\nu_s + \theta\delta(X_s)h_s)ds.$$

Note that

$$\begin{aligned}&E^{h,\nu}[\theta\{\int_0^T f(X_s,h_s)ds + \int_0^T \varphi(X_s,h_s)^* dW_s\} - \frac{1}{2}\int_0^T |\nu_s + \theta\delta(X_s)h_s|^2 ds] \\ &= E^{h,\nu}[\textstyle\int_0^T \{-\frac{\theta}{2}h_s^* Q_\theta(X_s)h_s + \theta h_s^*(g(X_s) + U(X_s))\}ds - \frac{1}{2}\int_0^T |\nu_s|^2 ds] \\ &\equiv E^{h,\nu}[\textstyle\int_0^T \Xi(X_s,h_s,\nu_s;\theta)ds]\end{aligned}$$

when setting $\Xi(x,h,\nu;\theta) = -\frac{\theta}{2}h^* Q_\theta(x)h + \theta h^* g(x) + \theta U(x) - \frac{1}{2}|\nu|^2$. Thus, (5.1) is written as

$$\tilde{J}(0,x;T) = \inf_{h.} \sup_{\nu.} E^{h,\nu}[\int_0^T \Xi(X_s,h_s,\nu_s;\theta)ds]. \tag{5.3}$$

Therefore, the Isaacs equation of this stochastic differential game with value $\tilde{J}$ defined by (5.3) and the controlled process X_t governed by (5.2) is now seen as

$$\frac{\partial v}{\partial t} + \frac{1}{2}\mathrm{tr}[\lambda\lambda^* D^2 v] + \sup_{\nu\in R^M} \inf_{h\in R^m}[\{\beta + \lambda(\nu + \theta\delta h)\}^* Dv + \Xi(x,h,\nu;\theta)] = 0.$$

Since

$$\sup_{\nu}[\nu^*\lambda^*(x)Dv - \frac{1}{2}|\nu|^2] = \frac{1}{2}(Dv)^*\lambda\lambda^*(x)Dv,$$

$$\inf_{h}[\theta\{\lambda\delta h\}^* Dv - \frac{\theta}{2}h^* Q_\theta h + \theta h^* g] = \frac{\theta}{2}(\delta^*\lambda^* Dv + g)^* Q_\theta^{-1}(\delta^*\lambda^* Dv + g),$$

$N_\theta = I + \theta\delta Q_\theta^{-1}\delta^*$ and $\beta_\theta = \beta + \theta\lambda\delta Q_\theta^{-1}g$, it can be written as

$$\frac{\partial v}{\partial t} + \frac{1}{2}\mathrm{tr}[\lambda\lambda^* D^2 v] + \beta_\theta^* Dv + \frac{1}{2}(Dv)^*\lambda N_\theta \lambda^* Dv + \frac{\theta}{2}g^* Q_\theta^{-1} g + \theta U = 0$$

which is identical to (1.8). Thus, we have a stochastic differential game interpretation of our problem (1.6). While, we do not need the proof of the verification theorem which shows the solution v to (1.8) is identical to the value, namely $\tilde{J}(0,x;T) = v(0,x;T)$. In the proof of our duality theorem to be seen later, we only need the expression of the H-J-B equation (3.16) of ergodic type as the one of the stochastic differential game which is the infinite horizon counterpart of (5.1). Here, we further give a comment to the value (5.1) of the game which suggests the reason why (5.1) has the same H-J-B equation as (1.8).

Remark 5.1. Set $\rho_T := \left.\frac{dP^{h,\nu}}{dP}\right|_{\mathcal{F}_T}$. Then, the value (5.1) has an expression such that

$$\begin{aligned}\tilde{J}(0,x;T) = \inf_{h.}[-\inf_{\nu.}\{E^{h,\nu}[-\theta\{\textstyle\int_0^T f(X_s,h_s)ds + \int_0^T \varphi(X_s,h_s)^* dW_s\}] \\ +E[\rho_T \log \rho_T]\}].\end{aligned} \tag{5.4}$$

Here, $E[\rho_T \log \rho_T]$ is a relative entropy of the probability measure $P^{h,\nu}$ with respect to P on $(\Omega, \mathcal{F}_T)$.

H-J-B equation of ergodic type:

$$\chi(\theta) = \frac{1}{2}\mathrm{tr}[\lambda\lambda^* D^2 w] + \beta_\theta^* Dw + \frac{1}{2}(Dw)^*\lambda N_\theta \lambda^* Dw + \frac{\theta}{2}g^* Q_\theta^{-1} g + \theta U$$

is seen to be identical to

$$\begin{aligned}\chi(\theta) &= \tfrac{1}{2}\mathrm{tr}[\lambda\lambda^* D^2 w] + \sup_\nu \inf_h[\{\beta + \lambda(\nu + \theta\delta h)\}^* Dw + \Xi(x,h,\nu;\theta)] \\ &= \tfrac{1}{2}\mathrm{tr}[\lambda\lambda^* D^2 w] + [\{\beta + \lambda(\tilde{\nu} + \theta\delta\tilde{h})\}^* Dw + \Xi(x,\tilde{h},\tilde{\nu};\theta)],\end{aligned} \tag{5.5}$$

where

$$\tilde{h}(x) = Q_\theta^{-1}(\delta^*\lambda^* Dw(x) + g(x)), \quad \tilde{\nu}(x) = \lambda^* Dw(x).$$

We assume that the probability space is the canonical one, namely P is defined on $(\Omega, \mathcal{F}, \mathcal{F}_t) = (W^N, \mathcal{B}(W^N); \mathcal{B}_{t+}(W^N))$. Owing to estimate (3.4) and assumption (2.1)-(2.4), we see that

$$|\tilde{h}(x)|,\ |\tilde{\nu}(x)| \le C(1+|x|), \quad C > 0 \tag{5.6}$$

and thus it can be seen that

$$E[e^{\int_0^T \{\tilde{\nu}(X_s)+\theta\delta(X_s)\tilde{h}(X_s)\}^* dW_s - \frac{1}{2}\int_0^T |\tilde{\nu}(X_s)+\theta\delta(X_s)\tilde{h}(X_s)|^2 ds}] = 1, \quad \forall T > 0.$$

Therefore, we can define a probability measure $\tilde{P}$ on a canonical space $(W^N, \mathcal{B}(W^N))$ such that

$$\left.\frac{d\tilde{P}}{dP^X}\right|_{\mathcal{F}_T} = e^{\int_0^T \{\tilde{\nu}(X_s)+\theta\delta(X_s)\tilde{h}(X_s)\}^* dW_s - \frac{1}{2}\int_0^T |\tilde{\nu}(X_s)+\theta\delta(X_s)\tilde{h}(X_s)|^2 ds},$$

where $P^X := P \circ X^{-1}$ and $X_.$ is the solution to stochastic differential equation (1.1). Then, X_t satisfies

$$dX_t = \{\beta(X_t) + \lambda(X_t)(\tilde{\nu}(X_t) + \theta\delta(X_t)\tilde{h}(X_t))\}dt + \lambda(X_t)d\tilde{W}_t, \quad \tilde{X}_0 = x, \tag{5.7}$$

with the Brownian motion process $\tilde{W}_t$ under $\tilde{P}$ defined by

$$\tilde{W}_t = W_t - \int_0^t \{\tilde{\nu}(X_s) + \theta\delta(X_s)\tilde{h}(X_s)\}^* ds.$$

6. Duality theorem

In this section we shall present large deviation estimates for the controlled semi-martingales given in section 1. For the proof of the duality theorem we prepare some useful formulae. Set

$$\begin{aligned} L^w &:= \tfrac{1}{2}\mathrm{tr}[\lambda\lambda^* D^2] + D_p H(x, Dw)^* D \\ &= \tfrac{1}{2}\mathrm{tr}[\lambda\lambda^* D^2] + [\beta + \lambda(\tilde{\nu} + \theta\delta\tilde{h})]^* D. \end{aligned}$$

Then, H-J-B equation ergodic type (3.16) is written as

$$\chi(\theta) = L^w w + \Xi(x, \tilde{h}, \tilde{\nu}; \theta) \tag{6.1}$$

and

$$\begin{aligned} \Xi(x, \tilde{h}, \tilde{\nu}; \theta) &= -\tfrac{\theta}{2}\tilde{h}^* Q_\theta \tilde{h} + \theta\tilde{h}^* g + \theta U - \tfrac{1}{2}|\tilde{\nu}|^2 \\ &= -\tfrac{\theta}{2}(\tilde{h} - Q_\theta^{-1} g)^* Q_\theta (\tilde{h} - Q_\theta^{-1} g) + \tfrac{\theta}{2} g^* Q_\theta^{-1} g + \theta U - \tfrac{1}{2}|\lambda^* Dw|^2 \\ &= -\tfrac{\theta}{2}(\delta^*\lambda^* Dw)^* Q_\theta^{-1}(\delta^*\lambda^* Dw) + \tfrac{\theta}{2} g^* Q_\theta^{-1} g + \theta U - \tfrac{1}{2}(Dw)^*\lambda\lambda^* Dw \\ &= -\tfrac{1}{2}(Dw)^*\lambda N_\theta \lambda^* Dw + \tfrac{\theta}{2} g^* Q_\theta^{-1} g + \theta U \end{aligned}$$

Further,

$$\chi'(\theta) = L^w w' + U + \frac{1}{2}(\lambda^* Dw + \delta(\delta^*\delta)^{-1} g)^* \frac{\partial N_\theta}{\partial\theta}(\lambda^* Dw + \delta(\delta^*\delta)^{-1} g)$$

and

$$\begin{aligned} &\tfrac{\theta}{2}(\lambda^* Dw + \delta(\delta^*\delta)^{-1} g)^* \tfrac{\partial N_\theta}{\partial\theta}(\lambda^* Dw + \delta(\delta^*\delta)^{-1} g) \\ &= \tfrac{\theta}{2}(\lambda^* Dw + \delta(\delta^*\delta)^{-1} g)^*(\delta Q_\theta^{-1}\delta^* + \theta(\delta Q_\theta^{-1}\delta^*)^2)(\lambda^* Dw + \delta(\delta^*\delta)^{-1} g) \\ &= \tfrac{\theta}{2}(\lambda^* Dw)^*(\delta Q_\theta^{-1}\delta^* + \theta(\delta Q_\theta^{-1}\delta^*)^2)\lambda^* Dw \\ &\qquad + \theta(\delta(\delta^*\delta)^{-1} g)^*(\delta Q_\theta^{-1}\delta^* + \theta(\delta Q_\theta^{-1}\delta^*)^2)\lambda^* Dw \\ &\qquad\qquad + \tfrac{\theta}{2}(\delta(\delta^*\delta)^{-1} g)^*(\delta Q_\theta^{-1}\delta^* + \theta(\delta Q_\theta^{-1}\delta^*)^2)\delta(\delta^*\delta)^{-1} g \\ &= \tfrac{1}{2}(\lambda^* Dw)^*(N_\theta - I)N_\theta\lambda^* Dw + g^*\theta Q_\theta^{-1}\delta^* N_\theta\lambda^* Dw \\ &\qquad\qquad + \tfrac{1}{2}(\theta Q_\theta^{-1}\delta^* + \theta^2 Q_\theta^{-1}\delta^*\delta Q_\theta^{-1}) g. \end{aligned}$$

Therefore we obtain

$$\chi(\theta) - \theta\chi'(\theta) = L^w(w - \theta w') - |\tilde{\nu} + \theta\delta^*\tilde{h}|^2 \tag{6.2}$$

by seeing that

$$(N_\theta\lambda^* Dw + \theta\delta Q_\theta^{-1} g)^*(N_\theta\lambda^* Dw + \theta\delta Q_\theta^{-1} g) = |\tilde{\nu} + \theta\delta^*\tilde{h}|^2. \tag{6.3}$$

We have the following duality theorem.

Theorem 6.1. *Assume the assumptions of Theorem 3.1. Then, for* $\kappa \in (\chi'(-\infty), \chi'(0-))$, *we have*

$$\lim_{T\to\infty} \frac{1}{T} \inf_{h\in\mathcal{A}(T)} \log P\left(\frac{1}{T}F_T(X_., h_.) \leq \kappa\right) = - \inf_{k\in(\chi'(-\infty),\kappa]} I(k) = -I(\kappa),$$

$$I(k) := \sup_{\theta<0}\{\theta k - \chi(\theta)\}.$$

Moreover, for $\theta(\kappa)$ *such that* $\chi'(\theta(\kappa)) = \kappa \in (\chi'(-\infty), \chi'(0-))$ *take the strategy* $\hat{h}_t^{(\theta(\kappa),T)}$ *defined in Proposition 2.2. Then,*

$$\lim_{T\to\infty} \frac{1}{T} \log P\left(\frac{1}{T}F_T(X_., h^{(\theta(\kappa),T)}) \leq \kappa\right) = - \inf_{k\in(\chi'(-\infty),\kappa]} I(k) = -I(\kappa).$$

Proof. Let us first give an upper estimate. For a given constant $\chi'(-\infty) < \kappa < \chi'(0-)$,

$$\begin{aligned}\log P(F_T(X_., h_.) \leq \kappa T) &= \log P(e^{\theta F_T(X_.,h_.)} \geq e^{\theta\kappa T}) \\ &\leq \log E[e^{\theta F_T(X_.,h_.) - \theta\kappa T}] \\ &= \log E[e^{\theta F_T(X_.,h_.)}] - \theta\kappa T.\end{aligned}$$

Therefore,

$$\begin{aligned}\inf\nolimits_{h_.} \log P((F_T(X_., h_.) \leq \kappa T) &\leq \inf\nolimits_{h_.} \log E[e^{\theta F_T(X_.,h_.)}] - \theta\kappa T \\ &= v(0, x; T) - \theta\kappa T,\end{aligned}$$

which implies

$$\varlimsup_{T\to\infty} \frac{1}{T} \inf_{h_.} \log P(F_T(X_., h_.) \leq \kappa T) \leq \chi(\theta) - \theta\kappa$$

for all $\theta < 0$. Hence, we obtain

$$\begin{aligned}\varlimsup_{T\to\infty} \tfrac{1}{T} \inf\nolimits_{h_.} \log P(F_T(X_., h_.) \leq \kappa T) &\leq \inf\nolimits_{\theta<0}\{\chi(\theta) - \theta\kappa\} \\ &= -I(\kappa) \\ &= -\inf\nolimits_{k\in(\chi'(-\infty),\kappa]} I(k).\end{aligned}$$

Now, let us prove the converse inequality. Take a constant κ and $\epsilon > 0$ such that $\chi'(0-) - \epsilon > \kappa - \epsilon > \chi'(-\infty)$. Then, there exists θ_ϵ such that

$$\inf_{\theta<0}\{\chi(\theta) - \theta(\kappa - \epsilon)\} = \chi(\theta_\epsilon) - \theta_\epsilon\chi'(\theta_\epsilon).$$

We write θ_ϵ as θ for simplicity in the following. Assuming the probability space $(\Omega, \mathcal{F}, P)$ is the canonical one, introduce a probability measure $\tilde{P}$ with the density

$$\left.\frac{d\tilde{P}}{dP}\right|_{\mathcal{F}_T} = e^{\int_0^T \{\tilde{\nu}(X_s)+\theta\delta(X_s)\tilde{h}(X_s)\}^* dW_s - \frac{1}{2}\int_0^T |\tilde{\nu}(X_s)+\theta\delta(X_s)\tilde{h}(X_s)|^2 ds}.$$

Then, as we have seen above, the solution X_t to (1.1) satisfies (5.7) with the Brownian motion process $\tilde{W}_t = W_t - \int_0^t \{\tilde{\nu}(X_s) + \theta\delta(X_s)\tilde{h}(X_s)\}ds$ under $\tilde{P}$. Let us first show the following lemma.

Lemma 6.1. *For each ϵ, by taking sufficiently large T*

$$\tilde{P}(\int_0^T f(X_s, h_s)ds + \int_0^T \varphi(X_s, h_s)^* dW_s > \kappa T) < \epsilon, \tag{6.4}$$

holds for any $h_\cdot \in \mathcal{A}_T$.

Proof. Set $\tilde{h}_s = \tilde{h}(X_s)$ and $\tilde{\nu}_s = \tilde{\nu}(X_s)$. Then, we see that

$$\begin{aligned}
&\theta\{\textstyle\int_0^T f(X_s, h_s)ds + \int_0^T \varphi(X_s, h_s)^* dW_s\} \\
&= \textstyle\int_0^T \Xi(X_s, \tilde{h}_s, \tilde{\nu}_s; \theta)ds + \theta \int_0^T \tilde{h}_s \delta(X_s)^* d\tilde{W}_s \\
&+\theta \textstyle\int_0^T (h_s - \tilde{h}_s)^*(\delta(X_s)^* - S^{1/2}(X_s)E_{m,M})d\tilde{W}_s + \frac{1}{2}\int_0^T |\tilde{\nu}_s + \theta\delta(X_s)\tilde{h}_s|^2 ds \\
&+\theta \textstyle\int_0^T (h_s - \tilde{h}_s)^* S^{1/2}(X_s)E_{m,M} d\tilde{W}_s - \frac{\theta}{2}\int_0^T (h_s - \tilde{h}_s)^* S(X_s)(h_s - \tilde{h}_s)ds,
\end{aligned} \tag{6.5}$$

where $E_{m,M}$ is an $m \times M$ matrix such that $(E_{m,M})_{ij} = \delta_{i,j}$, for $i, j = 1, \ldots, m$ and $(E_{m,M})_{i,j} = 0$, for $m + 1 \le j \le M$, $i = 1, \ldots, m$. Indeed,

$$\begin{aligned}
&\theta \int_0^T f(X_s, h_s)ds + \theta \int_0^T \varphi(X_s, h_s)^* dW_s \\
&= \theta \textstyle\int_0^T [f(X_s, h_s) + h_s^*\delta(X_s)^*\{\tilde{\nu}_s + \theta\delta(X_s)\tilde{h}_s\}]ds + \theta \int_0^T h_s^*\delta(X_s)^* d\tilde{W}_s \\
&= \theta \textstyle\int_0^T [f(X_s, \tilde{h}_s) + \tilde{h}_s^*\delta(X_s)^*\{\tilde{\nu}_s + \theta\delta(X_s)\tilde{h}_s\}]ds + \theta \int_0^T \tilde{h}_s^*\delta(X_s)^* d\tilde{W}_s \\
&\quad +\theta \textstyle\int_0^T [f(X_s, h_s) - f(X_s, \tilde{h}_s) + (h_s - \tilde{h}_s)^*\delta(X_s)^*\{\tilde{\nu}_s + \theta\delta(X_s)\tilde{h}_s\}]ds \\
&\quad +\theta \textstyle\int_0^T (h_s - \tilde{h}_s)^*\delta(X_s)^* d\tilde{W}_s \\
&= -\tfrac{\theta}{2} \textstyle\int_0^T \tilde{h}_s^*(S(X_s) - \theta\delta^*\delta(X_s))\tilde{h}_s ds + \theta \int_0^T \tilde{h}_s g(X_s)ds + \theta \int_0^T U(X_s)ds \\
&\quad -\tfrac{\theta}{2} \textstyle\int_0^T h_s^*(S(X_s) - \theta\delta^*\delta(X_s))h_s ds + \theta \int_0^T \{h_s g(X_s) + h_s^*\delta(X_s)\tilde{\nu}_s\}ds \\
&\quad +\tfrac{\theta}{2} \textstyle\int_0^T \tilde{h}_s^*(S(X_s) - \theta\delta^*\delta(X_s))\tilde{h}_s ds - \theta \int_0^T \{\tilde{h}_s g(X_s) + \tilde{h}_s^*\delta(X_s)^*\tilde{\nu}_s\}ds \\
&\quad +\theta \textstyle\int_0^T \tilde{h}_s\delta(X_s)^* d\tilde{W}_s + \frac{1}{2}\int_0^T |\tilde{\nu}_s + \theta\delta(X_s)\tilde{h}_s|^2 ds - \frac{1}{2}\int_0^T |\tilde{\nu}_s|^2 ds \\
&\quad +\theta \textstyle\int_0^T (h_s - \tilde{h}_s)^*\delta(X_s)^* d\tilde{W}_s - \frac{\theta^2}{2}\int_0^T (h_s - \tilde{h}_s)^*\delta^*\delta(X_s)(h_s - \tilde{h}_s)ds \\
&\quad \equiv I_T.
\end{aligned}$$

Since $\Xi(x, h, \nu; \theta) = -\frac{\theta}{2}h^*Q_\theta h + \theta h^* g + \theta U - \frac{1}{2}|\nu|^2$, $\tilde{h} = Q_\theta^{-1}(g + \delta^*\tilde{\nu})$, and $Q_\theta = S - \theta\delta^*\delta$, we obtain

$$\begin{aligned}
I_T = &\textstyle\int_0^T \Xi(X_s, \tilde{h}_s, \tilde{\nu}_s; \theta)ds + \theta \int_0^T \tilde{h}_s^*\delta(X_s)^* d\tilde{W}_s + \frac{1}{2}\int_0^T |\tilde{\nu}_s + \theta\delta(X_s)\tilde{h}_s|^2 ds \\
&+\theta \textstyle\int_0^t (h_s - \tilde{h}_s)^*\delta(X_s)^* d\tilde{W}_s - \frac{\theta}{2}\int_0^T (h_s - \tilde{h}_s)^* S(X_s)(h_s - \tilde{h}_s)ds
\end{aligned}$$

and see that (6.5) holds. Set

$$M_t^h = \int_0^t (h_s - \tilde{h}_s)^* S^{\frac{1}{2}}(X_s) E_{m,M} d\tilde{W}_s.$$

Then,

$$\begin{aligned} w(X_T) - w(X_0) &= \int_0^T L^w w(X_s) ds + \int_0^T (Dw)^* \lambda(X_s) d\tilde{W}_s \\ &= \chi(\theta)T - \int_0^T \Xi(X_s, \tilde{h}_s, \tilde{\nu}_s; \theta) ds + \int_0^T (Dw)^* \lambda(X_s) d\tilde{W}_s \end{aligned}$$

and

$$\begin{aligned} &\theta\{\int_0^T f(X_s, h_s) ds + \int_0^T \varphi(X_s, h_s)^* dW_s\} \\ &= \chi(\theta)T + w(X_0) - w(X_T) + \int_0^T \{(Dw)^* \lambda(X_s) + \theta \tilde{h}_s \delta(X_s)^*\} d\tilde{W}_s \\ &+\theta \int_0^T (h_s - \tilde{h}_s)^* (\delta(X_s)^* - S^{1/2}(X_s) E_{m,M}) d\tilde{W}_s + \tfrac{1}{2} \int_0^T |\tilde{\nu}_s + \theta \delta(X_s) \tilde{h}_s|^2 ds \\ &+\theta(M_T^h - \tfrac{1}{2}\langle M^h \rangle_T). \end{aligned} \tag{6.6}$$

Thanks to (6.2), we have

$$\begin{aligned} &(w - \theta w')(X_T) - (w - \theta w')(X_0) \\ &\quad = \int_0^T L^w (w - \theta w')(X_s) d\tilde{W}_s + \int_0^T D(w - \theta w')(X_s)^* \lambda(X_s) d\tilde{W}_s \\ &\quad = \tfrac{1}{2} \int_0^T |\tilde{\nu} + \theta \delta^* \tilde{h}|^2 (X_s) ds + (\chi(\theta) - \theta \chi'(\theta))T \\ &\qquad\qquad + \int_0^T D(w - \theta w')(X_s)^* \lambda(X_s) d\tilde{W}_s. \end{aligned} \tag{6.7}$$

Therefore,

$$\begin{aligned} &\theta\{\int_0^T f(X_s, h_s) ds + \int_0^T \varphi(X_s, h_s)^* dW_s\} \\ &= \theta\chi'(\theta)T + \theta w'(X_0) - \theta w'(X_T) + \theta \int_0^T \{(Dw')^* \lambda(X_s) + \tilde{h}_s^* \delta(X_s)^*\} d\tilde{W}_s \\ &\quad +\theta \int_0^T (h_s - \tilde{h}_s)^* (\delta(X_s)^* - S^{1/2}(X_s) E_{m,M}) d\tilde{W}_s + \theta(M_T^h - \tfrac{1}{2}\langle M^h \rangle_T). \end{aligned} \tag{6.8}$$

Now we can prove (6.4). Since $\kappa = \chi'(\theta) + \epsilon$,

$$\begin{aligned} &\tilde{P}(\int_0^T f(X_s, h_s) ds + \int_0^T \varphi(X_s, h_s)^* dW_s > \kappa T) \\ &\quad \le \tilde{P}(w'(X_0) - w'(X_T) > \tfrac{\epsilon T}{5}) + \tilde{P}(\int_0^T (Dw')^* \lambda(X_s) d\tilde{W}_s > \tfrac{\epsilon T}{5}) \\ &\qquad + \tilde{P}(\int_0^T \tilde{h}_s^* \delta(X_s)^* d\tilde{W}_s > \tfrac{\epsilon T}{5}) + \tilde{P}(M_T^h - \tfrac{1}{2}\langle M^h \rangle_T > \tfrac{\epsilon T}{5}) \\ &\qquad\quad + \tilde{P}(\int_0^T (h_s - \tilde{h}_s)^* (\delta(X_s)^* - S^{1/2}(X_s) E_{m,M}) d\tilde{W}_s > \tfrac{\epsilon T}{5}) \\ &\quad \equiv Q_1 + Q_2 + Q_3 + Q_4 + Q_5. \end{aligned}$$

Then, we have

$$Q_1 \le \frac{5^2}{\epsilon^2 T^2} \tilde{E}[|w'(X_T) - w'(X_0)|^2].$$

Noting that the law $P \circ \bar{X}_{\cdot}^{-1}$ defined in Remark in Section 3 is nothing but $\tilde{P}$ and that w' is at most polynomial growth, we see that $\tilde{E}[|w'(X_T) - w'(X_0)|^2]$ is bounded by a positive constant C independent of T thanks to Lemma 3.2. Thus, Q_1 is seen to be small enough when T is sufficiently large. Next,

$$\begin{aligned} Q_2 &\le \tfrac{5^2}{\epsilon^2 T^2}\tilde{E}[|\textstyle\int_0^T (Dw')^*\lambda(X_s)d\tilde{W}_s|^2] \\ &= \tfrac{5^2}{\epsilon^2 T^2}[\textstyle\int_0^T \tilde{E}(Dw')^*\lambda\lambda^* Dw'(X_s)]ds \end{aligned}$$

and $(Dw')^*\lambda\lambda^*Dw'$ is at most polynomial growth again we see that $\tilde{E}[(Dw')^*\lambda\lambda^*Dw'(X_s)]$ is bounced by a positive constant. Therefore, Q_2 is small enough when T is sufficiently large. As for Q_3 and Q_4, the same arguments apply since $\tilde{h}_s = \tilde{h}(X_s)$ and $h_s = h(X_s)$ with functions h, $\tilde{h}$ which are at most polynomial growth. Concerning Q_4, we have

$$Q_4 \le e^{-\epsilon T/5}\tilde{E}[e^{M_T^h - \frac{1}{2}\langle M^h\rangle_T}] \le e^{-\epsilon T/5}.$$

Hence we obtain (6.4). □

Proof of Theorem 6.1. The proof of Theorem 6.1 can proceed in a similar way to the proof of Theorem 2.4 in [28]. Indeed, set

$$\tilde{M}_t^\theta := \int_0^t \{\tilde{\nu}_s + \theta\delta(X_s)\tilde{h}_s\}d\tilde{W}_s$$

and

$$\begin{aligned} A_1 &= \{-\tilde{M}_T \ge -\epsilon T\} \\ A_2 &= \{-\tfrac{1}{2}\langle\tilde{M}^\theta\rangle_T \ge (\chi(\theta) - \theta\chi'(\theta) - \epsilon)T\} \\ A_3 &= \{F_T(X_., h_.) \le \kappa T\}. \end{aligned}$$

Thanks to (6.7) we have

$$\begin{aligned} &\tilde{P}(\tfrac{1}{2}\langle\tilde{M}^\theta\rangle_T + (\chi(\theta) - \theta\chi'(\theta))T > \epsilon T) \\ &= \tilde{P}((w - \theta w')(X_T) - (w - \theta w')(X_0) - \textstyle\int_0^T D(w - \theta w')^*\lambda(X_s)d\tilde{W}_s > \epsilon T) \end{aligned}$$

and similarly to the above, we see that

$$\tilde{P}(\frac{1}{2}\langle\tilde{M}^\theta\rangle_T + (\chi(\theta) - \theta\chi'(\theta))T > \epsilon T) < \epsilon$$

holds for sufficiently large T. Thus we see that $\tilde{P}(A_2^c) < \epsilon$. We have already seen that $\tilde{P}(A_3^c) < \epsilon$. It is similar to the above showing that $\tilde{P}(A_1^c) < \epsilon$. Thus,

$$\begin{aligned} P(F_T(X_., h_.) \le \kappa T) &= \tilde{E}[e^{-\tilde{M}_T^\theta - 1/2\langle\tilde{M}^\theta\rangle_T}; F_T(X_., h_.) \le \kappa T] \\ &\ge \tilde{E}[e^{-\tilde{M}_T^\theta - 1/2\langle\tilde{M}^\theta\rangle_T}; A_1 \cap A_2 \cap A_3] \\ &\ge e^{(\chi(\theta) - \theta\chi'(\theta) - 2\epsilon)T}\tilde{P}(A_1 \cap A_2 \cap A_3) \\ &\ge e^{(\chi(\theta) - \theta\chi'(\theta) - 2\epsilon)T}\{1 - \tilde{P}(A_1^c) - \tilde{P}(A_2^c) - \tilde{P}(A_3^c)\} \\ &\ge e^{(\chi(\theta) - \theta\chi'(\theta) - 2\epsilon)T}(1 - 3\epsilon). \end{aligned}$$

Therefore,

$$\begin{aligned}\underline{\lim}_{T\to\infty}\tfrac{1}{T}\inf_{h.}\log P(F_T(X.,h.)\le\kappa T) &\ge \chi(\theta)-\theta\chi'(\theta)-2\epsilon\\ &=\chi(\theta)-\theta(\kappa-\epsilon)-2\epsilon\\ &\ge \inf_{\theta<0}\{\chi(\theta)-\theta(\kappa-\epsilon)\}-2\epsilon\end{aligned}$$

for each ϵ. Since $\chi(\theta)$ is smooth and convex $\sup_{\theta<0}\{\theta\kappa-\chi(\theta)\}$ is convex, and thus continuous. Hence

$$\underline{\lim}_{T\to\infty}\frac{1}{T}\inf_{h.}\log P(F_T(X.,h.)\le\kappa T)\ge-\sup_{\theta<0}\{\theta\kappa-\chi(\theta)\}.$$

Now, for $\chi'(-\infty)<\kappa<\chi'(0-)$, take a strategy $h^{(\theta(\kappa),T)}$ defined in Proposition 2.2. Then,

$$\begin{aligned}\log P(F_T(X.,h^{(\theta(\kappa),T)})\le\kappa T) &\le \log E[e^{\theta F_T(X.,h^{(\theta(\kappa),T)})-\theta\kappa T}]\\ &=\inf_{h.}\log E[e^{\theta F_T(X.,h.)}]-\theta\kappa T\\ &=v(0,x;T)-\theta\kappa T.\end{aligned}$$

Therefore,

$$\overline{\lim}_{T\to\infty}\frac{1}{T}\log P(F_T(X.,h^{(\theta(\kappa),T)})\le\kappa T)\le\chi(\theta(\kappa))-\theta(\kappa)\kappa=-\sup_{\theta<0}\{\theta\kappa-\chi(\theta)\}.$$

On the other hand, it is obvious that

$$\begin{aligned}&\underline{\lim}_{T\to\infty}\tfrac{1}{T}\log P(F_T(X.,h^{(\theta(\kappa),T)})\le\kappa T)\\ &\qquad\ge\underline{\lim}_{T\to\infty}\tfrac{1}{T}\inf_{h.}\log P(F_T(X.,h.)\le\kappa T)\\ &\qquad=-\sup_{\theta<0}\{\theta\kappa-\chi(\theta)\}\end{aligned}$$

and hence we obtain our theorem. □

7. Linear Gaussian models

The linear Gaussian case can be independently discussed. Let as assume that

$$\begin{aligned}&\lambda(x)\equiv\lambda,\quad \beta(x)=Bx+b,\quad S(x)\equiv S,\quad g(x)=Ax+a,\\ &U(x)=\tfrac{1}{2}x^*Vx+m,\quad \delta(x)\equiv\delta,\end{aligned}$$

where λ, B, S, A, V, δ are constant matrices, b, a are constant vectors and m is a constant. Moreover we assume that

$$\delta^*\delta>0,\quad S>0,\quad V\ge0.$$

In this case, the solution $v(t,x)$ to H-J-B equation (1.8) has an explicit representation such that

$$v(t,x)=\frac{1}{2}x^*P(t)x+q(t)^*x+l(t),$$

where $P(t)$, $q(t)$ and $l(t)$ are the solutions to the following equations:

$$\dot{P}(t)+K_1^*P(t)+P(t)K_1+P(t)\lambda N_\theta\lambda^*P(t)+\theta A^*Q_\theta^{-1}A+\theta V=0,\quad P(T)=0,$$

$$\dot{q}(t)+(k_1+\lambda N_\theta\lambda^*P(t))^*q(t)+P(t)k_1+\theta A^*Q_\theta^{-1}a=0,\quad q(T)=0$$

$$\dot{l}(t)+\tfrac{1}{2}\mathrm{tr}[\lambda\lambda^*P(t)]+k_1^*q(t)+\tfrac{1}{2}q(t)^*\lambda N_\theta\lambda^*q(t)+\tfrac{\theta}{2}a^*Q_\theta^{-1}a+\theta m=0,\quad l(T)=0$$

where $K_1=B+\theta\lambda\delta Q_\theta^{-1}A$ and $k_1=b+\theta\lambda\delta Q_\theta^{-1}a$.

If we assume that
(I) $\lambda\lambda^*>0,\ \ A^*A>0,$
or

(II) B is stable,
then we can see that as $T\to\infty$,

$$P(t;T)\to\overline{P},\ q(t;T)\to\overline{q},\ \frac{l(t;T)}{T}\to\chi(\theta),$$

where $\overline{P}$ is the unique, negative semi-definite solution to the stationary Riccati equation:

$$K_1^*\overline{P}+\overline{P}K_1+\overline{P}\lambda N_\theta\lambda^*\overline{P}+\theta A^*Q_\theta^{-1}A+\theta V=0$$

such that $K_1+\lambda N_\theta\lambda^*\overline{P}$ is stable, $\overline{q}$ is the one of the algebraic equation:

$$(k_1+\lambda N_\theta\lambda^*\overline{P})^*\overline{q}+\overline{P}K_1+\theta A^*Q_\theta^{-1}a=0,$$

and $\chi(\theta)$ is the constant given by

$$\chi(\theta)=\frac{1}{2}\mathrm{tr}[\lambda\lambda^*\overline{P}]+k_1^*\overline{q}+\frac{1}{2}\overline{q}^*\lambda N_\theta\lambda^*\overline{q}+\frac{\theta}{2}a^*Q_\theta^{-1}a+\theta m.$$

Further, the pair of a function $w(x)=\frac{1}{2}x^*\overline{P}x+\overline{q}^*x$ and a constant $\chi(\theta)$ turns out to be a solution to the H-J-B equation of ergodic type. Indeed, let us assume (I). Then, setting $K:=-\theta\delta Q_\theta^{-1}A-\lambda^*(\lambda\lambda^*)^{-1}(B+I)$, we see that $K_1+\lambda K=-I$ and that it is stable. Thus, (K_1,λ) is stabilizable. Further, setting $K':=\sqrt{-\theta}\sqrt{Q_\theta^{-1}}\delta^*\lambda^*$ and $K'':=\frac{1}{\sqrt{-\theta}}\sqrt{Q_\theta}A(A^*A)^{-1}(-B-I)$, we see that $K_1^*+C^*(K'+K'')=-I$ and therefore (K_1^*,C^*) is stabilizable, namely (C,K_1) is detectable. Thus, according to general theory of Riccati equations ([5], [35]) we have the above convergence result. On the other hand, assume (II). Then, setting $\tilde{K}:=-\theta\delta Q_\theta^{-1}A$, we have $K_1+\lambda\tilde{K}=B$ and see that it is stable. Thus, (K_1,λ) turns out to be stabilizable. Further, we can see that $K_1^*+C^*K'=B^*$ and that (K_1^*,C^*) is stabilizable, namely (C,K_1) is detectable. Hence, in the same way as above we see that the above convergence result holds.

Differentiability of the solution to the H-J-B equation with respect to θ is also seen through independent arguments of the above general case but similarly to [16], and thus we can deduce the same statement as Theorem 6.1 under assumption (I) or (II).

8. Acknowledgements

Supported in part by grant-in-aid for Scientific Research 25400150, 23244015, 23340028 JSPS.

References

1. BENSOUSSAN, A. (1988). *Perturbation Methods in Optimal Control*, Wiley/ Gauthier-Villars
2. BENSOUSSAN, A. (1992). *Stochastic Control of Partially Observable Systems*, Cambridge University Press
3. BENSOUSSAN, A. AND J. FREHSE (1992). On Bellman equations of ergodic control in R^N, J. Reine Angew. Math., **429**, 125-160
4. BENSOUSSAN, A., FREHSE, J. AND NAGAI, H. (1998). Some results on risk-sensitive control with full information, *Appl. Math. Optim.* **37**, 1-41
5. BUCY, R.S. AND JOSEPH, P.D. (1987). *Filtering for stochastic processes with applications to guidance*, Chelsea, New York
6. BROWNE, S. (1999). Beating a moving target : optimal portfolio strategies for outperforming a stochastic benchmark, *Finance and Stochastics* **3** 275-294
7. BROWNE, S. (1999). The risk and rewards of minimizing shortfall probability, *J. Portfolio Management* **25** (4) 76-85
8. DAVIS, M. AND LLEO, S.(2008). Risk-sensitive benchmarked asset management, *Quantitative Finance* **8** 415-426
9. DEUSCHEL, J.-D. AND STROOCK, D.W. (1989) *Large deviations*, Academic Press, New York
10. DEMBO, A. AND ZEITOUNI, O. (1998). *Large deviations Techniques and applications*, Springer-Verlag, Berlin
11. DONSKER, M.D. AND VARADHAN, S.R.S (1975). Asymptotic evaluation of certain Markov process expectations for large time, I,II, *Comm. Pure and Appl. Math.* **28**, 1-47, 279-301
12. DONSKER, M.D. AND VARADHAN, S.R.S (1976). Asymptotic evaluation of certain Markov process expectations for large time, III, *Comm. Pure and Appl. Math.* **29**, 389-461
13. GÄRTIER, J. (1977). On large deviations from the invariant measure, *Th. Proba. Appl.* **22**, 24-39
14. HASMINSKII, R.Z. (1980). *Stochastic stability of differential equations*, SIJTHOFF & NOORDHOFF Maryland, USA
15. HATA H. AND IIDA, Y. (2006). A risk-sensitive stochastic control approach to an optimal investment problem with partial information, *Finance and Stochastics* **10** 395–426
16. HATA, H., NAGAI, H. AND SHEU, S.J. (2010) Asymptotics of the probability minimizing a "down-side" risk, Annals of Applied Probability **20**, 52-89
17. HATA, H. AND SEKINE, J.(2005). Solving long term optimal investment problems with Cox-Ingersoll-Ross interest rates, *Advances in Mathematical Economics* **8** 231–255
18. HATA, H. AND SEKINE, J. (2010). Explicit solution to a certain non-ELQG risk-sensitive stochastic control problem, *Appl. Math. Optim.* **62** 341-380
19. HATA, H. AND SHEU, S.J. (2009). Down-side risk" probability minimization problem for a multidimensional model with stochastic volatilty, preprint
20. ICHIHARA, N. (2011) Recurrence and transience o optimal feedback processes associated with Bellman equations of ergodic type. *SIAM Journal on Control and Opti-*

mization **49** 1938-1960

21. Ichihara, N. and Sheu, S.J. (2013) Large time behavior of solutions of Hamilton-Jacobi-Bellman equations with quadratic nonlinearity in gradients, *SIAM J. math. Anal.* **45** 279-306
22. Kuroda, K. and Nagai, H. (2002). Risk sensitive portfolio optimization infinite time horizon, *Stochastics and Stochastics Reports*, **73** 309–331
23. Ladyzhenskaya, O.A. and Ural'tseva, N.N. (1968). *Linear and Quasilinear Elliptic Equations*, Academic Press, New York
24. Nagai, H. (1996). Bellman equations of risk-sensitive control, *SIAM J. Control and Optimization* **34** 74-101
25. Nagai. H. (2003). Optimal strategies for risk-sensitive portfolio optimization problems for general factor models, *SIAM J. Cont. Optim.* **41** 1779–1800
26. Nagai, H. (2010). Asymptotics of the probability minimizing a "down-side" risk under partial information, *Quant. Finance.* **11**,789-803
27. Nagai, H. (2012) Downside risk minimization under prescribed consumption level, *Risk and Decision Analysis.* **3** 191-200
28. Nagai, H. (2012) Downside risk minimization via a large deviation approach. *Annals of Appl. Prob.* **22** 608-669
29. Nagai, H. (2013) *Downside risk minimization: Large deviation estimates for controlled semi-martingales.* "Real Options, Ambiguity, Risk and Insurance" , Ed. A. Bensoussan et al. , Studies in Probability, Optimization and Statistics vol. 5, IOS Press
30. Nagai, H. (2014). Robust estimates of certain large deviation probabilities for controlled semi-martingales, to appear in a volume of Banach Center Publications
31. Pham, H. (2003). A large deviations approach to optimal long term investment, *Finance and Stochastics* **7** 169–195
32. Pham, H. (2003). A risk-sensitive control dual approach to a large deviations control problem, *Systems and Control Letters* **49** 295–309
33. Stettner, L. (2004). Duality and risk sensitive portfolio optimization, *Mathematics of Finance, Ed. G. Yin and Q. Zhang, Contemporary Math.* **351** AMS 333-347
34. Stutzer, M. (2003). Portfolio choice with endogeous utility: a large deviations approach, *J. Econometrics* **116** 365-386
35. Wonham, W.M. (1979). *Linear Multivariable Control: A Geometric Approach*, Springer-Verlag

Chapter 23

A comparison theorem for backward SPDEs with jumps

Bernt Øksendal* †, Agnès Sulem‡ and Tusheng Zhang§

In honor of Professor M. Fukushima on occasion of his 80th birthday

In this paper, we obtain a comparison theorem for backward stochastic partial differential equation (SPDEs) with jumps. We apply it to introduce space-dependent convex risk measures as a model for risk in large systems of interacting components.

Contents

1. Introduction and framework

There are several papers dealing with comparison theorems for backward stochastic partial differential equations (BSPDEs). One of the first seems to be the paper [MY]. The results of that paper were subsequently extended (still for linear BSPDEs only) in the paper [DM]. Other related papers are [DQT] and also our own paper [ØSZ] (for reflected BSPDE).

The paper which seems to be closest to ours is [MYZ]. Here more general non-linear BSPDEs are considered, and a comparison theorem is proved for such equations by exploiting the relation between BSPDEs and coupled systems of forward-backward SDEs (FBSDEs).

Our paper also deals with quite general non-linear BSPDEs, but it differs from [MYZ] in several ways:

*The research leading to these results has received funding from the European Research Council under the European Community's Seventh Framework Programme (FP7/2007-2013) / ERC grant agreement no [228087]

†Center of Mathematics for Applications (CMA), Dept. of Mathematics, University of Oslo, P.O. Box 1053 Blindern, N–0316 Oslo, Norway, email: oksendal@math.uio.no and Norwegian School of Economics and Business Administration, Helleveien 30, N–5045 Bergen, Norway

‡INRIA Paris-Rocquencourt, Domaine de Voluceau, Rocquencourt, BP 105, Le Chesnay Cedex, 78153, France, Université Paris-Est, F-77455 Marne-la-Vallée, France and CMA, University of Oslo, email: agnes.sulem@inria.fr

§School of Mathematics, University of Manchester, Oxford Road, Manchester M13 9PL, United Kingdom and School of Mathematics and University of Science and Technology of China, Hefei, China, Tusheng.zhang@manchester.ac.uk

(i) First, our paper includes jumps.

(ii) Second, our BSPDEs are slightly different. They have stronger conditions on the second order term, but allow more general drift terms.

(ii) Third, our method is different, being based on an approximation technique.

Let $B_t = B_t(\omega), t \geq 0$ be a Brownian motion and $\tilde{N}(dt,dz) := N(dt,dz) - \nu(dz)dt$ an independent compensated Poisson random measure on a filtered probability space $(\Omega, \mathcal{F}, \{\mathcal{F}_t\}_{0\leq t\leq T}, P)$, where ν is the Lévy measure associated with the Poisson measure $N(\cdot,\cdot)$ on $[0,\infty)\times\mathbb{R}$. Let D be a bounded domain in $\mathbb{R}^d$. Denote by $A(t)$ the following second order differential operator on D equipped with the Dirichlet boundary condition:

$$A(t) = \sum_{i,j=1}^{d} \frac{\partial}{\partial x_i}(a_{ij}(t,x)\frac{\partial}{\partial x_j}),$$

where $a = (a_{ij}(t,x)) : [0,T]\times D \to \mathbb{R}^{d\times d}$ is a measurable, symmetric matrix-valued function. Set $L = L^2(\mathbb{R},\nu)$. Let $b(t,x,u,v,Z,r(\cdot))$ be a measurable mapping from $[0,T]\times \mathbb{R}\times\mathbb{R}\times\mathbb{R}^d\times\mathbb{R}\times L$ into $\mathbb{R}$. Let $\beta(t) = (\beta_1(t),...,\beta_d(t)), t\geq 0$ be a given progressively measurable $\mathbb{R}^d$-valued stochastic process. From now on, if $u(t,x)$ is a function of (t,x), we sometimes write $u(t)$ for the function $u(t,\cdot)$. Consider the solution of the following backward stochastic partial differential equation (BSPDE):

$$\begin{aligned} &du(t,x)\\ &= -A(t)u(t)dt - b(t,x,u(t,x),\nabla u(t,x),Z(t,x),r(t,x,\cdot))dt + Z(t,x)dB_t\\ &\quad - <\beta(t),\nabla Z(t,x)> dt + \int_{\mathbb{R}} r(t,x,z)\tilde{N}(dt,dz), t\in(0,T)\\ &u(T,x) = \phi(x) \quad a.s. \end{aligned} \tag{1.1}$$

Here ϕ is an $\mathcal{F}_T$-measurable $H := L^2(D)$-valued random variable. Let V be the Sobolev space $H_0^{1,2}(D)$ and V^* be its dual.

Definition 1.1. An adapted random field $u(t,x)$ is said to be the solution of the BSPDE (1.1) if

(i) $u \in D([0,T];H)\cap L^2(\Omega\times[0,T];V)$,

(ii) for $t>0$, u satisfies the following equation almost surely in V^*

$$\begin{aligned} &u(t,x)\\ &= \phi(x) + \int_t^T A(s)u(s)ds + \int_t^T b(s,x,u(s,x),\nabla u(s,x),Z(s,x),r(s,x,\cdot))ds\\ &- \int_t^T Z(s,x)dB_s + \int_t^T <\beta(s),\nabla Z(s,x)> ds - \int_t^T\int_{\mathbb{R}} r(s,x,z)\tilde{N}(ds,dz) \end{aligned} \tag{1.2}$$

Remark 1.1. Equation (1.2) is equivalent to that for any $\psi\in V$,

$$\begin{aligned} &< u(t,\cdot),\psi >\\ &=< \phi(\cdot),\psi > + \int_t^T < A(s)u(s),\psi > ds\\ &+ \int_t^T < b(s,\cdot,u(s,\cdot),\nabla u(s,\cdot),Z(s,\cdot),r(s,\cdot,\cdot)),\psi > ds \end{aligned}$$

$$- \int_t^T < Z(s,\cdot), \psi > dB_s - \int_t^T \int_D < \beta(s), \nabla\psi(x) > Z(s,x)dxds$$
$$- \int_t^T \int_{\mathbb{R}} < r(s,\cdot,z), \psi > \tilde{N}(ds,dz) \tag{1.3}$$

almost surely.

The aim of this paper is to prove a comparison theorem for the above BSPDEs with jumps.

2. Main result

Introduce the following assumptions:
(**A.1**). There exists $\delta_1 > 0$ and $0 < a < 1$ such that

$$\sum_{i,j=1}^{d} a_{ij}(t,x)z_i z_j \geq (\frac{1}{2a}|\beta|^2(t) + \delta_1)|z|^2, \ \forall z \in \mathbb{R}^d \text{ and } x \in D \tag{2.1}$$

(**A.2**). There exists $C > 0$ such that

$$|b(t,x,u_1,v_1,Z_1,r) - b(t,x,u_2,v_2,Z_2,r)| \leq C(|u_1 - u_2| + |Z_1 - Z_2| + |v_1 - v_2|) \tag{2.2}$$

(**A.3**).

$$b(t,x,u,v,Z,r_1) - b(t,x,u,v,Z,r_2) \leq \int_{\mathbb{R}} (r_1(z) - r_2(z))\lambda(t,z)\nu(dz), \tag{2.3}$$

where $0 \leq \lambda(t,z) \leq C(1 \wedge |z|)$.

For $i = 1,2$, consider BSPDEs:

$$\begin{aligned} &du_i(t,x) \\ &= -A(t)u_i(t)dt - b_i(t,x,u_i(t,x),\nabla u_i(t,x),Z_i(t,x),r_i(t,x,\cdot))dt \\ &\quad + Z_i(t,x)dB_t - < \beta(t), \nabla Z_i(t,x) > dt + \int_{\mathbb{R}} r_i(t,x,z)\tilde{N}(dt,dz), \ t \in (0,T) \\ &u_i(T,x) = \phi_i(x) \quad a.s. \end{aligned} \tag{2.4}$$

See e.g. [ØPZ] for information about BSPDEs with jumps.
The following theorem is the main result of this paper.

Theorem 2.1. *(Comparison theorem) Assume (A.1), (A.2) (A.3) hold for one of the coefficients b_i, say, b_2. If $\phi_1(x) \leq \phi_2(x)$ and*

$$\begin{aligned} &b_1(t,x,u_1(t,x),\nabla u_1(t,x),Z_1(t,x),r_1(t,x,\cdot)) \\ &\qquad \leq b_2(t,x,u_1(t,x),\nabla u_1(t,x),Z_1(t,x),r_1(t,x,\cdot)), \end{aligned}$$

then we have $u_1(t,x) \leq u_2(t,x), x \in D$, a.e. for every $t \in [0,T]$.

Proof. For $n \geq 1$, define functions $\psi_n(z)$, $f_n(x)$ as follows (see [DP1]).

$$\psi_n(z) = \begin{cases} 0 & \text{if } z \leq 0, \\ 2nz & \text{if } 0 \leq z \leq \frac{1}{n}, \\ 2 & \text{if } z > \frac{1}{n}. \end{cases} \tag{2.5}$$

$$f_n(x) = \begin{cases} 0 & \text{if } x \leq 0, \\ \int_0^x dy \int_0^y \psi_n(z)dz & \text{if } x > 0. \end{cases} \tag{2.6}$$

We have

$$f_n'(x) = \begin{cases} 0 & \text{if } x \leq 0, \\ nx^2 & \text{if } 0 \leq x \leq \frac{1}{n}, \\ 2x - \frac{1}{n} & \text{if } x > \frac{1}{n}. \end{cases} \tag{2.7}$$

Also $f_n(x) \uparrow (x^+)^2$ as $n \to \infty$. For $h \in K := L^2(D)$, set

$$F_n(h) = \int_D f_n(h(x))dx.$$

F_n has the following derivatives for $h_1, h_2 \in K$,

$$F_n'(h)(h_1) = \int_D f_n'(h(x))h_1(x)dx, \tag{2.8}$$

$$F_n''(h)(h_1, h_2) = \int_D f_n''(h(x))h_1(x)h_2(x)dx. \tag{2.9}$$

Applying Ito's formula we obtain

$$\begin{aligned}
&F_n(u_1(t) - u_2(t)) \\
&= F_n(\phi_1 - \phi_2) + \int_t^T F_n'(u_1(s) - u_2(s))(A(s)(u_1(s) - u_2(s)))ds \\
&+ \int_t^T F_n'(u_1(s) - u_2(s))(b_1(s, u_1(s), \nabla u_1(s), Z_1(s), r_1(s)) \\
&\qquad - b_2(s, u_2(s), \nabla u_2(s), Z_2(s), r_2(s)))ds \\
&+ \int_t^T F_n'(u_1(s) - u_2(s))(< \beta(s), \nabla Z_1(s) - \nabla Z_2(s) >)ds \\
&- \int_t^T F_n'(u_1(s) - u_2(s))(Z_1(s) - Z_2(s))dB_s \\
&- \frac{1}{2}\int_t^T F_n''(u_1(s) - u_2(s))(Z_1(s) - Z_2(s), Z_1(s) - Z_2(s))ds \\
&- \int_t^T \int_{\mathbb{R}} \Big\{ F_n(u_1(s-) - u_2(s-) + r_1(s, \cdot, z) - r_2(s, \cdot, z)) \\
&\qquad - F_n(u_1(s-) - u_2(s-)) \Big\} \tilde{N}(ds, dz) \\
&- \int_t^T \int_{\mathbb{R}} \Big\{ F_n(u_1(s) - u_2(s) + r_1(s, \cdot, z) - r_2(s, \cdot, z)) - F_n(u_1(s) - u_2(s)) \\
&\qquad - F_n'(u_1(s) - u_2(s))(r_1(s, \cdot, z) - r_2(s, \cdot, z)) \Big\} ds\nu(dz)
\end{aligned}$$

$$=: I_n^1 + I_n^2 + I_n^3 + I_n^4 + I_n^5 + I_n^6 + I_n^7 + I_n^8. \tag{2.10}$$

In view of the assumptions (A.1)–(A.3), we have

$$\begin{aligned}
I_n^2 &= \int_t^T F_n'(u_1(s) - u_2(s))(A(s)(u_1(s) - u_2(s)))ds \\
&= \int_t^T \int_D f_n'(u_1(s,x) - u_2(s,x))(\sum_{i,j=1}^d \frac{\partial}{\partial x_i}(a_{ij}(t,x)\frac{\partial}{\partial x_j}(u_1(s,x) - u_2(s,x))))dxds \\
&= -\int_t^T \int_D f_n''(u_1(s,x) - u_2(s,x)) \sum_{i,j=1}^d a_{ij}(t,x)\frac{\partial}{\partial x_i}(u_1(s,x) - u_2(s,x)) \\
&\quad \times \frac{\partial}{\partial x_j}(u_1(s,x) - u_2(s,x))dxds, \\
&\leq -\int_t^T (\frac{1}{2a}|\beta|^2(s) + \delta_1)ds \int_D f_n''(u_1(s,x) - u_2(s,x))|\nabla(u_1(s,x) - u_2(s,x))|^2 dx,
\end{aligned} \tag{2.11}$$

$$I_n^6 = -\frac{1}{2}\int_t^T \int_D f_n''(u_1(s,x) - u_2(s,x))|Z_1(s,x) - Z_2(s,x)|^2 dxds, \tag{2.12}$$

and

$$\begin{aligned}
I_n^8 &= -\int_t^T \int_{\mathbb{R}} \int_D \Big\{ f_n(u_1(s,x) - u_2(s,x) + r_1(s,x,z) - r_2(s,x,z)) \\
&\quad - f_n(u_1(s,x) - u_2(s,x)) - f_n'(u_1(s,x) - u_2(s,x)) \\
&\quad\quad \times (r_1(s,x,z) - r_2(s,x,z)) \Big\} dxds\nu(dz) \\
&= -\frac{1}{2}\int_t^T \int_{\mathbb{R}} \int_D f_n''(u_1(s,x) - u_2(s,x) + \theta(s,x,z) \\
&\quad \times (r_1(s,x,z) - r_2(s,x,z)))(r_1(s,x,z) - r_2(s,x,z))^2 dxds\nu(dz),
\end{aligned} \tag{2.13}$$

where $0 \leq \theta(s,x,z) \leq 1$.

We further write I_n^3 as

$$\begin{aligned}
&I_n^3 \\
&= \int_t^T \int_D f_n'(u_1(s,x) - u_2(s,x))(b_1(s,x,u_1(s,x),\nabla u_1(s,x),Z_1(s,x),r_1(s,x,\cdot)) \\
&\quad - b_2(s,x,u_2(s,x),\nabla u_2(s,x),Z_2(s,x),r_2(s,x,\cdot)))dxds \\
&= \int_t^T \int_D f_n'(u_1(s,x) - u_2(s,x))(b_1(s,x,u_1(s,x),\nabla u_1(s,x),Z_1(s,x),r_1(s,x,\cdot)) \\
&\quad - b_2(s,x,u_1(s,x),\nabla u_1(s,x),Z_1(s,x),r_1(s,x,\cdot)))dxds \\
&+ \int_t^T \int_D f_n'(u_1(s,x) - u_2(s,x))(b_2(s,x,u_1(s,x),\nabla u_1(s,x),Z_1(s,x),r_1(s,x,\cdot)) \\
&\quad - b_2(s,x,u_2(s,x),\nabla u_1(s,x),Z_1(s,x),r_1(s,x,\cdot)))dxds
\end{aligned}$$

$$+\int_t^T \int_D f_n'(u_1(s,x)-u_2(s,x))(b_2(s,x,u_2(s,x),\nabla u_1(s,x),Z_1(s,x),r_1(s,x,\cdot))$$
$$-b_2(s,x,u_2(s,x),\nabla u_2(s,x),Z_1(s,x),r_1(s,x,\cdot)))dxds$$
$$+\int_t^T \int_D f_n'(u_1(s,x)-u_2(s,x))(b_2(s,x,u_2(s,x),\nabla u_2(s,x),Z_1(s,x),r_1(s,x,\cdot))$$
$$-b_2(s,x,u_2(s,x),\nabla u_2(s,x),Z_2(s,x),r_1(s,x,\cdot)))dxds$$
$$+\int_t^T \int_D f_n'(u_1(s,x)-u_2(s,x))(b_2(s,x,u_2(s,x),\nabla u_2(s,x),Z_2(s,x),r_1(s,x,\cdot))$$
$$-b_2(s,x,u_2(s,x),\nabla u_2(s,x),Z_2(s,x),r_2(s,x,\cdot)))dxds$$
$$:= I_{n,1}^3 + I_{n,2}^3 + I_{n,3}^3 + I_{n,4}^3 + I_{n,5}^3. \tag{2.14}$$

Now, $I_{n,1}^3 \le 0$ by the assumption on b_1 and b_2, and

$$I_{n,2}^3 \le C\int_t^T \int_D ((u_1(s,x)-u_2(s,x))^+)^2 dxds, \tag{2.15}$$

by the Lipschtiz condition of b_2. Recalling the constant δ_1 in (A.1), we can find a constant C_{δ_1} such that

$$\begin{aligned}
I_{n,3}^3 \le& C\int_t^T ds \int_D dx f_n'(u_1(s,x)-u_2(s,x))|\nabla u_1(s,x)-\nabla u_2(s,x)| \\
\le& \delta_1 \int_t^T ds \int_D f_n''(u_1(s,x)-u_2(s,x))|\nabla u_1(s,x)-\nabla u_2(s,x)|^2 dx \\
&+ C_{\delta_1}\int_t^T ds \int_D \frac{f_n'(u_1(s,x)-u_2(s,x))^2}{f_n''(u_1(s,x)-u_2(s,x))} dx \\
\le& \delta_1 \int_t^T ds \int_D f_n''(u_1(s,x)-u_2(s,x))|\nabla u_1(s,x)-\nabla u_2(s,x)|^2 dx \\
&+ C\int_t^T \int_D ((u_1(s,x)-u_2(s,x))^+)^2 dxds,
\end{aligned} \tag{2.16}$$

where we have used the fact that there exists a constant C (independent of n) such that for $y \ge 0$,

$$\frac{f_n'(x)^2}{f_n''(x+y)} \le C(x^+)^2. \tag{2.17}$$

This can be easily checked using the definition of f_n. By a similar trick, for any $\delta_2 > 0$, we have

$$\begin{aligned}
I_{n,4}^3 \le& C\int_t^T \int_D f_n'(u_1(s,x)-u_2(s,x))|Z_1(s,x)-Z_2(s,x)|dxds \\
\le& \delta_2 \int_t^T ds \int_D f_n''(u_1(s,x)-u_2(s,x))|Z_1(s,x)-Z_2(s,x)|^2 dx \\
&+ C_{\delta_2}\int_t^T \int_D ((u_1(s,x)-u_2(s,x))^+)^2 dxds.
\end{aligned} \tag{2.18}$$

In view of the assumption (2.3), we have

$$\begin{aligned}
I^3_{n,5} \leq & \int_t^T ds \int_D dx f'_n(u_1(s,x)-u_2(s,x)) \int_{\mathbb{R}} (r_1(s,x,z)-r_2(s,x,z))\lambda(s,z)\nu(dz) \\
\leq & \int_t^T ds \int_D dx f'_n(u_1(s,x)-u_2(s,x)) \\
& \quad \times \int_{\mathbb{R}} (r_1(s,x,z)-r_1(s,x,z))\chi_{\{r_1(s,x,z)>r_2(s,x,z)\}}\lambda(s,z)\nu(dz) \\
\leq & \frac{1}{2}\int_t^T \int_{\mathbb{R}}\int_D f''_n(u_1(s,x)-u_2(s,x)+\theta(s,x,z)(r_1(s,x,z)-r_2(s,x,z))) \\
& \quad \times (r_1(s,x,z)-r_2(s,x,z))^2\chi_{\{r_1(s,x,z)>r_2(s,x,z)\}}dxds\nu(dz) \\
& + C\int_t^T \int_{\mathbb{R}} \lambda(s,z)^2 \int_D \frac{f'_n(u_1(s,x)-u_2(s,x))^2}{f''_n(u_1(s,x)-u_2(s,x)+\theta(s,x,z)(r_1(s,x,z)-r_2(s,x,z)))} \\
& \quad \times \chi_{\{r_1(s,x,z)>r_2(s,x,z)\}}dxds\nu(dz) \\
\leq & \frac{1}{2}\int_t^T \int_{\mathbb{R}}\int_D f''_n(u_1(s,x)-u_2(s,x)+\theta(s,x,z)(r_1(s,x,z)-r_2(s,x,z))) \\
& \quad \times (r_1(s,x,z)-r_2(s,x,z))^2 dxds\nu(dz) \\
& + C\int_{\mathbb{R}}(1\wedge|z|)^2\nu(dz)\int_t^T\int_D((u_1(s,x)-u_2(s,x))^+)^2dxds, \qquad (2.19)
\end{aligned}$$

where we have used (2.17) again.

Now, by integration by parts, for $0<a<1$,

$$\begin{aligned}
I^4_n = & \int_t^T \int_D < f'_n(u_1(s,x)-u_2(s,x))\beta(s), \nabla(Z_1(s,x)-Z_2(s,x)) > dxds \\
= & -\int_t^T \int_D < f''_n(u_1(s,x)-u_2(s,x))\nabla(u_1(s,x)-u_2(s,x)), \\
& \quad \beta(s)(Z_1(s,x)-Z_2(s,x)) > dxds \\
\leq & \frac{1}{2}a\int_t^T\int_D f''_n(u_1(s,x)-u_2(s,x))|Z_1(s,x)-Z_2(s,x)|^2dxds \\
& + \frac{1}{2a}\int_t^T \beta^2(s)\int_D f''_n(u_1(s,x)-u_2(s,x))|\nabla(u_1(s,x)-u_2(s,x))|^2dxds \qquad (2.20)
\end{aligned}$$

Now, choose $\delta_2>0$ sufficiently small to satisfy that

$$\delta_2 + \frac{1}{2}a \leq \frac{1}{2} \qquad (2.21)$$

Adding (2.11),(2.13), (2.16), (2.20), (2.14), (2.18) and (2.19) together and taking into account of (2.21) we deduce that

$$I^2_n + I^3_n + I^4_n + I^6_n + I^8_n \leq C\int_t^T\int_D((u_1(s,x)-u_2(s,x))^+)^2dxds \qquad (2.22)$$

Thus it follows from (2.10), (2.11) and (2.22) that

$$\begin{aligned}
&F_n(u_1(t)-u_2(t))\\
&\quad \leq F_n(\phi_1-\phi_2)+C\int_t^T\int_D((u_1(s,x)-u_2(s,x))^+)^2dxds\\
&\quad -\int_t^T F_n'(u_1(s)-u_2(s))(Z_1(s)-Z_2(s))dB_s\\
&\quad -\int_t^T\int_{\mathbb{R}}\Big\{F_n(u_1(s-)-u_2(s-)+r_1(s,\cdot,z)-r_2(s,\cdot,z))\\
&\qquad -F_n(u_1(s-)-u_2(s-))\Big\}\tilde{N}(ds,dz)
\end{aligned} \tag{2.23}$$

Take expectation and let $n\to\infty$ to get

$$E[\int_D((u_1(t,x)-u_2(t,x))^+)^2dx]\leq\int_t^T dsE[\int_D((u_1(s,x)-u_2(s,x))^+)^2dx] \tag{2.24}$$

Gronwall's inequality yields that

$$E[\int_D((u_1(t,x)-u_2(t,x))^+)^2dx]=0, \tag{2.25}$$

which completes the proof of the theorem. □

3. Application

Let $u(t,x)$ be the solution of a BSDE of the form (1.1) satisfying the conditions (A.1)-(A.3). Assume that b does not depend on u, i.e.

$$b(t,x,u,\nabla u,Z,r)=b(t,x,Z,r) \quad \text{for all} \quad t,x,u,Z,r. \tag{3.1}$$

Moreover, assume that $b(t,x,Z,r)$ is concave with respect to Z,r for all t,x. If we, for example, regard $\phi(x)$ as a financial standing at time $t=T$ and at the point x, we may as in [ØSZ] define the risk $\rho(\phi)(x)$ of ϕ at time $t=0$ and at the point x by

$$\rho(\phi)(x)=-u(0,x); \quad x\in\mathbb{R}^d. \tag{3.2}$$

Using the comparison theorem (Theorem 2.1) we can now verify that $\phi\to\rho(\phi)$ is a convex risk measure, in the sense that it satisfies the following conditions:
(3.3) (Convexity) $\rho(\lambda\phi_1+(1-\lambda)\phi_2)\leq\lambda\rho(\phi_1)+(1-\lambda)\rho(\phi_2)$ for all $\lambda\in[0,1]$ and all ϕ_1,ϕ_2.
(3.4) (Monotonicity) $\phi_1\leq\phi_2\Rightarrow\rho(\phi_1)\geq\rho(\phi_2)$.
(3.5) (Translation invariance) $\rho(\phi+a)=\rho(\phi)-a$ for all ϕ and all constants a.

Thus we have an extension of the convex risk measure concept (see e.g. [FS]) to a space-dependent situation. This might be of relevance in large systems of interacting components.

References

BBP. G. Barles, R. Buckdahn, , E. Pardoux: Backward stochastic differential equations and integral-partial differential equations. Stochastics and Stochastics Reports 60(1-2) (1997) 57-83.

DP1. C. Donai-Martin, E. Pardoux: White noise driven SPDEs with reflection. Probab. Theory Rel. Fields 95, 1-24 (1993).

DM. K. Du and Q. Meng: A revisit to W_2^n - theory of super-parabolic backward stochastic partial differential equations in R^d. Stochastic Processes and Their Applications 120 (2010), 1996 - 2015.

DQT. K.Du, J. Qui and S. Tang: L^p theory for superparabolic backward stochastic partial differential equationsin the whole space. Appl. Math. Optim. 65 (2012), 175-219

FS. H. Föllmer and A. Schied: Stochastic Finance, Third Edition. De Gruyter 2011.

MYZ. J. Ma, H. Yin and J. Zhang: On non-Markovian forward-backward SDEs and backward stochastic PDEs. Stochastic Processes and Their Applications 122 (2012) 3980-4004.

MY. J.Ma and J. Yong: On linear, degenerate backward stochastic partial differential equations. Probab. Theory Rel. Fields 113 (1999), 135-170,

ØSZ. B. Øksendal, A. Sulem and T. Zhang: Singular stochastic control of stochastic partial differential equations and backward SPDEs with reflection. Mathematics of Operations Research, DOI 10.1287 (2013).

ØPZ. B. Øksendal, F. Proske and T. Zhang: Backward stochastic partial differential equations with jumps and application to optimal control of random jump fields. Stochastics 77:5 (2005) 381-399.

Chapter 24

On a construction of a space-time diffusion process with boundary condition

Yoichi Oshima*

In this paper, we use the analytic and the stochastic calculus related to the lower bounded semi-Dirichlet forms to construct a diffusion process associated with a differential generator containing a boundary condition on smooth compact sets. For such differential generator, we give a condition under which the associated bilinear form becomes a semi-Dirichlet form. Once the condition is satisfied, then there corresponds a diffusion process. By using the stochastic calculus related to the semi-Dirichlet forms, we give a representation of the solution of a partial differential equation with boundary and terminal value conditions. Moreover a stochastic representation of the diffusion process is given.

Contents

1. Introduction

Let D be a bounded domain of $\mathbb{R}^d$ with $\mathcal{C}^2$-class boundary ∂D. Throughout in this paper, we assume that we are given a family of bounded functions $(a_{ij}(\tau,x))$ on $[0.\infty)\times D$ which is symmetric and uniformly elliptic relative to x, that is $a_{ij}(\tau,x)=a_{ji}(\tau,x)$ and there exist positive conatants $\lambda_1\le\lambda_2$ such that

$$\lambda_1\sum_{i=1}^d p_i^2 \le \sum_{i,j=1}^d a_{ij}(\tau,x)p_ip_j \le \lambda_2\sum_{i=1}^d p_i^2 \tag{1.1}$$

for all $(\tau,x)\in[0,\infty)\times D$ and $\mathbf{p}=(p_1,p_2,\ldots,p_d)\in\mathbb{R}^d$.

Let $\{b_i(\tau,x)\}_{1\le i\le d}$ and $\{\beta_i(\tau,x)\}_{1\le i\le d}$ be two families of bounded measurable functions on $[0,\infty)\times D$ and $[0,\infty)\times\partial D$, respectively. Furthermore assume that $\sum_{i=1}^d\beta_i(\tau,x)\frac{\partial}{\partial x_i}$ is a vector field on the tangent space of ∂D and the derivative of $\beta_i(\tau,x)$ along ∂D is uniformly bounded relative to $(\tau,x)\in[0,\infty)\times\partial D$. We consider that $\{a_{ij}(\tau,x)\}$, $\{b_i(\tau,x)\}$ and $\{\beta_i(\tau,x)\}$ are defined for $\tau\in(-\infty,\infty)$ by putting $a_{ij}(\tau,x)=a_{ij}(0,x)$, $b_i(\tau,x)=b_i(0,x)$ and $\beta_i(\tau,x)=\beta_i(0,x)$ for $\tau<0$.

*Department of Mathematics, Faculty of Engineering, Kumamoto University, Kumamoto 860-8555, Japan, oshima@gpo.kumamoto-u.ac.jp

Since D is of compact closure, its boundary ∂D is covered by a finite number of open sets $\{O^{(k)}\}_{1\le k\le n}$. Furthermore, for each $k \ge 1$, there exists an open set $U^{(k)}$ of $\mathbb{R}^d$ and a family of $\mathcal{C}^2$-functions $\Phi^{(k)}(\xi) = (\varphi_1^{(k)}(\xi), \varphi_2^{(k)}(\xi), \cdots, \varphi_d^{(k)}(\xi))$ of $\xi = (\xi_1, \xi_2, \ldots, \xi_d)$ such that any point of $U^{(k)}$ is mapped to a point of $O^{(k)}$ as one to one map and

$$\partial D \cap O^{(k)} = \{x = \Phi^{(k)}(\xi', 0) : (\xi', 0) \in U^{(k)}\},$$

where $\xi' = (\xi_1, \xi_2, \ldots, \xi_{d-1})$. Let $\{\delta^{(k)}(x)\}_{1\le k\le n}$ be a partition of unity subordinate to $\{O^{(k)}\}$, that is $\delta^{(k)}$ is a $\mathcal{C}^\infty$ function supported by $O^{(k)}$ and $\sum_{k=1}^n \delta^{(k)}(x) = 1$ on a neighbourhood of ∂D. Let $\mathcal{C}^1(\overline{D})$ be the family of functions φ on $\overline{D}$ which can be extended to a function of $\mathcal{C}^1(\mathbb{R}^d)$ and F be the Sobolev space of order 1 on $\overline{D}$, that is $F = \{\varphi \in L^2(\overline{D}; m) : \partial\varphi/\partial x_i \in L^2(\overline{D}; m) \text{ for all } 1 \le i \le d\}$, where $m(dx) = dx$ is the Lebesgue measure. Then F is equal to the closure of $\mathcal{C}^1(\overline{D})$ relative to the Dirichlet norm $\mathbf{D}_1$ (see [4] § 1.5), where $\mathbf{D}_1(\varphi, \psi) = \mathbf{D}(\varphi, \psi) + (\varphi, \psi)_H$ with $H = L^2(\overline{D}, m)$ and

$$\mathbf{D}(\varphi, \psi) = \int_D \sum_{i=1}^d \frac{\partial \varphi}{\partial x_i} \frac{\partial \psi}{\partial x_j}\, dx. \tag{1.2}$$

Define a bilinear form $E^{(\tau)}$ on F by

$$E^{(\tau)}(\varphi, \psi) = E^{(\tau),1}(\varphi, \psi) - E^{(\tau),2}(\varphi, \psi) \tag{1.3}$$

for

$$E^{(\tau),1}(\varphi, \psi) = \sum_{i,j=1}^d \int_D a_{ij}(\tau, x) \frac{\partial \varphi}{\partial x_i} \frac{\partial \psi}{\partial x_j} dx - \sum_{i=1}^d \int_D b_i(\tau, x) \frac{\partial \varphi}{\partial x_i} \psi(x) dx \tag{1.4}$$

$$E^{(\tau),2}(\varphi, \psi) = \sum_{i=1}^d \int_{\partial D} \beta_i(\tau, x) \frac{\partial \varphi}{\partial x_i} \psi(x) dS, \tag{1.5}$$

where dS is the surface element of ∂D.

The purpose of this paper is to construct a space-time diffusion process corresponding to the time dependent family of semi-Dirichlet forms $(E^{(\tau)}, F)$ with $\tau \ge 0$. The corresponding diffusion process is associated with a parabolic differential generator with oblique reflecting boundary condition (see Burdzy, Chen and Sylvester [1], Fukushima [3], Kim [6] and Tsuchiya [14]). Our construction is based upon the general theory of time dependent semi-Dirichlet forms given by Ma, Overbeck and Röckner [7], Ma and Röckner [8] and Oshima [12].

For a subspace F of H and a bilinear form $E(\varphi, \psi)$ defined for $\varphi, \psi \in F$, (E, F) is called a semi-Dirichlet form on H if it satisfies the following conditions:

$(E.1)$ (Lower boundedness): There exists a non-negative constant α_0 such that $E_{\alpha_0}(\varphi, \varphi) \ge 0$ for all $\varphi \in F$.

$(E.2)$ (Sector condition): There exists a constant K such that

$$|E(\varphi, \psi)| \le K E_{\alpha_0}(\varphi, \varphi)^{1/2} E_{\alpha_0}(\psi, \psi)^{1/2} \tag{1.6}$$

for all $\varphi, \psi \in F$.

$(E.3)$ (Closedness): F is a Hilbert space relative to the symmetric bilinear form $E^s_\alpha(\varphi,\psi) = \frac{1}{2}\left(E_\alpha(\varphi,\psi) + E_\alpha(\psi,\varphi)\right)$ for any $\alpha > \alpha_0$.

$(E.4)$ (Markov property): For any $\varphi \in F$ and non-negative constant a, $\varphi \wedge a \in F$ and $E(\varphi \wedge a, \varphi \wedge a) \le E(\varphi \wedge a, \varphi)$.

In Section 2, we shall show that $(E^{(\tau)}, F)$ given by (1.3) is a semi-Dirichlet form on H for any τ with common constants α_0 and K. Under our present assumption, we shall show that the symmetric Dirichlet form $(\mathbf{D}, H^1(\overline{D})$ is a reference form of $(E^{(\tau)}, F)$, that is, for any $\alpha > \alpha_0$, there exist positive constants $c_1 \le c_2$ such that

$$c_1\, \mathbf{D}_1(\varphi,\varphi) \le E_\alpha(\varphi,\varphi) \le c_2 \mathbf{D}_1(\varphi,\varphi)$$

for all $\varphi \in F$. In particular, F satisfies the condition $(\mathcal{E}.5)$ given in [11] which implies that $uv \in F_b$ for any $u, v \in F_b$, where F_b is the family of bounded functions of F. For any τ and $\delta > 0$, take a strictly positive δ-coexcessive function $\widehat{h}_\delta(\tau,\cdot) \in F$. As in [11] Theorem 2.4.8, we may assume that $\widehat{h}_\delta(x,\cdot) \ge c_\delta(\tau) > 0$ for some positive constant $c_\delta(\tau)$ which is uniformly bounded from below by a positive constant on each finite τ-interval. We shall also show that $(E^{(\tau)}, F)$ satisfies a condition which implies $(\mathcal{E}.6)$ in [11] relative to the time dependent bilinear form $(\mathcal{B}, \mathcal{F})$ given in the next section.

In Section 3, we construct the associated space-time process and give a probabilistic representation of a weak solution of a parabolic differential equation with boundary condition. Put $Z = R^1 \times \overline{D}$ and let ν be the measure on Z given by $d\nu(\tau,x) = d\tau \times dm(x)$. Let $\mathcal{H} = L^2(\mathbb{R}^1; H)$ be the family of functions $u(\tau,x)$ such that $u(\tau,\cdot) \in H$ for all $\tau \in \mathbb{R}^1$ and

$$\|u\|^2_{\mathcal{H}} = \int_{\mathbb{R}^1} \|u(\tau,\cdot)\|^2_H \, d\tau < \infty.$$

Define $(\mathcal{F}, \|\cdot\|_{\mathcal{F}})$ similarly by using F instead of H in the definition of $(\mathcal{H}, \|\cdot\|_{\mathcal{H}})$. Let F' be the dual space of F and consider that $F \subset H = H' \subset \mathcal{F}'$. For any $\varphi \in F$ and $\psi \in F'$, let (φ,ψ) be the canonical coupling between the elements of F' and F. Define $\mathcal{F}' = L^2(\mathbb{R}^1; F')$ similarly to $\mathcal{F}$. For $u \in \mathcal{F}$, let $\frac{\partial u}{\partial \tau}$ be the distribition sense derivative of $\tau \mapsto u(\tau,\cdot) \in F'$, that is $\int_{\mathbb{R}^1} \partial u(\tau,\cdot)/\partial\tau \xi(\tau)\, d\tau = \int_{\mathbb{R}^1} u(\tau,\cdot)\xi'(\tau)\, d\tau$ as an element of F' for any $\xi \in C_0^\infty(\mathbb{R}^1)$. Put $\mathcal{W} = \{u \in \mathcal{F} : \|u\|_{\mathcal{W}} < \infty\}$ with $\|u\|^2_{\mathcal{W}} = \|\partial u/\partial\tau\|^2_{\mathcal{F}'} + \|u\|^2_{\mathcal{F}}$ and let

$$\mathcal{E}(u,v) = \begin{cases} -\left(\frac{\partial u}{\partial \tau}, v\right) + \mathcal{B}(u,v) & u \in \mathcal{W}, v \in \mathcal{F} \\ \left(\frac{\partial v}{\partial \tau}, u\right) + \mathcal{B}(u,v) & u \in \mathcal{F}, v \in \mathcal{W}, \end{cases} \tag{1.7}$$

where

$$\left(\frac{\partial u}{\partial \tau}, v\right) = \int_{\mathbb{R}^1} \left(\frac{\partial u}{\partial \tau}, v(\tau,\cdot)\right) d\tau,$$
$$\mathcal{B}(u,v) = \int_{\mathbb{R}^1} E^{(\tau)}(u(\tau,\cdot), v(\tau,\cdot))\, d\tau.$$

We say that $(\mathcal{E}, \mathcal{F})$ a *time dependent semi-Dirichlet form.* Then there exists a resolvent $\{G_\alpha\}_{\alpha>0}$ such that $G_\alpha f \in \mathcal{W}$ for any $f \in \mathcal{H}$ and

$$\mathcal{E}_\alpha(G_\alpha f, v) = (f, v) \quad \text{for any } f \in \mathcal{F}'.$$

By choosing a suitable resolvent kernel $R_\alpha((\tau,x),dz)$ such that $R_\alpha f(\tau,x) = \int_{\overline{D}} R_\alpha((\tau,x),dz)f(z)$ is a q.c. modification of $G_\alpha f$ for any $f \in \mathcal{H}$ and $\alpha > \alpha_0$, there corresponds a diffusion process $Z_t = (\tau(t), X_t)$ on Z with resolvent $\{R_\alpha\}$. Then $\tau(t)$ is the uniform motion to the right, that is $\tau(t) = \tau(0) + t$.

We will show in Section 3 that the measure $d\tau \times dS(x)$ is a smooth measure of the space-time process. By using the local time associated with the smooth measure $d\tau \times dS$, we gve a probabilistic representation of the weak solution of

$$\frac{\partial u}{\partial \tau} = \sum_{i,j=1}^{d} \frac{\partial}{\partial x_i}\left(a_{ij}(\tau,x)\frac{\partial u}{\partial x_j}\right) + \sum_{i=1}^{d} b_i(\tau,x)\frac{\partial u}{\partial x_i} \tag{1.8}$$

on D with boundary condition

$$\frac{\partial u}{\partial \mathbf{n}} + \sum_{i=1}^{d} \beta_i(\tau,x)\frac{\partial u}{\partial x_i} = g(\tau,x) \tag{1.9}$$

on ∂D and terminal condition $u(T,x) = \varphi(x)$ for $x \in D$. Such representation are given by Burdzy, Chen and Sylvester [1] and M. Tsuchiya [15] for some space-time domains and boundary conditions.

For a positive quasi-continuous δ-coexcessive function $\widehat{h}_\delta \in \mathcal{F}$, define the energy $e^{(\delta)}(A)$ of an additive functional A_t by

$$e^{(\delta)}(A) = \frac{1}{2}\lim_{t\to 0}\frac{1}{t}\mathbb{E}_{\widehat{\nu}}(A_t^2), \tag{1.10}$$

where $d\widehat{\nu}(\tau,x) = \widehat{h}_\delta(\tau,x)d\nu$. Let $\overset{\circ}{\mathcal{M}}$ be the family of martingale additive functionals M such that $e^{(\delta)}(M) < \infty$ and $\mathcal{N}$ be the family of continuous additive functionals N such that $e^{(\delta)}(N) = 0$. In [11], only the weak sense Fukushima's decomposition is given explicitly in the time dependent case. Under our present setting, Fukushima's decomposition relative to the energy $e^{(\delta)}$ is also possible. In Section 3, we also mention the decomposition and give the characterizations of the martingale and zero energy parts of the decomposition. See also L. Ma, Z.M. Ma and W. Sun [9] for such decomposition.

2. Semi-Dirichlet form property of $E^{(\tau)}$

Let $(\mathbf{D}, F)$ be the Dirichlet form on H defined by (1.2). For a bilinear form (E, F), $\mathbf{D}$ is called a reference form of E if there exists a non-negative constant α_0 such that, for any $\alpha > \alpha_0$ we can choose a suitable positive constants $c_1(\alpha)$ and $c_2(\alpha)$ satisfying

$$c_1(\alpha)\mathbf{D}_1(\varphi,\varphi) \le E_\alpha(\varphi,\varphi) \le c_2(\alpha)\mathbf{D}_1(\varphi,\varphi) \tag{2.1}$$

for all $\varphi \in F$.

Note that $\mathbf{D}$ is a reference form of E if (2.1) holds for some $\alpha_1 > \alpha_0$. In fact, if (2.1) holds for some $\alpha_1 > \alpha_0$, then for any $\alpha \ge \alpha_1$, $E_\alpha(\varphi,\varphi) \ge E_{\alpha_1}(\varphi,\varphi) \ge c_1(\alpha_1)\mathbf{D}_1(\varphi,\varphi)$. On the other hand, if $\alpha_1 > \alpha > \alpha_0$, then

$$\begin{aligned} E_\alpha(\varphi,\varphi) &= E_{\alpha_0}(\varphi,\varphi) + (\alpha-\alpha_0)(\varphi,\varphi) \\ &\ge \frac{\alpha-\alpha_0}{\alpha_1-\alpha_0}\left(E_{\alpha_0}(\varphi,\varphi) + (\alpha_1-\alpha_0)(\varphi,\varphi)\right) \\ &\ge k_1(\alpha)\mathbf{D}_1(\varphi,\varphi) \end{aligned}$$

for $k_1(\alpha) = c_1(\alpha_1)(\alpha - \alpha_0)/(\alpha_1 - \alpha_0)$. Therefore, for any $\alpha > \alpha_0$, by putting $c_1(\alpha) = c_1(\alpha_1) \wedge k_1(\alpha)$ it holds that $c_1(\alpha)\mathbf{D}_1(\varphi,\varphi) \le E_\alpha(\varphi,\varphi)$. Similarly, the second inequality $E_\alpha(\varphi,\varphi) \le c_2(\alpha)\mathbf{D}_1(\varphi,\varphi)$ of (2.1) holds for $c_2(\alpha) = c_2(\alpha_1) \times ((\alpha - \alpha_0)/(\alpha_1 - \alpha_0)) \vee 1$.

Lemma 2.1. $(\mathbf{D}, F)$ *is a reference form of* $(E^{(\tau),1}, F)$ *for all* τ, *that is there exists a constants* $\alpha_1 \ge 0$ *such that, for any* $\alpha > \alpha_1$,

$$c_3\mathbf{D}_1(\varphi,\varphi) \le E_\alpha^{(\tau),1}(\varphi,\varphi) \le c_4\mathbf{D}_1(\varphi,\varphi) \tag{2.2}$$

for all $\varphi \in F$ *and some constants* c_3 *and* c_4 *independent of* τ.

Proof. Put $\lambda_1 = \sup\{\lambda(\tau) : \tau \ge 0\}$. Then, for any $\tau \ge 0$,

$$\begin{aligned}
E^{(\tau),1}(\varphi,\varphi) &= \int_D \sum_{i,j=1}^{d} a_{ij}(\tau,x)\frac{\partial\varphi}{\partial x_i}\frac{\partial\varphi}{\partial x_j}\,dx \\
&\quad - \int_D b_i(\tau,x)\frac{\partial\varphi}{\partial x_i}\,\psi(x)dx \\
&\ge \lambda_1\mathbf{D}(\varphi,\varphi) - \|b\|_\infty \mathbf{D}(\varphi,\varphi)^{1/2}\|\varphi\| \\
&\ge \lambda_1\mathbf{D}(\varphi,\varphi) - \frac{1}{2}\left(\lambda_1\mathbf{D}(\varphi,\varphi) + \frac{1}{\lambda_1}\|b\|_\infty^2\|\varphi\|^2\right).
\end{aligned}$$

Hence,

$$\frac{\lambda_1}{2}\mathbf{D}(\varphi,\varphi) \le E_{\alpha_1}^{(\tau),1}(\varphi,\varphi)$$

for $\alpha_1 = \frac{1}{2\lambda_1}\|b\|_\infty^2$. This implies for $\alpha > \alpha_1$ and $c_3 = (\lambda_1/2) \vee (\alpha - \alpha_1)$ that

$$c_3\mathbf{D}_1(\varphi,\varphi) \le \frac{\lambda_1}{2}\mathbf{D}(\varphi,\varphi) + (\alpha - \alpha_1)\|\varphi\|^2 \le E_\alpha^{(\tau),1}(\varphi,\varphi).$$

For any $\varphi, \psi \in F$,

$$\begin{aligned}
|E^{(\tau),1}(\varphi,\psi)| &\le \lambda_2\mathbf{D}(\varphi,\varphi)^{1/2}\mathbf{D}(\psi,\psi)^{1/2} + \|b\|_\infty\mathbf{D}(\varphi,\varphi)^{1/2}\|\psi\| \\
&\le \lambda_2\mathbf{D}(\varphi,\varphi)^{1/2}\left(\mathbf{D}(\psi,\psi)^{1/2} + \|b\|_\infty\|\psi\|\right).
\end{aligned}$$

Hence, noting that $\sqrt{a} + \sqrt{b} \le \sqrt{2(a+b)}$, we have

$$\begin{aligned}
|E_{\alpha_1}^{(\tau),1}(\varphi,\psi)| &\le \sqrt{2}\lambda_2\mathbf{D}(\varphi,\varphi)^{1/2}\mathbf{D}_{\|b\|_\infty}(\psi,\psi)^{1/2} + \alpha_1\|\varphi\|\,\|\psi\| \\
&\le (\sqrt{2}\lambda_2 + \alpha_1 + \|b\|_\infty)\mathbf{D}_1(\varphi,\varphi)^{1/2}\mathbf{D}_1(\psi,\psi)^{1/2}.
\end{aligned}$$

Hence

$$|E_{\alpha_1}^{(\tau),1}(\varphi,\psi)| \le c_4\mathbf{D}_1(\varphi,\varphi)^{1/2}\mathbf{D}_1(\psi,\psi)^{1/2} \tag{2.3}$$

for $c_4 = \sqrt{2}\lambda_2 + \alpha_1 + \|b\|_\infty$. As we noted before the statement of the lemma, this implies the assertion. □

Let $\{O^{(k)}\}_{1\le k\le n}$ be a covering of ∂D and $\{\delta^{(k)}(x)\}$ be a partition of unity subordinated to $\{O^{(k)}\}$ defined in Section 1. Let $\varphi \in \mathcal{C}^1(\overline{D})$. Then we may assume that φ is a restriction to D of a function of $\mathcal{C}^1(\mathbb{R}^d)$, which is also denoted by φ. Decompose φ as

$$\varphi(x) = \sum_{k=1}^{n} \varphi(x)\delta^{(k)}(x) + \varphi(x)\left(1 - \sum_{k=1}^{n} \delta^{(k)}(x)\right) \equiv \varphi_1(x) + \varphi_2(x).$$

Then $\varphi_2 \in \mathcal{C}_0^1(D)$ and hence $E^{(\tau)}(\varphi_2, \psi) = E^{(\tau),1}(\varphi_2, \psi)$ for any $\psi \in \mathcal{C}^1(\overline{D})$.

Since

$$E^{(\tau),2}(\varphi, \psi) = \sum_{k=1}^{d} E^{(\tau),2}(\varphi\delta^{(k)}, \psi),$$

by considering each term $E^{(\tau),2}(\varphi\delta_i, \psi)$ separately, we may assume that φ is supported by $O = O^{(k)}$ for fixed k. Put $\Phi = \Phi^{(k)}$, then by the mapping $\xi = \Phi(x)$ given in § 1.1, O is mapped onto $V \subset \mathbb{R}^d$ and $D \cap O$ is mapped onto $V \cap \{\xi = (\xi_1, \ldots, \xi_{d-1}, \xi_d) \in \mathbb{R}^d : \xi_d > 0\}$. Let $x = \Psi(\xi) = (\psi_1(\xi), \psi_2(\xi), \ldots, \psi_d(\xi))$ be the inverse map of Φ and put $\Gamma = (\gamma_{ij})$ for $\gamma_{ij} = \partial\psi_i/\partial\xi_j$. Then $dx = \rho(\xi)d\xi$ for $\rho(\xi) = \det(\Gamma)$. Furthermore the surface measure is given by $dS = \overline{S}(\xi', 0))d\xi'$, where $\xi' = (\xi_1, \xi_2, \ldots, \xi_{d-1})$ and $\overline{S}(\xi) = \left(\sum_{i=1}^{d} \Gamma_{id}^2(\xi)\right)^{1/2}$ by using the cofactor Γ_{id} of γ_{id} of Γ.

For any φ and $\psi \in \mathcal{C}^1(\overline{D})$, put $\overline{\varphi}(\xi) = \varphi(\Psi(\xi))$ and $\overline{\psi}(\xi) = \psi(\Psi(\xi))$, respectively. Since $\sum_{i=1}^{d} \beta_i(\tau, x)\partial\varphi/\partial x_i$ for $x \in \partial D$ is a derivative along the tangent space of ∂D, it is transformed by the mapping $x = \Psi(\xi)$ to a derivative on $\mathbb{R}^{d-1} \times \{0\}$, that is to

$$\sum_{j=1}^{d-1} \sum_{i=1}^{d} \beta_i(\tau, \Phi(\xi', 0))\gamma_{ij}^{-1}(\xi', 0) \frac{\partial\overline{\varphi}(\xi', 0)}{\partial\xi_j},$$

where (γ_{ij}^{-1}) is the inverse matrix of Γ. Therefore, by putting $\overline{\beta}_j(\tau, \xi') = \sum_{i=1}^{d} \beta_i(\tau, \Phi(\xi', 0))\gamma_{ij}^{-1}(\xi', 0)$ we can write $E^{(\tau),2}(\varphi, \psi)$ as

$$\begin{aligned} E^{(\tau),2}(\varphi, \psi) &= \int_{\partial D} \sum_{i=1}^{d} \beta_i(\tau, x)\frac{\partial\varphi}{\partial x_i}\psi(x)dS(x) \\ &= \int_{\mathbb{R}^{d-1}} \sum_{j=1}^{d-1} \overline{\beta}_j(\tau, \xi')\frac{\partial\overline{\varphi}(\xi', 0)}{\partial\xi_j}\overline{\psi}(\xi', 0)\overline{S}(\xi', 0)d\xi'. \end{aligned} \tag{2.4}$$

Lemma 2.2. *There exist positive constants c_5 and c_6 satisfying*

$$\left|E^{(\tau),2}(\varphi, \psi)\right| \leq c_5\mathbf{D}_1(\varphi, \varphi)^{1/2}\mathbf{D}_1(\psi, \psi)^{1/2} \tag{2.5}$$

$$E^{(\tau),2}(\varphi, \varphi) \geq -\epsilon\mathbf{D}(\varphi, \varphi) - \frac{c_6}{\epsilon}(\varphi, \varphi) \tag{2.6}$$

for all $\varphi, \psi \in F$ and $\epsilon > 0$.

Proof. Assume first that φ is supported by $O = O^{(k)}$ for sme k. Further, approximating φ by functions of $\mathcal{C}_0^2(\overline{D} \cap O)$, we may assume that $\varphi \in \mathcal{C}_0^2(\overline{D} \cap O)$ and hence we may consider that φ is a restriction of a function $\varphi \in \mathcal{C}_0^2(O)$ to $\overline{D}$. The corresponding function $\overline{\varphi}$ by the mapping Φ is also considered as a restriction of $\overline{\varphi} \in \mathcal{C}_0^2(V)$ to $\mathbb{R}_+^d = \{\xi = (\xi_1, \xi_2, \ldots, \xi_d) : \xi_d \geq 0\}$. Furthermore, we may assume that $\overline{\varphi} \in \mathcal{C}_0^2(\mathbb{R}^d)$ by putting $\overline{\varphi} = 0$ outside of V.

By (2.4), we can write

$$
\begin{aligned}
-E^{(\tau),2}(\varphi,\psi) &= \sum_{j=1}^{d-1}\int_{\mathbb{R}^{d-1}}\int_0^\infty \frac{\partial}{\partial\xi_d}\left(\overline{\beta}_j(\tau,\xi')\frac{\partial\overline{\varphi}}{\partial\xi_j}\,\overline{\psi}(\xi',\xi_d)\overline{S}(\xi',\xi_d)\right)d\xi' d\xi_d\\
&= \sum_{j=1}^{d-1}\int_{\mathbb{R}^d_+}\overline{\beta}_j(\tau,\xi')\frac{\partial^2\overline{\varphi}}{\partial\xi_j\partial\xi_d}\overline{\psi}(\xi)\overline{S}(\xi)d\xi\\
&\quad+\sum_{j=1}^{d-1}\int_{\mathbb{R}^d_+}\overline{\beta}_j(\xi')\,\frac{\partial\overline{\varphi}}{\partial\xi_j}\,\frac{\partial\overline{\psi}}{\partial\xi_d}\overline{S}(\xi)d\xi\\
&\quad+\sum_{j=1}^{d-1}\int_{\mathbb{R}^d_+}\overline{\beta}_j(\tau,\xi')\,\frac{\partial\overline{\varphi}}{\partial\xi_j}\,\overline{\psi}(\xi)\,\frac{\partial\overline{S}}{\partial\xi_d}\,d\xi\\
&= \mathrm{I}+\mathrm{II}+\mathrm{III}.
\end{aligned}
\tag{2.7}
$$

For the inward unit normal vector $\mathbf{n}$ of ∂D and the unit vector $\mathbf{d}$ of the direction of $(\gamma_{id})_{1\le i\le d}$, since $\rho(\xi)=\overline{S}(\xi)\left(\sum_{i=1}^d\gamma_{id}^2\right)^{1/2}(\mathbf{n},\mathbf{d})$ is positive and continuous, there exist positive constants k_1 and k_2 satisfying $k_1\overline{S}(\xi)\le\rho(\xi)\le k_2\overline{S}(\xi)$ for all $\xi\in V\cap\mathbb{R}^d_+$. Hence, by putting $\|\overline{\beta}\|_\infty=\max\{\|\overline{\beta}_i\|_\infty : 1\le i\le d\}$, it holds that

$$
\begin{aligned}
\mathrm{II} &\le \|\overline{\beta}\|_\infty\sum_{j=1}^{d-1}\left(\int_{\mathbb{R}^d_+}\left(\frac{\partial\overline{\varphi}}{\partial\xi_j}\right)^2\overline{S}(\xi)d\xi\right)^{1/2}\left(\int_{\mathbb{R}^d_+\cap V}\left(\frac{\partial\overline{\psi}}{\partial\xi_d}\right)^2\overline{S}(\xi)d\xi\right)^{1/2}\\
&\le \frac{1}{k_1}\|\overline{\beta}\|_\infty\mathbf{D}(\varphi,\varphi)^{1/2}\mathbf{D}(\psi,\psi)^{1/2}.
\end{aligned}
$$

Since $\frac{1}{\overline{S}(\xi)}\frac{\partial\overline{S}}{\partial\xi_d}$ is bounded by some constant k_3 on $V\cap\mathbb{R}^d_+$,

$$
\begin{aligned}
\mathrm{III} &\le k_3\|\overline{\beta}\|_\infty\sum_{j=1}^{d-1}\left(\int_{\mathbb{R}^d_+\cap V}\left(\frac{\partial\overline{\varphi}}{\partial\xi_j}\right)^2\overline{S}(\xi)d\xi\right)^{1/2}\left(\int_{\mathbb{R}^d_+\cap V}\overline{\psi}^2(\xi)\overline{S}(\xi)d\xi\right)^{1/2}\\
&\le \frac{k_3}{k_1}\|\overline{\beta}\|_\infty\mathbf{D}(\varphi,\varphi)^{1/2}\mathbf{D}_1(\psi,\psi)^{1/2}.
\end{aligned}
$$

For fixed ξ_d, since $\overline{\varphi}(\xi',\xi_d)$ has compact support in $V_{\xi_d}=\{\xi' : (\xi',\xi_d)\in V\cap\mathbb{R}^d_+\}$, by the integration by parts formula, I can be written as

$$
\begin{aligned}
\mathrm{I} &= -\sum_{j=1}^{d-1}\int_{\mathbb{R}^d_+\cap V}\frac{\partial\overline{\varphi}}{\partial\xi_d}\,\frac{\partial\overline{\psi}}{\partial\xi_j}\,\overline{\beta}_j(\xi')\overline{S}(\xi)d\xi\\
&\quad-\sum_{j=1}^{d-1}\int_{\mathbb{R}^d_+\cap V}\frac{\partial\overline{\varphi}}{\partial\xi_d}\,\overline{\psi}(\xi)\,\frac{\partial}{\partial\xi_j}\left(\overline{\beta}_j(\xi')\overline{S}(\xi)\right)d\xi.
\end{aligned}
$$

We may assume that k_3 also satisfies $(1/\overline{S}(\xi))|\partial\overline{S}/\partial\xi_j|<k_3$ on $\mathbb{R}^d_+\cap V$ for all $1\le j\le d-1$. Furthermore, there exists a constant k_4 such that $\|\partial\overline{\beta}_j/\partial\xi_j\|_\infty\le k_4$ for all $j\le d-1$. Hence, similarly to II and III, it holds that

$$
|\mathrm{I}|\le\frac{1}{k_1}\|\overline{\beta}\|_\infty\mathbf{D}(\varphi,\varphi)^{1/2}\mathbf{D}(\psi,\psi)^{1/2}
$$

$$+\frac{1}{k_1}(k_3\|\overline{\beta}\|_\infty + k_4)\mathbf{D}(\varphi,\varphi)^{1/2}\mathbf{D}_1(\psi,\psi)^{1/2}.$$

Therefore we obtained (2.5) for $\varphi \in \mathcal{C}^1(O \cap \overline{D})$.

For general $\varphi \in \mathcal{C}^1(\overline{D})$, put $\varphi_k = \varphi\delta^{(k)}$. Then, for each k, there exists a constant γ_k independent of φ such that (2.5) holds for φ_k instead of φ. Hence

$$|E^{(\tau),2}(\varphi,\psi)| \le \sum_{k=1}^{n} |E^{(\tau),2}(\varphi_k,\psi)| \le \sum_{k=1}^{n} \gamma_k \mathbf{D}_1(\varphi_k,\varphi_k)^{1/2}\mathbf{D}_1(\psi,\psi)^{1/2}.$$

Clearly $(\varphi_k,\varphi_k) \le (\varphi,\varphi)$. Furthermore, since

$$\begin{aligned}\left(\frac{\partial\varphi_k}{\partial x_i}\right)^2 &\le 2\left(\left(\frac{\partial\varphi}{\partial x_i}\right)^2 + \varphi^2(x)\left\|\frac{\partial\delta^{(k)}}{\partial\xi_i}\right\|_\infty\right)\\ &\le 2k_5\left(\left(\frac{\partial\varphi}{\partial x_i}\right)^2 + \varphi^2(x)\right)\end{aligned}$$

for $k_5 = 1 \vee \max\{\|\partial\delta^{(k)}/\partial\xi_i\|_\infty : 1 \le i \le n\}$, we obtain that

$$\begin{aligned}|E^{(\tau),2}(\varphi,\psi)| &\le 2k_5\sum_{k=1}^{n}\gamma_k\left(\mathbf{D}(\varphi,\varphi) + (d+1)(\varphi,\varphi)\right)^{1/2}\mathbf{D}_1(\psi,\psi)^{1/2}\\ &\le c_5\mathbf{D}_1(\varphi,\varphi)^{1/2}\mathbf{D}_1(\psi,\psi)^{1/2}\end{aligned}$$

for $c_5 = 2k_5(d+1)^{1/2}\sum_{k=1}^{n}\gamma_k$.

To show (2.6), take an arbitrary positive ϵ. For any $\varphi \in \mathcal{C}_0^2(O \cap \overline{D})$, similarly to (2.7),

$$\begin{aligned}E^{(\tau),2}(\varphi,\varphi) &= -\frac{1}{2}\sum_{j=1}^{d-1}\int_{\mathbb{R}^d_+}\frac{\partial}{\partial\xi_d}\left(\overline{\beta}_j(\xi')\frac{\partial\overline{\varphi}^2}{\partial\xi_j}\overline{S}(\xi)\right)d\xi\\ &= -\frac{1}{2}\sum_{j=1}^{d-1}\int_{\mathbb{R}^d_+}\overline{\beta}_j(\xi')\frac{\partial^2\overline{\varphi}^2}{\partial\xi_j\partial\xi_d}\overline{S}(\xi)d\xi\\ &\quad -\frac{1}{2}\sum_{j=1}^{d-1}\int_{\mathbb{R}^d_+}\overline{\beta}_j(\xi')\frac{\partial\overline{\varphi}^2}{\partial x_j}\frac{\partial\overline{S}}{\partial\xi_d}d\xi\\ &= \sum_{j=1}^{d-1}\int_{\mathbb{R}^d_+}\frac{\partial\overline{\varphi}}{\partial\xi_d}\overline{\varphi}(\xi)\frac{\partial(\overline{\beta}_j\overline{S})}{\partial\xi_j}d\xi\\ &\quad -\int_{\mathbb{R}^d_+}\sum_{j=1}^{d-1}\overline{\beta}_j(\xi')\frac{\partial\overline{\varphi}}{\partial x_j}\overline{\varphi}(\xi)\frac{\partial\overline{S}}{\partial\xi_d}d\xi\\ &\ge -\frac{1}{k_1}(k_3\|\overline{\beta}\|_\infty + k_4 + k_3)\mathbf{D}(\varphi,\varphi)^{1/2}\|\varphi\|\\ &\ge -\epsilon\mathbf{D}(\varphi,\varphi) - \frac{c_6}{\epsilon}(\varphi,\varphi)\end{aligned}$$

for $c_6 = 4(1/k_1)^2(k_3\|\overline{\beta}\|_\infty + k_4 + k_3)^2$. This implies the assertion for $\varphi \in \mathcal{C}^1(O \cap \overline{D})$. Extension to any $\varphi \in \mathcal{C}^1(\overline{D})$ is similar to the proof of (2.5). □

By virtue of (2.5), dS does not charge any set of zero $\mathbf{D}$-capacity. Hence the following result holds.

Corollary 2.1. *dS is a smooth measure relative to $(\mathbf{D}, F)$.*

Theorem 2.1. *For any $\tau \geq 0$, $(E^{(\tau)}, F)$ is a semi-Dirichlet form on $L^2(D; dx)$ with reference form $(\mathbf{D}, F)$. Furthermore, the constants α_0, K in $(E.1), (E.2)$ and $c_1(\alpha)$, $c_2(\alpha)$ in (2.1) can be chosen independently of τ.*

Proof. By virtue of (2.2), there exist constants $\alpha_1 \geq 0$ and $c_3 > 0$ such that $E_{\alpha_1}^{(\tau),1}(\varphi, \varphi) \geq c_3\mathbf{D}_1(\varphi, \varphi)$. For any ϵ such that $\epsilon < c_3$, by (2.6), it holds that

$$E_{\alpha_1}^{(\tau)}(\varphi, \varphi) = E_{\alpha_1}^{(\tau),1}(\varphi, \varphi) + E^{(\tau),2}(\varphi, \varphi) \geq c_3\mathbf{D}_1(\varphi, \varphi) - \epsilon\mathbf{D}(\varphi, \varphi) - \frac{c_6}{\epsilon}(\varphi, \varphi).$$

Hence

$$E_{\alpha_0}^{(\tau)}(\varphi, \varphi) \geq (c_3 - \epsilon)\mathbf{D}_1(\varphi, \varphi) \geq 0, \tag{2.8}$$

for $\alpha_0 = \alpha_0 + c_6/\epsilon$. This implies the lower boundedness of $(E^{(\tau)}, F)$.

By virtue of (2.8), (2.3) and (2.5), it follws that

$$\begin{aligned} |E^{(\tau)}(\varphi, \psi)| &\leq c_4\mathbf{D}_1(\varphi, \varphi)^{1/2}\mathbf{D}_1(\psi, \psi)^{1/2} + c_5\mathbf{D}_1(\varphi, \varphi)^{1/2}\mathbf{D}_1(\psi, \psi)^{1/2} \\ &\leq KE_{\alpha_0}^{(\tau)}(\varphi, \varphi)^{1/2}E_{\alpha_0}^{(\tau)}(\psi, \psi)^{1/2} \end{aligned}$$

for $K = (c_4 + c_5)/(c_3 - \epsilon)$. Hence the sector condition holds.

In the above proof, we have seen that

$$\begin{aligned} (c_3 - \epsilon)\mathbf{D}_1(\varphi, \varphi) \leq E_{\alpha_0}^{(\tau)}(\varphi, \varphi) &\leq (c_4 + c_5)\mathbf{D}_1(\varphi, \varphi) + \alpha_0(\varphi, \varphi) \\ &\leq ((c_4 + c_5) \vee \alpha_0)\, \mathbf{D}_1(\varphi, \varphi). \end{aligned}$$

This implies (2.1) for $\alpha = \alpha_0$ by putting $c_1(\alpha_0) = c_3 - \epsilon$ and $c_2(\alpha_0) = (c_4 + c_5) \vee \alpha_0$. As noted after (2.1), this implies the relation (2.1) for any $\alpha > \alpha_0$. Hence $(\mathbf{D}, F)$ is a reference form of $(E^{(\tau)}, F)$.

Since $(\mathbf{D}, H^1(\overline{D})$ is closed, the closedness of $(E^{(\tau)}, F)$ is clear. Furthermore, for any $\varphi \in F$ and positive constant a, $\varphi \wedge a \in F$ and

$$\begin{aligned} E^{(\tau)}(\varphi \wedge a, \varphi \wedge a) &= \int_{D\cap\{\varphi \leq a\}} \sum_{i,j=1}^{d} a_{ij}(\tau, x)\frac{\partial\varphi}{\partial x_i}\frac{\partial\varphi}{\partial x_j}dx \\ &\quad - \int_{D\cap\{\varphi \leq a\}} \sum_{i=1}^{d} b_i(\tau, x)\frac{\partial\varphi}{\partial x_i}\varphi(x)dx \\ &\quad - \int_{\partial D\cap\{\varphi \leq a\}} \sum_{i=1}^{d} \beta_i(\tau, x)\frac{\partial\varphi}{\partial x_i}\varphi(x)S(dx) \\ &= E^{(\tau)}(\varphi \wedge a, \varphi). \end{aligned}$$

Therefore, $(E^{(\tau)}, F)$ satisfies the Markov property. The independence of α_0 and K on τ is clear from the definition. □

Since $\mathcal{C}^1(\overline{D})$ is dense in F, $(E^{(\tau)}, F)$ is a regular semi-Dirichlet form.

Lemma 2.3. *A compact subset B of $\overline{D}$ is zero capacity relative to $E^{(\tau)}$ for some and hence all τ if and only if so is relative to* $\mathbf{D}$.

Proof. Suppose that $Cap^{(\tau),\alpha}(B) = 0$ for the α-capacity $Cap^{(\tau),\alpha}$ of $E^{(\tau)}$. Then there exists a decreasing sequence $\{B_n\}$ of relatively compact open sets such that $B_n \supset B$ and $\lim_{n\to\infty} Cap^{(\tau),\alpha}(B_n) = 0$. Let $e_{B_n}^{(\tau),\alpha}$ be a quasi-continuous modification of the α-equilibrium potential of B_n relative to $E^{(\tau)}$. Then

$$0 = \lim_{n\to\infty} Cap^{(\tau),\alpha}(B_n) \geq K_\alpha^2 \lim_{n\to\infty} E_\alpha^{(\tau)}(e_{B_n}^{(\tau),\alpha}, e_{B_n}^{(\tau),\alpha})$$

for some constant K_α. This implies that $\lim_{n\to\infty} \mathbf{D}(e_{B_n}^{(\tau),\alpha}, e_{B_n}^{(\tau),\alpha}) = 0$ by Lemma 2.2 and hence B is of zero capacity relative to $\mathbf{D}$.

Conversely, if B is of zero capacity relative to $\mathbf{D}$, then there exists a sequence $\{\varphi_n\} \subset F$ such that $\varphi_n \geq 1$ a.e. on a neighbourhood of B and $\lim_{n\to\infty} \mathbf{D}_1(\varphi_n, \varphi_n) = 0$. Since $E_\alpha^{(\tau)}(e_{B_n}^{(\tau),\alpha}, e_{B_n}^{(\tau),\alpha}) \leq K_\alpha^2 E_\alpha^{(\tau)}(\varphi_n, \varphi_n)$ for $\{B_n\}$ such that $B \subset B_n \subset \{x : \varphi_n(x) \geq 1\}$, Lemma 2.2 implies that $\lim_{n\to\infty} Cap^{(\tau),\alpha}(B_n) = 0$. □

Fixing a positive q.c. function $\ell \in F$, put $A^{(\tau),\ell}(\varphi, \psi) = E^{(\tau)}(\varphi, \psi\ell)$. This can be defined if $\varphi, \psi \in F$ satisfies $\psi\ell \in F$. Define a symmetric Dirichlet form $C^{(\tau),\ell}$ on $F^\ell = \{\varphi \in L^2(\overline{D}; \ell \cdot m) : \nabla\varphi \in L^2(\overline{D}; \ell \cdot m)\}$ by

$$C^{(\tau),\ell}(\varphi, \psi) = \sum_{i,j=1}^{d} \int_D a_{ij}(\tau, x) \frac{\partial \varphi}{\partial x_i} \frac{\partial \psi}{\partial x_j} \ell(x) dx.$$

If $\varphi \in F_b^\ell$, then $\varphi\ell \in L^2(D; m)$ and $\nabla(\varphi\ell) \in L^2(D; m)$. Hence $\varphi\ell \in F$ and $A^{(\delta),\ell}(\varphi, \psi) = E^{(\tau)}(\varphi, \psi\ell)$ is well defined and written as

$$\begin{aligned} A^{(\tau),\ell}(\varphi, \psi) = C^{(\tau),\ell}(\varphi, \psi) &+ \sum_{i,j=1}^{d} \int_D a_{ij}(\tau, x) \frac{\partial \varphi}{\partial x_i} \frac{\partial \ell}{\partial x_j} \psi(x) dx \\ &- \sum_{i=1}^{d} \int_D b_i(\tau, x) \frac{\partial \varphi}{\partial x_i} \psi(x)\ell(x) dx \\ &- \int_{\partial D} \sum_{i=1}^{d} \beta_i(\tau, x) \frac{\partial \varphi}{\partial x_i} \psi(x)\ell(x) dS(x). \end{aligned} \tag{2.9}$$

Note that $\mathcal{C}^1(\overline{D}) \subset F_b \cap F^\ell$. Furthermore, for any $\varphi, \psi \in F_b \cap F^\ell$, $A^{(\tau),\ell}(\varphi, \psi)$ satisfies the following condition which is used to show the condition $(\mathcal{E}.6)$ in Section 1.4 of [11] relative to the corresponding bilinear form $\mathcal{A}^{(\delta)}$ associated with the time dependent semi-Dirichlet form.

Lemma 2.4. *For any $\alpha > \alpha_0$, there exists a constant K_1 satisfying*

$$\begin{aligned} |A^{(\tau),\ell}(\varphi, \psi)| \leq K_1 &\left(\|\psi\|_\infty E_\alpha^{(\tau)}(\ell, \ell)^{1/2} + C_\alpha^{(\tau),\ell}(\psi, \psi)^{1/2} \right) \\ &\times \left(E_\alpha^{(\tau)}(\varphi, \varphi)^{1/2} + C_\alpha^{(\tau),\ell}(\varphi, \varphi)^{1/2} \right) \end{aligned} \tag{2.10}$$

for all $\varphi, \psi \in F_b \cap F^\ell$.

Proof. As the proof of equation (2.2),

$$\begin{aligned}\left|E^{(\tau),1}(\varphi,\psi\ell)\right| \leq\; & C^{(\tau),\ell}(\varphi,\varphi)^{1/2}C^{(\tau),\ell}(\psi,\psi)^{1/2} \\ & +\lambda_2\|\psi\|_\infty \mathbf{D}(\varphi,\varphi)^{1/2}\mathbf{D}(\ell,\ell)^{1/2} \\ & +\|b\|_\infty C^{(\tau),\ell}(\varphi,\varphi)^{1/2}\|\psi\|_{\ell\cdot m}.\end{aligned}$$

Hence, by Theorem 2.1, the right-hand side is dominated by

$$K_2\left(\|\psi\|_\infty E_\alpha^{(\tau)}(\ell,\ell)^{1/2}+C_1^{(\tau),\ell}(\psi,\psi)^{1/2}\right)\left(C^{(\tau),\ell}(\varphi,\varphi)^{1/2}+E_\alpha^{(\tau)}(\varphi,\varphi)^{1/2}\right)$$

for a suitable constant K_2. Decompose $E^{(\tau),2}(\varphi,\psi\ell)$ as (2.7) by taking $\psi\ell$ instead of ψ. Denote the corresponding terms I, II and III by I$'$, II$'$ and III$'$, respectively. Then, similarly to the proof of Lemma 2.2,

$$\begin{aligned}|\mathrm{II}'| &= \sum_{j=1}^{d-1}\int_{\mathbb{R}^d_+}\left|\overline{\beta}_j(\xi')\frac{\partial\overline{\varphi}}{\partial\xi_j}\frac{\partial(\overline{\psi\ell})}{\partial\xi_d}\overline{S}(\xi)\right|d\xi \\ &\leq \frac{1}{c_1}\|\beta\|_\infty C^{(\tau),\ell}(\varphi,\varphi)^{1/2}C^{(\tau),\ell}(\psi,\psi)^{1/2}+\frac{\|\psi\|_\infty}{k_1}\mathbf{D}(\varphi,\varphi)^{1/2}\mathbf{D}(\ell,\ell)^{1/2} \\ &\leq K_3\left(C^{(\tau),\ell}(\varphi,\varphi)^{1/2}+E_\alpha^{(\tau)}(\varphi,\varphi)^{1/2}\right) \\ &\quad\times\left(C^{(\tau),\ell}(\psi,\psi)^{1/2}+\|\psi\|_\infty E_\alpha^{(\tau)}(\ell,\ell)^{1/2}\right)\end{aligned}$$

for some constant K_3, where $\overline{\ell}(\xi)=\ell(\Psi(\xi))$. Similarly,

$$|\mathrm{III}'| \leq K_4 C^{(\tau),\ell}(\varphi,\varphi)^{1/2}\|\psi\|_{\ell\cdot m}.$$

Furthermore,

$$\begin{aligned}|\mathrm{I}'| \leq\; & \frac{\|\overline{\beta}\|_\infty}{k_1}C^{(\tau),\ell}(\varphi,\varphi)^{1/2}C^{(\tau),\ell}(\psi,\psi)^{1/2} \\ & +\frac{\|\overline{\beta}\|_\infty\|\psi\|_\infty}{k_1}\mathbf{D}(\varphi,\varphi)^{1/2}\mathbf{D}(\ell,\ell)^{1/2}+\frac{k_3}{k_1}C^{(\tau),\ell}(\varphi,\varphi)^{1/2}\|\psi\|_{\ell\cdot m} \\ \leq\; & K_5\left(C^{(\tau),\ell}(\varphi,\varphi)^{1/2}+E_\alpha^{(\tau)}(\varphi,\varphi)^{1/2}\right) \\ & \times\left(C_1^{(\tau),\ell}(\psi,\psi)^{1/2}+\|\psi\|_\infty E_\alpha^{(\tau)}(\ell,\ell)^{1/2}\right).\end{aligned}$$

Combining these inequalities we obtain (2.10). □

3. Associated space-time process and its application

By virtue of Theorem 2.1, there corresponds a time dependent semi-Dirichlet form$(\mathcal{E},\mathcal{F})$ on $\mathcal{H}$ defined by (1.7). Let $\mathbb{M}=(Z_t,\mathbb{P}_z)$ be the associated space-time Hunt process on Z, that is its resolvent $R_\alpha u$ is a q.c. function of $\mathcal{W}$ satisfying

$$\mathcal{E}_\alpha(R_\alpha u,v)=(u,v) \tag{3.1}$$

for all $\alpha > \alpha_0$ and $u \in \mathcal{H}$, where (u, v) is the inner product in $L^2(Z;\nu)$. Furthermore, there also exists a ν-dual resolvent $\widehat{R}_\alpha u \in \mathcal{W}$ satisfying $\mathcal{E}_\alpha(v, \widehat{R}_\alpha u) = (u, v)$ for $\alpha > \alpha_0$. By the resolvent equation, for any $\alpha > 0$, $R_\alpha u$ and $\widehat{R}_\alpha u$ can be extended to the operators on $L^\infty(Z;\nu)$ and $L^1(Z;\nu)$ respectively.

Let $\mathcal{L}$ be the generator corresponding to the resolvent $R_\alpha f$, that is $\mathcal{L}$ is defined for any function $u \in \mathcal{D}(\mathcal{L}) = \{R_\alpha f : f \in \mathcal{F}', \alpha > \alpha_0\}$ by $\mathcal{L}u = \alpha u - f$ for $u = R_\alpha f \in \mathcal{D}(\mathcal{L})$. Since $\tau(t) = \tau(0) + t$, the transition function $p_t f \in \mathcal{W}$ can be expressed as $p_t f(\tau, x) = \int_{\overline{D}} p(\tau, x; \tau + t, dy) f(\tau + t, y)$ by a time-inhomogeneous transition function $p(\tau, x; t, dy)$. For any function $\varphi \in L^2(\overline{D}; m)$, put $p(s, x; t, \varphi) = \int_{\overline{D}} p(s, x; t, dy)\varphi(y)$.

Lemma 3.1. *For any $\varphi \in L^2(\overline{D}; m)$ and $v \in \mathcal{F}$,*

$$-\left(\frac{\partial}{\partial \tau} p(\tau, x; t, \varphi), v(\tau, \cdot)\right) + E^{(\tau)}\left(p(\tau, \cdot; t, \varphi), v(\tau, \cdot)\right) = 0 \quad \text{a.e.}\ \tau < t. \tag{3.2}$$

Proof. Assume that φ and $v(\tau, \cdot)$ belong to $\mathcal{C}^1(\overline{D})$. For any $\xi(\tau) \in \mathcal{C}_0^1(\mathbb{R}^1)$, put $f(\tau, x) = \xi(\tau)\varphi(x)$. Then

$$\begin{aligned} \alpha R_\alpha f(\tau, x) &= \alpha \int_0^\infty e^{-\alpha t} p(\tau, x; \tau + t, \varphi)\xi(\tau + t)dt \\ &= \int_0^\infty e^{-\alpha t} \frac{\partial}{\partial t}\left(\xi(\tau + t)p(\tau, x; \tau + t, \varphi)\right) dt + \xi(\tau)\varphi(x). \end{aligned}$$

Hence, for $v(\tau, x) = \eta(\tau)\psi(x)$ with $\eta \in \mathcal{C}_0^1(\mathbb{R}^1)$ and $\psi \in \mathcal{C}^1(\overline{D})$,

$$\begin{aligned} 0 &= \mathcal{E}_\alpha(R_\alpha f, v) - (f, v) \\ &= -\int_0^\infty \int_{\mathbb{R}^1} e^{-\alpha t} \frac{\partial}{\partial \tau}\left(\xi(\tau + t)p(\tau, \cdot; \tau + t, \varphi), \psi\right) \eta(\tau) d\tau dt \\ &\quad + \int_0^\infty \int_{\mathbb{R}^1} e^{-\alpha t} \xi(\tau + t) E^{(\tau)}\left(p(\tau, \cdot; \tau + t, \varphi), \psi\right) \eta(\tau) d\tau dt \\ &\quad + \int_0^\infty \int_{\mathbb{R}^1} e^{-\alpha t} \left(\frac{\partial}{\partial t}\left(\xi(\tau + t)p(\tau, \cdot; \tau + t, \varphi)\right), \psi\right) \eta(\tau) d\tau dt. \end{aligned}$$

This implies that

$$\begin{aligned} &-\frac{\partial}{\partial \tau}\left(\xi(\tau + t)p(\tau, \cdot; \tau + t, \varphi), \psi\right) + \xi(\tau + t)E^{(\tau)}\left(p(\tau, \cdot; \tau + t, \varphi), \psi\right) \\ &+ \left(\frac{\partial}{\partial t}\left(\xi(\tau + t)p(\tau, \cdot; \tau + t, \varphi)\right), \psi\right) = 0. \end{aligned}$$

Therefore

$$-\left(\frac{\partial}{\partial \sigma} p(\sigma, \cdot; \tau + t, \varphi), \psi\right)\Big|_{\sigma = \tau} + E^{(\tau)}\left(p(\tau, \cdot; \tau + t, \varphi), \psi\right) = 0$$

which yields (3.2). □

Lemma 3.2. *The measure $d\eta(\tau, x) = d\tau \otimes dS(x)$ is a smooth measure relative to $(\mathcal{E}, \mathcal{F})$.*

Proof. For any $O = O^{(k)}$ and $f \in \mathcal{C}_0^1(O)$, similarly to the proof of Lemma 2.2,

$$\begin{aligned}\left|\int_{O\cap\partial D} f(x)dS(x)\right| &= \left|\int_{V\cap\{\xi:\xi_d=0\}} \overline{f}(\xi',0)\overline{S}(\xi',0)d\xi'\right| \\ &= \left|-\int_{\mathbb{R}^d_+} \frac{\partial}{\partial\xi_d}\left(\overline{f}(\xi)\overline{S}(\xi)\right)d\xi\right| \\ &\le k_1\mathbf{D}(f,f)^{1/2} + k_3\|f\| \\ &\le k_4\mathbf{D}_1(f,f)^{1/2}, \end{aligned} \tag{3.3}$$

for $k_4 = k_1 \vee k_3$. This implies that dS is a smooth measure relative to $\mathbf{D}$ and hence $\eta((a,b)\times B) = 0$ for any $0 \le a < b < \infty$ and Borel set B of zero $\mathbf{D}$-capacity.

Let B be a compact subset of $O \cap \partial D$ and $\{B_n\}$ be a decreasing sequence of relatively compact open sets of such that $B = \cap_n B_n$. Then, by [11] Theorem 6.2.4, the α-capacity $Cap^{(\alpha)}(\Lambda_n)$ and the α-equilibrium potential $e^\alpha_{\Lambda_n}$ of $\Lambda_n = (a,b)\times B_n$ satisfies

$$Cap^{(\alpha)}(\Lambda_n) \ge \mathcal{B}_\alpha(e^\alpha_{\Lambda_n}, e^\alpha_{\Lambda_n}) = \int_{\mathbb{R}^1} E^{(\tau)}_\alpha\left(e^\alpha_{\Lambda_n}(\widetilde{u},\cdot), e^\alpha_{\Lambda_n}(\tau,\cdot)\right)d\tau.$$

If $(a,b)\times B$ is of zero capacity, then there exists a sequence $\{B_n\}$ satisfying $\lim_{n\to\infty} Cap^{(\alpha)}(\Lambda_n) = 0$. Then, by choosing a subsequence if necessary, $\lim_{n\to\infty} E^{(\tau)}\left(e^\alpha_{\Lambda_n}(\tau,\cdot), e^\alpha_{\Lambda_n}(\tau,\cdot)\right) = 0$ a.e. τ. Since $e^\alpha_{\Lambda_n}(\tau,\cdot) \ge 1$ on B_n and $e^\alpha_{\Lambda_n}(\tau,\cdot) \in F$ for a.e. τ, this implies that B is of zero capacity relative to $(\mathbf{D}, F)$ and hence $\eta((a,b)\times B) = 0$.

By virtue of (3.3), for any $u(\tau,x) \in \mathcal{C}^1((a,b)\times O)$,

$$\left|\int_{\mathbb{R}^1}\int_{\partial D} u(\tau,x)dS(x)d\tau\right| \le k_4\int_{\mathbb{R}^1} E^{(\tau)}\left(u(\tau,\cdot),u(\tau,\cdot)\right)^{1/2}d\tau \le k_4\|u\|_{\mathcal{W}}.$$

As the proof of Lemma 2.2, this can be extended to all $u \in \mathcal{C}^1((a,b)\times\overline{D})$. Hence $d\tau\otimes dS$ is a smooth measure relative to $(\mathcal{E},\mathcal{F})$ by (6.2.21) and (6.2.27) of [11]. □

By virtue of Lemma 3.2, there exists a positive continuous additive functional L_t satisfying

$$\lim_{t\to 0}\mathbb{E}_{\widehat{h}\cdot\nu}\left(\int_0^t u(Z_t)dL_t\right) = \int_{\mathbb{R}^1\times\partial D}\widehat{h}(\tau,x)u((\tau,x)d\tau dS(x) \tag{3.4}$$

for any q.c. function u and α-coexcessive function $\widehat{h}$. The functional L_t is called the local time of ∂D.

For any fixed $T < \infty$ and $f(\tau,x) \in L^2([0,T]\times\overline{D};\nu)$,

$$R^T f(\tau,x) = \mathbb{E}_{(\tau,x)}\left(\int_\tau^T f(t,X_t)dt\right)$$

is the 0-order potential of $f(\tau,x)$. Hence, $R^T f$ is the unique function satisfying $\mathcal{E}(R^T f, v) = (f,v)$ for all $v \in \mathcal{F}$. Since v is arbitrary, this implies

$$\begin{aligned}&-\left(\frac{\partial}{\partial\tau}R^T f(\tau,\cdot), v(\tau,\cdot)\right) + E^{(\tau)}(R^T f(\tau,\cdot), v(\tau,\cdot)) \\ &= (f(\tau,\cdot), v(\tau,\cdot))\end{aligned} \tag{3.5}$$

for a.e. $0 \le \tau \le T$. Similarly, for $g(\tau, x) \in L^2([0,T] \times \partial D; \eta)$ with $d\eta(\tau, x) = d\tau dS(x)$, the potential of the measure $U^T(g \cdot \eta)$ given by

$$U^T(g \cdot \eta)(\tau, x) = \mathbb{E}_{(\tau,x)}\left(\int_\tau^T g(Z_t)dL_t\right)$$

is the unique function satisfying

$$\left(U^T(g \cdot \eta)(\tau, \cdot), \frac{\partial}{\partial \tau}v(\tau, \cdot)\right) + E^{(\tau)}(U^T(g \cdot \eta)(\tau, \cdot), v(\tau, \cdot))$$
$$= \langle g(\tau, \cdot) \cdot S, (\tau, \cdot)\rangle \tag{3.6}$$

for all $v \in \mathcal{W}$.

Theorem 3.1. *Let $f \in L^2([0,T] \times \overline{D}; \nu), g \in L^2([0,T] \times \partial D; \eta)$ and $\varphi \in L^2(\overline{D}; m)$. Then, the function $u(\tau, x)$ defined by*

$$u(\tau, x) = \mathbb{E}_{(\tau,x)}(\varphi(X_T)) + \mathbb{E}_{(\tau,x)}\left(\int_\tau^T f(t, X_t)dt\right)$$
$$+\mathbb{E}_{(\tau,x)}\left(\int_\tau^T g(t, X_t)dL_t\right) \tag{3.7}$$

is the unique function satisfying

$$\int_0^T \left(u(\tau, \cdot), \frac{\partial v}{\partial \tau}\right) d\tau + \int_0^T E^{(\tau)}(u(\tau, \cdot), v(\tau, \cdot))$$
$$= \int_0^T \int_D f(\tau, x)v(\tau, x)d\tau dx + \int_0^T \int_{\partial D} g(\tau, x)v(\tau, x)d\tau dS(x) \tag{3.8}$$

for all $v \in \mathcal{C}^1((0,T) \times \overline{D})$ such that $v(\tau, x) = 0$ for any τ outside of a compact subset of $(0,T)$.

Proof. Since $u(\tau, x) = p(\tau, x; T, \varphi) + R^T f(\tau, x) + U^T(g \cdot \eta)(\tau, x)$, (3.8) follows from (3.2), (3.5) and (3.6). □

Corollary 3.1. *The function $u(\tau, x)$ given by (3.7) is the unique function satisfying the following equations in the distribution sense.*

$$\frac{\partial u}{\partial \tau} + \sum_{i,j=1}^d \frac{\partial}{\partial x_j}\left(a_{ij}(\tau, x)\frac{\partial u}{\partial x_i}\right) + \sum_{i=1}^d b_i(\tau, x)\frac{\partial u}{\partial x_i} = -f(\tau, x) \tag{3.9}$$

for $(\tau, x) \in (0,T) \times D$,

$$\sum_{i=1}^d \left(\sum_{j=1}^d a_{ij}(\tau, x)n_j(\tau, x) + \beta_i(\tau, x)\right)\frac{\partial u}{\partial x_i} = g(\tau, x) \tag{3.10}$$

for $(\tau, x) \in (0,T) \times \partial D$ and $u(T, x) = \varphi(x)$ for $x \in \overline{D}$, where $\mathbf{n} = (n_1, n_2, \cdots, n_d)$ is the inward unit normal vector of ∂D.

Proof. Put $v(\tau,x) = \xi(\tau)\psi(x)$ in (3.8). Then the left-hand side of (3.8) can be written by Green's theorem as

$$
\begin{aligned}
&\int_0^T (u(\tau,\cdot),\psi)\,\xi'(\tau)d\tau + \int_0^T E^{(\tau)}\,(u(\tau,\cdot),\psi)\,\xi(\tau)d\tau \\
&= -\int_0^T \int_D \frac{\partial u}{\partial \tau}\psi(x)\xi(\tau)d\tau dx \\
&\quad - \sum_{i,j=1}^d \int_0^T \int_D \frac{\partial}{\partial x_j}\left(a_{ij}(\tau,x)\frac{\partial u}{\partial x_i}\right)\psi(x)\xi(\tau)d\tau dx \\
&\quad - \sum_{i=1}^d \int_0^T \int_D b_i(\tau,x)\frac{\partial u}{\partial x_i}\psi(x)\xi(\tau)d\tau dx \\
&\quad + \sum_{i=1}^d \int_0^T \int_{\partial D}\left(\sum_{j=1}^d a_{ij}n_j - \beta_i\right)(\tau,x)\,\frac{\partial u}{\partial x_i}\,\psi(x)\xi(\tau)d\tau dS(x).
\end{aligned}
$$

This compared with the right-hand side of (3.8) gives the result. □

For $\delta > \alpha_0$, let us fix a δ-coexcessive function $\widehat{h}_\delta \in \mathcal{F}$ which is bounded from below by a positive constant on any compact set of Z. See [12] Theorem 2.4.8 for the existence of such function. In [12], concerning to the stochastic calculus related to time dependent Dirichlet forms, only Fukushima's decomposition in the weak sense and its direct consequences are mentioned. But, most of the calculus similar to the time independent case as in Chapter 5 of [12] are possible. In particular, under our present settings, strong sense Fukushima's decomposition as Theorem 5.1.4 holds. Let us briefly mention the outline of it. The energy $e^{(\delta)}(A)$ of an additive functional A_t is defined by

$$
e^{(\delta)}(A) = \frac{1}{2}\lim_{\beta\to\infty}\beta^2\mathbb{E}_{\widehat{\nu}}\left(\int_0^\infty e^{-\beta t}\,A_t^2 dt\right), \tag{3.11}
$$

where $\widehat{\nu} = \widehat{h}_\delta \cdot \nu$. For a q.c. function $u \in \mathcal{W}$, put $A_t^{[u]} = u(Z_t) - u(Z_0)$. If $u \in \mathcal{W}_b$ satisfies $u\widehat{h} \in \mathcal{F}$, then the energy of $A^{[u]}$ is given by

$$
\begin{aligned}
e^{(\delta)}(A^{[u]}) &= \lim_{\beta\to\infty}\beta\left(u - \beta R_\beta u, u\widehat{h}_\delta\right) - \frac{1}{2}\lim_{\beta\to\infty}\beta\left(u^2 - \beta R_\beta u^2, \widehat{h}_\delta\right) \\
&= \mathcal{E}(u, u\widehat{h}_\delta) - \frac{1}{2}\mathcal{E}(u^2, \widehat{h}_\delta) \\
&= \mathcal{B}(u, u\widehat{h}_\delta) - \frac{1}{2}\mathcal{B}(u^2, \widehat{h}_\delta).
\end{aligned} \tag{3.12}
$$

In the previous section, for a strictly positive function $\ell \in F$, we introduced a Dirichlet form $(C^{(\tau),\ell}, F^\ell)$. By taking $\ell = \widehat{h}_\delta(\tau,\cdot)$ for each fixed τ, define a symmetric Dirichlet form $(C^{(\tau),\delta}, F^{(\tau),\delta})$ similarly by $F^{(\tau),\delta} = \{\varphi \in L^2(D : \widehat{h}_\delta(\tau,\cdot)\cdot m, \nabla\varphi \in L^2(D; \widehat{h}_\delta(\tau,\cdot)\cdot m)\}$ and

$$
C^{(\tau),\delta}(\varphi,\psi) = \int_D \sum_{i,j=1}^d a_{ij}(\tau,x)\,\frac{\partial\varphi}{\partial x_i}\frac{\partial\psi}{\partial x_j}\,\widehat{h}_\delta(\tau,x)\,dx.
$$

Let

$$
\mathcal{C}^\delta(u,v) = \int_{\mathbb{R}^1} C^{(\tau),\delta}\left(u(\tau,\cdot), v(\tau,\cdot)\right)\,d\tau
$$

and

$$\mathcal{A}^{\delta}(u,v) = \mathcal{B}(u, v\widehat{h}_{\delta}) = \int_{\mathbb{R}^1} A^{(\tau),\delta}\,(u(\tau,\cdot), v(\tau,\cdot))\, d\tau$$

where $A^{(\tau),\delta} = A^{(\tau),\ell}$ is defined by (2.9) for $\ell = \widehat{h}^{\delta}(\tau,\cdot)$. Then, by noting that $\widehat{h}_{\delta}$ is δ-coexcessive, it holds by (2.9) that

$$\mathcal{A}^{\delta}_{\alpha}(u,u) = \mathcal{C}^{\delta}_{\alpha-\delta/2}(u,u) + \frac{1}{2}\mathcal{E}_{\delta}(u^2, \widehat{h}_{\delta}) \geq \mathcal{C}^{\delta}_{\alpha-\delta/2}(u,u). \tag{3.13}$$

Let $\overset{\circ}{\mathcal{M}}$ and $\mathcal{N}$ be the families of martingale additive functionals of finite energy and continuous additive functional of zero energy relative to $\widehat{h}_{\delta}$. Then the following Fukushima's decomposition of $A^{[u]}_t = \widetilde{u}(Z_t) - \widetilde{u}(Z_0)$ holds.

Theorem 3.2. *For any $u \in \mathcal{W}_b$ such that $u\widehat{h}_{\delta} \in \mathcal{F}$, there exist uniquely $M^{[u]} \in \overset{\circ}{\mathcal{M}}$ and $N^{[u]} \in \mathcal{N}$ such that*

$$A^{[u]}_t = M^{[u]}_t + N^{[u]}_t. \tag{3.14}$$

Proof. Suppose that $u \in \mathcal{W}_b$ satisfies $u\widehat{h}_{\delta} \in \mathcal{F}$. Then $u(\tau,\cdot) \in F^{(\tau),\delta}$ for all τ. Hence (2.10) and (3.13) yields that

$$\begin{aligned}\mathcal{A}^{(\delta)}(u,v)^2 &\leq K_2\left(\|v\|^2_{\infty}\mathcal{B}_{\alpha}(\widehat{h}_{\delta},\widehat{h}_{\delta}) + \mathcal{C}^{\delta}_{\alpha}(v,v)\right)\left(\mathcal{B}_{\alpha}(u,u) + \mathcal{C}^{\delta}_{\alpha}(u,u)\right)\\ &\leq K_3\left(\|v\|^2_{\infty} + \mathcal{A}^{(\delta)}_{\alpha}(v,v)\right)\left(\mathcal{B}_{\alpha}(u,u) + \mathcal{A}^{(\delta)}_{\alpha}(u,u)\right)\end{aligned} \tag{3.15}$$

for some constant K_3 depending on $\widehat{h}_{\delta}$, where $\mathcal{C}^{\delta}_{\alpha}(u,v) = \mathcal{C}^{\delta}(u,v) + \alpha(u,v)_{\widehat{\nu}}$. This implies that the condition ($\mathcal{E}$.6) in [11] holds for $\mathcal{A}^{\delta}$. Hence any bounded function $u \in \mathcal{W}$ such that $u\widehat{h}_{\delta} \in \mathcal{F}$ is approximated by functions of the form $R_{\alpha}f$ relative to $\mathcal{B}_{\alpha} + \mathcal{A}^{\delta}_{\alpha}$ (see Theorem 1.4.8 of [11]). Then, as the proof of Theorem 5.1.4 in [11], we obtain the result. □

Since any function $w \in \mathcal{C}^1_0(\mathbb{R}^1 \times \overline{D})$ satisfies the condition of Theorem 3.2, the decomposition $A^{[w]}_t = M^{[w]}_t + N^{[w]}_t$ holds. The martingale part $M^{[w]}$ and the zero energy part $N^{[w]}$ are characterized as follows.

Corollary 3.2. *For any $w \in \mathcal{C}^1_0(\mathbb{R}^1 \times \overline{D})$, the martingale part $M^{[w]}$ satisfies*

$$\langle M^{[w]}\rangle_t = 2\int_0^t \sum_{i,j=1}^d a_{ij}(Z_s)\,\frac{\partial w}{\partial x_i}\frac{\partial w}{\partial x_j}(Z_s)\,ds. \tag{3.16}$$

Also the zero energy part $N^{[w]}$ satisfies,

$$\begin{aligned}\lim_{t\to 0}\frac{1}{t}\mathbb{E}_{v\cdot\nu}\left(N^{[w]}_t\right) &= \int_Z \frac{\partial w}{\partial \tau}\, v(\tau,x)d\nu(\tau,x)\\ &- \int_Z \sum_{i,j=1}^d a_{ij}(\tau,x)\,\frac{\partial w}{\partial x_i}\frac{\partial v}{\partial x_j}\,d\nu(\tau,x)\\ &+ \int_Z \sum_{i=1}^d b_i(\tau,x)\,\frac{\partial w}{\partial x_i}\,v(\tau,x)d\nu(\tau,x)\\ &+ \int\int_{\mathbb{R}^1\times\partial D}\sum_{i=1}^d \beta_i(\tau,x)\frac{\partial w}{\partial x_i}\,v(\tau,x)d\tau dS(x)\end{aligned} \tag{3.17}$$

for any $v \in \mathcal{F}_b$ with compact support.

Proof. Since v is bounded by a constant multiple of $\widehat{h}_\delta$, it holds that $\lim_{t\to 0}(1/t)\mathbb{E}_{v\cdot\nu}(N^{[w]}) = 0$. Hence

$$\begin{aligned}\int_Z v(z)d\mu_{\langle w\rangle}(z) &= \lim_{t\to 0}\frac{1}{t}\mathbb{E}_{v\cdot\nu}\left((w(Z_t)-w(Z_0))^2\right.\\ &= \lim_{t\to 0}\frac{1}{t}\left(2(w-p_tw, vw)_\nu - (w^2-p_tw^2, v)\right)\\ &= 2\mathcal{B}(w, wv) - \mathcal{B}(w^2, v)\\ &= 2\int_Z \sum_{i,j=1}^d a_{ij}(z)\frac{\partial w}{\partial x_i}\frac{\partial (wv)}{\partial x_j}\,d\nu(z)\\ &\quad - \int_Z \sum_{i,j=1}^d a_{ij}(z)\frac{\partial (w^2)}{\partial x_i}\frac{\partial v}{\partial x_j}\,d\nu(z)\\ &= 2\int_Z \sum_{i,j=1}^d a_{ij}(z)\frac{\partial w}{\partial x_i}\frac{\partial w}{\partial x_j}\,v(z)\,d\nu(z).\end{aligned}$$

Therefore $\mu_{\langle w}(dz) = 2\sum_{i,j=1}^d a_{ij}(z)\frac{\partial w}{\partial x_i}\frac{\partial w}{\partial x_j}\,d\nu(z)$ which implies (3.16). Similarly, (3.17) follows from

$$\lim_{t\to 0}\frac{1}{t}\,\mathbb{E}_{v\cdot\nu}(N^{[w]}) = \lim_{t\to 0}\frac{1}{t}\,\mathbb{E}_{v\cdot\nu}\left(w(Z_t)-w(Z_0)\right) = -\mathcal{E}(w,v).$$

□

References

1. K. Burdzy, Z.Q. Chen and J. Sylvester, The heat equation and reflected Brownian motion in time-dependent domains, *Ann. Probab.* **32** (2004), 775-804
2. E.B. Davies, *Heat kernels and spectral theory*, Cambridge University Press, 1989
3. M. Fukushima, Aconstruction of reflecting barrier Brownian motions for bounded domains, *Osaka J. Math.* **4** (1967), 183-215
4. M. Fukushima, On an L^p-estimate of resolvents of Markov processes, Publ. RIMS. Kyoto Univ. **13** (1977), 277-284
5. M. Fukushima, Y. Oshima and M. Takeda, *Dirichlet forms and symmetric Markov processes, 2nd edition*, De Druyter, 2010
6. J.H. Kim, Stochastic calculus related to non-symmetric Dirichlet forms, *Osaka J. Math.* **24** (1987), 331-371
7. Z.M. Ma, L. Overbeck and M. Röckner, Markov processes associated with semi-Dirichlet forms, *Osaka J. Math.* **32** (1995), 97-119
8. Z.M. Ma and M. Röckner, *Introduction to the theory of of (non-symmetric) Dirichlet forms*, Springer Verlag, Berlin-Heiderberg-New York, 1992
9. L. Ma, Z.M. Ma and W. Sun, Fukushima's decomposition for diffusions associated with semi-Dirichlet forms, Preprint (2011)
10. S. Mizohata, *The theory of partial differential equatins*, Cambridge University Press, London, 1973
11. Y. Oshima, On a construction of diffusion processes on moving domains, Potential Anal. **20** (2004), 1-31

12. Y. Oshima, *Semi-Dirichlet forms and Markov processes*, De Gruyter, 2013
13. K.T. Sturm, Analysis on local Dirichlet spaces, II. Upper Gaussian estimates for the fundamental solutions of parabolic equations, *Osaka J. Math.* **32** (1995), 275-312
14. M. Tsuchiya, On the oblique derivative problem for diffsion processes and diffusion equations with Hölder continuous coefficients, *Trans. Amer. Math. Soc.* **146** (1994), 257-281
15. M. Tsuchiya, Probabilistic representation of weak solutions to a parabolic equation with a mixed boundary condition, Preprint (2013)

Chapter 25

Lower bounded semi-Dirichlet forms associated with Lévy type operators

René L. Schilling* and Jian Wang†

Let $k : E \times E \to [0,\infty)$ be a non-negative measurable function on some locally compact separable metric space E. We provide some simple conditions such that the quadratic form with jump kernel k becomes a regular lower bounded (non-local,non-symmetric) semi-Dirichlet form. If $E = \mathbb{R}^n$ we identify the generator of the semi-Dirichlet form and its (formal) adjoint. In particular, we obtain a closed expression of the adjoint of the stable-like generator $-(-\Delta)^{\alpha(x)}$ in the sense of Bass. Our results complement a recent paper by Fukushima and Uemura [3] and establish the relation of these results with the symmetric principal value (SPV) approach due to Zhi-ming Ma and co-authors [5].

Contents

1. Introduction

Let (E, d, m) be a locally compact separable metric measure space. The reference measure m is a positive Radon measure with full topological support. Recently, Fukushima and Uemura [3] were able to construct a regular lower bounded semi-Dirichlet form and the corresponding jump-type Hunt process for a given jump kernel $k(x, y)$. A key ingredient in their construction are conditions that ensure that the symmetric part of the kernel, k_s, dominates the totally anti-symmetric part, k_a, where

$$k_s(x,y) := \tfrac{1}{2}\big(k(x,y) + k(y,x)\big) \quad \text{and} \quad k_a(x,y) := \tfrac{1}{2}\big(k(x,y) - k(y,x)\big).$$

For the readers' convenience let us briefly recall these assumptions, see [3, (2.1)–(2.4), Section 2],

$$x \mapsto \int_{y \neq x} \big(1 \wedge d(x,y)^2\big) k_s(x,y)\, m(dy) \in L^1_{\mathrm{loc}}(E, m), \tag{A0}$$

*Technische Universität Dresden, Institut für Mathematische Stochastik, 01062 Dresden, Germany, rene.schilling@tu-dresden.de
†School of Mathematics and Computer Science, Fujian Normal University, 350007 Fuzhou, China, jianwang@fjnu.edu.cn

$$C_1 := \sup_{x\in E} \int_{d(x,y)\geq 1} |k_a(x,y)|\, m(dy) < \infty, \tag{A1}$$

$$C_2 := \sup_{x\in E} \int_{d(x,y)<1} |k_a(x,y)|^\gamma\, m(dy) < \infty \quad \text{for some} \quad \gamma \in (0,1], \tag{A2}$$

$$C_3 := \sup_{x,y\in E,\, 0<d(x,y)\leq 1} \frac{|k_a(x,y)|^{2-\gamma}}{k_s(x,y)} < \infty \quad \text{for } \gamma \text{ from (A2)}. \tag{A3}$$

In this note we will simplify these conditions. If $E = \mathbb{R}^n$ we obtain the explicit expressions for the generator of the form and its formal adjoint in terms of Cauchy principal value integral (PV), which is related to the symmetric Cauchy principal value integral (SPV) in the sense of Ma *et al.* [5]. This is motivated by and improves more recent development on non-local Dirichlet forms and Lévy type operators, *e.g.* [3, 9].

2. Lower Bounded Semi-Dirichlet Forms

Let (E, d, m) be a locally compact separable metric measure space equipped with a positive Radon measure m, and $k(x, y)$ a non-negative Borel measurable function on the space $E \times E \setminus \Delta$, where Δ denotes the diagonal $\{(x, x) : x \in E\}$ in $E \times E$. The inner product and the norm in $L^2(E, m)$ are denoted by $\langle\cdot,\cdot\rangle_{L^2}$ and $\|\cdot\|_{L^2}$, respectively. As before, denote by k_s and k_a the symmetric part and the totally anti-symmetric part of k, respectively. Let $C_c^{\mathrm{lip}}(E)$ be the space of Lipschitz continuous functions on E with compact support. Throughout this section, we will assume (A0).

A (not necessarily symmetric) bilinear form $(\eta, \mathcal{F})$, $\mathcal{F} \subset L^2(E, m)$, is a lower bounded semi-Dirichlet form if the following conditions are satisfied: for some $\alpha > 0$ and $c > 0$

(1) $\eta(u, u) \geq -\alpha\langle u, u\rangle_{L^2}$ for all $u \in \mathcal{F}$;

(2) $\eta(u, v) \leq c\sqrt{\eta(u, u) + \alpha\langle u, u\rangle_{L^2}}\sqrt{\eta(v, v) + \alpha\langle v, v\rangle_{L^2}}$ for all $u, v \in \mathcal{F}$;

(3) $(\mathcal{F}, \eta(\cdot,\cdot) + \alpha\langle\cdot,\cdot\rangle_{L^2})$ is a complete subspace of $L^2(E, m)$;

(4) $u^+ \wedge 1 \in \mathcal{F}$ for all $u \in \mathcal{F}$ and $\eta(u^+ \wedge 1, u - u^+ \wedge 1) \geq 0$.

For further details we refer to [3, Section 1] and the references therein.

For each $n \in N$, we define the operator $L_n u$ for $u \in C_c^{\mathrm{lip}}(E)$ by

$$L_n u(x) := \int_{\{y\in E\,:\, d(x,y)>1/n\}} \big(u(y) - u(x)\big)k(x, y)\, m(dy), \quad x \in E,$$

and the quadratic form $\eta_n(u, v)$ for $u, v \in C_c^{\mathrm{lip}}(E)$ by

$$\eta_n(u, v) := -\langle L_n u, v\rangle_{L^2} = -\int_E L_n u(x)v(x)\, m(dx).$$

Due to (A0), all integrals appearing in the definition of L_n and η_n are absolutely convergent. Finally, set

$$\mathcal{E}(u, v) = \iint_{y\neq x} (u(x) - u(y))(v(x) - v(y))k_s(x, y)\, m(dx)\, m(dy),$$

$$\mathcal{F}^r = \Big\{u \in L^2(E, m) : u \text{ is Borel measurable and } \mathcal{E}(u, u) < \infty\Big\}.$$

The condition (A0) ensures that $(\mathcal{E}, \mathcal{F}^r)$ is a symmetric Dirichlet form on $L^2(E, m)$, and $\mathcal{F}^r$ contains the space $C_c^{\text{lip}}(E)$.

Set $\mathcal{E}_1(u,u) = \mathcal{E}(u,u) + \|u\|_{L^2}^2$, and write $\mathcal{F}^0$ for the $\mathcal{E}_1^{1/2}$-closure of $C_c^{\text{lip}}(E)$ in $\mathcal{F}^r$. In particular, $(\mathcal{E}, \mathcal{F}^0)$ is a regular symmetric Dirichlet form on $L^2(E, m)$, *cf.* [4, Example 1.2.4].

Our main result in this section is the following simple condition which guarantees that the limit of the forms $\eta_n(u,v)$, $n \to \infty$ exists, and defines a regular lower bounded semi-Dirichlet form. This generalizes and simplifies the earlier result by Fukushima and Uemura [3, Proposition 2.1 and Theorem 2.1].

Theorem 2.1. *Assume that* (A0) *is satisfied and that*

$$\sup_{x \in E} \int_{\{k_s(x,y)\neq 0\}} \frac{k_a(x,y)^2}{k_s(x,y)}\, m(dy) < \infty \tag{2.1}$$

holds. Then we have the following two statements.

(i) *For all* $u, v \in C_c^{\text{lip}}(E)$, *the limit* $\eta(u,v) = \lim_{n\to\infty} \eta_n(u,v)$ *exists. The form* $\eta(u,v)$ *has the following integral representation*

$$\eta(u,v) = \frac{1}{2}\,\mathcal{E}(u,v) + \iint_{y\neq x} (u(x) - u(y))\, v(y) k_a(x,y)\, m(dx)\, m(dy), \tag{2.2}$$

where the integral on the right hand side of (2.2) *is absolutely convergent.*

(ii) *The form* η *extends from* $C_c^{\text{lip}}(E) \times C_c^{\text{lip}}(E)$ *to* $\mathcal{F}^0 \times \mathcal{F}^0$ *such that the pair* $(\eta, \mathcal{F}^0)$ *is a regular lower bounded semi-Dirichlet form on* $L^2(E, m)$.

Let us briefly show that the conditions imposed by Fukushima and Uemura are more restrictive than (2.1). Indeed, if (A1)–(A3) hold, then we find for $x \in E$,

$$\begin{aligned}
&\int_{k_s(x,y)\neq 0} \frac{k_a(x,y)^2}{k_s(x,y)}\, m(dy) \\
&\le \int_{\substack{d(x,y)\le 1,\\ k_s(x,y)\neq 0}} \frac{|k_a(x,y)|^{2-\gamma}}{k_s(x,y)}\, |k_a(x,y)|^{\gamma}\, m(dy) + \int_{d(x,y)>1} k_s(x,y)\, m(dy) \\
&\le \left\{ \sup_{\substack{d(x,y)\le 1,\\ k_s(x,y)\neq 0}} \frac{|k_a(x,y)|^{2-\gamma}}{k_s(x,y)} \right\} \int_{d(x,y)\le 1} |k_a(x,y)|^{\gamma}\, m(dy) \\
&\quad + \int_{d(x,y)>1} k_s(x,y)\, m(dy) \\
&\le C_2 C_3 + C_1.
\end{aligned}$$

In the first inequality we have used that $|k_a(x,y)| \le k_s(x,y)$.

Sketch of the proof of Theorem 2.1. From the definition of η_n we find for all $u, v \in C_c^{\text{lip}}(E)$ that

$$\eta_n(u,v) = \frac{1}{2}\,\mathcal{E}_n(u,v) + \iint_{d(x,y)\ge 1/n} (u(x) - u(y)) v(y) k_a(x,y)\, m(dx)\, m(dy),$$

where

$$\mathcal{E}_n(u,v) = \iint_{d(x,y)\geq 1/n} (u(x)-u(y))(v(x)-v(y))k_s(x,y)\, m(dx)\, m(dy).$$

Because of (A0), $\mathcal{E}_n(u,v)$ converges to $\mathcal{E}(u,v)$ as $n\to\infty$.

To see the convergence of the non-symmetric part, we set for $x\in E$

$$h(x) := \int_{k_s(x,y)\neq 0} \frac{k_a(x,y)^2}{k_s(x,y)}\, m(dy).$$

An application of the Cauchy-Schwarz inequality and (2.1) show

$$\begin{aligned}
&\iint\limits_{d(x,y)\geq 1/n} |u(x)-u(y)||v(y)||k_a(x,y)|\, m(dx)\, m(dy)\\
&\quad = \iint\limits_{\substack{d(x,y)\geq 1/n,\\ k_s(x,y)\neq 0}} |u(x)-u(y)|k_s(x,y)^{1/2}\\
&\qquad\qquad \times |v(y)||k_a(x,y)|k_s(x,y)^{-1/2}\, m(dx)\, m(dy)\\
&\quad \leq \left[\ \iint\limits_{d(x,y)\geq 1/n} (u(x)-u(y))^2 k_s(x,y)\, m(dx)\, m(dy)\right]^{\frac12}\\
&\qquad\qquad \times \left[\ \iint\limits_{\substack{d(x,y)\geq 1/n,\\ k_s(x,y)\neq 0}} v(y)^2 \frac{k_a(x,y)^2}{k_s(x,y)}\, m(dx)\, m(dy)\right]^{\frac12}\\
&\quad \leq \Big[\mathcal{E}_n(u,u)\Big]^{\frac12}\left[\int v^2(y)h(y)\, m(dy)\right]^{\frac12}\\
&\quad \leq \Big[\mathcal{E}(u,u)\Big]^{\frac12} \|h\|_\infty^{\frac12}\|v\|_{L^2}.
\end{aligned}$$

This shows that the expression

$$\iint_{d(x,y)\geq 1/n} (u(x)-u(y))v(y)k_a(x,y)\, m(dx)\, m(dy)$$

converges absolutely as $n\to\infty$ and (i) follows. In order to see (ii), we use (i) and the argument used in the proof of [3, Theorem 2.1]. □

The semi-Dirichlet form $(\eta,\mathcal{F}^0)$ given by (2.2) is a coercive closed form in the sense of Ma-Röckner, *cf.* [7, Chapter I, Definition 2.4, page 16]. Then, by [7, Chapter I, Proposition 2.16, page 23], $(L,D(L))$ is the (pre-)generator of the form $(\eta,\mathcal{F}^0)$, where

$$D(L) = \{u\in\mathcal{F}^0 \mid v\mapsto \eta(u,v) \text{ is continuous w.r.t. } \|\cdot\|_{L^2} \text{ on } \mathcal{F}^0\}.$$

According to [7, Chapter I, Theorem 2.15, page 22], the generator $(L,D(L))$ is a linear operator mapping $D(L)$ into $L^2(E,m)$ such that

$$\eta(u,v) = -\langle Lu, v\rangle_{L^2}, \qquad u\in D(L),\ v\in\mathcal{F}^0. \tag{2.3}$$

However, for the semi-Dirichlet form $\eta(u,v)$ given by (2.2), it is in general difficult to find a closed expression for the generator $(L, D(L))$.

Note that any semi-Dirichlet form can be uniquely decomposed into three terms, which involve the integral in the sense of the symmetric Cauchy principal value (SPV), *cf.* [5, Definition 2.5, Theorems 2.6 and 4.1]. Motivated by this fact, we can obtain some information on $(L, D(L))$ if we assume that

$$\int_{y\neq x} (1\wedge d(x,y))|k_a(x,y)|\, m(dy) < \infty, \qquad x \in E. \tag{2.4}$$

Proposition 2.1. *Assume that* (A0), (2.1) *and* (2.4) *hold. Then,*

$$\eta(u,v) = -\langle Bu, v\rangle_{L^2}, \quad u \in C^*(E),\ v \in C_c^{\mathrm{lip}}(E),$$

where

$$C^*(E) := \left\{u \in C_c^{\mathrm{lip}}(E) \,:\, Bu \textit{ exists and belongs to } L^2(E,m)\right\}$$

and

$$\begin{aligned} Bu(x) :=& \mathrm{PV}\int_{y\neq x} \big(u(y)-u(x)\big)k_s(x,y)\, m(dy) \\ &+ \int_{y\neq x} \big(u(y)-u(x)\big)k_a(x,y)\, m(dy); \end{aligned}$$

$\mathrm{PV}\int \cdots dm$ *indicates the Cauchy principal value,* i.e. *for any $x \in E$, the limit*

$$\lim_{j\to\infty} \int_{\{y\in E\,:\, d(x,y)\geq 1/j\}} \big(u(y)-u(x)\big)k_s(x,y)\, m(dy).$$

Proof. The condition (2.4) ensures that the operator B is well defined for all x. According to the proof of Theorem 2.1, (A0) and (2.1) imply that for any $u \in C^*(E)$ and $v \in C_c^{\mathrm{lip}}(E)$, the form $\langle Bu, v\rangle_{L^2}$ is also well defined and finite.

On the other hand, under (A0) and (2.1), Theorem 2.1 shows that

$$\eta(u,v) = \frac{1}{2}\lim_{j\to\infty} \mathcal{E}_j(u,v) + \iint_{y\neq x} (u(x)-u(y))v(y)k_a(x,y)\, m(dx)\, m(dy),$$

with

$$\mathcal{E}_j(u,v) = \iint_{d(x,y)\geq 1/j} (u(x)-u(y))(v(x)-v(y))k_s(x,y)\, m(dx)\, m(dy).$$

Since $k_s(x,y) = k_s(y,x)$,

$$\begin{aligned} \mathcal{E}_j(u,v) =& \frac{1}{2}\iint_{d(x,y)\geq 1/j} (u(x)-u(y))v(x)k_s(x,y)\, m(dx)\, m(dy) \\ &- \frac{1}{2}\iint_{d(x,y)\geq 1/j} (u(x)-u(y))v(y)k_s(x,y)\, m(dx)\, m(dy) \\ =& \frac{1}{2}\iint_{d(x,y)\geq 1/j} (u(y)-u(x))v(y)k_s(y,x)\, m(dx)\, m(dy) \\ &- \frac{1}{2}\iint_{d(x,y)\geq 1/j} (u(x)-u(y))v(y)k_s(x,y)\, m(dx)\, m(dy) \end{aligned}$$

$$= \iint_{d(x,y)\geq 1/j} (u(y) - u(x))v(y)k_s(x,y)\, m(dx)\, m(dy),$$

and the claim follows by the dominated convergence theorem. □

To get an explicit expression for the generator associated with the semi-Dirichlet form $\eta(u,v)$, we now need to characterize the domain $C^*(E)$.

Theorem 2.2. *Assume that $E = \mathbb{R}^n$ is equipped with the Euclidean metric $d(x,y) = |x-y|$ and Lebesgue measure $m(dx) = dx$. Suppose that*

$$x \mapsto \int_{y\neq x} \big(1 \wedge |y-x|^2\big) k_s(x,y)\, dy \in L^2_{\text{loc}}(dx), \tag{2.5}$$

$$x \mapsto \int_{0<|z|\leq 1} |z||k_s(x,x+z) - k_s(x,x-z)|\, dz \in L^2_{\text{loc}}(dx) \tag{2.6}$$

and

$$x \mapsto \int_{\{y\in\mathbb{R}^n\,:\,|y-x|\geq 1\}} k_s(x,y)\, dy \in L^2(dx) \cup L^\infty(dx). \tag{2.7}$$

Let $(L, D(L))$ be a generator associated through (2.3) with the semi-Dirichlet form $\eta(u,v)$ given by (2.2). If (2.1) and (2.4) hold, then the set of twice differentiable functions with compact support is in the domain of L, i.e. $C_c^2(\mathbb{R}^n) \subset D(L)$, and on $C_c^2(\mathbb{R}^n)$ the operator L is of the following form

$$\begin{aligned} Lu(x) = &\int_{z\neq 0} \big(u(x+z) - u(x) - \nabla u(x)\cdot z\mathbf{1}_{\{|z|\leq 1\}}\big) k_s(x,x+z)\, dz \\ &+ \frac{1}{2}\nabla u(x) \cdot \int_{0<|z|\leq 1} z\big(k_s(x,x+z) - k_s(x,x-z)\big)\, dz \\ &+ \int_{y\neq x} \big(u(y) - u(x)\big) k_a(x,y)\, dy. \end{aligned} \tag{2.8}$$

Proof. *Step 1:* Note that the set $C_c^{\text{lip}}(\mathbb{R}^n)$ is dense in both $\mathcal{F}^0$ and $D(L)$ with respect to $\|\cdot\|_{L^2}$. According to Proposition 2.1, under (2.1), (2.4) and (2.5), the operator L has the following form

$$Lu(x) = \text{PV}\int_{y\neq x} \big(u(y) - u(x)\big) k_s(x,y)\, dy + \int_{y\neq x} \big(u(y) - u(x)\big) k_a(x,y)\, dy$$

on the set $C^*(\mathbb{R}^n) \cap D(L)$. Therefore, to get the required assertion it is sufficient to check that $C_c^2(\mathbb{R}^n) \subset C^*(\mathbb{R}^n) \cap D(L)$, and that the principal value integral is the same as (2.8).

First, since for any $\varepsilon \geq 0$,

$$\int_{\varepsilon\leq|z|\leq 1} z\,\big(k_s(x,x+z) + k_s(x,x-z)\big)\, dz = 0,$$

it holds for any $u \in C_c^2(\mathbb{R}^n)$ that for almost every $x \in \mathbb{R}^n$,

$$\begin{aligned}
&\mathrm{PV}\int_{y\neq x} \big(u(y)-u(x)\big)k_s(x,y)\,dy \\
&= \mathrm{PV}\int_{z\neq 0} \big(u(x+z)-u(x)\big)k_s(x,x+z)\,dz \\
&= \lim_{\varepsilon\to 0}\int_{|z|\geq\epsilon} \big(u(x+z)-u(x)\big)k_s(x,x+z)\,dz \\
&= \lim_{\varepsilon\to 0}\Bigg[\int_{|z|\geq\epsilon} \big(u(x+z)-u(x)\big)k_s(x,x+z)\,dz \\
&\qquad -\frac{1}{2}\int_{\varepsilon\leq|z|\leq 1} z\left(k_s(x,x+z)+k_s(x,x-z)\right)dz\cdot\nabla u(x)\Bigg] \qquad (2.9)\\
&= \lim_{\varepsilon\to 0}\Bigg[\int_{|z|\geq\epsilon} \big(u(x+z)-u(x)-\nabla u(x)\cdot z\mathbf{1}_{\{|z|\leq 1\}}\big)k_s(x,x+z)\,dz \\
&\qquad +\frac{1}{2}\int_{\varepsilon\leq|z|\leq 1} z\left(k_s(x,x+z)-k_s(x,x-z)\right)dz\cdot\nabla u(x)\Bigg] \\
&= \int_{z\neq 0} \big(u(x+z)-u(x)-\nabla u(x)\cdot z\mathbf{1}_{\{|z|\leq 1\}}\big)k_s(x,x+z)\,dz \\
&\qquad +\frac{1}{2}\nabla u(x)\cdot\int_{0<|z|\leq 1} z\big(k_s(x,x+z)-k_s(x,x-z)\big)\,dz,
\end{aligned}$$

where the last equality follows from (2.5), (2.6) and the dominated convergence theorem. Hence, (2.9) immediately yields that for any $u \in C_c^2(\mathbb{R}^n)$, $\mathrm{PV}\int_{y\neq x}\big(u(y)-u(x)\big)k_s(x,y)\,dy$ exists, and it also gives us the explicit expression (2.8) for L on $C_c^2(\mathbb{R}^n)$.

Step 2: Because of Step 1 and (2.3), to complete the proof we only need to verify that the operator L maps $C_c^2(\mathbb{R}^n)$ into $L^2(\mathbb{R}^n)$.

Let

$$\begin{aligned}
I_x &= \mathrm{PV}\int_{y\neq x} \big(u(y)-u(x)\big)k_s(x,y)\,dy \\
&= \int_{z\neq 0} \big(u(x+z)-u(x)-\nabla u(x)\cdot z\mathbf{1}_{\{|z|\leq 1\}}\big)k_s(x,x+z)\,dz \\
&\quad +\frac{1}{2}\nabla u(x)\cdot\int_{0<|z|\leq 1} z\big(k_s(x,x+z)-k_s(x,x-z)\big)\,dz
\end{aligned}$$

and $I\!I_x = \int_{y\neq x}\big(u(y)-u(x)\big)k_a(x,y)\,dy$. By the Cauchy-Schwarz inequality,

$$\begin{aligned}
\|I\!I_x\|^2_{L^2(dx)} &= \int\left(\int_{y\neq x}\big(u(y)-u(x)\big)k_a(x,y)\,dy\right)^2 dx \\
&= \int\left(\int_{k_s(x,y)\neq 0}\big(u(y)-u(x)\big)\sqrt{k_s(x,y)}\frac{k_a(x,y)}{\sqrt{k_s(x,y)}}\,dy\right)^2 dx \\
&\leq \int\left(\int\big(u(y)-u(x)\big)^2 k_s(x,y)\,dy\right)\left(\int_{k_s(x,y)\neq 0}\frac{k_a^2(x,y)}{k_s(x,y)}\,dy\right)dx \\
&\leq \left[\sup_{x\in\mathbb{R}^n}\int_{k_s(x,y)\neq 0}\frac{k_a^2(x,y)}{k_s(x,y)}\,dy\right]\mathcal{E}(u,u) < \infty,
\end{aligned}$$

where $\frac{1}{2}\mathcal{E}(u,u)$ is the symmetric part of $\eta(u,u)$ given by (2.2). On the other hand, for any $r>0$, it holds that

$$\|I_x\|^2_{L^2(dx)} \le \|\mathbf{1}_{B_{2r}(0)}(x)I_x\|^2_{L^2(dx)} + \|\mathbf{1}_{B^c_{2r}(0)}(x)I_x\|^2_{L^2(dx)}.$$

First,

$$\begin{aligned}
&\|\mathbf{1}_{B_{2r}(0)}(x)I_x\|^2_{L^2(dx)}\\
&= \int_{|x|\le 2r}\left(\mathrm{PV}\int \big(u(y)-u(x)\big)k_s(x,y)\,dy\right)^2 dx\\
&\le 8\left(\|u\|_\infty \vee \|\nabla^2 u\|_\infty\right)^2 \int_{|x|\le 2r}\left(\int \left(1\wedge |x-y|^2\right)k_s(x,y)\,dy\right)^2 dx\\
&\quad + \|\nabla u\|^2_\infty \int_{|x|\le 2r}\left(\int_{0<|z|\le 1} |z|\,|k_s(x,x+z)-k_s(x,x-z)|\,dz\right)^2 dx\\
&< \infty.
\end{aligned}$$

In the equality above we have used (2.9), and the inequality above follows from (2.5) and (2.6). Pick $r>1$ large enough such that $\operatorname{supp} u \subset B_r(0)$. We get that

$$\begin{aligned}
\|\mathbf{1}_{B^c_{2r}(0)}(x)I_x\|^2_{L^2(dx)} &= \int_{|x|\ge 2r}\left(\mathrm{PV}\int \big(u(y)-u(x)\big)k_s(x,y)\,dy\right)^2 dx\\
&= \int_{|x|\ge 2r}\left(\int u(y)k_s(x,y)\,dy\right)^2 dx\\
&= \int_{|x|\ge 2r}\left(\int_{|y|\le r} u(y)k_s(x,y)\,dy\right)^2 dx\\
&\le \|u\|^2_\infty \int\left(\int_{|x-y|>r} \mathbf{1}_{B_r(0)}(y)k_s(x,y)\,dy\right)^2 dx,
\end{aligned}$$

where the second equality follows again from (2.9).

If $x\mapsto \int_{|y-x|\ge 1} k_s(x,y)\,dy \in L^2(dx)$, then $\|\mathbf{1}_{B^c_{2r}(0)}(x)I_x\|^2_{L^2(dx)}<\infty$.

If $x\mapsto \int_{|y-x|\ge 1} k_s(x,y)\,dy \in L^\infty(dx)$, then, by the Cauchy-Schwarz inequality,

$$\begin{aligned}
&\|u\|^2_\infty \int\left(\int_{|x-y|>r} \mathbf{1}_{B_r(0)}(y)k_s(x,y)\,dy\right)^2 dx\\
&\le \|u\|^2_\infty \int\left(\int_{|x-y|>r} \mathbf{1}_{B_r(0)}(y)k_s(x,y)\,dy\right)\left(\int_{|x-y|>r} k_s(x,y)\,dy\right) dx\\
&\le \|u\|^2_\infty \left[\sup_{x\in\mathbb{R}^n}\int_{|x-y|>r} k_s(x,y)\,dy\right]\iint_{|x-y|>r} \mathbf{1}_{B_r(0)}(y)k_s(x,y)\,dy\,dx\\
&= \|u\|^2_\infty \left[\sup_{x\in\mathbb{R}^n}\int_{|x-y|>r} k_s(x,y)\,dy\right]\int \mathbf{1}_{B_r(0)}(x)\int_{|x-y|>r} k_s(x,y)\,dy\,dx\\
&< \infty,
\end{aligned}$$

where in the equality above we have used the symmetry of $k_s(x,y)\,dy\,dx$, and the last inequality follows from (A0). This also gives us that $\|\mathbf{1}_{B_{2r}^c(0)}(x)I_x\|^2_{L^2(dx)} < \infty$. The required assertion follows from all the conclusions above. □

3. The Adjoint of a Lévy Type Operator on $\mathbb{R}^n$

Assume that $E = \mathbb{R}^n$ is equipped with the Euclidean metric $d(x,y) = |x-y|$ and Lebesgue measure $m(dx) = dx$ as reference measure. If k is symmetric, then $k_a(x,y) = 0$ and Proposition 2.1 is identical with [8, Theorem 2.2]. As shown by Theorem 2.2, in this case the Cauchy principal value integral in the representation of L can be rewritten as an absolutely convergent integral if we introduce a regularizing term in the integrand, and so the expression of L becomes (2.8), which is a kind of symmetric Lévy type operator, see [10]. This observation enables us to consider the (formal) adjoint of general (not necessarily symmetric) Lévy type operators.

Let $C_c^\infty(\mathbb{R}^n)$ be the space of smooth functions with compact support on $\mathbb{R}^n$. For $f \in C_c^\infty(\mathbb{R}^n)$, define the following Lévy type operator

$$\begin{aligned} Lf(x) = &\int_{z\neq 0}\Big(f(x+z) - f(x) - \nabla f(x)\cdot z\mathbf{1}_{\{|z|\le 1\}}\Big)k(x,x+z)\,dz \\ &+ \frac{1}{2}\int_{0<|z|\le 1} z\,\big(k(x,x+z) - k(x,x-z)\big)\,dz\cdot\nabla f(x), \end{aligned} \tag{3.1}$$

where $\int_{z\neq 0}\big(1\wedge|z|^2\big)k(x,x+z)\,dz < \infty$ and

$$\int_{0<|z|\le 1} |z|\big|k(x,x+z) - k(x,x-z)\big|\,dz < \infty$$

for all $x \in \mathbb{R}^n$.

We will now present an explicit expression of the (formal) adjoint of the operator L. To state our result, we need a few assumptions. As before we write k_s and k_a for the symmetric and antisymmetric parts of j, *i.e.*

$$k_s(x,y) := \tfrac{1}{2}\big(k(x,y) + k(y,x)\big) \quad \text{and} \quad k_a(x,y) := \tfrac{1}{2}\big(k(x,y) - k(y,x)\big).$$

For $x, z \in \mathbb{R}^n$, we denote by

$$j^*(x,z) := \big|k(x,x+z) - k(x,x-z)\big| + \big|k(x+z,x) - k(x-z,x)\big|.$$

$$x \mapsto \int \big(1\wedge(y-x)^2\big)k_s(x,y)\,dy \in L^2_{\mathrm{loc}}(dx); \tag{H1}$$

$$x \mapsto \int_{\{y\in\mathbb{R}^n\,:\,|y-x|\ge 1\}} k_s(x,y)\,dy \in L^2(dx)\cup L^\infty(dx); \tag{H2}$$

$$x \mapsto \int_{0<|z|\le 1} |z|j^*(x,z)\,dz \in L^2_{\mathrm{loc}}(dx); \tag{H3}$$

$$\sup_{x\in\mathbb{R}^n}\int_{k_s(x,y)\neq 0}\frac{k_a(x,y)^2}{k_s(x,y)}\,dy < \infty; \tag{H4}$$

$$\sup_{x\in K}\sup_{\epsilon>0}\left|\int_{|y-x|\geq\varepsilon} k_a(x,y)\,dy\right| < \infty \quad \text{for every compact} \quad K\subset\mathbb{R}^n. \tag{H5}$$

Note that (H1) is just (2.5), which implies (A0). (H2) is (2.7), and (H4) is the same as (2.1). (H3) implies

$$x\mapsto\int_{0<|z|\leq 1}|z|\big|k_s(x,x+z)-k_s(x,x-z)\big|\,dz\in L^2_{\mathrm{loc}}(dx), \tag{3.2}$$

which is just (2.6). Although (H5) has no direct counterpart in Section 2, it is satisfied when (A1) and (A2) hold with $\gamma=1$ (as it is assumed in [3, Section 3]). As in the proof of Theorem 2.2, we can easily obtain that, under (H1)–(H3), the operator L given by (3.1) maps $C_c^\infty(\mathbb{R}^n)$ into $L^2(\mathbb{R}^n)$.

For any $f\in C_c^\infty(\mathbb{R}^n)$, define

$$\begin{aligned}\Lambda f(x):=&\int_{z\neq 0}\big(f(x+z)-f(x)-\nabla f(x)\cdot z\mathbf{1}_{\{|z|\leq 1\}}\big)\,k(x+z,x)\,dz\\&+\frac12\int_{0<|z|\leq 1}z\,\big(k(x+z,x)-k(x-z,x)\big)\,dz\cdot\nabla f(x).\end{aligned}\tag{3.3}$$

Theorem 3.1. *Assume that* (H1)–(H5) *hold.*

(i) *The operators* $(L,C_c^\infty(\mathbb{R}^n))$ *and* $(\Lambda,C_c^\infty(\mathbb{R}^n))$ *given by* (3.1) *and* (3.3), *respectively, are pre-generators corresponding to the semi-Dirichlet forms* $(\eta_L,\mathcal{F}^0\times\mathcal{F}^0)$ *and* $(\eta_\Lambda,\mathcal{F}^0\times\mathcal{F}^0)$, *generated by the kernels* $k(x,y)$ *and* $k(y,x)$ *on* $L^2(\mathbb{R}^n)$ *in the sense of* (2.3) *with* $D(L)=D(\Lambda)=C_c^\infty(\mathbb{R}^n)$.

(ii) *Let* $(L^*,C_c^\infty(\mathbb{R}^n))$ *be the dual operator for* L. *Then, for any function* $f\in C_c^\infty(\mathbb{R}^n)$

$$L^*f(x)=\Lambda f(x)+\kappa(x)f(x), \tag{3.4}$$

where $\kappa(x)$ *is a Borel measurable function on* $\mathbb{R}^n$ *such that the (signed) measure* $\kappa(x)\,dx$ *is the vague limit of the sequence of bounded (signed) measures* $\big\{\big(-2\int_{|x-y|>1/m}k_a(x,y)\,dy\big)\,dx\big\}_{m\in\mathbb{N}}$. *In particular, it holds that* $L^*(C_c^\infty(\mathbb{R}^n))\subset L^2(\mathbb{R}^n)$.

By (H1), (H3) and (H5), the operator L^* given by (3.4) is well defined on $C_c^\infty(\mathbb{R}^n)$. From the proof of Theorem 2.2, we know that, under (H1)–(H3) and (H5), the operator L^* maps $C_c^\infty(\mathbb{R}^n)$ into $L^2(\mathbb{R}^n)$. In particular, if j is symmetric, *i.e.* $k(x,y)=k(y,x)$ for all $x,y\in\mathbb{R}^n$, Theorem 3.1 shows that $L=\Lambda=L^*$ on $C_c^\infty(\mathbb{R}^n)$; that is, L defined by (3.1) is a symmetric Lévy type operator. For further details we refer to [8, Theorem 2.2] and [10, Theorem 1.2].

Proof of Theorem 3.1. The proof is divided into four steps. Throughout the proof we fix some $f,g\in C_c^\infty(\mathbb{R}^n)$. Note that the conditions on $k(x,y)$ in assumptions (H1)-(H5) are symmetric with respect to x, $y\in\mathbb{R}^n$. Therefore it is enough to prove part (i) for the operator L.

Step 1: For any $\varepsilon>0$, we define

$$\begin{aligned}L_\varepsilon f(x):=&\int_{|z|\geq\varepsilon}\Big(f(x+z)-f(x)-\nabla f(x)\cdot z\mathbf{1}_{\{|z|\leq 1\}}\Big)k(x,x+z)\,dz\\&+\frac12\int_{\varepsilon\leq|z|\leq 1}z\,\big(k(x,x+z)-k(x,x-z)\big)\,dz\cdot\nabla f(x)\end{aligned}$$

$$= \int_{|z|\geq\varepsilon} \Big(f(x+z) - f(x)\Big) k(x, x+z)\, dz$$
$$- \frac{1}{2} \int_{\varepsilon\leq|z|\leq 1} z\left(k(x,x+z) + k(x,x-z)\right) dz \cdot \nabla f(x).$$

Since for every $x \in \mathbb{R}^n$,

$$\int_{\varepsilon\leq|z|\leq 1} z\left(k(x,x+z) + k(x,x-z)\right) dz = 0,$$

we get

$$L_\varepsilon f(x) = \int_{|z|\geq\varepsilon} \Big(f(x+z) - f(x)\Big) k(x,x+z)\, dz;$$

due to (H1), $L_\varepsilon f \in L^2_{\text{loc}}(\mathbb{R}^n)$. Recall that $\langle f, g\rangle_{L^2} = \int f(x)g(x)\,dx$ denotes the inner product in $L^2(\mathbb{R}^n)$. We have

$$\begin{aligned} \langle L_\varepsilon f, g\rangle_{L^2} &= \int g(x) \int_{|z|\geq\varepsilon} (f(x+z) - f(x)) k(x,x+z)\, dz\, dx \\ &= \int g(x) \int_{|y-x|\geq\varepsilon} (f(y) - f(x)) k(x,y)\, dy\, dx. \end{aligned} \tag{3.5}$$

Note that under (H1)–(H3), for any $f \in C_c^\infty(\mathbb{R}^n)$, we can prove

$$\lim_{\varepsilon\to 0} \|L_\varepsilon f - Lf\|_{L^2} = 0, \tag{3.6}$$

using similar estimates as in the proof of Theorem 2.2. This along with (3.5), (H4) and Theorem 2.1 yield (i) for the operator L.

Step 2: We will now prove (ii). We begin with showing that the limit

$$A(f,g) := -\frac{1}{2} \lim_{\varepsilon\to 0} \Big[\langle L_\varepsilon f, g\rangle_{L^2} + \langle f, L_\varepsilon g\rangle_{L^2}\Big] \tag{3.7}$$

exists. Observe that

$$\begin{aligned} &\langle L_\varepsilon f, g\rangle_{L^2} + \langle f, L_\varepsilon g\rangle_{L^2} \\ &\quad= \Big[\int g(x) \int_{|y-x|\geq\varepsilon} (f(y) - f(x)) k(x,y)\, dy\, dx \\ &\qquad + \int f(x) \int_{|y-x|\geq\varepsilon} (g(y) - g(x)) k(x,y)\, dy\, dx\Big] \\ &\quad= \iint_{|y-x|\geq\varepsilon} \Big((f(y) - f(x))g(x) + (g(y) - g(x))f(x)\Big) k(x,y)\, dx\, dy \\ &\quad= \iint_{|y-x|\geq\varepsilon} \Big((f(y) - f(x))g(x) + (g(y) - g(x))f(x)\Big) k_s(x,y)\, dx\, dy \\ &\qquad + \iint_{|y-x|\geq\varepsilon} \Big((f(y) - f(x))g(x) + (g(y) - g(x))f(x)\Big) k_a(x,y)\, dx\, dy \\ &\quad=: I_1^\varepsilon + I_2^\varepsilon. \end{aligned}$$

If we change the roles of x and y in the expression of I_1^ε, we get

$$I_1^\varepsilon = -\iint_{|y-x|\ge\varepsilon} \Big(\big(f(y)-f(x)\big)g(y) + \big(g(y)-g(x)\big)f(y)\Big)k_s(x,y)\,dx\,dy,$$

which, if added to the original expression for I_1^ε, yields that

$$I_1^\varepsilon = -\iint_{|y-x|\ge\varepsilon} \big(f(y)-f(x)\big)\big(g(y)-g(x)\big)k_s(x,y)\,dx\,dy.$$

Because of (H1) we find

$$I_1 := \lim_{\varepsilon\to 0} I_1^\varepsilon = -\iint_{x\ne y} \big(f(y)-f(x)\big)\big(g(y)-g(x)\big)k_s(x,y)\,dx\,dy. \tag{3.8}$$

On the other hand, we see as in the proof of Theorem 2.1, that under (H1) and (H4), the limit $I_2 := \lim_{\varepsilon\to 0} I_2^\varepsilon$ exists and

$$I_2 = \iint \Big(\big(f(y)-f(x)\big)g(x) + \big(g(y)-g(x)\big)f(x)\Big)k_a(x,y)\,dx\,dy \tag{3.9}$$

with an absolutely convergent integral.

Step 3: According to [8, Theorem 2.2], the assumptions (H1) and (H3) imply that

$$-\frac{1}{2}\,I_1 = -\langle \widetilde{L}f, g\rangle_{L^2}, \tag{3.10}$$

where

$$\begin{aligned}\widetilde{L}f(x) := &\int_{z\ne 0} \Big(f(x+z)-f(x)-\nabla f(x)\cdot z\mathbf{1}_{\{|z|\le 1\}}\Big)k_s(x,x+z)\,dz\\ &+\frac{1}{2}\int_{0<|z|\le 1} z\,\big(k_s(x,x+z)-k_s(x,x-z)\big)\,dz\cdot\nabla f(x).\end{aligned}$$

With the same reasoning as above, the proof of Theorem 2.2 shows that, under (H1)—(H3), the operator $\widetilde{L}$ maps $C_c^\infty(\mathbb{R}^n)$ into $L^2(\mathbb{R}^n)$.

If we exchange the roles of x and y in the expression of I_2, we get because of the antisymmetry of k_a

$$I_2 = \iint_{x\ne y} \Big(\big(f(y)-f(x)\big)g(y) + \big(g(y)-g(x)\big)f(y)\Big)k_a(x,y)\,dx\,dy$$

and if we add this to the original expression for I_2, we see

$$I_2 = \iint_{x\ne y} \Big(f(y)g(y)-f(x)g(x)\Big)k_a(x,y)\,dx\,dy.$$

In the same way we find that

$$\begin{aligned}I_2^\varepsilon &= \iint_{|y-x|\ge\varepsilon} \Big(f(y)g(y)-f(x)g(x)\Big)k_a(x,y)\,dx\,dy\\ &= \iint_{|y-x|\ge\varepsilon} f(y)g(y)\,k_a(x,y)\,dx\,dy - \iint_{|y-x|\ge\varepsilon} f(x)g(x)\,k_a(x,y)\,dx\,dy\end{aligned}$$

$$= -\iint_{|y-x|\geq\varepsilon} f(x)g(x)\,k_a(x,y)\,dx\,dy - \iint_{|y-x|\geq\varepsilon} f(x)g(x)\,k_a(x,y)\,dx\,dy$$
$$= -2\int f(x)g(x)\Big[\int_{|y-x|\geq\varepsilon} k_a(x,y)\,dy\Big]dx.$$

Since $f, g \in C_c^\infty(\mathbb{R}^n)$ are arbitrary and since the limit $\lim_{\varepsilon\to 0} I_2^\varepsilon = I_2$ exists, we will see from Lemma 3.1 below that the vague limit of the family of (signed) measures $\big\{\big(-2\int_{|x-y|>\varepsilon} k_a(x,y)\,dy\big)\,dx\big\}_{\varepsilon>0}$, $\varepsilon\to 0$ exists and possesses a density function $\kappa(x)$ with respect to Lebesgue measure.

Thus, by (H5) again, for all $f, g \in C_c^\infty(\mathbb{R}^n)$,

$$I_2 = \int f(x)g(x)\,\kappa(x)\,dx = \langle \kappa f, g\rangle_{L^2}. \tag{3.11}$$

Step 4: Formally, $\Lambda + L = 2\tilde{L}$, and so the formal adjoint L^* of the operator L satisfies

$$\langle L^* f, g\rangle_{L^2} = \langle f, Lg\rangle_{L^2}, \qquad f, g \in C_c^\infty(\mathbb{R}^n).$$

On the other hand, according to (3.6), for any $f, g \in C_c^\infty(\mathbb{R}^n)$,

$$\begin{aligned} A(f,g) &= -\frac{1}{2}\lim_{\varepsilon\to 0}\Big[\langle L_\varepsilon f, g\rangle_{L^2} + \langle f, L_\varepsilon g\rangle_{L^2}\Big] \\ &= -\frac{1}{2}\Big[\langle Lf, g\rangle_{L^2} + \langle f, Lg\rangle_{L^2}\Big]. \end{aligned} \tag{3.12}$$

Combining (3.7)–(3.12), we have

$$\langle L^* f, g\rangle_{L^2} + \langle Lf, g\rangle_{L^2} = \langle f, Lg\rangle_{L^2} + \langle Lf, g\rangle_{L^2} = \Big(2\langle \widetilde{L} f, g\rangle_{L^2} + \langle \kappa f, g\rangle_{L^2}\Big).$$

Therefore,

$$L^* = (2\widetilde{L} + \kappa) - L$$

which is what we have claimed.

Step 5: Since κ is locally essentially bounded and $\Lambda(C_c^\infty(\mathbb{R}^n)) \subset L^2(\mathbb{R}^n)$, we see that L^* maps $C_c^\infty(\mathbb{R}^n)$ into $L^2(\mathbb{R}^n)$. □

Lemma 3.1. *Under the assumption* (H5), *$\kappa(x)$ given in the right side of the dual operator L^* defined by* (3.4) *is a locally essentially bounded measurable function such that $\kappa(x)\,dx$ is the vague limit of the sequence of (signed) measures* $\big\{\big(-2\int_{|x-y|>1/m} k_a(x,y)\,dy\big)\,dx\big\}_{m\in\mathbb{N}}$.

Proof. In the proof of Theorem 3.1 we have seen that the limit

$$\mu(fg) := \lim_{\varepsilon\to 0} I_2^\varepsilon = -2\lim_{\varepsilon\to 0}\int f(x)g(x)\Big[\int_{|y-x|\geq\varepsilon} k_a(x,y)\,dy\Big]dx$$

exists for all $f, g \in C_c^\infty(\mathbb{R}^n)$. Thus, μ is an element in $\mathcal{D}' = (C_c^\infty(\mathbb{R}^n))^*$. Since, under (H5), $|\int \varphi(x)\,\mu(dx)| \leq c_K\|\varphi\|_\infty$ for all continuous functions u with support in the compact set K, μ is a distribution of order zero, hence a (signed) Radon measure. Thus, it remains to show that μ is absolutely continuous with respect to Lebesgue measure dx.

For any $\varphi \in C_c^\infty(\mathbb{R}^n)$, we know from (H5) that for all $x \in \operatorname{supp}\varphi$,

$$\begin{aligned} f_\varphi(x) &:= \limsup_{m\to\infty} \left| \int_{|x-y|>1/m} k_a(x,y)\,dy \right| \\ &\le \sup_{x\in\operatorname{supp}\varphi}\ \sup_{m\in\mathbb{N}} \left| \int_{|x-y|>1/m} k_a(x,y)\,dy \right| \\ &\le C_\varphi < \infty. \end{aligned}$$

Therefore, by (a variant of) Fatou's lemma, we have

$$\begin{aligned} \left| \int \varphi(x)\,\mu(dx) \right| &\le \limsup_{m\to\infty} \int |\varphi(x)| \left| \int_{|x-y|>1/m} k_a(x,y)\,dy \right| dx \\ &\le \int |\varphi(x)| \limsup_{m\to\infty} \left| \int_{|x-y|>1/m} k_a(x,y)\,dy \right| dx \\ &\le \int |\varphi(x)| f_\varphi(x)\,dx. \end{aligned}$$

This proves that $\mu(dx) \ll dx$. Since $f_\varphi(x) \le C_\varphi$ on $\operatorname{supp}\varphi$, $\kappa := d\mu/dx$ is locally bounded. □

We have seen in Theorem 3.1 that, under the assumptions (H1)–(H5), the operator $(L, C_c^\infty(\mathbb{R}^n))$ given by (3.1) generates a semi-Dirichlet form $(\eta, \mathcal{F}^0 \times \mathcal{F}^0)$ on $L^2(\mathbb{R}^n)$. Therefore, there exists a unique sub-Markov semigroup $\{T_t\}_{t\ge 0}$ associated with the form $(\eta, \mathcal{F}^0 \times \mathcal{F}^0)$ on $L^2(\mathbb{R}^n)$. The dual semigroup $\{T_t^*\}_{t\ge0}$ is positivity preserving but it is, in general, not sub-Markovian, see [3, Section 1] and the references therein. The following result provides a sufficient condition for the sub-Markov property of the dual semigroup.

Corollary 3.1. *Let L, Λ and L^* be as in Theorem 3.1. If the killing term $\kappa(x)$ in the representation* (3.4) *of the dual operator L^* is non-positive, then the dual semigroup $\{T_t^*\}_{t\ge0}$ corresponding to the semi-Dirichlet form $(\eta, \mathcal{F}^0 \times \mathcal{F}^0)$ is sub-Markovian.*

Proof. We have seen in Theorem 3.1 that the operator $(\Lambda, C_c^\infty(\mathbb{R}^n))$ given by (3.3) is the pre-generator of the semi-Dirichlet form $(\eta_\Lambda, \mathcal{F}^0 \times \mathcal{F}^0)$ generated by the kernel $k(y,x)$ on $L^2(\mathbb{R}^n)$, *cf.* (2.3) with $D(\Lambda) = C_c^\infty(\mathbb{R}^n)$. Therefore, there is a unique sub-Markov semigroup $\{T_t^\Lambda\}_{t\ge0}$ associated with the form $(\eta_\Lambda, \mathcal{F}^0 \times \mathcal{F}^0)$ on $L^2(\mathbb{R}^n)$. It is well-known, *cf.* for example [6, Corollary, page 334], that $\{T_t^\Lambda\}_{t\ge0}$ is sub-Markovian if, and only if, the corresponding form satisfies

$$Nu \in \mathcal{F}^0 \quad \text{and} \quad \eta_\Lambda(Nu, u - Nu) \ge 0 \quad \text{for all } u \in \mathcal{F}^0$$

where $Nu = u^+ \wedge 1$ is the normal contraction. By definition, the form corresponding to $\{T_t^*\}_{t\ge0}$ is $\eta_{L^*}(u,v) = \eta_L(v,u)$, and the proof of Theorem 3.1 shows that $L^*u = \Lambda u + \kappa u$ and $\eta_{L^*}(u,v) = \eta_\Lambda(u,v) - \langle \kappa u, v\rangle_{L^2}$. Thus,

$$\begin{aligned} \eta_{L^*}(Nu, u - Nu) &= \eta_\Lambda(Nu, u-Nu) - \langle \kappa Nu, u - Nu\rangle_{L^2} \\ &\ge -\langle \kappa Nu, u - Nu\rangle_{L^2} \\ &= -\int \kappa(x)(u^+(x) \wedge 1)(u(x)-1)^+\,dx \ge 0. \end{aligned}$$

Therefore, $\{T_t^*\}_{t\ge0}$ is sub-Markovian. □

4. Example: Stable-like Processes

Let $E = \mathbb{R}^n$, and $m(dx) = dx$ be Lebesgue measure on $\mathbb{R}^n$. Consider the following integro-differential operator

$$Lu(x) = w(x) \int_{z \neq 0} \Big(u(x+z) - u(x) - \nabla u(x) \cdot z \mathbf{1}_{\{|z| \leq 1\}}(z) \Big) |z|^{-n-\alpha(x)}\, dz$$

for $u \in C_c^\infty(\mathbb{R}^n)$. The weight function $w(x)$ is chosen in such a way that

$$w(x) = \alpha(x) 2^{\alpha(x)-1} \frac{\Gamma\big(\frac{1}{2}\alpha(x) + \frac{1}{2}n\big)}{\pi^{n/2}\, \Gamma\big(1 - \frac{1}{2}\alpha(x)\big)},$$

and so $Le_\xi(x) = -|\xi|^{\alpha(x)} e_\xi(x)$, where $e_\xi(x) = e^{ix\cdot\xi}$, see *e.g.* [2, Exercise 18.23, page 184]. With this norming, L can be written as a pseudo-differential operator $-p(x, D)$ with the symbol $-|\xi|^{\alpha(x)}$,

$$Lu(x) = \int e^{ix\cdot\xi} |\xi|^{\alpha(x)} \widehat{u}(\xi)\, d\xi = -(-\Delta)^{\alpha(x)} u(x),$$

and this shows that $L = -(-\Delta)^{\alpha(x)}$ is a stable-like operator in the sense of Bass [1]. Note that, the stable-like operator is a special case of the Lévy type operator given by (3.1), and in this case, since for any x, $z \in \mathbb{R}^d$,

$$k(x, x+z) = w(x)|z|^{-n-\alpha(x)},$$

the second term on the right hand side of (3.1) vanishes automatically.

For $r > 0$, define

$$\beta(r) := \sup_{|x-y| \leq r} |\alpha(x) - \alpha(y)|.$$

Proposition 4.1. *Let $L = -(-\Delta)^{\alpha(x)}$ and suppose that there exist α_1, α_2 such that*

$$0 < \alpha_1 \leq \alpha(x) \leq \alpha_2 < 2 \quad \textit{for all} \quad x \in \mathbb{R}^n,$$

and

$$\int_0^1 \frac{\big(\beta(r)|\log r|\big)^2}{r^{1+\alpha_2}}\, dr < \infty.$$

(i) *The operator $(L, C_c^\infty(\mathbb{R}^n))$ generates a regular lower bounded semi-Dirichlet form on $L^2(\mathbb{R}^n)$ associated with the kernel*

$$k(x, y) = w(x)|x-y|^{-n-\alpha(x)}.$$

(ii) *If for all compact sets $K \subset \mathbb{R}^n$*

$$\sup_{x\in K} \sup_{\epsilon > 0} \left| \int_{|z|\geq\varepsilon} \left(\frac{w(x+z)}{|z|^{n+\alpha(x+z)}} - \frac{w(x)}{|z|^{n+\alpha(x)}} \right) dz \right| \leq c_K < \infty, \tag{4.1}$$

then the formal adjoint of L is given by

$$L^* f(x) = \int_{z\neq 0} \Big(f(x+z) - f(x) - \nabla f(x) \cdot z \boldsymbol{1}_{\{|z|\leq 1\}} \Big) \frac{w(x+z)}{|z|^{n+\alpha(x+z)}}\, dz$$

$$+\frac{1}{2}\int_{0<|z|\le 1} z\left(\frac{w(x+z)}{|z|^{n+\alpha(x+z)}}-\frac{w(x-z)}{|z|^{n+\alpha(x-z)}}\right)dz\cdot\nabla f(x)$$
$$+\kappa(x)f(x).$$

where $\kappa(x)$ is the density of a (signed) Radon measure on $\mathbb{R}^n$ such that $\kappa(x)\,dx$ is the vague limit of the sequence of (signed) measures

$$\left\{\left(\int_{|z|>1/m}\left(\frac{w(x+z)}{|z|^{n+\alpha(x+z)}}-\frac{w(x)}{|z|^{n+\alpha(x)}}\right)dz\right)dx\right\}_{m\in\mathbb{N}}.$$

Proof. We check the conditions of Theorems 2.1 and 3.1. Set

$$k(x,y)=w(x)|x-y|^{-n-\alpha(x)}.$$

Then,

$$k_s(x,y)=\frac{1}{2}\Big(w(x)|x-y|^{-n-\alpha(x)}+w(y)|x-y|^{-n-\alpha(y)}\Big),$$
$$k_a(x,y)=\frac{1}{2}\Big(w(x)|x-y|^{-n-\alpha(x)}-w(y)|x-y|^{-n-\alpha(y)}\Big).$$

From the definition of $w(x)$ it is easy to see that there exist constants $c_j>0, j=1,2,3$, such that for any $x,y\in\mathbb{R}^n$,

$$c_1\le w(x)\le c_2,\qquad |w(x)-w(y)|\le c_3|\alpha(x)-\alpha(y)|,$$

see [3, proof of Proposition 5.1].

(i) Since (A0) is obviously satisfied, we only have to verify (2.1) of Theorem 2.1. We have

$$\sup_{x\in\mathbb{R}^n}\int_{|x-y|\ge 1/e}\frac{k_a^2(x,y)}{k_s(x,y)}\,dy\le\sup_{x\in\mathbb{R}^n}\int_{|x-y|\ge 1/e}k_s(x,y)\,dy$$
$$\le c\int_{1/e}^{\infty}r^{-1-\alpha_1}\,dr<\infty.$$

To see

$$\sup_{x\in\mathbb{R}^n}\int_{|x-y|\le 1/e}\frac{k_a^2(x,y)}{k_s(x,y)}\,dy<\infty$$

we write

$$k_a(x,y)=\frac{1}{2}\Big[(w(x)-w(y))|x-y|^{-n-\alpha(x)}$$
$$+w(y)|x-y|^{-n}\left(|x-y|^{-\alpha(x)}-|x-y|^{-\alpha(y)}\right)\Big].$$

Then

$$\int_{|x-y|\le 1/e}\frac{(w(x)-w(y))^2\left(|x-y|^{-n-\alpha(x)}\right)^2}{w(x)|x-y|^{-n-\alpha(x)}+w(y)|x-y|^{-n-\alpha(y)}}\,dy$$
$$\le c\int_{|x-y|\le 1/e}\frac{(\alpha(x)-\alpha(y))^2}{|x-y|^{n+\alpha(x)}}\,dy$$

$$
\begin{aligned}
&\le c \int_{|x-y|\le 1/e} \frac{\beta^2(|x-y|)}{|x-y|^{n+\alpha_2}}\, dy \\
&= c \int_0^{1/e} \frac{\beta^2(r)}{r^{1+\alpha_2}}\, dr \\
&\le c \int_0^{1/e} \frac{\beta^2(r)|\log r|^2}{r^{1+\alpha_2}}\, dr < \infty.
\end{aligned}
$$

Since

$$
|x-y|^{-\alpha(x)} - |x-y|^{-\alpha(y)} = \int_{\alpha(y)}^{\alpha(x)} |x-y|^{-u} \log |x-y|^{-1}\, du,
$$

we obtain for all $|x-y| \le 1/e$,

$$
\begin{aligned}
&\left(|x-y|^{-\alpha(x)} - |x-y|^{-\alpha(y)}\right)^2 \\
&\qquad \le \left(\log |x-y|^{-1}\right)^2 \left(\alpha(x) - \alpha(y)\right)^2 |x-y|^{-2(\alpha(x)\vee\alpha(y))}.
\end{aligned}
$$

Therefore,

$$
\begin{aligned}
&\int_{|x-y|\le 1/e} \frac{w(y)^2\, |x-y|^{-2n} \left(|x-y|^{-\alpha(x)} - |x-y|^{-\alpha(y)}\right)^2}{w(x)|x-y|^{-n-\alpha(x)} + w(y)|x-y|^{-n-\alpha(y)}}\, dy \\
&\le c \int_{|x-y|\le 1/e} \frac{\left(\alpha(x)-\alpha(y)\right)^2 \left(\log |x-y|^{-1}\right)^2}{|x-y|^n} \\
&\qquad\qquad \times \frac{|x-y|^{-2(\alpha(x)\vee\alpha(y))}}{w(x)|x-y|^{-\alpha(x)} + w(y)|x-y|^{-\alpha(y)}}\, dy \\
&\le c \int_{|x-y|\le 1/e} \frac{\left(\alpha(x)-\alpha(y)\right)^2 \left(\log |x-y|^{-1}\right)^2}{|x-y|^n} |x-y|^{-(\alpha(x)\vee\alpha(y))}\, dy \\
&\le c \int_{|x-y|\le 1/e} \frac{\beta^2(|x-y|)\left(\log |x-y|^{-1}\right)^2}{|x-y|^{n+\alpha_2}}\, dy \\
&\le c \int_0^{1/e} \frac{(\beta(r)|\log r|)^2}{r^{1+\alpha_2}}\, dr,
\end{aligned}
$$

and (2.1) follows.

(ii) Clearly, the conditions (H1) and (H2) are satisfied. From part (i), we know that (H4) is also valid. Since (H5) is covered by (4.1), we only have to verify (H3). For all $x \in \mathbb{R}^n$,

$$
\begin{aligned}
&\int_{|x-y|\le 1/e} |x-y| \left| \frac{w(y)}{|x-y|^{n+\alpha(y)}} - \frac{w(x)}{|x-y|^{n+\alpha(x)}} \right| dy \\
&\qquad \le \int_{|x-y|\le 1/e} |x-y|\, |w(x)-w(y)|\, |x-y|^{-n-\alpha(y)}\, dy \\
&\qquad\quad + \int_{|x-y|\le 1/e} |x-y|\, w(x) \left| |x-y|^{-n-\alpha(y)} - |x-y|^{-n-\alpha(x)} \right| dy.
\end{aligned}
$$

With similar arguments as in the proof of part (i) we see that the right hand side of the inequality above is smaller than

$$c\int_0^{1/e}\frac{\beta(r)(1+|\log r|)}{r^{\alpha_2}}\,dr \le 2c\int_0^{1/e}\frac{\beta(r)\,|\log r|}{r^{\alpha_2}}\,dr < 2c\int_0^{1}\frac{\beta(r)\,|\log r|}{r^{\alpha_2}}\,dr.$$

Pick $\gamma<1/2$ such that $\alpha_2\le 1+2\gamma<2$. By the Cauchy-Schwarz inequality, we find

$$\int_0^1\frac{\beta(r)|\log r|}{r^{\alpha_2}}\,dr \le \left(\int_0^1\frac{1}{r^{2\gamma}}\,dr\right)^{1/2}\left(\int_0^1\frac{\beta(r)^2|\log r|^2}{r^{2\alpha_2-2\gamma}}\,dr\right)^{1/2} \le \frac{1}{\sqrt{1-2\gamma}}\left(\int_0^1\frac{\beta(r)^2|\log r|^2}{r^{1+\alpha_2}}\,dr\right)^{1/2}.$$

□

We close with this section with some comments on related results in [3, 9] and our Proposition 4.1.

Remark 4.1. (i) Assume that for $r\to 0$

$$\beta(r)\asymp r^{\beta},\quad \beta>\alpha_2/2,\qquad \text{or}\qquad \beta(r)\asymp r^{\alpha_2/2}|\log r|^{\varepsilon},\quad \varepsilon<-3/2.$$

Then Proposition 4.1(i) applies and shows that the operator $(L, C_c^\infty(\mathbb{R}^n))$ generates a regular lower bounded semi-Dirichlet form on $L^2(\mathbb{R}^n)$. This is, in particular, the case if the index function $\alpha(x)$ is locally Lipschitz continuous. Thus, Proposition 4.1(i) improves [3, Proposition 5.1] where the following assumptions are used: *there exist positive constants* α_1, α_2, M *and* δ *such that for* x, $y\in\mathbb{R}^n$,

$$0<\alpha_1\le\alpha(x)\le\alpha_2<2\quad \textit{with}\quad \alpha_2<1+\alpha_1/2$$

and

$$|\alpha(x)-\alpha(y)|\le M|x-y|^{\delta}\quad \textit{with}\quad 0<\frac12(2\alpha_2-\alpha_1)<\delta\le 1.$$

(ii) With essentially the same calculations as in the proof of Proposition 4.1(ii) we can see that

$$\int_0^1\frac{\beta(r)|\log r|}{r^{1+\alpha_2}}\,dr<\infty$$

guarantees that

$$\int\left(\frac{w(x+z)}{|z|^{n+\alpha(x+z)}}-\frac{w(x)}{|z|^{n+\alpha(x)}}\right)dz\quad\text{exists}$$

as a Lebesgue integral. This is, for example, the case if *there exist positive constants* α_1,α_2,δ *and* M *such that* $0<\alpha_1\le\alpha_2<1$, $\delta\in(\alpha_2,1]$, *and for all* $x,y\in\mathbb{R}^n$,

$$\alpha_1\le\alpha(x)\le\alpha_2\quad \textit{and}\quad |\alpha(x)-\alpha(y)|\le M|x-y|^{\delta}.$$

This shows that Proposition 4.1(ii) covers the conclusion in [9, Remark 4].

(iii) The assumption (4.1) ensures the existence of the killing term $\kappa(x)$ in the dual operator L^* given in Proposition 4.1(ii). We will claim that *under the conditions of Proposition 4.1 and if the index function* $\alpha(x)$ *also belongs to* $C_b^2(\mathbb{R}^n)$, *then* (4.1) *is fulfilled.* Indeed, in this case it is easily seen from the definition of $w(x)$ that $\alpha \in C_b^2(\mathbb{R}^n)$ entails $w \in C_b^2(\mathbb{R}^n)$. For any fixed $z \in \mathbb{R}^n$ with $z \neq 0$, set $f_z(x) := w(x)|z|^{-n-\alpha(x)}$. Then, for any $x,\ z \in \mathbb{R}^n$,

$$\nabla f_z(x) = |z|^{-n-\alpha(x)}\big[\nabla w(x) - \big(\log|z|\big) w(x)\nabla\alpha(x)\big],$$

and there is a constant $c_1 > 0$ such that for any $x,\ z \in \mathbb{R}^n$ with $0 < |z| < 1$,

$$\big|\nabla^2 f_z(x)\big| \le c_1 |z|^{-n-\alpha_2}\big(1 + \log^2|z|\big).$$

Thus, by Taylor's formula there exists $\theta \in (0,1)$ such that

$$f_z(x+y) - f_z(x) = \nabla f_z(x)\cdot y + \frac{1}{2} y^\top \nabla^2 f_z(x+\theta y) y \qquad \text{for all } x, y, z \in \mathbb{R}^n.$$

Therefore, for any $x \in \mathbb{R}^n$, by setting $y = z$ in the equality above,

$$\begin{aligned}
&\sup_{\epsilon>0}\left|\int_{|z|\ge\varepsilon}\Big(\frac{w(x+z)}{|z|^{n+\alpha(x+z)}} - \frac{w(x)}{|z|^{n+\alpha(x)}}\Big)\,dz\right| \\
&\le \sup_{\epsilon>0}\left|\int_{\varepsilon\le|z|<1}\big(f_z(x+z) - f_z(x)\big)\,dz\right| + \int_{|z|\ge1} f_z(x)\,dz \\
&\quad + \int_{|z|\ge1} f_z(x+z)\,dz \\
&\le c_3 + \sup_{\epsilon>0}\left|\int_{\varepsilon\le|z|<1}\Big(\nabla f_z(x)\cdot z + \frac{1}{2} z^\top \nabla^2 f_z(x+\theta z) z\Big)\,dz\right| \\
&= c_3 + \frac{1}{2}\sup_{\epsilon>0}\left|\int_{\varepsilon\le|z|<1}\Big(z^\top \nabla^2 f_z(x+\theta z) z\Big)\,dz\right| \\
&\le c_3 + c_4 \sup_{\epsilon>0}\int_{\varepsilon\le|z|<1} |z|^{-n-\alpha_2+2}\big(1+\log^2|z|\big)\,dz \\
&\le c_3 + c_5\int_0^1 \frac{1+\log^2 r}{r^{\alpha_2-1}}\,dr =: c_6 < \infty.
\end{aligned}$$

The desired assertion follows.

Acknowledgements

Financial support through DFG (grant Schi 419/5-2) and DAAD (PPP Kroatien) (for René L. Schilling), the National Natural Science Foundation of China (No. 11201073) and the Program for New Century Excellent Talents in Universities of Fujian (No. JA12053) (for Jian Wang) is gratefully acknowledged. We would like to thank Professors M. Fukushima and T. Uemura for helpful discussions on an earlier version of this paper.

References

1. Bass, R.: Uniqueness in law for pure jump Markov processes, *Probab. Theory Related Fields* **79** (1988), 271–287.
2. Berg, C. and Forst, G.: *Potential Theory on Locally Compact Abelian Groups*, Springer-Verlag, Ergebnisse Math. Grenzgeb. vol. **87**, Berlin 1975.
3. Fukushima, M. and Uemura, T.: Jump-type Hunt processes generated by lower bounded semi-Dirichlet forms, *Ann. Probab.* **40** (2012), 858–889.

4. Fukushima, M., Oshima, Y. and Takeda, M.: *Dirichlet Forms and Symmetric Markov Processes*, de Gruyter Studies in Math. vol. **19**, Walter de Gruyter, Berlin 1994.
5. Hu, Z.C., Ma. Z.M. and Sun, W.: Extensions of Lévy-Khintchine formula and Beurling-Deny formula in semi-Dirichlet forms setting, *J. Funct. Anal.* **239** (2006), 179–213.
6. Kunita, H.: Sub-Markov semi-groups in Banach lattices. In: *Proc. Intnl. Conf. on Functional Analysis and Related Topics (Tokyo 1969)*, Univ. of Tokyo Press, Tokyo 1970, 332–342.
7. Ma, Z.M. and Röckner, M.: *Introduction to the Theory of (Non-Symmetric) Dirichlet Forms*, Universitext, Springer-Verlag, Berlin 1992.
8. Schilling, R.L. and Uemura, T.: On the Feller property of Dirichlet forms generated by pseudo differential operators, *Tohoku Math. J.* **59** (2007), 401–422.
9. Uemura, T.: A remark on non-local operators with variable order, *Osaka J. Math.* **46** (2009), 503–514.
10. Wang, J.: Symmetric Lévy type operator, *Acta Math. Sin. Eng. Ser.* **25** (2009), 39–46.

Chapter 26

Ultracontractivity for non-symmetric Markovian semigroups

Ichiro Shigekawa* †

We discuss the ultracontractivity for non-symmetric Markovian semigroup. To do this, we make use of the symmetric bilinear form. For an application, we deal with the exponential convergence rate of the semigroup. We show that a non-symmetric semigroup converges faster than the symmetric semigroup. We give same examples of diffusions on Riemannian manifolds.

Contents

1. Introduction

Let $(M, \mathscr{B}, m)$ be a measure space. Suppose we are given a Markovian semigroup $\{T_t\}$ in $L^2(m)$. The semigroup $\{T_t\}$ is ultracontractive if T_t is bounded from L^1 to L^∞ for any $t > 0$. When semigroups are symmetric, several criterions for the ultracontractivity are already known. For example, for any constant $\mu > 0$, the followings are equivalent.

(i) There exists a constant c_1 such that

$$\|T_t f\|_\infty \le c_1 t^{-\mu/2} \|f\|_1, \quad \forall f \in L^1, \ \forall t > 0. \tag{1.1}$$

---

*Dedicated to Professor Masatoshi Fukushima on his 80th birthday

†Department of Mathematics, Kyoto University, Kyoto 606-8502, JAPAN, ichiro@math.kyoto-u.ac.jp

(ii) There exists a constant c_2 such that

$$\|f\|_2^{2+4/\mu} \le c_2\,\mathscr{E}(f,f)\,\|f\|_1^{4/\mu} \quad \forall f \in \mathrm{Dom}(\mathscr{E}) \cap L^1. \tag{1.2}$$

The inequality (1.2) is called the Nash inequality. If $\mu > 2$, then the following condition is also equivalent.

(iii) There exists a constant c_2 such that

$$\|f\|_{2\mu/(\mu-2)}^2 \le c_3\,\mathscr{E}(f,f), \quad \forall f \in \mathrm{Dom}(\mathscr{E}). \tag{1.3}$$

There are many applications of the ultracontractivity, for example, it bears the boundedness of fundamental solution. In this paper, we will extend it to the case of non-symmetric semigroup.

The organization of the paper is as follows. In Section 2, we give a general framework of non-symmetric semigroup in a Hilbert space. Even though the semigroup is non-symmetric, we associate a symmetric bilinear form to the semigroup. We also give a sufficient condition for the Markovian property of the associated bilinear form. In Section 3, we give a sufficient condition for the ultracontractivity of the semigroup. We also give a necessary condition. Using the ultracontractivity, we can deal with the exponential convergence rate of the semigroup. We show that a non-symmetric semigroup converges faster than the symmetric semigroup. Our method is to reduce the problem to the L^2 case which is shown in Hwang, Hwang-Ma and Sheu [4]. In Section 4, we treat a semigroup associated with a Dirichlet form that satisfies the sector condition. In Section 5, we discuss a semigroup with an invariant measure. In this case, we can give an estimate of $T_t f - m(f)$, where $m(f)$ stands for the integral with respect to the invariant measure m. This estimates is closely related to the spectral gap. In Section 6 and Section 7, we give some examples. In Section 6, we discuss a non-symmetric diffusion on a Riemannian manifold with boundary and in Section 7 we discuss a non-symmetric diffusion on a Riemannian manifold without boundary, in particular, in the case that the Riemannian manifold is compact.

Lastly the author would like to thank the anonymous referee for careful reading.

2. Non-symmetric semigroups

In this section, we give a general framework of non-symmetric semigroups.

2.1. *Symmetric bilinear forms of semigroups on a Hilbert space*

Our interest is in Markovian semigroups, but we start with a general setting in a Hilbert space H. We assume that H is a real Hilbert space and denote its inner product by $(\ ,\ )_H$. Let $\{T_t\}$ be a C_0 semigroup with a generator A. We assume that $A - \gamma$ is dissipative for some constant γ, i.e., $((A-\gamma)x, x)_H \le 0$ for all $x \in \mathrm{Dom}(A)$. All semigroups discussed in this paper are assumed to satisfy these conditions. We denote the dual semigroup of $\{T_t\}$ by $\{T_t^*\}$ and its generator by A^*.

If A is symmetric, there exists an associated symmetric bilinear form $\mathscr{E}$. In fact, we may take a closure of the following bilinear form:

$$\mathscr{E}(u,v) = -(Au, v), \quad u, v \in \mathrm{Dom}(A). \tag{2.1}$$

Here Dom stands for the domain. Even in non-symmetric case, we can introduce symmetric bilinear form $\tilde{\mathscr{E}}$ as follows:

$$\tilde{\mathscr{E}}(u,v) = -\frac{1}{2}\{(Au,v)+(A^*u,v)\}, \qquad \text{for } u,v \in \mathrm{Dom}(A)\cap\mathrm{Dom}(A^*). \tag{2.2}$$

Then we have

Proposition 2.1. $(\tilde{\mathscr{E}}, \mathrm{Dom}(A)\cap\mathrm{Dom}(A^*))$ *is closable.*

Proof. To show the closability, it suffices to show the following. (See, e.g., Ma-Röckner [5, p. 29, Propositio I.3.4]) If $\{u_n\} \subseteq \mathrm{Dom}(A)\cap\mathrm{Dom}(A^*)$ converges to 0 in H, then $\tilde{\mathscr{E}}(u_n,v)$ converges to 0 for any $v \in \mathrm{Dom}(A)\cap\mathrm{Dom}(A^*)$.

This can be easily seen from

$$2\tilde{\mathscr{E}}(u_n,v) = (Au_n,v)+(A^*u_n,v) = (u_n,A^*v)+(u_n,Av) \to 0.$$

□

So we can take a closure of $\tilde{\mathscr{E}}$, which we also denote by $\tilde{\mathscr{E}}$. The topology of $\mathrm{Dom}(\tilde{\mathscr{E}})$ is induced by the inner product $\tilde{\mathscr{E}}+(\gamma+1)(\ ,\ )_H$. To ensure that $\tilde{\mathscr{E}}$ is rich enough, we assume the following condition (A):

(A) $\mathrm{Dom}(A)\cap\mathrm{Dom}(A^*)$ is dense in $\mathrm{Dom}(A)$ and $\mathrm{Dom}(A^*)$, respectively.

We can easily show the following proposition by noting the topology of $\mathrm{Dom}(\tilde{\mathscr{E}})$ is weaker than the topology of $\mathrm{Dom}(A)$ with the graph norm of A.

Proposition 2.2. *Under the condition (A), we have* $\mathrm{Dom}(A)\cup\mathrm{Dom}(A^*) \subseteq \mathrm{Dom}(\tilde{\mathscr{E}})$. *We also have*

$$\tilde{\mathscr{E}}(u,v) = -\frac{1}{2}\{(Au,v)+(u,Av)\}, \quad \textit{for } u,v \in \mathrm{Dom}(A)$$

and

$$\tilde{\mathscr{E}}(u,v) = -\frac{1}{2}\{(u,A^*v)+(A^*u,v)\}, \quad \textit{for } u,v \in \mathrm{Dom}(A^*).$$

From now on, we always assume the condition (A) and the symmetric bilinear form $\tilde{\mathscr{E}}$ is defined as above.

2.2. *Convex set preserving*

Take any convex set C. We say that an operator S preserves C if $Su \in C$ for any $u \in C$. We say that the semigroup $\{T_t\}$ preserves C if T_t preserves C for all $t \in [0,\infty)$. Let us give a sufficient condition for this. We recall that for any u, there exists a unique shortest point v in C from u, i.e., $\|u-v\| = \inf\{\|u-w\|;\ w \in C\}$. We denote this shortest point v by Pu. Then we have

$$(u-Pu, v-Pu) \leq 0, \quad \forall v \in C. \tag{2.3}$$

By using this, we have the following

Theorem 2.1. *If $\{T_t\}$ and $\{T_t^*\}$ preserve C, then we have that $Pu \in \mathrm{Dom}(\tilde{\mathscr{E}})$ for any $u \in \mathrm{Dom}(\tilde{\mathscr{E}})$. Moreover we have*

$$\tilde{\mathscr{E}}(Pu, u - Pu) \geq 0. \tag{2.4}$$

Hence, the semigroup associated with $\tilde{\mathscr{E}}$ preserves C.

Proof. Take any $u \in \mathrm{Dom}(A) \cap \mathrm{Dom}(A^*)$. Since the semigroup $\{T_t\}$ preserves C, βG_β also preserves C. Here G_β is the resolvent operator. By the identity $AG_\beta = -(1 - \beta G_\beta)$, we have

$$\begin{aligned}(A\beta G_\beta Pu, u - \beta G_\beta Pu) &= -(\beta(1 - \beta G_\beta)Pu, u - \beta G_\beta Pu) \\ &= -\beta((1 - \beta G_\beta)Pu, u - Pu + Pu - \beta G_\beta Pu) \\ &= \beta(\beta G_\beta Pu - Pu, u - Pu) \\ &\quad - \beta((1 - \beta G_\beta)Pu, (1 - \beta G_\beta)Pu) \\ &\leq 0. \quad (\because (2.3))\end{aligned}$$

Hence

$$(A\beta G_\beta Pu, \beta G_\beta Pu) \geq (A\beta G_\beta Pu, u) = (\beta G_\beta Pu, A^*u).$$

From this, we have

$$\tilde{\mathscr{E}}_\gamma(\beta G_\beta Pu, \beta G_\beta Pu) \leq -(\beta G_\beta Pu, A^*u) + \gamma(\beta G_\beta Pu, \beta G_\beta Pu).$$

If we let $\beta \to \infty$, then the right hand side converges to $-(Pu, A^*u) + (Pu, Pu)$ and so we can see the left hand side is bounded. Since $\beta G_\beta Pu$ converges to Pu in H, it follows that $Pu \in \mathrm{Dom}(\tilde{\mathscr{E}})$. Furthermore, we have

$$\begin{aligned}\tilde{\mathscr{E}}_\gamma(Pu, Pu) &\leq \varliminf_{\beta\to\infty} \tilde{\mathscr{E}}_\gamma(\beta G_\beta Pu, \beta G_\beta Pu) \\ &\leq \lim_{\beta\to\infty} \{-(\beta G_\beta Pu, A^*u) + \gamma(\beta G_\beta Pu, \beta G_\beta Pu)\} \\ &= -(Pu, A^*u) + \gamma(Pu, Pu).\end{aligned}$$

By the same argument for A^*, we have

$$\tilde{\mathscr{E}}_\gamma(Pu, Pu) \leq -(Pu, Au) + \gamma(Pu, Pu).$$

Now, by adding them,

$$\begin{aligned}2\tilde{\mathscr{E}}_\gamma(Pu, Pu) &\leq -(Pu, A^*u) - (Pu, Au) + 2\gamma(Pu, Pu) \\ &= 2\tilde{\mathscr{E}}(Pu, u) + 2\gamma(Pu, Pu) \\ &= 2\tilde{\mathscr{E}}_\gamma(Pu, u) - 2\gamma(Pu, u) + 2\gamma(Pu, Pu) \\ &= 2\tilde{\mathscr{E}}_\gamma(Pu, u) + 2\gamma(Pu, Pu - u) \\ &\leq 2\tilde{\mathscr{E}}_\gamma(Pu, Pu)^{1/2}\tilde{\mathscr{E}}_\gamma(u, u)^{1/2} + 2\gamma(Pu, Pu - u).\end{aligned}$$

Thus

$$0 \geq \tilde{\mathscr{E}}_\gamma(Pu, Pu) - \tilde{\mathscr{E}}_\gamma(Pu, Pu)^{1/2}\tilde{\mathscr{E}}_\gamma(u, u)^{1/2} - \gamma(Pu, Pu - u)$$

$$\geq (\tilde{\mathscr{E}}_\gamma(Pu,Pu)^{1/2} - \tilde{\mathscr{E}}_\gamma(u,u)^{1/2}/2)^2 - \tilde{\mathscr{E}}_\gamma(u,u)/4 - \gamma(Pu, Pu-u).$$

This leads to

$$\tilde{\mathscr{E}}_\gamma(Pu,Pu)^{1/2} \leq \tilde{\mathscr{E}}_\gamma(u,u)^{1/2}/2 + \sqrt{\gamma(Pu,Pu-u) + \tilde{\mathscr{E}}_\gamma(u,u)/4}. \tag{2.5}$$

This identity holds for $u \in \mathrm{Dom}(A) \cap \mathrm{Dom}(A^*)$. Since $\mathrm{Dom}(A) \cap \mathrm{Dom}(A^*)$ is dense in $\mathrm{Dom}(\tilde{\mathscr{E}})$, for any $u \in \mathrm{Dom}(A) \cap \mathrm{Dom}(A^*)$, we can take a $\tilde{\mathscr{E}}_\gamma$-convergent sequence $\{u_n\}$ from $\mathrm{Dom}(A) \cap \mathrm{Dom}(A^*)$. Now (2.5) shows that $\{Pu_n\}$ is $\tilde{\mathscr{E}}_\gamma$-bounded. Hence we can take a $\tilde{\mathscr{E}}_\gamma$-weakly convergent subsequence $\{Pu_{n_j}\}$. By the above computation, we have had

$$\tilde{\mathscr{E}}_\gamma(Pu_{n_j}, Pu_{n_j}) \leq \tilde{\mathscr{E}}_\gamma(Pu_{n_j}, u_{n_j}) + \gamma(Pu_{n_j}, Pu_{n_j} - u_{n_j}).$$

Letting $j \to \infty$, we have

$$\begin{aligned}\tilde{\mathscr{E}}_\gamma(Pu,Pu) &\leq \varliminf_{n\to\infty} \tilde{\mathscr{E}}_\gamma(Pu_{n_j}, Pu_{n_j}) \\ &\leq \lim_{j\to\infty} \{\tilde{\mathscr{E}}_\gamma(Pu_{n_j}, u_{n_j}) + \gamma(Pu_{n_j}, Pu_{n_j} - u_{n_j})\} \\ &= \tilde{\mathscr{E}}_\gamma(Pu, u) + \gamma(Pu, Pu-u).\end{aligned}$$

Eventually we arrive at

$$\tilde{\mathscr{E}}(Pu, u - Pu) \geq 0.$$

It is known, see e.g., Ouhabaz [6], that this inequality implies that the semigroup associated with $\tilde{\mathscr{E}}$ preserves C. □

By using this, we can show that if $\{T_t\}$ and $\{T_t^*\}$ are Markovian, then the semigroup associated with $\tilde{\mathscr{E}}$ is Markovian.

3. Ultracontractive semigroups

Let (M, m) be a measure space and suppose a Markovian semigroup $\{T_t\}$ in $L^2(m)$ is given. We assume that the dual semigroup $\{T_t^*\}$ is also a Markovian semigroup. This implies $\{T_t\}$ satisfies the L^1 contraction property, i.e., for $f \geq 0$,

$$\int_M T_t f\, dm \leq \int_M f\, dm. \tag{3.1}$$

The existence of associated Markov process is not necessary. We denote the generator of $\{T_t\}$ by A:

$$T_t = e^{tA}.$$

The generator of $\{T_t^*\}$ is denoted by A^*. We assume the condition (A) in Section 2 for semigroups $\{T_t\}$ and $\{T_t^*\}$. As in Section 2, the associated symmetric bilinear form is denoted by $\tilde{\mathscr{E}}$, which satisfies Markovian property by Theorem 2.1.

Since $\{T_t\}$ is Markovian and L^1-contraction, it induces a C_0 semigroup in $L^p(m)$ for $p \in [1, \infty)$, which is denoted by $\{T_t\}$. The norm in $L^p(m)$ is denoted by $\|\cdot\|_p$ and the inner product in $L^2(m)$ is denoted by $(\ ,\ )_2$ or $(\ ,\ )$ for simplicity. The inner product $(\ ,\ )_2$ sometimes stands for the natural paring between L^p and L^q ($\frac{1}{p} + \frac{1}{q} = 1$).

We are interested in the ultracontractivity of semigroups. This notion is defined as follows:

Definition 3.1. The semigroup $\{T_t\}$ is ultracontractive if T_t is bounded from L^1 to L^∞ for any $t > 0$.

As was mentioned in Section 1, various criterions are well-known. In the sequel, we extend this result to the case that the semigroup is non-symmetric. We mainly follow the proof in the symmetric case (see, e.g., Davies [3]). Our tool is the symmetric bilinear form $\tilde{\mathscr{E}}$ that was introduced in the previous section.

In general, we denote the operator norm of A from L^p to L^q by $\|A\|_{p\to q}$. We define functions as follows:

$$a_t = \|T_t\|_{1\to\infty} = \|T_t^*\|_{1\to\infty}, \tag{3.2}$$
$$b_t = \|T_t\|_{1\to 2} = \|T_t^*\|_{2\to\infty}, \tag{3.3}$$
$$c_t = \|T_t\|_{2\to\infty} = \|T_t^*\|_{1\to 2}. \tag{3.4}$$

These are all increasing in t. Next proposition is easily obtained by the interpolation.

Proposition 3.1. *a_t, b_t, c_t satisfy*

$$a_{2t} \le b_t c_t, \tag{3.5}$$
$$b_t^2 \le a_t, \tag{3.6}$$
$$c_t^2 \le a_t. \tag{3.7}$$

Now we choose a parameter $\mu > 0$ and fix it. The ultracontractivity is formulated as follows: there exists a constant c_1 such that

$$\|T_t f\|_\infty \le c_1 t^{-\mu/2}\|f\|_1, \quad \forall f \in L^1,\ \forall t > 0. \tag{3.8}$$

The following sufficient condition is well-known, see, e.g., Coulhon [1, Proposition II.1].

Theorem 3.1. (3.8) *holds if there exists a constant c_2 such that*

$$\|f\|_2^{2+4/\mu} \le c_2\,\tilde{\mathscr{E}}(f,f)\,\|f\|_1^{4/\mu}, \quad \forall f \in \mathrm{Dom}(\tilde{\mathscr{E}}) \cap L^1(m). \tag{3.9}$$

To show the converse, we impose an additional condition.

Theorem 3.2. *Assume* (3.8) *and that for any $f \in \mathrm{Dom}(A^2)$, we have*

$$(A^2 f, f)_2 + (Af, Af)_2 \ge 0. \tag{3.10}$$

Then there exists a constant c_2 such that

$$\|f\|_2^{2+4/\mu} \le c_2\,\tilde{\mathscr{E}}(f,f)\,\|f\|_1^{4/\mu}, \quad \forall f \in \tilde{\mathscr{E}} \cap L^1. \tag{3.11}$$

Proof. Take any $f \in \mathrm{Dom}(A) \cap L^1$. By Proposition 3.1, we have

$$\|T_t\|_{1\to 2} \le c_1^{1/2} t^{-\mu/4}, \quad \|T_t^*\|_{2\to\infty} \le c_1^{1/2} t^{-\mu/4}$$

and

$$\|T_t^*T_t\|_{1\to\infty} \le c_1 t^{-\mu/2}.$$

Therefore

$$\begin{aligned} c_1 t^{-\mu/2}\|f\|_1^2 &\ge \|T_t^*T_t f\|_\infty \|f\|_1 \\ &\ge (T_t^*T_t f, f)_2 \\ &= (T_t f, T_t f)_2 \\ &= (f,f)_2 + \int_0^t \frac{d}{ds}(T_s f, T_s f)_2\, ds \\ &= (f,f)_2 + 2\int_0^t (AT_s f, T_s f)_2\, ds. \end{aligned}$$

We set $u(t) = (AT_t f, T_t f)_2$ and we will show that $u(t)$ is increasing in t. It is enough to show this when $f \in \mathrm{Dom}(A^2)$. Then, by (3.10), we have

$$u'(t) = (A^2 T_t f, T_t f)_2 + (AT_t f, AT_t f)_2 \ge 0$$

which implies that $u(t)$ is increasing.

Now, since $u(t)$ is increasing, we have

$$c_1 t^{-\mu/2}\|f\|_1^2 \ge (f,f)_2 + 2\int_0^t (Af,f)_2\, ds = (f,f)_2 + 2t(Af,f)_2$$

and

$$\|f\|_2^2 \le -2t(Af,f)_2 + c_1 t^{-\mu/2}\|f\|_1^2.$$

Now, by taking

$$t = \left(-\frac{\|f\|_1^2}{(Af,f)}\right)^{2/(\mu+2)},$$

we have

$$\|f\|_2^2 \le -(2+c_1)(Af,f)_2^{\mu/(\mu+2)}\|f\|_1^{4/(\mu+2)}.$$

Therefore

$$\begin{aligned} \|f\|_2^{2+4/\mu} &\le -(2+c_1)^{(\mu+2)/\mu}(Af,f)_2\|f\|_1^{4/\mu} \\ &= (2+c_1)^{(\mu+2)/\mu}\tilde{\mathscr{E}}(f,f)\|f\|_1^{4/\mu} \end{aligned}$$

which shows (3.11). □

The condition (3.10) above holds if A is normal. In fact

$$\begin{aligned} ((A+A^*)f,(A+A^*)f)_2 &= (Af,Af)_2 + (Af,A^*f)_2 \\ &\quad + (A^*f,Af)_2 + (A^*f,A^*f)_2 \\ &= (Af,Af)_2 + 2(Af,A^*f)_2 + (AA^*f,f)_2 \end{aligned}$$

$$= (Af, Af)_2 + 2(Af, A^*f)_2 + (A^*Af, f)_2$$
$$= 2(Af, Af)_2 + 2(A^2f, f)_2.$$

Let us consider the condition (3.10). If $f \in \mathrm{Dom}(A) \cap \mathrm{Dom}(A^*)$, then (3.10) can be rewritten as

$$0 \le (Af, A^*f)_2 + (Af, Af)_2 = -\tilde{\mathscr{E}}(Af, f) \le -\tilde{\mathscr{E}}(Af, f) - (Af, f)$$
$$= -\tilde{\mathscr{E}}_1(Af, f),$$

which means that A is dissipative with respect to $\tilde{\mathscr{E}}_1$. If A generate a semigroup in $\mathrm{Dom}(\tilde{\mathscr{E}})$, the semigroup must be contractive. So (3.10) corresponds to this kind of property.

When $\mu > 2$, we can give another kind of sufficient condition. Consider the following condition: There exists a constant $c_3 > 0$ such that for any $f \in \mathrm{Dom}(\tilde{\mathscr{E}})$, we have

$$\|f\|_{2\mu/(\mu-2)}^2 \le c_3\, \tilde{\mathscr{E}}(f, f), \quad \forall f \in \mathrm{Dom}(\tilde{\mathscr{E}}). \tag{3.12}$$

Recall that $\tilde{\mathscr{E}}$ is Markovian by Theorem 2.1. So $\tilde{\mathscr{E}}$ is symmetric and Markovian and the condition (3.12) is equivalent to (3.11).

We relax the condition (3.8) so that it holds only for $t > 0$ in the bounded interval $(0, 1]$ as follows:

Corollary 3.1. *Fix $\mu > 0$. If there exists a constant c_2 such that*

$$\|f\|_2^{2+4/\mu} \le c_2(\tilde{\mathscr{E}}(f) + \|f\|_2^2)\|f\|_1^{4/\mu}, \quad \forall f \in \mathrm{Dom}(\tilde{\mathscr{E}}) \cap L_+^1 \tag{3.13}$$

then, there exists a constant c_1 such that

$$\|T_t f\|_\infty \le c_1 t^{-\mu/2}\|f\|_1, \quad \forall f \in L^1,\ \forall t \in (0, 1]. \tag{3.14}$$

Proof. It is enough to consider $A - 1$ in place of A. Since the Dirichlet form associated to $A - 1$ is $\tilde{\mathscr{E}}(f) + \|f\|_2^2$, we have from Theorem 3.1

$$\|T_t f\|_\infty \le c_1' t^{-\mu/2} e^t \|f\|_1, \quad \forall t > 0. \tag{3.15}$$

Now (3.14) easily follows. □

The converse can be proved similarly as follows:

Corollary 3.2. *Fix $\mu > 0$. If there exists a constant c_2 such that*

$$\|T_t f\|_\infty \le c_1 t^{-\mu/2}\|f\|_1, \quad \forall f \in L^1,\ \forall t \in (0, 1] \tag{3.16}$$

and, in addition, there exists a constant $M > 0$ such that

$$((A - M)^2 f, f)_2 + ((A - M)f, (A - M)f)_2 \ge 0, \quad \forall f \in \mathrm{Dom}(A^2) \tag{3.17}$$

then there exists a constant c_2 such that

$$\|f\|_2^{2+4/\mu} \le c_2(\tilde{\mathscr{E}}(f) + \|f\|_2^2)\|f\|_1^{4/\mu}, \quad \forall f \in \mathrm{Dom}(\mathscr{E}) \cap L_+^1. \tag{3.18}$$

Proof. Consider $A - M$ in place of A. From (3.16), we have

$$e^{-Mt}\|T_t f\|_\infty \le c_1' t^{-\mu/2}\|f\|_1, \quad \forall t > 0.$$

Then we can show (3.18) by using Theorem 3.2. □

Corollary 3.1 and Corollary 3.2 give information of the behavior of semigroup as $t \to 0$. Let us consider the case $t \to \infty$, in particular, the exponential decay of the semigroup.

3.1. L^2 setting

We first summarize the fundamental facts in L^2 setting. Here we don't need the Markovian property of the semigroup.

Let us consider the following property:

$$\|f\|_2^2 \le \lambda^{-1}\tilde{\mathscr{E}}(f,f) \tag{3.19}$$

for some $\lambda > 0$. This condition is equivalent to the following:

$$\|T_t f\|_2^2 \le e^{-2\lambda t}\|f\|_2^2, \quad \forall t > 0. \tag{3.20}$$

In fact, since

$$\frac{d}{dt}\|T_t f\|_2^2 = 2(AT_t f, T_t f) = 2(AT_t f, T_t f) = -2\tilde{\mathscr{E}}(T_t f, T_t f),$$

(3.19) brings

$$\frac{d}{dt}\|T_t f\|_2^2 = -2\tilde{\mathscr{E}}(T_t f, T_t f) \le -2\lambda\|T_t f\|_2^2$$

which leads to (3.20). Conversely, if we assume (3.20), by differentiating it at $t = 0$, we have

$$-2\tilde{\mathscr{E}}(f,f) \le -2\lambda\|f\|_2^2.$$

We denote the maximal λ satisfying (3.19) by λ_{P}, i.e.,

$$\lambda_{\mathrm{P}} = \inf\left\{\frac{\tilde{\mathscr{E}}(f,f)}{\|f\|_2^2};\ f \ne 0\right\}. \tag{3.21}$$

This constants depends only on $\tilde{\mathscr{E}}$ and so it reflects the property of $\tilde{\mathscr{E}}$. But we connect this to the property of the semigroup $\{T_t\}$.

Next we define the constant $\lambda_{2\to2}$ by

$$\lambda_{2\to2} = -\lim_{t\to\infty}\frac{1}{t}\log\|T_t\|_{2\to2}. \tag{3.22}$$

It is well-known that

$$\lim_{n\to\infty}\|T_t^n\|_{2\to2}^{1/n} = e^{-t\lambda_{2\to2}}. \tag{3.23}$$

This is called the spectral radius because of the identity

$$e^{-t\lambda_{2\to2}} = \sup\{|\lambda|;\ \lambda \in \sigma(T_t)\}, \tag{3.24}$$

(see, e.g., Reed-Simon [8, Theorem VI.6]). Here $\sigma(T_t)$ stands for the spectrum of T_t.

If we assume (3.20), we have

$$\|T_t\|_{2\to 2} \le e^{-t\lambda_{\mathrm{P}}}$$

and hence we have

$$\lambda_{\mathrm{P}} \le \lambda_{2\to 2}. \tag{3.25}$$

We introduce another index λ_{B} as follows:

$$\lambda_{\mathrm{B}} = \inf \Re(\sigma(-A)). \tag{3.26}$$

We will see the relations among them.

Theorem 3.3. *We have the following inequality:*

$$\lambda_{\mathrm{P}} \le \lambda_{2\to 2} \le \lambda_{\mathrm{B}} \tag{3.27}$$

Proof. We have seen $\lambda_{\mathrm{P}} \le \lambda_{2\to 2}$.

Let us show $\lambda_{2\to 2} \le \lambda_{\mathrm{B}}$, but this is a general fact. It is known as spectral mapping theorem, which states

$$e^{t\sigma(A)} \subseteq \sigma(T_t)$$

(see e.g., Pazy [7, p. 45, Theorem 2.3], Davies [2, p. 42, Theorem 2.16].) This leads

$$\lambda_{2\to 2} \le \lambda_{\mathrm{B}}.$$

□

If the semigroup is symmetric, these three indices coincide as follows:

Theorem 3.4. *If the semigroup T_t is symmetric, we have*

$$\lambda_{\mathrm{P}} = \lambda_{2\to 2} = \lambda_{\mathrm{B}}. \tag{3.28}$$

This theorem can be extended to the case of normal operator as in the next theorem, and so we omit the proof.

Theorem 3.5. *If A is normal, we have*

$$\lambda_{\mathrm{P}} = \lambda_{2\to 2} = \lambda_{\mathrm{B}}. \tag{3.29}$$

Proof. We use the spectral decomposition. By using this, we have

$$A = \int_{\sigma(A)} \lambda dE_\lambda.$$

Here E_λ is the spectral resolution of A. The semigroup is expressed as

$$T_t = \int_{\sigma(A)} e^{t\lambda} dE_\lambda.$$

Hence

$$\begin{aligned}\|T_t f\|^2 &= \int_{\sigma(A)} |e^{t\lambda}|^2 d(E_\lambda f, f) \\ &= \int_{\sigma(A)} e^{2t\Re\lambda} d(E_\lambda f, f) \\ &\le \int_{\sigma(A)} e^{-2t\lambda_{\mathrm{B}}} d(E_\lambda f, f) \\ &= e^{-2t\lambda_{\mathrm{B}}} \|f\|^2.\end{aligned}$$

Here we used $\Re\lambda \le -\lambda_{\mathrm{B}}$ for $\lambda \in \sigma(A)$. This leads to $\lambda_{\mathrm{P}} \ge \lambda_{\mathrm{B}}$, which is the desired result. □

We denote the indices for $\tilde{\mathscr{E}}$ with $\tilde{}$, for instance $\tilde{\lambda}_{2\to 2}$, etc. Then the inequality $\lambda_{\mathrm{P}} \le \lambda_{2\to 2}$ can be written as

$$\tilde{\lambda}_{2\to 2} \le \lambda_{2\to 2},$$

which is regarded as a comparison between $\{T_t\}$ and the symmetric semigroup associated with $\tilde{\mathscr{E}}$. This inequality means the semigroup $\{T_t\}$ converges faster than symmetric one.

3.2. *Connection to the ultracontractivity*

We introduce another index $\lambda_{1\to\infty}$ as follows:

$$\lambda_{1\to\infty} = -\varlimsup_{t\to\infty} \frac{1}{t} \log \|T_t\|_{1\to\infty} \tag{3.30}$$

The index for $\tilde{\mathscr{E}}$ can be defined and we denote it by $\tilde{\lambda}_{1\to\infty}$.

Theorem 3.6. *Fix $\mu > 0$. We consider the following conditions.*

(i) There exists a constant c_1 such that

$$\|T_t f\|_\infty \le c_1 t^{-\mu/2} \|f\|_1, \quad \forall f \in L^1,\ \forall t \in (0, 1]. \tag{3.31}$$

(ii) There exists a constant c_2 such that

$$\|f\|_2^{2+(4/\mu)} \le c_2(\tilde{\mathscr{E}}(f, f) + \|f\|_2^2)\|f\|_1^{4/\mu}, \quad \forall f \in \mathrm{Dom}(\tilde{\mathscr{E}}) \cap L^1. \tag{3.32}$$

If we assume (i), then we have $\lambda_{1\to\infty} \ge \lambda_{2\to 2}$. Hence, if we assume (i) and $\lambda_{2\to 2} > 0$, then (3.8) *holds. If we assume (ii) and $\lambda_{\mathrm{P}} > 0$, then* (3.9) *in Theorem 3.1 holds.*

Proof. We first assume (i). Then, by virtue of Proposition 3.1, we have

$$\|T_t\|_{1\to\infty} \le \|T_1\|_{2\to\infty} \|T_{t-2}\|_{2\to 2} \|T_1\|_{1\to 2} \le \|T_1\|_{1\to\infty}^{1/2} \|T_{t-2}\|_{2\to 2} \|T_1\|_{1\to\infty}^{1/2},$$

which leads to $\lambda_{1\to\infty} \ge \lambda_{2\to 2}$. If, in addition, $\lambda_{2\to 2} > 0$, then, for any $\varepsilon > 0$, we have

$$\|T_t\|_{2\to 2} \le K \exp\{-(\lambda_{2\to 2} - \varepsilon)t\}, \quad \forall t > 0$$

for some $K > 0$. Hence

$$\|T_t\|_{1\to\infty} \le \|T_1\|_{1\to\infty} K \exp\{-(\lambda_{2\to 2} - \varepsilon)(t-2)\}.$$

Combining this with (3.31), we have

$$\|T_t f\|_\infty \le c_1' t^{-\mu/2} \|f\|_1, \quad \forall f \in L^1,\ \forall t > 0,$$

which is (3.8).

If we assume (ii) and $\lambda_{\mathrm{P}} > 0$, then (3.9) easily follows from the definition of $\lambda_{\mathrm{P}} > 0$. □

From now on, we assume that m is a *finite* measure.

Proposition 3.2. *Assume that m is a finite measure. Then, we have*

$$\lambda_{1\to\infty} \le \lambda_{2\to 2}. \tag{3.33}$$

If, in addition, $\lambda_{1\to\infty} > -\infty$, then the equality holds in (3.33).

Proof. Since m is a finite measure, the inclusions $\iota\colon L^2 \hookrightarrow L^1$ and $\iota^*\colon L^\infty \hookrightarrow L^2$ are bounded. Hence

$$\begin{aligned}\|T_t\|_{2\to 2} &= \|\iota^* T_t \iota\|_{2\to 2} \\ &\le \|\iota^*\|_{\infty\to 2} \|T_t\|_{1\to\infty} \|\iota\|_{2\to 1}.\end{aligned}$$

From this, $\lambda_{1\to\infty} \le \lambda_{2\to 2}$ follows.

If $\lambda_{1\to\infty} > -\infty$, then $\|T_T\|_{1\to\infty} < \infty$ for sufficiently large T. So

$$\begin{aligned}\|T_t\|_{1\to\infty} &\le \|T_T\|_{2\to\infty} \|T_{t-2T}\|_{2\to 2} \|T_T\|_{1\to 2} \\ &\le \|T_T\|_{1\to\infty} \|\iota\|_{2\to 1} \|T_{t-2T}\|_{2\to 2} \|\iota^*\|_{\infty\to 2} \|T_T\|_{1\to\infty}.\end{aligned}$$

Now $\lambda_{1\to\infty} \ge \lambda_{2\to 2}$ easily follows. □

Using this proposition, we have the following theorem.

Theorem 3.7. *Fix $\mu > 0$ and assume that m is a finite measure. If there exists a constant c_2 such that*

$$\|f\|_2^{2+(4/\mu)} \le c_2(\tilde{\mathscr{E}}(f,f) + \|f\|_2^2)\|f\|_1^{4/\mu}, \quad \forall f \in \mathrm{Dom}(\tilde{\mathscr{E}}) \cap L^1, \tag{3.34}$$

then $\lambda_{1\to\infty} = \lambda_{2\to 2}$, and hence we have,

$$\lambda_{1\to\infty} \ge \tilde{\lambda}_{1\to\infty}. \tag{3.35}$$

Proof. Assume (3.34). Then, by Corollary 3.1, (3.31) holds and hence we have $\lambda_{1\to\infty} > -\infty$. By using Proposition 3.2, we have $\lambda_{1\to\infty} = \lambda_{2\to 2}$. Similarly we have $\tilde{\lambda}_{1\to\infty} = \tilde{\lambda}_{2\to 2}$.

On the other hand, from Theorem 3.3 and Theorem 3.4, we have $\tilde{\lambda}_{2\to 2} \le \lambda_{2\to 2}$. Therefore

$$\tilde{\lambda}_{1\to\infty} = \tilde{\lambda}_{2\to 2} \le \lambda_{2\to 2} = \lambda_{1\to\infty},$$

which is the desired result. □

4. Dirichlet forms satisfying the sector condition

In this section, we discuss in the framework of non-symmetric Dirichlet form satisfying sector condition. Let $(M, \mathscr{B}, m)$ be a measure space and we are given a Dirichlet form $\mathscr{E}$ in $L^2(m)$. In this section, we do not assume that m is finite. For $\alpha \geq 0$, we define

$$\mathscr{E}_\alpha(f, g) = \mathscr{E}(f, g) + \alpha(f, g)_2. \tag{4.1}$$

We assume that $\mathscr{E}$ is positive definite (i.e., $\mathscr{E}(u, u) \geq 0$) and Markovian. We further assume that $\mathscr{E}$ satisfies the following weak sector condition: there exists a constant $K > 0$ such that

$$|\mathscr{E}(f, g)| \leq K\mathscr{E}_1(f, f)^{1/2}\mathscr{E}_1(f, f). \tag{4.2}$$

We define $\hat{\mathscr{E}}$ by

$$\hat{\mathscr{E}}(f, g) = \mathscr{E}(g, f), \tag{4.3}$$

which is assumed to be Markovian. The symmetric part of $\mathscr{E}$ is defined by

$$\tilde{\mathscr{E}} = \frac{1}{2}\{\mathscr{E} + \hat{\mathscr{E}}\}. \tag{4.4}$$

The semigroups associated with $\mathscr{E}$ and $\hat{\mathscr{E}}$ is denoted by $\{T_t\}$ and $\{T_t^*\}$, respectively. Their generators are denoted by A and A^*, respectively. All results in Section 3 hold by the same proof. Moreover Theorem 3.2 can be strengthened. In fact, we have the following.

Theorem 4.1. *Fix $\mu > 0$. The following two conditions are equivalent to each other.*

(i) There exists a constant c_1 such that

$$\|T_t f\|_\infty \leq c_1 t^{-\mu/2}\|f\|_1, \quad \forall f \in L^1, \forall t \in (0, 1]. \tag{4.5}$$

(ii) There exists a constant c_2 such that

$$\|f\|_2^{2+4/\mu} \leq c_2\,(\tilde{\mathscr{E}}(f, f) + \|f\|_2^2)\,\|f\|_1^{4/\mu}, \quad \forall f \in \mathrm{Dom}(\tilde{\mathscr{E}}) \cap L^1(m). \tag{4.6}$$

If $\mu > 2$ the following condition is also equivalent.

(iii) There exists a constant c_3 such that

$$\|f\|_{2\mu/(\mu-2)}^2 \leq c_3(\tilde{\mathscr{E}}(f, f) + \|f\|_2^2), \quad \forall \in \mathrm{Dom}(\tilde{\mathscr{E}}). \tag{4.7}$$

Proof. The implication (ii) $\Rightarrow$ (i) was shown in Theorem 3.1. Let us show the converse. $V = \mathrm{Dom}(\tilde{\mathscr{E}})$ is a Hilbert space with the inner product $\tilde{\mathscr{E}}_1 = \tilde{\mathscr{E}} + (\ ,\ )_2$. Take $\gamma > 0$ to be large enough. Then the semigroup $\{e^{-\gamma t}T_t\}$ is a C_0 semigroup in the Hilbert space V. (see, e.g. Tanabe [9, Theorem 3.6.1]) Moreover, (4.5) can be written as

$$\|e^{-\gamma t}T_t f\|_\infty \leq Ct^{-\mu/2}\|f\|_1, \quad \forall t > 0.$$

Take any $f \in \mathrm{Dom}(A) \cap L^1$. By Proposition 3.1,

$$\|e^{-\gamma t}T_t\|_{1\to 2} \leq C^{1/2}t^{-\mu/4}, \quad \forall t > 0,$$
$$\|e^{-\gamma t}T_t^*\|_{2\to\infty} \leq C^{1/2}t^{-\mu/4}, \quad \forall t > 0,$$

and so we have

$$\|e^{-2\gamma t}T_t^*T_t\|_{1\to\infty} \le Ct^{-\mu/2}.$$

Hence

$$\begin{aligned}
&Ct^{-\mu/2}\|f\|_1^2\\
&\quad\ge \|e^{-2\gamma t}T_t^*T_tf\|_\infty\|f\|_1\\
&\quad\ge (e^{-2\gamma t}T_t^*T_tf,f)\\
&\quad= (e^{-2\gamma t}T_tf,T_tf)\\
&\quad= (f,f)+\int_0^t\frac{d}{ds}(e^{-2\gamma s}T_sf,T_sf)\,ds\\
&\quad= (f,f)-2\int_0^t\{\tilde{\mathscr{E}}(e^{-\gamma t}T_sf,e^{-\gamma t}T_sf)+2\gamma(e^{-\gamma t}T_sf,e^{-\gamma t}T_sf)_2\}\,ds.
\end{aligned}$$

Since $e^{-\gamma t}T_t$ is bounded in V, we can take $K>0$ such that

$$\tilde{\mathscr{E}}(e^{-\gamma t}T_sf,e^{-\gamma t}T_sf) \le K\{\tilde{\mathscr{E}}(f,f)+(f,f)_2\}.$$

Therefore

$$Ct^{-\mu/2}\|f\|_1^2 \ge (f,f)-2(K+\gamma)\{\tilde{\mathscr{E}}(f,f)+(f,f)_2\}t.$$

Thus we have

$$\|f\|_2^2 \le 2(K+\gamma)\{\tilde{\mathscr{E}}(f,f)+(f,f)_2\}t+Ct^{-\mu/2}\|f\|_1^2.$$

Now we take

$$t=\left(\frac{\|f\|_1^2}{\tilde{\mathscr{E}}(f,f)+(f,f)_2}\right)^{2/(\mu+2)}$$

and get

$$\|f\|_2^2 \le (2K+2\gamma+C)\{\tilde{\mathscr{E}}(f,f)+(f,f)_2\}^{\mu/(\mu+2)}\|f\|_1^{4/(\mu+2)}.$$

Hence

$$\|f\|_2^{2+4/\mu} \le (2K+2\gamma+C)^{(\mu+2)/\mu}\{\tilde{\mathscr{E}}(f,f)+(f,f)_2\}\|f\|_1^{4/\mu},$$

which is a desired result.

The equivalence (ii) and (iii) is the known fact. □

5. Semigroup with an invariant measure

We go back to the framework in Section 3. Further we assume that m is an invariant probability measure for $\{T_t\}$ and $\{T_t\}$, i.e.,

$$\int_M T_tf\,dm=\int_M T_t^*f\,dm=\int_M f\,dm. \tag{5.1}$$

This is equivalent to $T_t 1 = T_t^* 1 = 1$. Further we have $1 \in \mathrm{Dom}(A) \cap \mathrm{Dom}(A^*)$ and $A1 = A^*1 = 0$. We also assume the uniqueness of the solution $Af = 0$. So 0 is an eigenvalue of A with multiplicity 1.

5.1. L^2 *theory*

First we summarize fundamental notions in L^2. The variance Var is defined by

$$\mathrm{Var}(f) = \int_M f(x)^2\, m(dx) - \left(\int_M f(x)\, m(dx)\right)^2. \tag{5.2}$$

The mean $m(f)$ is defined by

$$m(f) = \int_M f(x)\, m(dx).$$

Then we have $\mathrm{Var}(f) = \|f - m(f)\|_2^2$. Here $m(f)$ means a constant function. So we regard m as an operator in L^2.

For $\lambda > 0$, the following inequality is called **the Poincaré inequality**.

$$\mathrm{Var}(f) = \|f - m(f)\|_2^2 \leq \lambda^{-1} \tilde{\mathscr{E}}(f, f). \tag{5.3}$$

It is easy to show that this is equivalent to

$$\|T_t f - m(f)\|_2^2 \leq e^{-2\lambda t} \|f - m(f)\|_2^2, \quad \forall t > 0. \tag{5.4}$$

The largest λ satisfying (5.3) is denoted by γ_{P}, which we call the Poincaré constant.

$$\gamma_{\mathrm{P}} = \inf\left\{ \frac{\tilde{\mathscr{E}}(f, f)}{\|f - m(f)\|_2^2};\ f \neq m(f) \right\}. \tag{5.5}$$

Note that this constant depends only on $\tilde{\mathscr{E}}$. We also introduce the following index $\gamma_{2\to 2}$:

$$\gamma_{2\to 2} = -\lim_{t\to\infty} \frac{1}{t} \log \|T_t - m\|_{2\to 2}. \tag{5.6}$$

As was mentioned in Section 3, we have

$$\lim_{n\to\infty} \|(T_t - m)^n\|_{2\to 2}^{1/n} = e^{-\gamma_{2\to 2} t} \tag{5.7}$$

and

$$e^{-\gamma_{2\to 2} t} = \sup\{|\lambda|;\ \lambda \in \sigma(T_t - m)\}. \tag{5.8}$$

Therefore we have

$$\gamma_{\mathrm{P}} \leq \gamma_{2\to 2}. \tag{5.9}$$

Let us introduce another index, which we will call the spectral gap.

$$-\gamma_{\mathrm{SG}} = \sup \Re(\sigma(A) \setminus \{0\}). \tag{5.10}$$

Repeating the argument in Section 3, we have the following theorems.

Theorem 5.1. *We have*

$$\gamma_{\mathrm{P}} \leq \gamma_{2\to 2} \leq \gamma_{\mathrm{SG}}. \tag{5.11}$$

Theorem 5.2. *If A is normal, in particular, A is self-adjoint, we have*

$$\gamma_{\mathrm{P}} = \gamma_{2\to 2} = \gamma_{\mathrm{SG}}. \tag{5.12}$$

We can define indices associated with $\tilde{\mathscr{E}}$, which we denote with $\tilde{}$. For example

$$\tilde{\gamma}_{2\to 2} = -\lim \frac{1}{t} \log \|\tilde{T}_t - m\|_{2\to 2},$$

where $\tilde{T}_t$ is the semigroup associated with $\tilde{\mathscr{E}}$. Now the inequality $\gamma_{\mathrm{P}} \leq \gamma_{2\to 2}$ brings

$$\tilde{\gamma}_{2\to 2} \leq \gamma_{2\to 2}.$$

This inequality was essentially obtained in Huang, Huang-Ma and Sheu [4] but the formulation is little bit different.

5.2. *The ultracontractivity*

We now introduce the index related to the ultracontractivity. Define $\gamma_{1\to\infty}$ by

$$\gamma_{1\to\infty} = -\lim_{t\to\infty} \frac{1}{t} \log \|T_t - m\|_{1\to\infty}. \tag{5.13}$$

Set

$$a_t = \|T_t - m\|_{1\to\infty}.$$

Then

$$\begin{aligned} a_{t+s} &= \|T_{t+s} - m\|_{1\to\infty} = \|(T_t - m)\iota(T_s - m)\|_{1\to\infty} \\ &\leq \|T_t - m\|_{1\to\infty}\|\iota\|_{\infty\to 1}\|T_s - m\|_{1\to\infty} = a_t a_s. \end{aligned} \tag{5.14}$$

Here $\iota \colon L^\infty \to L^1$ denotes the inclusion map. This shows the existence of the limit in (5.13).

We have the following theorem by the same proof as in Proposition 3.2.

Proposition 5.1. *We have*

$$\gamma_{1\to\infty} \leq \gamma_{2\to 2}. \tag{5.15}$$

If $\gamma_{1\to\infty} > -\infty$, then the identity holds in (5.15).

In the case that the invariant measure exists, the ultracontractivity is formulated as follows: There exists a constant c_1 such that

$$\|T_t f - m(f)\|_\infty \leq c_1 t^{-\mu/2}\|f\|_1, \quad \forall t > 0. \tag{5.16}$$

Moreover the Nash inequality is formulated as follows: There exists a constant c_2 such that

$$\|f - m(f)\|_2^{2+(4/\mu)} \leq c_2 \tilde{\mathscr{E}}(f, f)\|f - m(f)\|_1^{4/\mu}, \quad \forall f \in \mathrm{Dom}(\tilde{\mathscr{E}}) \cap L^1. \tag{5.17}$$

It is clear that this induces the Poincaré inequality. In fact, noting $\|f\|_1 \le \|f\|_2$, we have

$$\|f - m(f)\|_2^{2+(4/\mu)} \le c_2 \tilde{\mathscr{E}}(f,f)\|f - m(f)\|_1^{4/\mu} \le c_2 \tilde{\mathscr{E}}(f,f)\|f - m(f)\|_2^{4/\mu}.$$

Now we have the following theorem:

Theorem 5.3. *Fix $\mu > 0$. We consider the following conditions.*

(i) There exists c_1 such that

$$\|T_t f - m(f)\|_\infty \le c_1 t^{-\mu/2}\|f\|_1, \quad \forall f \in L^1,\ \forall t > 0. \tag{5.18}$$

(ii) There exists c_2 such that

$$\|f - m(f)\|_2^{2+(4/\mu)} \le c_2 \tilde{\mathscr{E}}(f,f)\|f - m(f)\|_1^{4/\mu}, \quad \forall f \in \mathrm{Dom}(\tilde{\mathscr{E}}) \cap L^1. \tag{5.19}$$

If we assume (ii), then (i) follows. If we assume (i), then we have $\gamma_{1\to\infty} > 0$ and hence $\gamma_{2\to 2} = \gamma_{1\to\infty}$. If we assume (i) and the following condition, we have (ii):

$$(A^2 f, f)_2 + (Af, Af)_2 \ge 0 \quad \forall f \in \mathrm{Dom}(A^2). \tag{5.20}$$

Proof. First assume (ii). Take any $f \in \mathrm{Dom}(A) \cap L^1$ and set $u(t) = \|T_t f - m(f)\|_2^2$. By (5.19), we have

$$\begin{aligned} -\frac{du}{dt} &= -2(AT_t f, T_t f - m(f)) \\ &= -2(AT_t f, T_t f) \\ &= 2\tilde{\mathscr{E}}(T_t f, T_t f) \\ &\ge 2\|T_t f - m(f)\|_2^{2+4/\mu}/(c_2\|T_t f - m(f)\|_1^{4/\mu}) \\ &\ge 2u^{1+2/\mu}/(c_2\|f - m(f)\|_1^{4/\mu}). \end{aligned}$$

Hence

$$\frac{d}{dt}(u^{-2/\mu}) \ge \frac{4}{c_2\mu\|f - m(f)\|_1^{4/\mu}}$$

and

$$u^{-2/\mu}(t) \ge u(t)^{-2/\mu} - u(0)^{-2/\mu} \ge \frac{4t}{c_2\mu\|f - m(f)\|_1^{4/\mu}}.$$

Thus we have

$$\begin{aligned} \|T_t f - m(f)\|_2 = u(t)^{1/2} &\le (c_2\mu/4t)^{\mu/4}\|f - m(f)\|_1 \\ &\le (c_2\mu/4)^{\mu/4} t^{-\mu/4}\|f - m(f)\|_1. \end{aligned}$$

Similarly we have

$$\|T_t^* f - m(f)\|_2 \le (c_2\mu/4)^{\mu/4} t^{-\mu/4}\|f - m(f)\|_1.$$

Now, by using Proposition 3.1, we have (i).

Next we show (i) implies that $\gamma_{1\to\infty} > 0$. To show this, note that for sufficiently large T, there exists a constant $\kappa < 1$ such that $\|T_T f - m(f)\|_\infty \le \kappa \|f\|_1$. For any $k \in \mathbb{N}$, by using (5.14), we have

$$\|T_{kT} - m\|_{1\to\infty} \le \|T_t - m\|_{1\to\infty}^k \le \kappa^k.$$

Now $\gamma_{1\to\infty} > 0$ easily follows.

Next we assume (i) and (5.20). Take any $f \in \mathrm{Dom}(A) \cap L^1$. By Proposition 3.1, we have

$$\|T_t - m\|_{1\to 2} \le c_1^{1/2} t^{-\mu/4}, \quad \|T_t^* - m\|_{2\to\infty} \le c_1^{1/2} t^{-\mu/4}$$

amd

$$\|(T_t^* - m)(T_t - m)\|_{1\to\infty} \le c_1 t^{-\mu/2}.$$

Hence

$$\begin{aligned} c_1 t^{-\mu/2} \|f - m(f)\|_1^2 &\ge \|(T_t^* - m)(T_t - m) f\|_\infty \|f - m(f)\|_1 \\ &\ge ((T_t^* - m)(T_t - m) f, f - m(f))_2 \\ &= ((T_t - m) f, (T_t - m) f)_2 \\ &= (f - m(f), f - m(f))_2 \\ &\quad + \int_0^t \frac{d}{ds} ((T_s - m(f)) f, (T_s - m) f)_2 \, ds \\ &= \|f - m(f)\|_2^2 + 2 \int_0^t (A(T_s - m) f, (T_s - m) f)_2 \, ds \\ &= \|f - m(f)\|_2^2 + 2 \int_0^t (A T_s f, T_s f)_2 \, ds. \end{aligned}$$

Set $u(t) = (AT_t f, T_t f)_2$. We can show that $u(t)$ is increasing as in the proof of Theorem 3.2. So

$$\begin{aligned} c_1 t^{-\mu/2} \|f - m(f)\|_1^2 &\ge \|f - m(f)\|_2^2 + 2 \int_0^t (Af, f)_2 \, ds \\ &= \|f - m(f)\|_2^2 + 2t (Af, f)_2 \end{aligned}$$

and eventually we have

$$\|f - m(f)\|_2^2 \le -2t(Af, f)_2 + c_1 t^{-\mu/2} \|f - m(f)\|_1^2.$$

Now, taking

$$t = \left(-\frac{\|f - m(f)\|_1^2}{(Af, f)} \right)^{2/(\mu+2)}$$

we have

$$\|f - m(f)\|_2^2 \le -(2 + c_1)(Af, f)_2^{\mu/(\mu+2)} \|f - m(f)\|_1^{4/(\mu+2)}.$$

Therefore

$$\|f - m(f)\|_2^{2+4/\mu} \le -(2 + c_1)^{(\mu+2)/\mu}(Af, f)_2 \|f - m(f)\|_1^{4/\mu}$$
$$= (2 + c_1)^{(\mu+2)/\mu} \tilde{\mathscr{E}}(f, f)\|f - m(f)\|_1^{4/\mu},$$

which is (5.19). □

We can also have the following corollary similar to Corollary 3.1 and Corollary 3.2.

Corollary 5.1. *Fix $\mu > 0$. We consider the following conditions:*

(i) There esists a constant c_1 such that

$$\|T_t f - m(f)\|_\infty \le c_1 t^{-\mu/2}\|f\|_1, \quad \forall f \in L^1, \ \forall t \in (0, 1]. \tag{5.21}$$

(ii) There exists a constant c_2 such that

$$\|f - m(f)\|_2^{2+(4/\mu)} \le c_2(\tilde{\mathscr{E}}(f, f) + \|f\|_2^2)\|f - m(f)\|_1^{4/\mu}, \quad \forall f \in \mathrm{Dom}(\tilde{\mathscr{E}}) \cap L^1. \tag{5.22}$$

Then (ii) implies (i). Conversely, if we assume (i) and the following additional condition, then (ii) follows: There exists a constant $M > 0$ such that

$$((A - M)^2 f, f)_2 + ((A - M)f, (A - M)f)_2 \ge 0, \quad \forall f \in \mathrm{Dom}(A^2). \tag{5.23}$$

Proof. The condition (i) is equivalent to

$$\|T_t f\|_\infty \le c_3 t^{-\mu/2}\|f\|_1, \quad \forall f \in L^1, \ \forall t \in (0, 1].$$

In fact, assuming (5.21), we have

$$\|T_t f\|_\infty \le \|T_t f - m(f)\|_\infty + |m(f)|$$
$$\le c_1 t^{-\mu/2}\|f\|_1 + \|f\|_1 = (c_1 t^{-\mu/2} + 1)\|f\|_1.$$

The converse is similarly shown.

The condition (ii) is equivalent to

$$\|f\|_2^{2+(4/\mu)} \le c_4(\tilde{\mathscr{E}}(f, f) + \|f\|_2^2)\|f\|_1^{4/\mu}.$$

To see this, assume (5.22). Then we have

$$\|f\|_2^{2+(4/\mu)} \le 2^{1+(4/\mu)}(\|f - m(f)\|_2^{2+(4/\mu)} + |m(f)|^{2+(4/\mu)}$$
$$\le 2^{1+(4/\mu)}\{c_2(\tilde{\mathscr{E}}(f, f) + \|f\|_2^2)2^{4/\mu}\|f\|_1^{4/\mu} + \|f\|_2^2\|f\|_1^{4/\mu}\}.$$

The converse is similarly shown.

Now the corollary is reduced to Corollary 3.1 and Corollary 3.2. □

We are ready to state the main theorem.

Theorem 5.4. *Fix $\mu > 0$. Assume that there exists a constant c_2 such that*

$$\|f - m(f)\|_2^{2+(4/\mu)} \le c_2(\tilde{\mathscr{E}}(f,f) + \|f\|_2^2)\|f - m(f)\|_1^{4/\mu}, \quad \forall f \in \mathrm{Dom}(\tilde{\mathscr{E}}) \cap L^1. \tag{5.24}$$

Then $\gamma_{1\to\infty} = \gamma_{2\to 2}$, and hence we have

$$\gamma_{1\to\infty} \ge \tilde{\gamma}_{1\to\infty}. \tag{5.25}$$

If A is normal, we have $\gamma_{1\to\infty} = \tilde{\gamma}_{1\to\infty}$.

Proof. From the assumption we have $\gamma_{1\to\infty} > -\infty$ and $\tilde{\gamma}_{1\to\infty} > -\infty$. Hence, by Proposition 5.1, we have

$$\tilde{\gamma}_{1\to\infty} = \tilde{\gamma}_{2\to 2} \le \gamma_{2\to 2} = \gamma_{1\to\infty},$$

which is the desired result. If A is normal, then the equality holds by Theorem 5.2. □

This theorem shows that the semigroup converges to the mean faster than the symmetric one.

6. Non-symmetric diffusions on Riemannian manifolds with boundary

Let (M, g) be a C^∞ d-dimensional Riemannian manifold with boundary ∂M. We assume that M is compact. Let us consider the diffusion process on M with the generator

$$A = \frac{1}{2}\Delta + b. \tag{6.1}$$

Here Δ is the Laplace-Beltrami operator and b is a C^∞ vector field. Let m = vol be a normalized Riemannian volume. So our Hilbert space is $L^2(m)$. A is regarded as an operator on $L^2(m)$. Since M has a boundary, we impose the Dirichlet boundary condition.

The dual operator of A is

$$A^* = \frac{1}{2}\Delta - \nabla_b - \operatorname{div} b, \tag{6.2}$$

where $\operatorname{div} b$ is the divergence of b. Hence

$$\frac{1}{2}(A + A^*) = \frac{1}{2}\Delta - \frac{1}{2}\operatorname{div} b. \tag{6.3}$$

The semigroups generated by A and A^* is denoted by $\{T_t\}$ and $\{T_t^*\}$, respectively.

The Dirichlet form $\mathscr{E}$ associated with A is given by

$$\mathscr{E}(u,v) = -(Au, v) = \frac{1}{2}\int_M (\nabla u, \nabla v)\, dm - \int_M (\nabla_b u) v\, dm. \tag{6.4}$$

Associated symmetric bilinear form in the setting of Section 2 is given by

$$\tilde{\mathscr{E}}(u,v) = \frac{1}{2}\int_M (\nabla u, \nabla v)\, dm + \frac{1}{2}\int_M uv \operatorname{div} b\, dm. \tag{6.5}$$

We need that $\tilde{\mathscr{E}}$ is positive definite. So we assume

$$\operatorname{div} b \geq 0. \tag{6.6}$$

The domain of $\tilde{\mathscr{E}}$ is $H_0^1(M)$. If we assume this condition, then $\{T_t\}$ and $\{T_t^*\}$ become contraction Markovian semigroups.

By using Theorem 3.4 and Theorem 3.5, we have

Theorem 6.1. *We have $\tilde{\lambda}_{2\to 2} \leq \lambda_{2\to 2}$. If A is normal, then we have $\tilde{\lambda}_{2\to 2} = \lambda_{2\to 2}$.*

Since M is compact, the following Nash inequality holds:

$$\|f\|_2^{2+(4/d)} \leq c_2\tilde{\mathscr{E}}(f,f)\|f\|_1^{4/d}, \quad \forall f \in \mathrm{Dom}(\tilde{\mathscr{E}}) \cap L^1.$$

Now, by Theorem 3.7, we have $\tilde{\lambda}_{2\to 2} = \tilde{\lambda}_{1\to\infty}$ and $\lambda_{2\to 2} = \lambda_{1\to\infty}$. Thus we have the following theorem:

Theorem 6.2. *We have $\tilde{\lambda}_{1\to\infty} \leq \lambda_{1\to\infty}$. If A is normal, then we have $\tilde{\lambda}_{1\to\infty} = \lambda_{1\to\infty}$.*

The semigroup has a C^∞ density $q(t,x,y)$. From the definition, we have

$$\lambda_{1\to\infty} = -\lim_{t\to\infty} \frac{1}{t} \log \sup_{x,y\in M} q(t,x,y).$$

Similarly, the semigroup associated with $\tilde{\mathscr{E}}$ has a C^∞ density $p(t,x,y)$ and we have

$$\tilde{\lambda}_{1\to\infty} = -\lim_{t\to\infty} \frac{1}{t} \log \sup_{x,y\in M} p(t,x,y).$$

Theorem 6.2 says that $q(t,x,y)$ goes to 0 faster than $p(t,x,y)$ uniformly.

7. Non-symmetric diffusions on Riemannian manifolds without boundary

Let (M,g) be a C^∞ d-dimensional Riemannian manifold without boundary. We assume that M is compact. We consider a diffusion process with the generator

$$\frac{1}{2}\Delta + b. \tag{7.1}$$

Here Δ is the Laplace-Beltrami operator and b is a C^∞ vector field. The generator is regarded as a operator in $L^2(m)$ where m is the Riemannian volume $m = \mathrm{vol}$. The dual operator of the generator is

$$A^* = \frac{1}{2}\Delta - \nabla_b - \operatorname{div} b. \tag{7.2}$$

Since M is compact, there exists an invariant probability measure ν. Our first aim is to show that A generates a contraction semigroup in $L^2(\nu)$. To do this, we need some geometric notions.

We have two measures, the Riemannian volume m and the invariant measure ν. A dual operator depends on a reference measure. Suppose an operator S is given. We denote the dual operator with respect to m is denoted by S^* and the dual operator with respect to ν is denoted by S_ν^*. For the moment, the domain of S is taken to be $C^\infty(M)$. We use the Riemannian connection and we denote the exterior derivative by d. d^* denotes the dual operator with respect to m and d^*_ν denotes the dual operator with respect to ν. Since we are given a Riemannian metric g, the tangent space $T(M)_x$ is isomorphic to the cotangent space $T^*(M)_x$. We denote this isomorphism by

$$\flat\colon T(M)_x \longrightarrow T^*(M)_x$$

and its inverse by

$$\sharp\colon T^*(M)_x \longrightarrow T(M)_x.$$

We express these mapping as

$$T(M) \ni X \mapsto X^\flat \in T^*(M)$$

and

$$T^*(M) \ni \theta \mapsto \theta^\sharp \in T(M).$$

The inner product is defined in $T^*(M)_x$ and $T(M)_x$. We denote the both of them by $(\ ,\ )$.

b is a vector field. We denote the corresponding 1-form by $\omega_b = b^\sharp$. Then we have

$$Af = \frac{1}{2}\Delta f + bf = \frac{1}{2}\Delta f + (df, \omega_b). \tag{7.3}$$

Hence

$$\begin{aligned}(f, A^*g)_2 &= (Af, g)_2 \\ &= \int_M \left(\frac{1}{2}\Delta f\, g + (df, \omega_b)g\right) dm \\ &= \int_M \left(-\frac{1}{2}(df, dg) + (df, g\omega_b)\right) dm \\ &= \int_M (df, -\frac{1}{2}dg + g\omega_b) dm \\ &= \int_M f d^*\left(-\frac{1}{2}dg + g\omega_b\right) dm.\end{aligned}$$

This means that

$$A^*g = d^*\left(-\frac{1}{2}dg + g\omega_b\right). \tag{7.4}$$

ν has the density and we denote it by e^{-U}. Then, by noting that

$$\begin{aligned}\int_M (df)g\, d\nu = \int_M (df)ge^{-U} dm &= \int_M fe^U d^*(ge^{-U})e^{-U} dm \\ &= \int_M fe^U d^*(e^{-U}g) d\nu,\end{aligned}$$

we have

$$d^*_\nu = e^U d^* e^{-U}. \tag{7.5}$$

For 1-form θ, we have

$$d_\nu^*\theta = e^U d^*(e^{-U}\theta) = d^*\theta + e^U e^{-U}(dU, \theta) = d^*\theta + (dU, \theta). \tag{7.6}$$

Theorem 7.1. *The semigroup generated by* $A = \frac{1}{2}\Delta + b$ *has an invariant measure* $\nu = e^{-U}m$ *if and only if*

$$\omega_b = -\frac{1}{2}dU + \omega_{\tilde{b}} \tag{7.7}$$

for some vector field $\tilde{b}$ *with* $d_\nu^*\omega_{\tilde{b}} = 0$.

Proof. Since $\nu = e^{-U}m$ is an invariant measure, we have $A^*(e^{-U}) = 0$. Hence, by (7.4), we have

$$d^*\left(-\frac{1}{2}d(e^{-U}) + \omega_b e^{-U}\right) = 0. \tag{7.8}$$

Note also that

$$-\frac{1}{2}d(e^{-U}) + \omega_b e^{-U} = \left(\frac{1}{2}dU + \omega_b\right)e^{-U}.$$

Now we put

$$\omega_{\tilde{b}} = \frac{1}{2}dU + \omega_b$$
$$\eta = e^{-U}\omega_{\tilde{b}}.$$

Then we have

$$d^*\eta = 0. \tag{7.9}$$

Hence, by using (7.5), we have

$$d_\nu^*\omega_{\tilde{b}} = d_\nu^*(e^U\eta) = e^U d^* e^{-U}(e^U\eta) = e^U d^*\eta = 0. \quad (\because (7.9))$$

Conversely, assume that ω_b is expressed as (7.7) with $d_\nu^*\omega_{\tilde{b}} = 0$. Using $d_\nu^* = e^U d^* e^{-U}$, we have

$$d^*(e^{-U}\omega_{\tilde{b}}) = 0.$$

Hence

$$0 = d^*\left(e^{-U}\left(\frac{1}{2}dU + \omega_b\right)\right) = d^*\left(-\frac{1}{2}d(e^{-U}) + e^{-U}\omega_b\right),$$

which is (7.8). Thus we have $A^*(e^{-U}) = 0$ and ν is an invariant measure. □

Since M is compact, the invariant measure always exists. So A is expressed as in Theorem 7.1.

From now on, we may and do assume that ω_b is expressed as (7.7) with $d_\nu^*\omega_{\tilde{b}} = 0$. Then

$$d_\nu^* df = e^U d^*(e^{-U}df) = e^U e^{-U}(dU, df) + e^U e^{-U} d^*df = (dU, df) + d^*df$$

and

$$\begin{aligned}-\frac{1}{2}d_\nu^* df + (\omega_{\tilde{b}}, df) &= -\frac{1}{2}(dU, df) - \frac{1}{2}d^* df + (\omega_{\tilde{b}}, df)\\ &= -\frac{1}{2}d^* df + (-\frac{1}{2}dU + \omega_{\tilde{b}}, df)\\ &= -\frac{1}{2}d^* df + (\omega_b, df)\\ &= Af.\end{aligned}$$

Thus we have

$$Af = -\frac{1}{2}d^* df + (\omega_b, df) = -\frac{1}{2}d_\nu^* df + (\omega_{\tilde{b}}, df). \tag{7.10}$$

We can also see that

$$\begin{aligned}(Af, f)_\nu &= -\frac{1}{2}\int_M (d_\nu^* df, f)\, d\nu + \int_M (\omega_b, df) f\, d\nu\\ &= -\frac{1}{2}\int_M (df, df)\, d\nu + \frac{1}{2}\int_M (\omega_b, d(f^2))\, d\nu\\ &= -\frac{1}{2}\int_M (df, df)\, d\nu + \frac{1}{2}\int_M (d_\nu^*\omega_b) f^2\, d\nu\\ &= -\frac{1}{2}\int_M (df, df)\, d\nu \le 0.\end{aligned}$$

Let A_ν^* be the dual operator of A with respect to ν. We denote the inner product in $L^2(\nu)$ by $(\ ,\ )_\nu$. To get the expression of A_ν^*,

$$\begin{aligned}(Af, g)_\nu &= -\frac{1}{2}(d_\nu^* df, g)_\nu + ((\omega_{\tilde{b}}, df), g)_\nu\\ &= -\frac{1}{2}(f, d_\nu^* dg)_\nu + (\omega_{\tilde{b}}, d(fg) - f dg)_\nu\\ &= -\frac{1}{2}(f, d_\nu^* dg)_\nu - (f, (\omega_{\tilde{b}}, dg))_\nu.\end{aligned}$$

This means that

$$A_\nu^* g = -\frac{1}{2}d_\nu^* dg - (\omega_{\tilde{b}}, dg).$$

We can also show that $(A_\nu^* f, f)_\nu \le 0$.

Now we have obtained that A and A_ν^* generate a Markovian contraction semigroup in $L^2(\nu)$ and the associated symmetric Dirichlet form is given by

$$\tilde{\mathscr{E}}(f, g) = \frac{1}{2}\int_M (df, dg) d\nu.$$

We can see that A is self-adjoint if and only if $\tilde{b} = 0$. Hence $-\frac{1}{2}d_\nu^* d$ is the unique self-adjoint operator generating a semigroup with invariant measure ν. We summarize as follows:

Theorem 7.2. *The operator A generates a semigroup with invariant measure $\nu = e^{-U}m$ if and only if it is of the form*

$$A = -\frac{1}{2}d_\nu^* d + \omega_{\tilde{b}} \tag{7.11}$$

with $d_\nu^* \omega_{\tilde{b}} = 0$. *In this case,* $A_\nu^* = -\frac{1}{2} d_\nu^* d - \omega_{\tilde{b}}$ *and* A *is self-adjoint if and only if* $\omega_{\tilde{b}} = 0$.

The index $\gamma_{1\to\infty}$ is defined for the semigroup generated by A as (5.13). $\tilde{\gamma}_{1\to\infty}$ can be defined for symmetric one. Since M is compact, the following Nash inequality holds:

$$\|f - \nu(f)\|_2^{2+(4/d)} \leq c_2 \tilde{\mathscr{E}}(f,f) \|f - \nu(f)\|_1^{4/d}.$$

Now by Theorem 5.3, we have the following theorem.

Theorem 7.3. *We have* $\tilde{\gamma}_{1\to\infty} \leq \gamma_{1\to\infty}$. *If* A *is normal, we have* $\tilde{\gamma}_{1\to\infty} = \gamma_{1\to\infty}$.

T_t has a density $q(t,x,y)$ with respect to ν. From the definition, we have

$$\gamma_{1\to\infty} = -\lim_{t\to\infty} \frac{1}{t} \log \sup_{x,y\in M} |q(t,x,y) - 1|$$

Similarly associated symmetric semigroup has a density $p(t,x,y)$ and

$$\tilde{\gamma}_{1\to\infty} = -\lim_{t\to\infty} \frac{1}{t} \log \sup_{x,y\in M} |p(t,x,y) - 1|.$$

So the $\tilde{\gamma}_{1\to\infty} \leq \gamma_{1\to\infty}$ means that the density of semigroup converges to 1 faster than symmetric one uniformly.

References

1. T. Coulhon, Ultracontractivity and Nash type inequalities, *J. Funct. Anal.*, **141** (1996), no. 2, 510–539.
2. E. B. Davies, "*One-parameter semigroups*," Academic Press, London-New York, 1980.
3. E. B. Davies, "*Heat kernels and spectral theory*," Cambridge University Press, Cambridge, 1989.
4. C.-R. Hwang, S.-Y. Hwang-Ma and S.-J. Sheu, and Shuenn-Jyi Sheu, Accelerating diffusions, *Ann. Appl. Probab.*, **15** (2005), no. 2, 1433–1444.
5. Z.-M. Ma and M. Röckner, "*Introduction to the theory of (non-symmetric) Dirichlet forms*," Springer-Verlag, Berlin-Heidelberg-New York, 1992.
6. E. Ouhabaz, Invariance of closed convex sets and domination criteria for semigroups, *Potential Analysis*, **5** (1996), 611–625.
7. A. Pazy, "*Semigroups of linear operators and applications to partial differential equations*," Springer, New York-Berlin-Heidelberg, 1983.
8. M. Reed and B. Simon, "*Methods of modern mathematical physics. I. Functional analysis*," Academic Press, New York-London, 1972.
9. H. Tanabe, "*Equations of evolution*," Monographs and Studies in Mathematics, 6, Pitman, Boston, Mass.-London, 1979.

Chapter 27

Metric measure spaces with variable Ricci bounds and couplings of Brownian motions

Karl-Theodor Sturm *

The goal of this paper is twofold: we study metric measure spaces (X, d, m) with variable lower bounds for the Ricci curvature and we study pathwise coupling of Brownian motions. Given any lower semicontinuous function $\mathrm{k} : X \to \mathbb{R}$ we introduce the curvature-dimension condition $\mathrm{CD}(\mathrm{k}, \infty)$ which canonically extends the curvature-dimension condition $\mathrm{CD}(\mathrm{K}, \infty)$ of Lott-Sturm-Villani for constant $\mathrm{K} \in \mathbb{R}$. For infinitesimally Hilbertian spaces we prove

- its equivalence to an evolution-variation inequality $\mathrm{EVI}_{\mathrm{k}}$ which in turn extends the $\mathrm{EVI}_{\mathrm{K}}$-inequality of Ambrosio-Gigli-Savaré;
- its stability under convergence and its local-to-global property.

For metric measure spaces with uniform lower curvature bounds K we prove that for each pair of initial distributions μ_1, μ_2 on X there exists a coupling $B_t = (B_t^1, B_t^2)$, $t \geq 0$, of two Brownian motions on X with the given initial distributions such that a.s.

$$d\big(B^1_{s+t}, B^2_{s+t}\big) \leq e^{-\mathrm{K}t/2} \cdot d\big(B^1_s, B^2_s\big) \qquad (\forall s, t \geq 0).$$

Contents

*University of Bonn, Institute for Applied Mathematics, Endenicher Allee 60, 53115 Bonn, Germany

1. Heat Flow on Metric Measure Spaces

Throughout this paper, a *metric measure space* will be triple (X, d, m) where (X, d) is a complete, separable metric space and m is a measure on X equipped with its Borel σ-field $\mathcal{B}(X)$. To simplify the presentation, we also assume in addition that d is a length metric, that m has full topological support and that the following weak integrability property holds

$$\int_X e^{-C \cdot d^2(x',x)}\, dm(x) < \infty \tag{1.1}$$

(for some $x' \in X$ and $C \in \mathbb{R}$). $\mathcal{P}(X)$ will denote the space of Borel probability measures on X. There are two canonical ways to define the *heat flow* on a mms (X, d, m)

- either as the gradient flow for the *energy* $\mathcal{E}$ in the Hilbert space $L^2(X, m)$
- or as the gradient flow for the *entropy* Ent in the Wasserstein space $\mathcal{P}_2(X)$.

In the sequel, we will briefly present both approaches and we will illustrate that in great generality both approaches will coincide.

1.1. *Eulerian Approach via Energy*

Given any function $f : X \to \mathbb{R}$ we define its pre-energy by

$$\mathcal{E}^0(f) := \begin{cases} \int_X |Df|^2\, dm, & \text{if } f \text{ is bounded and Lipschitz} \\ +\infty, & \text{else} \end{cases}$$

and its *energy* – often called Cheeger energy – as the relaxation or lower semicontinuous envelop of the pre-energy:

$$\mathcal{E}(f) = \liminf_{g \to f \text{ in } L^2} \mathcal{E}^0(g).$$

Here $|Df|$ denotes the metric slope (or local Lipschitz constant) of a Lipschitz function defined as

$$|Df(x)| := \limsup_{y \to x} \frac{|f(y) - f(x)|}{d(y, x)}.$$

The energy $\mathcal{E}$ is a lower semicontinuous convex functional on $L^2(X, m)$. It is 2-homogeneous but not necessarily quadratic. Its domain $Dom(\mathcal{E}) := \{f \in L^2(X, m) : \mathcal{E}(f) < \infty\}$ is a dense linear subspace of $L^2(X, m)$ [3], Prop. 4.1. For each $f \in Dom(\mathcal{E})$ there exists the minimal weak upper gradient $|Df|_w$, a unique element of minimal norm in $\{g \in L^2(X, m) : \exists f_n \in \mathrm{Lip}_b(X), f_n \to f, |Df_n| \rightharpoonup g\}$. It satisfies $\mathcal{E}(f) = \int |Df|_w^2\, dm$.

Proposition 1.1. (Ambrosio et al. [3, Chapter 4])

- *The gradient flow for the energy $\mathcal{E}$ in $L^2(X, m)$ defines uniquely a continuous semigroup $(T_t)_{t \geq 0}$ of contractions in $L^2(X, m)$.*
- *For each $f \in L^2(X, m)$ the trajectory $t \mapsto T_t f$ is continuous in $t \in [0, \infty)$ and locally Lipschitz continuous in $t \in (0, \infty)$.*
- *The heat flow is mass preserving: for each $f \in L^2(X, m) \cap L^1(X, m)$*

$$\int T_t f\, dm = \int f\, dm.$$

- *It is contracting in L^p: for each $p \in [1,\infty]$ and each $f,g \in L^2(X,m) \cap L^p(X,m)$*

$$\|T_t f - T_t g\|_p \le \|f-g\|_p.$$

In more details, the heat flow $f_t = T_t f_0$ is defined via the differential inclusion

$$\frac{d}{dt} f_t \in -\partial^- \mathcal{E}(f_t)$$

where $\partial^- \mathcal{E}$ denotes the subdifferential of the convex function $\mathcal{E}$. The Laplacian is defined for those $f \in L^2$ with $\partial^- \mathcal{E}(f) \neq \emptyset$ as the the element of minimal L^2-norm within $-\partial^- \mathcal{E}(f)$.

In general, the heat semigroup – and equivalently the Laplacian – will neither be linear nor m-symmetric. The heat flow is linear if and only if the energy is a quadratic functional. And in this case, the heat semigroup will also be m-symmetric. Note that by this construction, the heat flow will be the solution to $\frac{d}{dt} f_t = \Delta f_t$ – whereas in parts of the literature it is regarded as the solution to $\frac{d}{dt} f_t = \frac{1}{2}\Delta f_t$.

Example 1.1. (Ohta et al. [17, 18]) Let (M,F,m) be a smooth Finsler space. Then the associated heat flow on M is linear if and only if the norm F on each tangent space is Hilbertian, or in other words, if and only the manifold is Riemannian.

Even if the heat semigroup on Finsler spaces is non-linear it shares many properties with the linear heat semigroup for Dirichlet forms, e.g. integrated Gaussian estimates à la Davies, pointwise comparison à la Cheeger-Yau and gradient estimates à la Bakry-Emery-Ledoux.

1.2. *Lagrangian Approach via Entropy*

For $p \in [1,\infty)$ we define the L^p-Wasserstein distance between $\mu_0, \mu_1 \in \mathcal{P}(X)$ by

$$W_p(\mu_0,\mu_1) = \left(\inf \int d(x,y)^p \, dq(x,y)\right)^{1/p}$$

and $W_\infty(\mu_0,\mu_1) = \inf \|d(.)\|_{L^\infty(X^2,q)}$ where in both cases the infimum is taken over all Borel probability measures q on $X \times X$ with marginals μ_0 and μ_1. Note that $W_\infty(\mu_0,\mu_1) = \lim_{p\to\infty} W_p(\mu_0,\mu_1)$.

The case $p=2$ will be of particular interest for us. We denote by $\mathcal{P}_2(X)$ the L^2-Wasserstein space over (X,d), i.e. the set of all Borel probability measures μ satisfying $\int_X d(x_0,x)^2 \mu(dx) < \infty$ for some, hence any, $x_0 \in X$.

Given a measure $\mu \in \mathcal{P}_2(X)$ we define its *relative entropy* or *Boltzmann entropy* by

$$\mathrm{Ent}(\mu) := \int \rho \log \rho \, dm\,,$$

if $\mu = \rho m$ is absolutely continuous w.r.t. m and $(\rho \log \rho)_+$ is integrable. Otherwise we set $\mathrm{Ent}(\mu) = +\infty$.

Definition 1.1. Given a number $\mathrm{K} \in \mathbb{R}$ we say that (X,d,m) satisfies the *curvature-dimension condition* $\mathrm{CD}(\mathrm{K},\infty)$ if the Boltzmann entropy is K-convex on the L^2-Wasserstein space $(\mathcal{P}_2(X), W_2)$.

Here a function S on a metric space (Y, d_Y) is called K-convex if every pair of points $y_0, y_1 \in Y$ can be joined by a (minimizing, constant speed) geodesic $(y_t)_{0\leq t\leq 1}$ in Y such

$$S(y_t) \leq (1-t)\, S(y_0) + t\, S(y_1) - \frac{\mathrm{K}}{2} t(1-t)\, d_Y(y_0, y_1)^2$$

for all $t \in [0,1]$. (The latter can equivalently be expressed by the fact that the function $u(t) = S(y_t)$ is upper semicontinuous in $t \in [0,1]$, continuous in $(0,1)$ and satisfies $u'' \geq \mathrm{K}|\dot{y}|^2$ weakly in $(0,1)$.) In the case $\mathrm{K} = 0$ it is the classical convexity. In the general case, K-convexity gives a precise meaning for weak solutions to the differential inequality $D^2 S \geq K$ on geodesic spaces.

Proposition 1.2. (Sturm, [24], Lott et al. [16] and Ambrosio et al. [2]) *The curvature-dimension condition* $\mathrm{CD}(\mathrm{K}, \infty)$ *has important stability and transformation properties:*

- *it is preserved under convergence of the underlying mms (with respect to mGH convergence or* $\mathbb{D}$*-convergence as well as with respect to the 'pointed versions' of these convergence concepts);*
- *it has the local-to-global property (provided the space is non-branching);*
- *it has the tensorization property (provided all the spaces are non-branching).*

Proposition 1.3. (Lott et al. [16] and Ambrosio et al. [2]) *The curvature-dimension condition* $\mathrm{CD}(\mathrm{K}, \infty)$ *for* $\mathrm{K} > 0$ *implies various functional inequalities, each of them with sharp constants,*

- *spectral gap estimate*
- *Talagrand inequality*
- *logarithmic Sobolev inequality.*

Proposition 1.4. (Ambrosio et al. [3]) *Assume that the condition* $\mathrm{CD}(\mathrm{K}, \infty)$ *holds true for some* $\mathrm{K} \in \mathbb{R}$.

(i) *For every* $\mu \in \mathcal{P}_2(X)$ *with* $\mathrm{Ent}(\mu) < \infty$ *there exists a unique gradient flow* $(P_t\mu)_{t\geq 0}$ *for the Boltzmann entropy* Ent *in the* L^2*-Wasserstein space* $(\mathcal{P}_2(X), W_2)$*, starting in* μ.

(ii) *For each* $\mu \in \mathcal{P}_2(X)$ *with* $\mathrm{Ent}(\mu) < \infty$ *and with* $f = \frac{d\mu}{dm} \in L^2(X, m)$ *the gradient flow of the entropy and the previously defined gradient flow of the energy coincide:*

$$P_t(f\, m) = (T_t f)\, m.$$

1.3. *Wasserstein Contraction, Gradient Estimates, and Bochner's Formula*

From now on, we will in addition always assume that the metric measure spaces (X, d, m) under consideration will be *"infinitesimally Hilbertian"* in the sense that the canonical energy (as introduced above) is *quadratic*:

$$\mathcal{E}(u+v) + \mathcal{E}(u-v) = 2\mathcal{E}(u) + 2\mathcal{E}(v) \qquad (\forall u, v \in Dom(\mathcal{E})).$$

To simplify the presentation, we also assume that every bounded function $f \in Dom(\mathcal{E})$ with $|Df|_w \leq 1$ admits a continuous representative ("property C"). Note that property

(ii) in the subsequent theorem already implies that the energy is quadratic. Moreover, each of properties (i) and (ii) imply the above mentioned property C.

Theorem 1.1. (Ambrosio et al. [2]) *For any mms (X, d, m) as above, the following properties are equivalent*

(i) *Curvature-dimension condition* $\mathrm{CD}(\mathrm{K}, \infty)$*;*

(ii) *Evolution-variation inequality* $\mathrm{EVI}_{\mathrm{K}}$*: for every $\mu_0 \in \mathcal{P}_2(X)$ there exists a curve $(\mu_t)_{t>0}$ in $Dom(\mathcal{E}nt)$ with $\lim_{t\to 0}\mu_t = \mu_0$ such that $\forall \nu \in \mathcal{P}_2(X), \forall t > 0$*

$$\frac{d^+}{dt}\frac{1}{2}W_2^2(\mu_t,\nu) + \frac{\mathrm{K}}{2}W_2^2(\mu_t,\nu) \le \mathcal{E}nt(\nu) - \mathrm{Ent}(\mu_t);$$

(iii) *L^2-Wasserstein contraction: $\forall f_1, f_2 \in L^2(X,m), \forall t \in \mathbb{R}_+$:*

$$W_2\Big(T_t(f_1\, m), T_t(f_2\, m)\Big) \le e^{-Kt}\, W_2\Big((f_1\, m), (f_2\, m)\Big);$$

(iv) *L^2-gradient estimate: $\forall u \in Dom(\mathcal{E})$ and $\forall t > 0$*

$$|DT_t u|_w^2 \le e^{-2\mathrm{K}t}T_t(|Du|_w^2);$$

(v) *Bochner's inequality or Bakry-Emery condition* $\mathrm{BE}(\mathrm{K}, \infty)$*: $\forall u \in Dom(\Delta)$ with $\Delta u \in Dom(\mathcal{E})$ and $\forall \phi \in Dom(\Delta) \cap L^\infty(X,m)$ with $\phi \ge 0$, $\Delta\phi \in L^\infty(X,m)$*

$$\int \left(\frac{1}{2}\Delta - \mathrm{K}\right) \phi \cdot |Du|_w^2\, dm \ge \int \phi \cdot \langle \nabla u, \nabla \Delta u\rangle\, dm. \tag{1.2}$$

Corollary 1.1. (Ambrosio et al. [3] and Rajala et al. [19]) *Assume that some/all of the above properties are satisfied. Then*

- *$(\mathcal{E}, Dom(\mathcal{E}))$ is a strongly local, quasi-regular Dirichlet form.*
- *Its carrè du champ operator coincides with the squared minimal weak upper gradient: $\Gamma(u) = |Du|_w^2$.*
- *The generator of the Dirichlet form is the negative Laplacian $-\Delta$, introduced as the single-valued subdifferential of $\mathcal{E}$. It is a linear self-adjoint operator.*
- *The $\mathrm{EVI}_{\mathrm{K}}$-curve in (ii) coincides with the heat flow: $\mu_t = P_t\mu_0$.*
- *For every $\mu_0, \mu_1 \in \mathcal{P}_2(X)$ with $\mu_0, \mu_1 \ll m$ the geodesic connecting them is unique.*

The previous theorem is of fundamental importance for our understanding of curvature bounds on metric measure spaces and it has plenty of applications. It also allows for various extensions and improvements. One direction of improvement is to combine the curvature bound K with a dimension bound N which leads to the curvature-dimension condition $\mathrm{CD}(\mathrm{K}, N)$ introduced in [25] and also studied (in slightly modified form) in [27], see also [5], [10]. Erbar-Kuwada-Sturm [12] succeeded to formulate proper versions (taking into account the additional information of the upper dimension bound $N < \infty$) of each of the properties in the previous theorem – among them $\mathrm{EVI}_{\mathrm{K},N}$ and $\mathrm{BE}(\mathrm{K}, N)$ inequalities – and to prove their equivalence. The $\mathrm{CD}(\mathrm{K}, N)$-condition for finite N also has the advantage that it allows to deduce curvature bounds under time change of the process and/or conformal transformation of the metric [26].

Another direction will be to consider variable curvature bounds instead of uniform bounds. This will be the topic of the final chapter 3 of this paper. There we will introduce and study appropriate modifications of the previous properties (i), (ii), (iv) and (v) for non-constant curvature bounds k : $X \to \mathbb{R}$. For a more refined approach – in a more 'regular' setting however – which even allows to define a *Ricci tensor*, see [26].

Yet another direction of improving the previous theorem consists in studying L^p-versions instead of L^2-versions. Using a remarkable self-improvement property of Bochner's inequality in the 'smooth' Γ_2-setting (i.e. assuming the existence of a nice algebra of functions), Bakry [6] deduced L^p versions of the gradient estimate (iv). Savaré [22] extended this argument to the non-smooth setting and combined it with Kuwada's duality argument [14, 15] to obtain an L^p-Wasserstein contraction estimate in the spirit of [20].

Theorem 1.2. (Savaré, [22]) *Assume that* (X,d,m) *satisfies* $CD(K,\infty)$. *Then for all* $p \in [1,\infty]$, *all* $\mu_1, \mu_2 \in \mathcal{P}(X)$ *and all* $t \in \mathbb{R}_+$

$$W_p(P_t\mu_1, P_t\mu_2) \le e^{-Kt}\, W_p(\mu_1, \mu_2).$$

This result will be the key ingredient for the construction of coupled pairs of Brownian motions on X which is the content of the subsequent chapter.

2. Coupled Pairs of Brownian Motions on X

2.1. *Brownian Motion on X*

As a consequence of the previous theorem (the classical $p = 2$ version suffices), the gradient flow $(P_t)_{t>0}$ extends to all of $\mathcal{P}_2(X)$ such that for fixed $t > 0$ the mapping $\mu \mapsto P_t\mu$ is Lipschitz continuous (w.r.t. the W_2-metric). Let us put $p_t(x,.) = P_t\delta_x$. Then the mapping $x \mapsto p_t(x,.)$ from X to $\mathcal{P}_2(X)$ is Lipschitz continuous, in particular, Borel measurable. Thus $p_t(.,.)$ is a Markov kernel on $(X, \mathcal{B}(X))$.

Uniqueness of the gradient flow implies $P_s(P_t\mu) = P_{s+t}\mu$ for all μ with finite entropy; by continuity in μ this extends to all of $\mathcal{P}_2(X)$. Thus $(p_t)_{t\ge 0}$ is a semigroup of Markov kernels on $(X, \mathcal{B}(X))$. Note that each of the measures $p_t(x,.)$ for $x \in X$ and $t > 0$ is absolutely continuous w.r.t. m.

Since the Dirichlet form $\mathcal{E}$ on $L^2(X,m)$ is quasi-regular and local, there exists an m-invariant continuous strong Markov process which (or more precisely, the transition semigroup of which) is m-equivalent to the semigroup $(p_t)_{t\ge 0}$. Since the latter is absolutely continuous w.r.t. m the continuous Markov process can be chosen to be equivalent to $(p_t)_{t\ge 0}$ ([13], Theorems 4.5.1 and 4.5.4).

This continuous stochastic process can be obtained as follows: Given the Markov semigroup $(p_t)_{t\in\mathbb{R}_+}$ on $(X, \mathcal{B}(X))$, an arbitrary initial distribution $\mu \in \mathcal{P}(X)$ and a finite subset $J = \{t_1, \ldots, t_r\}$ of $\mathbb{R}_+$ we define the *finite dimensional distribution* P_J^μ – a probability measure on X^r – as follows

$$P_J^\mu(B_1 \times \ldots B_r) = \int_X \int_{B_1} \cdots \int_{B_r} p_{t_r - t_{r-1}}(x_{r-1}, dx_r) \ldots p_{t_1}(x_0, dx_1)\, \mu(dx_0).$$

The probability measures $\{P_J^\mu : \ J \text{ finite } \subset \mathbb{R}_+\}$ constitute a *consistent family* which implies that their projective limit

$$\mathbf{P}^\mu_{\mathbb{R}_+} = \lim_{\leftarrow} P_J^\mu$$

exists: it is a probability measure on $(X^{\mathbb{R}_+}, \mathcal{B}(X)^{\mathbb{R}_+})$ with the property that

$$(\pi_J)_* \mathbf{P}^\mu_{\mathbb{R}_+} = P^\mu_J$$

for each finite $J \subset \mathbb{R}_+$ where π_J denotes the projection $\omega \mapsto (\omega(t_1), \ldots, \omega(t_r))$ from $X^{\mathbb{R}_+}$ onto X^r. The measures $\mathbf{P}^x_{\mathbb{R}_+} := \mathbf{P}^{\delta_x}_{\mathbb{R}_+}$ depend in a measurable way on x and

$$\mathbf{P}^\mu_{\mathbb{R}_+}(.) = \int_X \mathbf{P}^x_{\mathbb{R}_+}(.)\, \mu(dx)$$

for any $\mu \in \mathcal{P}(X)$ (cf. [8]).

There are various ways to construct a continuous modification out of this stochastic process. Firstly, since we already know that there exists a continuous modification we may refer to a result of Doob which states that in this case the subset $\mathcal{C}(\mathbb{R}_+, X)$ has full outer measure w.r.t. $\mathbf{P}^\mu_{\mathbb{R}_+}$. This allows to define a probability measure $\mathbf{P}^\mu$ on $\mathcal{C}(\mathbb{R}_+, X)$ equipped with the trace σ-field $\mathcal{B}(X)^{\mathbb{R}_+} \cap \mathcal{C}(\mathbb{R}_+, X)$ by

$$\mathbf{P}^\mu(A \cap \mathcal{C}(\mathbb{R}_+, X)) := \mathbf{P}^\mu_{\mathbb{R}_+}(A) \qquad (\forall A \in \mathcal{B}(X)^{\mathbb{R}_+}).$$

The process $(\pi_t)_{t\geq 0}$ on $(\mathcal{C}(\mathbb{R}_+, X), \mathcal{B}(X)^{\mathbb{R}_+} \cap \mathcal{C}(\mathbb{R}_+, X), \mathbf{P}^\mu)$ will do the job ([8], Theorem 63.2 and Lemma 63.8) and the process $(B_t)_{t\geq 0} := (\pi_{t/2})_{t\geq 0}$ then will be a Brownian motion on X with initial distribution μ.

Alternatively, we may restrict the given process $(\pi_t)_{t\in\mathbb{R}_+}$ on $(X^{\mathbb{R}_+}, \mathcal{B}(X)^{\mathbb{R}_+}, \mathbf{P}^\mu_{\mathbb{R}_+})$ to dyadic time instances, i.e. replace $\mathbb{R}_+$ by the set $\mathsf{D} := \{k \cdot 2^{-n} : k, n \in \mathbb{N}_0\}$. Then extend this process (with time parameter D) by local uniform continuity – outside of a zero set – to a process with time parameter $\mathbb{R}_+$. Indeed, for $\mathbf{P}^\mu_{\mathbb{R}_+}$-a.e. ω the limit

$$B_t(\omega) = \lim_{s\to t/2, s\in\mathsf{D}} \pi_s(\omega)$$

exists for each $t \in \mathbb{R}_+$ and the trajectory $t \mapsto B_t(\omega)$ is continuous. Moreover, for each $t \in \mathbb{R}_+$ the random variables $\pi_{t/2}$ and B_t coincide $\mathbf{P}^\mu_{\mathbb{R}_+}$-a.s. (Cf. [8], Lemma 63.5 and subsequent Remarks 1+2). Note that for the construction of the process $(B_t)_{t\in\mathbb{R}_+}$ the measures $\mathbf{P}^\mu_{\mathbb{R}_+}$ may be replaced by its projection onto the space X^{D}.

Proposition 2.1. *Given any initial distribution μ on X, the stochastic process $(B_t)_{t\in\mathbb{R}_+}$ with values in X, constructed as above on the probability space $(X^{\mathsf{D}}, \mathcal{B}(X)^{\mathsf{D}}, \mathbf{P}^\mu_{\mathsf{D}})$, is a Brownian motion on X with initial distribution μ.*

Recall that by convention a Brownian motion on X has generator $\frac{1}{2}\Delta$. Note that the link to the semigroup $(T_t)_{t\geq 0}$ – which we defined as the semigroup of selfadjoint operators on $L^2(X, m)$ with generator Δ – is given by

$$T_t u(x) = \mathbb{E}_x [u(B_{2t})]$$

for m-a.e. $x \in X$, each $t > 0$ and each Borel measurable L^2-function u on X.

2.2. *Coupled Semigroups in Discrete Time*

For the sequel, we have to introduce additional notation.

$$\mathcal{B}^u(X^2) := \bigcap_{\alpha \in \mathcal{P}(X^2)} \mathcal{B}^\alpha(X^2)$$

will denote the σ-field of *universally measurable* subsets of X^2. It is the intersection of all the $\mathcal{B}^\alpha(X^2)$ where α runs through the set $\mathcal{P}(X^2)$ of all Borel probability measures on X^2 and where $\mathcal{B}^\alpha(X^2)$ is the completion of the Borel σ-field on X^2 w.r.t. $\alpha \in \mathcal{P}(X^2)$. Moreover, $\mathsf{D} := \{k \cdot 2^{-n} : k, n \in \mathbb{N}_0\}$ will denote the set of nonnegative *dyadic* numbers whereas $\mathsf{D}_n := \{k \cdot 2^{-n} : k \in \mathbb{N}_0\}$ for fixed $n \in \mathbb{N}_0$.

Lemma 2.1. *For each $t \in \mathbb{R}_+$ there exists a Markov kernel q_t^* on $(X^2, \mathcal{B}^u(X^2))$ with the following properties:*

(i) *For each $(x,y) \in X^2$ the probability measure $q_t^*\big((x,y),.\big)$ is a coupling of the probability measures $p_t(x,.)$ and $p_t(y,.)$.*

(ii) *For each $(x,y) \in X^2$ and $q_t^*\big((x,y),.\big)$-a.e. $(x',y') \in X^2$*

$$d(x',y') \le e^{-Kt} \cdot d(x,y).$$

Proof. Applying Theorem 1.2 with $p = \infty$ to $\mu_1 = \delta_x, \mu_2 = \delta_y$ yields the existence of at least one probability measure $q_t^*((x,y),.)$ with properties (i) and (ii) for each $x, y \in X^2$. The class of all these measures is closed (for given x,y). Thus according to a classical measurable selection theorem (see e.g. [9], Thm. 6.9.2), we may choose the optimal coupling satisfying (i) and (ii) in such a way that the map

$$(x,y) \mapsto q_t^*((x,y),.), \qquad \big(X^2, \mathcal{B}^u(X^2)\big) \to \big(\mathcal{P}(X^2), \mathcal{B}\left(\mathcal{P}(X^2)\right)\big)$$

is measurable. (Note that the σ-field of universally measurable subsets of X^2 contains all Souslin subsets of X^2.) □

Lemma 2.2. *For each $n \in \mathbb{N}_0$ there exists a Markov semigroup $(q_t^{(n)})_{t \in \mathsf{D}_n}$ on $(X^2, \mathcal{B}^u(X^2))$ with the following properties:*

(i) *For each $t \in \mathsf{D}_n$ and each $(x,y) \in X^2$ the probability measure $q_t^{(n)}\big((x,y),.\big)$ is a coupling of the probability measures $p_t(x,.)$ and $p_t(y,.)$.*

(ii) *For each $t \in \mathsf{D}_n$, each $(x,y) \in X^2$ and $q_t^{(n)}\big((x,y),.\big)$-a.e. $(x',y') \in X^2$*

$$d(x',y') \le e^{-Kt} \cdot d(x,y).$$

Proof. Given $t \in \mathsf{D}_n$, say $t = k \cdot 2^{-n}$ we put

$$q_t^{(n)} := \underbrace{q_{2^{-n}}^* \circ \ldots \circ q_{2^{-n}}^*}_{k \text{ times}}.$$

This obviously defines for each $t \in \mathsf{D}_n$ a Markov kernel on $(X^2, \mathcal{B}^u(X^2))$ and $q_s^{(n)} \circ q_t^{(n)} = q_{s+t}^{(n)}$ for all $s, t \in \mathsf{D}_n$. Moreover, properties (i) and (ii) of the theorem are inherited (by iteration) from the corresponding properties (i) and (ii) for the kernel $q_{2^{-n}}^*$ (see previous lemma). Indeed, for each $i = 1, \ldots, k$, each $(x_{i-1}, y_{i-1}) \in X^2$ and $q^{(n)}_{2^{-n}}((x_{i-1}, y_{i-1}), .)$-a.e. $(x_i, y_i) \in X^2$

$$d(x_i, y_i) \le e^{-K2^{-n}} \cdot d(x_{i-1}, y_{i-1}).$$

This yields (ii) for the kernel q_t with $t = k2^{-n}$.

Moreover, for each bounded Borel function f on X

$$\begin{aligned}
&\int_{X^2} f(x_k)\, q^{(n)}_{k2^{-n}}((x_0, y_0), d(x_k, y_k)) \\
&= \int_{X^2} \left[\int_{X^2} f(x_k)\, q^*_{2^{-n}}((x_{k-1}, y_{k-1}), d(x_k, y_k)) \right] \\
&\qquad\qquad (q^*_{2^{-n}})^{k-1}((x_0, y_0), d(x_{k-1}, y_{k-1})) \\
&= \int_{X^2} \left[\int_{X^2} f(x_k)\, p_{2^{-n}}(x_{k-1}, dx_k) \right] (q^*_{2^{-n}})^{k-1}((x_0, y_0), d(x_{k-1}, y_{k-1})) \\
&= \ldots \\
&= \int_{X^2} \left[\int_{X^2} f(x_k)\, p_{2\cdot 2^{-n}}(x_{k-2}, dx_k) \right] (q^*_{2^{-n}})^{k-2}((x_0, y_0), d(x_{k-2}, y_{k-2})) \\
&= \ldots \\
&= \int_{X^2} \left[\int_{X^2} f(x_k)\, p_{(k-1)\cdot 2^{-n}}(x_1, dx_k) \right] q^*_{2^{-n}}((x_0, y_0), d(x_1, y_1)) \\
&= \int_X f(x_k)\, p_{k\cdot 2^{-n}}(x_0, dx_k).
\end{aligned}$$

Similarly,

$$\int_{X^2} f(y_k)\, q^{(n)}_{k2^{-n}}((x_0, y_0), d(x_k, y_k)) = \int_X f(y_k)\, p_{k\cdot 2^{-n}}(y_0, dy_k).$$

This proves property (i). □

Remark 2.1. For given $(x, y) \in X^2$ and $t \in \mathsf{D}$, say $t = k\,2^{-m}$, let us consider the set

$$\{q^{(n)}_t((x, y), .) : n \ge m\} \subset \mathcal{P}(X^2).$$

Being a subset of the set of all couplings of $p_t(x, .)$ and $p_t(y, .)$, this set is relatively compact, [27], Lemma 4.4. Thus there exists a coupling $q_t((x, y), .)$ and a subsequence $(n_l)_{l\in\mathbb{N}}$ such that

$$q_t = \lim_{l\to\infty} q^{n_l}_t$$

(in the sense of weak convergence).

2.3. *Coupled Stochastic Processes*

For the sequel, let us fix an initial distribution $\alpha \in \mathcal{P}(X^2)$ with marginals $\alpha_1 = (e_1)_*\alpha$ and $\alpha_2 = (e_2)_*\alpha$. To simplify notation, in the following lemma and its proof we drop α from the notation. For any $n \in \mathbb{N}_0$ and any finite subset J of D_n, say $J = \{t_1, \ldots, t_r\}$ with $t_1 < \ldots < t_r$, we define a probability measure $Q^{(n)}_J$ on $(X^2)^{|J|}$ by

$$Q^{(n)}_J(A_1 \times \ldots \times A_r) =$$

$$= \int_{X^2}\int_{A_1}\cdots\int_{A_r} q^{(n)}_{t_r-t_{r-1}}((x_{r-1},y_{r-1}),d(x_r,y_r))\ \ldots$$
$$\ldots\ q^{(n)}_{t_1}((x_0,y_0),d(x_1,y_1))\,\alpha(d(x_0,y_0)).$$

For fixed $n \in \mathbb{N}_0$, obviously $\{Q^{(n)}_J : J \text{ finite } \subset \mathsf{D}_n\}$ is a *consistent* family of probability measures.

Lemma 2.3. *For fixed finite $J \subset \mathsf{D}$, say $J \subset \mathsf{D}_m$, the family $\{Q^{(n)}_J : n \in \mathbb{N}_0,\ n \geq m\}$ is a tight family of probability measures on $(X^2)^{|J|}$.*

Proof. Let J be given as $J = \{t_1, \ldots, t_r\}$ with $t_i \in \mathsf{D}_m$. Put $P^{\alpha_1}_{t_i}(.) = \int_X p_{t_i}(x_0,.)\alpha_1(dx_0)$ and $P^{\alpha_2}_{t_i}(.) = \int_X p_{t_i}(y_0,.)\alpha_2(dy_0)$. The families $\{P^{\alpha_1}_{t_i} : i = 1, \ldots, r\}$ and $\{P^{\alpha_2}_{t_i} : i = 1, \ldots, r\}$ of probability measures on X are tight. Thus, given $\epsilon > 0$ there exist compact sets $A_1, A_2 \subset X$ such that

$$P^{\alpha_1}_{t_i}(X \setminus A_1) < \epsilon, \quad P^{\alpha_2}_{t_i}(X \setminus A_2) < \epsilon \qquad (\forall i).$$

Put $\vec{A} = (A_1 \times A_2)^r$. Then for all $n \in \mathbb{N}_0$

$$\begin{aligned} Q^{(n)}_J((X^2)^r \setminus \vec{A}) &\leq \sum_{i=1}^r Q^{(n)}_{t_i}(X^2 \setminus A_1 \times A_2) \\ &\leq \sum_{i=1}^r \left[Q^{(n)}_{t_i}((X \setminus A_1) \times X) + Q^{(n)}_{t_i}(X \times (X \setminus A_2))\right] \\ &= \sum_{i=1}^r \left[P^{\alpha_1}_{t_i}(X \setminus A_1) + P^{\alpha_2}_{t_i}(X \setminus A_2)\right] \\ &\leq 2r \cdot \epsilon. \end{aligned}$$

Since $\vec{A}$ is a compact subset in $(X^2)^{|J|}$ this proves the claim. □

Remark 2.2. Given $J = \{t_1, \ldots, t_r\}$ as above, define $\vec{e_1} : (X^2)^r \to X^r$ by $((x_1,y_1), \ldots, (x_r,y_r)) \mapsto (x_1, \ldots, x_r)$. Then

$$(\vec{e_1})_*\, Q^{(n)}_J = P^{\alpha_1}_J \qquad (\forall n)$$

where $P^{\alpha_1}_J$ is defined – as before – as a probability measure on X^r by

$$P^{\alpha_1}_J(B_1 \times \ldots B_r) = \int_X\int_{B_1}\cdots\int_{B_r} p_{t_r-t_{r-1}}(x_{r-1},dx_r)\ \ldots\ p_{t_1}(x_0,dx_1)\,\alpha_1(dx_0).$$

Similarly, $(\vec{e_2})_* Q^{(n)}_J = P^{\alpha_2}_J$.

Proposition 2.2. *(i) There exist a projective family $\{Q^\alpha_J : J \text{ finite } \subset \mathsf{D}\}$ of probability measures and a subsequence $(n_l)_{l\in\mathbb{N}}$ such that for each finite $J \subset \mathsf{D}$*

$$Q^{(n_l)}_J \to Q^\alpha_J \quad \textit{weakly on } (X^2)^{|J|}$$

as $l \to \infty$.

(ii) For each finite $J \subset \mathsf{D}$

$$(\overrightarrow{e_1})_* Q_J^\alpha = P_J^{\alpha_1} \qquad (\overrightarrow{e_2})_* Q_J^\alpha = P_J^{\alpha_2}$$

where $P_J^{\alpha_1}$ and $P_J^{\alpha_2}$ denote the finite dimensional distribution of the heat flow on X with initial distribution α_1 and α_2, resp.

Proof. (i) For fixed J the existence of a converging subsequence $Q_J^{(n_l)}, l \in \mathbb{N}$, follows from the tightness result of the previous lemma. A diagonal sequence argument allows to choose this subsequence jointly for all finite $J \subset \mathsf{D}$. The consistency condition is preserved under convergence.

(ii) is an immediate consequence of Remark 2.2. □

The previous proposition together with Kolmogorov's extension theorem yields

Corollary 2.1. *There exists a probability measure $\mathbf{Q}_\mathsf{D}^\alpha$ on $(X^2)^\mathsf{D}$ such that for all finite $J \subset \mathsf{D}$*

$$(\pi_J)_* \mathbf{Q}_\mathsf{D}^\alpha = Q_J^\alpha.$$

Moreover,

$$(\overrightarrow{e_1})_* \mathbf{Q}_\mathsf{D}^\alpha = \mathbf{P}_\mathsf{D}^{\alpha_1} \qquad (\overrightarrow{e_2})_* \mathbf{Q}_\mathsf{D}^\alpha = \mathbf{P}_\mathsf{D}^{\alpha_2}.$$

Now let $\pi_t = (\pi_t^1, \pi_t^2) : (X^2)^\mathsf{D} \to X^2$, $\omega \mapsto (\omega_1(t), \omega_2(t))$ be the coordinate process. Then under the measure $\mathbf{Q}_\mathsf{D}^\alpha$

- $(\pi_{t/2}^1)_{t\in\mathsf{D}}$ is a Brownian motion on X (restricted to dyadic time instances) with initial distribution α_1;
- $(\pi_{t/2}^2)_{t\in\mathsf{D}}$ is a Brownian motion on X (restricted to dyadic time instances) with initial distribution α_2.

It follows that under $\mathbf{Q}_\mathsf{D}^\alpha$ the process $(\pi_t^1)_{t\in\mathsf{D}}$ has the local uniform continuity property (63.6) from [8]. The same is true for the process $(\pi_t^2)_{t\in\mathsf{D}}$. Hence, also the joint process $\pi_t = (\pi_t^1, \pi_t^2), t \in \mathsf{D}$, satisfies an analogous property. Thus for $\mathbf{Q}_\mathsf{D}^\alpha$-a.e. ω the limit

$$B_t = \lim_{s\to t, s\in\mathsf{D}} \pi_{s/2}$$

exists for each $t \in \mathbb{R}_+$, it coincides with $\pi_{t/2}$ for each $t \in \mathsf{D}$ and the trajectory $t \mapsto B_t(\omega)$ is continuous ([8], Lemma 63.5).

Theorem 2.1. *Given any initial distribution α on X^2 then the stochastic process $(B_t)_{t\in\mathbb{R}_+}$ with values in X^2, constructed as above on the probability space $((X^2)^\mathsf{D}, \mathcal{B}(X^2)^\mathsf{D}, \mathbf{Q}_\mathsf{D}^\alpha)$, is a coupling of two Brownian motions $(B_t^1)_{t\in\mathbb{R}_+}$ and $(B_t^2)_{t\in\mathbb{R}_+}$ with values in X and with initial distributions α_1 and α_2, resp., and it satisfies*

$$d(B_{s+t}^1, B_{s+t}^2) \le e^{-Kt/2} \cdot d(B_s^1, B_s^2) \qquad (\forall s,t \ge 0) \tag{2.1}$$

for $\mathbf{Q}_\mathsf{D}^\alpha$-a.e. path.

Remark 2.3. The construction of coupled Brownian motions satisfying the pathwise estimate (2.1) is well-known in the case of smooth Riemannian manifold, cf. [28] and references therein. In this case, it is most naturally constructed using the Brownian motion on the frame bundle and estimates for the stochastic parallel transport.

3. Variable Curvature Bounds

For the remaining discussions, let us assume that (X,d,m) is a mms which satisfies all the previous requirements (complete separable length space, full support, integrability condition (1.1), infinitesimally Hilbertian, condition C) and which satisfies one of the equivalent properties of Theorem 1.1 for some (large negative) $\mathrm{K} \in \mathbb{R}$. In addition to this uniform bound K we want to impose another variable curvature bound k. Here and in the sequel k will denote a lower semicontinuous function on X, bounded from below by the number K and locally m-integrable.

3.1. *Bochner Inequality and Gradient Estimates*

Given such a function k we define the *Schrödinger semigroup* $(T_t^{2\mathrm{k}})_{t\geq 0}$ as the strongly continuous semigroup of operators on $L^2(X,m)$ with generator $\Delta - 2\mathrm{k}$. (If k is locally bounded, this operator can be understood in any of the possible ways. If k is merely L^1_{loc} and bounded from below, the operator should be regarded as the *form sum* of the Laplacian Δ and the multiplication operator $-2\mathrm{k}$.) See e.g. [23]. An explicit representation of this semigroup is given in terms of Brownian motion on X by means of the Feynman-Kac formula:

$$T_t^{2\mathrm{k}}u(x) = \mathbb{E}_x\left[e^{-\int_0^{2t}\mathrm{k}(B_s)ds}\cdot u(B_{2t})\right]$$

for m-a.e. $x \in X$, each $t > 0$ and each bounded Borel measurable L^2-function u on X. To see the strong continuity in L^p note that $\int_0^t \mathrm{k}(B_s)ds \to 0$ as $t \to 0$ $\mathbf{P}^x$-a.s. for m-a.e. $x \in X$ since k is locally integrable. Thus $M_t = \exp(-\int_0^{2t}\mathrm{k}(B_s)ds)$ is uniformly bounded from above by $e^{2|\mathrm{K}|t}$ and a.s. continuous at $t = 0$. Lebesgue's dominated convergence theorem therefore implies for each Borel $f \in L^p(X,m)$

$$\begin{aligned}\int_X \left|T_t^{2\mathrm{k}}f - T_t f\right|^p dm &\leq \int_X \mathbb{E}_x\left[|M_t - 1|\cdot f(B_{2t})\right]^p dm(x)\\ &\leq \int_X \left(\int_X \mathbb{E}_x\left[|M_t - 1|^{p'}\right]^{p/p'} p_t(y,dx)\right)\cdot |f|^p(y)\,dm(y)\\ &\to 0\end{aligned}$$

as $t \to 0$. Since in addition we know $\int_X \left|T_t f - f\right|^p dm \to 0$ this proves the strong continuity in L^p.

Theorem 3.1. *For any mms (X,d,m) as above, the following properties are equivalent*

(iv') *L^2-gradient estimate: $\forall u \in Dom(\mathcal{E})$ and $\forall t > 0$*

$$\Gamma(T_t u) \leq T_t^{2\mathrm{k}}(\Gamma(u)). \tag{3.1}$$

(v') *Bochner's inequality or Bakry-Emery condition* $\mathrm{BE}(\mathrm{k},\infty)$*: $\forall u \in Dom(\Delta)$ with $\Delta u \in Dom(\mathcal{E})$ and $\forall \phi \in Dom(\Delta)\cap L^\infty(X,m)$ with $\phi \geq 0$, $\Delta\phi \in L^\infty(X,m)$*

$$\int \left(\frac{1}{2}\Delta - \mathrm{k}\right)\phi\cdot\Gamma(u)\,dm \geq \int \phi\cdot\Gamma(u,\Delta u)\,dm. \tag{3.2}$$

Recall that according to Cor. 1 the square field operator $\Gamma(u)$ of a function $u \in Dom(\mathcal{E})$ coincides with the squared minimal weak upper gradient $|Du|_w^2$.

Proof. The proof follows the argumentation for constant curvature bounds. Put

$$f(s) = \int T_s^{2k}\phi \cdot \Gamma(T_{t-s}u)\, dm = \int \phi \cdot T_s^{2k}\Gamma(T_{t-s}u)\, dm$$

and assume (v'). Then

$$f'(s) = \int (\Delta - 2k)T_s^{2k}\phi \cdot \Gamma(T_{t-s}u)\, dm - 2\int T_s^{2k}\phi \cdot \Gamma(T_{t-s}u, \Delta T_{t-s}u))\, dm$$
$$\geq 0.$$

(The regularity issues which guarantee that these calculations are applicable have been discussed in detail in [2] and [12].) Thus $f(t) \geq f(0)$. This is the claim in (iv'). Conversely, assume (iv'). Then $f(s) \geq f(0)$ for all $s > 0$. Thus $f'(0) \geq 0$. That is,

$$\int (\Delta - 2k)\phi \cdot \Gamma(T_t u)\, dm - 2\int \phi \cdot \Gamma(T_t u, \Delta T_t u))\, dm \geq 0$$

for all $t > 0$. By density (or passing to $t \to 0$) this yields (v'). □

3.2. *Curvature-Dimension Condition and Evolution-Variation Inequality*

Let us introduce the adaption of the $\mathrm{CD}(\mathrm{K},\infty)$-condition to non-constant curvature bounds.

Definition 3.1. We say that (X, d, m) satisfies the *curvature-dimension condition* $\mathrm{CD}(\mathrm{k},\infty)$ with the curvature bound $\mathrm{k} : X \to \mathbb{R}$ if for every $\mu_0, \mu_1 \in \mathcal{P}_2(X) \cap Dom(\mathrm{Ent})$ there exists a geodesic $(\mu_t)_{t\in[0,1]}$ in $\mathcal{P}_2(X)$ connecting them and a probability measure $\Theta \in \Gamma(X)$ such that $\mu_t = (e_t)_*\Theta$ and

$$\mathrm{Ent}(\mu_t) \leq (1-t)\,\mathrm{Ent}(\mu_0) + t\,\mathrm{Ent}(\mu_1) - \int_0^1 \int_{\Gamma(X)} g(s,t)\cdot \mathrm{k}(\gamma_s)\cdot |\dot\gamma|^2\, d\Theta(\gamma)\, ds \qquad (3.3)$$

for all $t \in [0,1]$. Here $e_t : \gamma \mapsto \gamma(t)$ denotes the evaluation map, $\Gamma(X)$ denotes the space of (constant speed, minimizing) geodesics in X parametrized by $[0,1]$ and $|\dot\gamma|$ denotes the speed of such a geodesic. Moreover, $g(s,t) = \min\{s(1-t), t(1-s)\}$ is the Green function of the unit interval.

Note that for constant $\mathrm{k} = \mathrm{K}$

$$\int_0^1 \int_{\Gamma(X)} g(s,t)\cdot \mathrm{k}(\gamma_s)\cdot |\dot\gamma|^2\, d\Theta(\gamma)\, ds = K \cdot \frac{t(1-t)}{2} \cdot W_2^2(\mu_0,\mu_1).$$

Definition 3.2. We say that (X, d, m) satisfies the *evolution-variation inequality* EVI_k with the curvature bound $\mathrm{k} : X \to \mathbb{R}$ if for every $\mu_0 \in \mathcal{P}_2(X)$ there exists a curve $(\mu_t)_{t>0}$ in $Dom(\mathrm{Ent})$ with $\lim_{t\to 0}\mu_t = \mu_0$ and for each $t > 0$ and each $\nu \in \mathcal{P}_2(X)$ there exists a probability measure Θ_t on $\Gamma(X)$ such that $(e_s)_*\Theta_t$ for $s \in [0,1]$ defines a geodesic in $\mathcal{P}_2(X)$ connecting μ_t and ν and such that

$$\frac{d^+}{dt}\frac{1}{2}W_2^2(\mu_t,\nu) + \int_0^1 \int_{\Gamma(X)} (1-s)\,\mathrm{k}(\gamma_s)\,|\dot\gamma|^2\, d\Theta_t(\gamma)\, ds \leq \mathrm{Ent}(\nu) - \mathrm{Ent}(\mu_t). \qquad (3.4)$$

Note that for both (3.3) and (3.4) the measures Θ and Θ_t are indeed unique [19], cf. Cor. 1, provided one restricts the discussion to absolutely continuous measures μ_0, μ_1 or μ_t, ν, resp.

Theorem 3.2. *For any mms (X, d, m) as above, the following properties are equivalent*

(i') *The curvature-dimension condition* $\mathrm{CD}(\mathrm{k}, \infty)$;

(ii') *The evolution-variation inequality* $\mathrm{EVI}_{\mathrm{k}}$.

Proof. "(i')$\Longrightarrow$(ii')": Let us first note that for each geodesic $(\eta_r)_{r\in[0,1]}$ with the property (3.3) a straightforward calculation yields

$$\frac{d^+}{dr}\Big|_{r=0} \mathrm{Ent}(\eta_r) \leq \mathrm{Ent}(\eta_1) - \mathrm{Ent}(\eta_0) - \int_0^1 \int_{\Gamma(X)} (1-s)\cdot \mathrm{k}(\gamma_s)\cdot |\dot\gamma|^2 \, d\Theta(\gamma)\, ds. \tag{3.5}$$

We apply this to the geodesic connecting $\mu_t = \eta_0$ and $\nu = \eta_1$. Thanks to [1], Thm 6.3 and Prop. 6.6 we already know

$$\frac{d^+}{dt}\frac{1}{2}W_2^2(\mu_t, \nu) \leq \frac{d^+}{dr}\Big|_{r=0} \mathrm{Ent}(\eta_r).$$

(Originally, this was proven only for 'good' geodesics. However, according to [19] geodesics in $\mathcal{P}_2(X)$ are unique and thus 'good' - provided their endpoints are absolutely continuous measures.) Combining this with the previous estimate yields the claim.

"(ii')$\Longrightarrow$(i')": Let an arbitrary geodesic $(\mu_r)_{r\in[0,1]}$ be given, fix $r \in (0,1)$, and let $(\mu_r^t)_{t>0}$ be the solution to the $\mathrm{EVI}_{\mathrm{k}}$-flow starting at μ_r. Obviously, any $\mathrm{EVI}_{\mathrm{k}}$-flow is also an $\mathrm{EVI}_{\mathrm{K}}$-flow for any constant $\mathrm{K} \leq \mathrm{k}$ and thus coincides with the heat flow $(P_t\mu_r)_{t>0}$ starting in μ_r. Now apply the $\mathrm{EVI}_{\mathrm{k}}$-inequality (3.4) to μ_r^t and $\nu = \mu_0$ (or $\nu = \mu_1$, resp.). Then

$$\frac{d^+}{dt}\frac{1}{2}W_2^2(\mu_r^t, \mu_0) + \int_0^1 \int_{\Gamma(X)} (1-s)\, \mathrm{k}(\gamma_s)\, |\dot\gamma|^2 \, d\Theta_t^0(\gamma)\, ds \leq \mathrm{Ent}(\mu_0) - \mathrm{Ent}(\mu_r^t) \tag{3.6}$$

and

$$\frac{d^+}{dt}\frac{1}{2}W_2^2(\mu_r^t, \mu_1) + \int_0^1 \int_{\Gamma(X)} (1-s)\, \mathrm{k}(\gamma_s)\, |\dot\gamma|^2 \, d\Theta_t^1(\gamma)\, ds \leq \mathrm{Ent}(\mu_1) - \mathrm{Ent}(\mu_r^t) \tag{3.7}$$

where Θ_t^0 and Θ_t^1 denote probability measures on $\Gamma(X)$ such that $(e_s)_*\Theta_t^0, s \in [0,1]$ and $(e_s)_*\Theta_t^1, s \in [0,1]$ are geodesics connecting μ_r^t with μ_0 or μ_1, resp. Since $(\mu_r)_{r\in[0,1]}$ is a geodesic

$$\begin{aligned}(1-r)\, W_2^2(\mu_0, \mu_r^t) + r\, W_2^2(\mu_r^t, \mu_1) &\geq (1-r)r\, W_2^2(\mu_0, \mu_1)\\ &= (1-r)\, W_2^2(\mu_0, \mu_r) + r\, W_2^2(\mu_r, \mu_1)\end{aligned}$$

for all $t \geq 0$ and thus

$$(1-r)\cdot \frac{d^+}{dt}\frac{1}{2}W_2^2(\mu_r^t, \mu_0) + r\cdot \frac{d^+}{dt}\frac{1}{2}W_2^2(\mu_r^t, \mu_1) \geq 0.$$

Adding up (3.6) – multiplied by $(1-r)$ – and (3.7) – multiplied by r – and taking into account (3.2) therefore yields

$$\begin{aligned}&(1-r)\mathrm{Ent}(\mu_0)+r\mathrm{Ent}(\mu_1)-\mathrm{Ent}(\mu_r^t)\\ &\geq (1-r)\int_0^1\int_{\Gamma(X)}(1-s)\,\mathrm{k}(\gamma_s)\,|\dot\gamma|^2\,d\Theta_t^0(\gamma)\,ds\\ &\quad +r\int_0^1\int_{\Gamma(X)}(1-s)\,\mathrm{k}(\gamma_s)\,|\dot\gamma|^2\,d\Theta_t^1(\gamma)\,ds.\end{aligned}$$

Now let us consider the limit $t\to 0$. Lower semicontinuity of Ent allows to pass to the limit on the LHS of the previous estimate. Passing to the limit on the RHS is justified by the fact that

$$\Theta\mapsto\int_0^1\int_{\Gamma(X)}(1-s)\,\mathrm{k}(\gamma_s)\,|\dot\gamma|^2\,d\Theta(\gamma)\,ds$$

is a lower semicontinuous function on the space of probabilty measure on $\Gamma(X)$. Thus

$$\begin{aligned}&(1-r)\mathrm{Ent}(\mu_0)+r\mathrm{Ent}(\mu_1)-\mathrm{Ent}(\mu_r)\\ &\geq (1-r)\int_0^1\int_{\Gamma(X)}(1-s)\,\mathrm{k}(\gamma_s)\,|\dot\gamma|^2\,d\Theta^0(\gamma)\,ds\\ &\quad +r\int_0^1\int_{\Gamma(X)}(1-s)\,\mathrm{k}(\gamma_s)\,|\dot\gamma|^2\,d\Theta^1(\gamma)\,ds\end{aligned}\tag{3.8}$$

where Θ^0 and Θ^1 now denote probability measures on $\Gamma(X)$ which represent geodesics in $\mathcal{P}_2(X)$ connecting μ_r with μ_0 and μ_1, resp.

Now we will reparametrize and glue together the measures Θ^0 and Θ^1. First we define the measure $\hat\Theta^0:=(\Psi^0)_*\Theta^0$ on the space of geodesics $\gamma:[0,r]\to X$ and the measure $\hat\Theta^1=(\Psi^1)_*\Theta^1$ on the space of geodesics $\gamma:[r,1]\to X$ by means of the rescaling maps

$$(\Psi^0\gamma)_s=\gamma_{1-s/r},\qquad (\Psi^1\gamma)_s=\gamma_{(s-r)/(1-r)}.$$

Note that the evaluation map at time r for both measures leads to the same projection

$$(e_r)_*\Theta^0=\mu_r=(e_r)_*\Theta^1.$$

Disintegration w.r.t. μ_r yields Markov kernels $\vartheta^0(y,d\gamma)$ and $\vartheta^1(y,d\gamma)$ such that

$$\hat\Theta^i(d\gamma)=\int_X\vartheta^i(y,d\gamma)\,\mu_r(dy)$$

for $i=0,1$. Let $\Gamma_r(X)$ denote the space of constant speed curves $\gamma:[0,1]\to X$ for which $\gamma^0:=\gamma\big|_{[0,r]}$ and $\gamma^1:=\gamma\big|_{[r,1]}$ are geodesics. Define a probability measure Θ on $\Gamma_r(X)$ by

$$\Theta(d\gamma)=\int_X\vartheta^0(y,d\gamma^0)\,\vartheta^1(y,d\gamma^0)\,\mu_r(dy).$$

By construction, this measure Θ lives on the set of piecewise geodesic curves. Our claim, however, is that it is indeed supported by the set of geodesics. Let us consider the curve $\tilde\mu_s=(e_s)_*\Theta$, $s\in[0,1]$, which connects μ_0 and μ_1. Moreover, $W_2(\tilde\mu_0,\tilde\mu_r)=W_2(\mu_0,\mu_r)$

and $W_2(\tilde{\mu}_r,\tilde{\mu}_1) = W_2(\mu_r,\mu_1)$. Thus $(\tilde{\mu}_s)_{s\in[0,1]}$ is a geodesic in $\mathcal{P}_2(X)$ and therefore Θ is supported by the set of geodesics in X.

In terms of this probability measure Θ on $\Gamma(X)$, (3.8) can be rewritten as

$$(1-r)\mathrm{Ent}(\mu_0) + r\mathrm{Ent}(\mu_1) - \mathrm{Ent}(\mu_r) \geq \int_0^1 \int_{\Gamma(X)} g(r,s)\,\mathrm{k}(\gamma_s)\,|\dot{\gamma}|^2\,d\Theta(\gamma)\,ds.$$

This is the claim for a given $r \in [0,1]$. A priori the measure Θ might depend on r. Uniqueness of the optimal plan Θ, however, implies then that the claim holds for every r. □

3.3. *From Bakry-Emery to Curvature-Dimension*

The following action estimate in the spirit of [2], Theorem 4.16, is regarded as a key ingredient for proving that the Bakry-Emery condition $\mathrm{BE}(\mathrm{k},\infty)$ implies the curvature-dimension condition $\mathrm{CD}(\mathrm{k},\infty)$ and the evolution-variation inequality $\mathrm{EVI}_{\mathrm{k}}$.

Lemma 3.1. i) *Let $\rho_s = f_s\, m$, $s\in[0,1]$ be a regular curve (in the sense of [2], Def. 4.10) and put $\rho_{s,t} = P_{st}\rho_s$. Then for every Lipschitz function ϕ with bounded support and for all $t>0$*

$$\int \phi_1\,d\rho_{1,t} - \int \phi_0\,d\rho_0 - \int_0^1 |\dot{\rho}_s|^2\,ds + 2t\Big(\mathrm{Ent}(\rho_{1,t}) - \mathrm{Ent}(\rho_0)\Big)$$
$$\leq -\int_0^1\int_0^{st}\int_X T_r\left(2\mathrm{k}\cdot T^{2\mathrm{k}}_{st-r}\Gamma(\phi_s)\right)\,d\rho_s\,dr\,ds$$

where $\phi_s = Q_s\phi$, $s\in[0,1]$, denote the Hamilton-Jacobi flow induced by ϕ.

ii) *The same holds true for every geodesic $(\rho_s)_{s\in[0,1]}$ in $\mathcal{P}_2(X)$ with absolutely continuous measures in which case $\int_0^1|\dot{\rho}_s|^2\,ds = W_2^2(\rho_0,\rho_1)$.*

Proof. i) For $\mathrm{k}=0$, obviously the RHS vanishes and the estimate coincides with the result in [2], Thm 4.16. There is only one estimate which changes if one passes from $\mathrm{k}=0$ to $\mathrm{k}\neq 0$: In the previous case, the gradient estimate allows to estimate

$$-\int_0^1\int_X T_{st}\Gamma(\phi_s + t\,g^\epsilon_{s,t})\cdot f_s\,dm\,ds$$

in terms of

$$-\int_0^1\int_X \Gamma(T_{st}(\phi_s + t\,g^\epsilon_{s,t}))\cdot f_s\,dm\,ds. \tag{3.9}$$

(For the definition of $g^\epsilon_{s,t}$ we refer to [2].) In the case $\mathrm{k}\neq 0$, our new gradient estimate (3.1) will allow to estimate

$$-\int_0^1\int_X T^{2\mathrm{k}}_{st}\Gamma(\phi_s + t\,g^\epsilon_{s,t})\cdot f_s\,dm\,ds$$

in terms of the same expression (3.9) as before. Thus

$$\int \phi_1\,d\rho_{1,t} - \int \phi_0\,d\rho_0 - \int_0^1 |\dot{\rho}_s|^2\,ds + 2t(\mathrm{Ent}(\rho_{1,t}) - \mathrm{Ent}(\rho_0))$$

$$\leq -\int_0^1 \int_X T_{st}\Gamma(\phi_s + t\, g^\epsilon_{s,t}) \cdot f_s \, dm\, ds$$
$$+ \int_0^1 \int_X T^{2k}_{st}\Gamma(\phi_s + t\, g^\epsilon_{s,t}) \cdot f_s \, dm\, ds. \tag{3.10}$$

Duhamel's principle states that

$$T_{st} - T^{2k}_{st} = \int_0^{st} T_r\big(2k \cdot T^{2k}_{st-r}\big)\, dr$$

in the sense of operators. Thus the RHS of (3.10) can be estimated by

$$-\int_0^1 \int_0^{st} \int_X T_r \left(2k \cdot T^{2k}_{st-r}\Gamma(\phi_s + t\, g^\epsilon_{s,t})\right) \cdot f_s \, dm\, dr\, ds.$$

Thanks to the uniform estimates and convergence properties of $g^\epsilon_{s,t}$ established in [2], we obtain in the limit $\epsilon \to 0$

$$\int \phi_1 \, d\rho_{1,t} - \int \phi_0 \, d\rho_0 - \int_0^1 |\dot\rho_s|^2 \, ds + 2t(\mathrm{Ent}(\rho_{1,t}) - \mathrm{Ent}(\rho_0))$$
$$\leq -\int_0^1 \int_0^{st} \int_X T_r \left(2k \cdot T^{2k}_{st-r}\Gamma(\phi_s)\right) \cdot f_s \, dm\, dr\, ds$$

which is the first claim.

ii) Given any geodesic $(\rho_s)_{s\in[0,1]}$ with absolutely continuous measures $\rho_s = f_s\, m$, we approximate it by regular curves $(\rho^n_s)_{s\in[0,1]}$. The measures ρ^n_s will be absolutely continuous with densities f^n_s. We may choose the approximation such that $f^n_s(x) \to f_s(x)$ for a.e. $(x, s) \in X \times [0, 1]$ as $n \to \infty$. Put $\rho^n_{1,t} = \mu^n_t = P_t\mu^n_0$ and let $\phi_s = Q_s\phi$ be the Hamilton-Jacobi flow induced by any Lipschitz function ϕ with bounded support. Part i) of the lemma then yields

$$\int \phi_1 \, d\rho^n_{1,t} - \int \phi_0 \, d\rho^n_0 - \int_0^1 |\dot\rho^n_s|^2 \, ds + 2t\Big(\mathrm{Ent}(\rho^n_{1,t}) - \mathrm{Ent}(\rho^n_0)\Big)$$
$$\leq -\int_0^1 s \int_0^t \int_X T_{sr} \left(2k \cdot T^{2k}_{s(t-r)}\Gamma(\phi_s)\right) \cdot f^n_s \, dm\, dr\, ds.$$

Of course, $\int \phi_1 \, d\rho^n_{1,t} \to \int \phi_1 \, d\rho_{1,t}$ and $\int \phi_0 \, d\rho^n_0 \to \int \phi_0 \, d\rho_0$ as $n \to \infty$ as well as $\int_0^1 |\dot\rho^n_s|^2 \, ds \to W_2^2(\rho_0, \rho_1)$. Lower semicontinuity of the entropy implies $\mathrm{Ent}(\rho_{1,t}) \leq \liminf_{n\to\infty} \mathrm{Ent}(\rho^n_{1,t})$ and by construction of the regular curve we may achieve that $\mathrm{Ent}(\rho_0) \geq \limsup_{n\to\infty} \mathrm{Ent}(\rho^n_0)$. Passing to the limit $n \to \infty$ the previous estimate thus yields

$$\int \phi_1 \, d\rho_{1,t} - \int \phi_0 \, d\rho_0 - W_2^2(\rho_0, \rho_1) + 2t\Big(\mathrm{Ent}(\rho_{1,t}) - \mathrm{Ent}(\rho_0)\Big)$$
$$\leq -\limsup_{n\to\infty} \int_0^1 s \int_0^t \int_X T_{sr} \left(2k \cdot T^{2k}_{s(t-r)}\Gamma(\phi_s)\right) \cdot f^n_s \, dm\, dr\, ds$$
$$\leq -\int_0^1 s \int_0^t \int_X T_{sr} \left(2k \cdot T^{2k}_{s(t-r)}\Gamma(\phi_s)\right) \cdot f_s \, dm\, dr\, ds \tag{3.11}$$

where the last inequality follows by means of Fatou's lemma from lower boundedness of $T_{sr}\big(2\mathrm{k}\cdot T^{2\mathrm{k}}_{s(t-r)}\Gamma(\phi_s)\big)$ from and the a.e. convergence of $f^n_s \to f_s$ (together with $\int f^n_s\,dm = \int f_s\,dm$). □

Lemma 3.2. *Let ρ_0 and $\rho_1 \in \mathcal{P}_2(X) \cap Dom(\mathrm{Ent})$ be given (without restriction, we may assume that both measures have bounded densities), let $(\rho_s)_{s\in[0,1]}$ be the unique geodesic connecting them and put $\rho_{1,t} = P_t\rho_1$.*

Fix a Kantorovich potential ϕ_0 for the optimal transport from ρ_0 to ρ_1 and for each $t > 0$ let ϕ^t_0 be a Lipschitz function with bounded support such that $\int \phi^t_1\,d\rho_{1,t} - \int \phi^t_0\,d\rho_0 \geq W_2^2(\rho_0,\rho_{1,t}) - 2t^2$ (i.e. ϕ^t_0 is 'almost optimal' for the Kantorovich duality problem for ρ_0 and $\rho_{1,t}$) and

$$\Gamma(\phi^t_s)(x) \to \Gamma(\phi_s)(x) \qquad \text{in } L^1(X\times[0,1], dm\otimes ds)$$

as $t \to 0$ where $\phi^t_s = Q_s\phi^t_0$ denotes the Hamilton-Jacobi flow induced by ϕ^t_0. Then

$$\limsup_{t\to 0}\frac{1}{2t}\Big[W_2^2(\rho_0,\rho_{1,t}) - W_2^2(\rho_0,\rho_1)\Big]$$
$$\leq \mathrm{Ent}(\rho_0) - \mathrm{Ent}(\rho_1) - \int_0^1 s\int_X \Gamma(\phi_s)\cdot \mathrm{k}\cdot f_s\,dm\,ds.$$

Proof. Due to the CD(K, ∞)-condition we know that all the measures ρ_s will be absolutely continuous with uniformly bounded densities, say $f_s \leq C_0$.

Applying the estimate of the previous Lemma to the function ϕ^t_0 (inducing the transport almost towards the measure $\rho_{1,t}$) and dividing by $-2t$ gives

$$-\frac{1}{2t}\Big[W_2^2(\rho_0,\rho_{1,t}) - W_2^2(\rho_0,\rho_1)\Big] + t - \Big(\mathrm{Ent}(\rho_{1,t}) - \mathrm{Ent}(\rho_0)\Big)$$
$$\geq \frac{1}{2t}\int_0^1 s\int_0^t\int_X \Gamma(\phi^t_s)\cdot T^{\mathrm{k}}_{s(t-r)}\left(2\mathrm{k}\cdot T_{sr}f_s\right)\,dm\,dr\,ds$$
$$\geq \int_0^1 s\int_X \Gamma(\phi^t_s)\cdot\Big[\frac{1}{t}\int_0^t T^{2\mathrm{k}}_{s(t-r)}\big((\mathrm{k}\wedge C_1)\cdot T_{sr}f_s\big)\,dr\Big]\,dm\,ds$$

for any $C_1 \geq 0$. Strong continuity of the semigroups $(T_r)_{r>0}$ and $(T^{2\mathrm{k}}_r)_{r>0}$ in L^p implies that for any $p \in [1,\infty)$

$$\frac{1}{t}\int_0^t T^{2\mathrm{k}}_{s(t-r)}\big((\mathrm{k}\wedge C_1)\cdot T_{sr}f_s\big) \to (\mathrm{k}\wedge C_1)\cdot f_s \qquad \text{in } L^p$$

as $t \to 0$ and (after passing to a subsequence – which we drop from the notation) also a.e. on $X\times[0,1]$. Moreover, note that

$$\left|\frac{1}{t}\int_0^t T^{2\mathrm{k}}_{s(t-r)}\big((\mathrm{k}\wedge C_1)\cdot T_{sr}f_s\big)\right| \leq (|K|\vee C_1)\cdot C_0\cdot e^{2|K|}$$

for a.e. (x,s) and all $t \leq 1$. Together with the L^1-convergence $\Gamma(\phi^t_s) \to \Gamma(\phi_s)$ this gives that

$$\Gamma(\phi^t_s)\cdot\frac{1}{t}\int_0^t T^{2\mathrm{k}}_{s(t-r)}\big((\mathrm{k}\wedge C_1)\cdot T_{sr}f_s\big) \to \Gamma(\phi_s)\cdot(\mathrm{k}\wedge C_1)\cdot f_s$$

in $L^1(X \times [0,], dm \otimes ds)$ as $t \to 0$. Thus

$$-\limsup_{t\to 0} \frac{1}{2t}\Big[W_2^2(\rho_0, \rho_{1,t}) - W_2^2(\rho_0, \rho_1)\Big] - \Big(\mathrm{Ent}(\rho_1) - \mathrm{Ent}(\rho_0)\Big)$$
$$\geq \int_0^1 s \int_X \Gamma(\phi_s) \cdot \big(\mathrm{k} \wedge C_1\big) \cdot f_s \, dm \, ds \tag{3.12}$$

for all $C_1 \geq 0$. Monotone convergence allows to pass to the limit $C_1 \to \infty$ in (3.12). Thus

$$-\limsup_{t\to 0} \frac{1}{2t}\Big[W_2^2(\rho_0, \rho_{1,t}) - W_2^2(\rho_0, \rho_1)\Big] - \Big(\mathrm{Ent}(\rho_1) - \mathrm{Ent}(\rho_0)\Big)$$
$$\geq \int_0^1 s \int_X \Gamma(\phi_s) \cdot \mathrm{k} \cdot f_s \, dm \, ds$$

which is the claim. □

3.4. *Local-to-Global and Stability*

Recall our assumption that (X, d, m) is infinitesimally Hilbertian.

Theorem 3.3. *If the curvature-dimension condition* $\mathrm{CD}(\mathrm{k}, \infty)$ *holds locally then it also holds globally.*

Here we say that the $\mathrm{CD}(\mathrm{k}, \infty)$-condition holds locally on X if X can be covered by open sets U_i such that for each pair of measures μ_0, μ_1 supported in $\overline{U}_i$ a connecting geodesic $(\mu_t)_{t\in[0,1]}$ exists (which might leave the set $\overline{U}_i$) with property (3.3).

Proof. According to [19] we know that an optimal transport never charges branching geodesics. This allows to pass to a "pathwise version" of (3.3), cf. [24] ("localization in space"): for Θ-a.e. $\gamma \in \Gamma(X)$ and every $t \in [0, 1]$

$$\log \rho_t(\gamma_t) \leq (1 - t) \log \rho_0(\gamma_0) + t \log \rho_1(\gamma_1) - \int_0^1 g(s, t) \cdot \mathrm{k}(\gamma_s) \cdot |\dot{\gamma}|^2 \, ds. \tag{3.13}$$

This in turn is nothing but an integrated version of the fact that $t \mapsto \rho_t(\gamma_t)$ is upper semicontinuous on $[0, 1]$, continuous on $(0, 1)$ and satisfies the differential inequality $\frac{\partial^2}{\partial t^2} \log \rho_t(\gamma_t) \geq k(\gamma_t)|\dot{\gamma}|^2$ weakly on $(0, 1)$. This obviously allows for a "localization in time". The proof thus follows by a careful adaption of the arguments in [24]. □

As a consequence of the previous local-to-global result we obtain that the curvature-dimension condition $\mathrm{CD}(\mathrm{k}, \infty)$ is stable under tensorization.

Corollary 3.1. *Assume that mms* (X_i, d_i, m_i) *for* $i = 1, \ldots, n$ *are given, each of which satisfies a curvature-dimension condition* $\mathrm{CD}(\mathrm{k}_i, \infty)$ *for some lower semicontinuous, lower bounded function* $\mathrm{k}_i : X_i \to \mathbb{R}$. *Then the product space*

$$\big(X, d, m\big) = \Big(X_1 \times \ldots \times X_n, \sqrt{d_1^2 + \ldots + d_n^2}, m_1 \otimes \ldots m_n\Big)$$

satisfies the curvature-dimension condition $\mathrm{CD}(\mathrm{k}, \infty)$ *for the function*

$$\mathrm{k}(x_1, \ldots, x_n) := \min\{0, \mathrm{k}_1(x_1), \ldots, \mathrm{k}_n(x_n)\}.$$

Proof. Let us first prove the claim in the particular case where all the functions k_i are continuous. According to the previous theorem, it suffices to prove that the $\mathrm{CD}(\mathrm{k},\infty)$-condition holds locally on X. Given $x \in X$ and $\epsilon > 0$ there exists open neighborhoods $V_i \supset U_i$ of x_i and constants K_i such that V_i contains the convex hull of $\overline{U}_i$ (i.e. geodesics with endpoints in $\overline{U}_i$ do not leave V_i) and such that

$$\mathrm{k}_i(y) \geq \mathrm{K}_i \geq \mathrm{k}_i(z) - \epsilon$$

for $y, z \in V_i$. Put $\mathrm{K} := \min\{0, \mathrm{K}_1, \dots, \mathrm{K}_n\}$. Then for optimal transports in X_i with marginals μ_0^i, μ_1^i supported in $\overline{U}_i$ the 'classical' $\mathrm{CD}(\mathrm{K}_i,\infty)$-condition (with constant curvature bound K_i) applies. Due to the tensorization property of the latter [24], the $\mathrm{CD}(\mathrm{K},\infty)$-condition (with constant curvature bound K defined as above) applies to optimal transports in X with marginals μ_0, μ_1 supported in $\overline{U} = \overline{U}_1 \times \ldots \times \overline{U}_n$. Thus the $\mathrm{CD}(\mathrm{k}-\epsilon,\infty)$-condition holds locally on X. According to the previous local-to-global theorem this implies that the $\mathrm{CD}(\mathrm{k}-\epsilon,\infty)$-condition holds globally on X. Passing to the limit $\epsilon \to 0$ yields (via monotone convergence) the $\mathrm{CD}(\mathrm{k},\infty)$-condition globally on X. This proves the claim in the particular case of continuous k_i.

Now let us treat the case of general k_i. For each i there exists an monotone sequence of continuous functions $\mathrm{k}_i^l : X_i \to \mathbb{R}$ such that

$$\mathrm{k}_i^l(x) \nearrow \mathrm{k}_i(x) \qquad (\forall x \in X_i)$$

as $l \to \infty$. Applying the previous result to the functions $(\mathrm{k}_i^l)_{i=1\dots,n}$ for fixed l yields that X satisfies the $\mathrm{CD}(\mathrm{k}^l,\infty)$-condition globally on X with $\mathrm{k}^l(x_1,\dots,x_n) := \min\{0, \mathrm{k}_1^l(x_1),\dots,\mathrm{k}_n^l(x_n)\}$. In the limit $l \to \infty$, the $\mathrm{CD}(\mathrm{k},\infty)$-condition follows by monotone convergence. □

Remark 3.1. Obviously, also the $\mathrm{BE}(\mathrm{k},\infty)$-condition is stable under tensorization.

Our next goal is to analyze how the curvature-dimension condition $\mathrm{CD}(\mathrm{k},\infty)$ behaves under change of measure.

Given functions V and λ on X, we say that V is strongly λ-convex if for every geodesic $\gamma \in \Gamma(X)$ and every $t \in [0,1]$

$$V(\gamma_t) \leq (1-t)\,V(\gamma_0) + t\,V(\gamma_1) - \int_0^1 g(s,t)\cdot\lambda(\gamma_s)\cdot|\dot{\gamma}|^2\,ds. \tag{3.14}$$

Proposition 3.1. *If (X,d,m) satisfies the curvature-dimension condition $\mathrm{CD}(\mathrm{k},\infty)$ and if $V : X \to \mathbb{R}$ is strongly λ-convex then $(X,d,e^{-V}m)$ satisfies the curvature-dimension condition $\mathrm{CD}(\mathrm{k}+\lambda,\infty)$.*

Proof. Recall that the relative entropy Ent' w.r.t. $m' = e^{-V}m$ is given by

$$\mathrm{Ent}'(\mu) = \mathrm{Ent}(\mu) + \int V\,d\mu.$$

Integrating (3.14) w.r.t. the optimal path measure Θ from (3.3) leads to

$$\int V\,\mu_t \leq (1-t)\int V\,d\mu_0 + t\int V\,d\mu_1 - \int_0^1\int_{\Gamma(X)} g(s,t)\cdot\lambda(\gamma_s)\cdot|\dot{\gamma}|^2\,d\Theta(\gamma)\,ds. \tag{3.15}$$

Adding this to (3.3) yields the claim. □

Finally, we want to study whether the $\mathrm{CD}(\mathrm{k},\infty)$-condition is stable under convergence. The precise formulation of this question already requires some care: the curvature bounds for the approximating spaces and for the limit space will be functions defined on different spaces. To avoid additional complications, we will restrict the discussion in the sequel to *normalized* mms, i.e. mms (X,d,m) with $m(X)=1$.

Recall that a sequence of normalized mms (X_n,d_n,m_n), $n\in\mathbb{N}$, converges to a mms (X,d,m) in L^2-transportation distance $\mathbb{D}$ if and only if there exists a metric space (X^*,d^*) and isometric embeddings $\iota_n : X_n \hookrightarrow X^*$, $\iota : X \hookrightarrow X^*$ such that the push forward measures converge w.r.t. L^2-Wasserstein distance W_2^* on (X^*,d^*):

$$W_2^*\Big((\iota_n)_* m_n, \iota_* m\Big) \to 0.$$

Definition 3.3. We say that a function $\mathrm{k} : X \to \mathbb{R}$ is *asymptotically dominated* by a sequence of functions $\mathrm{k}_n : X_n \to \mathbb{R}$ if for each $\epsilon > 0$ there exists $n' \in \mathbb{N}$, isometric embeddings $(\iota_n)_{n\geq n'}, \iota$ into a common metric space (X^*,d^*) as above and a lower semicontinuous function $\mathrm{k}^* : X^* \to \mathbb{R}$ such that $\forall n \geq n'$

$$\mathrm{k} \leq \mathrm{k}^* \circ \iota \quad \text{on } X, \qquad \mathrm{k}^* \circ \iota_n \leq \mathrm{k}_n + \epsilon \quad \text{on } X_n. \tag{3.16}$$

Theorem 3.4. *For each $n \in \mathbb{N}$, let (X_n,d_n,m_n) be a normalized mms which satisfies the curvature-dimension condition $\mathrm{CD}(\mathrm{k}_n,\infty)$ for some function $\mathrm{k}_n : X_n \to \mathbb{R}$. Assume that for $n\to\infty$ the sequence of spaces (X_n,d_n,m_n) converges in $\mathbb{D}$-distance to some normalized mms (X,d,m) and assume that the function $\mathrm{k} : X \to \mathbb{R}$ is asymptotically dominated by the sequence of functions $\mathrm{k}_n : X_n \to \mathbb{R}$. Then (X,d,m) satisfies the curvature-dimension condition $\mathrm{CD}(\mathrm{k},\infty)$.*

In brief words: The curvature-dimension condition $\mathrm{CD}(\mathrm{k},\infty)$ is stable under convergence.

Proof. Our proof follows the argumentation in [24]. Given the mms (X_n,d_n,m_n), $n\in\mathbb{N}$, and (X,d,m) as above we may assume without restriction that they are already isometrically embedded as subsets into some space (X^*,d^*). Let q_n be an W_2^*-optimal coupling of m_n and m. Its disintegration kernel allows to map each probability measure μ which is absolutely continuous w.r.t. m onto a probability measure μ^n which is absolutely continuous w.r.t. m_n in such a way that $\mathrm{Ent}_n(\mu^n) \leq \mathrm{Ent}(\mu)$.

Given two probability measures μ_0,μ_1 (supported on X) with finite entropy w.r.t. m we thus obtain in a canonical way corresponding probability measures μ_0^n,μ_1^n (supported on X_n) with finite entropy w.r.t. m_n. The curvature-dimension condition $\mathrm{CD}(\mathrm{k}_n,\infty)$ for the space (X_n,d_n,m_n) yields the existence of a probability measure Θ^n on $\Gamma(X_n)\subset\Gamma(X^*)$ which induces a geodesic $\mu_t^n = (e_t)_*\Theta^n$ connecting μ_0^n,μ_1^n and which satisfies

$$\begin{aligned}\mathrm{Ent}_n(\mu_t^n) &\leq (1-t)\,\mathrm{Ent}_n(\mu_0^n) + t\,\mathrm{Ent}_n(\mu_1^n)\\ &\quad - \int_0^1\int_{\Gamma(X_n)} g(s,t)\cdot \mathrm{k}_n(\gamma_s)\cdot|\dot\gamma|^2\,d\Theta^n(\gamma)\,ds\end{aligned}$$

for each t. Tightness implies the existence of a converging subsequence – again denoted by $(\Theta^n)_{n\geq n'}$ – and a limit measure $\Theta\in\mathcal{P}(\Gamma(X^*))$. Lower semicontinuity of $n\mapsto \mathrm{Ent}_n(\mu_t^n)$ (w.r.t. both measures involved) then provides the estimate

$$\mathrm{Ent}(\mu_t) \leq \liminf_{n\to\infty} \mathrm{Ent}_n(\mu_t^n).$$

Moreover, by construction we have $\mathrm{Ent}(\mu_t) = \lim_{n\to\infty} \mathrm{Ent}_n(\mu_t^n)$ for $t=0$ and $t=1$. It remains to prove

$$\liminf_{n\to\infty} \int_0^1 \int_{\Gamma(X_n)} g(s,t)\cdot \mathrm{k}_n(\gamma_s)\cdot |\dot\gamma|^2\, d\Theta^n(\gamma)\, ds$$
$$\geq \int_0^1 \int_{\Gamma(X)} g(s,t)\cdot \mathrm{k}(\gamma_s)\cdot |\dot\gamma|^2\, d\Theta(\gamma)\, ds.$$

Using the embedding into X^* and the estimates between $\mathrm{k}_n, \mathrm{k}^*$ and k it suffices to prove

$$\liminf_{n\to\infty} \int_0^1 \int_{\Gamma(X^*)} g(s,t)\cdot \mathrm{k}^*(\gamma_s)\cdot |\dot\gamma|^2\, d\Theta^n(\gamma)\, ds$$
$$\geq \int_0^1 \int_{\Gamma(X^*)} g(s,t)\cdot \mathrm{k}^*(\gamma_s)\cdot |\dot\gamma|^2\, d\Theta(\gamma)\, ds.$$

This finally follows from the weak convergence $\Theta^n \to \Theta$ and from the fact that

$$\Theta \mapsto \int_0^1 \int_{\Gamma(X^*)} g(s,t)\cdot \mathrm{k}^*(\gamma_s)\cdot |\dot\gamma|^2\, d\Theta(\gamma)\, ds$$

is a lower semicontinuous function on $\mathcal{P}(\Gamma(X^*))$ which in turn follows from the lower semicontinuity of k^*. □

References

1. Ambrosio, L., Gigli, N., Mondino, A. and Rajala, T., Riemannian Ricci curvature lower bounds in metric measure spaces with σ-finite measures, Preprint arXiv:1207.4924, 2012.
2. Ambrosio, L., Gigli, N. and Savaré, G., Bakry-Émery curvature-dimension condition and Riemannian Ricci curvature bounds, Preprint arXiv:1209.5786, 2012.
3. Ambrosio, L., Gigli, N. and Savaré, G., Calculus and heat flow in metric measure spaces and applications to spaces with Ricci bounds from below, Invent. Math. 2013.
4. Ambrosio, L., Gigli, N. and Savaré, G., Metric measure spaces with Riemannian Ricci curvature bounded from below, Preprint arXiv:1109.0222, 2011.
5. Bacher, K. and Sturm, K.T., Localization and tensorization properties of the curvature-dimension condition for metric measure spaces, J. Funct. Anal. **259** (2010), 28–56.
6. Bakry, D., L'hypercontractivité et son utilisation en théorie des semigroupes, pp. 1-114 in *Lectures on probability theory, Saint-Flour 1992*, Lecture Notes in Math., 1581, Springer, 1994.
7. Bakry, D. and Émery, M., Diffusions hypercontractives, Séminaire de probabilités, XIX, 1983/84, Lecture Notes in Math. **1123**, 177–206, Springer, Berlin 1985.
8. Bauer, H., *Probability theory and elements of measure theory*, 2nd ed. Academic Press 1982.
9. Bogachev, V.I., *Measure theory. Vol. I, II.* Springer 2007.
10. Cavalletti, F. and Sturm, K.T., Local curvature-dimension condition implies measure-contraction property, J. Funct. Anal. **262** (2012), 5110–5127.
11. Erbar, M., The heat equation on manifolds as a gradient flow in the Wasserstein space, Ann. Inst. Henri Poincaré Probab. Stat., **46** (2010), 1–23.

12. Erbar, M., Kuwada, K. and Sturm, K.T. On the equivalence of the entropic curvature-dimension condition and Bochner's inequality on metric measure spaces, Preprint arXiv:1303.4382(2013).
13. Fukushima, M., Oshima, Y. and Takeda, M., *Dirichlet forms and symmetric Markov processes.* De Gruyter 2011.
14. Kuwada, K., Duality on gradient estimates and Wasserstein controls, J. Funct. Anal. **258** (2010), 3758 - 3774.
15. Kuwada, K., Space-time Wasserstein control and Bakry-Ledoux type gradient estimates, Preprint 2013.
16. Lott, J. and Villani, C. Ricci curvature for metric-measure spaces via optimal transport, Annals of Math. **169** (2009), 903-991.
17. Ohta, S.-I. and Sturm, K.T., Heat flow on Finsler manifolds, Comm. Pure Appl. Math. **62** (2009), 1386–1433.
18. Ohta, S.-I. and Sturm, K.T., Bochner–Weitzenböck formula and Li-Yau esitmate on Finsler manifolds, Adv. Math **252** (2014), 429 - 448.
19. Rajala, T. and Sturm, K.T., Non-branching geodesics and optimal maps in strong CD(K, ∞)-spaces , Preprint arXiv:1207.6754, to appear in Calc. Var. PDEs.
20. Renesse, M.-K. von and Sturm, K.T., Transport inequalities, gradient estimates, entropy, and Ricci curvature, Comm. Pure Appl. Math. **58** (2005), 923–940.
21. Revuz,D. and Yor, M., *Continuous martingales and Brownian motion.* Springer 1999.
22. Savaré, G., Self-improvement of the Bakry-Émery condition and Wasserstein contraction of the heat flow in RCD(K,∞) metric measure spaces, to appear in Disc. Cont. Dyn. Sist. A. (2013) 1-23
23. Stollmann, P. and Voigt, J., Perturbation of Dirichlet forms by measures, Potential Analysis **5** (1996), 109 - 138.
24. Sturm, K.T., On the geometry of metric measure spaces. I, Acta Math. **196** (2006), 65–131.
25. Sturm, K.T., On the geometry of metric measure spaces. II, Acta Math. **196** (2006), 133–177.
26. Sturm, K.T., Ricci Tensor for Diffusion Operators and Curvature-Dimension Inequalities under Conformal Transformations and Time Changes, Preprint Arxiv, 2014.
27. Villani, C. *Optimal transport, old and new.* Springer 2009.
28. Wang, Feng-Yu, On estimation of the logarithmic Sobolev constant and gradient estimates of heat semigroups, Probab. Theory Related Fields **108** (1997), 87–101.

Chapter 28

Intrinsic ultracontractivity and semi-small perturbation for skew product diffusion operators

Matsuyo Tomisaki*

We consider a diffusion operator $\mathcal{L}$ associated with a skew product of a one-dimensional generalized diffusion process R and the spherical Brownian motion Θ with respect to a positive continuous additive functional $\mathbf{f}(t)$ of R. We give sufficient conditions for [SSP] (i.e., 1 is a semismall perturbation of the operator $\mathcal{L}$) and [IU] (i.e., the associated heat kernel is intrinsically ultracontractive) in terms of characteristics of R and $\mathbf{f}(t)$.

Contents

1. Introduction

Let s^{R} be a continuous strictly increasing function on an open interval $I = (l_1, l_2)$, and m^{R} be a right continuous strictly increasing function on I, where $0 \le l_1 < l_2 \le \infty$. We denote by $\mathrm{R} = [R_t, P_r^{\mathrm{R}}]$ a one-dimensional generalized diffusion process on I with scale function s^{R}, speed measure m^{R} and no killing measure (ODGDP with $(s^{\mathrm{R}}, m^{\mathrm{R}}, 0)$ for brief). Let $\Theta = [\Theta_t, P_\theta^{\Theta}]$ be the spherical Brownian motion on $S^{d-1} \subset \mathbb{R}^d$ with generator $(1/2)\Delta^{\Theta}$, Δ^{Θ} being the spherical Laplacian on S^{d-1}. In this paper we study [IU] and [SSP] for the generator of the skew product $\mathrm{X} = [X_t = (R_t, \Theta_{\mathbf{f}(t)}), P_{(r,\theta)}^{\mathrm{X}} = P_r^{\mathrm{R}} \otimes P_\theta^{\Theta},\ (r,\theta) \in I \times S^{d-1}]$ with respect to a positive continuous additive functional (PCAF for brief) $\mathbf{f}(t)$ of the ODGDP R.

We consider the following symmetric bilinear form $(\mathcal{E}^{\mathrm{R}}, \mathcal{F}^{\mathrm{R}})$.

$$\mathcal{E}^{\mathrm{R}}(u,v) = \int_I \frac{du}{ds^{\mathrm{R}}} \frac{dv}{ds^{\mathrm{R}}} \, ds^{\mathrm{R}},$$

$$\mathcal{F}^{\mathrm{R}} = \{u \in L^2(I, m^{\mathrm{R}}) : u \text{ is absolutely continuous on } I \text{ with respect to } ds^{\mathrm{R}} \text{ and } \mathcal{E}^{\mathrm{R}}(u,u) < \infty\}.$$

We set $\mathcal{C}^{\mathrm{R}} = \{u \circ s^{\mathrm{R}} : u \in C_0^\infty(J)\}$, where $J = s^{\mathrm{R}}(I)$ and $C_0^\infty(A)$ is the set of all infinitely

*Nara Women's University, Kita-Uoya Nishimachi, Nara, 630-8506 Japan, e-mail: tomisaki@cc.nara-wu.ac.jp

differentiable functions on A with compact support. Then $(\mathcal{E}^{\rm R}, \mathcal{F}^{\rm R})$ is a regular, strongly local, irreducible Dirichlet form on $L^2(I, m^{\rm R})$ possessing $\mathcal{C}^{\rm R}$ as its core and corresponding to the ODGDP R (see [1], [6]). We denote by $p_t^{\rm R}$ the semigroup, that is,

$$p_t^{\rm R} f(r) = E^{P_r^{\rm R}}[f(R_t)] = \int_I p^{\rm R}(t,r,\xi) f(\xi)\, dm^{\rm R}(\xi), \qquad t>0,\ r\in I,$$

for $f \in C_b(I)$, where $C_b(A)$ is the set of all bounded continuous functions on a set A, E^P stands for the expectation with respect to the probability measure P, and $p^{\rm R}(t,r,\xi)$ denotes the transition probability density of R with respect to $m^{\rm R}$.

We denote by $(\mathcal{E}^{\Theta}, \mathcal{F}^{\Theta})$ the Dirichlet form on $L^2(S^{d-1}, m^{\Theta})$ corresponding to Θ, where m^{Θ} stands for the standard spherical element of S^{d-1}, so that $|S^{d-1}| = \int_{S^{d-1}} dm^{\Theta}$ is the total area of S^{d-1}. We denote by p_t^{Θ} the semigroup, that is,

$$p_t^{\Theta} f(\theta) = E^{P_\theta^{\Theta}}[f(\Theta_t)] = \int_{S^{d-1}} p^{\Theta}(t,\theta,\varphi) f(\varphi)\, dm^{\Theta}(\varphi),\ t>0,\ \theta\in S^{d-1},$$

for $f \in C_b(S^{d-1})$, where $p^{\Theta}(t,\theta,\varphi)$ stands for the transition probability density of Θ with respect to m^{Θ}. It is known that $p^{\Theta}(t,\theta,\varphi)$ is represented by spherical harmonics S_n^l, that is,

$$p^{\Theta}(t,\theta,\varphi) = \sum_{n=0}^{\infty} e^{-\gamma_n t} \sum_{l=1}^{\kappa(n)} S_n^l(\theta) S_n^l(\varphi), \tag{1.1}$$

where $\gamma_n = \frac{1}{2} n(n+d-2)$, and $\kappa(n) = (2n+d-2)\cdot(n+d-3)!/n!\,(d-2)!$ which is the number of spherical harmonics of weight n (see [3], [7]). In the following we put $\kappa_n(\theta,\varphi) = \sum_{l=1}^{\kappa(n)} S_n^l(\theta) S_n^l(\varphi)$, so that

$$p^{\Theta}(t,\theta,\varphi) = \sum_{n=0}^{\infty} e^{-\gamma_n t} \kappa_n(\theta,\varphi). \tag{1.2}$$

We note that $\kappa_0(\theta,\varphi) = |S^{d-1}|^{-1}$ and

$$\kappa_n(\theta,\varphi) = \frac{(2n+d-2)\Gamma(d/2)}{2\pi(d-2)} C_n^{(d-2)/2}[(\theta,\varphi)], \quad d\ge 3,\ n\ge 1,$$

$$\kappa_n(\theta,\theta) = \frac{(2n+d-2)\Gamma(d/2)\Gamma(n+d-2)}{2\pi\Gamma(d-1)n!}, \quad d\ge 2,\ n\ge 1,$$

where C_n^{ν} is the Gegenbauer polynomial, and (θ,φ) is the inner product of θ and φ of S^{d-1}.

We turn to a skew product of $\mathrm{R} = [R_t, P_r^{\rm R}]$ and $\Theta = [\Theta_t, P_\theta^{\Theta}]$. It is known that there exists the local time $l^{\rm R}(t,r)$ of R which is continuous in $(t,r)\in[0,\infty)\times I$ and satisfies $\int_0^t 1_A(R_u)\,du = \int_A l^{\rm R}(t,r)\, dm^{\rm R}(r)$, $t>0$, for every measurable set $A\subset I$, where 1_A is the indicator for a set A (see [7]). Let ν be a Radon measure on I and assume that the support of ν coincides with I. We set

$$\mathbf{f}(t) = \int_I l^{\rm R}(t,r)\, d\nu(r).$$

We also set $M = I \times S^{d-1}$ and denote by m^{X} the product measure $m^{\mathrm{R}} \otimes m^{\Theta}$. The skew product of the ODGDP R and the spherical Brownian motion Θ with respect to the PCAF $\mathbf{f}(t)$ is given by $\mathrm{X} = [X_t = (R_t, \Theta_{\mathbf{f}(t)}), P^{\mathrm{X}}_{(r,\theta)} = P^{\mathrm{R}}_r \otimes P^{\Theta}_{\theta},\ (r,\theta) \in M]$. We set

$$\mathcal{E}^{\mathrm{X}}(f,g) = \int_{S^{d-1}} \mathcal{E}^{\mathrm{R}}(f(\cdot,\theta), g(\cdot,\theta))\, dm^{\Theta}(\theta) + \int_I \mathcal{E}^{\Theta}(f(r,\cdot), g(r,\cdot))\, d\nu(r),$$

for $f, g \in \mathcal{C}^{\mathrm{X}}$, where $\mathcal{C}^{\mathrm{X}} = \{f(s^{\mathrm{R}}(r), \theta) : f \in C_0^{\infty}(J \times S^{d-1})\}$ and $J = s^{\mathrm{R}}(I)$. Then by means of Theorems 1.1 and 7.2 of [5] and Theorems 3.1 and 5.1 of [11], we immediately obtain the following result. We should note that R is not necessarily conservative.

Proposition 1.1. *The form* $(\mathcal{E}^{\mathrm{X}}, \mathcal{C}^{\mathrm{X}})$ *is closable on* $L^2(M, m^{\mathrm{X}})$. *The closure* $(\mathcal{E}^{\mathrm{X}}, \mathcal{F}^{\mathrm{X}})$ *is a regular Dirichlet form, it is corresponding to the skew product* X, *and the semigroup* $\{p^{\mathrm{X}}_t,\ t > 0\}$ *is irreducible.*

The semigroup p^{X}_t is represented by

$$p^{\mathrm{X}}_t f(r,\theta) = E^{P^{\mathrm{R}}_r \otimes P^{\Theta}_{\theta}}[f(R_t, \Theta_{\mathbf{f}(t)})], \qquad t > 0,\ (r,\theta) \in M, \tag{1.3}$$

for $f \in C_b(M)$. Combining this with (1.1), we obtain the following.

Proposition 1.2. *There exists the transition probability density* $p^{\mathrm{X}}(t,x,y)$ *with respect to* m^{X}, *that is,*

$$p^{\mathrm{X}}_t f(x) = \int_M p^{\mathrm{X}}(t,x,y) f(y)\, dm^{\mathrm{X}}(y), \qquad t > 0,\ x \in M,$$

for $f \in C_b(M)$. $p^{\mathrm{X}}(t,x,y)$ *is continuous in* (t,x,y), *and it is represented by*

$$p^{\mathrm{X}}(t,(r,\theta),(\xi,\varphi)) = \sum_{n=0}^{\infty} p^{(n)}(t,r,\xi)\kappa_n(\theta,\varphi), \tag{1.4}$$

where $p^{(n)}(t,r,\xi)$ *is the transition probability density with respect to* m^{R} *corresponding to an ODGDP* $\mathrm{R}^{(n)}$ *with* $(s^{\mathrm{R}}, m^{\mathrm{R}}, \gamma_n \nu)$, *and the series converges uniformly in* $(t,(r,\theta),(\xi,\varphi)) \in K$ *for any compact set* $K \subset (0,\infty) \times M \times M$.

We prove this proposition in Section 3. The α-Green function $G^{\mathrm{X}}(\alpha,x,y)$ corresponding to the skew product X is given by

$$G^{\mathrm{X}}(\alpha,x,y) = \int_0^{\infty} e^{-\alpha t} p^{\mathrm{X}}(t,x,y)\, dt,$$

for $\alpha > 0$, and $x, y \in M$. There exists the 0-Green function $G^{\mathrm{X}}(0,x,y)$ if and only if $s^{\mathrm{R}}(l_1) > -\infty$ or $s^{\mathrm{R}}(l_2) < \infty$ (see Theorem 1 of [14]). Therefore Proposition 1.3, Theorem 1.1, Remark 1.1, and Corollary 1.1 hold true for $\alpha \geq 0$, whenever $s^{\mathrm{R}}(l_1) > -\infty$ or $s^{\mathrm{R}}(l_2) < \infty$. The following result is an immediate consequence of Proposition 1.2.

Proposition 1.3. *Let* $\alpha > 0$. $G^{\mathrm{X}}(\alpha,x,y)$ *is continuous in* (α,x,y), *and it is represented by*

$$G^{\mathrm{X}}(\alpha,(r,\theta),(\xi,\varphi)) = \sum_{n=0}^{\infty} G^{(n)}(\alpha,r,\xi)\kappa_n(\theta,\varphi), \tag{1.5}$$

where $G^{(n)}(\alpha, r, \xi)$ *is the* α*-Green function corresponding to an ODGDP* $\mathrm{R}^{(n)}$*, that is,*

$$G^{(n)}(\alpha, r, \xi) = \int_0^\infty e^{-\alpha t} p^{(n)}(t, r, \xi)\, dt. \tag{1.6}$$

Further the series in (1.5) converges uniformly in $((r, \theta), (\xi, \varphi)) \in K$ *for any compact set* $K \subset M \times M$.

Now we state our main results. Let us fix a point $c_o \in I$ arbitrarily and set

$$J_{\mu,\lambda}(r) = \int_{(c_o, r]} d\mu(\xi) \int_{(c_o, \xi]} d\lambda(\eta),$$

for $r \in I$, where μ and λ are Borel measures on I, and the integral $\int_{(a,b]}$ is read as $-\int_{(b,a]}$ if $a > b$. The first result is concerned with [SSP], that is, 1 is a semismall perturbation of $L^{\mathrm{X}} + \alpha$ on M, where L^{X} is the selfadjoint operator in $L^2(M, m^{\mathrm{X}})$ associated with the Dirichlet form $(\mathcal{E}^{\mathrm{X}}, \mathcal{F}^{\mathrm{X}})$.

Theorem 1.1. *Let* $\alpha > 0$. *Assume* $\min\{J_{s^{\mathrm{R}}, m^{\mathrm{R}}}(l_i),\ J_{m^{\mathrm{R}}, s^{\mathrm{R}}}(l_i)\} < \infty$ *for* $i = 1, 2$.

(i) *Assume one of the following conditions* (S1) *and* (S2).

(S1) $|s^{\mathrm{R}}(l_1)| < \infty$ *and* ν *satisfies*

$$\int_{(l_1, c_o)} ds^{\mathrm{R}}(\xi) \int_{(\xi, c_o)} d\nu(\eta) = \infty, \tag{1.7}$$

$$\int_{(l_1, c_o)} d\nu(\xi) = \infty. \tag{1.8}$$

(S2) $|s^{\mathrm{R}}(l_1)| = \infty$ *and* ν *satisfies*

$$\int_{(l_1, c_o)} |s^{\mathrm{R}}(\xi)|\, d\nu(\xi) = \infty. \tag{1.9}$$

Then for any $\varepsilon > 0$ *there is a compact set* $K \subset M$ *such that*

$$\lim_{|x| \to l_1} G^{\mathrm{X}}(\alpha, x, y)^{-1} \int_{M \setminus K} G^{\mathrm{X}}(\alpha, x, z) G^{\mathrm{X}}(\alpha, z, y) dm^{\mathrm{X}}(z) < \varepsilon, \tag{1.10}$$

for $y \in M \setminus K$.

(ii) *Assume one of the following conditions* (S3) *and* (S4).

(S3) $|s^{\mathrm{R}}(l_2)| < \infty$ *and* ν *satisfies*

$$\int_{(c_o, l_2)} ds^{\mathrm{R}}(\xi) \int_{(c_o, \xi)} d\nu(\eta) = \infty, \tag{1.11}$$

$$\int_{(c_o, l_2)} d\nu(\xi) = \infty. \tag{1.12}$$

(S4) $|s^{\mathrm{R}}(l_2)| = \infty$ *and* ν *satisfies*

$$\int_{(c_o,l_2)} |s^{\mathrm{R}}(\xi)|\, d\nu(\xi) = \infty. \tag{1.13}$$

Then for any $\varepsilon > 0$ *there is a compact set* $K \subset M$ *satisfying (1.10) with* l_1 *replaced by* l_2, *for* $y \in M \setminus K$.

Remark 1.1. If (S2)[resp. (S4)] is satisfied, then (1.10) is equivalent to

$$\int_{M\setminus K} G^{\mathrm{X}}(\alpha, x_0, z) G^{\mathrm{X}}(\alpha, z, y) dm^{\mathrm{X}}(z) < \varepsilon G^{\mathrm{X}}(\alpha, x_0, y), \tag{1.14}$$

where $x_0 = (l_1, \theta)$ [resp. (l_2, θ)] and both hand sides of (1.14) are independent of $\theta \in S^{d-1}$.

Theorem 1.1 and Remark 1.1 are proved in Section 3.

We next observe an eigenfunction expansion for $p^{\mathrm{X}}(t, x, y)$. It is well known that $p^{\mathrm{R}}(t, \xi, \eta)$ has the following eigenfunction expansion (1.15) whenever $\min\{J_{s^{\mathrm{R}},m^{\mathrm{R}}}(l_i),\ J_{m^{\mathrm{R}},s^{\mathrm{R}}}(l_i)\} < \infty$ for $i = 1, 2$ ([7], see also Proposition 2.1).

$$p^{\mathrm{R}}(t, \xi, \eta) = \sum_{j=0}^{\infty} e^{-\lambda_j^{\mathrm{R}} t} \phi_j^{\mathrm{R}}(\xi) \phi_j^{\mathrm{R}}(\eta), \qquad t > 0,\ \xi, \eta \in I, \tag{1.15}$$

where $\{\phi_j^{\mathrm{R}}\}_{j=0}^{\infty}$ is a complete orthonormal base of eigenfunctions with eigenvalues $0 \le \lambda_0^{\mathrm{R}} < \lambda_1^{\mathrm{R}} \le \lambda_2^{\mathrm{R}} \le \cdots$ repeated according to multiplicity, and the series converges uniformly on any compact set of $(0, \infty) \times I \times I$.

We introduce the following conditions.

$(E1)$ For each $n \ge 1$, there exists a positive function $h^{(n)}$ satisfying (2.3) with s, m and k replaced by s^{R}, m^{R} and $\gamma_n \nu$, respectively. Further $\min\{J_{s^{(n)},m^{(n)}}(l_i),\ J_{m^{(n)},s^{(n)}}(l_i)\} < \infty$ for $i = 1, 2$, where

$$s^{(n)}(\xi) = \int_{(c_o,\xi]} h^{(n)}(\eta)^{-2} ds^{\mathrm{R}}(\eta), \quad m^{(n)}(\xi) = \int_{(c_o,\xi]} h^{(n)}(\eta)^2 dm^{\mathrm{R}}(\eta).$$

$(E2)$ There exists a positive constant B_0 such that

$$\int_I u(\xi)\, d\nu(\xi) \ge B_0 \int_I u(\xi)\, dm^{\mathrm{R}}(\xi), \qquad u \in C_0(I). \tag{1.16}$$

Here $C_0(I)$ is the set of all continuous functions on I with compact support.

Under one of the conditions (Ej), $j = 1, 2$, we obtain the following two results.

Proposition 1.4. *Assume* $\min\{J_{s^{\mathrm{R}},m^{\mathrm{R}}}(l_i),\ J_{m^{\mathrm{R}},s^{\mathrm{R}}}(l_i)\} < \infty$ *for* $i = 1, 2$, *and one of* (Ej), $j = 1, 2$. *Then there is a complete orthonormal base of eigenfunctions* $\{\phi_j^{\mathrm{X}}\}_{j=0}^{\infty}$ *with eigenvalues* $0 \le \lambda_0^{\mathrm{X}} < \lambda_1^{\mathrm{X}} \le \lambda_2^{\mathrm{X}} \le \cdots$ *repeated according to multiplicity, so that*

$$p^{\mathrm{X}}(t, x, y) = \sum_{j=0}^{\infty} e^{-\lambda_j^{\mathrm{X}} t} \phi_j^{\mathrm{X}}(x) \phi_j^{\mathrm{X}}(y), \qquad t > 0,\ x, y \in M, \tag{1.17}$$

where the series converges uniformly on any compact set of $(0,\infty)\times M\times M$. *Further* $\lambda_0^{\mathrm{X}}=\lambda_0^{\mathrm{R}}$ *and* $\phi_0^{\mathrm{X}}(x)=\phi_0^{\mathrm{R}}(|x|)/\sqrt{|S^{d-1}|}$.

Theorem 1.2. *Assume* $\min\{J_{s^{\mathrm{R}},m^{\mathrm{R}}}(l_i),\ J_{m^{\mathrm{R}},s^{\mathrm{R}}}(l_i)\}<\infty$ *for* $i=1,2$, *and one of* (Ej), $j=1,2$. *Then there exists the limit*

$$\lim_{|y|\to l_i} p^{\mathrm{X}}(t,x,y)/\phi_0^{\mathrm{X}}(y)=q(t,x,(l_i,\theta)),\quad t>0,\ x\in M, \tag{1.18}$$

for $y=(r,\theta)\in M$. *The convergence is uniform in* $(t,x,\theta)\in K$ *for any compact set* $K\subset(0,\infty)\times M\times S^{d-1}$. *The limit* $q(t,x,(l_i,\theta))$ *is given by the following. If* $|s^{\mathrm{R}}(l_1)|<\infty$,

$$q(t,(\xi,\varphi),(l_1,\theta))=|S^{d-1}|^{1/2}\phi_0^{\mathrm{R}+}(l_1)^{-1}\sum_{n=0}^{\infty}p^{(n)}{}_2^{+}(t,l_1,\xi)\kappa_n(\theta,\varphi). \tag{1.19}$$

In particular, if (1.7) and (1.8) are satisfied,

$$q(t,(\xi,\varphi),(l_1,\theta))=|S^{d-1}|^{-1/2}\phi_0^{\mathrm{R}+}(l_1)^{-1}p^{\mathrm{R}}{}_2^{+}(t,l_1,\xi). \tag{1.20}$$

If $|s^{\mathrm{R}}(l_1)|=\infty$,

$$q(t,(\xi,\varphi),(l_1,\theta))=|S^{d-1}|^{1/2}\phi_0^{\mathrm{R}}(l_1)^{-1}\sum_{n=0}^{\infty}p^{(n)}(t,l_1,\xi)\kappa_n(\theta,\varphi). \tag{1.21}$$

In particular, if (1.9) is satisfied,

$$q(t,(\xi,\varphi),(l_1,\theta))=|S^{d-1}|^{-1/2}\phi_0^{\mathrm{R}}(l_1)^{-1}p^{\mathrm{R}}(t,l_1,\xi). \tag{1.22}$$

If $|s^{\mathrm{R}}(l_2)|<\infty$,

$$q(t,(\xi,\varphi),(l_2,\theta))=|S^{d-1}|^{1/2}\phi_0^{\mathrm{R}-}(l_2)^{-1}\sum_{n=0}^{\infty}p^{(n)}{}_2^{-}(t,l_2,\xi)\kappa_n(\theta,\varphi). \tag{1.23}$$

In particular, if (1.11) and (1.12) are satisfied,

$$q(t,(\xi,\varphi),(l_1,\theta))=|S^{d-1}|^{-1/2}\phi_0^{\mathrm{R}-}(l_2)^{-1}p^{\mathrm{R}}{}_2^{-}(t,l_2,\xi). \tag{1.24}$$

If $|s^{\mathrm{R}}(l_2)|=\infty$,

$$q(t,(\xi,\varphi),(l_2,\theta))=|S^{d-1}|^{1/2}\phi_0^{\mathrm{R}}(l_2)^{-1}\sum_{n=0}^{\infty}p^{(n)}(t,l_2,\xi)\kappa_n(\theta,\varphi). \tag{1.25}$$

In particular, if (1.13) is satisfied,

$$q(t,(\xi,\varphi),(l_2,\theta))=|S^{d-1}|^{-1/2}\phi_0^{\mathrm{R}}(l_2)^{-1}p^{\mathrm{R}}(t,l_2,\xi). \tag{1.26}$$

The series of the right hand sides of (1.19), (1.21), (1.23) and (1.25) converge uniformly in $(t,\xi,\theta,\varphi)\in K$ *for any compact set* $K\subset(0,\infty)\times I\times S^{d-1}\times S^{d-1}$.

In Section 4, we prove Proposition 1.4 and Theorem 1.2. We there explain precise meaning of $\phi_0^{\mathrm{R}+}(l_1)$, ${p^{(n)}}_2^+(t, l_1, \xi)$, etc.

Finally we consider a sufficient condition for $p^{\mathrm{X}}(t, x, y)$ satisfying the following condition [IU] (i.e., intrinsic ultracontractivity).

[IU] For any $t > 0$, there exists a constant $C_t > 0$ such that

$$p^{\mathrm{X}}(t, x, y) \le C_t\ \phi_0^{\mathrm{X}}(x)\phi_0^{\mathrm{X}}(y), \quad x, y \in M.$$

In [16] we gave a sufficient condition for $p^{\mathrm{R}}(t, r, \xi)$ satisfying [IU]. We explain results in [16]. We may assume that $m^{\mathrm{R}}(l_1) < 0 < m^{\mathrm{R}}(l_2)$ and $s^{\mathrm{R}}(l_1) < 0 < s^{\mathrm{R}}(l_2)$ without loss of generality. We fix points $c_i \in I$, $i = 1, 2$, such that $m^{\mathrm{R}}(c_1) < 0 < m^{\mathrm{R}}(c_2)$ and $s^{\mathrm{R}}(c_1) < 0 < s^{\mathrm{R}}(c_2)$. Let (Aj), $j = 1, 2, 3, 4$, be the following conditions:

$$(A1) \qquad \int_{(l_1,c_1)} |m^{\mathrm{R}}(\xi)|\, ds^{\mathrm{R}}(\xi) < \infty \quad \text{and} \quad \int_{(l_1,c_1)} \frac{\mu_1(\xi)}{|m^{\mathrm{R}}(\xi)|}\, dm^{\mathrm{R}}(\xi) < \infty,$$

where $\mu_1(\xi) = \sup_{l_1 < \eta \le \xi} |m^{\mathrm{R}}(\eta)|\{s^{\mathrm{R}}(\eta) - s^{\mathrm{R}}(l_1)\}$.

$$(A2) \qquad \int_{(l_1,c_1)} |m^{\mathrm{R}}(\xi)|\, ds^{\mathrm{R}}(\xi) = \infty \quad \text{and} \quad \int_{(l_1,c_1)} \frac{\nu_1(\xi)}{|s^{\mathrm{R}}(\xi)|}\, ds^{\mathrm{R}}(\xi) < \infty,$$

where $\nu_1(\xi) = \sup_{l_1 < \eta \le \xi} |s^{\mathrm{R}}(\eta)|\{m^{\mathrm{R}}(\eta) - m^{\mathrm{R}}(l_1)\}$.

$$(A3) \qquad \int_{(c_2,l_2)} m^{\mathrm{R}}(\xi)\, ds^{\mathrm{R}}(\xi) < \infty \quad \text{and} \quad \int_{(c_2,l_2)} \frac{\mu_2(\xi)}{m^{\mathrm{R}}(\xi)}\, dm^{\mathrm{R}}(\xi) < \infty,$$

where $\mu_2(\xi) = \sup_{\xi \le \eta < l_2} m^{\mathrm{R}}(\eta)\{s^{\mathrm{R}}(l_2) - s^{\mathrm{R}}(\eta)\}$.

$$(A4) \qquad \int_{(c_2,l_2)} m^{\mathrm{R}}(\xi)\, ds^{\mathrm{R}}(\xi) = \infty \quad \text{and} \quad \int_{(c_2,l_2)} \frac{\nu_2(\xi)}{s^{\mathrm{R}}(\xi)}\, ds^{\mathrm{R}}(\xi) < \infty,$$

where $\nu_2(\xi) = \sup_{\xi \le \eta < l_2} s^{\mathrm{R}}(\eta)\{m^{\mathrm{R}}(l_2) - m^{\mathrm{R}}(\eta)\}$.

Theorem 1.3. (Tomisaki, [16]) *Assume* $\min\{J_{s^{\mathrm{R}},m^{\mathrm{R}}}(l_i),\ J_{m^{\mathrm{R}},s^{\mathrm{R}}}(l_i)\} < \infty$ *for* $i = 1, 2$, *and if* $\max\{J_{s^{\mathrm{R}},m^{\mathrm{R}}}(l_1),\ J_{m^{\mathrm{R}},s^{\mathrm{R}}}(l_1)\} = \infty$ *[resp.* $\max\{J_{s^{\mathrm{R}},m^{\mathrm{R}}}(l_2),\ J_{m^{\mathrm{R}},s^{\mathrm{R}}}(l_2)\} = \infty$*], assume* $(A1)$ *or* $(A2)$ *[resp.* $(A3)$ *or* $(A4)$*]. Then the following (1.27) holds.*

$$\sup_{r,\xi \in I} p^{\mathrm{R}}(t, r, \xi)/\phi_0^{\mathrm{R}}(r)\phi_0^{\mathrm{R}}(\xi) < \infty, \quad t > 0. \tag{1.27}$$

By using this theorem, we show [IU] for $p^{\mathrm{X}}(t, x, y)$ under some additional condition on ν.

Theorem 1.4. *Assume* $\min\{J_{s^{\mathrm{R}},m^{\mathrm{R}}}(l_i),\ J_{m^{\mathrm{R}},s^{\mathrm{R}}}(l_i)\} < \infty$ *for* $i = 1, 2$, *and if* $\max\{J_{s^{\mathrm{R}},m^{\mathrm{R}}}(l_1),\ J_{m^{\mathrm{R}},s^{\mathrm{R}}}(l_1)\} = \infty$ *[resp.* $\max\{J_{s^{\mathrm{R}},m^{\mathrm{R}}}(l_2),\ J_{m^{\mathrm{R}},s^{\mathrm{R}}}(l_2)\} = \infty$*], assume* $(A1)$ *or* $(A2)$ *[resp.* $(A3)$ *or* $(A4)$*]. Further assume* $(E2)$*. Then [IU] holds for* $p^{\mathrm{X}}(t, x, y)$*, that is,*

$$\sup_{x,y \in M} p^{\mathrm{X}}(t, x, y)/\phi_0^{\mathrm{X}}(x)\phi_0^{\mathrm{X}}(y) < \infty, \quad t > 0. \tag{1.28}$$

In Section 4 we show Theorem 1.4.

We should note the results due to Murata [9]. He showed that [IU] implies the following [SP] (i.e., small perturbation) (see Theorem 1.1 of [9]).

[SP] 1 is a small perturbation of $L^X + \alpha$ on M, i.e., for any $\varepsilon > 0$ there exists a compact subset K of M such that

$$\int_{M\setminus K} G^X(\alpha, x, z) G^X(\alpha, z, y)\, dm^X(z) \le \varepsilon G^X(\alpha, x, y), \qquad x, y \in M \setminus K.$$

The condition [SP] also implies the following condition [SSP] (i.e., semismall perturbation).

[SSP] 1 is a semismall perturbation of $L^X + \alpha$ on M, i.e., for any $\varepsilon > 0$ there exists a compact subset K of M such that

$$\int_{M\setminus K} G^X(\alpha, x^0, z) G^X(\alpha, z, y) dm^X(z) \le \varepsilon G^X(\alpha, x^0, y), \qquad y \in M \setminus K,$$

where x^0 is a reference point fixed in M.

Therefore we obtain the following as an immediate consequence of Theorem 1.4.

Corollary 1.1. *Under the same assumption as Theorem 1.4, [SP] and [SSP] hold.*

[SSP] of Corollary 1.1 is different from that of Theorem 1.1 in that a reference point x^0 is taken from inside of M. The assumption of Corollary 1.1 does not imply that of Theorem 1.1, and vice versa.

The organization of this paper is as follows. In Section 2 we summarize some results on ODGDPs. In Sections 3 and 4 we prove our main results. Section 5 is devoted to the following Laplace-Beltrami operator. Let $M = (0, \infty) \times S^{d-1}$, and

$$\Delta = f^{1-d} \partial/\partial r \left(f^{d-1} \partial/\partial r \right) + f^{-2} \Delta^{\Theta},$$

where f is a nonnegative smooth function on $[0, \infty)$ such that $f > 0$ on $(0, \infty)$, $f(0) = 0$, $f'(0) = 1$ and $f''(0) = 0$. This operator was recently treated by Murata (see [10]). He established a necessary and sufficient condition for [SSP] (1 is a semismall perturbation of $-\Delta + 1$ on M). We give a sufficient condition for the corresponding heat kernel satisfying [IU].

2. One-dimensional generalized diffusion processes

Let m be a right continuous nondecreasing function on an open interval $I = (l_1, l_2)$, where $-\infty \le l_1 < l_2 \le \infty$, s be a continuous increasing function on I, and k be a right continuous nondecreasing function on I. We assume that the support of the measure $dm(\xi)$ on I induced by $m(\xi)$ is equal to I. For a function u on I, we set $u(l_i) = \lim_{\xi \to l_i, \xi \in I} u(\xi)$ if there exists the limit, for $i = 1, 2$. We set $I^* = I \cup \{\xi; \xi = l_i \text{ with } |m(l_i)| + |s(l_i)| + |k(l_i)| < \infty, \ i = 1, 2\}$. Following [4], we call the boundary point l_i to be

(s, m, k)-regular if $J_{s,m+k}(l_i) < \infty$ and $J_{m+k,s}(l_i) < \infty$,
(s, m, k)-exit if $J_{s,m+k}(l_i) < \infty$ and $J_{m+k,s}(l_i) = \infty$,
(s, m, k)-entrance if $J_{s,m+k}(l_i) = \infty$ and $J_{m+k,s}(l_i) < \infty$,
(s, m, k)-natural if $J_{s,m+k}(l_i) = \infty$ and $J_{m+k,s}(l_i) = \infty$.

Let $\mathcal{G}$ be a one-dimensional generalized diffusion operator with scale function s, speed measure m, and killing measure k (ODGDO with (s,m,k) for brief). In the following, for a measurable functions u on I, $u^+(\xi)$ stands for the right derivative with respect to $s(\xi)$, that is, $u^+(\xi) = \lim_{\varepsilon\downarrow 0}\{u(\xi+\varepsilon)-u(\xi)\}/\{s(\xi+\varepsilon)-s(\xi)\}$, provided it exists.

Following McKean [8] (see also Section 4.11 of [7]), we can define the elementary solution of the following equation

$$\frac{\partial}{\partial t}p(t,\xi,\eta) = \mathcal{G}p(t,\xi,\eta), \quad t>0,\ \xi,\eta\in I,$$

where $\mathcal{G}$ is applied to ξ or η. It is known that $p(t,\xi,\eta)$ satisfies the following properties:

$$\begin{aligned}
&0 < p(t,\xi,\eta) = p(t,\eta,\xi) \text{ is continuous on } (0,\infty)\times I\times I,\\
&p(s+t,\xi,\eta) = \int_I p(s,\xi,\zeta)p(t,\zeta,\eta)\,dm(\zeta), \quad s,t>0,\ \xi,\eta\in I,\\
&p(t,l_i,\eta) = 0, \quad t>0,\ \eta\in I, \quad \text{if } l_i \text{ is not entrance},\\
&p_2^+(t,l_i,\eta) = 0, \quad t>0,\ \eta\in I, \quad \text{if } l_i \text{ is entrance},
\end{aligned}$$

where $p_2^+(t,\xi,\eta) = \lim_{\varepsilon\downarrow 0}\{p(t,\xi+\varepsilon,\eta)-p(t,\xi,\eta)\}/\{s(\xi+\varepsilon)-s(\xi)\}$. It is also known that there exists an ODGDP with (s,m,k) $\mathbb{D} = [x(t):\ t\geq 0,\ P_\xi : \xi\in I^*]$ such that

$$P_\xi(x(t)\in E) = \int_E p(t,\xi,\eta)\,dm(\eta), \quad t>0,\ \xi\in I^*.$$

By this reason, $p(t,\xi,\eta)$ is sometimes called the transition probability density with respect to m. We should note that the absorbing boundary condition is posed at the regular boundary point l_i.

We next define the α-Green function $G(\alpha,\xi,\eta)$ corresponding to the ODGDO $\mathcal{G}$ with (s,m,k). For $\alpha\geq 0$ and $i=1,2$, let $g_i(\cdot,\alpha)$ be a function on I satisfying the following properties.

(g.1) $g_i(\xi,\alpha)$ is positive and continuous in ξ.
(g.2) $g_1(\xi,\alpha)$ is nondecreasing in ξ and $g_2(\xi,\alpha)$ is nonincreasing in ξ.
(g.3) If $|s(l_i)|<\infty$, then $g_i(l_i,\alpha)=0$.
(g.4) $g_i(\xi,\alpha)$ satisfies

$$\begin{aligned}
g_i(\xi,\alpha) =& g_i(c_o,\alpha) + g_i^+(c_o,\alpha)\{s(\xi)-s(c_o)\}\\
&+\int_{(c_o,\xi]}\{s(\xi)-s(\eta)\}g_i(\eta,\alpha)\{\alpha dm(\eta)+dk(\eta)\}, \quad \xi\in I.
\end{aligned} \tag{2.1}$$

Here $g_i^+(\xi,\alpha) = \lim_{\varepsilon\downarrow 0}\{g_i(\xi+\varepsilon,\alpha)-g_i(\xi,\alpha)\}/\{s(\xi+\varepsilon)-s(\xi)\}$, $i=1,2$. It is known that there exist functions $g_i(\cdot,\alpha)$, $i=1,2$, satisfying the properties (g.1)–(g.4) (see Section 4.6 of [7]).

We set $W(\alpha) = g_1^+(\xi,\alpha)g_2(\xi,\alpha) - g_1(\xi,\alpha)g_2^+(\xi,\alpha)$. Note that $W(\alpha)$ is a positive number independent of $\xi\in I$. We put

$$G(\alpha,\xi,\eta) = G(\alpha,\eta,\xi) = W(\alpha)^{-1}g_1(\xi,\alpha)g_2(\eta,\alpha), \tag{2.2}$$

for $\alpha > 0,\ \xi, \eta \in I,\ \xi \le \eta$. It is known that

$$G(\alpha, \xi, \eta) = \int_0^\infty e^{-\alpha t} p(t, \xi, \eta)\, dt, \quad \alpha > 0,\ \xi, \eta \in I^*.$$

It is easy to see that, if $k \neq 0$, then there exists $G(0, \xi, \eta)$. We also note that, in the case $k = 0$, there exists $G(0, \xi, \eta)$ if and only if $|s(l_i)| < \infty$ for $i = 1$ or 2 (see [15]).

For $\beta \ge 0$ let $\mathcal{H}_\beta$ be the set of all positive functions h satisfying

$$\begin{aligned} h(\xi) =& h(c_o) + h^+(c_o)\{s(\xi) - s(c_o)\} \\ &+ \int_{(c_o,\xi]} \{s(\xi) - s(\eta)\} h(\eta) \{\beta dm(\eta) + dk(\eta)\}, \quad \xi \in I. \end{aligned} \tag{2.3}$$

For $h \in \mathcal{H}_\beta$, we set

$$\begin{aligned} &g_{h,i}(\xi, \alpha) = g_i(\xi, \alpha)/h(\xi), \quad i = 1, 2,\ \alpha \ge 0, \\ &s_h(\xi) = \int_{(c_o,\xi]} h(\eta)^{-2}\, ds(\eta), \quad m_h(\xi) = \int_{(c_o,\xi]} h(\eta)^2\, dm(\eta). \end{aligned}$$

Let $\mathcal{G}_h^*$ be the ODGDO with $(s_h, m_h, 0)$, and $G_h^*(\gamma, \xi, \eta)$ and $p_h^*(t, \xi, \eta)$ be the γ-Green function and the transition probability density corresponding to $\mathcal{G}_h^*$, respectively. Then we obtain the following.

$$\begin{aligned} &G_h^*(\alpha - \beta, \xi, \eta) = G_h^*(\alpha - \beta, \eta, \xi) \\ &= W(\alpha) g_{h,1}(\xi, \alpha) g_{h,2}(\eta, \alpha) = G(\alpha, \xi, \eta)/h(\xi)h(\eta), \end{aligned}$$

for $\alpha \ge \beta \ge 0$ and $l_1 < \xi \le \eta < l_2$, and

$$p_h^*(t, \xi, \eta) = e^{-\beta t} p(t, \xi, \eta)/h(\xi)h(\eta), \tag{2.4}$$

for $t > 0,\ \xi, \eta \in I$ (see [16]).

We note an eigenfunction expansion for $p(t, \xi, \eta)$. The following result is well known.

Proposition 2.1. (Itô et al. [7]) *Assume that both of l_1 and l_2 are not (s, m, k)-natural. Then there is a complete orthonormal base of eigenfunctions $\{\phi_j\}_{j=0}^\infty$ with eigenvalues $0 \le \lambda_0 < \lambda_1 \le \lambda_2 \le \cdots$ repeated according to multiplicity, so that*

$$p(t, \xi, \eta) = \sum_{j=0}^\infty e^{-\lambda_j t} \phi_j(\xi) \phi_j(\eta), \qquad t > 0,\ \xi, \eta \in I, \tag{2.5}$$

where the series converges uniformly on any compact set of $(0, \infty) \times I \times I$.

We may also have an eigenfunction expansion for $p(t, \xi, \eta)$ in the case that l_1 or l_2 is (s, m, k)-natural. We introduce the following conditions.

$(R1)$ There exists a function $h \in \mathcal{H}_\beta$ such that both of l_1 and l_2 are not $(s_h, m_h, 0)$-natural.

$(R2)$ There exists a positive constant k_0 such that

$$\int_I u(\xi)\, dk(\xi) \ge k_0 \int_I u(\xi)\, dm(\xi), \qquad u \in C_0(I). \tag{2.6}$$

Proposition 2.2. *Assume that both of l_1 and l_2 are not $(s, m, 0)$-natural. Further assume that k is not null measure and one of the conditions (Rj), $j = 1, 2$, is satisfied. Then we obtain the same conclusion as Proposition 2.1.*

Proof. The assertion of the proposition under the condition $(R1)$ is obtained in the proof of Theorem 2.12 of [16]).

Suppose the condition $(R2)$. In this case, we show the following.

$$\int_I p(t, \xi, \xi)\, dm(\xi) < \infty, \quad t > 0. \tag{2.7}$$

Then by means of a standard fact we have a representation (2.5) for a complete orthonormal base of eigenfunctions $\{\phi_j\}_{j=0}^{\infty}$ with eigenvalues $0 \le \lambda_0 \le \lambda_1 \le \lambda_2 \le \cdots$ repeated according to multiplicity, and the series converges uniformly on any compact set of $(0, \infty) \times I \times I$. Since the semigroup $\{p_t, t > 0\}$ is irreducible, λ_0 has multiplicity one and the ϕ_0 is positive.

Now we give a proof of (2.7). We note the representation of p_t by Kac's formula. For the ODGDP $\mathbb{D} = [x(t), P_\xi : \xi \in I^*]$ with (s, m, k), we denote by $\mathbb{D}^o = [x(t), P_\xi^o : \xi \in I^*]$ an ODGDP with $(s, m, 0)$. Namely, $\mathbb{D}^o$ has the same scale function s and the same speed measure m as those of $\mathbb{D}$, but the killing measure of $\mathbb{D}^o$ is null. Let $l^o(t, \xi)$ be the local time of $\mathbb{D}^o$ which is continuous in $(t, \xi) \in [0, \infty) \times I$ and satisfies $\int_0^t 1_A(x(u))\, du = \int_A l^o(t, \xi)\, dm(\xi)$, $t > 0$, for every measurable set $A \subset I$ (see [7]). We set

$$\mathbf{k}(t) = \int_I l^o(t, \xi)\, dk(\xi).$$

Then the semigroup $\{p_t, t > 0\}$ of $\mathbb{D}$ has the following representation.

$$\begin{aligned} p_t f(\xi) &= E^{P_\xi}[f(x(t))] = E^{P_\xi^o}[f(x(t))e^{-\mathbf{k}(t)}] \\ &= E^{P_\xi^o}[f(x(t))e^{-\mathbf{k}(t)}; t < \sigma_{l_1} \wedge \sigma_{l_2}], \quad t > 0,\ \xi \in I, \end{aligned}$$

for $f \in C_b(I)$, where $\sigma_a = \inf\{t \ge 0 : x(t) = a\}$ and $a \wedge b = \min\{a, b\}$. For $t > 0$, the local time $l^o(t, \xi)$ is continuous in ξ and $l^o(t, \xi) = 0$ for $\xi \notin [\min_{0 \le u \le t} x(u), \max_{0 \le u \le t} x(u)]$ (see [7]). Combining this with (2.6), we get

$$\begin{aligned} \mathbf{k}(t) &= \int_I l^o(t, \xi)\, dk(\xi) \ge k_0 \int_I l^o(t, \xi)\, dm(\xi) \\ &= k_0(t \wedge \sigma_{l_1} \wedge \sigma_{l_2}),\ t > 0,\ P_\xi^o - a.e.,\ \xi \in I, \end{aligned}$$

and hence

$$\begin{aligned} p_t f(\xi) &= E^{P_\xi^o}[f(x(t))e^{-\mathbf{k}(t)}; t < \sigma_{l_1} \wedge \sigma_{l_2}] \\ &\le E^{P_\xi^o}[f(x(t))e^{-k_0 t}; t < \sigma_{l_1} \wedge \sigma_{l_2}] = e^{-k_0 t} E^{P_\xi^o}[f(x(t))], \end{aligned}$$

for nonnegative $f \in C_b(I)$. Let denote by $p^o(t, \xi, \eta)$ the transition probability density of $\mathbb{D}^o$ with respect to m. Since $p(t, \xi, \eta)$ and $p^o(t, \xi, \eta)$ are continuous in (t, ξ, η), we get

$$p(t, \xi, \eta) \le e^{-k_0 t} p^o(t, \xi, \eta), \quad t > 0,\ \xi, \eta \in I.$$

By the assumption that both of l_1 and l_2 are not $(s, m, 0)$-natural,

$$A^o := \int_I p^o(t, \xi, \xi)\, dm(\xi) < \infty.$$

Therefore we obtain

$$\int_I p(t, \xi, \xi)\, dm(\xi) \le e^{-k_o t} A^o,$$

which shows (2.7).

Thus the proof is completed. □

We give a sufficient condition for $(R1)$. Let us consider the following conditions.

$(Q1)$ l_1 or l_2 is not (s, m, k)-natural.

$(Q2)$ Both of l_1 and l_2 are (s, m, k)-natural, but l_1 or l_2 is not $(s, m, 0)$-entrance.

$(Q3)$ Both of l_1 and l_2 are (s, m, k)-natural and $(s, m, 0)$-entrance. There is a function $g_1(\xi, \alpha)$ or $g_2(\xi, \alpha)$ satisfying one of the following.

$$\int_{(c_o, l_2)} g_1(\xi, \alpha)^2\, dm(\xi) < \infty. \tag{2.8}$$

$$\int_{(c_o, l_2)} g_1(\xi, \alpha)^{-2}\, ds(\xi) \int_{(c_o, \xi)} g_1(\eta, \alpha)^2\, dm(\eta) < \infty. \tag{2.9}$$

$$\int_{(l_1, c_o)} g_2(\xi, \alpha)^2\, dm(\xi) < \infty. \tag{2.10}$$

$$\int_{(l_1, c_o)} g_2(\xi, \alpha)^{-2}\, ds(\xi) \int_{(\xi, c_o)} g_2(\eta, \alpha)^2\, dm(\eta) < \infty. \tag{2.11}$$

Proposition 2.3. *Assume that both of l_1 and l_2 are not $(s, m, 0)$-natural, and one of the conditions (Qj), $j = 1, 2, 3$. Then $(R1)$ holds.*

Proof. Assume $(Q1)$. We may assume that l_1 is (s, m, k)-natural and l_2 is not (s, m, k)-natural. Since l_2 is not (s, m, k)-natural,

$$l_2 \text{ is not } (s_h, m_h, 0)\text{-natural}, \tag{2.12}$$

for any $h \in \mathcal{H}_\alpha$ (see Theorem 1.1 of [13]). If l_1 is not $(s, m, 0)$-entrance, we set $h(\xi) = g_2(\xi, \alpha)$. Since h is nonincreasing on I and l_1 is $(s, m, 0)$-regular or -exit, $J_{s_h, m_h}(l_1) \le J_{s,m}(l_1) < \infty$. Since $h(l_1) = \infty$, by means of Theorem 1.1 of [13], we get

$$l_1 \text{ is } (s_h, m_h, 0)\text{-regular or -exit according to } |m_h(l_1)| < \infty \text{ or } = \infty. \tag{2.13}$$

If l_1 is $(s, m, 0)$-entrance, we set $h(\xi) = g_1(\xi, \alpha)$. Since h is nondecreasing on I and l_1 is $(s, m, 0)$-entrance, $J_{m_h, s_h}(l_1) \le J_{m,s}(l_1) < \infty$. Since $h(l_1) = 0$, by means of Theorem 1.1 of [13], we get

$$l_1 \text{ is } (s_h, m_h, 0)\text{-entrance}. \tag{2.14}$$

Assume $(Q2)$. We may assume that l_2 is not $(s, m, 0)$-entrance. If l_1 is not $(s, m, 0)$-entrance, we set $h(x) = g_1(x, \alpha) + g_2(x, \alpha)$. Since h is nonincreasing on (l_1, c_1) for some

$c_1 \in I$ and l_1 is $(s,m,0)$-regular or -exit, $J_{s_h,m_h}(l_1) \le J_{s,m}(l_1) < \infty$. Since $h(l_1) = \infty$, we get (2.13). Since h is nondecreasing on (c_2, l_2) for some $c_2 \in I$, and l_2 is $(s,m,0)$-regular or -exit, $J_{s_h,m_h}(l_2) \le J_{s,m}(l_2) < \infty$. Since $h(l_2) = \infty$, we have (2.13) with l_2 in place of l_1. If l_1 is $(s,m,0)$-entrance, we set $h(x) = g_1(x,\alpha)$. Since h is nondecreasing on I, l_1 is $(s,m,0)$-entrance, and l_2 is $(s,m,0)$-regular or -exit, $J_{m_h,s_h}(l_1) \le J_{m,s}(l_1) < \infty$ and $J_{s_h,m_h}(l_2) \le J_{s,m}(l_2) < \infty$. Since $h(l_1) = 0$ and $h(l_2) = \infty$, we get (2.14), and (2.13) with l_2 in place of l_1.

Assume $(Q3)$. It is enough to consider the case that there is a function $g_1(\xi,\alpha)$ satisfying (2.8) or (2.9). We put $h(\xi) = g_1(\xi,\alpha)$. Since $J_{m_h,s_h}(l_1) \le J_{m,s}(l_1) < \infty$, we have (2.14). On the other hand, we have the following three cases for the state of l_2 (see Theorem 1.1 of [13]).

l_2 is $(s_h, m_h, 0)$-regular if $|m_h(l_2)| < \infty$.
l_2 is $(s_h, m_h, 0)$-exit if $|m_h(l_2)| = \infty$ and $J_{s_h,m_h}(l_2) < \infty$.
l_2 is $(s_h, m_h, 0)$-natural if $|m_h(l_2)| = \infty$ and $J_{s_h,m_h}(l_2) = \infty$.

By means of (2.8) or (2.9), we have (2.13) with l_2 in place of l_1. □

3. Semismall perturbation

Let $\mathrm{X} = [X_t = (R_t, \Theta_{\mathbf{f}(t)}), P^{\mathrm{X}}_{(r,\theta)} = P^{\mathrm{R}}_r \otimes P^{\Theta}_\theta,\ (r,\theta) \in M]$ be the skew product of the ODGDP R and the spherical Brownian motion Θ with respect to the PCAF $\mathbf{f}(t)$ as in Section 1. In this section, we show Proposition 1.2 and Theorem 1.1.

Proof of Proposition 1.2. The semigroup p^{X}_t of the skew product X is represented by (1.3). By virtue of (1.2) we obtain the following

$$\begin{aligned} p^{\mathrm{X}}_t f(r,\theta) &= \int_{S^{d-1}} E^{P^{\mathrm{R}}_r}\left[f(R_t,\varphi)\, p^{\Theta}(\mathbf{f}(t),\theta,\varphi)\right] dm^{\Theta}(\varphi) \\ &= \sum_{n=0}^{\infty} \int_{S^{d-1}} E^{P^{\mathrm{R}}_r}\left[f(R_t,\varphi) e^{-\gamma_n \mathbf{f}(t)}\right] \kappa_n(\theta,\varphi)\, dm^{\Theta}(\varphi). \end{aligned} \tag{3.1}$$

We also note that

$$\begin{aligned} E^{P^{\mathrm{R}}_r}\left[f(R_t,\varphi) e^{-\gamma_n \mathbf{f}(t)}\right] &= \int_I f(\xi,\varphi) E^{P^{\mathrm{R}}_r}\left[R_t \in d\xi, e^{-\gamma_n \mathbf{f}(t)}\right] \\ &= \int_I f(\xi,\varphi) p^{(n)}(t,r,\xi)\, dm^{\mathrm{R}}(\xi), \end{aligned} \tag{3.2}$$

where $p^{(n)}(t,r,\xi)$ is the transition probability density with respect to m^{R} corresponding to an ODGDP $\mathrm{R}^{(n)}$ with $(s^{\mathrm{R}}, m^{\mathrm{R}}, \gamma_n \nu)$. Obviously $p^{(0)}(t,r,\xi) = p^{\mathrm{R}}(t,r,\xi)$. (3.1) coupled with (3.2) shows that there exists the transition probability density $p^{\mathrm{X}}(t,x,y)$ with respect to m^{X}, which is given by (1.4). Further the series converges uniformly in $(t,(r,\theta),(\xi,\varphi)) \in K$ for any compact set $K \subset (0,\infty) \times M \times M$. Since each $p^{(n)}(t,r,\xi)$ is continuous in (t,r,ξ) and $\kappa_n(\theta,\varphi)$ is continuous in (θ,φ), $p^{\mathrm{X}}(t,x,y)$ is continuous in (t,x,y). □

It follows from (1.6) and (3.2) that

$$p^{(n)}(t,r,\xi) \le p^{\mathrm{R}}(t,r,\xi), \quad t > 0,\ \xi,\eta \in I, \tag{3.3}$$

$$G^{(n)}(\alpha,r,\xi) \le G^{\mathrm{R}}(\alpha,r,\xi), \quad \alpha > 0,\ \xi,\eta \in I. \tag{3.4}$$

Noting $\int_{S^{d-1}} p^{\Theta}(t,\theta,\varphi)\,dm^{\Theta}(\theta) = \int_{S^{d-1}} p^{\Theta}(t,\theta,\varphi)\,dm^{\Theta}(\varphi) = 1$, we also obtain

$$\begin{aligned}&\int_{S^{d-1}} G^{\mathrm{X}}(\alpha,(r,\theta),(\xi,\varphi))\,dm^{\Theta}(\theta)\\&=\int_{S^{d-1}} G^{\mathrm{X}}(\alpha,(r,\theta),(\xi,\varphi))\,dm^{\Theta}(\varphi) = G^{\mathrm{R}}(\alpha,r,\xi),\end{aligned} \tag{3.5}$$

and

$$\begin{aligned}&\int_{S^{d-1}} p^{\mathrm{X}}(t,(r,\theta),(\xi,\varphi))\,dm^{\Theta}(\theta)\\&=\int_{S^{d-1}} p^{\mathrm{X}}(t,(r,\theta),(\xi,\varphi))\,dm^{\Theta}(\varphi) = p^{\mathrm{R}}(t,r,\xi).\end{aligned}$$

We next show that there exist some limits of $p^{\mathrm{X}}(t,x,y)$ and $G^{\mathrm{X}}(\alpha,x,y)$ as $|x| \to l_i$ $(i=1,2)$ for $x \in M$. For each $n = 0,1,2,\cdots$, let $g_i^{(n)}(\xi,\alpha)$, $i=1,2$ be functions satisfying (g.1)-(g.4) with s^{R}, m^{R} and $\gamma_n\nu$ in place of s, m and k, respectively. We set $W^{(n)}(\alpha) = g_1^{(n)+}(\xi,\alpha)g_2^{(n)}(\xi,\alpha) - g_1^{(n)}(\xi,\alpha)g_2^{(n)+}(\xi,\alpha)$. Note that there exist the following limits for $n=0,1,2,\cdots$, and (3.6), (3.8), (3.10) and (3.12) converge uniformly in $(t,\xi) \in K$ for any compact $K \subset (0,\infty)\times I$. If $|s^{\mathrm{R}}(l_1)| < \infty$,

$$p^{(n)}{}_2^{+}(t,l_1,\xi) = \lim_{r\to l_1} p^{(n)}(t,r,\xi)/\{s^{\mathrm{R}}(r) - s^{\mathrm{R}}(l_1)\}, \tag{3.6}$$

$$g_1^{(n)+}(l_1,\alpha) = \lim_{r\to l_1} g_1^{(n)}(r,\alpha)/\{s^{\mathrm{R}}(r) - s^{\mathrm{R}}(l_1)\}. \tag{3.7}$$

If $|s^{\mathrm{R}}(l_1)| = \infty$,

$$p^{(n)}(t,l_1,\xi) = \lim_{r\to l_1} p^{(n)}(t,r,\xi), \tag{3.8}$$

$$g_1^{(n)}(l_1,\alpha) = \lim_{r\to l_1} g_1^{(n)}(r,\alpha). \tag{3.9}$$

If $|s^{\mathrm{R}}(l_2)| < \infty$,

$$p^{(n)}{}_2^{-}(t,l_2,\xi) = \lim_{r\to l_2} p^{(n)}(t,r,\xi)/\{s^{\mathrm{R}}(r) - s^{\mathrm{R}}(l_2)\}, \tag{3.10}$$

$$g_2^{(n)-}(l_2,\alpha) = \lim_{r\to l_1} g_2^{(n)}(r,\alpha)/\{s^{\mathrm{R}}(r) - s^{\mathrm{R}}(l_2)\}. \tag{3.11}$$

If $|s^{\mathrm{R}}(l_2)| = \infty$,

$$p^{(n)}(t,l_2,\xi) = \lim_{r\to l_2} p^{(n)}(t,r,\xi), \tag{3.12}$$

$$g_2^{(n)}(l_2,\alpha) = \lim_{r\to l_2} g_2^{(n)}(r,\alpha). \tag{3.13}$$

Theorem 3.1. *Assume* $\min\{J_{s^{\mathrm{R}},m^{\mathrm{R}}}(l_1),\ J_{m^{\mathrm{R}},s^{\mathrm{R}}}(l_1)\} < \infty$.
(i) *If* $|s^{\mathrm{R}}(l_1)| < \infty$, *then there exist the following limits.*

$$\lim_{r\to l_1} p^{\mathrm{X}}(t,(r,\theta),(\xi,\varphi))/\{s^{\mathrm{R}}(r) - s^{\mathrm{R}}(l_1)\}$$

$$= \sum_{n=0}^{\infty} p^{(n)}{}_2^{+}(t, l_1, \xi)\kappa_n(\theta, \varphi), \tag{3.14}$$

$$\lim_{r \to l_1} G^{\mathrm{X}}(\alpha, (r, \theta), (\xi, \varphi)/\{s^{\mathrm{R}}(r) - s^{\mathrm{R}}(l_1)\}$$

$$= \sum_{n=0}^{\infty} W^{(n)}(\alpha)^{-1} g_1^{(n)+}(l_1, \alpha) g_2^{(n)}(\xi, \alpha)\kappa_n(\theta, \varphi), \tag{3.15}$$

for $t > 0$, $\alpha \geq 0$, $\xi \in I$, *and* $\theta, \varphi \in S^{d-1}$. *The convergence of both hand sides of (3.14) are uniform in* $(t, \xi, \theta, \varphi) \in K$ *for any compact* $K \subset (0, \infty) \times I \times S^{d-1} \times S^{d-1}$. *The convergence of both hand sides of (3.15) are uniform in* $(\xi, \theta, \varphi) \in K$ *for any compact* $K \subset I \times S^{d-1} \times S^{d-1}$. *Especially if* ν *satisfies (1.7) and (1.8), then the right hand sides of (3.14) and (3.15) are equal to* $|S^{d-1}|^{-1} p^{\mathrm{R}}{}_2^{+}(t, l_1, \xi)$ (> 0) *and* $|S^{d-1}|^{-1} W^{\mathrm{R}}(\alpha)^{-1} g_1^{\mathrm{R}+}(l_1, \alpha) g_2^{\mathrm{R}}(\xi, \alpha)$ (> 0), *respectively.*
(ii) *If* $|s^{\mathrm{R}}(l_1)| = \infty$, *then there exist the following limits.*

$$\lim_{r \to l_1} p^{\mathrm{X}}(t, (r, \theta), (\xi, \varphi)) = \sum_{n=0}^{\infty} p^{(n)}(t, l_1, \xi)\kappa_n(\theta, \varphi), \tag{3.16}$$

$$\lim_{r \to l_1} G^{\mathrm{X}}(\alpha, (r, \theta), (\xi, \varphi)) = \sum_{n=0}^{\infty} G^{(n)}(\alpha, l_1, \xi)\kappa_n(\theta, \varphi), \tag{3.17}$$

for $t > 0$, $\alpha > 0$, $\xi \in I$, *and* $\theta, \varphi \in S^{d-1}$. *The convergence of both hand sides of (3.16) are uniform in* $(t, \xi, \theta, \varphi) \in K$ *for any compact* $K \subset (0, \infty) \times I \times S^{d-1} \times S^{d-1}$. *The convergence of both hand sides of (3.17) are uniform in* $(\xi, \theta, \varphi) \in K$ *for any compact* $K \subset I \times S^{d-1} \times S^{d-1}$. *Especially if* ν *satisfies (1.9), then the right hand sides of (3.16) and (3.17) are equal to* $|S^{d-1}|^{-1} p^{\mathrm{R}}(t, l_1, \xi)$ (> 0) *and* $|S^{d-1}|^{-1} G^{\mathrm{R}}(\alpha, l_1, \xi)$ (> 0), *respectively.*

Theorem 3.2. *Assume* $\min\{J_{s^{\mathrm{R}}, m^{\mathrm{R}}}(l_2),\ J_{m^{\mathrm{R}}, s^{\mathrm{R}}}(l_2)\} < \infty$.
(i) *If* $|s^{\mathrm{R}}(l_2)| < \infty$, *then there exist the following limits.*

$$\lim_{r \to l_2} p^{\mathrm{X}}(t, (r, \theta), (\xi, \varphi))/\{s^{\mathrm{R}}(l_2) - s^{\mathrm{R}}(r)\}$$

$$= -\sum_{n=0}^{\infty} p^{(n)}{}_2^{-}(t, l_2, \xi)\kappa_n(\theta, \varphi), \tag{3.18}$$

$$\lim_{r \to l_2} G^{\mathrm{X}}(\alpha, (r, \theta), (\xi, \varphi))/\{s^{\mathrm{R}}(l_2) - s^{\mathrm{R}}(r)\}$$

$$= -\sum_{n=1}^{\infty} W^{(n)}(\alpha)^{-1} g_1^{(n)}(\xi, \alpha) g_2^{(n)-}(l_2, \alpha)\kappa_n(\theta, \varphi), \tag{3.19}$$

for $t > 0$, $\alpha \geq 0$, $\xi \in I$, *and* $\theta, \varphi \in S^{d-1}$. *The convergence of both hand sides of (3.18) are uniform in* $(t, \xi, \theta, \varphi) \in K$ *for any compact* $K \subset (0, \infty) \times I \times S^{d-1} \times S^{d-1}$. *The convergence of both hand sides of (3.19) are uniform in* $(\xi, \theta, \varphi) \in K$ *for any compact* $K \subset I \times S^{d-1} \times S^{d-1}$. *Especially if* ν *satisfies (1.11) and (1.12), then the right hand sides of (3.18) and (3.19) are equal to* $-|S^{d-1}|^{-1} p^{\mathrm{R}}{}_2^{-}(t, l_2, \xi)$ (> 0) *and* $-|S^{d-1}|^{-1} W^{\mathrm{R}}(\alpha)^{-1} g_1^{\mathrm{R}}(\xi, \alpha) g_2^{\mathrm{R}-}(l_2, \alpha)$ (> 0), *respectively.*

(ii) *If* $|s^{\mathrm{R}}(l_2)| = \infty$, *then there exist the following limits.*

$$\lim_{r\to l_2} p^{\mathrm{X}}(t,(r,\theta),(\xi,\varphi)) = \sum_{n=0}^{\infty} p^{(n)}(t,l_2,\xi)\kappa_n(\theta,\varphi), \tag{3.20}$$

$$\lim_{r\to l_2} G^{\mathrm{X}}(\alpha,(r,\theta),(\xi,\varphi)) = \sum_{n=0}^{\infty} G^{(n)}(\alpha,l_2,\xi)\kappa_n(\theta,\varphi), \tag{3.21}$$

for $t>0,\ \alpha>0,\ \xi\in I,\ \theta,\varphi\in S^{d-1}$. *The convergence of both hand sides of (3.20) are uniform in* $(t,\xi,\theta,\varphi)\in K$ *for any compact* $K\subset(0,\infty)\times I\times S^{d-1}\times S^{d-1}$. *The convergence of both hand sides of (3.21) are uniform in* $(\xi,\theta,\varphi)\in K$ *for any compact* $K\subset I\times S^{d-1}\times S^{d-1}$. *Especially if* ν *satisfies (1.13), then the right hand sides of (3.20) and (3.21) are equal to* $|S^{d-1}|^{-1}p^{\mathrm{R}}(t,l_2,\xi)\ (>0)$ *and* $|S^{d-1}|^{-1}G^{\mathrm{R}}(\alpha,l_2,\xi)\ (>0)$, *respectively.*

In order to show above theorems we prepare some lemmas. Let σ_a be the first hitting time to the point a for an ODGDP R with $(s^{\mathrm{R}},m^{\mathrm{R}},0)$, that is, $\sigma_a=\min\{t\ge 0: R(t)=a\}$. Then it is known that

$$E^{P_a^{\mathrm{R}}}\left[e^{-\alpha\sigma_b-\gamma_n\mathbf{f}(\sigma_b)}\right] = g_1^{(n)}(a,\alpha)/g_1^{(n)}(b,\alpha), \tag{3.22}$$

$$E^{P_b^{\mathrm{R}}}\left[e^{-\alpha\sigma_a-\gamma_n\mathbf{f}(\sigma_a)}\right] = g_2^{(n)}(b,\alpha)/g_2^{(n)}(a,\alpha), \tag{3.23}$$

for $\alpha\ge 0$ and $l_1<a<b<l_2$ ([7]).

Noting (g.4) for $\mathrm{R}^{(n)}$, we easily obtain the following (3.24) and (3.25).

$$\begin{aligned} &g_i^{(n)}(\xi,\alpha)\\ &= g_i^{(n)}(a,\alpha)\sum_{j=0}^{\infty}U_j(\xi;a;\alpha,\gamma_n) + g_i^{(n)+}(a,\alpha)\sum_{j=0}^{\infty}V_j(\xi;a;\alpha,\gamma_n), \end{aligned} \tag{3.24}$$

$$\begin{aligned} &g_i^{(n)+}(\xi,\alpha)\\ &= g_i^{(n)}(a,\alpha)\sum_{j=0}^{\infty}u_j(\xi;a;\alpha,\gamma_n) + g_i^{(n)+}(a,\alpha)\sum_{j=0}^{\infty}v_j(\xi;a;\alpha,\gamma_n), \end{aligned} \tag{3.25}$$

for $i=1,2,\ \alpha\ge 0$ and $a,\xi\in I$, where

$$\begin{aligned} U_0(\xi;a;\alpha,\beta) &=1,\\ U_j(\xi;a;\alpha,\beta) &= \int_{(a,\xi]} ds^{\mathrm{R}}(\eta)\int_{(a,\eta]} U_{j-1}(\zeta;a;\alpha,\beta)\{\alpha dm^{\mathrm{R}}(\zeta)+\beta d\nu(\zeta)\},\ j\ge 1,\\ V_0(\xi;a;\alpha,\beta) &= \int_{(a,\xi]} ds^{\mathrm{R}}(\eta),\\ V_j(\xi;a;\alpha,\beta) &= \int_{(a,\xi]} ds^{\mathrm{R}}(\eta)\int_{(a,\eta]} V_{j-1}(\zeta;a;\alpha,\beta)\{\alpha dm^{\mathrm{R}}(\zeta)+\beta d\nu(\zeta)\},\ j\ge 1,\\ u_0(\xi;a;\alpha,\beta) &= \int_{(a,\xi]}\{\alpha dm^{\mathrm{R}}(\eta)+\beta d\nu(\eta)\},\\ u_j(\xi;a;\alpha,\beta) &= \int_{(a,\xi]}\{\alpha dm^{\mathrm{R}}(\eta)+\beta d\nu(\eta)\}\int_{(a,\eta]} u_{j-1}(\zeta;a;\alpha,\beta)\,ds^{\mathrm{R}}(\zeta),\ j\ge 1,\\ v_0(\xi;a;\alpha,\beta) &=1, \end{aligned}$$

$$v_j(\xi;a;\alpha,\beta) = \int_{(a,\xi]} \{\alpha dm^{\mathrm{R}}(\eta) + \beta d\nu(\eta)\} \int_{(a,\eta]} v_{j-1}(\zeta;a;\alpha,\beta) ds^{\mathrm{R}}(\zeta), \ j \geq 1.$$

We put $d_o = d/2$ or $d_o = (d+1)/2$ according to d being even or odd, and $C_o = \sum_{n=1}^{\infty}(\gamma_n)^{-d_o}\kappa_n$. Note that C_o is a positive number depending only on d.

Lemma 3.1. *Let $l_1 < a < b < l_2$. Then*

$$\sum_{n=1}^{\infty} \kappa_n E^{P_a^{\mathrm{R}}}\left[e^{-\gamma_n \mathbf{f}(\sigma_b)}\right] \leq C_o/U_{d_o}(b;a;0,1). \tag{3.26}$$

Further if $|s^{\mathrm{R}}(l_1)| < \infty$, then

$$\sum_{n=1}^{\infty} \kappa_n \sup_{l_1 < r \leq a} \left\{s^{\mathrm{R}}(r) - s^{\mathrm{R}}(l_1)\right\}^{-1} E^{P_r^{\mathrm{R}}}\left[e^{-\gamma_n \mathbf{f}(\sigma_b)}\right] \leq C_o/V_{d_o}(b;a;0,1). \tag{3.27}$$

Proof. By means of (3.22) and (3.24),

$$\begin{aligned} E^{P_a^{\mathrm{R}}}\left[e^{-\gamma_n \mathbf{f}(\sigma_b)}\right] &= g_1^{(n)}(a,0)/g_1^{(n)}(b,0) \\ &\leq \left\{\sum_{j=0}^{\infty}(\gamma_n)^j U_j(b;a;0,1)\right\}^{-1} \leq \left\{(\gamma_n)^{d_o} U_{d_o}(b;a;0,1)\right\}^{-1}. \end{aligned}$$

Therefore we obtain (3.26).

Assume $|s^{\mathrm{R}}(l_1)| < \infty$. By means of (g.3) and (2.1) for $\mathrm{R}^{(n)}$, $g_1^{(n)}(l_1,0) = 0$ and $g_1^{(n)}(\xi,0)$ is convex with respect to $s^{\mathrm{R}}(\xi)$. Therefore we get

$$\begin{aligned} &\{s^{\mathrm{R}}(r) - s^{\mathrm{R}}(l_1)\}^{-1} E^{P_r^{\mathrm{R}}}\left[e^{-\gamma_n \mathbf{f}(\sigma_b)}\right] \\ &= g_1^{(n)}(r,0)/\{s^{\mathrm{R}}(r) - s^{\mathrm{R}}(l_1)\} g_1^{(n)}(b,0) \\ &\leq g_1^{(n)+}(r,0)/g_1^{(n)}(b,0) \leq g_1^{(n)+}(a,0)/g_1^{(n)}(b,0). \end{aligned}$$

By means of (3.24) again,

$$\begin{aligned} &g_1^{(n)+}(a,0)/g_1^{(n)}(b,0) \\ &\leq \left\{\sum_{j=0}^{\infty}(\gamma_n)^j V_j(b;a;0,1)\right\}^{-1} \leq \left\{(\gamma_n)^{d_o} V_{d_o}(b;a;0,1)\right\}^{-1}. \end{aligned}$$

Therefore we obtain (3.27). □

In the same way as above we obtain the following result.

Lemma 3.2. *Let $l_1 < b < a < l_2$. Then*

$$\sum_{n=1}^{\infty} \kappa_n E^{P_a^{\mathrm{R}}}\left[e^{-\gamma_n \mathbf{f}(\sigma_b)}\right] \leq C_o/U_{d_o}(b;a;0,1). \tag{3.28}$$

Further if $|s^{\mathrm{R}}(l_2)| < \infty$, *then*

$$\sum_{n=1}^{\infty} \kappa_n \sup_{a \le r < l_2} \left\{ s^{\mathrm{R}}(l_2) - s^{\mathrm{R}}(r) \right\}^{-1} E^{P_r^{\mathrm{R}}} \left[e^{-\gamma_n \mathbf{f}(\sigma_b)} \right] \le C_o / |V_{d_o}(b; a; 0, 1)|. \tag{3.29}$$

Next we note the following.

Lemma 3.3. *Let* $\alpha \ge 0$ *and* $l_1 < a < b < c < l_2$. *Then it holds true that*

$$\sum_{n=1}^{\infty} \kappa_n \sup_{t>0,\ l_1 < r \le a,\ b \le \xi \le c} p^{(n)}(t, r, \xi) < \infty, \tag{3.30}$$

$$\sum_{n=1}^{\infty} \kappa_n \sup_{l_1 < r \le a,\ b \le \xi \le c} G^{(n)}(\alpha, r, \xi) < \infty. \tag{3.31}$$

Further if $|s^{\mathrm{R}}(l_1)| < \infty$, *then*

$$\sum_{n=1}^{\infty} \kappa_n \sup_{t>0,\ l_1 < r \le a,\ b \le \xi \le c} p^{(n)}(t, r, \xi) / \left\{ s^{\mathrm{R}}(r) - s^{\mathrm{R}}(l_1) \right\} < \infty, \tag{3.32}$$

$$\sum_{n=1}^{\infty} \kappa_n \sup_{l_1 < r \le a,\ b \le \xi \le c} G^{(n)}(\alpha, r, \xi) / \left\{ s^{\mathrm{R}}(r) - s^{\mathrm{R}}(l_1) \right\} < \infty. \tag{3.33}$$

Proof. We take a point $a_o \in (a, b)$. Then it follows from (2.2) and (3.22) for $\mathrm{R}^{(n)}$ that

$$p^{(n)}(t, r, \xi) = \int_0^t P_r^{\mathrm{R}^{(n)}}(\sigma_{a_o} \in du) p^{(n)}(t - u, a_o, \xi),$$

for $l_1 < r \le a$ and $b \le \xi \le c$. Noting the fact $\Lambda = \sup_{u>0,\ b \le \xi \le c} p^{\mathrm{R}}(u, a_o, \xi) < \infty$ (see [8]), we have

$$\begin{aligned} p^{(n)}(t, r, \xi) &\le \Lambda P_r^{\mathrm{R}^{(n)}}(\sigma_{a_o} < t) \\ &\le \Lambda e E^{P_r^{\mathrm{R}^{(n)}}} \left[e^{-\sigma_{a_o}/t}; \sigma_{a_o} < t \right] \le \Lambda e E^{P_r^{\mathrm{R}}} \left[e^{-\gamma_n \mathbf{f}(\sigma_{a_o})} \right], \end{aligned} \tag{3.34}$$

for $t > 0,\ l_1 < r \le a$ and $b \le \xi \le c$. Combining (3.34) with (3.26), we obtain

$$\sum_{n=1}^{\infty} \kappa_n \sup_{t>0,\ l_1 < r \le a,\ b \le \xi \le c} p^{(n)}(t, r, \xi) \le C_o \Lambda e / U_{d_o}(a_o; a; 0, 1) < \infty, \tag{3.35}$$

which shows (3.30) and (3.31) with $\alpha > 0$. If $|s^{\mathrm{R}}(l_1)| < \infty$, then (3.34) coupled with (3.27) implies

$$\sum_{n=1}^{\infty} \kappa_n \sup_{t>0,\ l_1 < r \le a,\ b \le \xi \le c} \frac{p^{(n)}(t, r, \xi)}{s^{\mathrm{R}}(r) - s^{\mathrm{R}}(l_1)} \le C_o \Lambda e / V_{d_o}(a_o; a; 0, 1) < \infty,$$

from which we get (3.32) and (3.33) with $\alpha > 0$.

We show (3.31) and (3.33) with $\alpha = 0$. By means of (3.25),

$$G^{(n)}(0, r, \xi)$$

$$
\begin{aligned}
&= g_1^{(n)}(r,0)g_2^{(n)}(\xi,0)/\{g_1^{(n)+}(\xi,0)g_2^{(n)}(\xi,0) - g_1^{(n)}(\xi,0)g_2^{(n)+}(\xi,0)\} \\
&\le g_1^{(n)}(r,0)/g_1^{(n)+}(\xi,0) \le g_1^{(n)}(a,0)/g_1^{(n)+}(b,0) \\
&\le \left\{\sum_{j=0}^{\infty}(\gamma_n)^j u_j(b;a;0,1)\right\}^{-1} \le \{(\gamma_n)^{d_o}u_{d_o}(b;a;0,1)\}^{-1},
\end{aligned}
$$

for $l_1 < r \le a$ and $b \le \xi \le c$. Therefore

$$
\begin{aligned}
&\sum_{n=1}^{\infty}\kappa_n \sup_{l_1<r\le a,\ b\le\xi\le c} G^{(n)}(0,r,\xi) \\
&\le \sum_{n=1}^{\infty}\kappa_n(\gamma_n)^{-d_o}/u_{d_o}(b;a;0,1) \le C_o/u_{d_o}(b;a;0,1),
\end{aligned}
$$

which shows (3.31) with $\alpha = 0$. If $|s^{\mathrm{R}}(l_1)| < \infty$, then, by the same reason as in the proof of Lemma 3.1,

$$
\begin{aligned}
&\sup_{l_1<r\le a,\ b\le\xi\le c} G^{(n)}(0,r,\xi)/\{s^{\mathrm{R}}(r) - s^{\mathrm{R}}(l_1)\} \\
&\le g_1^{(n)+}(a,0)g_2^{(n)}(b,0)/\{g_1^{(n)+}(b,0)g_2^{(n)}(b,0) - g_1^{(n)}(b,0)g_2^{(n)+}(b,0)\} \\
&\le g_1^{(n)+}(a,0)/g_1^{(n)+}(b,0) \\
&\le \left\{\sum_{j=0}^{\infty}(\gamma_n)^j v_j(b;a;0,1)\right\}^{-1} \le \{(\gamma_n)^{d_o}v_{d_o}(b;a;0,1)\}^{-1}.
\end{aligned}
$$

Therefore

$$
\sum_{n=1}^{\infty}\kappa_n \sup_{l_1<r\le a,\ b\le\xi\le c} G^{(n)}(0,r,\xi)/\left\{s^{\mathrm{R}}(r) - s^{\mathrm{R}}(l_1)\right\} \le C_o/v_{d_o}(b;a;0,1),
$$

which shows (3.33) with $\alpha = 0$. □

In the same way as above we obtain the following.

Lemma 3.4. *Let $\alpha \ge 0$ and $l_1 < a < b < c < l_2$. Then it holds true that*

$$
\sum_{n=1}^{\infty}\kappa_n \sup_{t>0,\ c\le r<l_2,\ a\le\xi\le b} p^{(n)}(t,r,\xi) < \infty, \tag{3.36}
$$

$$
\sum_{n=1}^{\infty}\kappa_n \sup_{c\le r<l_2,\ a\le\xi\le b} G^{(n)}(\alpha,r,\xi) < \infty. \tag{3.37}
$$

Further if $|s^{\mathrm{R}}(l_2)| < \infty$, then

$$
\sum_{n=1}^{\infty}\kappa_n \sup_{t>0,\ c\le r<l_2,\ a\le\xi\le b} p^{(n)}(t,r,\xi)/\left\{s^{\mathrm{R}}(l_2) - s^{\mathrm{R}}(r)\right\} < \infty, \tag{3.38}
$$

$$
\sum_{n=1}^{\infty}\kappa_n \sup_{c\le r<l_2,\ a\le\xi\le b} G^{(n)}(\alpha,r,\xi)/\left\{s^{\mathrm{R}}(l_2) - s^{\mathrm{R}}(r)\right\} < \infty. \tag{3.39}
$$

We are ready to show Theorems 3.1 and 3.2. We only give

Proof of Theorem 3.1. Assume $\min\{J_{s^{\mathrm{R}},m^{\mathrm{R}}}(l_1),\ J_{m^{\mathrm{R}},s^{\mathrm{R}}}(l_1)\} < \infty$, so that l_1 is not $(s^{\mathrm{R}}, m^{\mathrm{R}}, 0)$-natural.

(i) Assume $|s^{\mathrm{R}}(l_1)| < \infty$. Then, by means of (1.4), (1.5), (3.6), (3.7), (3.32), and (3.33), we obtain (3.14) and (3.15). The formula (3.32) also shows that the convergence of the left hand side of (3.14) is uniform in $(t, \xi, \theta, \varphi) \in K$ and the series of the right hand sides of (3.14) converges uniformly in $(t, \xi, \theta, \varphi) \in K$ for any compact $K \subset (0, \infty) \times I \times S^{d-1} \times S^{d-1}$. The formula (3.33) shows that the convergence of the left hand side of (3.15) is uniform in $(\xi, \theta, \varphi) \in K$ and the series of the right hand sides of (3.15) converges uniformly in $(\xi, \theta, \varphi) \in K$ for any compact $K \subset I \times S^{d-1} \times S^{d-1}$. We note that, under the assumptions (1.7) and (1.8),

$$p^{(n)}{}_2^+(t, l_1, \xi) = 0, \ g_1^{(n)+}(l_1, \alpha) = 0, \quad n = 1, 2, \cdots. \tag{3.40}$$

Indeed, the condition (1.8) coupled with $|s^{\mathrm{R}}(l_1)| < \infty$ implies

$$\int_{(l_1, c_o)} d\nu(\xi) \int_{(\xi, c_o)} ds^{\mathrm{R}}(\eta) = \infty. \tag{3.41}$$

This coupled with (1.7) means that l_1 is $(s^{\mathrm{R}}, m^{\mathrm{R}}, \gamma_n \nu)$-natural, and hence we have (3.40) (see [7], [8]). Thus we obtain the statement(i) of the theorem.

(ii) Assume $|s^{\mathrm{R}}(l_1)| = \infty$. Then, by means of (1.4), (1.5), (3.8), (3.9), (3.30), and (3.31), we obtain (3.16) and (3.17). The formula (3.30) shows that the convergence of the left hand side of (3.16) is uniform in $(t, \xi, \theta, \varphi) \in K$ and the series of the right hand sides of (3.16) converges uniformly in $(t, \xi, \theta, \varphi) \in K$ for any compact $K \subset (0, \infty) \times I \times S^{d-1} \times S^{d-1}$. The formula (3.31) shows that the convergence of the left hand side of (3.17) is uniform in $(\xi, \theta, \varphi) \in K$ and the series of the right hand sides of (3.17) converges uniformly in $(\xi, \theta, \varphi) \in K$ for any compact $K \subset I \times S^{d-1} \times S^{d-1}$. We note that, under the assumption (1.9),

$$p^{(n)}(t, l_1, \xi) = 0, \ g_1^{(n)}(l_1, \alpha) = 0, \quad n = 1, 2, \cdots. \tag{3.42}$$

Indeed, $|s^{\mathrm{R}}(l_1)| = \infty$ and (1.9) imply (1.7) and (3.41), respectively. Therefore l_1 is $(s^{\mathrm{R}}, m^{\mathrm{R}}, \gamma_n \nu)$-natural, and hence we have (3.42) (see [7], [8]). Thus we obtain the statement(ii) of the theorem. □

We turn to [SSP] in the sense of Theorem 1.1. First we show [SP] for the α-Green function $G^{(n)}(\alpha, x, y)$ of ODGDP $\mathrm{R}^{(n)}$. In [16] we characterized [SP] for ODGDOs in terms of the state of boundary points l_i $(i = 1, 2)$. By virtue of Theorem 2.6 of [16], we see that, if both of l_1 and l_2 are not $(s^{\mathrm{R}}, m^{\mathrm{R}}, \gamma_n \nu)$-natural, then for any $\varepsilon > 0$ there exist two points $a, b \in I$ satisfying (3.43) below. Note that a and b generally depend on n. We also note that l_1 or l_2 may be $(s^{\mathrm{R}}, m^{\mathrm{R}}, \gamma_n \nu)$-natural as we saw in the proof of Proposition 2.2, and hence we cannot apply Theorem 2.6 of [16]. However the following result shows that [SP] holds for all $\mathrm{R}^{(n)}$, and a and b do not depend on n, whenever $J_{s^{\mathrm{R}},m^{\mathrm{R}}}(l_i) < \infty$ or $J_{m^{\mathrm{R}},s^{\mathrm{R}}}(l_i) < \infty$ for $i = 1, 2$.

Proposition 3.1. *Assume* $\min\{J_{s^{\mathrm{R}},m^{\mathrm{R}}}(l_i),\ J_{m^{\mathrm{R}},s^{\mathrm{R}}}(l_i)\} < \infty$ *for* $i = 1, 2$. *Then for any*

$\varepsilon > 0$ there exist two points $a, b \in I$ such that

$$\int_{I\setminus[a,b]} G^{(n)}(\alpha, r, \eta)G^{(n)}(\alpha, \eta, \xi)dm^{\mathrm{R}}(\eta) < \varepsilon G^{(n)}(\alpha, r, \xi), \tag{3.43}$$

for $r, \xi \in I \setminus [a, b]$, and $n = 0, 1, 2, \cdots$.

Proof. The assumption $\min\{J_{s^{\mathrm{R}},m^{\mathrm{R}}}(l_i),\ J_{m^{\mathrm{R}},s^{\mathrm{R}}}(l_i)\} < \infty$, $i = 1, 2$, implies that both of l_1 and l_2 are not $(s^{\mathrm{R}}, m^{\mathrm{R}}, 0)$-natural. Therefore, by virtue of Theorem 2.6 of [16], for any $\varepsilon > 0$ there exist two points $a, b \in I$ such that

$$\int_{I\setminus[a,b]} G^{\mathrm{R}}(\alpha, \eta, \eta)dm^{\mathrm{R}}(\eta) < \varepsilon. \tag{3.44}$$

By means of (g.2) and (3.4), we have

$$\begin{aligned}
&\int_{I\setminus[a,b]} G^{(n)}(\alpha, r, \eta)G^{(n)}(\alpha, \eta, \xi)dm^{\mathrm{R}}(\eta)\\
&\le G^{(n)}(\alpha, r, \xi)\int_{I\setminus[a,b]} G^{(n)}(\alpha, \eta, \eta)dm^{\mathrm{R}}(\eta)\\
&\le G^{(n)}(\alpha, r, \xi)\int_{I\setminus[a,b]} G^{\mathrm{R}}(\alpha, \eta, \eta)dm^{\mathrm{R}}(\eta),
\end{aligned}$$

for $r, \xi \in I \setminus [a, b]$ and $n = 0, 1, 2, \cdots$. Combining this with (3.44) we obtain the assertion of the proposition. □

We next note the following.

Proposition 3.2. *Assume* $\min\{J_{s^{\mathrm{R}},m^{\mathrm{R}}}(l_i),\ J_{m^{\mathrm{R}},s^{\mathrm{R}}}(l_i)\} < \infty$ *for $i = 1, 2$. Then for any $\varepsilon > 0$ there is a compact set $K \subset I$ such that*

$$\int_{(I\setminus K)\times S^{d-1}} G^{\mathrm{X}}(\alpha, x, z)G^{\mathrm{X}}(\alpha, z, y)dm^{\mathrm{X}}(z) < \varepsilon G^{\mathrm{X}}(\alpha, x, y), \tag{3.45}$$

for $x = (r, \theta),\ y = (\xi, \varphi) \in (I \setminus K) \times S^{d-1}$ satisfying $\theta = \varphi$.

Remark 3.1. We should note that (3.45) only holds for $x = (r, \theta),\ y = (\xi, \varphi) \in (I \setminus K) \times S^{d-1}$ satisfying $\theta = \varphi$. Therefore it does not yield [SP].

Proof. Let us fix an $\varepsilon > 0$ arbitrarily. We take two points $a, b \in I$ satisfying (3.43) for $r, \xi \in I \setminus [a, b]$, and $n = 0, 1, 2, \cdots$. By using properties of spherical harmonics S_n^l and (1.5), we see

$$\begin{aligned}
&\int_{(I\setminus[a,b])\times S^{d-1}} G^{\mathrm{X}}(\alpha, (r, \theta), (\eta, \psi))G^{\mathrm{X}}(\alpha, (\eta, \psi), (\xi, \theta))\, dm^{\mathrm{R}}(\eta)dm^{\Theta}(\psi)\\
&= \int_{(I\setminus[a,b])\times S^{d-1}} \left\{\sum_{n=0}^{\infty} G^{(n)}(\alpha, r, \eta)\kappa_n(\theta, \psi)\right\}\\
&\qquad \times \left\{\sum_{n=0}^{\infty} G^{(n)}(\alpha, \eta, \xi)\kappa_n(\psi, \theta)\right\} dm^{\mathrm{R}}(\eta)dm^{\Theta}(\psi)
\end{aligned}$$

$$= \sum_{n=0}^{\infty} \kappa_n(\theta,\theta) \int_{I\setminus[a,b]} G^{(n)}(\alpha,r,\eta)G^{(n)}(\alpha,\eta,\xi)\, dm^{\mathrm{R}}(\eta)$$
$$< \varepsilon \sum_{n=0}^{\infty} \kappa_n G^{(n)}(\alpha,r,\xi) = \varepsilon G^{\mathrm{X}}(\alpha,(r,\theta),(\xi,\theta)),$$

for $(r,\theta),(\xi,\theta) \in (I\setminus[a,b]) \times S^{d-1}$. Thus we obtain the proposition. □

Now we show Theorem 1.1.

Proof of Theorem 1.1. We only show the statement (i). Assume $\min\{J_{s^{\mathrm{R}},m^{\mathrm{R}}}(l_i), J_{m^{\mathrm{R}},s^{\mathrm{R}}}(l_i)\} < \infty$ for $i = 1,2$, which implies that both of l_1 and l_2 are not $(s^{\mathrm{R}}, m^{\mathrm{R}}, 0)$-natural. Therefore, by virtue of Theorem 2.6 of [16], for any $\varepsilon > 0$ there exists a compact set $K \subset I$ such that

$$\int_{I\setminus K} G^{\mathrm{R}}(\alpha,\eta,\eta) dm^{\mathrm{R}}(\eta) < \varepsilon.$$

Assume (S1). By means of Theorem 3.1, there exists the limit

$$\lim_{r\to l_1} G^{\mathrm{X}}(\alpha,(r,\theta),(\xi,\varphi))/\{s^{\mathrm{R}}(r) - s^{\mathrm{R}}(l_1)\}$$
$$= |S^{d-1}|^{-1} W^{\mathrm{R}}(\alpha)^{-1} g_1^{\mathrm{R}+}(l_1,\alpha) g_2^{\mathrm{R}}(\xi,\alpha) > 0.$$

We also have, by virtue of (3.5),

$$\int_I dm^{\mathrm{R}}(\eta) \int_{S^{d-1}} G^{\mathrm{X}}(\alpha,(\eta,\psi),(\xi,\varphi)) dm^{\Theta}(\psi)$$
$$= \int_I G^{\mathrm{R}}(\alpha,\eta,\xi)\, dm^{\mathrm{R}}(\eta) < \infty.$$

Combining these facts with (3.33) and by using the dominated convergence theorem, we get

$$\lim_{r\to l_1} \{s^{\mathrm{R}}(r) - s^{\mathrm{R}}(l_1)\}^{-1}$$
$$\times \int_{(I\setminus K)\times S^{d-1}} G^{\mathrm{X}}(\alpha,(r,\theta),(\eta,\psi)) G^{\mathrm{X}}(\alpha,(\eta,\psi),(\xi,\varphi))\, dm^{\mathrm{R}}(\eta)\, dm^{\Theta}(\psi)$$
$$= |S^{d-1}|^{-1} W^{\mathrm{R}}(\alpha)^{-1} g_1^{\mathrm{R}+}(l_1,\alpha)$$
$$\times \int_{(I\setminus K)\times S^{d-1}} g_2^{\mathrm{R}}(\eta,\alpha) G^{\mathrm{X}}(\alpha,(\eta,\psi),(\xi,\varphi))\, dm^{\mathrm{R}}(\eta)\, dm^{\Theta}(\psi)$$
$$= |S^{d-1}|^{-1} W^{\mathrm{R}}(\alpha)^{-1} g_1^{\mathrm{R}+}(l_1,\alpha) \int_{I\setminus K} g_2^{\mathrm{R}}(\eta,\alpha) G^{\mathrm{R}}(\alpha,\eta,\xi)\, dm^{\mathrm{R}}(\eta),$$

from which

$$\lim_{r\to l_1} G^{\mathrm{X}}(\alpha,(r,\theta),(\xi,\varphi))^{-1}$$
$$\times \int_{(I\setminus K)\times S^{d-1}} G^{\mathrm{X}}(\alpha,(r,\theta),(\eta,\psi)) G^{\mathrm{X}}(\alpha,(\eta,\psi),(\xi,\varphi))\, dm^{\mathrm{R}}(\eta)\, dm^{\Theta}(\psi)$$

$$= g_2^{\mathrm{R}}(\xi,\alpha)^{-1}\int_{I\setminus K} g_2^{\mathrm{R}}(\eta,\alpha)G^{\mathrm{R}}(\alpha,\eta,\xi)\,dm^{\mathrm{R}}(\eta)$$
$$\le \int_{I\setminus K} G^{\mathrm{R}}(\alpha,\eta,\eta)\,dm^{\mathrm{R}}(\eta) < \varepsilon.$$

Thus (1.10) follows.

Assume (S2). By means of Theorem 3.1, there exists the limit

$$\lim_{r\to l_1} G^{\mathrm{X}}(\alpha,(r,\theta),(\eta,\psi)) = |S^{d-1}|^{-1}G^{\mathrm{R}}(\alpha,l_1,\eta) > 0.$$

Noting (3.31), in the same way as above we have

$$\lim_{r\to l_1}\int_{(I\setminus K)\times S^{d-1}} G^{\mathrm{X}}(\alpha,(r,\theta),(\eta,\psi))G^{\mathrm{X}}(\alpha,(\eta,\psi),(\xi,\varphi))\,dm^{\mathrm{R}}(\eta)\,dm^{\Theta}(\psi)$$
$$= |S^{d-1}|^{-1}\int_{(I\setminus K)\times S^{d-1}} G^{\mathrm{R}}(\alpha,l_1,\eta)G^{\mathrm{X}}(\alpha,(\eta,\psi),(\xi,\varphi))\,dm^{\mathrm{R}}(\eta)\,dm^{\Theta}(\psi)$$
$$= |S^{d-1}|^{-1}\int_{I\setminus K} G^{\mathrm{R}}(\alpha,l_1,\eta)G^{\mathrm{R}}(\alpha,\eta,\xi)\,dm^{\mathrm{R}}(\eta)$$
$$\le |S^{d-1}|^{-1}G^{\mathrm{R}}(\alpha,l_1,\xi)\int_{I\setminus K} G^{\mathrm{R}}(\alpha,\eta,\eta)\,dm^{\mathrm{R}}(\eta)$$
$$< \varepsilon|S^{d-1}|^{-1}G^{\mathrm{R}}(\alpha,l_1,\xi),$$

from which (1.10) and Remark 1.1 follow. □

4. Intrinsic ultracontractivity

In this section, we show Proposition 1.4, Theorem 1.2, and Theorem 1.4.

Proof of Proposition 1.4. Assume $\min\{J_{s^{\mathrm{R}},m^{\mathrm{R}}}(l_i),\ J_{m^{\mathrm{R}},s^{\mathrm{R}}}(l_i)\} < \infty$ for $i = 1,2$, and one of the conditions (Ej), $j = 1,2$. Since both of l_1 and l_2 are not $(s^{\mathrm{R}}, m^{\mathrm{R}}, 0)$-natural, we have (1.15) by virtue of Proposition 2.2. For each $n \ge 1$, $(E1)$ and $(E2)$ imply $(R1)$ and $(R2)$, respectively, where (s,m,k) is replaced by $(s^{\mathrm{R}}, m^{\mathrm{R}}, \gamma_n\nu)$. Therefore $p^{(n)}(t,r,\xi)$ has the eigenfunction expansion, that is,

$$p^{(n)}(t,r,\xi) = \sum_{j=0}^{\infty} e^{-\lambda_j^{(n)}t}\phi_j^{(n)}(r)\phi_j^{(n)}(\xi),$$

where $\{\phi_j^{(n)}\}_{j=0}^{\infty}$ is a complete orthonormal base of eigenfunctions with eigenvalues $0 \le \lambda_0^{(n)} < \lambda_1^{(n)} \le \lambda_2^{(n)} \le \cdots$ repeated according to multiplicity, and the series converges uniformly in $(t,r,\xi) \in K$ for any compact set $K \subset (0,\infty)\times I\times I$. Combining this with (1.4), we get

$$p^{\mathrm{X}}(t,(r,\theta),(\xi,\varphi)) = \sum_{n=0}^{\infty} \kappa_n(\theta,\varphi)\sum_{j=0}^{\infty} e^{-\lambda_j^{(n)}t}\phi_j^{(n)}(r)\phi_j^{(n)}(\xi).$$

The series converges uniformly in $(t,(r,\theta),(\xi,\varphi)) \in K$ for any compact set $K \subset (0,\infty)\times M\times M$. We note that $\lambda_0^{(0)} = \lambda_0^{\mathrm{R}} < \lambda_0^{(n)}$, $n = 1,2,\cdots$. Thus we get (1.17) with $0 \le \lambda_0^{\mathrm{X}} =$

$\lambda_0^{\mathrm{R}} < \lambda_1^{\mathrm{X}} \leq \lambda_2^{\mathrm{X}} \leq \cdots$.

We show $\phi_0^{\mathrm{X}}(x) = \phi_0^{\mathrm{R}}(|x|)/\sqrt{|S^{d-1}|}$.

$$\begin{aligned}
&\sum_{j=0}^{\infty} e^{-\lambda_j^{\mathrm{X}} t} \phi_j^{\mathrm{X}}(x) \int_M \phi_j^{\mathrm{X}}(y)\phi_0^{\mathrm{R}}(|y|)\, dm^{\mathrm{X}}(y) \\
&= p_t^{\mathrm{X}}\phi_0^{\mathrm{R}}(x) = \int_M p^{\mathrm{X}}(t,x,y)\phi_0^{\mathrm{R}}(|y|)\, dm^{\mathrm{X}}(y) \\
&= \int_{S^{d-1}} E^{P_r^{\mathrm{R}}}\left[\phi_0^{\mathrm{R}}(R_t) p^{\Theta}(\mathbf{f}(t),\theta,\varphi)\right] dm^{\Theta}(\varphi) \\
&= E^{P_r^{\mathrm{R}}}\left[\phi_0^{\mathrm{R}}(R_t)\right] = \sum_{j=0}^{\infty} e^{-\lambda_j^{\mathrm{R}} t}\phi_j^{\mathrm{R}}(r) \int_I \phi_j^{\mathrm{R}}(\xi)\phi_0^{\mathrm{R}}(\xi)\, dm^{\mathrm{R}}(\xi) \\
&= e^{-\lambda_0^{\mathrm{R}} t}\phi_0^{\mathrm{R}}(r).
\end{aligned}$$

Since $\lambda_0^{\mathrm{X}} = \lambda_0^{\mathrm{R}}$, we have $C\phi_0^{\mathrm{X}}(x) = \phi_0^{\mathrm{R}}(r)$, where $C = \int_M \phi_0^{\mathrm{X}}(y)\phi_0^{\mathrm{R}}(|y|)\, dm^{\mathrm{X}}(y)$. Noting $C = \int_M C^{-1}\phi_0^{\mathrm{R}}(|y|)^2\, dm^{\mathrm{X}}(y) = C^{-1}|S^{d-1}|$, we get $C = |S^{d-1}|^{1/2}$. Thus the proof is completed. □

Theorem 1.2 immediately follows from the fact $\phi^{\mathrm{X}}(x) = |S^{d-1}|^{-1/2} \times \phi^{\mathrm{R}}(|x|)$ and Theorems 3.1 and 3.2.

Proof of Theorem 1.4. Under the assumption of the theorem, (1.27) holds, that is,

$$\Lambda_t = \sup_{r,\xi\in I}\left\{p^{\mathrm{R}}(t,r,\xi)/\phi_0^{\mathrm{R}}(r)\phi_0^{\mathrm{R}}(\xi)\right\} < \infty, \quad t > 0.$$

For $t > 0$, the local time $l^{\mathrm{R}}(t,\xi)$ is continuous in ξ and $l^{\mathrm{R}}(t,\xi) = 0$ for $\xi \notin [\min_{0\leq u\leq t} R(u), \max_{0\leq u\leq t} R(u)]$ (see [7]). Therefore

$$\begin{aligned}
\mathbf{f}(t) &= \int_I l^{\mathrm{R}}(t,\xi)\, d\nu(\xi) \\
&\geq B_0 \int_I l^{\mathrm{R}}(t,\xi)\, dm^{\mathrm{R}}(\xi) = B_0(t\wedge\sigma_{l_1}\wedge\sigma_{l_2}),\ t>0,\ P_r^{\mathrm{R}} - a.e.,\ r\in I.
\end{aligned}$$

Since $p^{\Theta}(t,\theta,\varphi) \leq p^{\Theta}(t,\theta,\theta) = A_t^{\Theta} = \sum_{n=0}^{\infty} e^{-\gamma_n t}\kappa_n$,

$$\begin{aligned}
&E^{P_r^{\mathrm{R}}}\left[|f(R_t,\varphi)|p^{\Theta}(\mathbf{f}(t),\theta,\varphi)\right] \\
&\leq E^{P_r^{\mathrm{R}}}\left[|f(R_t,\varphi)|p^{\Theta}(\mathbf{f}(t),\theta,\theta)\right] \\
&= E^{P_r^{\mathrm{R}}}\left[|f(R_t,\varphi)|p^{\Theta}(\mathbf{f}(t),\theta,\theta); t<\sigma_{l_1}\wedge\sigma_{l_2}\right] \\
&\leq E^{P_r^{\mathrm{R}}}\left[|f(R_t,\varphi)|p^{\Theta}(B_0 t,\theta,\theta); t<\sigma_{l_1}\wedge\sigma_{l_2}\right] \\
&\leq A_{B_0 t}^{\Theta} E^{P_r^{\mathrm{R}}}\left[|f(R_t,\varphi)|\right].
\end{aligned}$$

Therefore for any $f \in L^2(M, m^{\mathrm{X}})$,

$$|p_t^{\mathrm{X}} f(r,\theta)| \leq \int_{S^{d-1}} E^{P_r^{\mathrm{R}}}\left[|f(R_t,\varphi)|p^{\Theta}(\mathbf{f}(t),\theta,\varphi)\right] dm^{\Theta}(\varphi)$$

$$
\begin{aligned}
&\leq A^{\Theta}_{B_0 t}\int_{S^{d-1}} E^{P^{\mathrm{R}}_r}\left[|f(R_t,\varphi)|\right] dm^{\Theta}(\varphi)\\
&= A^{\Theta}_{B_0 t}\int_{I\times S^{d-1}} |f(\xi,\varphi)| p^{\mathrm{R}}(t,r,\xi)\, dm^{\mathrm{R}}(\xi) dm^{\Theta}(\varphi)\\
&\leq A^{\Theta}_{B_0 t}\Lambda_t\int_{I\times S^{d-1}} |f(\xi,\varphi)|\phi^{\mathrm{R}}_0(r)\phi^{\mathrm{R}}_0(\xi)\, dm^{\mathrm{R}}(\xi) dm^{\Theta}(\varphi)\\
&\leq |S^{d-1}|^{1/2} A^{\Theta}_{B_0 t}\Lambda_t\phi^{\mathrm{R}}_0(r)\|f\|_{L^2(M,m^{\mathrm{X}})}.
\end{aligned}
$$

Since $\phi^{\mathrm{X}}_0(x)=\phi^{\mathrm{R}}_0(|x|)/\sqrt{|S^{d-1}|}$, we get

$$
|p^{\mathrm{X}}_t f(x)|\leq C_t\|f\|_{L^2(M,m^{\mathrm{X}})}\phi^{\mathrm{X}}_0(x), \qquad x\in M,
$$

for all $f\in L^2(M,m^{\mathrm{X}})$, where C_t is a positive constant depending on t. This shows (1.28) (see [2]). □

5. Example

In this section, we consider the Laplace-Beltrami operator Δ treated by Murata ([10]). Let $I=(0,\infty)$, $M=I\times S^{d-1}$, and

$$
\Delta = f^{1-d}\partial/\partial r\left(f^{d-1}\partial/\partial r\right)+f^{-2}\Delta^{\Theta},
$$

where f is a nonnegative smooth function on $[0,\infty)$ such that $f>0$ on I, $f(0)=0$, $f'(0)=1$ and $f''(0)=0$. We put

$$
\begin{aligned}
&ds^{\mathrm{R}}(\xi)=f^{1-d}(\xi)\,d\xi, \quad dm^{\mathrm{R}}(\xi)=f^{d-1}(\xi)\,d\xi,\\
&d\nu(\xi)=f^{-2}(\xi)\,dm^{\mathrm{R}}(\xi)=f^{d-3}\,d\xi.
\end{aligned}
$$

Let R be a diffusion process on I with $(s^{\mathrm{R}},m^{\mathrm{R}},0)$, and X be the skew product of R and the spherical Brownian motion Θ with respect to the PCAF $\mathbf{f}(t)$, where $\mathbf{f}(t)=\int_I l^{\mathrm{R}}(t,r)\,d\nu(r)$. Then Δ is the generator of X. We denote the corresponding items for R or X by $p^{\mathrm{R}}(t,r,\xi)$, $p^{\mathrm{X}}(t,x,y)$, etc.

Since $\lim_{\xi\to 0} f(\xi)/\xi=1$, we have

$$
J_{s^{\mathrm{R}},m^{\mathrm{R}}}(0)=\int_0^1 f^{1-d}(\xi)\,d\xi\int_\xi^1 f^{d-1}(\eta)\,d\eta=\infty, \tag{5.1}
$$

$$
J_{m^{\mathrm{R}},s^{\mathrm{R}}}(0)=\int_0^1 f^{d-1}(\xi)\,d\xi\int_\xi^1 f^{1-d}(\eta)\,d\eta<\infty, \tag{5.2}
$$

$$
J_{\nu,s^{\mathrm{R}}}(0)=\int_0^1 f^{d-3}(\xi)\,d\xi\int_\xi^1 f^{1-d}(\eta)\,d\eta=\infty. \tag{5.3}
$$

Therefore $|s^{\mathrm{R}}(0)|=\infty$ and (1.9) is satisfied. We next note that

$$
\begin{aligned}
&\int_1^\infty dm^{\mathrm{R}}(\xi)=\int_1^\infty f^{d-1}(\xi)\,d\xi=\infty, \quad \text{or}\\
&\int_1^\infty ds^{\mathrm{R}}(\xi)=\int_1^\infty f^{1-d}(\xi)\,d\xi=\infty,
\end{aligned}
$$

since $\int_1^\infty f^{d-1}(\xi)\,d\xi \int_1^\infty f^{1-d}(\xi)\,d\xi \ \geq\ \left(\int_1^\infty d\xi\right)^2 \ =\ \infty$. We also note $J_{s^{\mathrm{R}},m^{\mathrm{R}}}(\infty) + J_{m^{\mathrm{R}},s^{\mathrm{R}}}(\infty) = \int_1^\infty f^{d-1}(\xi)\,d\xi \int_1^\infty f^{1-d}(\xi)\,d\xi = \infty$, and hence

$$J_{s^{\mathrm{R}},m^{\mathrm{R}}}(\infty) = \infty \quad \text{or} \quad J_{m^{\mathrm{R}},s^{\mathrm{R}}}(\infty) = \infty.$$

Noting Lemmas 3.2 and 3.3 of [10], we see the following. If $J_{m^{\mathrm{R}},s^{\mathrm{R}}}(\infty) < \infty$, then $s^{\mathrm{R}}(\infty) = \infty$ and

$$J_{\nu,s^{\mathrm{R}}}(\infty) = \int_1^\infty f^{d-3}(\xi)\,d\xi \int_1^\xi f^{1-d}(\eta)\,d\eta = \infty, \tag{5.4}$$

that is, (1.13) is satisfied. If $J_{s^{\mathrm{R}},m^{\mathrm{R}}}(\infty) < \infty$, then

$$J_{s^{\mathrm{R}},\nu}(\infty) = \int_1^\infty f^{1-d}(\xi) \int_1^\xi f^{d-3}(\eta)\,d\eta < \infty, \tag{5.5}$$

that is, (1.11) is not satisfied.

The following result with $x_0 = (0,\theta)$ is obtained by Murata (see Theorem 1.1 of [10]). We can derive it from Theorem 1.1 and Remark 1.1.

Proposition 5.1. (Murata, [10]) *Assume $J_{m^{\mathrm{R}},s^{\mathrm{R}}}(\infty) < \infty$ or $J_{s^{\mathrm{R}},m^{\mathrm{R}}}(\infty) < \infty$. Then for any $\varepsilon > 0$ there is a compact set $K \subset M$ such that*

$$\int_{M\setminus K} G^{\mathrm{X}}(\alpha, x_0, z) G^{\mathrm{X}}(\alpha, z, y)\, dm^{\mathrm{X}}(z) < \varepsilon G^{\mathrm{X}}(\alpha, x_0, y), \tag{5.6}$$

for $y \in M \setminus K$, where $x_0 = (0,\theta)$. Further, if $J_{m^{\mathrm{R}},s^{\mathrm{R}}}(\infty) < \infty$, then (5.6) also holds for $x_0 = (\infty,\theta)$. Both hand sides of (5.6) are independent of $\theta \in S^{d-1}$.

We next note the following.

Lemma 5.1. *If $J_{s^{\mathrm{R}},m^{\mathrm{R}}}(\infty) < \infty$ or $J_{m^{\mathrm{R}},s^{\mathrm{R}}}(\infty) < \infty$, then the condition $(E1)$ is satisfied.*

Proof. By virtue of (5.1) and (5.2), the end point 0 is $(s^{\mathrm{R}}, m^{\mathrm{R}}, 0)$-entrance.

Assume that $J_{s^{\mathrm{R}},m^{\mathrm{R}}}(\infty) < \infty$, from which the end point ∞ is not $(s^{\mathrm{R}}, m^{\mathrm{R}}, 0)$-natural. By means of (5.5), $J_{s^{\mathrm{R}},\nu}(\infty) < \infty$ and hence $J_{s^{\mathrm{R}},m^{\mathrm{R}}+\nu}(\infty) < \infty$, that is, the end point ∞ is not $(s^{\mathrm{R}}, m^{\mathrm{R}}, \gamma_n\nu)$-natural. Therefore $(Q1)$ with $(s^{\mathrm{R}}, m^{\mathrm{R}}, \gamma_n\nu)$ in place of (s,m,k) is satisfied. By virtue of Proposition 2.3, $(R1)$ with $(s^{\mathrm{R}}, m^{\mathrm{R}}, \gamma_n\nu)$ in place of (s,m,k) holds. Thus we see that the condition $(E1)$ is satisfied.

Assume that $J_{m^{\mathrm{R}},s^{\mathrm{R}}}(\infty) < \infty$. Then $J_{s^{\mathrm{R}},m^{\mathrm{R}}}(\infty) = \infty$ and hence $J_{s^{\mathrm{R}},m^{\mathrm{R}}+\nu}(\infty) = \infty$. We also have $J_{m^{\mathrm{R}}+\nu,s^{\mathrm{R}}}(\infty) = \infty$ by means of (5.4). Therefore the end point ∞ is $(s^{\mathrm{R}}, m^{\mathrm{R}}, 0)$-entrance and $(s^{\mathrm{R}}, m^{\mathrm{R}}, \gamma_n\nu)$-natural. Further $J_{s^{\mathrm{R}},m^{\mathrm{R}}+\nu}(0) = J_{m^{\mathrm{R}}+\nu,s^{\mathrm{R}}}(0) = \infty$ by (5.1) and (5.3), that is, the end point 0 is $(s^{\mathrm{R}}, m^{\mathrm{R}}, \gamma_n\nu)$-natural. Fix an $\alpha > 0$ arbitrarily and let $g_2(\xi,\alpha)$ be a function satisfying (g.1)-(g.4) with $(s^{\mathrm{R}}, m^{\mathrm{R}}, \gamma_n\nu)$ in place of (s,m,k). We show that $g_2(\xi,\alpha)$ satisfies (2.11). For a positive constant A, we put $h(\xi,\alpha;A) = \xi^{-\delta} K_\beta(\sqrt{\alpha}\,A^{d-1}\xi)$, where $K_\beta(\xi)$ is a modified Bessel function, $\delta = (d-2)/2$, and $\beta = \sqrt{\gamma_n A^{2d} + \delta^2}$. Note that $h(\xi) = h(\xi,\alpha;A)$ is positive and decreasing on I, and satisfies

$$h''(\xi) + \frac{d-1}{\xi} h'(\xi) - (\alpha A^{2(d-1)} + \gamma_n A^{2d}\xi^{-2}) h(\xi) = 0.$$

We take a constant $C > 1$ satisfying $\sqrt{\gamma_n C^{2d} + \delta^2} - \sqrt{\gamma_n C^{-2d} + \delta^2} < 1$. Since $\lim_{\xi \to 0} f(\xi)/\xi = 1$, there is a point $a \in I$ such that $C^{-1}\xi \le f(\xi) \le C\xi$ for $0 < \xi \le a$. Therefore there exist positive constants Λ_i $(i = 1, 2)$ such that

$$\Lambda_1 h(\xi, \alpha; C^{-1}) \le g_2(\xi, \alpha) \le \Lambda_2 h(\xi, \alpha; C), \quad 0 < \xi \le a.$$

From this, we get

$$\begin{aligned} &\int_0^a g_2(\xi,\alpha)^{-2} ds^{\mathrm{R}}(\xi) \int_\xi^a g_2(\eta,\alpha)^2 dm^{\mathrm{R}}(\eta) \\ &\le (\Lambda_2/\Lambda_1)^2 C^{2(d-1)} \int_0^a h(\xi,\alpha;C^{-1})^{-2} \xi^{1-d}\, d\xi \int_\xi^a h(\eta,\alpha;C)^2 \eta^{d-1}\, d\eta. \end{aligned}$$

The last integral converges if and only if so does the following one.

$$\int_0^a \left(\xi^{-\delta-\beta_1}\right)^{-2} \xi^{1-d}\, d\xi \int_\xi^a \left(\eta^{-\delta-\beta_2}\right)^2 \eta^{d-1}\, d\eta, \tag{5.7}$$

where $\beta_1 = \sqrt{\gamma_n C^{-2d} + \delta^2}$ and $\beta_2 = \sqrt{C^{2d} + \delta^2}$. Since $\beta_2 - \beta_1 < 1$, the integral (5.7) converges. Thus $g_2(\xi, \alpha)$ satisfies (2.11) with $(s^{\mathrm{R}}, m^{\mathrm{R}}, \gamma_n \nu)$ in place of (s, m, k). Therefore $(Q3)$ with $(s^{\mathrm{R}}, m^{\mathrm{R}}, \gamma_n \nu)$ in place of (s, m, k) is satisfied. By virtue of Proposition 2.3, $(R1)$ with $(s^{\mathrm{R}}, m^{\mathrm{R}}, \gamma_n \nu)$ in place of (s, m, k) holds. Thus we see that the condition $(E1)$ is satisfied. □

The following proposition follows from Proposition 1.4 and Lemma 5.1.

Proposition 5.2. (Murata, [10] and Pinchover, [12]) *Assume $J_{m^{\mathrm{R}}, s^{\mathrm{R}}}(\infty) < \infty$ or $J_{s^{\mathrm{R}}, m^{\mathrm{R}}}(\infty) < \infty$. Then there is a complete orthonormal base of eigenfunctions $\{\phi_j^{\mathrm{X}}\}_{j=0}^\infty$ with eigenvalues $0 \le \lambda_0^{\mathrm{X}} < \lambda_1^{\mathrm{X}} \le \lambda_2^{\mathrm{X}} \le \cdots$ repeated according to multiplicity, so that*

$$p^{\mathrm{X}}(t,x,y) = \sum_{j=0}^\infty e^{-\lambda_j^{\mathrm{X}} t} \phi_j^{\mathrm{X}}(x) \phi_j^{\mathrm{X}}(y), \qquad t > 0,\ x, y \in M, \tag{5.8}$$

where the series converges uniformly on any compact set of $(0,\infty) \times M \times M$. Further $\lambda_0^{\mathrm{X}} = \lambda_0^{\mathrm{R}}$ and $\phi_0^{\mathrm{X}}(x) = \phi_0^{\mathrm{R}}(|x|)/\sqrt{|S^{d-1}|}$.

The representation (5.8) is a known result ([12]). If $J_{m^{\mathrm{R}}, s^{\mathrm{R}}}(\infty) < \infty$, then by a standard fact on ODGDPs we get $\lambda_0^{\mathrm{R}} = 0$ and $\phi_0^{\mathrm{R}}(\xi) = (\int_0^\infty f^{d-1}(\xi)\, d\xi)^{-1/2}$, so that

$$\lambda_0^{\mathrm{X}} = 0, \quad \phi_0^{\mathrm{X}}(x) = \left\{ |S^{d-1}| \int_0^\infty f^{d-1}(\xi)\, d\xi \right\}^{-1/2}.$$

This fact is pointed out in [10].

The following result is also obtained by Murata (see Theorem 1.3 of [10]). We can obtain it by means of Theorem 1.2.

Proposition 5.3. (Murata, [10]) *Assume $J_{m^{\mathrm{R}}, s^{\mathrm{R}}}(\infty) < \infty$ or $J_{s^{\mathrm{R}}, m^{\mathrm{R}}}(\infty) < \infty$. Then there exists the limit*

$$\lim_{|y| \to l_i} p^{\mathrm{X}}(t,x,y)/\phi_0^{\mathrm{X}}(y) = q(t, x, (l_i, \theta)), \quad t > 0,\ x \in M,$$

for $y = (r, \theta) \in M$*, where* $l_i = 0$ *or* ∞*. The convergence is uniform in* $(t, x, \theta) \in K$ *for any compact set* $K \subset (0, \infty) \times M \times S^{d-1}$*. The limit* $q(t, x, (l_i, \theta))$ *is given as follows.*

$$q(t, (\xi, \varphi), (0, \theta)) = |S^{d-1}|^{-1/2}\phi_0^{\mathrm{R}}(0)^{-1}p^{\mathrm{R}}(t, 0, \xi).$$

If $J_{m^{\mathrm{R}}, s^{\mathrm{R}}}(\infty) < \infty$*,*

$$q(t, (\xi, \varphi), (\infty, \theta)) = |S^{d-1}|^{-1/2}\phi_0^{\mathrm{R}}(\infty)^{-1}p^{\mathrm{R}}(t, \infty, \xi).$$

If $J_{s^{\mathrm{R}}, m^{\mathrm{R}}}(\infty) < \infty$*,*

$$q(t, (\xi, \varphi), (\infty, \theta)) = |S^{d-1}|^{1/2}\phi_0^{\mathrm{R}^-}(\infty)^{-1}\sum_{n=0}^{\infty} p^{(n)}{}_2^-(t, \infty, \xi)\kappa_n(\theta, \varphi),$$

where the series converges uniformly in $(t, \xi, \theta, \varphi) \in K$ *for any compact* $K \subset (0, \infty) \times I \times S^{d-1} \times S^{d-1}$*.*

Finally we observe a sufficient condition for [IU]. We note that there exists no positive number B_0 satisfying (1.16) in the case $J_{s^{\mathrm{R}}, m^{\mathrm{R}}}(\infty) < \infty$. Indeed, $J_{s^{\mathrm{R}}, m^{\mathrm{R}}}(\infty) < \infty$ implies $\int_1^\infty f^{1-d}(\xi)\, d\xi < \infty$, so that, $\limsup_{\xi \to \infty} f(\xi) = \infty$. Since $C = \sup_{\xi \in I} f(\xi) < \infty$ implies

$$\int_I h(\xi)\, d\nu(\xi) \geq C^{-2} \int_I h(\xi)\, dm^{\mathrm{R}}(\xi), \quad h \in C_0(I),$$

we obtain the following result from Theorem 1.4 and Corollary 1.1.

Proposition 5.4. *Assume* $J_{m^{\mathrm{R}}, s^{\mathrm{R}}}(\infty) < \infty$*,* $\sup_{\xi \in I} f(\xi) < \infty$*, and*

$$\int_2^\infty \frac{f^{1-d}(r)}{\int_1^r f^{1-d}(\xi)\, d\xi} \sup_{r \leq \xi < \infty} \left(\int_1^\xi f^{1-d}(\eta)\, d\eta\right)\left(\int_\xi^\infty f^{d-1}(\eta)\, d\eta\right) dr < \infty.$$

Then [IU] and [SP] hold, that is,

$$\sup_{x, y \in M} p^{\mathrm{X}}(t, x, y) < \infty, \quad t > 0,$$

and for any $\varepsilon > 0$ *there exists a compact subset* K *of* M *such that*

$$\int_{M \setminus K} G^{\mathrm{X}}(\alpha, x, z)G^{\mathrm{X}}(\alpha, z, y)\, dm^{\mathrm{X}}(z) \leq \varepsilon G^{\mathrm{X}}(\alpha, x, y), \qquad x, y \in M \setminus K.$$

Acknowledgment The author would like to thank Professor M. Murata for his valuable comments.

References

1. Z. Q. Chen and M. Fukushima, *Symmetric Markov Processes, Time Change and Boundary Theory, Princeton University Press*, Princeton University Press, 2012.
2. E. B. Davies and B. Simon, *Ultracontractivity and the heat kernel for Schrödinger operators and the Dirichlet Laplacians*, J. Funct. Anal. **59** (1984), 335–395.
3. A. Erdélyi, *Higher transcendental functions* **1,2**, New York, 1953.

4. W. Feller, *The parabolic differential equations and the associated semi-groups of transformations*, Ann. of Math., **55** (1952), 468–519.
5. M. Fukushima and Y. Oshima, *On the Skew of Symmetric Diffusion Processes*, Forum Math.**1** (1989), 103–142.
6. M. Fukushima, Y. Oshima and M. Takeda, *Dirichlet Forms and Symmetric Markov Processes*, Walter de Gruyter, Berlin, 1994
7. K. Itô and H. P. McKean, Jr., *Diffusion Processes and their Sample Paths*, Springer-Verlag, New York, 1974.
8. H. P. McKean, Jr., *Elementary solutions for certain parabolic differential equations*, Trans. Amer. Math. Soc. **82** (1956), 519–548.
9. M. Murata, *Integral representations of nonnegative solutions for parabolic equations and elliptic Martin boundaries*, J. Funct. Anal. **245** (2007), 177–212.
10. M. Murata, *Nonnegative solutions of the heat equation on rotationally symmetric Riemannian manifolds and semismall perturbations*, Rev. Mat. Iberoamericana **27** (2011), 885–907.
11. H. Okura, *A new approach to the skew product of symmetric Markov processes*, Memoirs of the Faculty of Engineering and Design, Kyoto Institute of Technology **46** (1998), 1–12.
12. Y. Pinchover, *Maximum and anti-maximum principles and eigenfunctions estimates via perturbation theory of positive solutions of elliptic equations*, Math. Ann. **314** (1999), 555–590.
13. T. Takemura, *State of boundaries for harmonic transforms of one-dimensional generalized diffusion processes*, Ann. Reports of Graduate School of Humanities and Sciences, Nara Women's University **25** (2010), 285–294.
14. T. Takemura and M. Tomisaki, *Recurrence/transience criteria for skew product diffusion processes*, Proceedings of the Japan Academy **87**, Ser.A (2011), 119-122.
15. M. Tomisaki, *Asymptotic behavior of elementary solutions of transient generalized diffusion equations*, J. Math. Soc. Japan **40** (1988), 561–581.
16. M. Tomisaki, *Intrinsic ultracontractivity and small perturbation for one dimensional generalized diffusion operators*, J. Funct. Anal. **251** (2007), 289–324.

Author Index